ACCESSING Autodesk®
ARCHITECTURAL
DESKTOP 2005

ACCESSING Autodesk® ARCHITECTURAL DESKTOP 2005

WILLIAM G. WYATT

autodesk Press

THOMSON
DELMAR LEARNING

Australia • Canada • Mexico • Singapore • Spain • United Kingdom • United States

Accessing Autodesk® Architectural Desktop 2005

William G. Wyatt

Vice President, Technology and Trades SBU:
Alar Elken

Editorial Director:
Sandy Clark

Senior Acquisitions Editor:
James DeVoe

Senior Development Editor:
John Fisher

Marketing Director:
Dave Garza

Channel Manager:
Dennie Williams

Marketing Coordinator:
Casey Bruno

Production Director:
Mary Ellen Black

Production Manager:
Andrew Crouth

Production Editor
Stacy Masucci

Art/Design Specialist:
Mary Beth Vought

Technology Project Manager:
Kevin Smith

Technology Project Specialist:
Linda Verde

COPYRIGHT © 2005 Thomson Delmar Learning. Thomson, the Star Logo, and Delmar Learning are trademarks used herein under license.

Printed in Canada

1 2 3 4 5 XX 06 05 04

For more information contact
Thomson Delmar Learning
Executive Woods
5 Maxwell Drive, PO Box 8007,
Clifton Park, NY 12065-8007
Or find us on the World Wide Web at
www.delmarlearning.com

ALL RIGHTS RESERVED. No part of this work covered by the copyright hereon may be reproduced in any form or by any means—graphic, electronic, or mechanical, including photocopying, recording, taping, Web distribution, or information storage and retrieval systems—without the written permission of the publisher.

For permission to use material from the text or product, contact us by
Tel.   (800) 730-2214
Fax   (800) 730-2215
www.thomsonrights.com

Library of Congress Cataloging-in-Publication Data:

ISBN: 1-4018-8352-4

NOTICE TO THE READER

Publisher does not warrant or guarantee any of the products described herein or perform any independent analysis in connection with any of the product information contained herein. Publisher does not assume, and expressly disclaims, any obligation to obtain and include information other than that provided to it by the manufacturer.

The reader is expressly warned to consider and adopt all safety precautions that might be indicated by the activities herein and to avoid all potential hazards. By following the instructions contained herein, the reader willingly assumes all risks in connection with such instructions.

The publisher makes no representation or warranties of any kind, including but not limited to, the warranties of fitness for particular purpose or merchantability, nor are any such representations implied with respect to the material set forth herein, and the publisher takes no responsibility with respect to such material. The publisher shall not be liable for any special, consequential, or exemplary damages resulting, in whole or part, from the readers' use of, or reliance upon, this material.

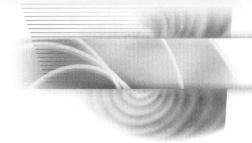

CONTENTS

PREFACE
FEATURES OF THIS EDITION xxii
    Style Conventions xxii
    How to Use This Book xxiii
    Command Access Tables xxiii
    Organizing Tutorial Directories xxiv
    Online Companion xxiv
    We Want to Hear from You xxv
    About the Author xxv
    Dedication xxv
    Acknowledgements xxv

CHAPTER 1 INTRODUCTION TO ARCHITECTURAL DESKTOP
OVERVIEW 1
OBJECTIVES 1
ADVANTAGES OF OBJECT TECHNOLOGY 2
    Selecting Installation Options 2
STARTING ARCHITECTURAL DESKTOP 2
    Selecting Templates 3
ARCHITECTURAL DESKTOP SCREEN LAYOUT 4
    Tools of the Tool Palette 7
    Customizing with the Content Browser 11
    Using the Properties Palette 16
    Accessing Design Resources 17
    Resources of the Style Manager 18
    Inserting Multi-View Blocks and Masking Blocks 18
    Inserting Documentation 19
    Creating Detail Compponents 20
    Inserting Structural Compponents 21
    Accessing Shortcut Menus 22
    Using Toolbars to Access Commands 23
CONFIGURING DRAWING SETUP 25
    Setting Units 26
    Setting Scale 26
    Defining the Layer Standard 27
    Controlling Display in Drawing Setup 27
TOOLS OF THE LAYER MANAGER 28
    Creating Layer Snapshots 29

Steps to Creating a Layer Snapshot 29
USING LAYOUTS 30
DEFINING DISPLAY CONTROL FOR OBJECTS AND VIEWPORTS 31
    Selecting Display Representations 32
    Contents of Display Representation Sets 33
    Applying Display Configurations to Viewports 34
VIEWING OBJECTS WITH THE OBJECT VIEWER 37
    Editing Objects with Object Display 39
    Materials 40
    Display Properties 40
    General Properties 41
UPDATING THE DISPLAY OF OBJECTS 42
CREATING PROJECTS 42
    Using the Project Browser to Create the Project 43
    Creating Project Content with the Project Navigator 46
    Using Project Components to Develop Construction Documents 47
    Steps to Attaching an Element Drawing to a Construct File 52
    Steps to Creating a Model Space View Within a View Drawing 54
    Keeping the Project Navigator Up to Date 55
SUMMARY 57
REVIEW QUESTIONS 58
TUTORIAL 1.1 CREATING A NEW DRAWING 59
TUTORIAL 1.2 VIEWING THE DISPLAY REPRESENTATION
OF OBJECTS AND SNAPSHOTS 60
TUTORIAL 1.3 CREATING A PROJECT 65

CHAPTER 2 CREATING FLOOR PLANS

INTRODUCTION 71
OBJECTIVES 71
DRAWING WALLS 72
    Creating Walls with the WallAdd Command 73
    Using Add Selected 74
    Changing Walls with the Apply Tool Properties To Command 75
CONTENTS OF THE WALL PROPERTIES PALETTE 76
    Setting Justification for Precision Drawing 80
    Creating Straight Walls 81
    Designing Curved Walls 82
    Using the Offset Option to Move the Wall Handle 86
SETTING UP FOR DRAWING AND EDITING WALLS 87
    Using the Node Object Snap 88
    Grips of the Wall 88
    Using Grips to Edit Walls 89
    Steps to Copying a Wall Using Grips 90
    Setting Justification 90

USING EDIT JUSTIFICATION	91
CHANGING WALL DIRECTION WITH WALLREVERSE	92
CONNECTING WALLS WITH AUTOSNAP	93
Using Wall Cleanup	94
Setting the Cleanup Radius	94
Using Priority to Control Cleanup	96
Creating Cleanup Group Definitions	96
Steps to Creating and Applying a Cleanup Group	97
Managing Wall Defect Markers	97
MODIFYING WALLS	99
Creating L and T Wall Intersections	99
Using the Wall Cleanup T command	100
Merging Walls	101
Joining Walls	102
Using WallOffsetCopy to Draw Interior Walls	103
Using WallOffsetMove	106
Using the Offset Set From Command	107
Editing Walls with AutoCAD Editing Commands	110
SUMMARY	111
REVIEW QUESTIONS	112
TUTORIAL 2.1 DRAWING WALLS AND USING CLOSE	113
TUTORIAL 2.2 DRAWING CURVED WALLS AND USING ORTHO CLOSE	117
TUTORIAL 2.3 DRAWING WALLS USING WALL OFFSET COPY COMMAND	122
TUTORIAL 2.4 EDITING WALLS USING EDIT JUSTIFICATION, REVERSE, WALL OFFSET, JOIN, AND MERGE WALL	130
TUTORIAL 2.5 EDITING WALLS USING TRIM, MATCH, AND ADD SELECTED COMMANDS	136
PROJECT	140

CHAPTER 3 ADVANCED WALL FEATURES

INTRODUCTION	141
OBJECTIVES	141
ACCESSING WALL STYLES	141
CREATING AND EDITING WALL STYLES	142
Creating a New Wall Style Name	146
Steps to Creating a New Wall Style	148
Creating a New Style from an Existing Style	148
Defining the Properties of a Wall Style	149
Steps to Overriding the Drawing Default Display Representation for a Wall Style	169
Assigning Materials to Wall Components	170
Steps to Creating Material Definitions for Walls Using the Content Browser	171
Exporting and Importing Wall Styles	173

Steps to Importing a Wall Style 174
Purging Wall Styles 174
Identifying Wall Styles Using Inspect 175
EXTENDING A WALL WITH ROOFLINE/FLOORLINE 176
Extending the Wall with FloorLine 177
CREATING WALL ENDCAPS 178
Creating Custom Wall Endcaps 179
Steps to Creating an Endcap 180
Modifying Endcaps 182
USING EDIT IN PLACE AND OVERRIDE ENDCAP STYLE 183
Applying Calculate Automatically to Change the Endcap 185
Overriding the Priority of Components in a Wall Style 187
CREATING WALL MODIFIERS AND STYLES 188
Inserting and Editing Wall Modifiers in the Workspace 192
Creating Wall Modifier Styles 196
Steps to Converting a Polyline to a Wall Modifier for a Selected Wall 197
CREATING WALL SWEEPS USING PROFILES 198
CREATING THE SWEEP 200
Steps to Creating a Sweep 201
MODIFYING SWEPT WALLS 202
Using Edit In Place with Wall Sweep 202
ADDING MASS ELEMENTS USING BODY MODIFIERS 204
CREATING ADDITIONAL FLOORS 206
CREATING A FOUNDATION PLAN 209
PLOTTING SHEETS 209
Steps to Plotting 210
SUMMARY 215
REVIEW QUESTIONS 215
TUTORIAL 3.1 CREATING A WALL STYLE 216
TUTORIAL 3.2 CREATING AND EDITING AN ENDCAP 222
TUTORIAL 3.3 CREATING WALL MODIFIERS 229
TUTORIAL 3.4 CREATING THE BASEMENT FLOOR PLAN 235
TUTORIAL 3.5 CREATING THE FOUNDATION PLAN 244
TUTORIAL 3.6 PLOTTING FLOOR PLANS 251
PROJECT 254

CHAPTER 4 PLACING DOORS AND WINDOWS

INTRODUCTION 255
Objectives 255
PLACING DOORS IN WALLS 256
Properties of a Door 257
Setting the Swing 262
Locating the Insertion Point of the Door 262
Defining Door Properties 269

CREATING AND USING DOOR STYLES 272
    Creating a New Door Style 273
EDITING A DOOR STYLE 273
    Shifting a Door Within the Wall 284
    Shifting the Door Along a Wall 286
PLACING WINDOWS IN WALLS 287
    Defining Window Properties 290
    Editing a Window with Grips 292
THE WINDOWS PALETTE 293
CREATING A NEW WINDOW STYLE 293
CREATING MUNTINS FOR WINDOWS 300
    Adding Shutters to a Window Style 304
CREATING OPENINGS 305
    Defining Opening Properties 308
    Adding an Opening with Precision 308
    Modifying Openings 309
    Adding and Editing the Profile of DoorS, Windows, and Openings 310
APPLYING TOOL PROPERTIES 311
    Steps to Using Apply Tool Properties for Doors 312
IMPORTING AND EXPORTING DOOR AND WINDOW STYLES 313
    Steps to Importing Door Styles 313
SUMMARY 314
REVIEW QUESTIONS: 315
TUTORIAL 4.1 INSERTING DOORS 315
TUTORIAL 4.2 INSERTING WINDOWS AND
CREATING WINDOW STYLES 323
TUTORIAL 4.3 CREATING AN AEC PROFILE AND
USING THE PROFILE IN A DOOR STYLE 329
PROJECTS 332

CHAPTER 5 DOOR AND WINDOW ASSEMBLIES

INTRODUCTION 335
OBJECTIVES 335
ADDING DOOR/WINDOW ASSEMBLIES 335
COMPONENTS OF A DOOR/WINDOW ASSEMBLY STYLE 340
    Defining the Components Using Design Rules 345
MODIFYING DOOR/WINDOW ASSEMBLIES 352
    Changing the Door/Window Assembly Style in the Drawing 352
    Saving In Place Edit Changes to the Style 355
    Steps to Creating an Override Assignment 358
    Editing the Profile of the Frame or Mullion 359
    Steps to Adding a Profile Override 360
    Steps to Editing a Modified Profile Using In Place Edit 361
    Steps to Creating a Frame Override to the Edge Assignment 362

Steps to Creating an Override of Division Assignment..364
Steps to Creating a Miter..365
Steps to Using Interference Add to Trim the Assembly...366
SUMMARY..368
REVIEW QUESTIONS..368
TUTORIAL 5.1 CREATING AND EDITING DOOR/WINDOW
ASSEMBLIES FOR A FRONT ENTRANCE UNIT...369
PROJECT...376

CHAPTER 6 CREATING ROOFS AND ROOF SLABS

INTRODUCTION..377
OBJECTIVES..377
CREATING A ROOF WITH ROOFADD..378
    Creating a Hip Roof...383
DEFINING ROOF PROPERTIES..383
CREATING A GABLE ROOF..387
    Steps to Creating a Gable Roof..387
EDITING AN EXISTING ROOF TO CREATE GABLES..388
    Using Edit Edges/Faces to Edit Roof Planes..388
    Steps to Editing Roof Edges and Faces..388
    Using Grips to Edit a Roof...390
    Steps to Creating a Gable Using Grips..390
CREATING A SHED ROOF..391
CREATING A GAMBREL ROOF USING DOUBLE SLOPED ROOF....................391
CREATING A FLAT ROOF..392
CONVERTING POLYLINES OR WALLS TO ROOFS...392
CREATING ROOF SLABS..393
    Using the Direction Property to Create Roof Slabs..399
    Creating the Roof Slab from a Wall, Roof, or Polyline...400
MODIFYING ROOF SLABS..403
TOOLS FOR EDITING ROOF SLABS..406
    Trimming a Roof Slab..406
    Extending a Roof Slab...409
    Mitering Roof Slabs...410
    Cutting a Roof Slab...412
    Adding a Roof Slab Vertex..413
    Removing the Vertex of the Roof Slab...415
    Creating Holes in a Roof..416
    Using Boolean Add/Subtract/Detach to Combine Mass Elements.......................418
    Creating a Roof Dormer..420
DETERMINING ROOF INTERSECTIONS..422
    Creating a Mass Group..422
    Steps to Cutting and Editing Roof Slabs..424
CREATING ROOF SLAB STYLES..425
CREATING A NEW ROOF EDGE STYLE...429

ASSIGNING ROOF SLAB EDGE STYLES TO SLAB EDGES 436
USING ADD EDGE PROFILE IN THE WORKSPACE 438
    Steps to Adding an Edge Profile 439
CHANGING THE PROFILE WITH EDIT EDGE PROFILE IN PLACE 441
EXTENDING WALLS TO THE ROOF 441
SUMMARY 441
REVIEW QUESTIONS 442
TUTORIAL 6.1 CREATING A HIP ROOF 443
TUTORIAL 6.2 CREATING A GABLE ROOF 446
TUTORIAL 6.3 CREATING AND EDITING A GAMBREL ROOF 450
TUTORIAL 6.4 CREATING DORMERS AND ROOF SLABS 455
PROJECT 463

CHAPTER 7 CREATING SLABS FOR FLOORS AND CEILINGS

INTRODUCTION 465
OBJECTIVES 465
ADDING AND MODIFYING A FLOOR SLAB 465
MODIFYING THE SLAB USING PROPERTIES AND GRIPS 472
    Using Grips 474
ACCESSING SLAB STYLES 475
ATTACHING A SLAB EDGE STYLE TO A SLAB 476
    Editing the Slab Edge Styles of a Slab 478
    Adding an Edge Profile in the Workspace 479
CHANGING THE PROFILE WITH EDIT EDGE PROFILE IN PLACE 481
SUMMARY 482
REVIEW QUESTIONS 482
TUTORIAL 7.1 CREATING A CATHEDRAL CEILING 483
TUTORIAL 7.2 CREATING A FLAT ROOF WITH A SLAB 489
PROJECT 494

CHAPTER 8 STAIRS AND RAILINGS

INTRODUCTION 497
OBJECTIVES 497
CREATING THE STAIR CONSTRUCT 498
    Steps to Creating a Stair Construct 498
USING THE STAIRADD COMMAND 499
    Properties of a Stair 501
SPECIFYING FLIGHT POINTS TO CREATE STAIR SHAPES 508
    Creating a Multi-Landing Stair 508
    Creating a U-Shaped Stair 509
    Creating Spiral Stairs 512
    Editing a Stair Using Grips 514
    Grips of the Multi-Landing Stair 514
    Grips of the U-Shaped Stair 515

Grips of the Spiral Stair	515
CUSTOMIZING THE EDGE OF A STAIR	**517**
Projecting Stairs to Walls	518
Removing Stair Customization	519
Creating a Polyline from the Edge of the Stair	520
CREATING STAIR STYLES	**521**
Using the Style Manager to Create Stair Styles	522
ANCHORING A STAIR TO A LANDING	**533**
CREATING A STAIR TOWER	**534**
CREATING RAILINGS	**536**
ADDING A RAILING	**538**
DEFINING RAILING STYLES	**538**
Steps to Importing and Editing a Railing Style in the Style Manager	540
Components of a Railing Style	540
MODIFYING THE RAILING	**551**
Editing Post Locations	551
Reversing the Railing	554
Anchoring a Railing	555
Adding a Profile for Railing Components	556
Changing the Profile with Edit In Place	557
DISPLAYING THE STAIR IN MULTIPLE LEVELS	**558**
SUMMARY	**561**
REVIEW QUESTIONS	**561**
TUTORIAL 8.1 CREATING STRAIGHT STAIRS	
WITH VERTICAL ORIENTATIONS	**562**
TUTORIAL 8.2 CREATING A U-SHAPED STAIR AND A STAIR TOWER	**567**
TUTORIAL 8.3 CREATING STAIR AND RAILING STYLES	**571**
PROJECT	**583**

CHAPTER 9 USING AND CREATING SYMBOLS

INTRODUCTION	**585**
OBJECTIVES	**585**
SETTING THE SCALE AND LAYER FOR SYMBOLS AND ANNOTATION	**586**
USING THE CONTENT BROWSER	**587**
Steps to Creating a Tool Palette and Adding Content	588
USING THE DESIGNCENTER	**591**
INSERTING SYMBOLS	**593**
Steps to Inserting Symbols into a Drawing from the Content Browser	595
PROPERTIES OF MULTI-VIEW BLOCKS	**595**
Editing Multi-View Blocks using Properties	597
EDITING THE MULTI-VIEW BLOCK DEFINITION PROPERTIES	**599**
CREATING A MULTI-VIEW BLOCK	**602**
Steps to Creating and Defining a New Multi-View Block	604
Defining the Properties of a New Multi-View Block	604

IMPORTING AND EXPORTING MULTI-VIEW BLOCKS	606
Steps to Exporting Multi-View Blocks	606
INSERTING NEW MULTI-VIEW BLOCKS	607
IN-PLACE EDITING OF VIEW BLOCK OFFSETS	608
Steps to In-Place Edit view block Insertion Offsets	608
Steps to In-Place Edit of Attribute location	609
INSERTING LAYOUTS WITH MULTIPLE FIXTURES	610
Viewing Multi-View Blocks with Reflected Display Representation	612
CREATING MASKING BLOCKS	613
PROPERTIES OF THE MASKING BLOCK	614
USING PROPERTIES TO EDIT EXISTING MASKING BLOCKS	615
ATTACHING OBJECTS TO MASKING BLOCKS	615
Detaching Masking Blocks	616
CREATING MASKING BLOCKS	617
Steps to Creating a Masking Block	617
Setting Definition Properties of a New Mask Block	618
INSERTING NEW MULTI-VIEW AND MASKING BLOCKS	620
CREATING SYMBOLS FOR THE DESIGNCENTER	621
SUMMARY	626
REVIEW QUESTIONS	627
TUTORIAL 9.1 INSERTING MULTI-VIEW BLOCKS WITH PRECISION	628
TUTORIAL 9.2 CREATING AND MODIFYING MULTI-VIEW BLOCKS	635
TUTORIAL 9.3 ACCESSING MASK BLOCKS FROM THE DESIGNCENTER AND CREATING MASK BLOCKS	642
PROJECT	647

CHAPTER 10 ANNOTATING AND DOCUMENTING THE DRAWING

INTRODUCTION	649
OBJECTIVES	649
PLACING ANNOTATION ON A DRAWING	650
Placing Text	650
Placing Leaders in the Drawing	651
CREATING DIMENSIONS	653
Inserting AEC Dimension (2) Dimensions	655
Creating AEC Manual Dimensions	656
Using the Style Manager to Create an AEC Dimension Style	657
Editing AEC Dimensions	664
Editing AEC Dimensions with the Shortcut Menu	666
Angular Dimensions	672
Creating Radial Dimensions	674
PLACING STRAIGHT CUT LINES	675
REVISION CLOUDS	676
INSERTING CHASES	678

- Steps to Inserting a Chase using the Content browser 681
- CONTENTS OF THE MISCELLANEOUS FOLDER 681
- CREATING TAGS AND SCHEDULES FOR OBJECTS 682
  - Properties of Objects and Object Styles 684
  - Placing Door Tags 686
  - Steps to Placing a Door Tag 687
  - Placing Window Tags 688
  - Steps to Placing a Window Tag 689
  - Room and Finish Tags 689
  - Steps to Creating a Space from Walls within a Project 695
  - Object Tags 698
  - Wall Tags 699
  - Steps to Placing Wall Tags 700
- EDITING TAGS AND SCHEDULE DATA 701
- ADDING A SCHEDULE TABLE 701
- USING SCHEDULE TABLE STYLES 704
- RENUMBERING TAGS 712
- UPDATING A SCHEDULE 714
- EDITING THE CELLS OF A SCHEDULE 715
- USING BROWSE PROPERTY DATA 715
- CHANGING THE SELECTION SET OF A SCHEDULE 718
  - Adding Objects to an Existing Schedule 718
  - Removing Objects from an Existing Schedule 719
  - Reselecting Objects for an Existing Schedule 720
  - Finding Objects in the Drawing Listed in the Schedule 721
- EXPORTING SCHEDULE DATA 722
  - Steps to Exporting Schedules to Data Files 724
  - Creating schedules for project Data 725
  - Steps to Creating a Comprehensive Project-Based Schedule 725
  - Inserting Properties without Tags 726
  - Property Set Definitions 727
- INSERTING DATA FIELDS IN A SCHEDULE 731
- SUMMARY 731
- REVIEW QUESTIONS 732
- TUTORIAL 10.1 INSERTING AEC DIMENSIONS 733
- TUTORIAL 10.2 ADDING SPACE TAGS FOR ROOMS 740
- TUTORIAL 10.3 PLACING TAGS FOR WINDOWS AND CREATING WINDOW SCHEDULES 743
- TUTORIAL 10.4 PLACING ALPHABETIC TAGS AND QUANTITIES FOR DOOR SCHEDULES 753
- TUTORIAL 10.5 CREATING TAGS AND SCHEDULES FOR A PROJECT 758
- PROJECT 766

CHAPTER 11 CREATING ELEVATIONS, SECTIONS, AND DETAILS

INTRODUCTION 769
OBJECTIVES 769
CREATING THE MODEL FOR ELEVATIONS AND SECTIONS 770
    Steps to Creating a Model 770
CREATING BUILDING ELEVATIONS AND SECTIONS 772
    Steps to Creating an Elevation 775
    Editing the Properties Palette of the building elevation/Section line 778
REFRESHING AND REGENERATING ELEVATION/SECTIONS 781
    CONTROLLING Display of an Elevation or Section 785
    Creating Elevation Styles to Control Elevation Display 786
    Steps to Creating and Editing an Elevation Style 790
EDITING THE LINEWORK OF THE ELEVATION 791
MERGING LINES TO THE ELEVATION 794
    Modifying the Material Hatch Pattern 794
CREATING SECTIONS OF THE MODEL USING CALLOUTS 796
    Steps to Creating a Section 796
    Creating a Horizontal and vertical Sections 800
    Steps to Creating a Horizontal Section 800
    Steps to Creating a Vertical Section 802
    Creating a Live Section 802
    Creating 2D Sections with Hidden Line Projection 804
    Creating a Napkin Sketch View 805
    Using the Detail Component Manager 808
    Steps to Creating Welded wire fabric 813
    Steps to Creating Gravel Boundary filling component 814
    Steps to Creating Masonry Units 816
KEYNOTING 826
    Reference Keynoting 827
    Sheet Keynotes 831
    Editing Legends 833
SUMMARY 835
REVIEW QUESTIONS 836
TUTORIAL 11.1 CREATING A MODEL VIEW USING
THE PROJECT NAVIGATOR 837
TUTORIAL 11.2 CREATING ELEVATIONS AND ELEVATION STYLES 841
TUTORIAL 11.3 CREATING A 2D SECTION AND A LIVE SECTION 852
TUTORIAL 11.4 CREATING DETAILS AND KEYNOTING 856
PROJECT 868

CHAPTER 12 CREATING MASS MODELS, SPACES, AND BOUNDARIES

INTRODUCTION 871
OBJECTIVES 871
CREATING MASS MODELS 872
    Inserting Mass Elements 873
    Selecting Mass Element Shapes 875
    Steps to Inserting a Mass Element (Gable) 879
    Creating Mass Elements Using Extrusion and Revolution................................880
    Creating Mass Elements Using Drape 882
MODIFYING MASS ELEMENTS 883
    Trimming Mass Elements 885
    Splitting Mass Elements 886
    Editing Faces 887
CREATING STYLE DEFINITIONS FOR MASS ELEMENTS 889
CREATING COMPLEX MODELS 891
CREATING GROUPS FOR MASS ELEMENTS 891
ADDING MASS ELEMENTS TO A GROUP 893
    Attaching Elements to a Group 893
    Detaching Elements from a Mass Group 895
    Boolean Operations with Mass Elements 895
    Subtracting Mass Elements 896
    Using the Intersect Operation 897
    Using the Model Explorer 898
    Mouse Operations in the Model Explorer 901
    Using the Object Viewer Shortcut Menu 901
    Using the Tree View Shortcut Menus 903
    Creating Mass Elements with the Model Explorer 903
    Creating Floorplate Slices and Boundaries 905
    Converting the Slice to a Polyline 908
    Converting AEC Objects to Mass Elements 909
SUMMARY 911
REVIEW QUESTIONS 912
TUTORIAL 12.1 CREATING MASS ELEMENT COMPONENTS
FOR A FIREPLACE 913
TUTORIAL 12.2 CREATING MASS ELEMENTS FOR A TERRAIN 915
PROJECTS 918

CHAPTER 13 DRAWING COMMERCIAL STRUCTURES

INTRODUCTION 923
OBJECTIVES 923
INSERTING STRUCTURAL MEMBERS 924
    Using the Structural Member Catalog 924
    Creating a Structural Member Style 927

Steps to Creating a Structural Member Style 927
ADDING STRUCTURAL MEMBERS 928
    Inserting Columns 929
    Steps to Adding a Column 932
    Inserting Beams 933
    Steps to Adding a Beam 935
    Inserting a Brace 935
    Steps to Adding a Brace 937
    Modifying Ends with Trim Planes 938
    Customizing Structural Members Using Styles 939
    Converting Linework to Structural Members 943
    Creating a Rigid Frame 947
    Creating Complex Columns 948
CREATING CUSTOM SHAPES 948
    Steps to Creating a Custom Shape 949
    Creating Structural Members in the Style Wizard 951
CREATING BAR JOISTS 953
    Steps to Creating a Bar Joist 953
CREATING STRUCTURAL GRIDS 953
    Specifying the Size of the Column Grid 959
    Steps to Dynamically Sizing a Rectangular Column Grid 960
    Steps to Dynamically Sizing a Radial Column Grid 961
    Clipping the Column Grid 963
    Defining the Column for the Grid 965
REFINING THE COLUMN GRID 965
    Adding and Removing Column Grid Lines 966
    Converting Existing lines to Column Grids 970
    Dimensioning the Column Grid 973
CREATING CEILING GRIDS 974
    Steps for Attaching a Ceiling Grid to a space object 978
CHANGING THE CEILING GRID 979
    Adding and Removing Ceiling Grid Lines 980
    Adding Boundaries and Holes 982
USING LAYOUT CURVES TO PLACE COLUMNS 986
VIEWING THE MODEL 988
    Creating a Walkthrough 990
SUMMARY 992
REVIEW QUESTIONS 993
TUTORIAL 13.1 CREATING DECK FLOOR FRAMING 993
TUTORIAL 13.2 CREATING STRUCTURAL PLANS 998
TUTORIAL 13.3 DIMENSIONING THE COLUMN GRID 1004
TUTORIAL 13.4 CREATING AND MODIFYING A CEILING GRID 1009
PROJECT 1012

INDEX 1015

PREFACE

Accessing Autodesk Architectural Desktop, 3rd edition, is a comprehensive presentation of the tools included in Autodesk® Architectural Desktop™ 2005. The format of the text includes the introduction of tools followed by an explanation of the options of the command and how it is used in development of drawings. A tool access table is provided for each command presented. The text includes screen captures of the dialog boxes associated with each tool. Tutorials are included at the end of each chapter to show you step by step how to use the tools of the chapter. In addition, the tutorials provide a written guide for you to refer to when doing the independent projects located at the end of each chapter. Projects and review questions are also provided for each chapter to reinforce your knowledge of the software. The Autodesk Architectural Desktop 2005 software is state-of-the-art software for architectural design. Recently, programs of study in architectural technology and architecture have integrated computer-aided design within the curriculum. With this in mind, *Accessing Autodesk Architectural Desktop*, 3rd edition, has been written in order to provide detailed and systematic explanations of the application of the tools to create architectural working drawings. The book includes techniques and tutorials that apply the software to the creation of drawings for residential and commercial buildings.

The first eleven chapters of the book provide the necessary instruction regarding the use of Autodesk Architectural Desktop 2005 for creating architectural working drawings. The beginning chapters will step you through the placement and editing of doors, walls, windows, and roofs. The tutorials track the development of working drawings from the beginning of a residence. Specific tutorials are included to demonstrate how the software is used to create floor plans, foundation plans, elevations, and sections. Each tutorial focuses on how to access a command and how to use the command to create architectural working drawings. This format allows the reader to immediately begin drawing floor plans.

Chapter 12 explains how to develop mass models using the Model Explorer to represent the shape of a building. Tutorials include instruction in creating mass models, developing the mass elements and space boundaries, and generating walls. Chapter 13 explains how the tools of Autodesk Architectural Desktop 2005 can be used to create drawings for commercial buildings. This chapter includes the structural catalog, layout curves, ceiling grids, camera views, and walkthroughs.

FEATURES OF THIS EDITION

Included in Accessing Autodesk Architectural Desktop are tutorials, which provide step-by-step instruction in the use of Autodesk Architectural Desktop 2005 to develop architectural working drawings for residential and commercial buildings. The tutorials, command access tables, and the contents of the CD together help the reader identify the appropriate commands and techniques to develop the working drawings. The features of this edition are summarized below:

- Command access tables are provided for each new command introduced. They describe in table format how to access the commands from the menu bar, command line, tool palette, and shortcut menu.
- Tutorials are included to introduce and reinforce the options of the commands. They include both residential and commercial building applications.
- The CD shipped with the book includes the drawings and solutions for the tutorials and the following appendices:

    Appendix A, "Plotting Sheets," includes a description of the Autodesk DWF Viewer.

    Appendix B, "Miscellaneous Folder of the Documentation Tool Catalog," includes an explanation regarding additional dimensioning and documentation tools located in the Miscellaneous Folder of the Documentation Tool Catalog.

    Appendix C, "Creating Curtain Walls" includes step-by-step procedures for creating curtain walls.

- Development of complex roofs, roof slabs, roof intersections, and dormers for residential and commercial buildings are included.
- Supplemental content for Chapters 12 and 13 are enclosed on the CD.

STYLE CONVENTIONS

Throughout the book you are requested to select commands and respond to the command in the command line or on screen with the mouse. The text style conventions are used systematically to enhance the understanding and recall of the commands. The style conventions of the text are as follows:

Element	Example
Commands	**WallAdd**
Menu	**Format>Style Manager**
Dialog box elements	The **Edit** button
Command line prompts	Elevation line start point:
Keyboard input	Press ENTER to end the command.
User Input	Type **OFFICE—CAD** in the field.
File and directory names	*C:\Documents and Settings\All users\Application Data\Autodesk*

HOW TO USE THIS BOOK

The design of each chapter of the text is to introduce the commands of Autodesk Architectural Desktop to allow you to gain the skills and understanding of the software to create architectural working drawings. Each chapter includes an introduction and objectives. Read carefully the objectives of the chapter to determine the commands and types of drawings created in the chapter. When you encounter a tutorial marker, stop and perform the tutorial located at the end of the chapter. The purpose of the tutorial is not to finish quickly but to gain hands-on experience in using the commands to create drawings. You may find it most helpful to repeat the tutorial after completing the chapter to gain recognition of the commands on the tool palette and of the content of the Properties palette. The book includes screen captures of the tool palettes and Properties palettes to allow you to study the commands when you might not have access to the software. However, having access to the software as you read the book greatly enhances the learning process. Summary and review questions are provided at the end of each chapter, as are additional project exercises.

COMMAND ACCESS TABLES

The commands included in the text are summarized and presented in command access tables. Included in each table is a list of methods to access the command from the menu bar, command prompt, tool palette, and shortcut menu. Commands to create objects are usually selected from tool palettes, while commands to edit objects are selected from shortcut menus of the selected object.

The command access table for the **Wall** tool (**WallAdd** command) is shown below as an example.

Command prompt	WALLADD
Tool palette	Select the Wall tool from the Design palette as shown in Figure 1.1

ORGANIZING TUTORIAL DIRECTORIES

Drawing files for all tutorials are located on the CD shipped with the book. The drawing files are sorted by chapter and located in the *ADT Tutor* folder on the CD. Create a student folder and copy the *ADT Tutor* and all its subdirectories from the CD to your student folder. After copying the folder to your computer, select the *ADT Tutor* folder of your *ADT Student* directory in Windows Explorer, right-click, and choose **Properties**. Clear the Read-only attribute of the folder contents. This will enable you to open tutorial drawings or projects from the *ADT Tutor* files of your student directory and save your modified drawing files there. Project folders will also be located in your *ADT Tutor* folder.

 STOP. Create your student folder now.

ONLINE COMPANION

If you have access to the Internet, you can access the Online Companion. Additional resources and links to other sites are available. Access the Online Companion at:

http://www.autodeskpress.com/resources/olcs/index.aspj. After you access the web address, click on the OLC for Accessing Autodesk Architectural Desktop 2005.

This address also provides other resources available from Autodesk Press.

An e.resource CD with PowerPoint® presentations for each chapter is available for the instructor. The ISBN is 1-4018-8353-2.

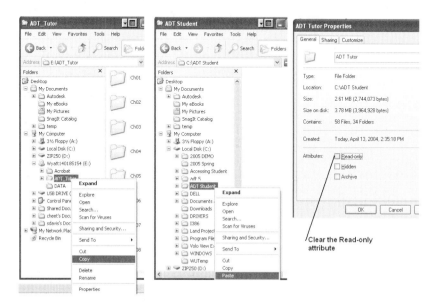

Figure P.1 *Create a student directory for the ADT Tutor*

WE WANT TO HEAR FROM YOU

We welcome your comments and suggestions regarding the contents of this text. Your input will result in the improvement of future publications. Please forward your comments and questions to:

The CADD Team
C/O Autodesk Press
Executive Woods
5 Maxwell Drive
Clifton Park, NY 12065-8007

ABOUT THE AUTHOR

William G. Wyatt, Sr., is an instructor at John Tyler Community College in Chester, Virginia. He has taught architectural drafting and related technical courses in the Architectural Engineering Technology program since 1972. He earned his Doctor of Education from Virginia Tech and his Master of Science and Bachelor of Science degrees in Industrial Technology from Eastern Kentucky University. He earned his Associate in Applied Science in Architectural Technology from John Tyler Community College. He is a certified Architectural and Building Construction Technician and Autodesk Certified Instructor for the Autodesk Training Center at Tidewater Community College.

DEDICATION

There are so many to thank in my circle of gratitude. I must always start with my late parents, Leslie and Catherine Wyatt, whose love has always been a cornerstone of stability in my life. I know they would have enjoyed this book.

A special thank you to our children, Leslie Wyatt, who is a rising junior at Mary Washington College, Sarah Wyatt, who will be a sophomore at Virginia Tech, and Will Wyatt, Jr., who will be starting at Virginia Commonwealth University in the Fall 2004. I wish them well as they continue their academic endeavors and turn their dreams into realities.

In addition, a heartfelt thanks to my father-in-law and mother-in-law, Kermit and Helen Hedahl, for their acceptance and understanding of the time needed to complete this project. Most importantly, to my wife Bevin Hedahl Wyatt, who provides continuous encouragement, support, and love. She has done more than her share to make this project possible.

ACKNOWLEDGEMENTS

The author would like to thank and acknowledge the team of reviewers who reviewed the chapters and provided guidance and direction to the work. A special thanks to the following reviewers, who reviewed the chapters in detail:

Paul Adams, Denver Technical College, Denver, Colorado

Deanna Blickham, Moberly Area Community College, Moberly, Missouri

David Braun, Spokane Community College, Spokane, Washington

Paul N. Champigny, New England Institute of Technology, Warwick, Rhode Island

James Freygang, Ivy Tech State College, South Bend, Indiana

Lynn A. Gurnett, York County Technical College, Wells, Maine

Donald W. Hain, Orleans/Niagara BOCES, Sanborn, New York

Christopher LeBlanc, Porter & Chester Institute, Chicopee, Massachusetts

Jeff Levy, Pulaski County High School, Dublin, Virginia

Joseph M. Liston, University of Arkansas Westark, Fort Smith, Arkansas

Jeff Porter, Porter & Chester Institute, Watertown, Connecticut

Charles T. Walling, Silicon Valley College, Walnut Creek, California

A special thanks goes to David Byrnes of Design Department at Emily Carr Institute of Art & Design in Vancouver, British Columbia, and the CAD Training Centre at the British Columbia Institute of Technology, Burnaby, British Columbia for his careful editing of the manuscript. His technical expertise has allowed this text to include practical details and tips that will benefit the reader. In addition, a special thanks to Lloyd Cannaday of Patrick Henry Community College, for performing initial technical editing and promoting the use of Autodesk Architectural Desktop in the Virginia Community Colleges.

I would like to thank Ronald A. Williams of Ronald A. Williams, LTD, Autodesk Education Representative of Virginia and David Butts of CADRE Systems, Inc., for their support and encouragement to this project.

The author would like to acknowledge and thank the following staff from Delmar Publishers:

Sandy Clark, Editorial Director; James DeVoe, Senior Acquisitions Editor; John Fisher, Senior Developmental Editor; Stacy Masucci, Production Editor; and Mary Beth Vought, Art & Design Specialist.

The author would like to acknowledge and thank the following people:

Copyediting: Gail Taylor

Composition: Vince Potenza of SoundLightMind Media Design and Development

Introduction to Architectural Desktop

OVERVIEW

Autodesk® Architectural Desktop™ is intended to assist the designer in developing architectural working drawings as well as preliminary schematics and computer models of a building. The first release of Autodesk Architectural Desktop was based upon AutoCAD Release 14 in 1998. Autodesk Architectural Desktop Release 2005 utilizes the new features of AutoCAD 2005.

Autodesk Architectural Desktop creates components of a building as 3D objects. Objects consist of walls, doors, windows, stairs, and roofs. Because 3D objects are used to develop the drawings, the drawings can be put together to create a three-dimensional model of a building. In addition, Architectural Desktop includes tools for creating schedules that allow you to extract a comprehensive list of doors, windows, furniture, rooms, and wall types from the drawing and models.

OBJECTIVES

After completing this chapter, you will be able to

- Describe the advantages of object technology
- Describe how to start Architectural Desktop and identify the components of the Architectural Desktop workspace
- Identify the purpose and resources of the **Content Browser**™, **DesignCenter**™, **Detail Component Manager**, **Display Manager**, and **Style Manager**
- Identify the purpose of ADT templates and Drawing Setup
- Identify the display options of tool palettes and the Properties palette
- Describe how to use shortcut menus for command access
- Identify how display configurations are used to control object display

- Describe how to use the Object Viewer and the **ObjectDisplay** command
- Describe how to use the **Project Browser** and **Project Navigator** to create a project

ADVANTAGES OF OBJECT TECHNOLOGY

Using Architectural Desktop requires a basic knowledge of AutoCAD because AutoCAD is integrated into the Architectural Desktop. Doors and windows as well as the walls are inserted in a drawing through ObjectARX® technology. These objects consist of several entities that have associative interaction with other objects when inserted. When doors and windows are placed in the wall, they behave as magnets within the wall, and the wall is correctly edited for the window or door. Using ObjectARX technology reduces the need to edit the drawing. Architectural Desktop objects are three-dimensional and consist of several components. For instance, a door object consists of a door panel, frame, stop, and swing components.

Another advantage of object technology is the use of the Object Viewer. The Object Viewer can be accessed from the shortcut menu of any object, and is a separate window designed for viewing one or more objects. The object can be viewed in orthographic, isometric, and perspective views. The navigation of the Object Viewer allows the object to be viewed dynamically from any vantage point, as shown in Figure 1.33. The Object Viewer can be used to view interior spaces or individual object components in perspective as shaded or rendered images. Rendered views of objects can be saved as bitmap image files and exported to other documents.

SELECTING INSTALLATION OPTIONS

There are two installation options available when you install Architectural Desktop. The content of this book is based upon the Standard installation, which installs the Imperial and Metric content in the DesignCenter and most other application features. The Custom installation allows you to select Metric D A CH content. It allows you to choose the following features: CAD Standards, Database, Dictionaries, Drawing Encryption, Fonts, New Features Workshop, Migrate Custom Settings, Portable License Utility, Reference Manager, Samples, Tutorials, and VBA Support.

STARTING ARCHITECTURAL DESKTOP

Start Architectural Desktop by choosing the **Start** button of Windows XP/Windows 2000 and selecting **All Programs>Autodesk>Autodesk Architectural Desktop 2005** or by double-clicking on the **Autodesk Architectural Desktop 2005** icon on the desktop. Once Autodesk Architectural Desktop is launched, it opens to Drawing 1 based upon the *Aec Model (Imperial Stb).dwt* template. When you select **QNew** from the **Standard** toolbar, a new drawing is created based upon the *Aec Model (Imperial Stb).dwt* template. If you select **File >New** from the menu bar, the **Select Template** dialog box will open, allowing you to specify a template.

Tip: Control of the display of the **Startup** dialog box is set in the **Startup** option in the **System** tab of the **Options** dialog box. When Startup is set to **Do not show a Startup dialog**, the **Select Template** dialog box is displayed. When Startup is set to **Show Startup dialog**, the **Create New Drawing** dialog box is displayed.

New drawings in Architectural Desktop should utilize one of the templates as listed in Table 1.1. Drawings developed as part of a project are printed or plotted from drawings created based upon the sheet templates listed in the *C:\Documents and Settings\All Users\Application Data\Autodesk\ADT 2005\enu\Template* folder.

Template	Purpose
Aec Model (Imperial Ctb)	Create ADT drawings using object styles based on imperial units and use color dependent plotting style tables for plotting.
Aec Model (Imperial Stb)	Create ADT drawings using object styles based on imperial units and use style dependent plotting style tables for plotting.
Aec Model (Metric Ctb)	Create ADT drawings using object styles based on metric units and use color dependent plotting style tables for plotting
Aec Model (Metric Stb)	Create ADT drawings using object styles based on metric units and use style dependent plotting style tables for plotting.

Table 1.1 *Release 2005 templates*

Start a new drawing using imperial units by selecting either **Aec Model (Imperial Ctb).dwt** or **AecModel (Imperial Stb).dwt**. There are no model space graphic entities in the templates.

SELECTING TEMPLATES

Templates specifically designed for Architectural Desktop include preset values for annotation, dimensioning, grid, layer standards, limits, snap, and units. The settings for the imperial and metric templates are shown in Table 1.2.

Setting	Imperial	Metric
Annotation Plot Size	3/32"	3.5 mm
Grid	10'	3000 mm
Layer Standards	AecLayerStd.dwg	AecLayerStd.dwg
Layer Key Style	AIA (256 Color)	BS1192 Descriptive (256)
Limits	0,0 to 288',192'	0,0 to 53600,41400
Scale Factor	96	1:100 mm
Snap	1/16	1 mm
Units	Architectural	Decimal
Units Precision	1/32 Precision	0.00 Precision

Table 1.2 *Settings of the Architectural Desktop templates*

When you create a new drawing using an AEC template, the default drawing name is Drawing1 because it has not been saved to a directory. To save a drawing with a specific name, choose **SaveAs** from the **File** menu and type a new drawing name in the file name edit box.

ARCHITECTURAL DESKTOP SCREEN LAYOUT

The Autodesk Architectural Desktop 2005 screen is shown in Figure 1.1. Components unique to Architectural Desktop are ADT tool palettes, **Navigation** toolbar, **Shapes toolbar**, and the Drawing Window status bar. The screen includes a menu bar at the top and four docked toolbars. The docked toolbars are from left to right: **Standard**, **Navigation**, **Shapes**, and **Layer Properties**. The tool palettes shown in Figure 1.1 are the primary source for accessing commands for Autodesk Architectural Desktop. The Design palette is shown in Figure 1.1; however, you can select other palette tabs to display additional palettes for placing various styles of architectural objects. When you select a tool from the tool palette, information regarding the settings of the current tool is displayed in the Properties palette. If you select an existing object, its properties are displayed and can be edited as shown in the Properties palette on the right in Figure 1.1.

The drawing window includes the Model and Work tabs. The Work tab shown in Figure 1.1 is a layout that displays the building in isometric view at left and in plan view on the right. The Drawing Window status bar located above the command window, shown in Figure 1.2, includes context menus; on the left is the **Drawing** menu and on the right are the **Viewport Scale**, **Annotation Scale**, and **Display**

Introduction to Architectural Desktop 5

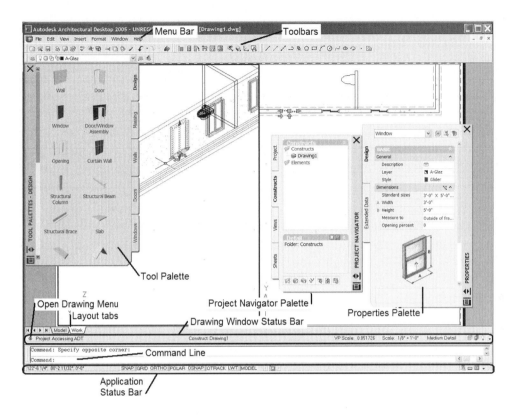

Figure 1.1 ADT 2005 screen layout

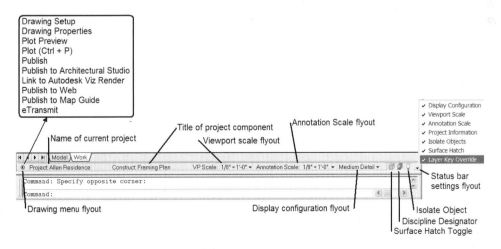

Figure 1.2 Drawing Window status bar

Configuration flyouts. The **Drawing** menu includes the following: **Drawing Setup, Drawing Properties, Plot Preview, Plot, Publish to Architectural Studio, Link to Autodesk VIZ Render, Publish to Web**, and **eTransmit**. You can select **VIZ Render** from the **Drawing** menu to render the drawing contents. The **Display Configuration** menu includes a flyout to select a display configuration from the list of display configurations. (The display configurations will be discussed later in this chapter.) If a floating viewport is active, the zoom scale factor of the viewport is displayed in the lower right corner of the Drawing Window status bar. The name and type of drawing associated with a project are displayed in the middle of the Drawing Window status bar.

The Application status bar located below the command line shown in Figure 1.3 includes toggles to control the display of the workspace. The **Maximize Viewport** toggle can be selected to display a full screen view of the viewport. The alternate **Miminize Viewport** toggle shown in Figure 1.3 can be selected to return the screen view of the viewport to include other inactive viewports. The **Drawing Window Status Bar** toggle located in the lower right corner is selected to turn on or off the display of the Drawing Window status bar. All toolbars, properties palettes, and dialog boxes are removed from the workspace when you select the **Clean Screen Mode** toggle. The **Clean Screen Mode** and **Maximize Viewport** toggles have been turned on as shown in Figure 1.4. The **Plotting Status** button is displayed during background plotting. Select this button to open the **Plot and Publish Details** dialog box and view the details of the current plot.

The menu bar includes **File, Edit, View, Insert, Format, Window,** and **Help** menus. The content of each menu is shown in Figures 1.5 and 1.6. The following additional menus are listed in the **Window>Pulldowns** cascade menu shown in Figure 1.6: **Design Pulldown, Document Pulldown, CAD Manager Pulldown, 3D Solids Pulldown,** and **Customize**. The **Design** and **Document** menus include content similar to the **Design** and **Documentation** menus of Release 3.3. The **Customize** menu option allows you to modify the pulldown menus from the menu groups.

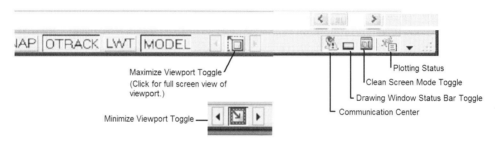

Figure 1.3 *Application status bar*

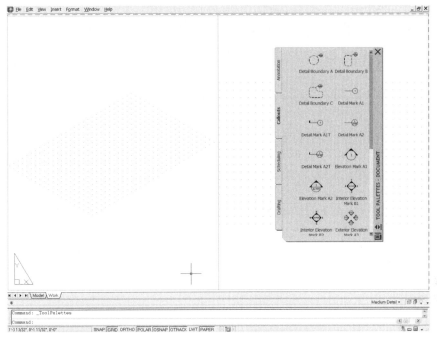

Figure 1.4 *Clean Screen Mode and Maximize Viewport applied to workspace*

Tip: Select the **Profile** tab of the **Options** dialog box to create a profile to save the display of toolbars, tool palettes, menus, Properties palette, **Project Navigator**, and the support path to the tool palettes.

TOOLS OF THE TOOL PALETTE

The tools of the tool palette include the basic tools for creating Architectural Desktop objects. The tools include commands for adding doors, windows, walls, and the roof. To modify existing objects, select the object and choose commands from the shortcut menu to edit the object. Tool palettes can be displayed as described in Table 1.3, by selecting **Tool Palettes** from the **Window** menu or by typing CTRL + 3 in the command line.

Command	CTRL + 3
Menu bar	Window>Tool Palettes

Table 1.3 *Accessing Tool Palettes*

Commands are selected by clicking the icon of the tool palette. When the command is executed, the Properties palette opens; review the settings in the Properties palette

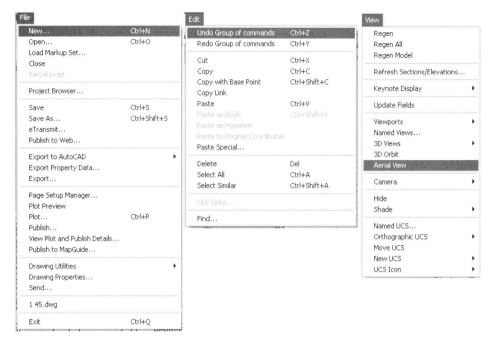

Figure 1.5 File, Edit and View menus

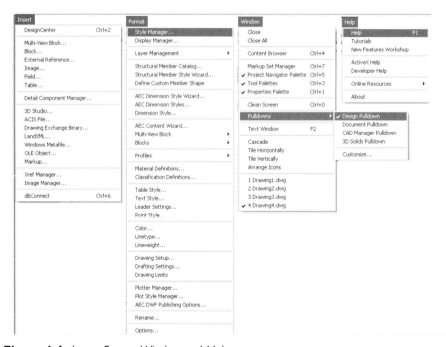

Figure 1.6 Insert, Format, Window and Help menus

Introduction to Architectural Desktop

and respond to the prompts in the command line. The tool palettes can be customized to include additional tools needed by the designer. You can drag styles of objects created in the **Style Manager** to the tool palettes. You can click and drag content such as plumbing fixtures and furniture from the DesignCenter and **Content Browser** onto tool palettes. The **Content Browser** includes catalogs of tools that can be placed on existing or new tool palettes. Tool palettes are grouped into three palette sets: Design, Document, and Detailing. The content of the palette sets shown in Figure 1.7 are summarized below:

> **Design** – Includes Design, Massing, Walls, Doors, Windows palettes. The Design palette includes tools for placing walls, doors, windows, window assemblies, openings, curtain walls, structural columns, structural beams, structural braces, slabs, roof slabs, roofs, stairs, railings, structural column grids, ceiling grids, and spaces. The Walls, Doors, and Windows palettes include sample styles of objects.
>
> **Document** – Includes tool palettes for dimensioning, annotating, tagging, and placing schedules. The Callout palette includes tools for placing detail, elevation, and section marks with options for creating view drawings.
>
> **Detailing** – Includes detail tools from each of 15 Master Format divisions. Tools on these palettes allow you to place section views or plan views of building components as necessary for the development of details.

The title bar of the tool palettes includes the following toggles: **Close**, **Auto-hide**, and **Properties**, as shown in Figure 1.8. The **Close** button will close the tool palette from

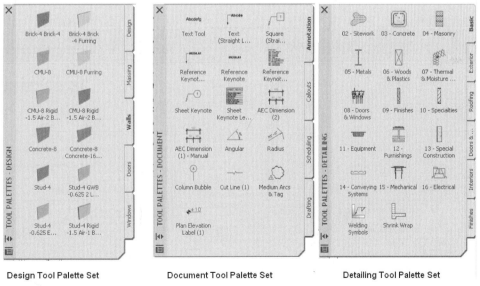

Figure 1.7 *Tool palette sets*

display on the screen. The **Auto-hide** toggle on the title bar will collapse the tool palette. When **Auto-hide** is toggled ON, only the title bar of the tool palette will be displayed in the drawing after a tool has been selected. The tool palette will expand to a full palette when you move the pointer over the title bar. You can move a tool palette on the graphics screen by clicking on the title bar and dragging the cursor to a different location. If you position the pointer over the edges of the tool palette, the resize arrow will be displayed, as shown at right in Figure 1.8. You can click and drag the resize arrow to resize the tool palette. If you resize the tool palette, the tabs may become obstructed, as shown at left in Figure 1.8. You can select the edge of the stacked palettes as shown in Figure 1.8 to display a list of palettes in the shortcut menu, from which a palette can be selected.

The Properties button of the tool palettes title bar shown in Figure 1.8 will open a menu at right, allowing you to set the properties of tool palettes. The options of this menu are described below:

> **Move** – Displays a four-headed arrow on the cursor that allows you to move the palette.
>
> **Size** – Allows you to move the pointer over the tool palette, right click, and drag the pointer to specify the size of the palette.
>
> **Close** – Closes the tool palette.
>
> **Allow Docking** – Toggled ON, the tool palette can be docked at the left or right side of the graphics window. A docked tool palette has a grab bar at the top. If you left-click on the grab bar and then move your pointer to the middle of the screen, the tool palette will undock.

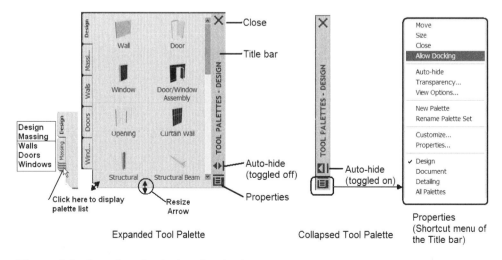

Figure 1.8 *Controlling the display of tool palettes*

Auto-hide – If checked, the tool palettes will collapse when the cursor is moved from the tool palettes. When **Auto-hide** is toggled OFF, the tool palette remains expanded.

Transparency – If selected, a **Transparency** dialog box will open as shown in Figure 1.9. This dialog box allows you to adjust the level of transparency by moving the slide bar from less to more. The dialog box also has a check box to turn off transparency. When transparency is toggled ON, you can view your drawing through the tool palette, as shown at right in Figure 1.9.

View Options – Opens a **View Options** dialog box as shown in Figure 1.10. This dialog box allows you to increase the icon image size by sliding the **Image Size** to the right. The format of the icon and text can be defined by checking one of the following options: **Icon only**, **Icon with text**, or **List view**. Selecting **Icon only** will cause the display of only the icons on the tool palette without the text. The **Icon with text** option displays the icon with the name of the command centered below the icon. The **List view** option displays the icon with the name of the command listed to the right of the icon. The lower portion of the dialog box allows you to specify the format for either the current palette or all palettes. A horizontal tool palette can be created when you resize the tool palette if **List view** is not selected in the **View Options** dialog box.

New Palette – Creates a new palette.

Rename Palette Set – Allows you to change the name of a tool palette.

Customize – Opens the **Customize** dialog box as shown in Figure 1.11. The tool palettes are listed in the left pane. The palette sets and the associated palettes are listed in the right pane. You can click and drag a palette from the left pane to a palette set of the right pane. The shortcut menu of each pane allows you to rename, delete, and create new palettes or palette sets.

Properties – Opens the Palette Properties dialog box. The dialog box indicates if the palette is shared from a catalog, and its name and description.

CUSTOMIZING WITH THE CONTENT BROWSER

The **Content Browser** stores the content of the tools and tool palettes displayed in Architectural Desktop. The tools displayed on the tool palettes are defined by the settings in the **Content Browser**. The **Content Browser** can be used to share tools among users. Tools can also be accessed from Web sites. A tool catalog stores pointers to the physical tool catalogs instead of storing the tool definitions in a file. Tool catalogs may include the following:

Tools – Used to place or edit objects. A tool can be used to place a wall in the drawing.

Tool palette – Multiple tools displayed as a group because they are used for similar tasks. The tool palette is considered a single object.

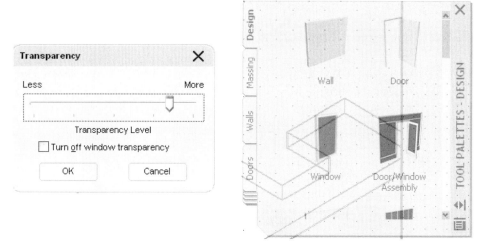

Figure 1.9 *Transparency of tool palettes turned on*

Tool package – A collection of tools that you can select for your tool palette.

Access the **Content Browser** as shown in Table 1.4.

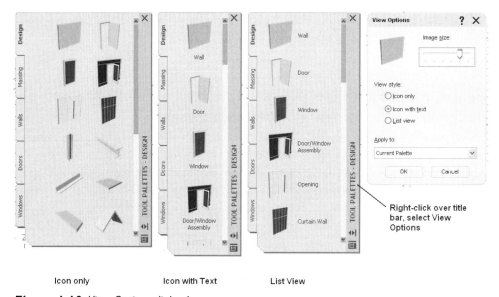

Figure 1.10 *View Options dialog box*

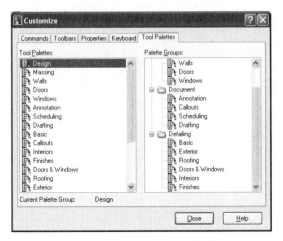

Figure 1.11 *Customize palette dialog box*

Command prompt	CONTENTBROWSER
Toolbar	Select Content Browser from the Navigation toolbar shown in Figure 1.22

Table 1.4 *Content Browser command access*

When you access the **Content Browser**, it opens as shown in Figure 1.12.

The **Content Browser** shown in Figure 1.12 consists of two panes. The left pane consists of navigation controls at the top and bottom. When a catalog is opened, the categories will be displayed in the left pane of Figure 1.13. The right pane displays the contents of tool catalog categories. The Autodesk Architectural Desktop Stock catalog has been opened as shown in Figure 1.13.

If you move the pointer over the categories listed in the left pane, the subcategories of the category will expand as shown in Figure 1.13. You can select the subcategory by clicking on its title while expanded. The Mass Elements subcategory is shown opened in Figure 1.14.

The right pane displays the tools of the subcategory. You can return to the previous state of the **Content Browser** by selecting the **Back** arrow shown at the top of Figure 1.14.

You can open additional windows of the **Content Browser** if you right-click over the **Content Browser** and choose **New Window**. In addition, you can move the point-

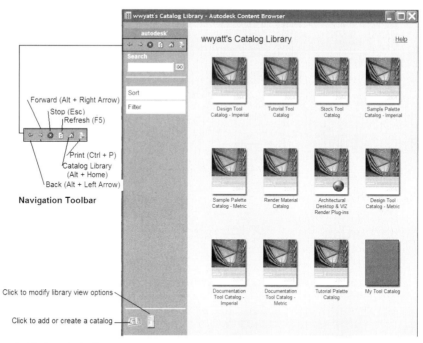

Figure 1.12 Autodesk Content Browser

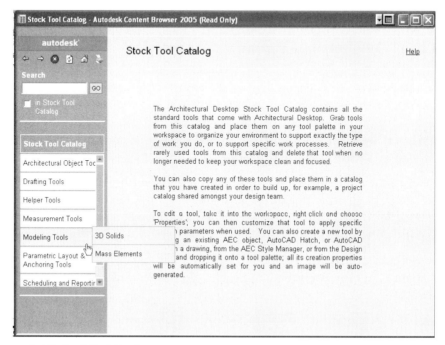

Figure 1.13 Stock Tool Catalog

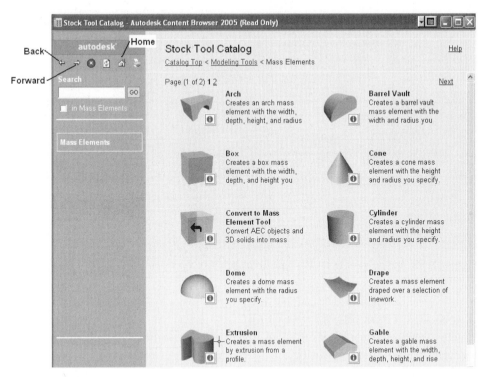

Figure 1.14 *Mass Elements tool catalog*

er over the **Content Browser** and press CTRL + N to open an additional window of the **Content Browser**. The My Tool Catalog shown in Figure 1.15 is an empty catalog located in the **Content Browser**. You can i-drop™ tools from other catalogs or the palettes of your drawing into the My Tool Catalog. The buttons at the bottom in the left pane are used to create new palettes, packages, or categories. After opening an additional **Content Browser** window, you can navigate to other catalogs in the new window and drag content into your catalog. The i-drop shown in the right pane allows you to drag a tool or a tool palette into the current workspace. In Figure 1.15, the **Basic Legend** tool was i-dropped from the Autodesk Architectural Desktop Stock Tool Catalog shown on the left to the My Tool Catalog shown at right. The My Tool Catalog is provided to create custom tool palettes.

Tip: The Sample Palette Catalog-Imperial and Sample Palette Catalog-Metric of the **Content Browser** consist of palette sets included when the software is installed. Therefore, you can click and drag the i-drop for a palette from the **Content Browser** to restore tool content of the original palettes.

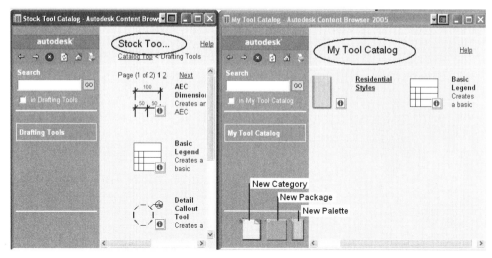

Figure 1.15 *Adding tools to the My Tool Catalog*

USING THE PROPERTIES PALETTE

When you choose a tool from the tool palette, the Properties palette opens to display the current settings for the new object. The current settings for a tool are inherited from the last insertion of that tool. Open the Properties palette as shown in Table 1.5. The Properties palette is used to set the properties of new objects and to modify existing objects. The palette includes two tabs, **Design** and **Extended Data**. The **Design** tab of a wall object is shown in Figure 1.16.

Command prompt	PROPERTIES or CTRL + 1
Shortcut menu	Select an object, right-click, and choose Properties

Table 1.5 *Accessing the Properties palette*

The most basic properties are listed in the Basic section at the top. If you are drawing a wall, the Basic section lists the general information such as style, dimensions of the wall, and location, while the Advanced section includes cleanups, style overrides, and worksheets. Depending on the object created, the Properties palette may include graphical illustrations of features of the object. Selecting the **Illustration** toggle as shown in Figure 1.16 will turn off/on the display of the illustration feature. Each category of the Properties palette can be collapsed by selecting the **Close** toggle, or it can be opened by selecting the **Open** toggle, as shown in Figure 1.16. The entire Basic or Advanced category can be opened or closed by selecting the **Open** or **Close** category toggles. The **Extended Data** tab of the Properties palette provides space

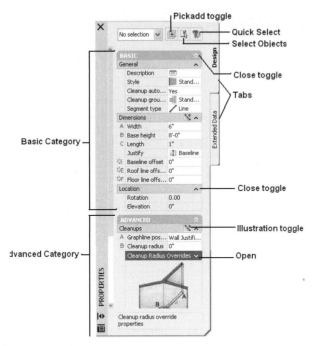

Figure 1.16 *Properties palette*

for recording documentation files that support the object and may include property data for schedules.

ACCESSING DESIGN RESOURCES

The resources for creating a drawing are located in the **Style Manager**, **Content Browser**, DesignCenter, **Detail Component Manager**, and **Structural Catalog**. Throughout your work with Architectural Desktop, you will open each of these resources to insert objects or annotation. Wall, door, and window objects are inserted into the drawing using the tools of the Design tool palette set. Each object is defined in a style definition. You can access the styles of objects in the **Style Manager**, which allows you to create, edit, import, and export styles. Therefore, to create complex walls that include brick veneer and concrete masonry units, you can create a wall style that includes brick and concrete masonry unit wall components. The wall style definition controls the shape and appearance of the wall. Refer to Table 1.6 to access the **Style Manager**. Wall styles can be dragged from the **Style Manager**, shown at right in Figure 1.17, to a tool palette or to the **Content Browser**.

Command prompt	AECSTYLEMANAGER
Menu bar	Format>Style Manager

Table 1.6 *Accessing the Style Manager*

RESOURCES OF THE STYLE MANAGER

Doors, walls, windows, and other objects included in Architectural Desktop are defined by styles. The style definition for windows includes the type of window, frame size, glass thickness, and typical sizes available. The styles included in an Architectural Desktop drawing are accessed by the **Style Manager**. The **Style Manager**, shown in Figure 1.17, organizes object styles into Architectural Objects, Documentation Objects, and Multi-Purpose Objects folders. The Architectural Objects folder includes such styles as doors, windows, stairs, and railings. The Documentation Objects folder includes such styles as elevation styles, section styles, and schedule tables. The Multi-Purpose Objects folder includes layer key styles, profiles, and material definitions. Styles are not included in the template—therefore the **Style Manager** is used to copy a style from other drawings to the current drawing. Styles can be created and edited in the current drawing or imported from and exported to other drawings. The use of the **Style Manager** will be introduced in Chapter 3. Additional styles are located in the Imperial and Metric folders of the *C:\Documents and Settings\All Users\Application Data\Autodesk\ADT 2005\enu\Styles* directory.

Tip: The settings in Windows Explorer may suppress the display of selected hidden files and directories. To view hidden files and folders, launch Windows Explorer, and select **Tools>Folder Options** from the menu bar. Select the **View** tab, and select **Show hidden files and folders**.

INSERTING MULTI-VIEW BLOCKS AND MASKING BLOCKS

Furniture, appliances, and plumbing fixtures are inserted in the drawing from the DesignCenter or the design catalogs of the **Content Browser**. These objects are multi-view and masking blocks located in the *Custom Applications>Design* folders in the DesignCenter, as shown in Figure 1.18. The Custom view of the DesignCenter displays a preview image of a selected symbol. Refer to Table 1.7 to access the DesignCenter.

Command prompt	CTRL + 2
Toolbar	Select DesignCenter from the Navigation toolbar

Table 1.7 *Accessing the DesignCenter*

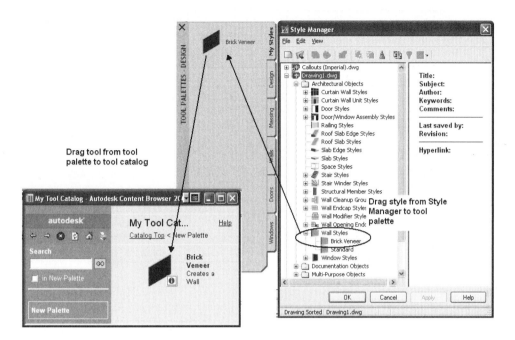

Figure 1.17 *Copying styles from the Style Manager to tool palettes*

These extensive libraries of symbols allow you to easily place symbols in a drawing from the DesignCenter. The contents of the Design directories based on the Standard installation are listed in Table 1.8; the toilet shown in Figure 1.18 is selected from the *Imperial\Design\Mechanical\Plumbing Fixtures\Bath* folder. The DesignCenter allows you to preview the block prior to inserting it in the drawing. The Plumbing Fixtures folder of the Imperial Design directory is shown in Figure 1.18.

These components are also located in the Design Tool Catalog-Imperial and Design Tool Catalog-Metric catalogs of the **Content Browser**. Refer to Table 1.4 to access the **Content Browser**.

To insert content from the DesignCenter, double-click on the component or click and drag the component into the drawing or onto a custom tool palette. To insert tools from a tool catalog, click on the i-drop of the tool and drag the i-drop into the drawing or onto a tool palette.

INSERTING DOCUMENTATION

Dimensions, schedules, and symbols can be placed in the drawing from the Documentation folder of the DesignCenter or the Documentation Tool Catalog-Metric and the Documentation Tool Catalog-Imperial catalogs of the **Content**

Imperial>Design	Metric>Design	Metric D A CH>Design
Conveying	Bathroom	Furniture
Electrical	Domestic Furniture	Office
Equipment	Electrical Services	Site
Furnishing	Kitchen Fittings	Symbols Austria
General	Office Furniture	
Mechanical	Pipe and Ducted Services	
Site	Site	
Special Construction		
Specialties		

Table 1.8 *DesignCenter folders of symbols*

Browser. The folders include break marks, callouts, dimensions, keynotes, schedules and tags.

CREATING DETAIL COMPONENTS

Components for details such as brick, concrete masonry units, and 2 × 4s can be selected from the **Detail Component Manager**. Refer to Table 1.9 to access the **Detail Component Manager**. The Detail tool palette set allows access to the **Detail Component Manager** and its content. (The **Detail Component Manager** will be presented in Chapter 11.)

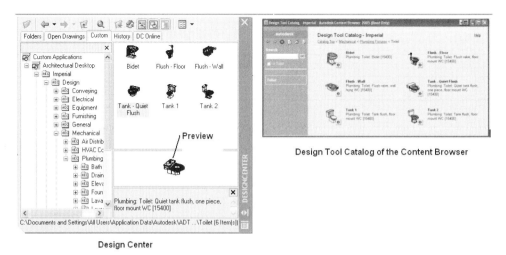

Figure 1.18 *DesignCenter and Design Tool Catalog of the Content Browser*

Command prompt	AECDTLCOMPMANAGER
Menu bar	Insert>Detail Component Manager
Toolbar	Select Detail Component Manager from the Navigation toolbar

Table 1.9 *Accessing the Detail Component Manager*

INSERTING STRUCTURAL COMPONENTS

Structural components such as precast concrete, steel, and wood components can be inserted in the drawing from the Structural Member Catalog. Refer to Table 1.10 to access the Structural Member Catalog. (The Structural Member Catalog shown in Figure 1.20 is presented in Chapter 13.)

Command prompt	AECSMEMBERCATALOG
Menu bar	Format>Structural Member Catalog

Table 1.10 *Accessing the Structural Member Catalog*

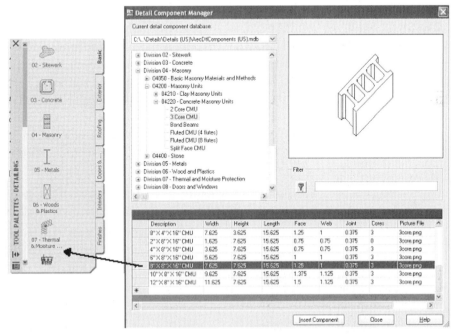

Figure 1.19 *Detail Component Manager*

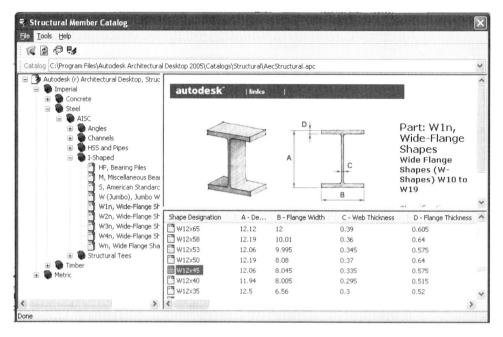

Figure 1.20 *Structural Member Catalog*

ACCESSING SHORTCUT MENUS

Shortcut menus are menus displayed when you right-click with the mouse. **Shortcut menus are the primary method for accessing commands to change existing Architectural Desktop objects.** Therefore, after placing a wall, you can select the wall and right-click—the shortcut menu options allow you to edit the wall or the style of the wall. Shown in Figure 1.21 is the shortcut menu displayed after a wall is selected. Notice that the content of this menu includes commands for editing the wall. The cascaded menu includes basic modify commands that are common to all entities and objects.

The menu displayed varies according to cursor location. The content of the shortcut menu reflects the context of the location. Shortcut menus are displayed in the following locations:

- Drawing area after selecting one or more objects
- Drawing area with no objects selected
- Drawing area during a command
- Inside the command window

- Within the DesignCenter
- Within area of text in the Multi-line text editor
- Over a tool palette or toolbar
- Over the Model or layout tabs
- Over the status bar

If you right-click over the drawing area when no object or entity has been selected, the shortcut menu includes zoom and pan options as shown at left in Figure 1.21. The shortcut menu also includes the following commands to control display of objects:

> **Isolate Objects>Isolate Objects (AecIsolateObjects)** – Turns off the display of all objects except selection.
>
> **Isolate Objects>Hide Objects (AecHideObjects)** – Turns off the display of selected objects.
>
> **Isolate Objects>End Objects Isolation (AecUnisolateObjects)** – Turns off control of Isolate Objects and Hide Objects.

USING TOOLBARS TO ACCESS COMMANDS

The toolbars displayed when the software is installed are **Standard**, **Navigation**, **Layer Properties**, and **Shapes**. These toolbars, shown in Figure 1.22, are similar to the basic AutoCAD 2005 toolbars. The commands of the toolbars are described in

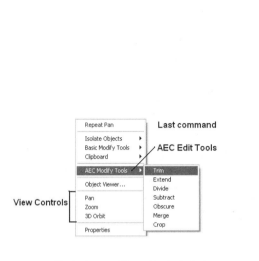

Figure 1.21 Accessing the shortcut menu

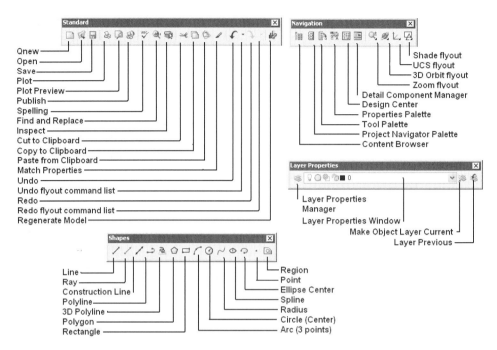

Figure 1.22 *Toolbars of Architectural Desktop*

Figure 1.22. The following additional toolbars are included in the ADT menu group: **3D Orbit, Dimension, Inquiry, Modify, Object Snap, Properties, Refedit, Shading, UCS, UCS II, Views,** and **Zoom**. The toolbars of the ADT menu group are listed in the **Customize** dialog box that opens when you access the **Toolbar** command. Additional toolbars of the ADT menu group can be displayed by moving the pointer over an existing toolbar, right-clicking, and choosing from the shortcut menu list.

 Tip: The settings for the display of toolbars are saved in profiles. Profiles are listed on the Profile tab of the **Options** dialog box. Right-click over the command window and choose **Options** from the shortcut menu to open the **Options** dialog box.

Standard

The **Standard** toolbar has been altered from the basic AutoCAD 2005 **Standard** toolbar. The **Standard** toolbar of Autodesk Architectural Desktop 2005 does not include **Zoom, Pan, Properties, DesignCenter, Tool Palettes,** and **Help**. However, some of these commands are located on the **Navigation** toolbar described below.

Navigation

The **Navigation** toolbar provides access to the **Content Browser, Project Navigator** palette, tool palettes, Properties palette, DesignCenter, and **Detail Component**

Manager. The toolbar also includes flyouts for **Zoom**, **3D Orbit**, **UCS**, and **Shade** commands. The commands of the **Navigation** toolbar are shown in Figure 1.22.

Layer Properties

The **Layer Properties** toolbar provides access to the commands related to layers as shown in Figure 1.22. The **Layer Properties** window displays the name and state of the current layer. This toolbar also includes the **Layer Properties Manager** command, which opens the **Layer Manager** for creating and editing layers. The **Layer Manager** includes additional layer commands to access the layer key styles.

Shapes

The **Shapes** toolbar is similar to the AutoCAD 2005 **Draw** toolbar; it provides access to commands for creating entities in the drawing. The commands for modifying geometry are accessed by selecting the object, right-clicking, and choosing from the shortcut menu. Additional commands for editing AutoCAD entities are located on the **Modify** toolbar.

CONFIGURING DRAWING SETUP

The **Drawing Setup** command allows you to set the units, scale, layering, and display. Access the **Drawing Setup** command as shown in Table 1.11.

Menu bar	Format>Drawing Setup
Command prompt	AECDWGSETUP
Drawing Window status bar	Select Drawing menu>Drawing Setup

Table 1.11 *Drawing Setup command access*

Selecting the **Drawing Setup** command opens the **Drawing Setup** dialog box shown in Figure 1.23. This dialog box consists of the **Units, Scale, Layering,** and **Display** tabs. The settings of the **Drawing Setup** dialog box control units and scale of annotation added to the drawing using Architectural Desktop annotation tools. Symbols inserted from the AEC DesignCenter will be scaled according to the scale factor specified in the **Scale** tab of the **Drawing Setup** dialog box. The layer standard used by Architectural Desktop is set in the **Layering** tab of the **Drawing Setup** dialog box. Symbols inserted in the drawing from the DesignCenter will be placed on layers according to the layer standard specified in the **Drawing Setup** dialog box. The **Units** tab shown in Figure 1.23 allows you to set the drawing units.

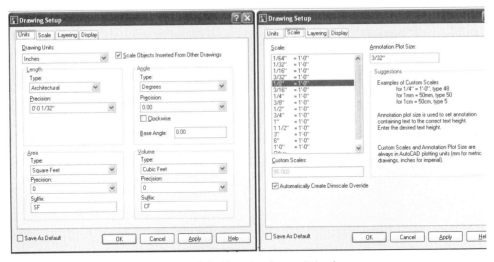

Figure 1.23 *Units and Scale tabs of the Drawing Setup dialog box*

SETTING UNITS

The **Units** tab allows you to set the units of the Architectural Desktop symbols. This tab is also used to set the units to metric. The **Units** tab of the **Drawing Setup** dialog box includes a **Drawing Units** drop-down list and **Linear**, **Area**, **Angular**, and **Volume** sections, as shown in Figure 1.23.

SETTING SCALE

The **Scale** tab allows you to specify the scale factor of the drawing. Symbols and annotation inserted in the drawing will be scaled according to the scale factor specified. The **Drawing Scale** listed in Figure 1.23 indicates the drawing is set for 1/8"=1'-0"; the resulting scale factor is shown inactive below the scale list. The scale factor is set to 96 for the 1/8"=1'-0" scale. For example, if a receptacle symbol is placed in a drawing with the scale set to 1/4"=1'-0", it is drawn with a 6" diameter circle. This symbol, if placed in a drawing with the 1/8"=1'-0" scale, would be drawn with a 12" diameter circle because the scale factor has changed from 48 to 96. The scale of a drawing sets the scale factor, which is used as a multiplier for selected symbols and annotation. The scale factor is the ratio between the size of the AutoCAD entity and its size when printed on paper. The typical architectural scales and the associated scale factors are listed in the **Scale** tab of the **Drawing Setup** dialog box shown in Figure 1.23.

 Tip The Drawing Setup scale can also be set in the Drawing Window status bar.

DEFINING THE LAYER STANDARD

The **Layering** tab shown in Figure 1.24 allows you to select a predefined layer standard for the drawing. It includes a **Layer Standards/Key File to Auto Import** edit field and a **Default Layer Standard Layer Key Style** list. Using the AecLayerStd file and the AIA (256 color)* current drawing will place architectural features such as walls, doors, and windows on the layer as defined in the AIA standard. This standard can be verified in this tab. The tutorials included in this book utilize the AIA (256 Colors) layer key style of the *AecLayerStd.dwg* drawing.

CONTROLLING DISPLAY IN DRAWING SETUP

The **Display** tab, as shown in Figure 1.24, lists the objects in the left window of Architectural Desktop. The display representations available for each object are in the right window. The display representations shown in Figure 1.24 are for a door object.

The options of the **Display** tab are described below:

> **Object Type** – Lists the objects of Architectural Desktop.
>
> **Display Representations** – Lists the display representations for the objects listed in the **Object Type** list.
>
> **Drawing Default Display Configuration** – Lists the display representation that will be applied to viewports as a default for the current drawing.

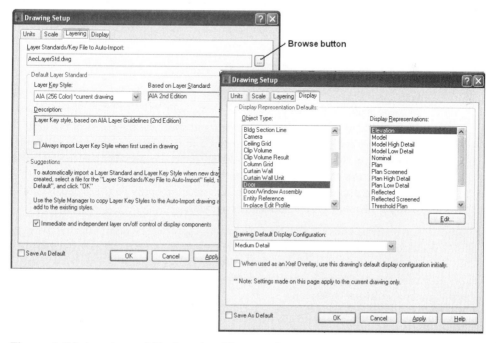

Figure 1.24 *Layering and Display tabs of Drawing Setup*

TOOLS OF THE LAYER MANAGER

The **Layer Manager** of Architectural Desktop is included in the **Layer Properties** toolbar rather than the **Layer** command of AutoCAD 2005. The **Layer Manager** includes options to change layer standards and layer key styles and to create layer snapshots or groups. The **Layer Manager** functions similarly to the AutoCAD **Layer** command; however, the additional features improve layer controls. Additional layer controls are also available in the Express Tools, which can be added by selecting AutoCAD Express Tools from the **Install Supplemental Tools** section of Disk 1. Access the **Layer Manager** as shown in Table 1.12.

Menu bar	Format>Layer Management>Layer Properties Manager
Command prompt	LAYER
Toolbar	Select Layer Properties Manager from the Layer Properties toolbar

Table 1.12 *Layer Manager command access*

The **Layer Manager** shown in Figure 1.25 consists of a menu bar, toolbar, left, and right panes. The commands of the toolbar are shown in the Figure 1.25. The right pane lists the layers of the drawing and allows you to control such properties as visibility, freeze, color, linetype, and lineweight.

The Layer Manager includes the following functions:

> **Groups** – Named groups of layers can be created as a set of layers. Select the All layer group in the left pane shown in Figure 1.25, right-click, choose **New Group Filter** from the shortcut menu, and overtype the name of the filter group. To assign layers to the group, click and drag layers from the right pane to the group name of the left pane. If you select a group name in the left pane, right-click, and choose **Select Layers>Add**, you can choose entities from the workspace that will add the layer of the selected entity to the group. The All group shown in Figure 1.25 includes all layers.
>
> **Layer Snapshots** – Can be created that capture the state (visibility, color, lock, linetype, lineweight) of the layers defined by the snapshot. A layer snapshot can be created that freezes all layers except five layers. This layer snapshot can be restored later in the drawing to recall the "freeze all layers but the five layers" selection.
>
> **Layer Key Styles** – Styles define the layer naming and properties that conform to the standard. The AIA Layer Key style specifies a standard layer naming convention. Architectural Desktop is programmed to place objects on

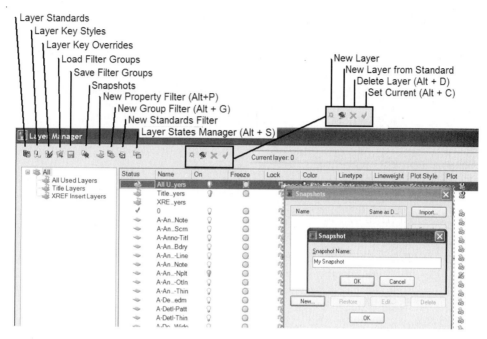

Figure 1.25 *Layer Manager dialog box*

layers as specified in the layer key. Therefore, when you place a door in the drawing, it will be placed on a layer that conforms to the AIA layer naming convention. The layer name in the drawing is A-Door, which is mapped to the layer key name Door.

CREATING LAYER SNAPSHOTS

Layer snapshots allow you to save the display setting of layers as a snapshot name. Layer snapshots can be used to create special plan views of a floor plan such as Electrical, Heating, Ventilating, Air Conditioning, Lighting, Plumbing, and Structural. Each of these plans requires the display of only selected architectural features. The layer snapshot is created to save the layer settings, which can be recalled later. The named snapshot can be exported as a file, which can then be imported into other drawings.

STEPS TO CREATING A LAYER SNAPSHOT

1. Choose **Layer Properties Manager** from the **Layer Properties** toolbar.
2. Change properties of layers, that is, freeze/thaw the desired layers in the layer list.
3. Choose the **Snapshots** button in the **Layer Manager** to open the **Snapshots** dialog box as shown in Figure 1.25.

4. Choose the **New** button and type a name for the snapshot, and then select **OK** to dismiss the **Snapshot** dialog box.

5. Select **OK** to dismiss the **Snapshots** dialog box.

After a snapshot has been created, the remaining buttons of the **Snapshots** dialog box are activated. The purpose of the buttons of the **Snapshots** dialog box are as follows:

Import – Opens snapshot files that can be used in the current drawing.

Export – Creates a snapshot file that saves the layer settings in a file. The snapshot file has an SSL extension.

New – Allows you to create a name for the snapshot in the Snapshot dialog box.

Restore – Allows you to select the name of a snapshot from the snapshot list and then select the **Restore** button to execute the layer settings.

Edit – Opens the **Snapshot Edit** dialog box, which can be used to change the color, linetype, lineweight, and visibility of the layers included in the snapshot. Additional layers can be added or removed from the layer snapshot definition in the Snapshot Edit dialog box.

Delete – Removes the snapshot names from the snapshot list.

 STOP. Do Tutorial 1.1, "Creating a New Drawing," at the end of the chapter.

USING LAYOUTS

The ADT templates include layout tabs specifically designed for the display of Architectural Desktop objects. The layout tabs included with the templates are Model and Work. The layout tabs at the lower left corner of the drawing area in Figure 1.1 provide quick access to model space or paper space viewports. The Model tab is designed for you to work in model space. The Work layout tab includes viewports that allow you to work in either model space or paper space. The viewports of the Work layout tab include a left viewport set to the SW Isometric view and a right viewport set to the Plan or Top view. You can toggle between model space and paper space within a layout by moving the mouse over the space you want, and then double-clicking. You can also toggle between paper space and model space by selecting MODEL/PAPER on the status bar. The status bar toggle will also display the name of the current space that is active. When you toggle on model space, you set the TILE-MODE system variable to 1; this variable is set to 0 when you work in paper space.

Drawings can be plotted from the Model or Work layout tabs or from Sheets in the **Project Navigator**. Specifying a display configuration for a viewport controls how objects are displayed to create a type of working drawing.

DEFINING DISPLAY CONTROL FOR OBJECTS AND VIEWPORTS

Display control in Architectural Desktop consists of three levels: display representations, display representation sets, and display configurations. These levels can be viewed and edited in the **Display Manager** dialog box. The **AecDisplayManager** command opens the **Display Manager** dialog box. Access the **AecDisplayManager** command as shown in Table 1.13.

Menu bar	Format>Display Manager
Command prompt	AECDISPLAYMANAGER

Table 1.13 *Display Manager command access*

If this command is accessed while in paper space, you will be prompted to select a viewport, and the **Display Manager** will open presenting information relative to the selected viewport. If model space is current when you select this command, the **Display Manager** will open presenting information relative to the active viewport.

The **Display Manager** dialog box shown in Figure 1.26 has combined all three levels of the display system into one dialog box. The **Display Manager** consists of three folders: Configurations, Sets, and Representations by Object. Each of these folders can

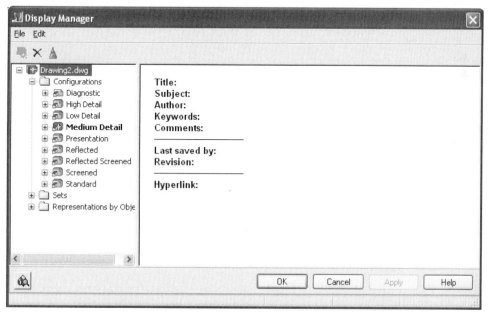

Figure 1.26 *Display configurations*

be opened to view the available options in the right pane. Selecting the (+) plus sign of the Configurations folder within the tree in the left pane will open the folder and display the available configurations. In Figure 1.26 the Work configuration is shown bold because it is the configuration applied to the current viewport.

SELECTING DISPLAY REPRESENTATIONS

Display representations control how objects are displayed. There are 36 display representations defined for the objects of Architectural Desktop. One or more display representations available can be applied to define the display of the object. The display representations define how much detail to display. Objects consist of several display components; the number of display components varies according to the detail desired for the display representation. Each component of an object can be controlled for visibility, layer, color, and linetype in the display representation. You can view the properties of a display representation in the **Display Manager** as shown in Figure 1.27.

In Figure 1.27 the (+) symbol of the Representations by Object folder has been selected to display a list of all objects. If you scroll down the list of objects and select an object in the left pane, the right pane of the **Display Manager** will display a list of the display representations for that object. The display representations for the door object are listed in Figure 1.27. The list of Sets in which the display representations can be used is listed across the top of the right pane. The display representation for an object varies according to the desired representation of that object in the drawing. For example, a door object has the following display representations: Elevation,

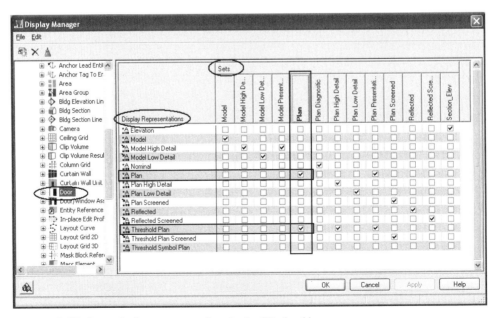

Figure 1.27 Door display representations in the Display Manager

Model, Model High Detail, Model Low Detail, Nominal, Plan, Plan High Detail, Plan Low Detail, Reflected Screened, Threshold Plan, Threshold Plan Screened, and Threshold Symbol Plan. Each of these representations will turn on or off the display of specific components of the object. A door when shown for a floor plan could use Plan set, which uses the Plan and Threshold Plan representations as shown in Figure 1.27. Certain components of the door are visible in the Plan representation, while others common to other display representations are not included. Each of these representations will turn on or off the display of specific components of the object.

A door object consists of the following components when the **Plan** display representation is used: Door panel, Frame, Stop, Swing, Direction, Door Panel Above Cut Plane, Frame Above Cut Plane, Stop Above Cut Plane, Swing Above Cut Plane, Door Panel Below Cut Plane, Frame Below Cut Plane, Stop Below Cut Plane, and Swing Below Cut Plane. When the Nominal representation is assigned, the Stop, Stop Above Cut Plane, and Stop Below Cut Plane components are not included, and a simple representation of the door is displayed. When you select a display configuration for a viewport, a set of display representations is applied to the objects in the viewport. The door shown in Figure 1.28 is displayed according to the display configuration for the viewport, which ultimately defines the display representations to apply to the door. The door shown at left in Figure 1.28 is displayed by hidden lines as defined by the Screened display representation. The Reflected Screened display configuration assigns the Screened display representation to objects in the viewport.

CONTENTS OF DISPLAY REPRESENTATION SETS

A display representation set groups display representations to obtain an appropriate object representation for a drawing. The display representation sets are shown across the top of Figure 1.27. Notice in Figure 1.27 the Plan and Threshold display representations are checked for the Plan set. You can view the available sets by expanding the Sets folder of the **Display Manager** as shown in Figure 1.29. In this figure the left pane list the 13 sets. The Plan set is selected in the left pane and the **Display Representation Control** tab is selected in the right pane. The column in the right pane

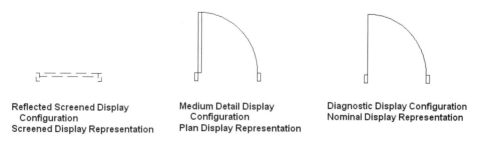

Reflected Screened Display
Configuration
Screened Display Representation

Medium Detail Display
Configuration
Plan Display Representation

Diagnostic Display Configuration
Nominal Display Representation

Figure 1.28 *Display configurations applied to the display of a door*

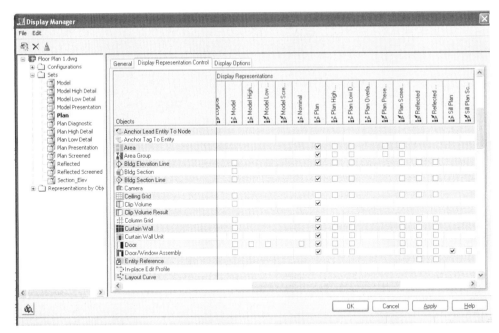

Figure 1.29 Sets folder of the Display Manager

lists all objects and the top row lists all display representations available. If a box is checked, that object and its display representation is defined for the **Plan** set.

A display representation set therefore controls the visibility of the components according to the needs of the drawing.

APPLYING DISPLAY CONFIGURATIONS TO VIEWPORTS

The third level of display control uses a display configuration to apply one or more display representation sets to a viewport. The current display configuration assigned to a viewport is listed in the lower right corner of the Drawing Window status bar as shown in Figure 1.2. You can select from this menu to change the display configuration for the viewport. If you create additional layouts, you should set a display configuration current for the new layout.

However, to view the content of a display configuration, select **Format>Display Manager**, and expand the Configurations tab of the left pane as shown in Figure 1.30. The right pane of this tab describes the display configuration.

The purpose of each tab in the right pane is described below.

> **General** – Allows you to edit the name and description of the display configuration.

Introduction to Architectural Desktop 35

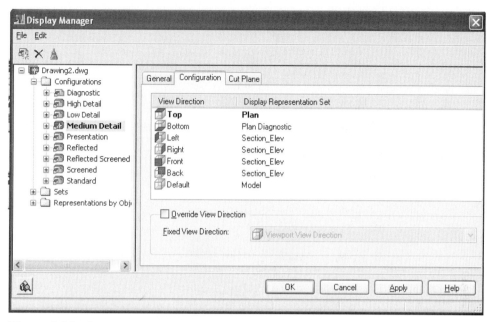

Figure 1.30 *Configuration of the Medium Detail display configuration*

Configuration – Allows you to assign display representation sets to define the display representation set for a given view direction. In Figure 1.30, the **Configuration** tab is shown for the **Medium Detail** display configuration, and the view directions with their respective display representation sets are shown in the right pane. Therefore, when this display configuration is used and the objects are viewed from the top, the Plan display representation set will be applied. The **Section_Elev** display representation set will be applied when the design is viewed from the left, right, front, or back. The **Model** display representation set will be used when a pictorial view is applied.

Cut Plane – This tab allows you to change the cutting plane height and the range of display. You can modify the cut plane height to view each floor of a model.

You can change the display representation set assigned to a view direction by selecting the **View Direction** of the **Configuration** tab, clicking on the current name in the **Display Representation Set** column, and selecting from the list as shown in Figure 1.31.

The display configuration shown in Figure 1.32 is **View Directional Dependent**. A **View Directional Dependent** display configuration will control the display of an object dependent upon the view. You can check the **Override View Direction** check box of the **Configuration** tab and assign a **Fixed View Direction** for the viewport by selecting from the **Fixed View Direction** list as shown in Figure 1.32.

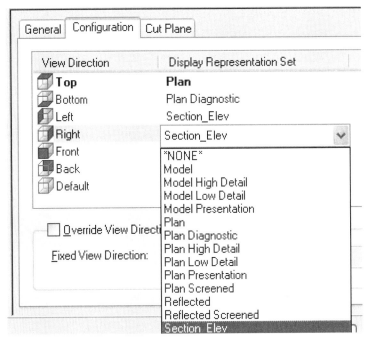

Figure 1.31 *Selecting the display representation set for a view direction*

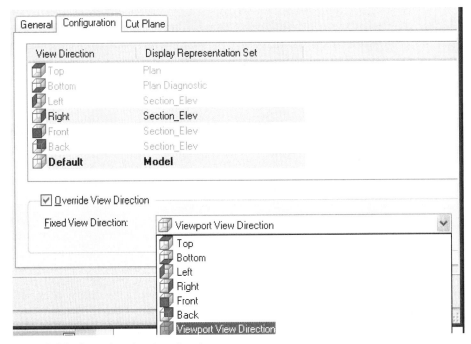

Figure 1.32 *Overriding the view direction*

Although display configurations can be edited, the display configurations available in the template are usually adequate for creating architectural working drawings. In essence, the beginner does not need to edit the display configurations if the Aec Model (Imperial Ctb) or Aec Model (Imperial Stb) template is used.

 Note: Display configurations can cause objects to be visible in one viewport but not necessarily in all viewports. In later tutorials, such display settings will become obvious, and later chapters will discuss how display configurations can be used to create the correct display of objects for working drawings.

VIEWING OBJECTS WITH THE OBJECT VIEWER

One or more objects can be selected and viewed in the Object Viewer window. The Object Viewer allows you to view the objects using display configurations from various view directions. The Object Viewer shown in Figure 1.33 includes a listing of the commands on the toolbars of the Object Viewer. Access the Object Viewer as shown in Table 1.14.

| Shortcut menu | Select an AEC object, right-click, and choose Object Viewer |

Table 1.14 *Object Viewer command access*

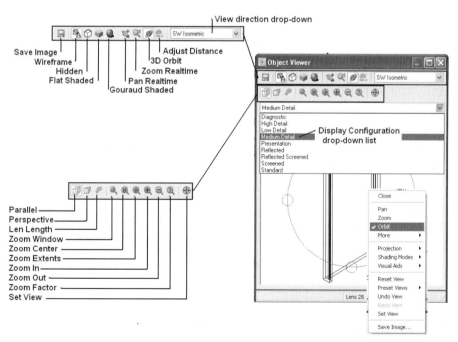

Figure 1.33 *Object Viewer display of a door*

When the Object Viewer opens, the object is displayed in the window. The shortcut menu in the graphics window of the viewer includes an option for perspective projection. The viewer image can also be saved as a *.png, *.jpg, *.bmp, or *tif file.

The Object Viewer allows you to edit the type of view displayed by selecting buttons in the toolbars. The flyout in the upper right corner allows you to change the view to one of the following views: Top, Bottom, Front, Back, Left, Right, Southwest, Southeast, Northeast, Northwest. The drop-down list below the view buttons allows you to select a display configuration for viewing the object, as shown in Figure 1.33. You can view the object using **Pan**, **Zoom**, or **Orbit** when you move your mouse over the Object Viewer. The view option can be selected from the **Object Viewer** toolbar. The **Orbit** option is the default viewing method when the Object Viewer opens. If the Orbit mode is selected, you can view the object from virtually any rotation and tilt angle. If you move your mouse over the view and right-click, a shortcut menu allows you to choose other viewing options, as shown in Figure 1.33.

The menu options of this shortcut menu are summarized below:

> **Close** – Closes the Object Viewer.
>
> **Pan** – Executes **Realtime Pan** on the image in the Object Viewer.
>
> **Zoom** – Executes **Realtime Zoom** on the image in the Object Viewer.
>
> **Orbit** – Allows you to click with the mouse to rotate and tilt the object. The Orbit mode of viewing the object is toggled ON by default.
>
> **More** – Allows you to select **Zoom Window**, **Zoom Extents**, **Zoom Center**, **Zoom In**, **Zoom Out**.
>
> **Projection** – Allows you to view the object in orthographic views (parallel) or in perspective.
>
> **Shading Modes** – Includes the following display methods: Wireframe, Hidden, Flat Shaded, Gouraud Shaded.
>
> **Visual Aids** – Allows you to select a Compass or Grid to aid in visualizing the object.
>
> **Reset View** – Returns view of the object to the original view.
>
> **Preset Views** – Allows you to select one of the following views of the object: Top, Bottom, Left, Right, Front, Back, Southwest, Southeast, Northeast, Northwest.
>
> **Undo View** – Performs an **Undo** to the view changes in the Object Viewer.
>
> **Redo View** – Returns to view prior to executing the Undo.
>
> **Set View** – Not active for the Object Viewer. If an object in the drawing is selected and then the Object Viewer is opened, the **Set View** option allows

the view in the drawing area to be equal to that displayed in the Object Viewer.

Save Image – Opens the **Save Image File** dialog box, which allows the file to be save in the following formats: *.png, *.jpg, *.bmp, or *.tif file.

EDITING OBJECTS WITH OBJECT DISPLAY

The display of each object can also be controlled as an exception to that defined in the style or display representation. This level of individual control is accessed by **ObjectDisplay**. Access **ObjectDisplay** as shown in Table 1.15.

Shortcut menu	Select an object, right-click, and choose Edit Object Display
Command prompt	OBJECTDISPLAY, select an object.

Table 1.15 *ObjectDisplay command access*

When you select an object, right-click, and select **Edit Object Display**, the **Object Display** dialog box opens as shown in Figure 1.34. The **Object Display** dialog box consists of **Materials**, **Display Properties**, and **General Properties** tabs. The **Materials** tab shown in Figure 1.34 is for a door, and you can add or edit the materials assigned for the components. The tabs of the **Object Display** dialog box are described below.

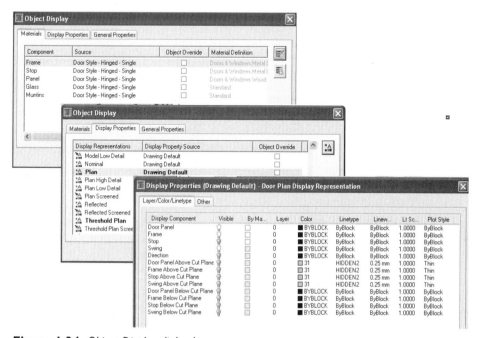

Figure 1.34 *Object Display dialog box*

MATERIALS

Component – The **Component** column lists the components for the object based upon the current display configuration.

Source – The **Source** column lists the style source in which the material is assigned.

Object Override – The **Object Override** check box, if checked, allows you to change the material assigned from the default assigned in the style. When this check box is clear, the materials defined in style are listed in the **Material Definition** column and may not be edited.

Material Definition – The **Material Definition** column lists the current material defined in the style or new materials defined by an override.

Edit Material – The **Edit Material** button, when **Object Override** is checked, allows you to change the material definition and the display representation used for the material.

Add New Material – The **Add New Material** button, when **Object Override** is checked, allows you to add a new material name.

DISPLAY PROPERTIES

The **Display Properties** tab is used to view the current properties and toggle ON an override for the display. The display representations of a door object are shown in Figure 1.34.

The options of the **Display Properties** tab are listed below.

Display Representations – The **Display Representations** column lists the available display representations for the object. The display representations shown in bold are the current display representations.

Display Property Source – The **Display Property Source** column lists the source for the display representation. If a door is being edited, this column would state either **Drawing Default** or **Door Override** as the source.

Object Override – The Object Override check box when clear uses the default display representation. If this box is checked, an exception or override can be defined for the display representation. When you check this box, the **Display Properties** dialog box opens.

Properties – The **Properties** button opens the **Display Properties** dialog box shown in Figure 1.34, which lists how the components of the object will be displayed. Figure 1.34 shows the display properties of a door for the Plan display representation.

This dialog box for a door consists of the **Layer/Color/Linetype** and **Other** tabs. Additional tabs may be included depending upon the object; however, the **Layer/Color/Linetype** tab is standard because it allows you to turn off/on the display of individual components of an object. The options of the **Layer/Color/Linetype** tab are described below.

> **Display Component** – The **Display Component** column lists the components of the objects for the display representation.
>
> **Visible** – The **Visible** light bulb icon is used to turn ON/OFF the display of the component.
>
> **By Material** – The **By Material** check box, when checked, sets the display of the component by the material definition.
>
> **Layer** – The **Layer** column lists the layer of the component.
>
> **Color** – The **Color** column lists the color assigned to the component. Click in this column to open the **Select Color** dialog box and edit the color for the component.
>
> **Linetype** – The **Linetype** column lists the linetype defined for the component. Click in this column to open the **Select Linetype** dialog box and edit the linetype for the component.
>
> **Lineweight** – The **Lineweight** column lists the lineweight for the component. Click in this column to open the **Lineweight** dialog box and edit the lineweight.
>
> **LtScale** – The **LtScale** column lists the linetype scale for the component. Click in this column and overtype a value for the linetype scale.
>
> **Plot Style** – The **Plot Style** column lists the plot style for the component. Click in this column to open the **Select Plot Style** dialog box and edit the plot style.

GENERAL PROPERTIES

The **General Properties** tab allows you to change basic properties as described below.

> **Color** – The default bylayer color is displayed; select the **Color** button and choose from the **Select Color** dialog box to change color assignment.
>
> **Layer** – The default layer assigned by the layer key is displayed; select the **Layer** button and choose from the **Select Layer** dialog box to change the layer assignment.
>
> **Linetype** – The default bylayer linetype is displayed; select the **Linetype** button and choose from the **Select Linetype** dialog box to change the linetype.
>
> **Linetype Scale** – The default linetype scale is listed; edit the field by overtyping the new linetype scale.
>
> **Lineweight** – The default **ByLayer** lineweight is displayed; select the **Lineweight** button and choose from the **Select Lineweight** dialog box to change lineweight.

Plot Style – The default ByLayer plot style is displayed; select the **Plot Style** button and choose from the **Select Plot Style** dialog box to change styles.

UPDATING THE DISPLAY OF OBJECTS

Objects may be incorrectly displayed during a drawing session. If objects viewed in a pictorial view are displayed in two dimensions, the display can be corrected by the **ObjRelUpdate** command. Access the **ObjRelUpdate** command as shown in Table 1.16.

Command prompt	OBJRELUPDATE
Toolbar	Select Regenerate Model from the Standard toolbar

Table 1.16 *Regenerate Model command access*

When you select the **ObjRelUpdate** command, respond in the command line as shown below:

    Command: _objrelupdate

    Select entities to update or RETURN for all.

    Select objects: ENTER *(Press* ENTER *to select all objects.)*

    7 found

    6 were not in current space.

    Select objects: *(Press* ENTER *to end the command.)*

Regenerating the model may be necessary when objects are displayed from reference files.

Tip: Select **2D Wireframe** from the **Navigation t**oolbar to remove the display distortion resulting from applying shade and hidden views.

STOP Do Tutorial 1.2, "Viewing the Display Representation of Objects and Snapshots," at the end of the chapter.

CREATING PROJECTS

Architectural Desktop includes a **Project Browser** and **Project Navigator** to create and define projects. Drawings for single story buildings can easily be developed without the use of projects. However, the drawings for multi-story buildings should be developed using a project because project tools provide organization and collaboration of the levels and divisions within the building. Project tools allow you to insert one or more levels to create a digital model of the building, which can be used for the development of elevations and section drawings. The project tools also include techniques for placing view drawings onto sheets for plotting or printing. These model management tools automate the design, data, and model presentation. The **Project Navigator** allows you to manage the use of external reference files and extract data across several drawings. Drawings are developed and classified as Elements, Constructs, Views, and Sheets in the **Project Navigator**. Drawings can be developed specifically for a project, or existing drawings can be converted to components of a project.

USING THE PROJECT BROWSER TO CREATE THE PROJECT

The first step in developing a model is to create a project name in the **Project Browser**. Select **File>Project Browser** from the menu bar (see Table 1.17 for command access) to open the **Project Browser**. A drawing must be open in the workspace to open the **Project Browser**. This drawing does not have to be a part of the project.

Menu bar	File>Project Browser
Command prompt	PROJECTBROWSER
Toolbar	Select Launch Project Browser on the Project Navigator palette

Table 1.17 *Project Browser access*

When you open the **Project Browser**, the left pane lists the current project and other projects in the Project Selector, as shown in Figure 1.35. The right pane consists of a project bulletin board connected to an HTML page. The contents of the bulletin board are specified in the HTML file, and its location is specified in the **Add Project** dialog box. The current path to the project can be identified if you select the drop-down list of the Project Selector. The default path to the My Projects folder shown in Figure 1.35 is *C:\Documents and Settings\Login\My Documents\Autodesk\My Projects*. Therefore you should edit this path to **your student folder** from the drop-down list before creating a project.

To create a new project, select the **New Project** button located in the lower left corner of the **Project Browser** to open the **Add Project** dialog box shown in Figure 1.36. To create a new project you must enter a name; a project number is optional. The project number and name appear in the Project Header shown in Figure 1.35. When you create a project, it becomes the current project. Only one project may be current for a given drawing. The remainder of the **Add Project** dialog box allows you to add additional details regarding the project. The paths to the default templates for the Sheet,

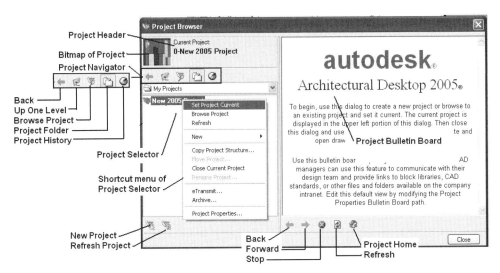

Figure 1.35 Navigating the Project Browser

View, Construct, and Element drawings are listed in the **Add Project** dialog box. The default templates are used for creating new drawings within the project.

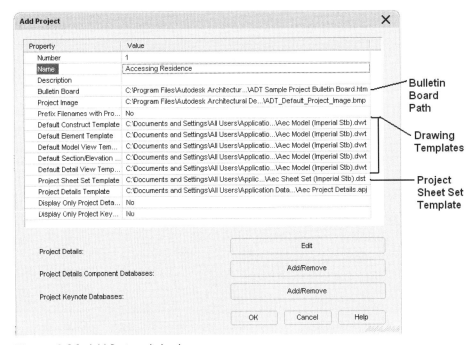

Figure 1.36 Add Project dialog box

The content of the **Add Project** dialog box can be revised by selecting the name of the project in the left pane of the **Project Browser**, right-clicking, and choosing **Project Properties**. The **Project Properties** command opens the **Modify Project** dialog box, which includes the content of the **Add Project** dialog box.

The default templates and project settings are specified in the **AEC Project Defaults** tab of the **Options** dialog box shown in Figure 1.37. The **AEC Project Defaults** tab specifies the default locations for the project, construct drawing, element drawing, model view, section/elevation, detail view, sheet set template, project details template, project bulletin board, and default project image. The tab also includes a **Create** button, which opens the **Create Sheet Set-Begin** dialog box to create custom sheet sets.

To set existing projects current, double-click on the project name in the Project Selector of the **Project Browser**. Additional tools for archiving, copying, closing, and browsing a project are included in the shortcut menu of the selected project in the Project Selector of the **Project Browser** as shown in Figure 1.35. The Archiving option allows you to create a zip or executable file that includes all support files for a project. The Archive option should be used to move your project to another computer. The **Browse Project** button of the **Project Browser** toolbar shown in Figure 1.35 can be selected to locate projects within the computer.

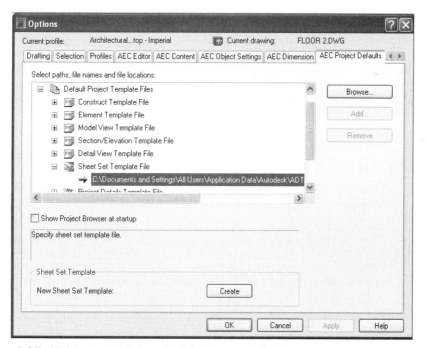

Figure 1.37 *AEC Project Defaults tab of the Options dialog box*

When a project is created, a directory is created within the folder listed in the Project Selector. The project folder includes a project file (*.apj extension), sheet set file (*.dst extension), and the following subdirectories: **Construct**, **Element**, **View**, and **Sheets**. The project file (*.apj) is created, which contains such data as project name, number, levels, templates, and project detail text information. Drawing files created in the project are saved to one of the subdirectories of the project directory. The folders are empty when the project is created; however, as the project develops, the drawing files of the project are automatically placed in the respective folders. An XML file is created with each drawing file created or used in the project. The XML file retains such supporting data as level, divisions, viewport scale, and display configuration. Therefore, the XML file is essential to the function of the drawing file in the project. Do **not** delete or move files within the project with the Windows Explorer.

CREATING PROJECT CONTENT WITH THE PROJECT NAVIGATOR

The **Project Browser** is used to create the project name and data supporting the project. However, the **Project Navigator** is used to create the drawings of the project. You can return to the **Project Browser** from the **Project Navigator** by selecting the **Launch Project Browser** button in the lower left corner of the **Project Navigator** as shown in Figure 1.38. Access the **Project Navigator** by selecting **Window>Project Navigator Palette** from the menu bar; refer to Table 1.18 for command access.

Command prompt	PROJECTNAVIGATORTOGGLE or CTRL +5
Menu bar	Select Window>Project Navigator Palette
Toolbar	Select Project Navigator for the Navigation toolbar

Table 1.18 *Project Navigator Palette access*

The **Project Navigator** includes four tabs: **Project, Constructs, Views**, and **Sheets**, as shown in Figure 1.38 and described below.

> **Project** – The **Project** tab defines the divisions and levels of the project. The floor elevations are specified in the level definition as shown at right.
>
> **Constructs** – Drawings classified as elements and constructs are the building blocks of floor plans and the model view drawings are created from this tab.
>
> **Views** – Views are created in this tab by attaching as construct drawings as reference files. The View drawing identifies level, divisions, and construct drawings.
>
> **Sheets** – Drawings classified as Sheets are created on this tab. The sheet drawing includes the titleblock and border for plotting. View drawings are attached to the sheet to create the document for plotting.

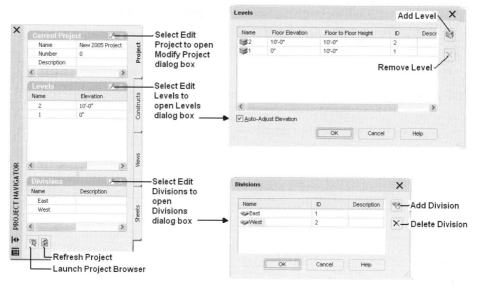

Figure 1.38 *Creating levels and divisions in the Project Navigator*

USING PROJECT COMPONENTS TO DEVELOP CONSTRUCTION DOCUMENTS

Project content is created to ultimately create the working drawings. The development of a floor plan is described in Figure 1.39. As shown at left, the first step is to define the levels and divisions of a project. After specifying levels, you create a floor plan as a construct drawing, which includes the walls, doors, and windows. The construct drawing is a unique drawing developed as part of the project and associated with a level and division. This construct is attached to a view drawing. The view completes the plan when dimensions, notes, and schedules are added. A view drawing is created for each floor plan. The floor plan view drawing is then referenced into a sheet drawing to place the drawing in a viewport within the sheet. The sheet drawing includes a titleblock and border as defined in the sheet template. The **Sheets** tab of the **Project Navigator** provides the functions of the Sheet Set Manager of AutoCAD 2005 drawings.

Elevations and sections are created by attaching each floor plan or other construct drawings to a model view drawing as shown in Figure 1.39. The model drawing includes all levels and divisions of the project. Elevation and section drawings are developed from the model as view drawings using the elevation and section tools of the Callout tool palette. The elevation and section view drawings are placed on a sheet created in the **Sheets** tab of the **Project Navigator**.

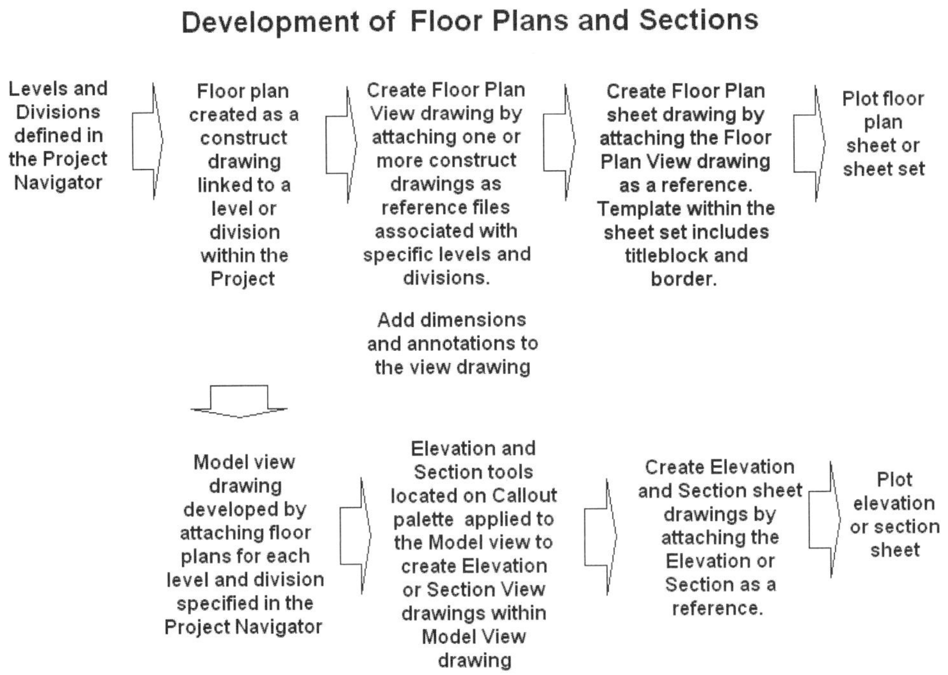

Development of Floor Plans and Sections

Levels and Divisions defined in the Project Navigator → **Floor plan created as a construct drawing linked to a level or division within the Project** → **Create Floor Plan View drawing by attaching one or more construct drawings as reference files associated with specific levels and divisions.** → **Create Floor Plan sheet drawing by attaching the Floor Plan View drawing as a reference. Template within the sheet set includes titleblock and border.** → **Plot floor plan sheet or sheet set**

Add dimensions and annotations to the view drawing

Model view drawing developed by attaching floor plans for each level and division specified in the Project Navigator → **Elevation and Section tools located on Callout palette applied to the Model view to create Elevation or Section View drawings within Model View drawing** → **Create Elevation and Section sheet drawings by attaching the Elevation or Section as a reference.** → **Plot elevation or section sheet**

Figure 1.39 *Development of plans and elevations using a project*

Defining Divisions and Levels of the Project

The **Project** tab lists the project data, levels, and divisions of the project. The level definition defines the floor height and floor elevations, while divisions specify wings or horizontal divisions of the building.

The **Edit Project**, **Edit Levels**, and **Edit Divisions** buttons located at right on the **Project Navigator** allow you to specify the content of the project. Selecting the **Edit Project** button of the **Project** tab opens the **Modify Project** dialog box, which allows you to edit the project data from the **Project Navigator**.

Levels of the project are defined by selecting the **Edit Levels** button, shown in Figure 1.38, to open the **Levels** dialog box. The **Levels** dialog box shown in Figure 1.38 includes level name, floor elevation, and floor to floor height. Each name of the level must be unique. Click in the columns of the **Levels** dialog box to edit the floor elevation and floor to floor height. Drawing files created for a level are inserted as reference files based on the floor elevation value.

Additional levels can be created by selecting the **Add Level** button, or you can select the name of a level, right-click, and choose **Add Level Above** or **Add Level Below**. The shortcut menu of a level name also includes **Copy Level** and **Copy Level and**

Contents options for creating a new level from an existing level. The **Copy Level and Contents** option can be used to copy the first floor plan of a house to the second level; the second level can then be edited to retain the bearing walls common to each level. The **Auto-Adjust Elevation** check box located in the lower left corner, if checked, will assign the floor elevation of the new level equal to the sum of the previous floor level and floor elevation. When you add a level between levels, the floor elevation of the levels above the new level is automatically adjusted. The **Floor Elevation** and **Floor to Floor Height** values may be edited at any point during the development of drawings for the project.

The **Divisions** section includes the **Edit Divisions** button, shown in Figure 1.38, which opens the **Divisions** dialog box to create divisions. Select the **Add Division** button to open the **Divisions** dialog box and define division names and descriptions. To edit the **Divisions** dialog box, click in the **Name**, **ID**, and **Description** columns.

Creating Constructs

The **Constructs** tab of the **Project Navigator** shown in Figure 1.40 allows you to create construct and element drawings. The **Constructs** tab includes a toolbar, shown at the bottom of the tab in Figure 1.40, which includes the commands for creating categories, elements, and constructs. The concept of the project is to create small unique drawings, as elements and constructs, which can be inserted as references into view drawings to create the desired plan or model. Elements and constructs are defined below:

> **Elements** – Drawings created as elements are the simplest components of the building that are repeated in the design. An element could be a drawing of a bathroom layout that is repeated several times in the final drawing. Element drawings are attached to construct drawings.
>
> **Constructs** – Drawings of unique content in the building are constructs. A construct can be a floor plan that consists of Architectural Desktop Objects. Element drawings can be referenced into construct drawings.

Prior to creating a new drawing, you can create a category within the **Construct** tab to organize the drawings. Categories can be created to classify the drawings as architectural, plans, details, or site work. Element or Construct categories are created by selecting the **Element** or **Construct** folder, and then selecting the **Add Category** command from the toolbar of the **Constructs** tab shown in Figure 1.40. To create a new drawing in a category, select the category name, and then select the **Add Element** or the **Add Construct** command from the toolbar on the **Constructs** tab.

When you select the **Add Element** command, the **Add Element** dialog box opens as shown in Figure 1.41. The template for the element drawing can be selected by selecting in the **Drawing Template** edit field as shown in Figure 1.41.

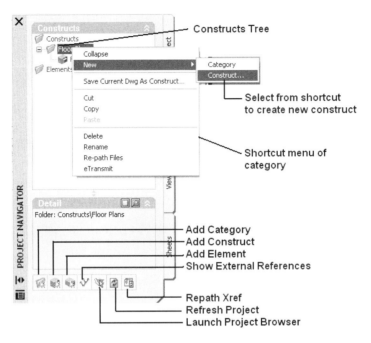

Figure 1.40 *Constructs tab*

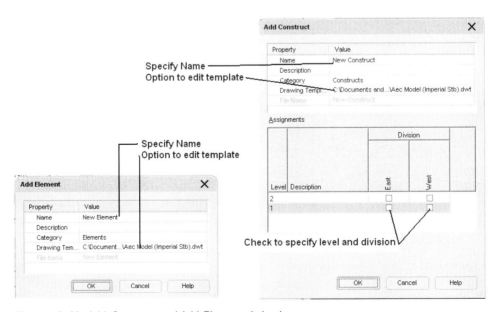

Figure 1.41 *Add Construct and Add Element dialog boxes*

When you select the **Add Construct** command, the **Add Construct** dialog box opens as shown at right in Figure 1.41. The **Add Construct** dialog box lists the levels and divisions defined for the project. If you select the check box to assign a division to a level, the construct drawing for the division will be inserted as an external reference file for the level specified in the view drawings. The drawing is attached and recorded in the project data files. You can double-click on a drawing listed in the **Construct** tab to open the new construct drawing.

Elements are the simplest drawing unit of the model. The drawing classified as an element can be inserted in several locations of the construct drawing. When you add an element drawing, its name will appear in the tree of the **Project Navigator** (see Figure 1.42). To open a new element, select the name of the element in the tree of the **Project Navigator**, right-click, and choose **Open**. Content of the element drawing is created by adding content from Autodesk Architectural Desktop. Close and save the element drawing before attaching the drawing to a construct drawing.

The content of other files can also be attached as a block, reference file, or overlay file to the construct file. The procedure to attach an element drawing to a construct file is shown below:

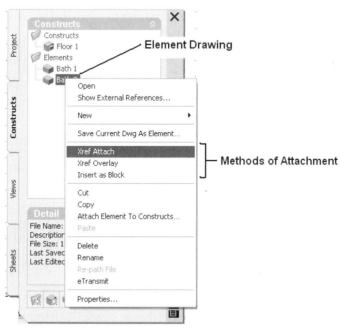

Figure 1.42 *Adding an element to a construct drawing*

STEPS TO ATTACHING AN ELEMENT DRAWING TO A CONSTRUCT FILE

1. Open the **Project Navigator**, select the construct drawing, right-click, and choose **Open**
2. Select the name of the element to insert in the **Project Navigator** tree, right-click, and choose one of the following: **Xref Attach**, **Xref Overlay**, or **Insert as Block**, as shown in Figure 1.42.

Drawings that were created as elements can be converted to constructs by selecting the name of the element in the **Project Navigator** and dragging it to the Construct category of the tree of the **Constructs** tab. When you drag an element to the Construct category, the **Add Construct** dialog box opens, allowing you to specify the levels and divisions of the construct. Constructs can also be converted to element drawings by dragging them into the Element category. When construct drawings are converted to elements, the level and division data is lost.

Existing drawings can be converted to element or construct drawings in the **Project Navigator**. To convert a drawing, open the drawing and the **Project Navigator**. Select the category of the construct or element in the **Project Navigator**, right-click, and choose **Save Current Dwg As Element** or **Save Current Dwg As Construct**. The **Add Element** or **Add Construct** dialog box opens, allowing you to specify the name and other data. When existing drawings are converted to element or construct drawings, an accompanying XML file is also created. The existing drawing is not copied, moved, or linked when converted to an element or construct. Therefore an existing floor plan can be converted to a construct drawing to take advantage of the **Project Navigator** management tools.

Creating Views

Views are created to specify which elements and constructs are used to define the drawing outcome of the model. View drawings can be created as General Views, Section/Elevation Views or Detail Views as shown in the shortcut menu in Figure 1.43. You could create a general view that includes only constructs of the first floor or a composite view of all floors. A view is created by selecting the **Views** tab as shown in Figure 1.43. A category can be created by selecting the **Add Category** button from the toolbar at the bottom of the **Views** tab. You can create a view by selecting the **Add View** command from the toolbar at the bottom of the tab. When you select **Add View**, the **Add View** dialog box opens, which prompts you to specify the type as General, Section/Elevation, or Detail. After you specify the General type, the **Add General View** dialog box opens as shown in Figure 1.44. The name, description, and template are defined on the **General** page.

The **Context** page, shown in Figure 1.44, lists the levels and divisions of the project. The check placed in the box for Division 1 of Level 1 indicates that the content specified for level **1** and division **1** will be included in the view. If you are creating a model

Introduction to Architectural Desktop 53

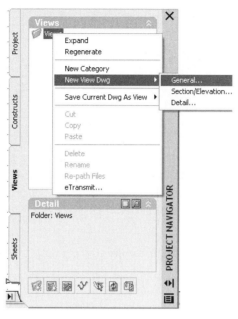

Figure 1.43 *Creating a view drawing*

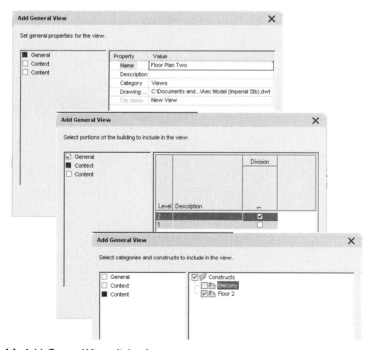

Figure 1.44 *Add General View dialog box*

view of the composite components, all levels should be checked for all divisions. At least one division must be checked for a division to create a view.

The constructs included in the view are specified in the **Content** page. The **Content** page shown in Figure 1.44 includes two construct drawings. The check to the left of the construct indicates that the construct will be displayed. The **Balcony element** check box is clear; therefore it will not be displayed.

Creating Model Space Views Within a View Drawing

A model space view can be created within an existing view drawing. The saved view specifies the display configuration, layer snapshot, and scale. A model space view could be created for detail and plan view drawings.

STEPS TO CREATING A MODEL SPACE VIEW WITHIN A VIEW DRAWING

1. Open a view drawing in the **Project Navigator**.
2. Select the view drawing name in the **View** tab of the **Project Navigator**, right-click, and choose **New Model Space View** from the shortcut menu to open the **Add Model Space View** dialog box shown in Figure 1.45.
3. Edit the name, display configuration, layer snapshot, and scale edit fields. Choose the **Define View Window** button to return to the workspace and select from point **p1** to **p2** shown in Figure 1.45 to specify the content of the view. Select **OK** to close the **Add Model Space View** dialog box.
4. Verify that the model space view is listed below the view drawing in the **Project Navigator**, as shown at right in Figure 1.45.

Creating Sheets

Prior to creating a new sheet, select the **Sheet Set** title in the **Project Navigator**, right-click, and choose **Properties** to open the **Sheet Set Properties** dialog box. Select the **Browse** button shown at right in Figure 1.46 to open the **Select Layout as Sheet Template** dialog box. The templates can be specified in the **Properties** of each subset of the sheet set in the **Project Navigator**. To create a new sheet, select a sheet category, right-click, and choose **New>Sheet** from the shortcut menu. When you create a new sheet, the **New Sheet** dialog box opens, as shown in Figure 1.47. Type the sheet number and name in the **New Sheet** dialog box and select **OK** to close the **New Sheet** dialog box.

After a sheet has been created, double-click on the sheet name to open the drawing. Select the **Views** tab of the **Project Navigator**, select a view, and drag the view attached to the pointer into the workspace of the sheet as shown in Figure 1.48. A phantom view of the view will be attached to the pointer as you drag the view onto the sheet. You can right-click to edit the scale of view prior to specifying the location of the view.

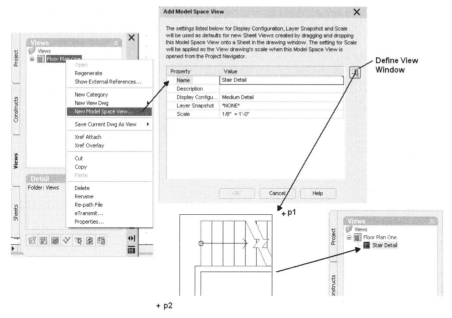

Figure 1.45 *Creating model space views in a view drawing.*

KEEPING THE PROJECT NAVIGATOR UP TO DATE

During the process of defining the components of a project in a networked environment, the project tree of the **Project Navigator** may not be displayed up to date.

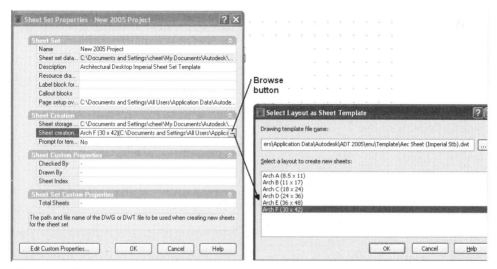

Figure 1.46 *Selecting a sheet template*

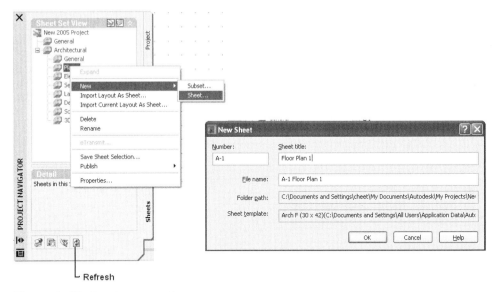

Figure 1.47 *Creating a new sheet*

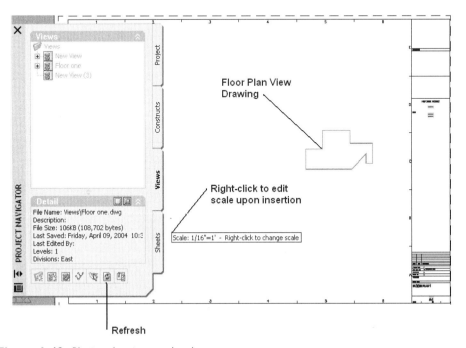

Figure 1.48 *Placing the view on the sheet*

Select the **Refresh Project** command located on the bottom toolbar of each tab to recalculate the settings of the project and display the current drawing with the latest information. The **Repathing Xref** command, also located on the toolbar of each tab, recalculates the path to the files referenced in the project. The **Repathing Xref** command should be selected when the names or locations of project files have changed. Because projects are developed as reference files, select the **Xref Manager** icon of the Drawing Window status bar to reload the files after changes have been made.

Tip: To move a project to another computer, select **Archive** from the shortcut menu of the **Project Browser**. When you archive a project, you can create a zip or executable file, which can be expanded in a different location. When you open the project on the new computer, an AutoCAD warning box will prompt you to repath the project. Select **Yes** to repath the project to the drive and directory information of the different computer.

STOP. Do Tutorial 1.3, "Creating A Project," at the end of the chapter.

SUMMARY

1. Architectural Desktop objects are created by selecting tools from the tool palette.
2. The shortcut menu of a selected Architectural Desktop object includes the commands for changing the object.
3. Doors, windows, and stairs are inserted in the drawing as objects.
4. Objects inserted in a drawing interact with other objects to reduce the need to edit the wall.
5. Toolbars for the Architectural Desktop are loaded from the ADT menu group through the **TOOLBAR** command.
6. Profiles can be created to retain the display of selected toolbars.
7. The resource files for Architectural Desktop are located in the *C:\Documents and Settings\All Users\Application Data\Autodesk\ADT 2005\enu* directory.
8. Commands can be selected from tool palettes. The **Content Browser** is used to create and customize tool palettes.
9. The name of the project associated with a drawing is listed on the Drawing Window status bar.
10. Projects are created using the **Project Browser**, whereas the **Project Navigator** is used to create drawings assigned to the project.
11. Units, scale, layer, and display settings are defined in the **Drawing Setup** dialog box.
12. Layer snapshots and groups are created in the **Layer Manager**.
13. The display of objects can be refreshed using the **ObjRelUpdate** command.

14. Templates for creating a floor plan using imperial units are Aec Model (Imperial Stb) and Aec Model (Imperial Ctb).
15. A display representation specifies which components of an object are displayed.
16. A display representation set identifies one or more display representations that are used to display objects.
17. Display configurations identify one or more display representation sets to govern the display of objects in a viewport.
18. Display configurations include specifying which display representation to be used according to the specified view direction in a viewport.
19. The **Edit Object Display** command allows you to override the display as defined by the display representation or object style.
20. One or more objects can be viewed in the Object Viewer.

REVIEW QUESTIONS

1. Doors and windows are inserted in a drawing as _____.
2. The menu group that includes the Architectural Desktop toolbars is _____.
3. The major categories of objects in the **Style Manager** are _____, _____, _____.
4. Resource files that include additional styles are located in the _____ directory.
5. Tool palettes are organized into the following palette sets: _____, _____, _____.
6. Several entities that have associative interactions when inserted are _____.
7. The _____ menu is located in the lower left corner of the Drawing Window status bar.
8. Describe the procedure to display hidden files in Windows XP.
9. Obtain a perspective view or one or more objects by selecting _____.
10. The _____ drawing is attached to a view drawing in a project.
11. What procedure can be used to modify a door or window using shortcut menus?
12. Profiles are created and set current through the _____ command.
13. The command used to set the layer standard and scale of a drawing is _____.
14. View drawings are attached to a _____ drawing, which is designed for plotting.

15. The command used to edit a display representation set is the _____.
16. The limits of the Architectural building model and view (Imperial-ctb) template are set to _____.
17. The annotation plot size is _____ as defined in the Architectural building model and view (Imperial-ctb) or Architectural building model and view (Imperial-stb) templates.
18. Options of a command used in placing an object are set in the _____ dialog box.
19. The feature of the **Layer Manager** that allows you to save layer states and recall them later in the drawing is _____.
20. The directories created for the components of a project are _____, _____, _____, _____.

TUTORIAL 1.1 CREATING A NEW DRAWING

1. Launch Architectural Desktop by selecting the shortcut from the desktop.
2. Choose **QNew** from the **Standard** toolbar.
3. Select the Model tab.
4. If tool palettes are not displayed, choose **Window>Tool Palettes** from the menu bar.
5. Right-click over the title bar of tool palette, and choose **View Options** to open the **View Options** dialog box. Select **Icon with Text** view style and choose **All Palettes** from the **Apply to** drop-down list. Select **OK** to close the **View Options** dialog box.
6. Right-click over the title bar of the tool palette, and choose **New Palette**. Overtype **Chapter 1** as the name of the new palette. Select the **Content Browser** from the **Navigation** toolbar. Open the Stock Tool Catalog>Drafting Tools category. Select and drag the i-drop of the Hatch and Gradient tool onto the new palette.
7. If the Properties palette is not displayed, choose **Window>Properties Palette**.
8. Choose the **Drawing** menu button on the left of the Drawing Window status bar, and select **Drawing Setup** from the menu.
9. Choose the **Units** tab, and verify the units are set to inches.
10. Choose the **Scale** tab, and change the scale to 1/4"=1'-0", verify that Annotation plot size = 3/32", and check **Automatically Create Dimstyle Override**.

11. Choose the **Layering** tab, and verify that the **Layer Standards/Key File to Auto-Import** is AecLayerStd.dwg and the **Layer Key Style** is AIA (256)*current drawing as shown in Figure 1.24.

12. Choose the **Display** tab and verify that the **Drawing Default Display Configuration** is set to **Medium Detail**.

13. Choose **OK** to dismiss the **Drawing Setup** dialog box.

14. Choose **File>SaveAs** from the menu bar.

15. Edit the **Save in** list to your student directory, type **Lab 1-1** in the **File name** edit field and choose **Save** to close the **Save Drawing As** dialog box.

TUTORIAL 1.2 VIEWING THE DISPLAY REPRESENTATION OF OBJECTS AND SNAPSHOTS

1. Open *Ex 1-2* from your *ADT Student\ADT Tutor\Ch1* directory.

2. Choose **File>SaveAs** from the menu bar and save the drawing as **Lab 1-2** in your student directory.

3. Choose the Model tab, and verify that the display configuration is set to **Medium Detail** in the Drawing Window status bar. Select the Work layout tab, double-click in the right viewport, and verify that the display configuration is **Medium Detail**.

4. To view the building with the reflected display configuration, verify that the right viewport is current. Choose the **Reflected** display configuration from the **Display Configuration** flyout menu of the Drawing Window status bar. The viewport should display the ceiling grid as shown in Figure 1T.1.

5. To view the settings of the Reflected display configuration, choose **Format>Display Manager** from the menu bar. Choose the (+) sign of the Configurations folder. Note that the Reflected display configuration is shown bold.

6. Choose the **Reflected** display configuration in the left pane.

7. Choose the **Configuration** tab of the right pane to display the View Directions as shown in Figure 1T.2.

8. Choose the (+) of the Sets folder in the left pane, and choose Reflected in the left pane. Choose the **Display Representation Control** tab in the right pane.

9. Scroll down the list of objects to **Ceiling Grid** and scroll to the right to the **Reflected** display representation as shown in Figure 1T.3. (If the scroll bar is not displayed in the dialog box, stretch the dialog box down by moving the pointer to lower edge of the dialog box; when the two-way arrow is displayed, click and drag down.)

Introduction to Architectural Desktop 61

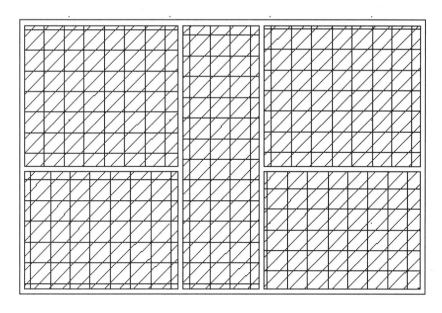

Figure 1T.1 *Reflected display configuration of Work layout tab*

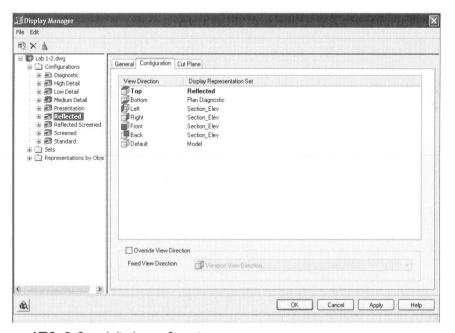

Figure 1T.2 *Reflected display configuration*

10. Expand the **Representations by Object** in the left pane by choosing the (+), and then choose the **Ceiling Grid** object to view the display representations of the Ceiling Grid object as shown in Figure 1T.4.

11. Choose the **OK** button to close the **Display Manager**.

12. Choose the **Layer Properties Manager** command from the **Layer Properties** toolbar, choose the **A-Area-Spce** layer, and choose the **Sunshine** icon to freeze this layer in all viewports. Freezing the A-Area-Spce layer will remove the display of the hatch lines representing the space in plan view.

13. To create a layer snapshot, choose the **Snapshots** button on the **Layer Manager** toolbar, shown in Figure 1T.5, select the **New** button, and type **Floor** as the name of the layer snapshot. Choose **OK** to close the **Snapshot** dialog box. Choose **OK** to dismiss the **Snapshots** dialog box.

14. Choose **OK** to close the **Layer Manager** and display the drawing without the spaces.

15. Choose the **Layer Properties Manager** command from **Layer Properties** toolbar; choose the **A-Area-Spce** layer, and thaw the layer in all viewports by selecting the **Snowflake** icon.

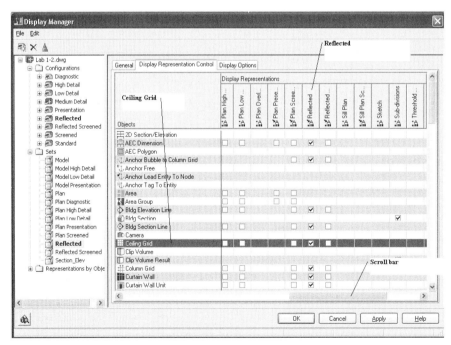

Figure 1T.3 *Reflected representation set includes Reflected representation for Ceiling Grid*

Introduction to Architectural Desktop 63

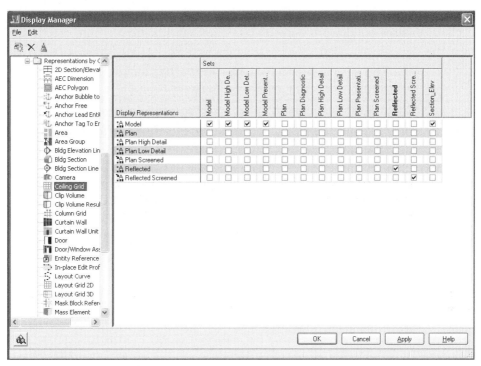

Figure 1T.4 *Ceiling Grid display representations*

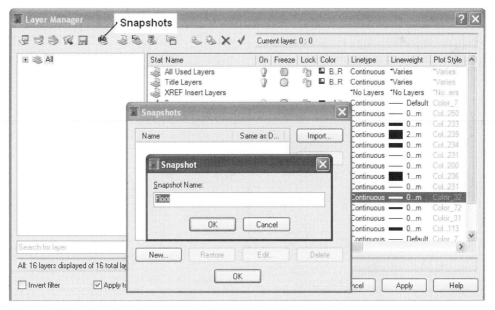

Figure 1T.5 *Creating a layer snapshot in the Layer Manager*

16. Choose **OK** to dismiss the **Layer Manager** and view the drawing with the spaces.

17. To view the results of the layer snapshot, choose the **Layer Properties Manager** command from the **Layer Properties** toolbar, and choose the **Snapshots** button from the **Layer Manager** toolbar.

18. Choose **Floor** in the snapshot list, and choose the **Restore** button. Choose **OK** to dismiss the **Snapshots** dialog box. Choose **OK** to close the **Layer Manager**.

19. To view the building in the Object Viewer, verify that the right viewport of the Work layout tab is active, select the entire building using a window selection set, select from **p1** to **p2**, as shown in Figure 1T.6, right-click, and choose **Object Viewer** from the shortcut menu.

20. Choose **SW Isometric** from the **View** flyout of the Object Viewer. Select **Maximize** on the Object Viewer title bar.

21. Verify that **Reflected** is selected from the **Configuration** list of the Object Viewer.

22. Move the cursor to the graphics screen of the Object Viewer, right-click, and choose **Projection>Perspective** from the shortcut menu.

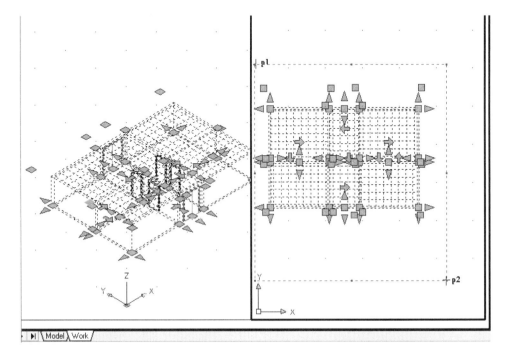

Figure 1T.6 *Building selected for view in Object Viewer*

23. Move the cursor to the graphic screen of the Object Viewer, right-click, and choose **Shading Modes>Flat Shaded**; the building should be displayed in the viewer as shown in Figure 1T.7.

24. Close the Object Viewer, and then choose **File>Save** from the menu bar.

TUTORIAL 1.3 CREATING A PROJECT

1. Choose **QNew** from the **Standard** toolbar. Choose **File>Project Browser** from the menu bar to open the **Project Browser**.

2. Choose the drop-down list of the Project Selector, and set the path to your *ADT Student\ADT Tutor\Ch 1*. Note below the path to your student folder for future reference.

3. Select the **New Project** button shown in Figure 1T.8 of the **Project Browser** to open the **Add Project** dialog box.

4. Type **1** in the **Number** edit field and **Ex 1-3** in the **Name** edit field. Select the **Browse** button to edit the Bulletin Board file path to your *student directory\Ch1\Access Project Bulletin Board.htm* as shown in Figure 1T.8. Select **OK** to close the **Add Project** dialog box.

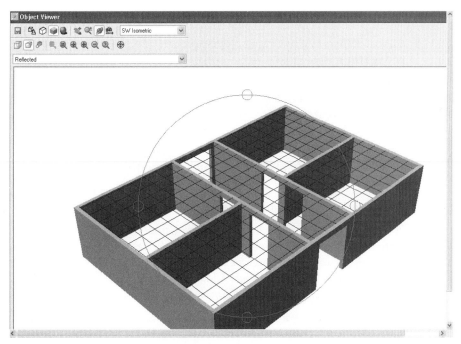

Figure 1T.7 *Object Viewer display of the building*

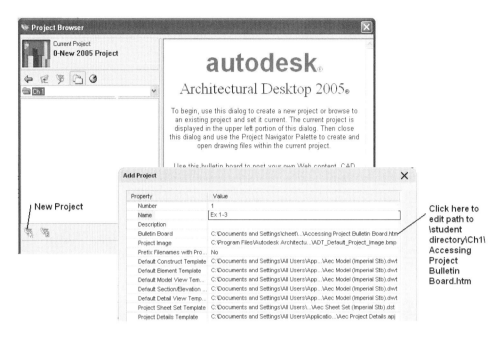

Figure 1T.8 *Creating a project in the Project Browser*

5. Choose **Close** to close the **Project Browser** and open the **Project Navigator** in the workspace.
6. Select the **Project** tab of the **Project Navigator**, and choose **Edit Levels**. Edit level 1 as follows: Name = **1**, Floor Elevation = **0**, Floor to Floor Height = **9'**. Clear the **Auto-Adjust Elevation** check box.
7. To create an additional level, choose **Add Level** button as shown in Figure 1T.9. Edit the properties of level 2 as follows: Name = **2**, Floor Elevation = **9'**, Floor to Floor Height = **8'**. Choose **OK** to close the **Levels** dialog box.
8. If the AutoCAD warning dialog box opens, choose **Yes** to regenerate the views and dismiss the AutoCAD warning dialog box.
9. To create a construct drawing, choose the **Constructs** tab. Choose the Constructs category, right-click, and choose **New>Construct** to open the **Add Construct** dialog box. Edit the **Add Construct** dialog box as follows: Name = **Floor 1** and check level 1 for division 1 as shown in Figure 1T.10. Choose **OK** to close the **Add Construct** dialog box.
10. To create a view drawing, choose the **Views** tab. Choose the Views category, right-click, and choose **New View Dwg>General** to open the **Add General View** dialog box.

Introduction to Architectural Desktop   67

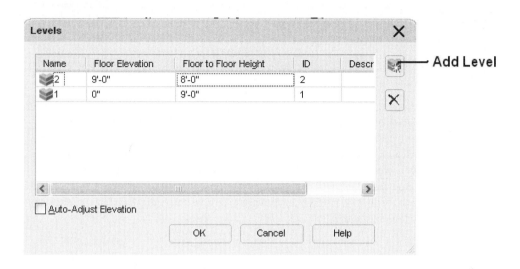

Figure 1T.9 *Defining levels for the project*

11. Type **Floor Plan 1** in the **Name** edit field of the **General** page of the **Add General View** dialog box as shown in Figure 1T.11. Choose the **Next** but-

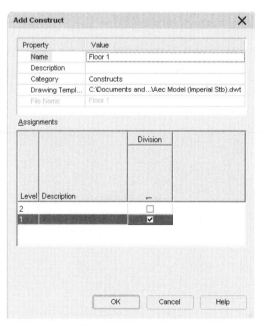

Figure 1T.10 *Creating a construct drawing*

ton to open the **Context** page: check Level 1 for Division 1. Click **Next** to open the **Content** page. Verify that **Floor 1** construct is checked for this construct. Choose **Finish** to close the **Add General View** dialog box. Double-click on the icon of **Floor Plan 1** to open the drawing. Verify that View: Floor Plan 1 is the drawing title listed in the Drawing Window status bar.

12. To create a sheet, choose the **Sheets** tab of the **Project Navigator**. Choose the Plans category, right-click, and choose **New>Sheet** to open the **New Sheet** dialog box. Type **A-2** in the **Number** field and **Level 1 Floor Plan** in the **Sheet title** edit field as shown in Figure 1T.12. Select **OK** to close the **New Sheet** dialog box. Double-click on the icon for A-2 Level 1 Floor Plan to open the sheet. The titleblock and border should be displayed at right. The Sheet A-2 Level 1 Floor Plan drawing title should be displayed in the Drawing Window status bar.

13. Select the **Project** tab of the **Project Navigator**. Select the **Launch Project Browser** button to open the **Project Browser**. Choose **Ex 1-3** project in the Project Selector, right-click, and choose **Close Current Project**. Save and close all drawings.

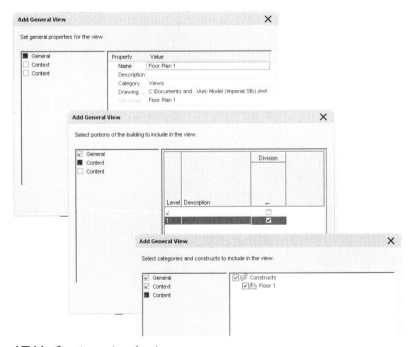

Figure 1T.11 *Creating a view drawing*

Figure 1T.12 *Creating a sheet drawing*

CHAPTER 2

Creating Floor Plans

INTRODUCTION

This chapter introduces the commands to draw and modify walls using the Standard wall style. The focus of this chapter is the placement of the Standard wall and the tools available for precise placement of walls for the creation of a floor plan. Options of the **WallAdd** command and how to use properties for editing will be explained. The tools to edit walls with precision such as **Justification**, **WallOffsetCopy**, **WallOffsetMove**, **Offset Set From**, **Cleanup 'L'**, **Cleanup 'T'**, grips, and editing wall defect markers will be presented.

OBJECTIVES

After completing this chapter, you will be able to

- Draw straight or curved walls using the **WallAdd** command
- Set justification, offset, width, and height properties of walls
- Use the Close and Ortho Close options to close a polygon shape
- Use the Offset option to shift the handle for the wall relative to the justification line
- Use grips to identify the justification of a wall
- Use object snaps to place walls with precision
- Set the cleanup radius to avoid wall error markers
- Use the **WallOffsetCopy**, **WallOffsetMove**, and **Offset Set From** commands for adding walls
- Merge walls with **Join** and **Wall Merge** commands

- Change display configurations to display wall graph lines
- Reverse the direction of a wall using the **Reverse** command
- Use **Add Selected** to execute the **WallAdd** command
- Use the Match option to match wall properties

DRAWING WALLS

Drawing walls is the beginning point for the creation of floor plans for residential or commercial structures. Walls created in Architectural Desktop are objects that have three-dimensional properties of width, height, and length. Therefore the wall can be viewed in plan to create a floor plan or viewed as an isometric to create a pictorial drawing of the building (see Figure 2.1). Walls placed in a drawing have attributes defined by their style. The simplest style of wall is the Standard wall style that consists of only two wall lines. The style of the wall is listed in the Basic section of the Properties palette shown in Figure 2.3. Additional styles of walls can be created that consist of several wall lines to represent more complex walls. The use of Wall styles to create more complex walls will be discussed in the Chapter 3.

Because walls are inserted as three-dimensional objects using object technology, the placement of doors and windows in a wall will cause the wall to be edited correctly to

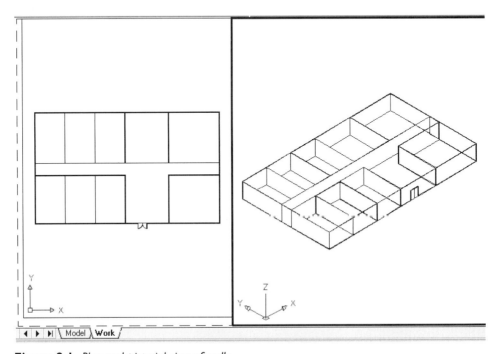

Figure 2.1 *Plan and pictorial view of walls*

display the other objects in the wall. Wall anchors exist that lock other objects such as doors to the anchor. Therefore you can select a door, select its grips, and slide it down the wall without moving the door out of the wall. When an object is created by Architectural Desktop, it is automatically placed on the correct layer. The entities created to represent a wall are automatically placed on the A-Wall layer if the AIA (256 color) Layer Key Style is used. The default color for the A-Wall layer is color 113 (a hue of green).

CREATING WALLS WITH THE WALLADD COMMAND

The **Wall** tool (**WallAdd** command) is accessed from the Design tool palette as shown in Figure 2.2. Access the **WallAdd** command as shown in Table 2.1.

Command prompt	WALLADD
Tool palette	Select the Wall tool from the Design palette as shown in Figure 2.2

Table 2.1 *WallAdd command access*

When you select the **Wall** tool from the Design tool palette, the Properties palette opens as shown in Figure 2.3, and you are prompted for a start point to begin the wall. Prior to specifying the start point you should review the settings in the Properties palette. The settings of a wall are retained from the last instance of the command. The start point and end point of the wall can be specified just as you would specify the length of an AutoCAD line by using absolute coordinates, direct distance entry, and relative coordinates or by clicking with the mouse. Once the wall is begun, the dis-

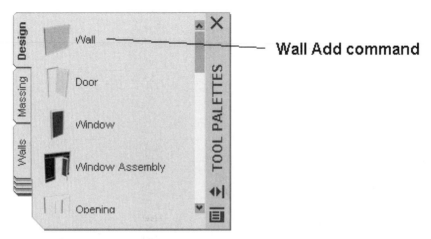

Figure 2.2 *Wall tool on the Design palette*

tance from the start point to the current cursor location is dynamically displayed in the **Length** edit field of the Properties palette as shown in Figure 2.3.

 Tip: To display the length of the wall dynamically in the Auto Track tooltip as shown in Figure 2.3, toggle ON POLAR on the status bar and toggle ON **Display Auto Track tooltip** in the **Drafting** tab of the **Options** dialog box.

USING ADD SELECTED

The **WallAddSelected** command allows you to select a wall, right-click, and choose **Add Selected**, and then begin drawing a wall. The properties of the new wall are inherited from the selected wall. Access **WallAddSelected** as shown in Table 2.2.

Command prompt	WALLADDSELECTED
Shortcut menu	Select a wall, right-click, and choose Add Selected

Table 2.2 *Add Selected command access*

When you select a wall and then right-click and choose **Add Selected** from the shortcut menu, you are prompted to begin drawing the wall as shown in the following command line sequence. **Add Selected** can be applied to insert other AEC objects. Therefore, if you select a door, right-click, and choose **Add Selected**, the **DoorAdd** command is active.

Command: WallAddSelected

Start point or [STyle/Group/WIdth/Height/OFfset/Justify/Match/Arc]:

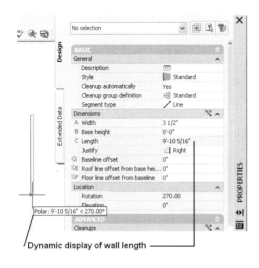
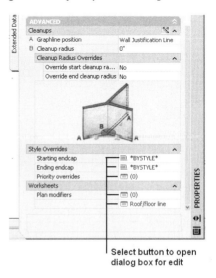

Figure 2.3 *Properties palette*

In a similar manner, if you are currently using the **WallAdd** command drawing walls, you can match the properties of an existing wall by the Match option of the **WallAdd** command. The Match option is selected by typing **M** in the command line. The following command line sequence demonstrates the use of the Match option.

>Command: WallAdd
>
>Start point or
>[STyle/Group/WIdth/Height/OFfset/Justify/Match/Arc]:
>**m** ENTER
>
>Select a wall to match: *(Select an existing wall to match its properties.)*
>
>Match [Style/Group/Width/Height/Justify] <All>:
>ENTER *(Press ENTER to select all properties.)*
>
>Start point or
>[STyle/Group/WIdth/Height/OFfset/Justify/Match/Arc]:
>*(Specify the start of the wall.)*
>
>End point or
>[STyle/Group/WIdth/Height/OFfset/Justify/Match/Arc]:
>*(Specify the end of the wall.)*
>
>End point or
>[STyle/Group/WIdth/Height/OFfset/Justify/Match/Arc/Undo]: ENTER
>*(Press ENTER to end the command.)*

The Match option listed in the command line allows you to select another wall in the drawing and match its properties. When this option is executed, you can match all the properties of another wall or select any one of the following properties: STyle, Group, WIdth, Height, and Justify. To select only a certain property, enter the option in the command line. The command line is used to select the options of Match.

CHANGING WALLS WITH THE APPLY TOOL PROPERTIES TO COMMAND

The **Apply Tool Properties to** command is accessed from the shortcut menu of the **Wall** tool on the Design tool palette. This command allows you to change one or more walls to the settings as specified in the Properties palette or to convert linework to walls. Access the **Apply Tool Properties to** command as shown in Table 2.3.

| Shortcut menu | Select the Wall tool of the Design palette, right-click, and choose Apply Tool Properties to>Wall or Apply Tool Properties to>Linework |

Table 2.3 *Apply Tool Properties To command access*

When you select the **Wall** tool, right-click, and choose **Apply Tool Properties to**, a submenu appears, and you can select either Wall or Linework options. You can select one or more walls, and then edit the Properties palette; the selected walls will match the properties specified in the Properties palette. If you select the linework option and then select lines, arc, or circles, the linework will be converted to walls.

CONTENTS OF THE WALL PROPERTIES PALETTE

The Properties palette consists of Basic and Advanced sections to describe a wall being placed. The Properties palette is changed by clicking in the edit field to the right of the property, and then typing a new value or selecting from the options of the drop-down list. The options of the Properties palette shown in Figure 2.3 are as follows:

Design – Basic

General

Description – Opens a **Description** dialog box to type a description.

Style – Displays the current wall style; the drop-down list includes all styles that have been used in the drawing.

Cleanup Automatically – Allows you to select Yes/No to automatically clean up the intersection of walls.

Cleanup group definition – Lists the current cleanup group name. The drop-down list includes cleanup styles that have been inserted in the drawing.

Segment type – This toggle includes the Line option to draw straight walls and the Arc option to draw curved walls.

Dimensions

A-Width – Allows you to set the width of the wall.

B-Base Height – Specifies the height of the wall.

C-Length – The dynamic display of the length of wall. This field can be used to change the length of the wall. The start point for the wall remains anchored, while the second point location moves. The angle of the wall remains the same while edited. The start and end points remain anchored for an arc wall segment. When an arc wall segment is placed, the length and radius are displayed.

Justify – Displays the current justification. The drop-down list displays the following justifications for the wall: Baseline, Center, Left, and Right.

Baseline Offset – Allows you to set a distance to offset the handle for placing the wall.

E-Roof line offset from base height – Allows you to extend the wall above or below the base height. Additional editing of roofline can be set in the **Roof and Floor Line** dialog box as shown in Figure 2.7.

F-Floor line offset from baseline – This option allows you to extend the wall above or below the bottom of the wall baseline. Additional editing of the roof line can be set in the **Roof and Floor Line** dialog box as shown in Figure 2.7.

Location

Rotation – Specifies the rotation of the wall relative to 3:00 o'clock position.

Elevation – Specifies the elevation of the wall relative to the World Coordinate System

Additional information – The **Additional information** button, displayed when you edit existing walls, opens the **Location** dialog box, which describes the location in three-dimensional space relative to the World Coordinate System or the User Coordinate System as shown in Figure 2.4. The Normal values, Z=1, X=0, and Y=0, will extrude the walls perpendicular to the XY plane as shown in Figure 2.5. Normal values can be edited as shown in Figure 2.5 to change the direction of wall extrusion.

Design – Advanced

Cleanups

A-Graphline position – Allows you to specify the graphline positions as either Wall Justification Line or Wall Center Line.

B-Cleanup radius – The radius specified to clean up wall intersections.

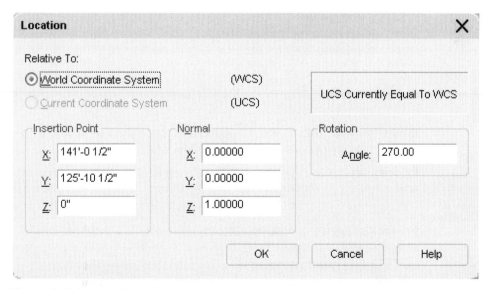

Figure 2.4 *Location dialog box*

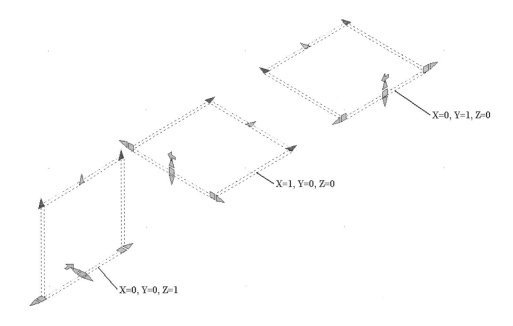

Figure 2.5 *Walls extruded in the direction of their normal values*

Cleanup Radius Overrides
> **Override start cleanup radius** – The Yes option allows you to specify a value for the cleanup radius at the start of the wall.
>
> **Override end cleanup radius** – The Yes option allows you to specify in the Properties palette a value for the cleanup radius at the end of the wall.

Style Overrides
> **Starting Endcap** – Allows you to specify an endcap style as an exception to the wall style definition at the start point of the wall.
>
> **Ending endcap** – Allows you to specify an endcap style as an exception to the wall style definition at the end point of the wall.
>
> **Priority overrides** – Allows you to specify a wall component priority value as an exception to those defined in the wall style.

Worksheets
> **Plan Modifiers** – The **Plan Modifiers** button opens the **Wall Modifiers** dialog box, shown in Figure 2.6, to assign a wall modifier style to the wall. Wall modifiers are representations of materials used to finish a wall such as pilasters or other projections not common throughout the length of the wall. Chapter 3 will include a discussion regarding wall modifiers.

Creating Floor Plans 79

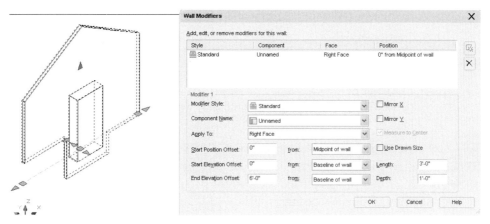

Figure 2.6 *Wall Modifiers dialog box*

Roof/Floor Line – Opens the **Roof and Floor Line** dialog box, shown in Figure 2.7, to create extensions to represent steps in the rooflines and floor lines of the wall. This feature will extend the wall to the roof for the gable wall.

The Properties palette can be edited prior to or after the start point for a wall is selected. Although most properties can be set in the Properties palette, you can also select options from the command line. You can select options by entering the characters capitalized in the options list of the command prompt. Therefore, to select the OFfset option, you would enter **OF**. The options of the **WallAdd** command are shown in the sequence below:

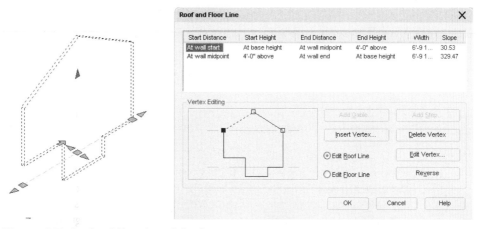

Figure 2.7 *Roof and Floor Line dialog box*

Command: WallAdd

Start point or
[STyle/Group/WIdth/Height/OFfset/Justify/Match/Arc]:
(Specify the start point.)

End point or
[STyle/Group/WIdth/Height/OFfset/Justify/Match/Arc]:
(Specify the end point.)

SETTING JUSTIFICATION FOR PRECISION DRAWING

When you draw a wall, a graph line is drawn in which the wall components follow as shown in Figure 2.8. This graph line becomes the handle for drawing the wall; it begins and ends at the coordinates specified for the wall. The grips of the wall are located on the graph line. The graph line is similar to an axis or control line in which the wall is displayed. The graph line is displayed dynamically as you draw the wall; however, the line is not visible after the wall segment is drawn.

When you select an existing wall, the graph line is displayed as well as the grips of the wall as shown in Figure 2.8. The graph line is shown as a dashed line. A directional arrow is shown indicating the direction the wall was drawn. The walls in Figure 2.8 were drawn from left to right. The justification of a wall sets the location of the graph line relative to the wall width. A wall drawn with right justification has its handle on

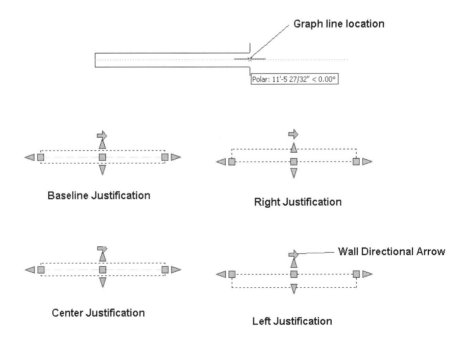

Figure 2.8 *Wall justifications*

the right when you orient yourself at the start point and face toward the end of the wall.

Types of justifications available are Right, Left, Center, and Baseline. The figures shown below illustrate each type of justification and its handle position.

The wall drawn in Figure 2.8 used the Standard wall style. This wall style positions the graph line in the center of the wall width for both Center and Baseline justifications. The Standard style consists of only two lines, which are offset one-half of the wall width value on each side of the Baseline. The value entered for the wall width is balanced one-half on each side of the Baseline. Therefore, the Standard wall style is unique because both its Baseline and Center justifications result in locating the wall handle in the center of the wall width.

Complex walls created through the use of wall styles can consist of wall lines offset a specific distance from the Baseline. Complex walls created using wall styles will be presented in Chapter 3.

CREATING STRAIGHT WALLS

Walls can be drawn straight or curved. The Segment property located in the General category of the Basic section of the Properties palette is used to toggle to Line or Arc segments. The Line toggle is the default wall type that is retained if arc walls were previously drawn. Therefore segment type will be **Line** unless you toggle **Arc** in the **Segment** edit field of the Basic section of the Properties palette. To draw a straight wall, select the **Wall** tool from the Design palette, edit the Properties palette, and then specify the beginning and ending points of the wall. The following command sequence demonstrates how to draw the straight wall shown in Figure 2.9 using Direct Distance Entry.

(Verify that ORTHO is toggled ON on the status bar.)

*(Select the **Wall** tool from the Design palette, and verify Style = Standard in the Properties palette.)*

Command: WallAdd

Start point or
[STyle/Group/WIdth/Height/OFfset/Justify/Match/Arc]:
(Click a location for p1 in the graphics screen.)

(Move the mouse to the right.)

End point or
[STyle/Group/WIdth/Height/OFfset/Justify/Match/Arc]: **20'** ENTER *(p2 is specified.)*

End point or
[STyle/Group/WIdth/Height/OFfset/Justify/Match/Arc/Undo]: ENTER
(Press ENTER to end the command.)

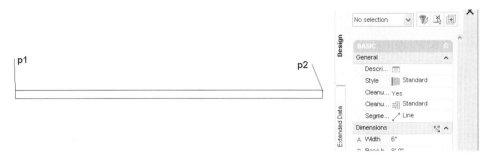

Figure 2.9 *Creating a straight wall*

The endpoint can be established by any of the following four coordinate entry methods: absolute coordinate, relative coordinate, polar coordinate, or direct distance entry. In the example above, direct distance entry was used to draw the wall 20' long.

Drawing walls in Architectural Desktop is very similar to drawing a line in AutoCAD. Notice in the command sequence that the Undo option is added to command options once the endpoint of the wall is established. If you are drawing a series of wall segments, you can undo each segment by continuing to select the Undo option in the **WallAdd** command prompt. To exit the command, press ENTER; the null response will cause the **WallAdd** command to be terminated.

DESIGNING CURVED WALLS

Curved walls are drawn by selecting **Arc** in the **Segment** edit field of the Basic section in the Properties palette. Three points establish the curve of the wall: the beginning point, a point along the curve, and the end point of the curved wall. The following command sequence illustrates the command line prompts and entries to create the curved wall shown in Figure 2.10.

*(Select the **Wall** tool from the Design palette and select the Arc Segment type in the Properties palette.)*

Command: WallAdd

Start point or
[STyle/Group/WIdth/Height/OFfset/Justify/Match/Arc]:

Start point or
[STyle/Group/WIdth/Height/OFfset/Justify/Match/Line]:
(Click a location in the graphics screen to locate p1 as shown in Figure 2.10.)

Mid point or
[STyle/Group/WIdth/Height/OFfset/Justify/Match/Line]:

@-5',3'6" *(Specifies p2 as shown in Figure 2.10.)*

End point or
[STyle/Group/WIdth/Height/OFfset/Justify/Match/Line]:
@-10',-3'6" *(Specifies p3 as shown in Figure 2.10.)*

Mid point or
[STyle/Group/WIdth/Height/OFfset/Justify/Match/Line/Undo]: ENTER

(Ends the WallAdd command.)

In the command sequence above, the start point is established by clicking a random location. The Mid point prompt allows you to establish a point (**p2**) for the curved wall to pass through. This point and all succeeding points can be established with any of the AutoCAD coordinate entry methods. Once the midpoint is established, additional points can be specified to continue drawing curved wall segments. Drawing a curved wall is similar to drawing an arc through three points with the Arc command of AutoCAD.

Using the Close Option

The **WallAdd** command includes options of Close and Ortho close. You can access these options from the command line or from shortcut menu displayed while drawing the wall. The Close option is used to draw a final wall segment from the endpoint of the last wall segment back to the start point of the first wall segment of the wall segment series as shown in Figure 2.11. This option is similar to the Close option when AutoCAD lines are drawn. The Close option of the AutoCAD **Line** command allows you to draw several line segments and enter a C for Close to close the last line segment to the beginning of the first line segment. The Close option for the wall becomes active after two wall segments have been drawn. Select the Close option by typing C in the command line or choosing **Close** from the shortcut menu displayed after two segments are drawn. The following command sequence was used to close the last wall segment.

Start point or
[STyle/Group/WIdth/Height/OFfset/Justify/Match/Arc]:
(Select p1 as shown in Figure 2.11.)

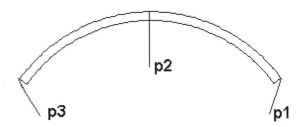

Figure 2.10 *Points specified for an arc wall*

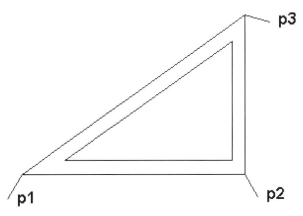

Figure 2.11 *Polygon drawn using the Close option*

End point or
[STyle/Group/WIdth/Height/OFfset/Justify/Match/Arc]:
(Select p2 as shown in Figure 2.11.)

End point or
[STyle/Group/WIdth/Height/OFfset/Justify/Match/Arc/Undo]: *(Select p3 as shown in Figure 2.11.)*

End point or

[STyle/Group/WIdth/Height/OFfset/Justify/Match/Arc/Undo/Close/ORtho close]: *(Right-click and choose Close from the shortcut menu.)*

(Final wall segment drawn to p1.)

 STOP. Do Tutorial 2.1, "Drawing Walls and Using Close," at the end of the chapter.

Using the Ortho Close Option

The Ortho close option draws the final wall segment of a series of walls to connect at an angle of 90 degrees to the beginning of the first wall segment. Execute this option by selecting **Ortho close** from the shortcut menu or typing **OR** in the command line. If ORTHO is toggled ON on the status bar, this option allows you to draw a rectangle after establishing the length and width distances by the first two wall segments. The following command sequence was used to create a rectangle using the Ortho close option. Note in the following command sequence, when you type **OR** you are prompted to select a "Point on wall in direction of close." This point sets the direction in which to complete the wall segment.

Start point or
[STyle/Group/WIdth/Height/OFfset/Justify/Match/Arc]:
(Select p1 as shown in Figure 2.12.)

End point or
[STyle/Group/WIdth/Height/OFfset/Justify/Match/Arc]:
(Select p2 as shown in Figure 2.12.)

End point or
[STyle/Group/WIdth/Height/OFfset/Justify/Match/Arc/Undo]: *(Select p3 as shown in Figure 2.12.)*

End point or
STyle/Group/WIdth/Height/OFfset/Justify/Match/Arc/Undo/
Close/Ortho close]: *(Right-click and choose **Ortho close** from the shortcut menu.)* OR

Point on wall in direction of close: *(Select a location near p4 as shown in Figure 2.12.)*

The Ortho close option is not active until two wall segments are drawn. In the command sequence above, when Ortho close is selected, you are prompted "Point on wall in direction of close:" In response to this prompt set ON AutoCAD ORTHO (F8 toggle) and select a point on the drawing area perpendicular to the last wall segment and in the direction of the beginning of the first wall segment. Because ORTHO is on, the next wall segment will be drawn perpendicular to the last and the Ortho close option will close the polygon to form a rectangle.

If the first two wall segments were not perpendicular to each other when Ortho close was used, the final wall segment will still close to the first wall segment at an angle of 90 degrees. In each example shown in Figure 2.13, walls were drawn through points **p1**, **p2**, and **p3**. Point **p4** was specified in response to the prompt "Point on wall in direction of close:" to close the polygon.

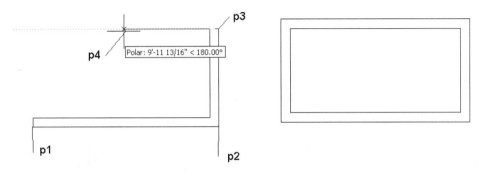

Figure 2.12 *Identifying a point for Ortho close to complete the rectangle*

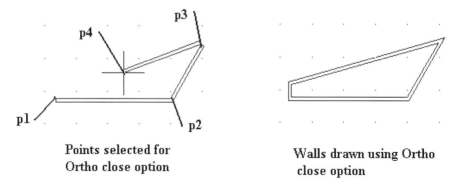

Points selected for Ortho close option

Walls drawn using Ortho close option

Figure 2.13 *Ortho close used to complete the walls of a polygon*

 STOP. Do Tutorial 2.2, "Drawing Curved Walls and Using Ortho Close," at the end of the chapter.

USING THE OFFSET OPTION TO MOVE THE WALL HANDLE

The **Baseline Offset** edit field of the Properties palette allows you to enter a distance to offset the handle for the wall relative to the graph line of the wall. If you enter an offset of 2" using a right wall justification, the handle for the wall will be located 2" beyond the right wall face, as shown on left in Figure 2.14. The wall shown is 8" wide and is being drawn from left to right.

If a negative value is entered for the offset distance, the handle for the same wall as shown in Figure 2.14 will move up 2" from the right face, as shown on the right in Figure 2.14. The ability to shift the handle relative to the justification of a wall is useful in creating a foundation plan. In residential construction drawings, the masonry walls are offset 3/4" from a wood stud partition line, as shown in Figure 2.15. Therefore, the outside of stud surface of the first floor is offset 3/4" from the outside of the concrete masonry unit of the foundation, as shown in Figure 2.15. Based on this arrangement of the wall elements, the masonry foundation wall would be created with a 7 5/8" wall width and an offset of 3/4". These settings would allow you to trace the first floor plan geometry to create a foundation plan using the Endpoint Object Snap mode.

In commercial construction the offset distance could be set according to the distance a wall is offset from the column centerline. This would allow you to trace the column line to draw a wall offset the desired distance between the column line and the wall.

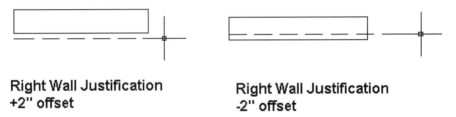

Figure 2.14 *Wall created with positive and negative baseline offset*

SETTING UP FOR DRAWING AND EDITING WALLS

When a new file is created to draw a floor plan, the Aec Model (Imperial-Ctb) or the Aec Model (Imperial-Stb) template is used. The file can be created as a new Construct of a project or created independent of a project. The Work layout tab is best suited to drawing the floor plan because it consists of two viewports with the **Medium Detail** display configuration. The **Node** object snap mode should be selected when you draw walls or manipulate mass elements. The **Node** object snap allows you to snap to the justification line of walls.

Representation of walls should be kept simple, and the width of a wall component set equal to the wall's structural component. If you are drawing 2 × 4 wood frame walls, the wall width for the wall should be set to 3 1/2", the actual width of a 2 × 4. Setting the width to the actual size of the structural component allows you to snap to the lines representing the wall for specific location dimensions. Wall components could

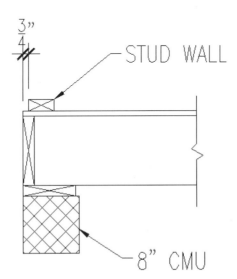

Figure 2.15 *Wall detail of wood wall and masonry foundation construction*

be added to represent the sheathing or gypsum board in a wall style. More complex walls such as brick veneer frame or brick veneer with concrete masonry units should be drawn using Wall styles. Complex Wall styles, discussed in Chapter 3, consist of multiple wall components that can be turned ON or OFF for plan representation.

USING THE NODE OBJECT SNAP

The **Node** object snap mode can be used to snap to the endpoint and midpoint of the wall along its graph line. The points selected using the **Node** object snap mode are located on the bottom and top of the wall along the graph line as shown in Figure 2.16. Therefore, the **Node** mode can be used to connect existing walls or when placing other objects such as windows and doors.

GRIPS OF THE WALL

You can use AutoCAD grips to lengthen, move, or increase the height or width of a wall. The function of the grips of a wall are described in Figure 2.17. The grips are located at the ends and midpoint of the wall justification line. When you select a grip, it becomes hot; movement of the pointer will stretch the grip to new location. Hot grips allow you to directly manipulate the size and location of a wall. Movement of a hot grip displays one or more dynamic dimensions. When a dynamic dimension is displayed **red**, you can type a value in the keyboard, and the wall will stretch to the dimension specified. When a grip is hot, you can press TAB to cycle through and highlight other dynamic dimensions, as shown in the lower right corner of Figure 2.17.

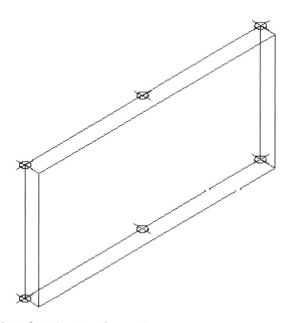

Figure 2.16 Node Object Snap locations for a wall

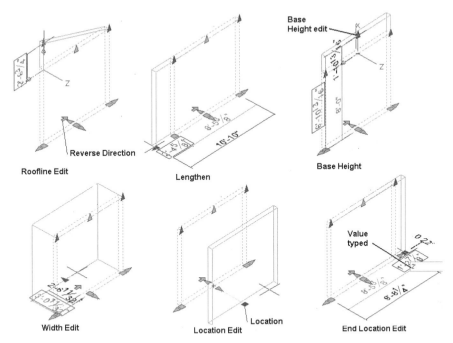

Figure 2.17 *Grip editing of walls*

USING GRIPS TO EDIT WALLS

The location grips of a wall can be used to modify the length of a wall or to copy the wall. All grip editing operations of Move, Mirror, Rotate, Scale, and Stretch are executable on the grips of a wall. However, the Move and Stretch options when used with **Copy** are most useful in creating floor plans. Each location grip is located at the endpoint and midpoint of the wall as shown in Figure 2.17. Grips displayed on the wall are warm, or hot. When a wall is selected, its grips are warm. When you select a warm grip it changes to red color and is ready for edit. Grip display and wall selection can be terminated by pressing ESC.

> **Warm Grips** – Displayed for a wall when you click on the wall. This highlights the wall, indicating that it is selected for editing, and displays the grips.
>
> **Hot Grips** – When warm grips are selected, they become hot grips, causing the grip box to fill with red. Once grips are hot, the grip point can be edited with the operations of Move, Mirror, Rotate, Scale, and Stretch. Grip options can be selected from the command window or you can right-click and choose from the shortcut menu.
>
> **Shift to Add** – Hold SHIFT down prior to clicking on the first grip to make more than one grip hot.

STEPS TO COPYING A WALL USING GRIPS
1. Select the wall to display the grips.
2. Select the midpoint grip to make it hot.
3. Right-click and choose **Copy** from the shortcut menu.
4. Move the pointer in the direction for the new wall.
5. Type the distance for the Copy Stretch in the dynamic dimension.
6. Press ESC twice to clear the grips and end grip editing.

SETTING JUSTIFICATION

Prior to beginning the floor plan, identify which wall justification best suits your needs. If you are working from a sketch that identifies the overall dimensions to the outer wall surface, the left or right wall justification should be used. The justification of the wall sets the location of the handle for placing the wall. This handle is located at the bottom of the wall or at the zero Z coordinate. Therefore, as the wall is drawn, the bottom of the wall is placed at Z=0 for all wall vertices. If a wall is drawn with center justification, the handle for the wall is located at the bottom of the wall and is centered between the sides of the wall.

When drawing a floor plan, you often start with a basic shape such as a rectangle. The overall dimensions of the basic shape are set by the outer wall surfaces. To draw walls dimensioned to the outer surfaces, set the justification to right and draw the walls from left to right or in a counterclockwise direction. When the justification is set to right and the polygon is drawn in a counterclockwise direction, the handles for the wall are on the outer wall lines. The same effect is obtained if the walls are drawn with left justification and drawn in a clockwise direction.

If you are drawing a wood frame structure, the overall dimensions of the structure should usually be a multiple of 16" for framing efficiency. Because studs and joists are usually placed 16" o.c., the building length should be a multiple of 16". Therefore, the overall dimensions of the building should not be a random distance but rather a distance that is a specific multiple of 16". Setting the SNAP and GRID distance to 16" will aid in controlling wall lengths to multiples of 16". With SNAP, GRID, and the appropriate wall justification, the outer wall surfaces will be snapped to dimensions that are multiples of 16 inches as the mouse pointer is moved. In Figure 2.18, the GRID and SNAP were set to 16". All the dimensions are multiples of 16" and are placed relative to the justification line. The right justification of the walls is shown by the location of the grips. The exterior walls were drawn counterclockwise, and the interior walls were drawn from top to bottom, because the directional arrows are pointing down, or in a counterclockwise direction. Notice that when interior walls intersect the exterior walls, the exterior walls are not cut into separate segments.

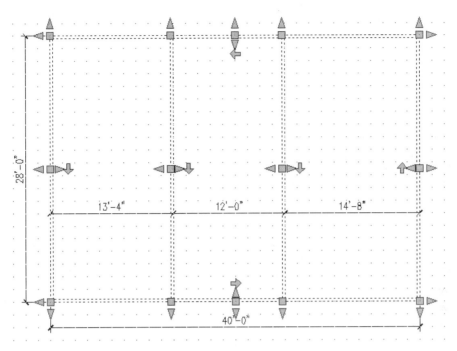

Figure 2.18 *Wall justification displayed with grips*

USING EDIT JUSTIFICATION

You can edit the justification of a wall in the Basic section of the Properties palette. However, the **Edit Justification** command can be used to specifically change the justification. Access this command as shown in Table 2.4. When this command is executed, trigger grips are displayed on the wall, which indicate the justification options as shown in Figure 2.19.

The current justification is shaded. If you move your pointer over a justification, a tip will display the name of the justification and dynamically display how the wall will change if this justification is selected. The justification can be changed by clicking the justification trigger grip. After you click the justification trigger grip, the markers are cleared and only the wall is displayed.

Command prompt	WALLEDITJUSTIFICATION
Shortcut menu	Select the wall, right-click, and choose Edit Justification

Table 2.4 *WallEditJustification command access*

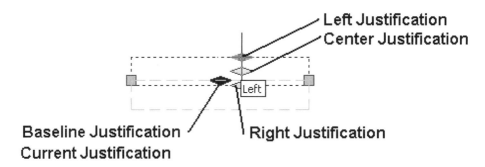

Figure 2.19 *Justification trigger grips of a wall*

CHANGING WALL DIRECTION WITH WALLREVERSE

The Reverse (WallReverse) command allows you to convert the direction in which a wall is drawn. This command is most useful in flipping the thickness of a wall to the other side of the justification line.

Access the WallReverse command as shown in Table 2.5.

Command prompt	WALLREVERSE
Shortcut menu	Select a Wall, right-click, and choose Reverse

Table 2.5 *WallReverse command access*

When you select a wall and then right-click and choose **Reverse**, the shortcut menu includes **In Place** and **Baseline** options. The **In Place** option will reverse the direction and retain wall location as shown at right in Figure 2.20. The **Baseline** option reverses the justification line; therefore when the wall is drawn with left or right justification, the thickness of the wall will flip to the other side. The **WallReverse Baseline** option does not move the justification line; it simply reverses the direction in which the wall is drawn as shown at right in Figure 2.20. Notice that the wall directional indicator is reversed from the top walls. The wall directional indicator is a trigger grip, which when selected will apply the **Reverse In Place** command. If you hold down CTRL when you select the trigger grip, the **Reverse Baseline** command is applied to the wall.

In summary, to draw a rectangular shape and place the graph line on the outside corners of the building use the Standard wall style with

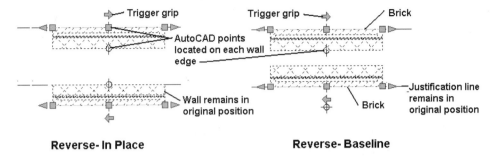

Figure 2.20 *Effects of the WallReverse command on existing walls*

Right justified walls, drawn in a counterclockwise direction

Or

Left justified walls, drawn in a clockwise direction.

CONNECTING WALLS WITH AUTOSNAP

Wall endpoints automatically connect to adjoining wall graph lines when the Autosnap is toggled ON. Autosnap is controlled by toggling ON **Autosnap New Wall Baselines** in the **AEC Object Settings** tab of the **Options** dialog box (see Figure 2.21). The Autosnap feature detects other graph lines of walls within a specified radius of the current wall graph line and will tie the two walls together. The Autosnap radius is set to 6" by default. The Autosnap feature can be toggled ON/OFF for grip editing by checking **Autosnap Grip Edited Wall Baselines**.

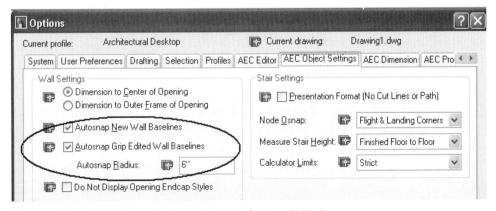

Figure 2.21 *AEC Objects Settings tab of the Options dialog box*

If the **Autosnap** toggle were turned OFF, walls would only intersect with other walls if they had connecting graph lines. If the AutoSnap radius is set to 6", a wall graph line will tie to other surrounding walls with graph lines within the 6" radius.

 Tip: The Diagnostic display configuration displays the wall graph line, wall directional arrows, and cleanup radii to help you edit wall intersections, as shown in Figure 2.23.

USING WALL CLEANUP

Autosnap assists you in connecting the graph lines of intersecting walls. The **Cleanup** properties of a wall allow you to control whether walls clean up when they intersect. If you are drawing walls representing the same type of construction, the walls should clean up. However, if you are drawing a floor plan with masonry and wood frame walls, the walls should not clean up or merge together. Control over the merge and cleanup of walls is governed in the cleanup settings of the wall and additional controls can be defined in the wall style definition. Because Autosnap is toggled ON, the cleanup radius of walls can be set to zero. The method of setting the cleanup radius is discussed in the "Setting the Cleanup Radius" section.

Cleanup automatically can be set to **Yes** or **No** in the Basic section of the Properties palette as shown in Figure 2.22. When cleanup is turned on by selecting **Yes**, walls with the same cleanup group definition and with connecting graph lines will cleanup as shown in Figure 2.22. Therefore walls should clean up if they share the same Standard wall style and Standard Cleanup Group Definition with connecting graph lines. The name of the Cleanup Group Definition of a wall is listed in the Basic section of Properties palette as shown in Figure 2.22. (Cleanup Group Definitions are discussed later in this chapter.) The cleanup setting of the wall is changed by editing the cleanup setting in the Basic section of the Properties palette of the selected wall.

If **Cleanup automatically** is set to **No** in the Basic section of the Properties palette, walls will not clean up as they are drawn. The default for cleanup is **Yes**; therefore each time the **WallAdd** command is selected, the cleanup is set to **Yes** unless you change it.

SETTING THE CLEANUP RADIUS

The **Cleanup** radius of the wall governs which walls are selected for cleanup. Walls will clean up with other walls if the cleanup radius crosses the justification line. This radius is centered by default along the Justification wall line. However the Advanced section of the Properties palette includes a **Graphline position** option, which allows you to change Graphline position to either the **Wall Center Line** or the **Wall Justification Line**. The **Cleanup Radius** is set in the Advanced section of the Properties palette as shown in Figure 2.23. The default cleanup radius for the Standard Wall is 0. If you need to edit the cleanup radius due to wall cleanup errors, set the cleanup to one-half the wall width.

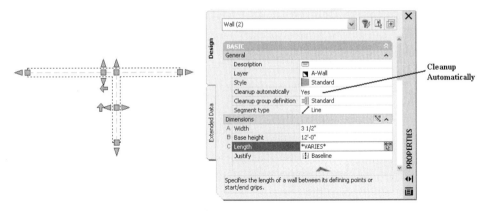

Figure 2.22 *Cleanup settings in the Properties palette of a wall*

If you select the Diagnostic display configuration on the Drawing Window status bar, you can view the graph line and cleanup radii of the walls. The Diagnostic display configuration also includes a wall directional arrow or triangle, which points to the end point of the wall. The width of the wall is not shown in the Diagnostic display configuration. Compare the wall cleanup radii shown on the left in Figure 2.23 with the actual wall cleanup shown on the right.

The **Cleanup** radius can be set as an exception or **Override** for the start or end of the wall segment. This allows you to edit the value of the cleanup radius for a specific end of the wall. The override is set in the Advanced section of the Properties palette.

Figure 2.23 *Cleanup radii displayed for walls using Diagnostic display configuration*

USING PRIORITY TO CONTROL CLEANUP

Complex walls created using Wall styles can have more than one wall component. The wall components clean up based upon the priority value assigned to the component. A wall can be created that consists of a concrete masonry unit component and a brick component. Each of the components can be assigned a priority value. Wall components with the same numerical priority value will clean up. Components with **the lowest value priority override** the display of components with **higher priority values**. In Figure 2.24 the brick component has a priority of 410 and the concrete masonry unit component has a priority of 400. Therefore, the concrete masonry units clean up together. Additional information regarding setting priorities of wall components is presented in Chapter 3.

CREATING CLEANUP GROUP DEFINITIONS

Cleanup Group Definitions can be defined to control wall cleanup. If walls have different cleanup group definitions, they will not clean up. The walls can be of the same style, but if they have different cleanup group definitions, the cleanup will not occur. Cleanup groups are created to control the merging of walls during automatic cleanup. Regardless of style name, walls can merge with other walls, and the intersections are cleaned up according to wall component priority number. The priority number of wall components controls how the walls clean up. Walls with components of equal priority numbers merge together if they share the same cleanup group definition.

Walls with different cleanup group definitions will not blend together regardless of wall component priority number. When cleanup group definitions are assigned to walls in the Properties palette, the walls that share the same cleanup group definition will merge

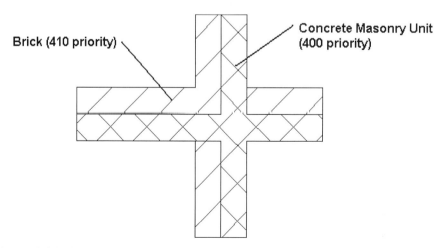

Figure 2.24 *Control of wall cleanup using priority*

based upon the priority number of the wall component. Walls that do not share the same cleanup group will not merge regardless of the wall component priority number.

Cleanup Groups are created in the **Style Manager**. Wall Cleanup Group Definitions are located in the Architectural Objects folder of the left pane in the **Style Manager**. Shown below are the steps to creating a cleanup group definition.

STEPS TO CREATING AND APPLYING A CLEANUP GROUP

1. Select **Format>Style Manager** from the menu bar.
2. Select the (+) to expand the Architectural Objects folder in the left pane.
3. Select Wall Cleanup Group Definitions in the left pane.
4. Select the **New Style** command on the **Style Manager** toolbar, and type **Mystyle** as the name of the new style as shown in Figure 2.25.
5. Select **OK** to dismiss the **Style Manager**.
6. Verify that **Medium Detail** is the current display configuration, select a wall in the current drawing, right-click, and choose **Properties** from the shortcut menu.
7. Click in the edit field of the **Cleanup Group Definitions** in the Properties palette. Select **Mystyle** from the drop-down list as shown in Figure 2.26. Walls do not merge as shown at right in Figure 2.26 because they have different cleanup group definitions.

MANAGING WALL DEFECT MARKERS

A defect marker, as shown on the left in Figure 2.27, will be displayed in substitution for a wall if a wall fails to clean up. Wall defect markers are displayed at the midpoint

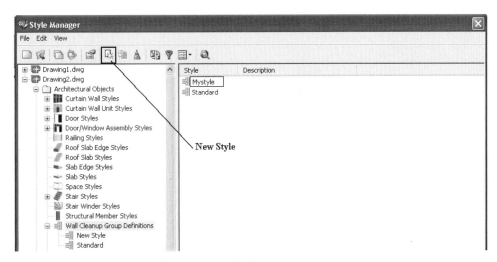

Figure 2.25 *Creating Wall Cleanup Group Definitions*

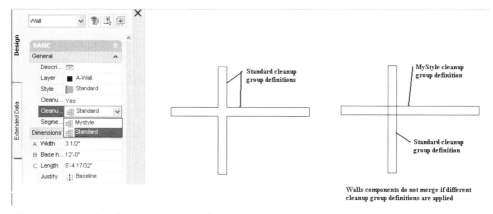

Figure 2.26 New Cleanup Group Definition in the Properties palette

of the wall and are colored red to alert you to the problem. The wall error marker is the **Defect Warning** component of the wall object and is placed on the G_Anno_Nplt layer, which has the no plot layer property. Because the wall defect marker has three grips, you can select a grip and stretch the wall. Access the Wall Graph display representation as shown in Table 2.6.

Command prompt	WALLGRAPHDISPLAYTOGGLE
Shortcut menu	Select a wall, right-click, and choose Cleanups>Toggle Wall Graph Display

Table 2.6 Wall Graph Display command access

When you toggle ON the wall graph display, the justification line is shown as well as other wall components. When defect markers occur, turn ON the Diagnostic display configuration on the Drawing Window status bar or turn ON the Graph display representation. The Diagnostic display configuration and Graph display representations allow you to visually verify if graph lines intersect. The Graph display

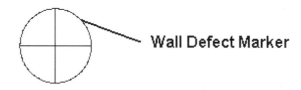

Figure 2.27 Wall defect marker

representation also displays the cleanup radius of the walls. Table 2.7 lists the reasons and solutions for editing defect markers.

Reason	Solution
Multiple walls intersecting at angles.	Select the walls and add a cleanup radius to each wall less than the wall width.
The wall is coincident with another wall.	Select the defect marker, right-click, and choose Basic Modify Tools>Erase from the shortcut menu.
Cleanup radius may be too large.	Select the wall and reduce the cleanup radius to the wall width in the Properties palette.

Table 2.7 *Reasons Wall Defect Markers are formed*

MODIFYING WALLS

When you place walls, it is necessary to modify the length and location of the walls. Walls are placed within the basic footprint of the building. Interior walls can be drawn or copied across the entire width or length of the building and later trimmed to length. In the following sections, interior walls will be created using the **WallOffsetCopy**, **WallOffsetMove**, and **Wall Offset Set From** commands. These commands allow you to copy and move walls with precision. Although most basic editing commands of AutoCAD can be applied to the walls, the **WallCleanupL** and **WallCleanupT** Architectural Desktop commands allow you to trim walls to create 'L' and 'T' intersections. Short segments of walls can be merged with the **Join** and **Wall Merge** commands.

CREATING L AND T WALL INTERSECTIONS

The **WallCleanupT** and **WallCleanupL** commands located in the **Cleanups** shortcut menu of a selected wall are used to form T and L wall intersections. Access the **WallCleanupL** command as shown in Table 2.8.

Command prompt	WALLCLEANUPL
Shortcut menu	Select a wall, right-click, and choose Cleanups>Apply 'L' Cleanup

Table 2.8 *Apply 'L' Cleanup command access*

When you choose **Apply 'L' Cleanup** from the shortcut menu, you are prompted as shown in the following command sequence to select the second wall component for the cleanup. The command **extends** or **trims** the wall endpoints that are closest to each other. Therefore you could select walls **p1** and **p2** as shown in Figure 2.28, right-click, and choose **Cleanups>Apply 'L' Cleanup**, and the wall segments closest to each other are removed to create the L intersection. Wall segment B is shorter than wall segment A; therefore segment B is trimmed as shown in Figure 2.28. If the walls do not cross, when the **Apply 'L' Cleanup** command is applied, the walls will extend to create an L intersection.

(Select a wall for editing at p1, right-click, and choose **Cleanups>Apply 'L' Cleanup**.*)*

Command: WallCleanupL

Select the second wall: *(Select the wall at p2.)*

(Wall cleaned up as shown in Figure 2.28.)

USING THE WALL CLEANUP T COMMAND

The **WallCleanupT** command deletes the shorter wall segment, which crosses the boundary wall. In Figure 2.28 wall segments A and B cross the boundary. A is removed because it is the shorter. Access the **WallCleanupT** as shown in Table 2.9.

Command prompt	WALLCLEANUPT
Shortcut menu	Select a wall, right-click, and choose Cleanups>Apply 'T' Cleanup

Table 2.9 *Apply 'T' Cleanup command access*

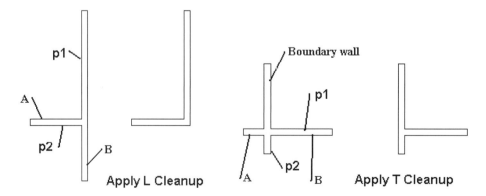

Figure 2.28 *Editing walls with Wall Cleanup L and Wall Cleanup T*

The following command line sequence created a T intersection using the **WallCleanupT** command as shown in Figure 2.28.

> (Select a wall for editing at p1, right-click and choose **Cleanups>Apply 'T' Cleanup**.)
>
> Command: WallCleanupT
>
> Select boundary wall: (Select the wall at p2 shown in Figure 2.28.)
>
> (Wall cleaned up as shown in Figure 2.28.)

MERGING WALLS

The **WallMergeAdd** command is used to merge two intersecting walls. The **WallMergeRemove** command is used to remove the merger of walls. The **WallMergeAdd** command is useful for creating short walls or projections of a wall, or for cleaning up walls that intersect perpendicular to the current wall. If several wall components are merged together, each wall component can also be removed from the merge using this command. Access the **WallMergeAdd** command as shown in Table 2.10.

Command prompt	WALLMERGEADD
Shortcut menu	Select a wall, right-click, and choose Cleanups>Add Wall Merge Condition

Table 2.10 *Wall Merge command access*

When you select a wall and then select **Cleanups>Add Wall Merge Condition** from the shortcut menu, you are prompted to select the walls to merge with the selected wall. Walls selected in response to this prompt will merge together. The command sequence of the **WallMergeAdd** command is shown below. Multiple components can be merged while remaining in the **WallMergeAdd** command.

> Command: WallMergeAdd
>
> (Select wall as shown at p1 in Figure 2.29, right-click and choose **Cleanups>Add Wall Merge Condition**.)
>
> Select walls to merge with: (Select the wall at p2 as shown in Figure 2.29.) 1 found
>
> Select walls to merge with: (Press ENTER to end the command.)
>
> (Walls merged as shown at right in Figure 2.29.)

If the results of the merger are not satisfactory, you can remove the merger using the **WallMergeRemove** command. Access the **WallMergeRemove** command as shown in Table 2.11.

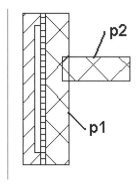 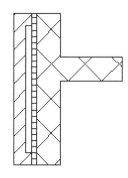

Figure 2.29 *Walls selected for Wall Merge*

Command prompt	WALLMERGEREMOVE
Shortcut menu	Select the wall, right-click, and choose Cleanups>Remove All Wall Merge Conditions

Table 2.11 *WallMergeRemove command access*

When you select a wall that has been merged, right-click, and choose **Cleanups>Remove All Wall Merge Conditions**, the command line will report the results as shown in the following command line sequence.

    Command: WallMergeRemove

    (1)   wall merge conditions removed.

Therefore, when this command is accessed from the shortcut menu, the actions are executed on the selected wall.

 Note: The commands **WallMergeAdd** and **WallMergeRemove** can also be accessed by entering **WALLMERGE** in the command line. The options of this command when entered in the command line are shown below.

    Command: wallmerge

    Wall Merge [Add/Remove/removeAll]: (Type **Add** to Wall Merge Add or **R** to Wall Merge Remove. Note that only the "A" is capitalized in Add and removeAll; therefore if you type **A**, wall merge will be applied to the selected walls.)

JOINING WALLS

When constructing a floor plan, you can draw short segments of walls that connect to other short walls as shown in Figure 2.30. The two segments when unselected appear as a single wall because of the cleanup function. However, each segment is treated uniquely when you place doors or windows in the wall. If you insert a door centered in the wall, the door will be centered in a wall segment. Therefore, the two wall segments should be joined together or joined to make one wall. The **Join** command will convert the two walls to one. Access the **WallJoin** command as shown in Table 2.12.

Command prompt	WALLJOIN
Shortcut menu	Select a wall segment, right-click, and choose Join

Table 2.12 *WallJoin command access*

(Select a wall segment, right-click, and choose **Join** from the shortcut menu.)

Command: WallJoin

Select second wall: *(Select wall at p1 as shown in Figure 2.30.)*

(Walls segments joined.)

USING WALLOFFSETCOPY TO DRAW INTERIOR WALLS

The **OFFSET** command of AutoCAD is frequently used to copy an existing entity a specified distance. The AutoCAD **OFFSET** command copies a wall; however, it does not compensate for the wall width as shown at A in Figure 2.31. The **Wall Offset** command performs a similar function; however, the wall width is considered in placing the wall. The shortcut menu of a wall has the following three wall offset options: **WallOffsetCopy**, **WallOffsetMove**, and **WallOffsetSet**. The **WallOffsetCopy** command copies the selected wall a specified distance from the wall. The specified distance can include the wall width as shown at B in Figure 2.31, or the specified distance can place walls exclusive of the wall width as shown at C in Figure 2.31. The wall shown at B in Figure 2.31 is offset by selecting the wall at **p1**, right-clicking and choosing **Offset>Copy**. A red line is displayed near the wall surface as you move your

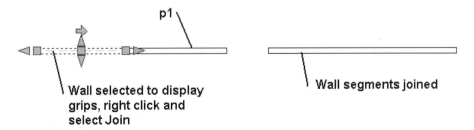

Figure 2.30 *Wall selected for Join command*

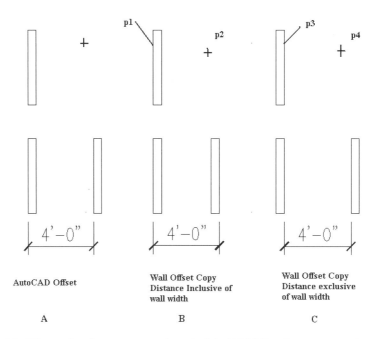

Figure 2.31 *Effects of wall component selected of the WallOffsetCopy command*

pointer; click near surface **p1** to specify it as the reference surface, and then move the pointer to a point near **p2** to specify the direction and type the offset distance in the dynamic dimension field of the workspace. The offset wall is positioned to set the overall dimension measured to the outer surface of the wall at **p1**. The wall surface selected becomes the reference point for the dimension.

The wall offset shown at C in Figure 2.31 is created by selecting the wall at **p3**, right-clicking and choosing **Offset>Copy**. As you move your pointer near the wall, a red line is displayed. Select the surface at **p3** to specify the reference surface, and then click near **p4** and type the offset distance in the dynamic dimension edit field. The offset distance becomes the interior dimension between the two walls. Selecting a wall by the surface nearest to the direction of the offset will create another wall relative to the surface selected and exclusive of the wall width.

The **WallOffsetCopy** command allows you to use direct manipulation and type the distance for the offset. The value typed will be displayed in the cursor tip. Access the **WallOffsetCopy** command as shown in Table 2.13.

Command prompt	WALLOFFSETCOPY
Shortcut menu	Select a wall, right-click, and choose Offset>Copy

Table 2.13 *WallOffsetCopy command access*

When you select a wall, right-click, and choose **Offset>Copy**, you are prompted as shown in the following command line sequence.

 Note: The tooltip described below specifying the component is displayed if you toggle ON **Display AutoSnap tooltip** in the **Drafting** tab of the **Options** dialog box. If the tip is not displayed, toggle Off OSNAP, to avoid the display of the Osnap mode.

(Select a wall as shown at pl in Figure 2.32.)

Command: WallOffsetCopy

Select the component to offset from:
(Move the pointer over the wall component to reference the offset; the tip displays the name of the wall component under the pointer, unless OSNAP is toggled ON; click to select the point p2 as shown in Figure 2.32.)

Select a point to offset to. **6'** ENTER
(Move the pointer down to set the direction; distance is dynamically displayed; then type the offset distance in the workspace as shown in Figure 2.33.)

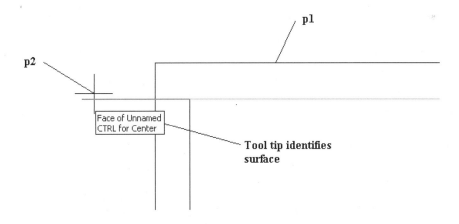

Figure 2.32 *Selection of wall face*

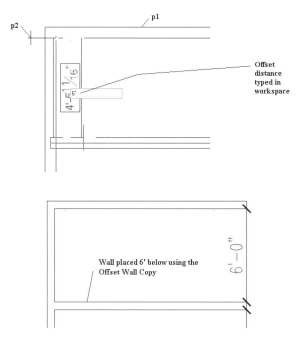

Figure 2.33 *Type offset distance in workspace*

Select a point to offset to. ENTER
(Press the ENTER to end the command.)

 STOP. Do Tutorial 2.3, "Drawing Walls Using Wall Offset Copy Command," at the end of the chapter.

USING WALLOFFSETMOVE

The **WallOffsetMove** command will allow you to move an existing wall offset from a specified point on the selected wall. Access the **WallOffsetMove** command as shown in Table 2.14.

Command prompt	WALLOFFSETMOVE
Shortcut	Select a wall, right-click, and choose Offset>Move

Table 2.14 *WallOffsetMove command access*

When you select a wall, and then select **Offset>Move** from the shortcut menu, you can move walls as shown in the following command line sequence.

> *(Select a wall as shown at p1 in Figure 2.34, right-click, and choose* **Offset>Move**.*)*
>
> Command: WallOffsetMove
>
> Select the component to offset from:
>
> *(Move the pointer over the wall component to reference the offset, the tip displays the name of the wall component under the pointer; click to select the point as shown in Figure 2.34.)*
>
> Select a point to offset to. **5'** ENTER
>
> *(Move the pointer down to set the direction, distance is dynamically displayed; then type the offset distance in the workspace as shown in Figure 2.35.)*
>
> *(Wall is moved 5' as specified.)*

USING THE OFFSET SET FROM COMMAND

The **WallOffsetSet** command will allow you to move an existing wall from a specified point on the selected wall and relative to existing walls in the drawing. This com-

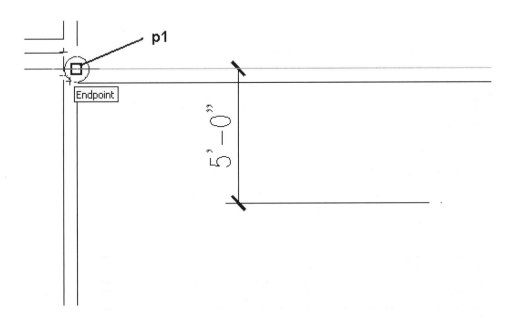

Figure 2.34 *Wall component selected for WallOffsetMove*

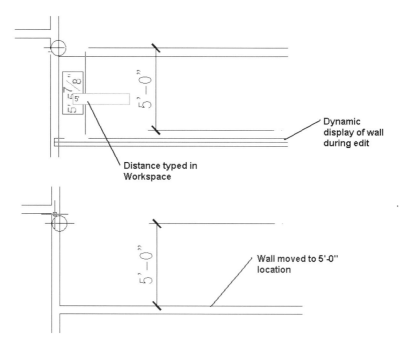

Figure 2.35 Move distance specified in workspace

mand allows you to measure distances from walls and enter a new distance. Access the **Offset Set From** command as shown in Table 2.15.

Command prompt	WALLOFFSETSET
Shortcut	Select a wall, right-click, and choose Offset>Set From

Table 2.15 WallOffsetSet command access

When you select a wall, and then select **Offset>Set From** from the shortcut menu, you can move walls as shown in the following command line sequence and Figure 2.36.

(Toggle ON running Endpoint Object Snap.)

(Select a wall as shown at p1 in Figure 2.36, right-click, and choose **Offset>Set From**.*)*

Command: WallOffsetSet

Select the component to offset from: *(Move the pointer over the wall component to reference the offset, click to select the point at p2 as shown in Figure 2.37.)*

Creating Floor Plans 109

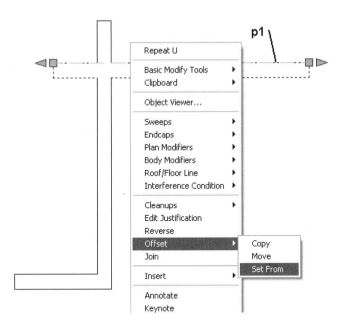

Figure 2.36 *Wall face specified for Offset Set From command*

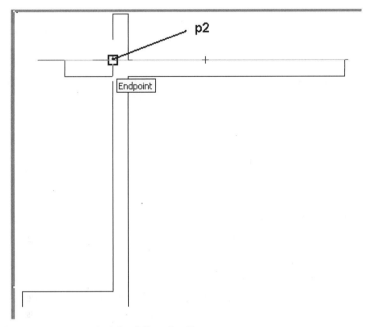

Figure 2.37 *Distance specified for Offset Set From*

Select a point or wall component. *(Move the pointer down and select point p3 to measure the distance between the two walls. Distance is displayed as a dynamic dimension as shown at left in Figure 2.38.)*

Enter new distance between the selected points. **5'** ENTER *(Value typed in the workspace.)*

(Selected wall surface is moved to a point 5' from the wall at p3 as shown at right in Figure 2.38.)

The AutoCAD **Offset** command can be used to offset walls. However, it does not compensate for wall widths or allow you to specify a reference point for the offset distance. Therefore when you use the AutoCAD **Offset** command to create a clear distance between walls, the thickness of the wall must be added to the clear distance between the walls.

Tip: To create a tub enclosure using 3 1/2" walls that has a 5' inside clear dimension between partitions, make the offset distance 5'-3 1/2" when using the AutoCAD **Offset** command.

EDITING WALLS WITH AUTOCAD EDITING COMMANDS

Most all other AutoCAD commands can be used to edit the walls of a drawing. Walls can be edited with the following AutoCAD commands: **Align, Array, Break, Copy, Chamfer, Fillet, Erase, Extend, Mirror, Move, Offset, Rotate, Scale, Stretch,** and **Trim**.

When the **Extend** and **Trim** commands are used, the graph line of the wall serves as the boundary or cutting edge. Wall cleanup radii may display a wall as connected to an adjacent wall, yet the graph line may not be connected. When you select a wall, the grips are displayed on the graph line as shown in Figure 2.39. The **Align** and **Scale** commands do not change the thickness of the wall.

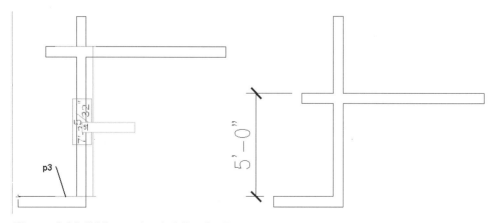

Figure 2.38 *Wall moved with Offset Set From*

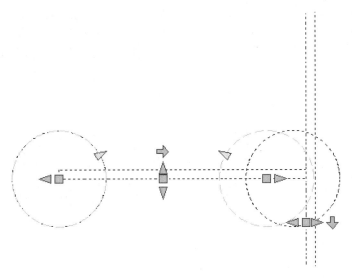

Figure 2.39 *Wall components extended beyond graph line*

 Warning: The **Explode** command should not be used on walls. The object definition is lost when you explode Architectural Desktop objects; therefore the objects will not interact with other objects.

 STOP. Do Tutorial 2.4, "Editing Walls Using Edit Justification, Reverse, Wall Offset, Join, and Merge Wall," and Tutorial 2.5, "Editing Walls Using Trim, Match, and Add Selected Commands," at the end of the chapter.

SUMMARY

1. Walls are created with the **WallAdd** command as objects with properties of width, height, and justification.
2. Draw curved walls by toggling ON **Arc** in the Properties palette and identifying a start point, a point on the curve, and the endpoint of the curve.
3. If the Close option is used, the last wall segment is drawn to connect to the beginning of the first wall segment.
4. The Ortho close option of the **WallAdd** command can be used to create walls in the shape of a rectangle.
5. The Baseline Offset option shifts the justification handle off the justification line a specified distance.
6. The handle for placing a wall is located at the Z=0 location along the justification line.

7. Object snap nodes are located at the beginning, middle, and end of the justification line along the top and bottom of the wall.
8. Grips of a wall are located at the beginning, middle, and end on the bottom of the justification line.
9. The Cleanup Radius value determines how close to an existing wall a new wall can begin and be automatically connected to the wall.
10. AutoCAD edit commands such as **Chamfer**, **Fillet**, **Extend**, **Offset**, **Stretch** and **Trim** can be used to create and modify walls.
11. The grips of a wall allow easy editing of the wall to copy, stretch, and move walls.
12. Wall Modifiers can be added to a wall to form a projection such as a pilaster.

REVIEW QUESTIONS

1. The option of the **WallAdd** command that allows you to move the wall handle a specified distance from the wall is _____.
2. The single object snap mode that allows you to snap to the midpoint and endpoints of the wall justification line is _____.
3. The Z coordinate of the wall justification line is _____.
4. The grips of a wall are located at _____, _____, and _____.
5. The nodes of a wall are located at the _____.
6. The width of a wall is changed by _____.
7. The length of a wall can be changed by the _____.
8. The handle of the wall is located on the _____ of the wall.
9. What is the purpose of the wall directional indicator?
10. Wall modifiers are defined for a wall with the _____ of the wall.
11. Walls with different cleanup group definition _____ blend together regardless of wall component priority number.
12. Wall components with low priority numbers will _____ the display of wall components with high priority numbers.
13. Create a curved wall by establishing three points: _____, _____, _____.
14. The _____ option draws the final wall segment of a polygon to connect at an angle of 90° to the beginning of the first wall segment.
15. The _____ command will display a dialog box for setting Autosnap.
16. Often you can remove wall error markers by editing the _____ of the wall.

17. The following commands of the **Modify** toolbar can be used to modify walls: _____, _____, _____.

18. The _____ command allows you to change the direction in which a wall has been drawn.

TUTORIAL 2.1 DRAWING WALLS AND USING CLOSE

1. Select **QNew** from the **Standard** toolbar.
2. Choose **File>SaveAs** from the menu bar and save the drawing as **Lab 2-1** in your student directory.
3. Select the Model tab.
4. Select the **Drawing** menu from the Drawing Window status bar and choose **Drawing Setup**.
5. Select the **Layering** tab of the **Drawing Setup** dialog box. Verify that the Layer Standards/Key File to Auto-Import: is **AecLayerStd** and the Layer Key Style is **AIA (256 color)** as shown in Figure 2T.1. Select OK to dismiss the **Drawing Setup** dialog box.
6. Select **Zoom Window** from the **Zoom** flyout as shown in Figure 2T.2.
7. Enter the following coordinates for the **Zoom** command as shown in the following command line sequence.

    Command: ZOOM

    Specify corner of window, enter a scale factor (nX or nXP), or

    [All/Center/Dynamic/Extents/Previous/Scale/Window] <real time>: **w**

    Specify first corner: **150',50'** ENTER

    Specify opposite corner: **200',100'** ENTER

8. Verify that tool palettes are displayed. If tool palettes are not displayed, select **Window>Tool Palettes** from the menu bar.
9. Verify that POLAR, OSNAP, and OTRACK are toggled ON on the status bar.
10. Move the pointer over OSNAP button on the status bar, right-click, and choose **Settings**. Edit the **Drafting Settings**, **Object Snap** tab as follows: clear all Osnap modes except **Endpoint** and **Node**. Select the **Options** button on the **Object Snap** tab, toggle ON **Display polar tracking vector**, **Display full-screen tracking vector**, **Display Auto Track tooltip**, and **Display AutoSnap tooltip**. Select **OK** to dismiss the **Options** dialog box and select **OK** to dismiss the **Drafting Settings** dialog box.
11. Select the **Wall** tool from the Design tool palette.

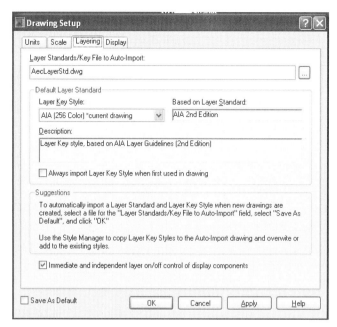

Figure 2T.1 *Drawing Setup dialog box*

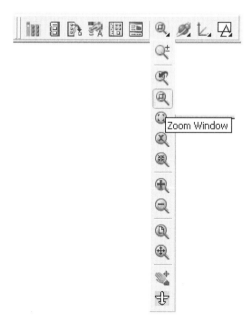

Figure 2T.2 *Zoom Window*

12. Edit the Properties palette as follows: Basic: Style = Standard, Cleanup automatically = Yes, Cleanup group definition = Standard, Segment type = Line, A-Width = 3 1/2, Base height = 8', Justify = Right, Baseline offset = 0, Roof line offset from base height = 0, and Floor line offset from baseline = 0 as shown in Figure 2T.3.

13. Begin drawing the walls shown in Figure 2T.4 using direct distance entry. Type the distances in the command line as shown in the following command sequence.

>Command: WallAdd
>
>Start point or
>[STyle/Group/WIdth/Height/OFfset/Justify/Match/Arc]:
>**170',75'** ENTER *(Point p1 established as shown in Figure 2T.4.)*
>
>*(Move the pointer down, verify polar angle in Auto Track tooltip=270.)*
>
>End point or
>[STyle/Group/WIdth/Height/OFfset/Justify/Match/Arc]:
>**12'** ENTER *(p2)*
>
>*(Move the pointer right, verify polar angle in Auto Track tooltip=0.)*
>
>End point or
>[STyle/Group/WIdth/Height/OFfset/Justify/Match/Arc]:
>**12'** ENTER *(p3)*
>
>*(Move the pointer down, verify polar angle in Auto Track tooltip=270.)*
>
>End point or
>[STyle/Group/WIdth/Height/OFfset/Justify/Match/Arc/Undo]:
>**2'** ENTER *(p4)*
>
>*(Move the pointer right, verify polar angle in Auto Track tooltip=0.)*
>
>End point or
>[STyle/Group/WIdth/Height/OFfset/Justify/Match/Arc/Undo/ Close/ORtho]: **12'** ENTER *(p5)*
>
>*(Move the pointer up, verify polar angle in Auto Track tooltip=90.)*
>
>End point or
>[STyle/Group/WIdth/Height/OFfset/Justify/Match/Arc/Undo/ Close/ORtho]: **2'** ENTER *(p6)*
>
>*(Move the pointer right, verify polar angle in Auto Track tooltip = 0.)*
>
>End point or
>[STyle/Group/WIdth/Height/OFfset/Justify/Match/Arc/Undo/ Close/ORtho]: **12'** ENTER *(p7)*

(Move the pointer up, verify polar angle in Auto Track tooltip=90.)

End point or
[STyle/Group/WIdth/Height/OFfset/Justify/Match/Arc/Undo/Close/ORtho]: **24'** ENTER *(p8)*

(Move the pointer left, verify polar angle in Auto Track tooltip=180.)

End point or
[STyle/Group/WIdth/Height/OFfset/Justify/Match/Arc/Undo/Close/ORtho]: **24'** ENTER *(p9)*

(Right-click and choose Close from shortcut menu.)

[STyle/Group/WIdth/Height/OFfset/Justify/Match/Arc/Undo/Close/ORtho]: **c** ENTER

(Wall segments drawn as shown below.)

14. Select the Work layout, and double-click in the left viewport. Select **Zoom Extents** from the **Navigation** toolbar. Double-click in the right viewport. Select **Zoom Extents** from the **Navigation** toolbar. Select 1/4" = 1'-0" from the **VP Scale** flyout of the Drawing Window status bar.

15. Save the drawing and select **File>Close** from the menu bar to close the drawing.

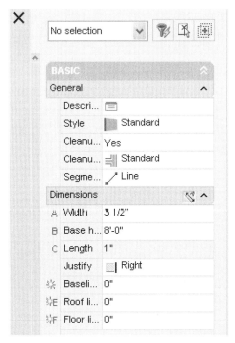

Figure 2T.3 *Wall Properties palette*

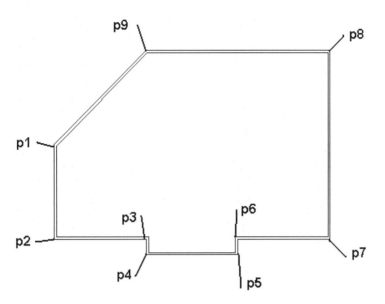

Figure 2T.4 *Close option used to draw the last wall segment*

TUTORIAL 2.2 DRAWING CURVED WALLS AND USING ORTHO CLOSE

1. Select **File>New** from the menu bar to select a template for a new drawing.

2. Select Aec Model (Imperial Stb).dwt from the template list of the **Select template** dialog box. Choose **Open** to dismiss the **Select template** dialog box.

3. Choose **File>SaveAs** from the menu bar and save the drawing as **Lab 2-2** in your student directory.

4. Select the Model tab.

5. Select the **Drawing** menu from the Drawing Window status bar and select **Drawing Setup**.

6. Select the **Layering** tab of the **Drawing Setup** dialog box. Verify that the Layer Standards/Key File to Auto-Import is **AecLayerStd** and the Layer Key Style is **AIA (256 color)**. Select **OK** to close the **Drawing Setup** dialog box.

7. Select **Zoom Window** from the **Zoom** flyout of the **Navigation** toolbar.

8. Enter the following coordinates as shown in the following command line sequence.

    Command: ZOOM

Specify corner of window, enter a scale factor (nX or nXP), or [All/Center/Dynamic/Extents/Previous/Scale/Window] <real time>: **_w**

Specify first corner: **150',25'** ENTER

Specify opposite corner: **200',100'** ENTER

9. Verify the tool palettes are displayed. If tool palettes are not displayed, select **Window>Tool Palettes** from the menu bar.
10. Verify that POLAR, OSNAP, and OTRACK are toggled ON on the status bar.
11. Move the pointer over OSNAP button on the status bar, right-click, and choose **Settings**. Edit the **Drafting Settings**, **Object Snap** tab as follows: clear all Osnap modes except **Endpoint** and **Node**. Select the **Options** button on the **Object Snap** tab, toggle ON **Display polar tracking vector**, **Display full screen tracking vector**, **Display Auto Track tooltip**, and **Display AutoSnap tooltip**. Select **OK** to dismiss the **Options** dialog box and select **OK** to dismiss the **Drafting Settings** dialog box.
12. Select the **Wall** tool from the Design tool palette.
13. Edit the Properties palette as follows: Basic Style = Standard, Cleanup automatically= Yes, Cleanup group definition = Standard, Segment= Line, A-Width = 3 1/2, Base height = 8', Justify = Right, Baseline offset = 0, Roof line offset from base height = 0, and Floor line offset from baseline = 0 as shown in Figure 2T.5.
14. Begin drawing the walls shown in Figure 2T.6 using direct distance entry and typing the distances in the command line as shown in the following command sequence.

Command: WallAdd

Start point or [STyle/Group/WIdth/Height/OFfset/Justify/Match/Arc]: **160',30'** ENTER *(Point p1 established as shown in Figure 2T.6.)*

*(Move the pointer to the right to display Auto Track tooltip polar angle =0 and notice the length is changing in the **Length** field of the Properties palette.)*

End point or [STyle/Group/WIdth/Height/OFfset/Justify/Match/Arc]: **40'** ENTER *(p2)*

(Toggle Segment Type to Arc in the Properties palette.)

Mid point or [STyle/Group/WIdth/Height/OFfset/Justify/Match/Arc]: *(Arc segment toggled on.)*

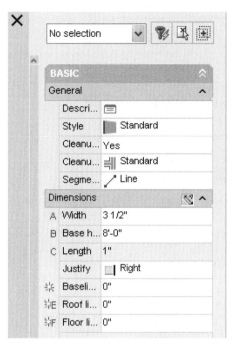

Figure 2T.5 *Wall Properties palette*

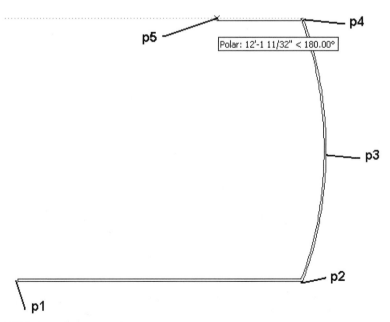

Figure 2T.6 *Specify location for wall segment*

End point or
[STyle/Group/WIdth/Height/OFfset/Justify/Match/Arc]:
@3',12' ENTER *(p3)*

End point or
[STyle/Group/WIdth/Height/OFfset/Justify/Match/Arc/Undo]:
@-3',24' ENTER *(p4)*

*(Right-click and choose **Ortho close** from the shortcut menu.)*

Mid point or
[STyle/Group/WIdth/Height/OFfset/Justify/Match/Arc/Undo/Close/ORtho]: OR

Point on wall in direction of close: *(Move the pointer to the left until the Autotrack tooltip polar angle =180, and then select a location near point p5 shown in Figure 2T.6.)*

(Wall segments drawn as shown in Figure 2T.7.)

15. Select the curved wall segment to display the grips and note the radius is displayed in the **D-Radius** field of the Properties palette as shown in Figure 2T.8.
16. Select **Zoom Extents** from the **Zoom** flyout of the **Navigation** toolbar. To view the Presentation display configuration, select Presentation from the **Display Configuration** flyout of the Drawing Window status bar.
17. Select **View>SE Isometric View** from the **Navigation** toolbar as shown in Figure 2T.9.

Figure 2T.7 *Wall segments drawn with arc*

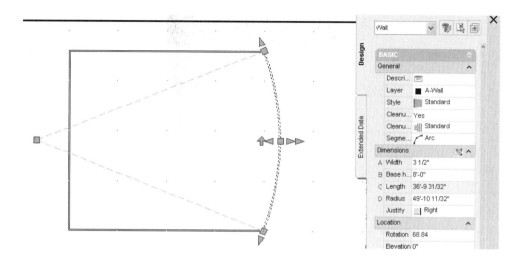

Figure 2T.8 *Wall Properties palette displays radius of arc wall*

Note the pictorial view of walls as shown in Figure 2T.10.

18. Save the drawing, and select **File>Close** from the menu bar to close the drawing.

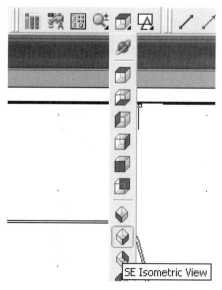

Figure 2T.9 *SE Isometric View from the View flyout*

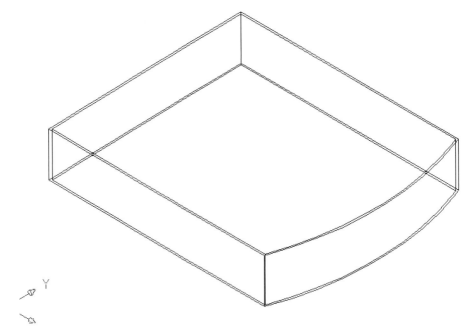

Figure 2T.10 *Pictorial view of walls*

TUTORIAL 2.3 DRAWING WALLS USING WALL OFFSET COPY COMMAND

1. Open Autodesk Architectural Desktop 2005, and select **QNew** from the Standard toolbar, then select **File>Project Browser** from the menu bar. Use the Project Selector drop-down list to navigate to your *ADT Student\ADT Tutor\Ch2* directory as shown in Figure 2T.11. Double-click on **Ex 2-3** to set this project current. Select **Close** to dismiss the **Project Browser**. (If your student folder does not include the project, refer to "Organizing Tutorial Directories" in the Preface.) Select **Yes** to the AutoCad message box to the repath the project.

2. Select the **Constructs** tab of the **Project Navigator**. Double-click on **Floor 1** to open the Floor 1 drawing. The Floor 1 drawing is empty and ready for you to start the floor plan.

3. Select the Work layout tab, and double-click in the right viewport. Select **Maximize Viewport** on the Application status bar. Select **Zoom Extents** from the **Navigation** toolbar.

4. Verify that Scale is 1/8"=1'-0" in the Drawing Window status bar.

5. If tool palettes are not displayed, select **Window>Tool Palettes** from the menu bar.

Creating Floor Plans 123

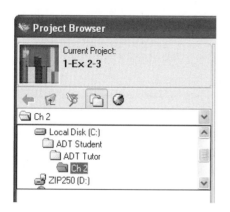

Chapter 2 folder of the ADT Tutor in your ADT Student folder

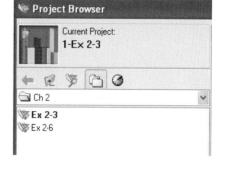

Projects of Chapter 2

Figure 2T.11 *Selecting the Ex 2-3 project*

6. Verify that POLAR, OSNAP, and OTRACK are toggled ON on the status bar.

7. Move the pointer over the OSNAP button on the status bar, right-click, and choose **Settings**. Edit the **Drafting Settings, Object Snap** tab as follows: clear all Osnap modes except **Endpoint** and **Node**. Select the **Options** button of the **Object Snap** tab, toggle ON **Display polar tracking vector, Display full screen tracking vector, Display Auto Track Tooltip,** and **Display AutoSnap tooltip**. Select **OK** to dismiss the **Options** dialog box and select **OK** to dismiss the **Drafting Settings** dialog box.

8. Select the **Wall** tool from the Design tool palette.

9. Edit the Properties palette as follows: Basic Style = Standard, Cleanup automatically= Yes, Cleanup group definition = Standard, Segment = Line, A-Width = 3.5, Base height= 8', Justify = Right, Baseline offset = 0, Roof line offset from base height = 0, and Floor line offset from baseline = 0.

10. Begin drawing the walls shown in Figure 2T.13 using direct distance entry and typing the distances in the command line as shown in the following command sequence and Figure 2T.12.

    Command: WallAdd

    Start point or
    [STyle/Group/WIdth/Height/OFfset/Justify/Match/Arc]:
    75',40' ENTER *(Point p1 established as shown in Figure 2T.12.)*

    (Move the pointer right to Auto Track tooltip polar angle =0°.)

    End point or

[STyle/Group/WIdth/Height/OFfset/Justify/Match/Arc]: **68'3** ENTER *(p2)*

(Move the pointer up to Auto Track tooltip polar angle =90°.)

End point or
[STyle/Group/WIdth/Height/OFfset/Justify/Match/Arc]: **28'** ENTER *(p3)*

*(Right-click and choose **Ortho close** from the shortcut menu.)*

End point or
[STyle/Group/WIdth/Height/OFfset/Justify/Match/Arc/Undo/Close/ORtho]: OR *(**Ortho close** option selected from shortcut menu.)*

Point on wall in direction of close: *(Move the pointer left to Auto Track tooltip polar angle =180° and select a point near p4 as shown in Figure 2T.12.)*

(Walls drawn to create the rectangle as shown in Figure 2T.13.)

11. Select **Zoom Window** from the **Zoom** flyout on the **Navigation** toolbar. Click a location near **p1** and **p2** as shown in Figure 2T.14.

12. Select the vertical wall at **p3** shown in Figure 2T.14, right-click, and choose **Offset>Copy** from the shortcut menu. Respond as shown in the following command line sequence to create interior walls using the **WallOffsetCopy** command.

   (Select wall at p3 as shown in Figure 2T.14.)

   Command: WallOffsetCopy

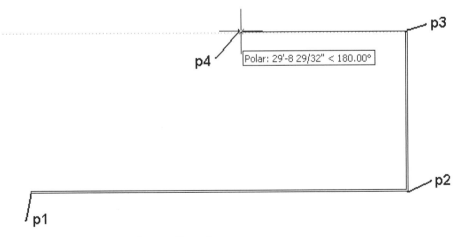

Figure 2T.12 *Location of points for wall segments*

Figure 2T.13 Rectangular polygon drawn with Ortho close option

Select the component to offset from: *(Select the endpoint of the wall as shown in Figure 2T.15.)*

(Move the pointer to the right with a polar angle of 0°.)

Select a point to offset to. **31'** ENTER *(Type distance, dynamically displayed in the autotrack tool tip, as shown in Figure 2T.16.)*

Select a point to offset to. **'z** ENTER *(Selects transparent zoom command.)*

>>Specify corner of window, enter a scale factor (nX or nXP), or [All/Center/Dynamic/Extents/Previous/Scale/Window] <real time>: **p** *(Selects previous option.)*

(Move the pointer to the right with a polar angle of 0°.)

Select a point to offset to. **12'** ENTER *(Type distance in command line, dynamically displayed in the Auto Track tooltip.)*

(Move the pointer to the right with a polar angle of 0°.)

Select a point to offset to. **8'** ENTER *(Type distance in command line, dynamically displayed in the Auto Track tooltip.)*

(Move the pointer to the right with a polar angle of 0.)

Select a point to offset to. **3'** ENTER *(Type distance in command line, dynamically displayed in the Auto Track tooltip.)*

Select a point to offset to. ENTER *(Press ENTER to end the command.)*

13. Select **Zoom Extents** from the **Zoom** flyout of the **Navigation** toolbar. Select Wall A as shown in Figure 2T.17.

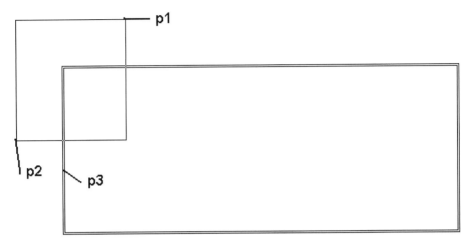

Figure 2T.14 Points specified for Zoom Window

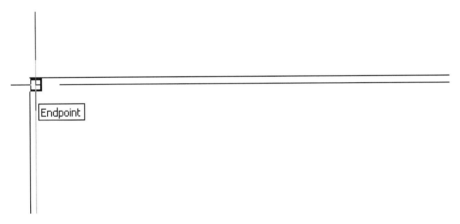

Figure 2T.15 Wall component specified for WallOffsetCopy

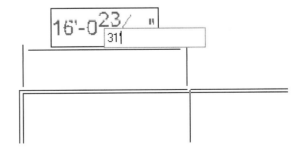

Figure 2T.16 Distance specified for WallOffsetCopy

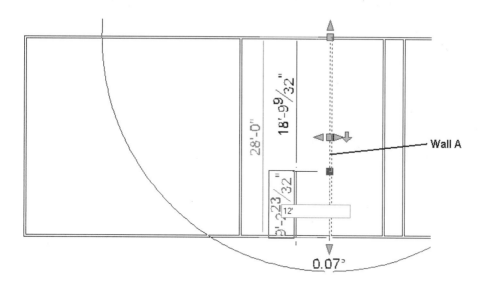

Figure 2T.17 *Wall selected for grip edit*

14. To edit the length of wall using grips, select the bottom square grip of Wall A to make it hot, move the pointer up, and type **12'** ENTER, dynamically displayed in tooltip, to stretch the wall up as shown in Figure 2T.17.
15. Press ESC to end grip edit and selection of wall.
16. To verify that **Autosnap** is toggled ON in the **Options** dialog box, right-click over the command line, and choose **Options**.
17. Select the **AEC Object Settings** tab and verify the following are checked ON as shown in Figure 2T.18: **Autosnap New Wall Baselines**, **Autosnap Grip Edited Wall Baselines**, and **Autosnap Radius = 6"**.
18. Select **OK** to dismiss the **Options** dialog box.
19. To view the justification line of the walls, click the **Display Configuration** menu of the Drawing Window status bar and select **Diagnostic** as shown in Figure 2T.19.
20. To view the cleanup radius, select all the walls in the drawing, and edit the Cleanup radius to **6"** in the Advanced section of the Properties palette as shown in Figure 2T.20.
21. Move the pointer over the workspace, and press ESC to view the cleanup radii as shown in Figure 2T.21.

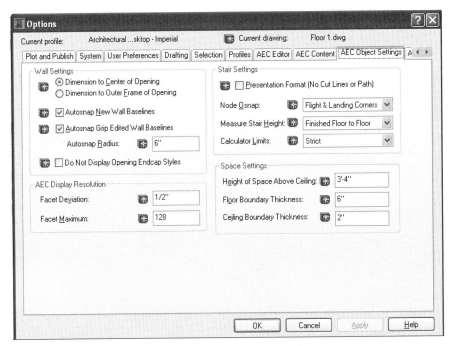

Figure 2T.18 *AEC Object Settings tab of the Options dialog box*

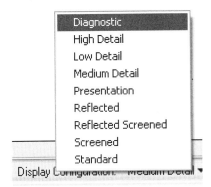

Figure 2T.19 *Diagnostic display configuration specified*

22. Select all the walls in the drawing, and set the Cleanup radius equal to **0** in the Advanced section of the Properties palette.

23. Move the pointer over the workspace, and press ESC to view the change in the cleanup radii.

Creating Floor Plans 129

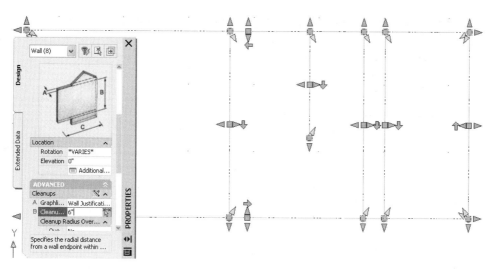

Figure 2T.20 *Walls selected for change of cleanup radius*

24. Click the **Display Configuration** menu of the Drawing Window status bar and select **Medium Detail**.
25. Use **WallOffsetCopy** to add walls **A** and **B** as shown in Figure 2T.22.
26. Save and close the drawing.
27. Select **File>Project Browser** from the menu bar. Select the current project, right-click, and choose **Close Current Project**.

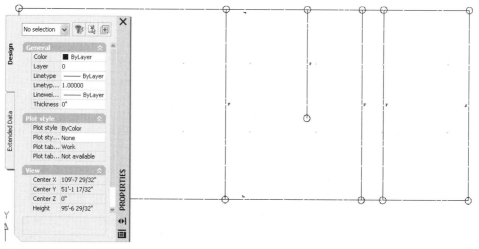

Figure 2T.21 *Cleanup radius displayed and graph line of walls*

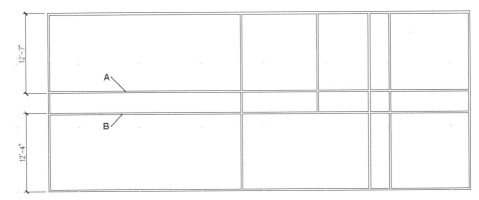

Figure 2T.22 *Dimensions for additional walls*

TUTORIAL 2.4 EDITING WALLS USING EDIT JUSTIFICATION, REVERSE, WALL OFFSET, JOIN, AND MERGE WALL

1. Select **Open** from the **Standard** toolbar.
2. Select your ADT Student\ADT Tutor\Ch2\Ex2-4 from the **Select File** dialog box, and select **Open** to close the **Select File** dialog box.
3. Choose **File>SaveAs** from the menu bar and save the drawing as **Lab 2-4** in your student directory.
4. Select the Model tab.
5. If tool palettes are not displayed, select **Window>Tool Palettes** from the menu bar.
6. Verify that POLAR, OSNAP, and OTRACK are toggled ON on the status bar.
7. Move the pointer over the OSNAP button on the status bar, right-click, and choose **Settings**. Edit the **Drafting Settings**, **Object Snap** tab as follows: clear all Osnap modes except **Endpoint** and **Node**. Select the **Options** button of the **Object Snap** tab, and toggle ON **Display polar tracking vector**, **Display full screen tracking vector**, **Display Auto Track tooltip**, and **Display AutoSnap tooltip**. Select **OK** to dismiss the **Options** dialog box and select **OK** to dismiss the **Drafting Settings** dialog box.
8. Select Wall A as shown in Figure 2T.23, right-click and choose **Edit Justification** from the shortcut menu. Move the pointer over the gray justification marker to identify the justification of the wall as shown in Figure 2T.23.
9. Move the pointer over the Left justification marker to display the tip and then click on the Left marker to change the justification as shown in Figure 2T.24.

Creating Floor Plans 131

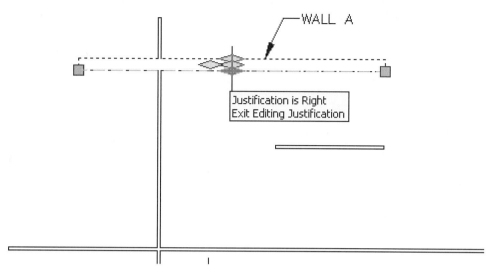

Figure 2T.23 *Justification markers displayed for the Edit Justification command*

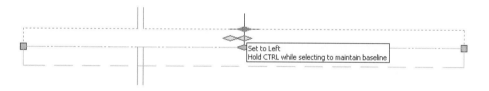

Figure 2T.24 *Dynamic display of wall with left justification*

10. Right-click and choose **Reverse>Baseline** from the shortcut menu.

11. Select Wall A shown in Figure 2T.23 and verify the wall has left justification and the directional trigger grip is reversed to point left. Press ESC to clear grip display of the wall.

12. Select Wall B shown in Figure 2T.25, right-click, and choose **Offset>Set From**. Respond to the following command prompts as shown below to move it **24'** from Wall C.

    Command: WallOffsetSet

    Select the component to offset from: *(Select the left face of Wall B as shown in Figure 2T.25.)*

    Select a point or wall component. *(Move the cursor to the right wall face of wall C as shown in Figure 2T.25.)*

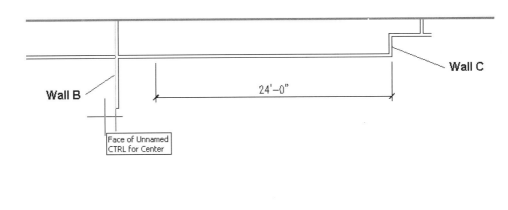

Figure 2T.25 *Offset component selected for Offset Set From command*

Enter new distance between the selected points. 24' ENTER
(Type distance in the tooltip box on screen, and press ENTER, and the wall moves 24' from wall C as shown in Figure 2T.26.)

13. Select wall D as shown in Figure 2T.27, right-click, and choose **Join** from the shortcut menu. Respond to the command line prompts as shown below.

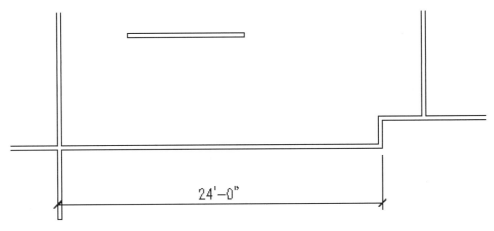

Figure 2T.26 *Wall moved using the Offset Set From command*

*(Select a wall segment, right-click, and choose **Join** from the shortcut menu.)*

Command: WallJoin

Select second wall: *(Select adjoining wall segment at p1 in Figure 2T.27.)*

Merged walls successfully

14. Select the wall at **p1** in Figure 2T.27 to verify the wall is joined. Press ESC to deselect the wall.

15. Select the wall at **p2** as shown in Figure 2T.28, right-click, and choose **Cleanups>Apply 'T' cleanup** from the shortcut menu. Respond to the command line prompts as shown below.

    (Wall selected at p2 as shown in Figure 2T.28.)

    Command: WallCleanupT

    Select boundary wall: *(Select wall at p3 as shown in Figure 2T.28.)*

    (Wall trimmed as shown in Figure 2T.28.)

16. Select the wall at **p1** as shown in Figure 2T.29, right-click, and choose **Cleanups>Apply 'L' cleanup** from the shortcut menu. Respond to the command line prompts as shown below.

    (Wall selected at p1 as shown in Figure 2T.29.)

    Command: WallCleanupL

    Select the second wall: *(Select wall at p2 as shown in Figure 2T.29.)*

    (Wall trimmed as shown in Figure 2T.29.)

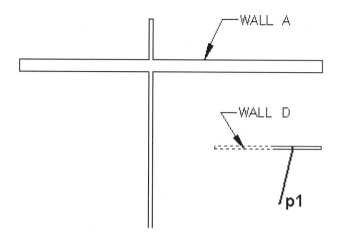

Figure 2T.27 *Wall segment selected for Join operation*

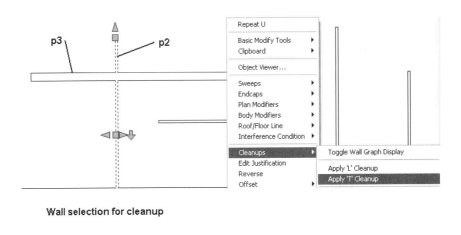

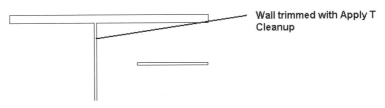

Figure 2T.28 *Wall selected for Wall Cleanup T*

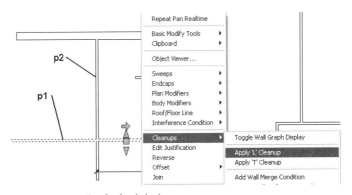

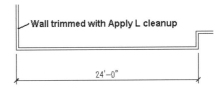

Figure 2T.29 *Wall trimmed using Apply 'L' Cleanup*

17. Select Wall E as shown in Figure 2T.30, right-click, and choose **Cleanups>Add Wall Merge Condition** from the shortcut menu. Respond to the command line prompts as shown below.

    (Wall E selected at p1 in Figure 2T.30; right-click and choose **Cleanups>Add Wall Merge Condition**.*)*

    Command: WallMergeAdd

    Select walls to merge with: **1 found** *(Select wall at p2 as shown in Figure 2T.30.)*

    Select walls to merge with: ENTER *(Press ENTER to end selection.)*

    Merged with 1 wall(s).

18. Select the wall at **p3**, as shown in Figure 2T.30, right-click and choose **Cleanups>Add Wall Merge Condition** from the shortcut menu. Respond to the command line prompts as shown below.

    (Wall selected at p3 in Figure 2T.30, right-click and choose **Cleanups>Add Wall Merge Condition**.*)*

    Command: WallMergeAdd

    Select walls to merge with: **1 found** *(Select wall at p1 as shown in Figure 2T.30.)*

    Select walls to merge with: **1 found** *(Select wall at p2 as shown in Figure 2T.30.)*

    Select walls to merge with: ENTER *(Press ENTER to end selection.)*

    Merged with 1 wall(s).

    (Wall E merged as shown in Figure 2T.30.)

19. Save and close your drawing.

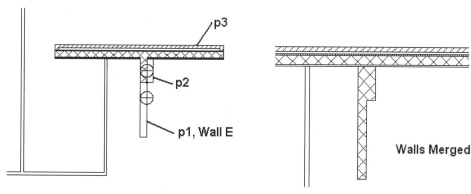

Figure 2T.30 *Wall selected for Merge Wall*

TUTORIAL 2.5 EDITING WALLS USING TRIM, MATCH, AND ADD SELECTED COMMANDS

1. Select **Open** from the **Standard** toolbar.

2. Select your ADT Student\ADT Tutor\Ch2\Ex2-5 from the **Select File** dialog box, and select **OK** to close the **Select File** dialog box.

3. Choose **File>SaveAs** from the menu bar and save the drawing as **Lab 2-5** in your student directory.

4. Select the Work tab.

5. If tool palettes are not displayed, select **Window>Tool Palettes** from the menu bar.

6. Verify that POLAR, OSNAP, and OTRACK are toggled ON on the status bar.

7. Move the pointer over OSNAP button on the status bar, right-click, and choose **Settings**. Edit the **Drafting Settings**, **Object Snap** tab as follows: clear all Osnap modes except **Endpoint** and **Node**. Select the **Options** button of the **Object Snap** tab, toggle ON **Display polar tracking vector**, **Display full screen tracking vector**, **Display Auto Track tooltip**, and **Display AutoSnap tooltip**. Select **OK** to dismiss the **Options** dialog box, and select **OK** to dismiss the **Drafting Settings** dialog box.

8. Right-click over the graphics screen, and choose **Basic Modify Tools>Extend** from the shortcut menu. Respond to the following command line prompts as shown below.

    (Extend command selected from graphics screen shortcut.)

    Command: _extend

    Current settings: Projection=UCS, Edge=None

    Select boundary edges ...

    Select objects: **1 found** *(Select wall at p1 shown in Figure 2T.31.)*

    Select objects: ENTER *(Press ENTER to end selection of boundaries.)*

    Select object to extend or shift-select to trim or [Project/Edge/Undo]: *(Select wall at p2 shown in Figure 2T.31.)*

    Select object to extend or shift-select to trim or [Project/Edge/Undo]: ENTER *(Press ENTER to end selection.)*

    (Wall extended as shown in Figure 2T.31.)

9. Select **Wall** from the Design tool palette and respond to the following command line prompts to create a new wall, which matches the properties of an existing wall.

    Command: WallAdd

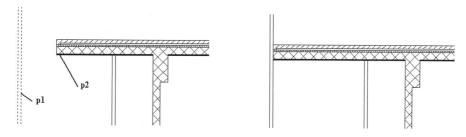

Figure 2T.31 Wall selected for the Extend command

Start point or
[STyle/Group/WIdth/Height/OFfset/Justify/Match/Arc]: **m** ENTER
(Match option selected.)

Select a wall to match: *(Select the wall at p1 in Figure 2T.32.)*

Match [Style/Group/Width/Height/Justify] <All>: ENTER *(Press ENTER to match all properties.)*

** Width does not apply to this wall style. ** *(Width is predefined in the wall style.)*

Start point or
[STyle/Group/WIdth/Height/OFfset/Justify/Match/Arc]: *(Select the endpoint of an existing wall at p2.)*

(Move the cursor to the left at a polar angle of 180°)

End point or
[STyle/Group/WIdth/Height/OFfset/Justify/Match/Arc]:
10' ENTER

(Wall drawn shown in Figure 2T.32.)

End point or
[STyle/Group/WIdth/Height/OFfset/Justify/Match/Arc]: ENTER
(Press ENTER to end the command.)

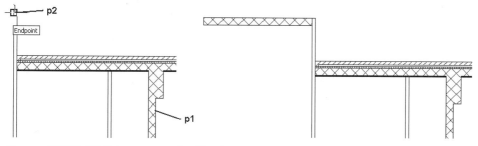

Figure 2T.32 Wall drawn using the Match option

10. Select the wall at **p3** shown in Figure 2T.33, right-click, and choose **Add Selected** from the shortcut menu. Respond to the command line prompts as shown below to add a wall.

(Add Selected chosen from shortcut menu of selected wall.)

Command: WallAddSelected

Start point or [STyle/Group/WIdth/Height/OFfset/Justify/Match/Arc]: *(Edit the Floor line offset from baseline to –24 in the Properties palette to extend the wall below the floor.)*

Start point or [STyle/Group/WIdth/Height/OFfset/Justify/Match/Arc]: *(Select Node object snap of wall shown in Figure 2T.33.)*

(Move the cursor down at a polar angle of 270°.)

End point or [STyle/Group/WIdth/Height/OFfset/Justify/Match/Arc]: **10'** ENTER *(Wall drawn 10' long as shown in 2T.33.)*

End point or [STyle/Group/WIdth/Height/OFfset/Justify/Match/Arc/Undo]: ENTER *(Press ENTER to end the command.)*

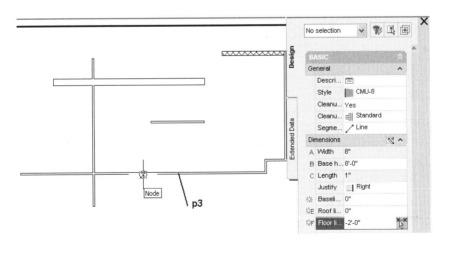

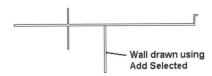

Figure 2T.33 *Wall drawn using Add Selected*

11. Select **SW Isometric** from the **View** flyout of the **Navigation** toolbar to view the walls as shown in Figure 2T.34.
12. Save and close the drawing.

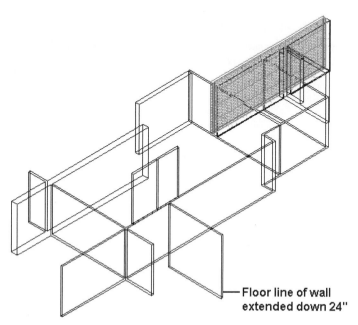

Figure 2T.34 *SW Isometric view of walls*

PROJECT

Exercise 2.1 Drawing the Walls of a Floor Plan

1. Open Autodesk Architectural Desktop 2005, and select **File>Project Browser** from the menu bar. Use the Project Selector drop-down list to navigate to your *ADT Student\Adt Tutor\Ch 2* directory. Double-click on **Ex 2-6** to set the project current. Select **Close** to dismiss the **Project Browser**.

2. Select the **Constructs** tab of the **Project Navigator**. Double-click on **Floor 1** to open the Floor 1 drawing.

3. Add additional walls as dimensioned in Figure 2T.35.

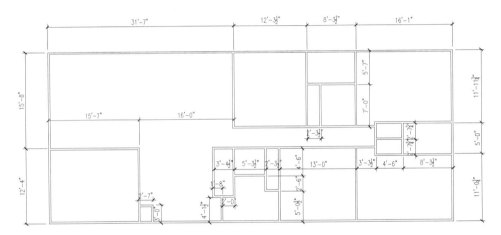

Figure 2T.35 *Wall placement dimensions*

CHAPTER 3

Advanced Wall Features

INTRODUCTION

Autodesk Architectural Desktop provides additional tools to enhance the creation of walls that consist of multiple wall components. The use of wall styles, wall groups, wall endcaps, wall surface modifiers, and body modifiers increases the speed of design and the detail of wall representation.

OBJECTIVES

After completing this chapter, you will be able to

- Create and modify a wall style
- Change the style of a wall
- Control the display properties of wall components to include materials
- Create wall endcap styles, wall modifier styles, and body modifiers to modify the properties of walls
- Use Edit In Place to edit wall endcaps, wall modifiers, and body modifiers
- Use the **Project Navigator** to develop additional floor plans and a foundation plan

ACCESSING WALL STYLES

Although the Standard wall style may be adequate in many design applications, it consists of only two wall lines. Other wall styles can be used to create wall representations for complex walls that consist of several wall components. Wall styles allow the designer to create a wall style for each wall type. Wall styles created in the drawing are listed in the **Style** drop-down list of the Properties palette. Other defined wall styles can be selected from the Walls tool palette as shown in Figure 3.1. Additional styles can be i-dropped into the drawing or onto tool palettes from the Walls category of the Design Tool Catalog-Imperial or Design Tool Catalog-Metric as shown in Figure 3.1. The Walls category consists of Brick, Casework, CMU, Concrete, and Stud subcategories.

When you click on a style from the Wall palette, the **WallAdd** command begins with the style preset in the Properties palette. After you place a wall style, the style is added to the styles listed in the **Style Manager** and in the **Style** drop-down list of the Properties palette as shown in Figure 3.2. Therefore the Properties palette retains a running list of the styles used in the drawing. If you want to change the style of a wall, select the wall and select a style from the **Style** list in the Properties palette. You can use the **Style Manager** to bring additional styles into the drawing that are not listed on the Walls tool palette. When you import or create a new style, it is listed in the Style list of the Properties palette.

CREATING AND EDITING WALL STYLES

The **Style Manager** is used to create, edit, import, and export wall styles in drawings. Unused styles of the drawing can be purged or deleted from the drawing. Additional styles can be i-dropped from the Design Tool Catalog or imported within the **Style Manager** from the *C:\Documents & Settings\All Users\Application Data\Autodesk\ADT 2005\enu\Styles* directory. Access the **Style Manager** as shown in Table 3.1.

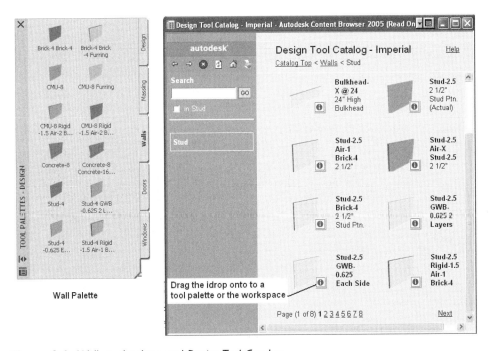

Figure 3.1 *Walls tool palette and Design Tool Catalog*

Advanced Wall Features

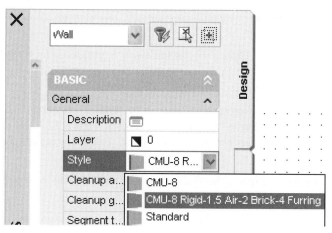

Figure 3.2 *Styles listed in the Style list of the Properties palette*

Menu bar	Format>Style Manager
Command prompt	WALLSTYLE
Command prompt	AECSTYLEMANAGER

Table 3.1 *Style Manager access*

When the **Style Manager** command is executed, the **Style Manager** dialog box opens as shown in Figure 3.3. It contains a menu bar and a toolbar. The commands of the toolbar are shown in Figure 3.3 and described below:

> **New Drawing** – Creates a new drawing file in which you can create new styles rather than create the styles in the current or existing drawings.
>
> **Open Drawing** – Opens the **Open Drawing** dialog box, which allows you to select drawings that are not currently opened. **Open Drawing** allows you to open other drawings and access their styles in the **Style Manager**. You can access predefined ADT styles from the imperial and metric folders of the *C:\Document and Settings\All Users\Application Data\Autodesk\ADT 2005\enu\Styles* folder.
>
> **Copy** – Copies a selected wall style or wall style folder to the clipboard.
>
> **Paste** – Pastes from the clipboard a wall style or wall style folder into another drawing listed in the tree of the left pane.
>
> **Edit Style** – Opens the **Wall Style Properties** dialog box, which allows you to define the detail of the selected wall style.
>
> **New Style** – Creates an edit field to type a new name for the wall style.

Set From – Specifies geometry for style definitions such as profiles and wall endcaps.

Purge Styles – Opens the **Select Wall Style** dialog box, which lists the wall styles not used in the drawing. When a wall style is purged from the drawing, its definition is deleted from the drawing.

Toggle View – Toggles between sorting styles by drawing or sorting styles by style type. The **Style Manager** on the left in Figure 3.4 is set to sorting styles by drawing, while the **Style Manager** on the right is sorting styles by style type.

Filter Style Type – Limits the display of styles in the tree view to a style selected by the filter as shown by Figure 3.5.

Details – Controls the format for displaying the contents in the right pane. Options for Icon Format include Small Icons, Large Icons, List, and Details.

When the **AecWallStyle** command opens the **Style Manager**, all styles of the drawing are listed in one of the following three categories: Architectural Objects, Documentation Objects, and Multi-Purpose Objects folders. Wall styles are included in the Wall Styles folder of the Architectural Objects folder. When the **WallStyle** command opens the **Style Manager**, only the Wall Styles are listed. When you

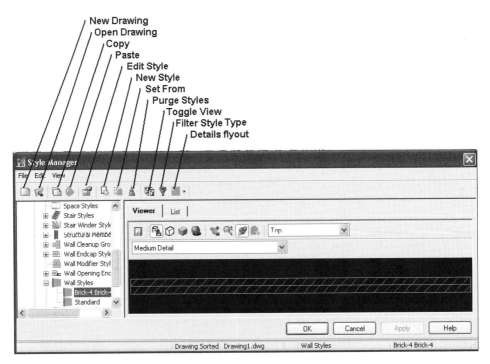

Figure 3.3 *Style Manager dialog box*

Advanced Wall Features 145

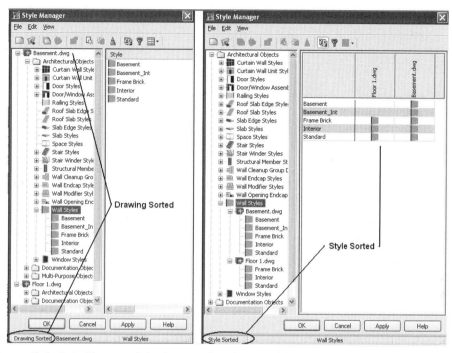

Figure 3.4 *Style Manager dialog boxes with styles sorted by drawing and by type*

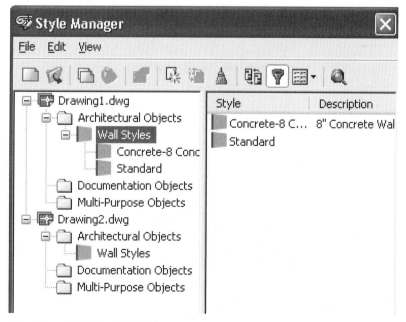

Figure 3.5 *Filter applied to Style Manager*

select the (+) sign, the style type expands, and the styles used in the drawing are shown in the left pane. When you select a wall style in the left pane, the description of the style is displayed in the right pane. The details of the selected style can be viewed by the **Viewer** or **List** tab. The **Viewer** tab of a selected wall is shown in Figure 3.6.

The **Viewer** tab includes a viewing toolbar and display representations list as shown in Figure 3.6. The **Viewing** toolbar allows you to view the wall style based upon the specified display representation and from the following view directions: Top, Bottom, Left, Right, Front, Back, SW Isometric, SE Isometric, NE Isometric, and NW Isometric. After you specify the view direction, the wall can be shaded using the selections of the **Shade** toolbar inside the **Style Manager**. The shortcut menu inside the **Viewer** tab includes options to change the view direction and shading of the wall style.

The **List** tab will display the properties of the wall style in text format as shown in Figure 3.7.

CREATING A NEW WALL STYLE NAME

The first step to create a new wall style is to establish a new name using the **New Style** button of the **Style Manager** as shown in Figure 3.8. Names for wall styles should be kept short and descriptive. Spaces are permitted in the wall style name; however, the (") inch symbol is not accepted. Once a name has been created, it will be listed in the

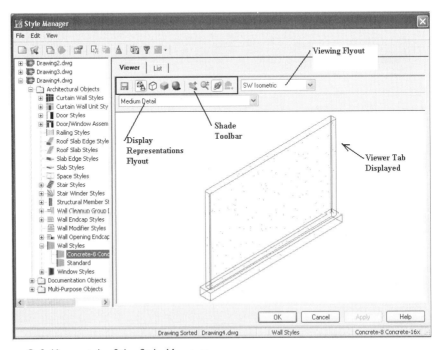

Figure 3.6 *Viewer tab of the Style Manager*

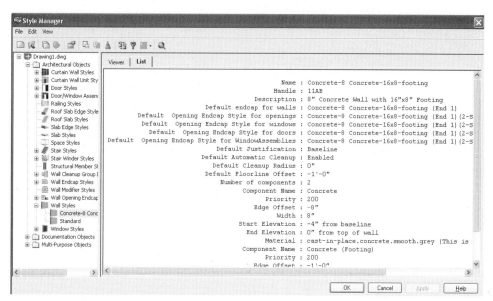

Figure 3.7 List tab of the Style Manager

style list of the left pane or tree view. When you select the **New Style** button on the **Style Manager** toolbar, the default name "New Style" is assigned and will be highlighted. You can overtype to enter a desired name. Specific properties of a wall style are specified when you select the **Edit Style** button on the **Style Manager** toolbar as shown in Figure 3.9.

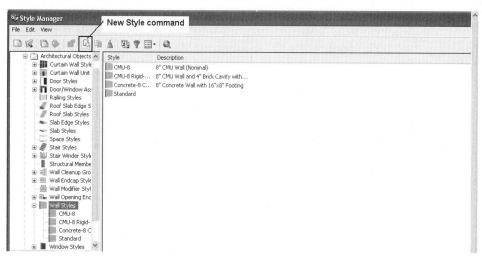

Figure 3.8 New Style command of the Style Manager

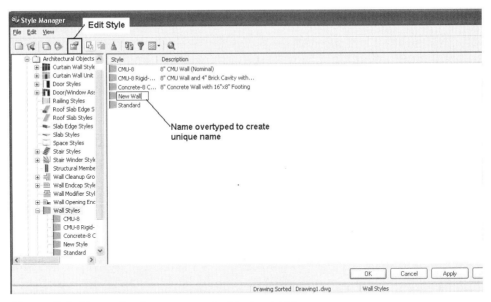

Figure 3.9 *New wall style name overtyped in the right pane*

STEPS TO CREATING A NEW WALL STYLE

1. Select **Format>Style Manager**.
2. Select a (+) sign of the Architectural Objects folder in tree view of the left pane.
3. Select the (+) sign of the Wall Styles folder.
4. Select the **Wall Styles** category, and then select the **New Style** button from the **Style Manager** toolbar as shown in Figure 3.8.
5. Overtype a new name in the right pane as shown in Figure 3.9.

Once a new name has been created and selected, the **Edit** button of the **Wall Styles** dialog box is used to define specific properties of the new style. Editing the wall style allows you to define the number and size of wall components, endcap styles, components, materials, classifications, and display representation of the wall components.

CREATING A NEW STYLE FROM AN EXISTING STYLE

The **CopyAndAssignStyle** command allows you to select an object and create a new style from the style definition of the selected object. The new style definition will be assigned to the selected object. The **CopyAndAssignStyle** command can be accessed from the shortcut menu of a selected wall as shown in Table 3.2. This command, when applied to a wall, creates a copy of the wall style assigned to the selected wall and opens the **Wall Style Properties** dialog box. The name and other properties of the new style can then be changed in the **Wall Style Properties** dialog box.

Command prompt	AECCOPYANDASSIGNSTYLE
Shortcut menu	Select a wall, right-click, and choose Copy Wall Style and Assign Style

Table 3.2 *AecCopyAndAssignStyle command access*

DEFINING THE PROPERTIES OF A WALL STYLE

Editing a wall style can occur prior to or during the use of the wall style. The current definition for the wall style is applied to all instances of the wall in the drawing. Therefore, if a wall style is defined initially to consist of three wall lines with a total width of 12" and the style is later edited to a width of 10", all previous walls placed in the drawing using this style will automatically be updated to 10"-wide walls. Using wall styles allows you to globally edit all the walls of an entire building by changing the original wall style definition.

To define a wall style, select the **Edit Style** button of the **Style Manager** as shown in Figure 3.9. Selecting the **Edit Style** button opens the **Wall Style Properties** dialog box, which consists of the following tabs: **General**, **Components**, **Materials**, **Endcaps/Opening Endcaps**, **Classifications**, and **Display Properties**. Each tab is used to define the characteristics of a wall style. A description of the options in each tab follows:

General Tab

The **General** tab for the Concrete-8 Concrete 16x8 footing wall style is shown in Figure 3.10.

> **Name** – Editing the name in essence renames the wall style. You cannot delete all of the text in the **Name** edit field—at least one alphanumeric character must remain.
>
> **Description** – You can edit the description by clicking in the **Description** edit field and typing text. Because new wall styles are created without a description, the **Description** edit field allows you to add the descriptive information. It can include specific sizes and material types that distinguish the wall style. The text in the **Description** edit field is optional.
>
> **Keynote** – Keynotes may be defined for the wall style. Choosing the **Select Keynote** button will open the **Select Keynote** dialog box, which allows you to specify keynotes from the 16 CSI divisions. Keynote commands of the Annotation palette can be used to add the keynote annotation when you select the wall.

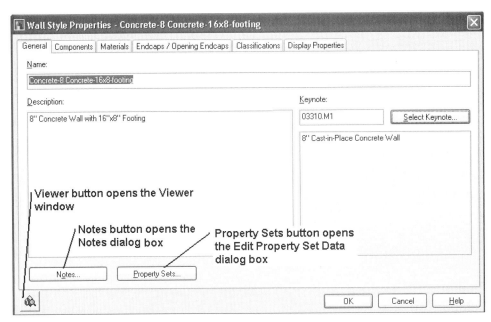

Figure 3.10 *General tab of the Wall Style Properties dialog box*

Notes – Selecting the **Notes** button shown in Figure 3.10 opens the **Notes** dialog box, which includes two tabs, **Notes** and **Reference Docs**, shown in Figure 3.11. To add text to the **Notes** tab, click in the edit field and begin typing. Information regarding construction notes or costs for a given wall style can be entered in the **Notes** tab. The **Select All** button in the lower right corner of the **Notes** dialog box allows you to select all the text in the **Notes** dialog box. Once the text is selected, right-click to display a shortcut menu that includes **Undo**, **Cut**, **Copy**, **Paste**, and **Delete** operations. Choosing these shortcut menu options allows the selected text to be manipulated through Windows operations and the Clipboard. The **Reference Docs** tab shown at right in Figure 3.11 allows you to list other files that can contain details of the construction or text files that describe the wall. Selecting the **Add** button shown in the lower left corner opens the **Select Reference Document** dialog box, which allows you to select a file and directory path for reference files. Selecting the **Edit** button allows you to edit the path shown for the reference document file or the description of the file. The **Description** field would be useful for describing the content of the reference document. The **Delete** button is used to delete a file from the list. Using a reference file list allows you to cross-reference drawings while working on a design file.

Property Sets – The **Property Sets** button opens the **Edit Property Set Data** dialog box. This dialog box allows you to edit the property sets associated with a wall style. Property data is used to develop schedules from the drawing and will be presented in Chapter 10.

Viewer – The **Viewer** button located in the lower left corner of the **General** tab as shown in Figure 3.10 opens the **Viewer** and allows you to preview the graphics of the wall style. A view of the wall can be either an orthographic or a selected pictorial view. The view shown in Figure 3.12 is a pictorial view of the Concrete-8 Concrete 16x8 footing wall style.

Components Tab

The components of a wall style represent physical components of the wall such as brick veneer, sheathing, and drywall. The **Components** tab consists of two panes. The component shown in the left pane is a dynamic view of the wall style; therefore, as you edit the component changes are reflected in the left pane. The shortcut menu of the left pane includes zoom and view direction options. The left pane shown in Figure 3.13 is the left side view of the wall. The right pane includes a list of the wall components and their properties. The properties of the components are set in the right pane and displayed in the left pane. The five buttons shown on the right margin of the tab allow you to add or remove components. The options of this tab are described below.

> **Add Component** – Selecting the **Add Component** button will create a new wall component identical to the component highlighted in the component list.
>
> **Remove Component** – The **Remove Component** button will remove the highlighted wall component from the list. Remove components by first selecting the Index number of the wall component and then selecting the **Remove Component** button.

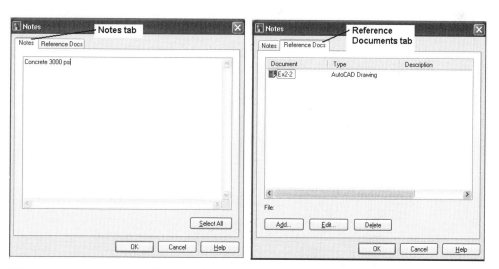

Figure 3.11 *Notes and Reference tabs of the Notes dialog box*

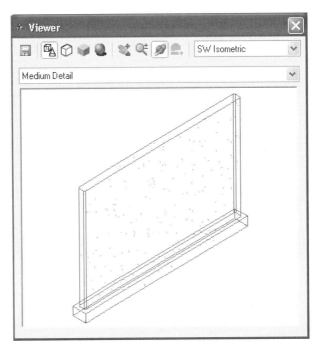

Figure 3.12 *Wall displayed in the Viewer*

Move Component Up In List – The **Move Component Up In List** button will move a selected wall component up within the list of components.

Move Component Down In List – The **Move Component Down In List** button will move a selected wall component down within the list of components.

Wall Style Browser – The **Wall Style Browser** button opens the **Wall Style Components Browser** dialog box, shown in Figure 3.14. The **Wall Style Components Browser** displays the wall components included in the wall styles of the drawing. The **Open** button on the toolbar allows you to open other drawings and import their wall components. The left pane of the **Wall Style Components Browser** lists the wall styles. As shown in Figure 3.14, a component of other wall styles can be selected in the left pane and the definition of a wall component from that style copied, using the shortcut menu, and pasted into the current style.

Index – The **Index** number identifies each wall component. The components of the Concrete-8 Concrete 16x8 footing wall style are shown in Figure 3.13. The Concrete-8 Concrete 16x8 footing wall style consists of (1) Concrete and (2) Concrete (footing) wall components.

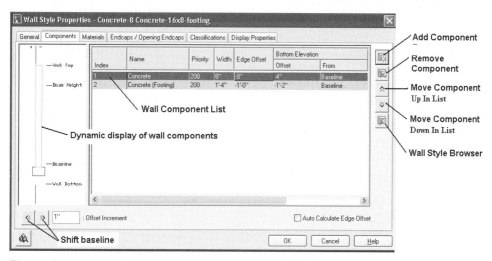

Figure 3.13 Components tab of the Wall Style Properties dialog box

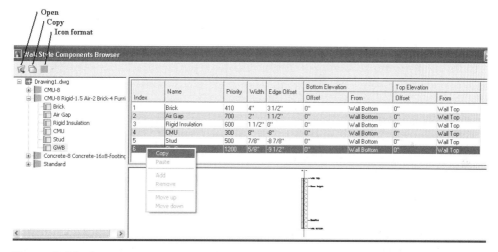

Figure 3.14 Components from wall styles listed in the Wall Style Components Browser

Name – Click in the **Name** edit field to type the name of a wall component. Spaces are permitted in the name; however, the name should be kept short and descriptive.

Priority – The **Priority** edit field allows you to assign a priority number to control how the wall component will clean up with other walls. Wall components with low priority numbers will override the display of wall components with higher priority numbers. The walls shown in Figure 3.15 were drawn with

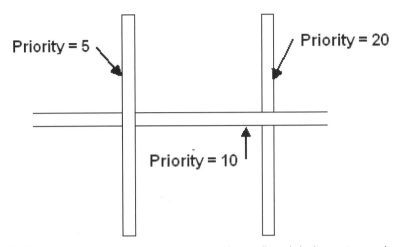

Figure 3.15 Walls with low priority numbers overriding walls with high priority numbers

priorities of 5, 10, and 20. The horizontal wall and the left wall have low component priority; therefore they override the display of the wall component with priority of 20.

Assigning priority numbers to wall components would be useful for controlling the display of components of fire-rated walls. Fire-rated walls are displayed without penetrations from other non-rated walls. Therefore, in the example shown in Figure 3.15, the right vertical wall could correctly represent a non-rated wall that intersects the horizontal fire-rated wall. Assigning priority to wall components controls the cleanup behavior of a wall when wall components intersect other walls.

The use of priorities to control the display of the CMU component is applied to the wall shown in Figure 3.16. The CMU is assigned the lowest priority number, 300. The Brick has a priority number of 810 and the Air Gap has a priority of 700; therefore the CMU display will override the lines that represent the brick and air gap. Because the CMU priority number is the lowest, the CMU display will override that of the Brick and Air Gap components, with equal priority numbers merging together.

> **Width** – The **Width** of a wall component is the distance between the faces of the component. The width is defined for the component by selecting the drop-down list as shown in Figure 3.17. The drop-down list defines a formula, which may consist of a fixed value, **Base Width**, operator, and operand. A fixed value is set by typing a value in the fixed value edit field and setting the **Base Width** value to 0 as shown in Figure 3.17.

Selecting the **Base Width** or other values from the drop-down list sets a variable value in the formula. The variable value is applied to the operator and operand in the for-

Advanced Wall Features 155

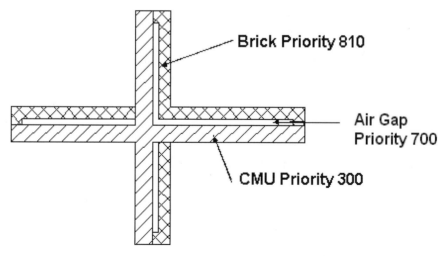

Figure 3.16 *The CMU-6 Air-2 Brick-4 wall style*

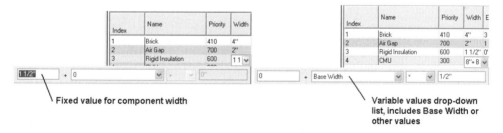

Figure 3.17 *Setting component width*

mula. The **Base Width** value is the width extracted from the **Width** edit field of the Properties palette of the wall. The operator and operand are then applied to the base value. The operator can be add, subtract, divide, or multiply. The **Base Width** is multiplied by as shown in Figure 3.17.

Positive or negative distances can be entered in the **Width** edit field as shown in Figure 3.18. The results of the width dimension setting is displayed in the left pane.

> **Edge Offset** – The **Edge Offset** column is used to define the location of a wall component relative to the **Baseline**. To set the **Edge Offset** distance, click in the **Edge Offset** column and select options from the drop-down list. The edge offset can be set to a fixed value or variable offsets. Figure 3.19 includes the settings for fixed and variable values.

Typing a value in the fixed offset field and setting the variable offset to zero will specify a fixed offset as shown at left in Figure 3.19. A variable offset is specified by select-

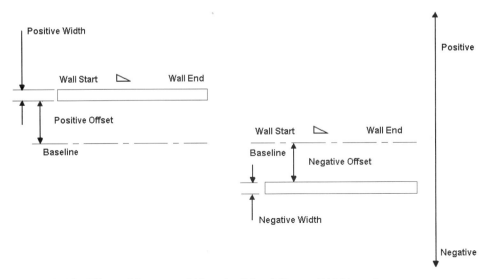

Figure 3.18 *Effects of Positive and Negative Edge Offset and Width settings*

ing the **Base Width** option as shown in the drop-down list. The **Base Width** value is extracted from the A-Width of the wall in the Properties palette. The **Base Width** is applied to the operator and operand as shown in Figure 3.19. The Operator options include add, subtract, multiply, and divide. The operand is a number applied to the operator and the base width. This method of specifying the edge-offset distance is used in the Standard wall style. The Standard wall style specifies the offset as BW*−1/2. This statement will multiply −1/2 times the base width to specify the offset. Therefore, if you were drawing with the Standard wall style and set the Width to 6" in the Properties palette, the offset distance would be −3, (6 × −1/2 = −3).

The variable offset method can also specify a fixed value, which is added to arithmetic operations of the **Base Width**. This method adds the fixed value to the **Base Width** to determine the arithmetic formula as shown at right in Figure 3.19.

The wall drawn in Figure 3.20 includes a brick and CMU wall components with zero edge offsets. The brick is displayed above the baseline because it has a +4 width and the CMU is displayed below the baseline because it has a −8 width value. The wall drawn in Figure 3.20 was drawn from left to right with baseline justification.

> **Bottom Elevation** – The **Bottom Elevation** column allows you to set the elevation of the bottom of each component. The bottom of the wall component will move up if the distance is assigned positive from the Wall Bottom and Baseline, or negative from the Wall Top and Base Height. Setting the Bottom Elevation above the wall bottom allows you to create a cavity wall, as shown in Figure 3.21.

Advanced Wall Features

Figure 3.19 *Fixed and variable settings*

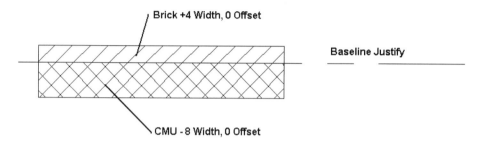

Figure 3.20 *Components with 4 and −8 wall widths and zero offset*

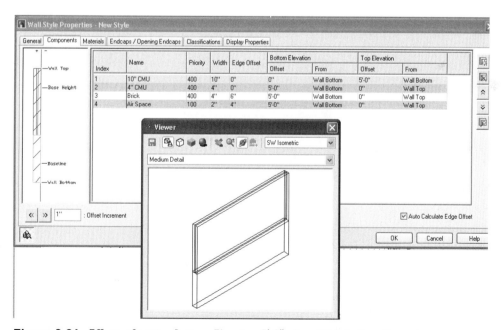

Figure 3.21 *Effects of setting Bottom Elevation 5'-0" above Wall Bottom for cavity components*

In Figure 3.21, the brick, air space, and 4" CMU components have a Bottom Elevation of 5' from Wall Bottom and a Top Elevation of zero from Wall Top. The 10" CMU wall has a Bottom Elevation of zero and Top Elevation of 5' from Wall Bottom. The top elevation restricts the height of this component. Controlling the elevation of wall components allows the wall style to simulate actual construction. A footing can be created as a wall component with a bottom elevation below the baseline of the wall. This component can be projected down by the **FloorLine** command presented in this chapter.

 Caution: Setting the bottom elevation with a positive value relative to the wall top or base height will cause the component to disappear. The component is defined but not visible.

Top Elevation – The **Top Elevation** column allows you to set the top elevation of a wall component. The top of the wall component will move down if the distance is assigned a positive value from the Baseline and Wall Bottom, or negative from the Wall Top and Base Height. This allows a wall component to have a top elevation different from other components. In Figure 3.22 the 4" wall component has a top elevation of 4' above the wall bottom. This 4" wall component could represent brick veneer placed in front of a masonry wall component.

The top of a wall component can be moved down by assigning a positive elevation relative to the baseline or wall bottom. A negative elevation distance will move the top elevation of a component down if the height of a wall component is set relative to the base height or wall top. The top elevation of a wall component such as brick veneer can be designed to adjust with the **RoofLine** command to cover the wood framing between floors.

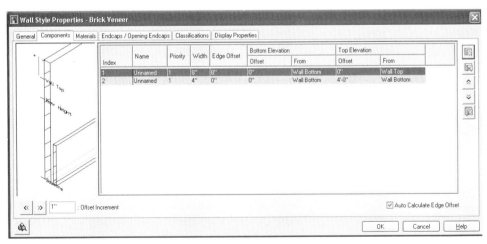

Figure 3.22 *Effects of one component Top Elevation Offset 4'-0"*

Caution: Setting the top elevation with a negative value relative to the wall bottom or baseline will cause the component to disappear. The component is defined but not visible.

Offset Increment – The **Offset Increment** text box allows you to specify a distance to incrementally shift the baseline. The baseline is shifted by selecting the **Shift Baseline** arrows as shown in Figure 3.13. When you select a wall component, you can click on the left directional arrow, and the component will step over 1" from the baseline. The result of the step is displayed in the offset list of the wall component.

Auto Calculate Edge Offset – The **Auto Calculate Edge Offset** check box, if selected, will adjust the offset equal to the previous width to avoid the overlap of components. In Figure 3.23, wall component Index 4 was the original component. Additional components 1, 2, and 3 were created with Auto Increment. The edge offset was increased by the width of 6" each time an additional component was added. The last component added was component 1.

Materials Tab

The Materials tab allows you to specify the name and definition of materials applied to the wall. The options of the Materials tab shown in Figure 3.24 are described below.

Component – The **Component** field lists the names of the wall components to apply materials.

Material Definition – The **Material Definition** column lists the material definitions for the component. If you click in this column, you can select from the list the material definitions included in the drawing. The Standard material definition is the only material definition listed unless materials definitions have been imported or other objects with material definitions have been inserted into the drawing. You can select **Open** from the **Style Manager** toolbar and import material definitions from the Material Definitions (Imperial).dwg of the *C:\Documents & Settings\All Users\Application Data\Autodesk\ADT 2005\enu\Styles\Imperial* directory. The materials are located in the Multi-

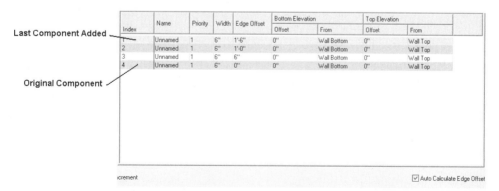

Figure 3.23 *Wall components added using Auto Calculate Edge Offset*

Purpose Objects folder of the Material Definition (Imperial).dwg file. The Material Definition (Imperial) file includes material representations for brick, concrete, and concrete masonry units.

Material definitions can also be created from materials i-dropped from the Autodesk Render Material Catalog in the **Content Browser**.

Edit Material – The **Edit Material** button opens the **Material Definition Properties** dialog box, to specify how material definitions are applied for surface and section rendering.

Add New Material – The **Add New Material** button opens the **New Material** dialog box, which allows you to specify a name for the new material.

Endcaps/Opening Endcaps Tab

The **Endcaps/Opening Endcaps** tab shown in Figure 3.26 allows you to edit how endcaps of a wall will be capped at the end of the wall or at an opening. The endcap of the brick and concrete masonry wall shown in Figure 3.25 has an endcap to return the brick to close the cavity. The **Endcaps/Opening Endcaps** tab shown in Figure 3.26 allows you to define an endcap style to be used for the ends of walls and the openings in a wall.

Endcap styles allow you to specify how the wall components terminate at the end of a wall. In Figure 3.26, the Concrete-8 Concrete 16x8 footing endcap style is assigned to cap the ends of the wall and the openings. Creating endcap styles will be discussed

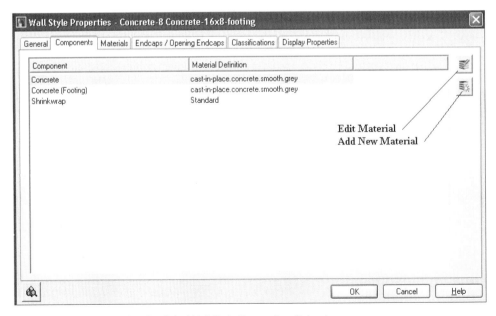

Figure 3.24 *Materials tab of the Wall Style Properties dialog box*

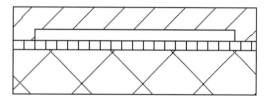

Figure 3.25 Brick and Concrete Masonry Wall

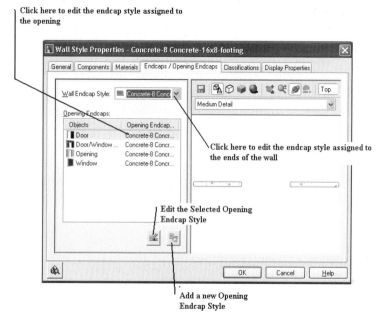

Figure 3.26 Endcaps/Opening Endcaps tab of the Wall Style Properties dialog box

later in this chapter. To change the endcap style assigned, click in the endcap list and select from the list of endcap styles in the drawing. The left pane of this tab include the following buttons:

> **Edit the Selected Opening Endcap Style** – This button opens the **Opening Endcap Style** dialog box, which allows you to edit the existing opening endcap style.

> **Add a New Opening Endcap Style** – Selecting this button opens the **Opening Endcap Style** dialog box to create a new opening endcap style.

The **Opening Endcap Style** dialog box has two tabs: **General** and **Design Rules**. The **General** tab allows you to change the name, description, and add notes. The **Design**

Rules tab allows you to assign endcaps to the start, end, sill, and head locations of the opening as shown in Figure 3.27. The wall opening endcap style specified in Figure 3.27 applies the jamb endcap named Concrete 8(End 1) to each jamb of the opening. You can specify different endcap styles for each of the four edges of the opening in the opening endcap style.

You can create and edit an opening endcap style from the **Style Manager** by selecting a style located in the Architectural Objects\Wall Opening Endcaps folder.

Classifications Tab

The **Classifications** tab allows you to assign a classification definition and classification to the wall style. **Classification Definitions** are created in the **Style Manager**. A **Classification Definition** includes classifications that create categories within an object style. You could create a Building Status classification definition that would consist of New and Existing classifications. Therefore a wall style could then be assigned the New or Existing classification. A display representation set could then be edited to show or hide all objects with the New classification. Classifications also allow the development of schedules restricted to one classification of an object style. A wall schedule could be developed which includes one classification within a wall style. Classifications allow you to sort objects with the same object style according to their classification. The **Classifications** tab shown in Figure 3.28 includes a Fire Rating classification definition and the 1 HR fire rating classification, which is selected from the drop-down list.

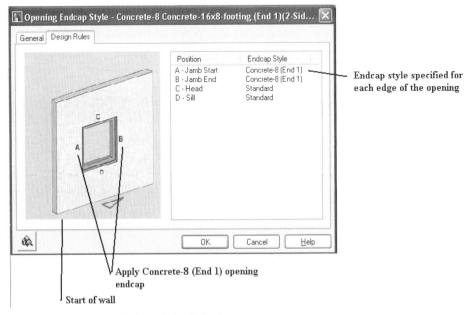

Figure 3.27 *Opening Endcap Style dialog box*

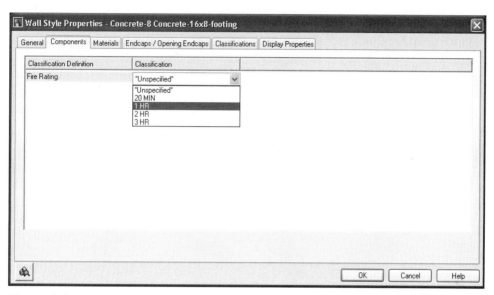

Figure 3.28 *Classifications tab of the Wall Style Properties dialog box*

If no classifications are defined in the **Style Manager**, this tab will not contain any information.

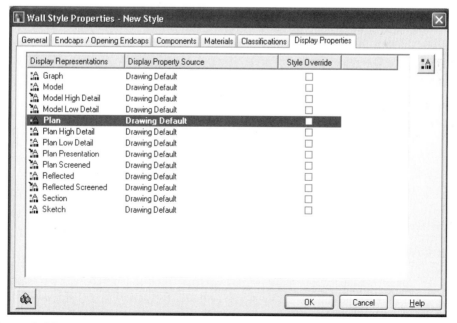

Figure 3.29 *Display Properties tab of the Wall Style Properties dialog box*

Display Properties Tab

The **Display Properties** tab shown in Figure 3.29 allows you to specify wall display according to the display representation. The available display representations of a wall are listed as shown in Figure 3.29. The display method for the wall can be specified for each of the display representations. The display representation that is current in the drawing is shown as bold. Therefore the **Display Properties** tab shown in Figure 3.29 allows you to edit the display for the Plan display representation.

The Plan display representation can be controlled by a **Style Override** or the **Drawing Default**. If you check the **Style Override** check box, the **Display Properties** dialog box opens, as shown at left in Figure 3.30. A **Style Override** lists each of the wall components defined by the wall style in the **Display Properties** dialog box. The **Display Properties** for the **Drawing Default** display of the wall is shown at right in Figure 3.30. This dialog box does not list each unique wall component of the wall. The wall components can be represented based upon the material assignment using either the **Drawing Default** or **Style Override**. The **Display Properties** dialog box allows you to specify if the wall component will be represented according to the assigned material or specified with unique layer, color, linetype, lineweight, and linetype scale.

If the wall is displayed as a **Style Override** without materials, hatching can be added or color and linetype altered to represent the materials of the wall. You can turn hatching on or off by controlling the display properties associated with the wall style. The display of selected wall components can be turned off. The display system will define how the first 20 components of a wall are displayed. The display of components with Index numbers greater than 20 are controlled according to the material assignment. Editing this tab controls the display properties of the wall as defined in the style. To create a **Style Override** for a wall style, you must override the **Drawing Default** display properties of the wall style.

The options of the tab shown in Figure 3.29 are described below.

> **Display Representations** – The **Display Representations** column lists the display representations available for the wall. The **Plan** display representation shown bold in Figure 3.29 is the current display representation set used in the active viewport.
>
> **Display Property Source** – The **Display Property Source** column lists object display categories for the current viewport. The display property source is either controlled by the **Drawing Default** global settings or by a **Style Override**.
>
> **Style Override** – The **Style Override** check box indicates if an override has been defined in the display properties. If this box is clear, the **Drawing Default** is controlling the display of the wall style. When you check this box, the **Display Properties** dialog box opens as shown in Figure 3.30.

Edit Display Properties – Selecting the **Edit Display Properties** button without checking the **Style Override** check box opens the **Display Properties** dialog box as shown at right in Figure 3.30.

The **Display Properties** dialog box consists of the tabs described below to define the display. The **Display Properties** dialog box shown in Figure 3.30 is from the **Wall Plan Display Representation**, which consists of four tabs to define the override of the style.

Layer/Color/Linetype Tab

The **Layer/Color/Linetype** tab allows you to specify the properties of the display components. The display components listed in this tab vary according to the display representation and wall style. The components of this tab are described below.

The **Layer/Color/Linetype** tab shown in Figure 3.30 is for a wall that consists of only two wall components: concrete and concrete footing. The **Layer/Color/Linetype** tab provides control of visibility, layer, color, linetype, line weight, and linetype scale for each component of the wall object. If a component is controlled "By Material," the layer, color, linetype, lineweight options are disabled as shown at right in Figure 3.30. The components of the wall object are described below.

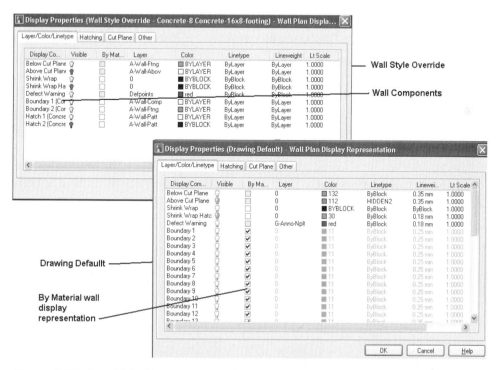

Figure 3.30 *Layer/Color/Linetype settings for a wall component*

Below Cut Plane – The **Below Cut Plane** component consists of the entity representation of the wall located at an elevation below the cutting plane. The default cutting plane elevation is 3'-6" for the Medium Detail display configuration defined in the **Display Manager**. You can alter the global height of the cutting plane by editing the **Display Manager** or overriding the cutting plane in the **Cut Plane** tab of the wall style. Turning on visibility turns on entities to represent objects below the cutting plane.

Above Cut Plane – The **Above Cut Plane** components consist of entities representing the object at an elevation above the cutting plane. In the dialog box shown in Figure 3.30, the visibility of components above the cutting plane is turned off.

Shrink Wrap – The **Shrink Wrap** component is displayed when the **Wall Interference** command is applied to a wall that intersects with other objects. Shrink-wrap entities add emphasis to the intersection of objects.

Defect Warning – The **Defect Warning** component is the wall defect marker displayed when walls are drawn either too short or coincident with other walls.

Boundary – The **Boundary** consists of the entities that represent the width of the wall components. In Figure 3.30, Boundary 1 (Concrete) and Boundary 2 (Concrete Footing) are the wall components created as Index 1 and Index 2. You can turn the display of each component off by turning off the visibility of the component. Display can be controlled for 20 boundaries.

Hatch – The **Hatch** component applies a hatch pattern for each wall component. The Hatch 1 (Concrete) and Hatch 2 (Concrete Footing) display components will apply the specified hatch pattern to each of the wall components. The hatch pattern applied is specified in the **Hatching** tab. To display hatching of a component, turn ON the light bulb in the **Visible** column.

The display of each component is controlled by editing the column for visibility, material, layer, color, linetype, lineweight, linetype scale, and plot style. The **Below Cut Plan** component is assigned to the A-Wall_Ftng layer. This layer has the hidden linetype property, and therefore the footing lines are displayed hidden.

Hatching Tab

The **Hatching** tab allows you to turn ON or OFF the hatching of the wall component. If **By Material** representation is turned OFF, the **Hatching** tab shown in Figure 3.31 can be used to represent the material. To change the hatch properties, click in the column and edit the hatch pattern, hatch scale factor, angle, orientation, and x offset, and y offset.

The type of hatch pattern, scale, angle, and orientation for the component can be set by the **Hatching** tab. You can edit the hatch pattern by selecting the **Pattern** column for the desired hatch. The default pattern is **User Single**, which will fill the hatch area with parallel lines. You can select other hatch patterns by selecting the **User Single** option of the **Pattern** column to display the **Hatch Pattern** dialog box as shown in Figure 3.32.

The **Hatch Pattern** dialog box allows you to select the pattern source from the pattern **Type** drop-down list, which includes Predefined, User-Defined, Custom, or Solid Fill. The Predefined option allows access to the AutoCAD hatch patterns. The User-Defined option, the default pattern of Architectural Desktop, will apply a single or double line hatch. The Custom option allows you to select custom hatch patterns from other hatch files, and the Solid Fill option will fill the hatch area with solid shading.

The hatch scale, angle, and orientation can be edited directly in the **Hatching** tab of the dialog box. To edit the scale of the hatch pattern, select the current scale; the scale value will be highlighted, and you can overtype the desired value for the scale. The orientation of the hatch can be set to **Object** or **Global**. The **Global** option will apply the angle of the hatch to all wall segments as if the wall were one continuous area, as shown below on the right. The **Object** orientation of hatch will orient each hatch according to the orientation of the wall. The **Object** orientation is shown on the left in Figure 3.33.

Cut Plane Tab

The **Cut Plane** tab allows you to set the cut plane height for the wall style. Cut plane elevations can be defined for the display configuration in the **Display Manager** or within the wall style. The options of the **Cut Plane** tab shown in Figure 3.34 are described below.

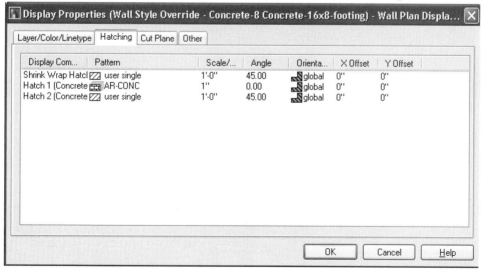

Figure 3.31 *Hatching tab for wall components*

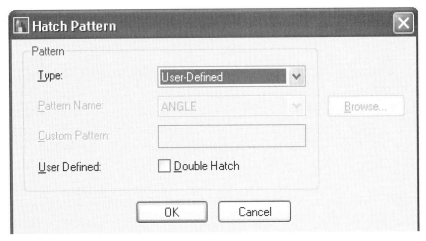

Figure 3.32 *Setting hatch pattern for wall components*

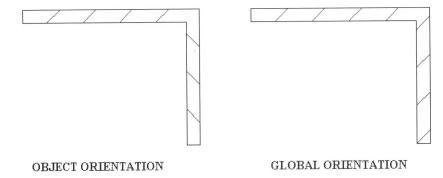

Figure 3.33 *Object and Global hatch orientation*

Override Display Configuration Cut Plane – This check box toggles control of the height of the cut plane to the wall style. Set the height of the cut plane by typing the height in the **Cut Plane Height** edit field.

Cut Plane Height – The **Cut Plane Height** is specified in the edit field.

Automatically Choose Above and Below Cut Plane Heights – This check box displays objects above and below the cut plane height.

Manual Above and Below Cut Plane Heights – This check box allows you to control the cut plane height uniquely by wall component. Selecting this check box allows you to add control to the display of each component.

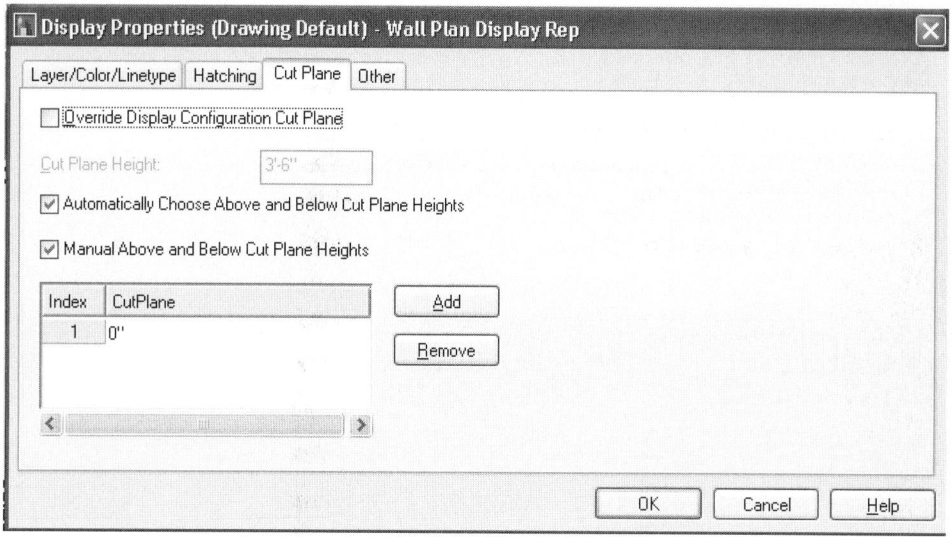

Figure 3.34 *Cut Plane tab*

Other Tab

The **Other** tab, shown in Figure 3.35, allows you to control the display of lines relative to the cutting plane. This tab allows you additional control of the wall components based upon their position relative to the cutting plane line.

This tab includes check boxes for turning on the visibility of openings, door frames, window frames, and inner wall lines that are cut by the cutting plane. Miter lines can also be checked for each of the wall components.

A foundation vent is shown in Figure 3.36. The wall is displayed incorrectly because the footing and wall should not break at the vent location. If the **Display Inner Lines Below** is toggled ON and **Hide Lines Below Openings at Cut Plane** is toggled OFF, the wall will be correctly displayed as shown at the far right of Figure 3.36.

STEPS TO OVERRIDING THE DRAWING DEFAULT DISPLAY REPRESENTATION FOR A WALL STYLE

1. Select a wall, right-click, and choose **Edit Wall Style** from the shortcut menu.

2. Select the **Display Properties** tab.

3. Select the current display representation set, and check the **Style Override** box.

4. Edit the **Wall Properties** dialog box, select the **Layer/Color/Linetype** tab, and set visibility, color, and hatching of the wall components.

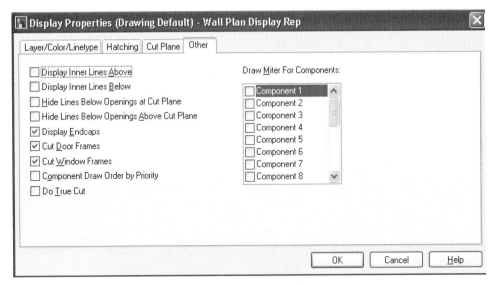

Figure 3.35 *Other tab of the Display Properties dialog box*

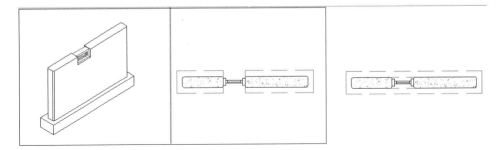

Figure 3.36 *Control of lines beneath openings to create accurate foundation vent display*

5. Select the **Hatching** tab to edit the style of hatching for wall components.
6. Select the **Cut Plane** tab to edit cut plane exceptions.

ASSIGNING MATERIALS TO WALL COMPONENTS

Material definitions are assigned to each wall component in the **Materials** tab of the wall style. The material definitions of a drawing are listed in the Material Definition styles in the **Style Manager**. The material definitions can be imported from the Material Definition (Imperial).dwg file located in the *C:\Documents and Settings\All Users\Application Data\Autodesk\ADT 2005\enu\Styles* directory. Material definitions can be created on the fly within the wall style using the render materials from the **Content Browser** as outlined in the following steps.

Advanced Wall Features 171

STEPS TO CREATING MATERIAL DEFINITIONS FOR WALLS USING THE CONTENT BROWSER

1. Select the **Content Browser** from the **Navigation** toolbar.
2. Open the Render Material Catalog, and expand a category of the catalog as shown in Figure 3.37.
3. Drag the i-drop of a render material from the catalog to the workspace.
4. Select **Format>Style Manager** from the menu bar. Expand the Architectural Objects folder and Wall Styles folder. Double-click on a wall style to open the **Wall Style Properties** dialog box.
5. Select the **Materials** tab, and select the **Add New Material** button as shown in Figure 3.38.
6. Type **Stone** in the name edit field of the new material definition in the **New Material** dialog box. Select **OK** to dismiss the **New Material** dialog box.
7. Select the **Edit Material** button of the **Materials** tab of the **Wall Style Properties** dialog box to open the **Material Definition Properties – Stone** dialog box shown in the middle dialog box of Figure 3.38.
8. Select the **Edit Display Properties** button of the **Display Properties** tab of the **Material Definition Properties – Stone** dialog box to edit the properties of the new material in the **Display Properties (Drawing Default) – Material Definition General Medium Detail Display Representation** dialog box. This dialog box allows you to edit the way materials are represented for a display configuration as shown in Figure 3.38.
9. Select the **Other** tab as shown in Figure 3.38. The render material i-dropped from the **Content Browser** will be listed in the drop-down list of materials as shown in Figure 3.38. The render material can be assigned for surface rendering and sectioning.
10. Select OK to dismiss all dialog boxes.

The **MaterialList** command allows you to determine the volume of a material used in a wall. If you assign materials to wall components, the command will list the material definition and the total volume of the material contained in the wall as shown in the following command sequence.

Command: materiallist

Select Objects: Specify opposite corner: 2 found (*Select two concrete walls.*)

Select Objects: ENTER

Material Name	Volume
Concrete.Cast-in-Place.Flat.Grey	154 CF

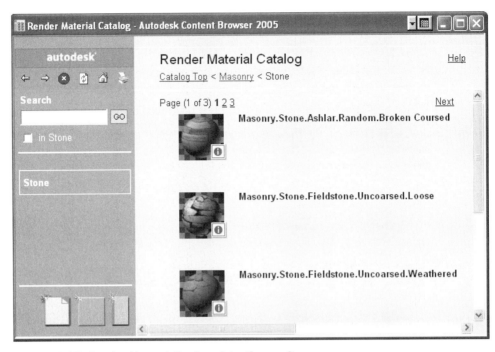

Figure 3.37 *Render Material Catalog of the Content Browser*

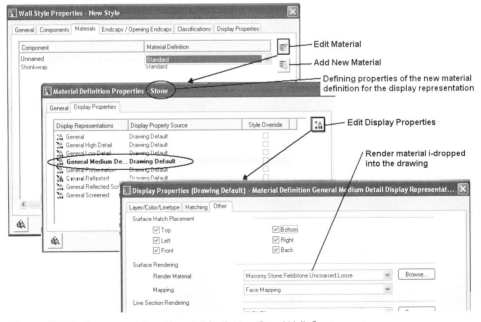

Figure 3.38 *Creating a New Material Definition for a Wall Component*

EXPORTING AND IMPORTING WALL STYLES

Wall styles can be i-dropped from the Design Tool Catalog-Imperial or Design Tool Catalog-Metric. The wall styles that have been i-dropped or selected from the Walls tool palette will be displayed within the Wall Style folder of the **Style Manager**. The wall styles of the Design Tool Catalog-Imperial and Design Tool Catalog-Metric are also located in the *C:\Documents and Settings\All Users\Application Data\Autodesk\ADT 2005\enu\Styles* directory.

You can import or export wall styles from other drawings in the **Style Manager**. The commands of the **Style Manager** toolbar as shown in Figure 3.3 are useful in importing and exporting styles. The **New Drawing**, **Open Drawing**, **Copy**, and **Paste** commands are used to import and export wall styles. The **New Drawing** button opens the **New Drawing** dialog box to create a new file for placing wall styles. The **Open Drawing** button opens the **Open Drawing** dialog box, allowing you to open an existing file. When you open a drawing within the **Style Manager**, its style contents are displayed in the left pane. The **Copy** and **Paste** commands of the toolbar allow you to copy a style from one drawing and paste it in another drawing in the left pane. If you have created a style in a drawing that is opened, the drawing will be listed in the left pane of the **Style Manager**. Therefore the **New Drawing** and **Open Drawing** buttons will open or create files listed in the left pane. To copy or export a file to another drawing, you copy the style to the clipboard and paste it into other drawings within the left pane. In addition to using the **Copy** and **Paste** buttons, you can click on a style and drag it to another drawing name and release it in the left pane over the file name. To select all styles from the list, select the first style; hold down SHIFT and select the last desired wall style. To select more than one wall style from the list of wall styles, hold CTRL down and select the desired wall styles. *Using* SHIFT *and* CTRL *works only on the right pane of the Style Manager.*

 Note: Copy and **Paste** are included in the shortcut menus of both the left and right panes of the **Style Manager** to allow you to copy and paste styles to other drawings.

If wall styles of a drawing are selected for export to a file that contains the same named wall styles, the **Import/Export – Duplicate Names Found** dialog box will open as shown in Figure 3.39. This dialog box displays the duplicate names. There are three options: **Leave Existing**, **Overwrite Existing**, and **Rename to Unique**.

The **Leave Existing** option will not replace the wall style of the destination drawing with that of the source file. The **Overwrite Existing** option revises the wall style definition of the destination file with that of the wall style of the source file. The **Rename to Unique** option will append a number such as 2, 3, or 4 to the wall style name from the source file to distinguish it from the wall style of the destination drawing.

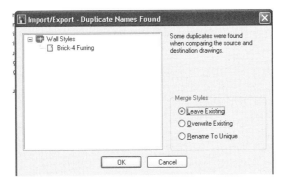

Figure 3.39 Import/Export-Duplicate Names Found dialog box

In summary, the steps to importing walls styles are as follows:

STEPS TO IMPORTING A WALL STYLE
1. Select **Format>Style Manager** from the menu bar.
2. Select the **Open** button on the **Style Manager** toolbar.
3. Select the source drawing that includes the desired wall style from the **Open Drawing** dialog box.
4. Select the source drawing in the left pane, and expand the Architectural Objects\Wall Styles folder.
5. Select the name of the desired wall style of the source file in the left pane.
6. Select **Copy** from the **Style Manager** toolbar (wall style copied to the clipboard).
7. Select name of the target file or the current drawing in the left pane; choose **Paste** from the **Style Manager** toolbar.

The wall style is imported to the target or current drawing.

The procedure to import wall styles can be applied to exporting styles to other drawings by copying the wall style to the clipboard in the **Style Manager** of the current drawing and pasting the style into the target drawing.

PURGING WALL STYLES
Wall styles can also be purged from a drawing to decrease the list of wall styles in the **Style Manager**. Only wall styles that have not been used in the drawing can be deleted. Wall styles can be purged by opening the **Style Manager**, expanding the Wall Styles folder of the current drawing in the left pane, then selecting the names of the wall styles, right-clicking, and choosing **Purge** from the shortcut menu. The **Select Wall Styles** dialog box opens as shown in Figure 3.40.

This dialog box lists all the wall styles of the drawing that have not been used. You can deselect a wall style for purging by clearing the check box for that wall style in the

Advanced Wall Features   175

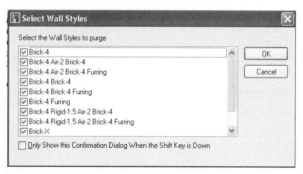

Figure 3.40 *Wall styles listed in the Select Wall Styles dialog box*

Select Wall Styles dialog box. Select the OK button, and all unused wall styles will be deleted from the drawing.

IDENTIFYING WALL STYLES USING INSPECT

The **Inspect** command allows you to hover the pointer over an object, and the tooltip will display the style and layer properties of the object. Access the **Inspect** command as shown in Table 3.3 and in Figure 3.41.

Command prompt	INSPECT
Toolbar	Select Inspect from the Standard toolbar

Table 3.3 *Inspect command access*

 Tip: The Inspect feature requires that **Display Auto Snap tooltip** be enabled in the **Drafting** tab of the **Options** dialog box. If object snap markers are displayed, turn off the OSNAP toggle in the status bar.

 STOP. Do Tutorial 3.1, "Creating a Wall Style," at the end of the chapter.

Figure 3.41 *Accessing Inspect*

EXTENDING A WALL WITH ROOFLINE/FLOORLINE

The **RoofLine** and **FloorLine** commands allow you to extend the wall beyond the base height and baseline. The commands can be useful in creating walls that extend to other objects such as gables or stairs. The Offset option of the command allows you extend a wall component uniformly a specified distance along its length. Access the **RoofLine** command as shown in Table 3.4.

Command prompt	ROOFLINE
Shortcut menu	Select a wall, right-click, and choose Roof/Floor Line>Modify Roof Line

Table 3.4 *RoofLine command access*

The options of the **RoofLine** command are as described below:

Offset – Allows you to specify the distance the wall is projected up.

Project to polyline – Allows you to select the walls and project the walls to the intersection of a selected polyline. The polyline is used as the boundary instead of the roof object.

Generate polyline – Creates a polyline from the wall; the polyline is created at the top of the wall.

Auto project – Projects the wall to the roof, roof slab, or stair.

Reset – Removes previous modifications to the roof line or floor line.

The Offset option extends the wall the specified offset distance toward the roof. The following command line sequence was used to extend the brick veneer up beyond the base height as shown in Figure 3.42. The **RoofLine** command allows you to adjust the brick veneer or other components to cover the floor system when the height of the floor system is determined.

(Select a wall, right-click, and choose **Roof/Floor Line>Modify Roof Line**.)

Command: RoofLine

RoofLine [Offset/Project to polyline/Generate polyline/Auto project/Reset]: **o** ENTER *(Specify the Offset option.)*

Enter offset <0>: **12** ENTER

RoofLine [Offset/Project to polyline/Generate polyline/Auto project/Reset]: ENTER *(Press ENTER to end the command.)*

EXTENDING THE WALL WITH FLOORLINE

The **FloorLine** command can be used to extend the wall up or down about the baseline of the wall. In Figure 3.42 the wall was extended down below the baseline to create the footing. The floor line extends the wall down to create the wall bottom. Access the **FloorLine** command as shown in Table 3.5.

Command prompt	FLOORLINE
Shortcut menu	Select a wall, right-click, and choose Roof/Floor Line>Modify Floor Line

Table 3.5 *FloorLine command access*

The base height of the wall is the distance between the baseline and the base height. The wall is inserted at Z=0 or on its baseline. The following command line sequence was used to extend the wall below the baseline to represent the footing.

(Select a wall, right-click, and choose **Roof/Floor Line>Modify Floor Line**.)

Command: FloorLine

FloorLine [Offset/Project to polyline/Generate polyline/Auto project/Reset]: **o** ENTER *(Selects the Offset option.)*

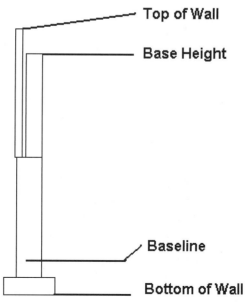

Figure 3.42 *Wall components extended with RoofLine and FloorLine*

Enter offset <1'-0">: **-16** ENTER *(Specifies the offset distance down.)*

FloorLine [Offset/Project to polyline/Generate polyline/Auto project/Reset]: ENTER *(Press ENTER to end the command.)*

CREATING WALL ENDCAPS

When openings are created in a wall or when a wall terminates, the treatment of the ends is controlled through Endcap styles. Wall Endcap styles enclose the end of each end of the wall, and Wall Opening Endcap styles enclose the edges of a wall when doors and windows are placed in the openings. The Wall Opening Endcap styles can be assigned to the sill, start jamb, end jamb, and header of the opening. The Wall Endcap and Wall Opening Endcaps are defined with the wall style. The Standard wall style uses a Standard wall endcap style and a Standard opening endcap style. The Standard endcap styles are simple straight lines connecting the two faces of the wall. Each wall style used in the drawing will include an endcap style. Therefore, when you import a wall style, the associated endcap will be imported.

Additional wall styles can be imported from the wall style files of the *C:\Documents and Settings\All Users\Application Data\Autodesk\ADT 2005\enu\Styles\Imperial* directory.

 Tip: Wall styles can be i-dropped from the Architectural Desktop Design Tool Catalog-Imperial>Walls of the Content Browser.

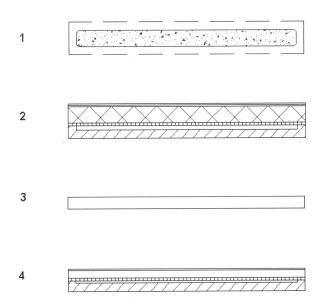

Figure 3.43 *Walls with Endcap styles*

The following wall styles files are available in the Imperial directory: Wall Styles-Brick (Imperial), Wall Styles-Casework (Imperial), Wall Styles-CMU (Imperial), Wall Styles-Concrete (Imperial), and Wall Styles-Stud (Imperial). As you import the style, the wall endcap style and the wall opening endcap styles also come into the drawing. Examples of wall styles with wall endcaps are shown in Figure 3.43.

The first wall style uses the Concrete-8 Concrete 16x8 footing wall endcap style. This endcap style creates an extension of the footing beyond the wall. The footing is created as a separate wall component, and the endcap extends the footing wall beyond the concrete wall component. The CMU-8 Rigid-1.5 Air-2 Brick-4 Furring endcap wall style is applied to the second wall style listed. This endcap style closes the cavity of the 4" brick wall component. The third wall style is the Standard wall style, which uses the simple straight line to connect the wall faces. The fourth wall style uses the Stud-4 Rigid-1.5 air-1 Brick-4 wall endcap style.

These four examples are applications of using an endcap style to enhance the wall design. Endcaps are created from polylines and assigned to the components of a wall. An endcap can be used with various wall styles; however, walls with multiple components should be capped with endcaps linked to each wall component.

Access **Endcap Styles** as shown in Table 3.6.

Menu bar	Format>Style Manager
Command prompt	WALLENDCAPSTYLE

Table 3.6 *WallEndCapStyle command access*

When you open the **Style Manager**, expand the Wall Endcap Styles folder as shown in Figure 3.44.

The Wall Endcap Styles and the Wall Opening Endcap Styles are shown for a drawing in Figure 3.44. Wall endcap styles are listed in the **Style Manager** for walls of the current drawing.

CREATING CUSTOM WALL ENDCAPS

To create a new endcap, first draw a polyline in the shape of the endcap. The selected polyline is assigned to a wall component in the wall style definition. The polyline does not have to be drawn the actual width of the wall. The shape of the polyline can be applied to walls of different widths. However, the ends of the polyline should align because the shape is applied to the end of the wall. Shown in Figure 3.45 are two polylines that were used to form endcaps. Adjacent to the endcaps are examples of walls created that use the endcap style. The polyline that capped the wall with unequal length polylines in essence bent the end of the wall when the endcap style was attached.

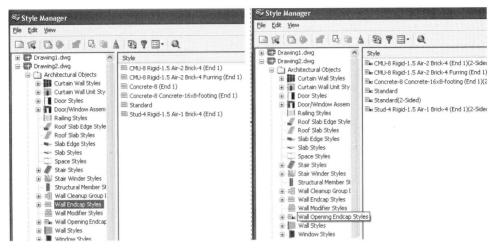

Figure 3.44 *Endcap styles of the Style Manager dialog box*

 Tip: Draw the polyline in a counterclockwise direction to retain its orientation when applied as endcap geometry in a wall style.

Once the endcap style is created, you assign it to a wall style by editing the wall style in the **Style Manager**.

The wall shown at left in Figure 3.47 consists of only one component. The following steps describe how to create a wall endcap style to cap the ends of this wall.

STEPS TO CREATING AN ENDCAP

1. Create a new drawing using the Aec Model (Imperial Stb) template.

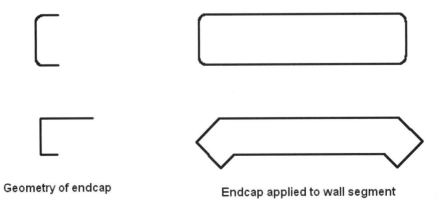

Figure 3.45 *Applications of endcaps to walls*

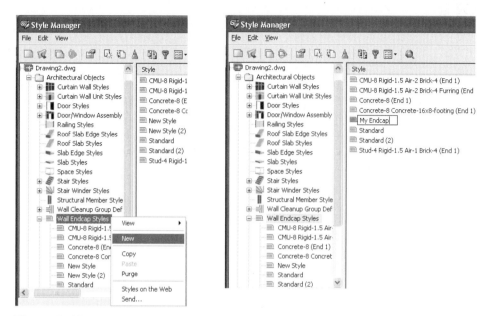

Figure 3.46 *Shortcut menu for Wall Endcap Styles in the Style Manager*

2. Draw a short segment of a wall, which includes only one wall component.
3. Draw a polyline in the shape of the line shown at **p1** in Figure 3.47.
4. Select **Format>Style Manager** from the menu bar.
5. Expand the Architectural Objects folder and the Wall Endcap Styles folder in the left pane.
6. Select the Wall Endcap Styles folder in the left pane, right-click, and choose New from the shortcut menu as shown in Figure 3.46.
7. Overtype **My Endcap** in the right pane of the **Style Manager**.
8. Select the new **My Endcap** style in the right pane, and choose **Set From** on the **Style Manager** toolbar as shown in Figure 3.46. The **Style Manager** temporarily closes, and you are prompted in the command line as follows:

   Command: _AecStyleManager

   Select a polyline: *(Select the polyline at p1 as shown in Figure 3.47.)*

   Enter component index for this segment <1>: ENTER *(Press ENTER to accept wall component with Index= 1.)*

   Add another component? [Yes/No] <No>: **N** ENTER *(Press ENTER to end component add.)*

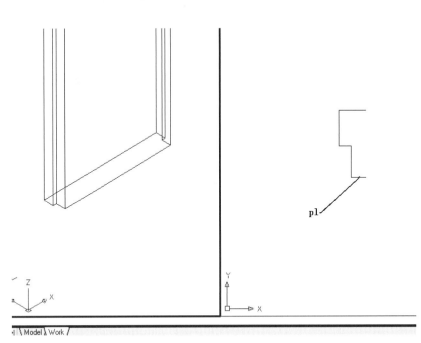

Figure 3.47 Creating an endcap style

Enter return offset <0">: ENTER (Press ENTER to attach without offset.)

9. **Style Manager** reopens; expand **Wall Styles** in the left pane.
10. Select the wall drawn in step 2, right-click, and select **Edit Wall Style** from the shortcut menu.
11. Select the **Endcaps/Opening Endcaps** tab, click the **Wall Endcap Style** list, and select the **My Endcap** style. Select **OK** to close the **Wall Style Properties-Standard** dialog box.

(Wall endcap geometry is applied to wall as shown at left in Figure 3.47.)

If a wall consists of two wall components, draw two polylines, one for each wall component. The index number of the wall component is linked to the polyline selected in the process of defining wall endcap styles.

MODIFYING ENDCAPS

If an existing endcap style needs to be modified, the **WallEndCap** command can be used to insert the polyline of the endcap style in the drawing for editing. This command is selected as shown in Table 3.7.

Command prompt	WALLENDCAP

Table 3.7 *WallEndcap command access*

When this command is typed in the command line, the command has two options: Define and as Pline. You must select the as Pline option to insert the endcap in the drawing. The Define option can be used to create an endcap style, whereas the as Pline option will insert the selected endcap as polyline geometry in the drawing. When the endcap is inserted as a polyline, the geometry can be edited and saved as an endcap. You can save the endcap using a new name. However, if the revised geometry is saved with the same endcap name as that of the original endcap, all endcaps with the same name will be updated to the new geometry definition. The **Edit In Place** command described below allows you to edit the polyline of the endcap while in place in the wall.

USING EDIT IN PLACE AND OVERRIDE ENDCAP STYLE

The endcap of a wall can be edited in the workspace using the options of the **Endcaps** shortcut menu. When you select a wall and right-click, the shortcut menu includes the following three options for editing wall endcaps: **Edit In Place**, **Override Endcap Style**, and **Calculate Automatically**. The **Edit In Place** option provides access to commands for editing the endcap and using grips to edit the endcap. Access **Edit In Place** as shown in Table 3.8.

Shortcut menu	Select wall, right-click, and choose Endcaps>Edit In Place
Command prompt	WALLENDCAPEDIT

Table 3.8 *WallEndcapEdit command access*

When you select a wall, right-click, and choose **Endcaps>Edit In Place**, you are prompted to select near the endcap to specify the endcap for editing. After you select a point near the endcap, the grips of the endcap are displayed as shown in Figure 3.48. You can select the grips and stretch the grips of the polyline to change the shape of the endcap. When the grips are displayed, right-click, and the shortcut menu includes the following additional options for editing the polyline:

> **Add Vertex** – Executes the **InplaceEditAddVertex** command, and you are prompted to select a point to add in the workspace. The existing polyline will be extended to include the vertex.
>
> **Remove Vertex** – Executes the **InplaceEditRemoveVertex** command, and you are prompted to select a point to remove in the workspace.

Hide Edge – Executes the **InplaceEditHideEdge** command. You are prompted to select an edge to hide in the workspace. The selected polyline segments are turned off.

Show Edge – Executes the **InplaceEditShowEdge** command. You are prompted to select an edge to display of polyline segments that were turned off with the **InplaceEditHideEdge** command.

Replace Endcap – Executes the **InplaceEditReplaceEndcap** command, and you are prompted to select an existing polyline in the drawing to be used as the new endcap.

Remove Endcap – Executes the **InplaceEditRemoveEndcap** command, removing the endcap of the wall.

Save All Changes – Saves the editing of the endcap with the **InplaceEditSaveAll** command.

Save As New Endcap Style – Saves the edited polyline as an endcap with a different name. The new name is typed in the **New Endcap Style** dialog box.

Discard All Changes – Exits the editing of the polyline without saving any changes.

When you select options from the shortcut menu, the grips are cleared and you are prompted to edit the polyline. Prior to saving or discarding the edit, you can select the

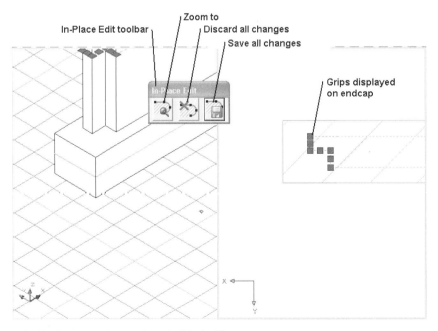

Figure 3.48 *Editing endcap style with Edit In Place*

endcap to display the grips and edit the grips or make additional selections from the shortcut menu.

In addition to the shortcut menu, the **In-Place Edit** toolbar is displayed during the edit, which includes the following commands: **Zoom To**, **Discard All Changes**, and **Save All Changes**, as shown in Figure 3.48. This toolbar remains on screen until you select **Discard All Changes** or **Save All Changes**.

Using Override Endcap Style

The **Override Endcap Styles** command allows you to select from a list of endcaps to use at the end of a wall. The default endcap is defined in the **Endcaps/Opening Endcaps** tab of the **Wall Style Properties** dialog box. The **Override Endcap Style** command allows you to change the endcap used at one end of the wall. Access the **Override Endcap Style** command as shown in Table 3.9.

Shortcut menu	Select a wall, right-click, and choose Endcaps>Override Endcap Style
Command prompt	WALLAPPLYENDCAP

Table 3.9 *Endcap Style command access*

When you select a wall, right-click, and choose **Endcaps>Override Endcap Style**, you are prompted to select a point near the endcap for edit. Selecting the point specifies the wall endcap for edit, and opens the **Select an Endcap Style** dialog box, as shown in Figure 3.49. Select a style from the dialog box, and the endcap is changed.

APPLYING CALCULATE AUTOMATICALLY TO CHANGE THE ENDCAP

The **Calculate Automatically** command allows you to select a polyline from the workspace and apply the polyline as an override or to create a wall endcap style as the default for the wall. This command automatically determines the component of the wall style for attachment to the wall. Access **Calculate Automatically** as shown in Table 3.10.

Shortcut menu	Select a wall, then select Endcaps>Calculate Automatically
Command prompt	WALLAUTOENDCAP

Table 3.10 *Automatically calculate endcap from polyline command access*

When you select a wall, right-click, and choose **Endcaps>Calculate Automatically**, you are prompted in the command line as shown in the following command line sequence.

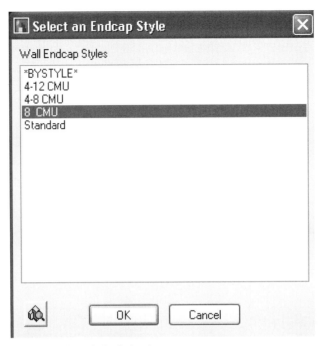

Figure 3.49 *Select an Endcap Style dialog box*

Command: WallAutoEndCap

Select polylines: *(Select a polyline as shown in Figure 3.50.)*
 1 found

Select polylines: ENTER *(Press ENTER to end selection of polylines.)*

Erase selected polyline(s)? [Yes/No] <No>: *(Press ENTER to select the No option to retain the polyline; select Yes to erase the polyline from the workspace.)*

Modify the current endcap style '8 CMU'? [Yes/No] <Yes>: **n** *(Select No to retain the current endcap style definition and create an override. Select Yes to edit the endcap style definition.)*

Apply the new wall endcap style to this end as [Wallstyledefault/Override] **w** ENTER

<Wallstyledefault>: ENTER *(Select Wallstyledefault to apply the endcap in the style definition or select Override to change the endcap for this wall end as an exception.)*

Before you select **Calculate Automatically,** you should draw a polyline in the workspace that you want to use as an endcap. The first prompt of the command is to select the polyline for the new endcap application. The command line sequence provides you

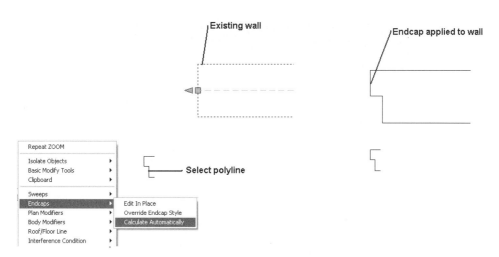

Figure 3.50 *Creating an endcap style using Calculate Automatically*

with the option to retain the polyline or erase it from the workspace. The polyline can be used to redefine the definition of current endcap style for the wall style or to apply the polyline as an override to the endcap style. The application of the polyline can be to the wall style definition or as an override to the wall style for the selected end of the wall.

When wall endcaps are edited for a wall, the results are displayed in the Advanced section of the Properties palette as shown in Figure 3.51. You can select from the endcap drop-down list in the Advanced section of the Properties palette to create an override for the endcap. The endcaps listed are the endcaps used by the walls inserted in the drawing or created in the **Style Manager**.

OVERRIDING THE PRIORITY OF COMPONENTS IN A WALL STYLE

The Wall Style Overrides are listed in the Advanced section of the Properties palette. The priority of the wall components can be edited as an exception to the priorities defined in the wall style. When you select the **Priority Overrides** button of the Advanced section of the Properties palette, the **Priority Overrides** dialog box opens as shown in Figure 3.52. This dialog box will be empty if no overrides have been set. To create an override, select the **Add Priority Override** button at right as shown in Figure 3.52. Priority overrides can be removed by selecting the component and then selecting the **Remove Priority Override** button. When you add an override, you define the name of the component, location on the wall, and priority value. If you click in the Component column, a drop-down list will display the names of the wall components defined in the wall style. The **Override** column allows you to specify the loca-

Figure 3.51 *Endcap style listed in style overrides section of the Properties palette*

tion of the override at the start or end of the wall. The **Priority** column allows you to type a number to specify the priority for the wall component.

 STOP. Do Tutorial 3.2, "Creating and Editing an Endcap," at the end of the chapter.

CREATING WALL MODIFIERS AND STYLES

Wall modifiers can be created to add projections to a wall at the beginning, end, or at a specified distance from the beginning, end, or midpoint of a wall. Wall modifiers can consist of rectangular projections or polyline shapes saved as wall modifier styles. You can create pilasters or other decorative projections of the wall using wall modifier styles. Wall modifiers are placed on existing walls by editing the properties of the wall in the Advanced section of the Properties palette. If wall modifiers are created through the Properties palette, the **Wall Modifiers** dialog box allows you to specify the style and location of the modifier as shown in Figure 3.53.

Wall modifiers can also be added in the workspace from the shortcut menu of a selected wall. The shortcut menu of a selected wall includes the following wall modifier options: **Add**, **Convert Polyline to Wall Modifier**, and **Remove**. Access the **WallModifierAdd** command as shown in Table 3.11.

Advanced Wall Features | 189

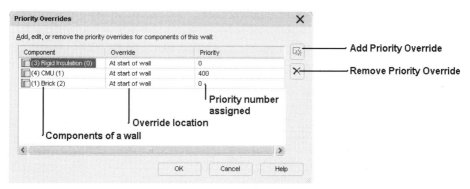

Figure 3.52 *Priority Overrides dialog box*

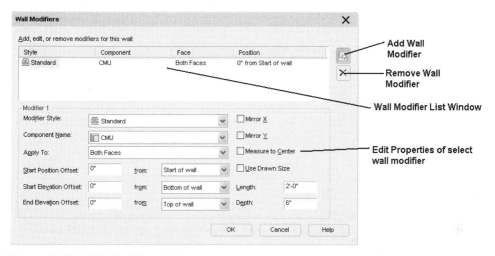

Figure 3.53 *Wall Modifiers dialog box*

Command prompt	WallModifierAdd
Shortcut menu	Select a wall, right-click, and choose Plan Modifiers>Add
Properties	Click in the Plan Modifiers field of the Advanced section of the Properties palette

Table 3.11 *WallModifierAdd command access*

If you insert a wall modifier by selecting the **Plan Modifiers** button on the Properties palette, the **Wall Modifiers** dialog box will open as shown in Figure 3.53. This dialog box allows you to specify the style, wall component, and location of the wall modifier. The top portion of the **Wall Modifiers** dialog box lists the wall modifiers that are currently attached to the selected wall. The bottom section of the dialog box is designed to edit the modifier that is selected in the top portion. The options of the **Wall Modifiers** dialog box are described below.

> **Add Wall Modifier** – The **Add Wall Modifier** button activates the dialog box allowing you to assign a wall modifier style to the selected wall. The Standard wall modifier style will be used as the default unless other wall modifier styles are selected and defined in the drawing.
>
> **Remove Wall Modifier** – If you select a wall modifier and then select the **Remove Wall Modifier** button, the wall modifier will be removed.
>
> **Wall Modifier List Window** – The **Wall Modifier** list window lists the wall modifiers that are defined for the selected wall.
>
> The lower portion of the dialog box is used to edit properties and conditions of attachment to the wall. The purpose of each of the edit fields of the lower portion of the **Wall Modifiers** dialog box is described below.
>
> **Modifier Style** – The **Modifier Style** drop-down list shows defined wall modifier styles that have been created in the drawing. Selecting a wall modifier style from the list will apply the predefined modifier style to the wall.
>
> **Component Name** – The **Component Name** drop-down list shows the wall components of the selected wall. Wall modifier styles can be applied to each wall component of a wall style.
>
> **Apply To** – The **Apply To** drop-down list allows the wall modifier style to be applied to the Left, Right, or Both Faces of the wall.
>
> **Start Position Offset** – The **Start Position Offset** edit box allows you to set the position horizontally along the wall to locate the wall modifier. The distance entered can be a positive or negative number. The from list allows the start position to be established relative to the wall start, wall end, or wall midpoint. Setting the distance to zero and measuring the distance relative to the wall midpoint will locate the projection in the middle of the wall.
>
> **Start Elevation Offset** – The **Start Elevation Offset** edit box establishes the elevation of the bottom of the projection. The bottom of the projection can be defined to start at the bottom of the wall or at some distance from the bottom. The **Start Elevation Offset** distance and the **from** list allow the distance to be defined from the wall top, wall base height, wall baseline, or wall bottom. The wall top and wall bottom are the top and bottom surfaces of the wall component being modified. Because wall components can be defined with different elevations in the wall style definition, the wall baseline, and base height options are provided. Wall baseline is the bottom of the baseline justification line, and the base height is the wall height from the baseline. Establish the distance between the baseline and the wall base height by setting the Height in the Properties palette.

End Elevation Offset – The **End Elevation Offset** section establishes the elevation of the top of the projection. The **End Elevation Offset** distance and the **from** list define the distance from the wall top, wall base height, wall baseline, or wall bottom for the top of the projection. The top of the projection is defined in the same manner as the bottom. Note that if the start elevation offset and end elevation offset are set to the same distance and reference surface, the modifier will become invisible because the projection is starting and ending at the same elevation.

Setting a distance in the start elevation offset from the wall bottom will lift the projection up from the bottom. Editing the start position offset, start elevation offset, and end elevation offset would allow the creation of the wall modifiers as shown in Figure 3.54.

A projection can be positioned relative to the beginning, midpoint, and end of the wall. In Figure 3.55, the start position offset is set to zero relative to the end of wall. This projection has a start elevation of 1'-0" from the wall bottom and an end elevation offset of a negative 1'-0".

Use Drawn Size – The **Use Drawn Size** check box turns OFF the size established in the **Length** and **Depth** edit fields. If this box is checked, the wall modifier will be created according to the actual size of the polyline geometry used to create the wall modifier style.

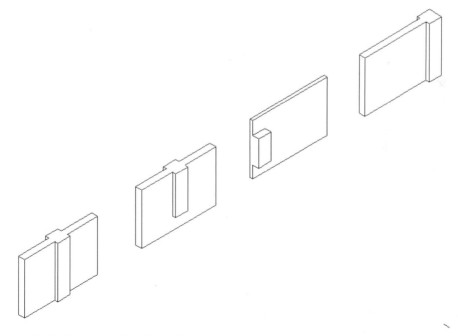

Figure 3.54 *Examples of wall modifier styles*

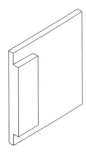

Figure 3.55 *Start and End Elevation set for wall modifier style*

Length – The **Length** edit field allows you to establish the length of the wall modifier. The distance entered in the **Length** edit box can be the same as the wall length.

Depth – The **Depth** edit field sets the distance the wall modifier will project from the wall.

Mirror X – The **Mirror X** check box is used to mirror the wall modifier in the X direction.

Mirror Y – The **Mirror Y** check box is used to mirror the wall modifier in the Y direction.

Measure to Center – The **Measure to Center** check box will position the wall modifier relative to its center along the wall.

INSERTING AND EDITING WALL MODIFIERS IN THE WORKSPACE

Wall modifiers can be added and removed, and polylines converted to modifiers in the workspace. The shortcut menu of a selected wall allows you add the wall modifier in the workspace. Adding a wall modifier in the workspace allows you to select points in the workspace to begin and end the wall modifier. When you select a wall, right-click, and then choose **Plan Modifiers>Add**, you are prompted as follows:

Command: WallModifierAdd

Select start point: *(Select a point near p1 as shown at left in Figure 3.56.)*

(Move the pointer to the right near p2, and verify polar tracking tooltip angle=0.)

Select end point: **24** ENTER *(Width of wall modifier specified.)*

Select the side to draw the modifier: *(Move the pointer down and select a point near p3, polar tracking tooltip=270 as shown in Figure 3.56.)*

Enter wall modifier depth <8 5/8">: **6** ENTER *(Specify distance for projection.)*

*(***Add Wall Modifier** *dialog box opens; select wall modifier style and elevations, then select* **OK** *to dismiss the* **Add Wall Modifier** *dialog box.)*

(Wall modifier created as shown at right in Figure 3.56.)

Removing a Wall Modifier

A wall modifier can be removed from a wall by accessing the **WallModifierRemove** command as shown in Table 3.12.

Command prompt	WALLMODIFIERREMOVE
Shortcut menu	Select the wall, right-click, and choose Plan Modifiers>Remove

Table 3.12 *WallModifierRemove command access*

When you access **WallModifierRemove** from the shortcut menu, you are prompted to select the wall modifier as shown below.

Command: WallModifierRemove

Select a modifier: *(Select a wall modifier at p1 in Figure 3.57.)*

Convert removed modifier to a polyline? [Yes/No] <No>: **N** ENTER
(Select N to not retain the polyline geometry in the workspace.)

(Wall modifier is removed as shown in Figure 3.57.)

Points located for Wall Modifiers   Edit Add Wall Modifier   Wall Modifier Added

Figure 3.56 *Creating a wall modifier*

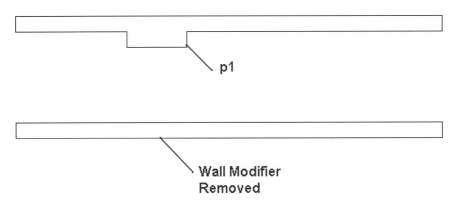

Figure 3.57 *Wall modifier removed from wall*

Edit In Place

A wall modifier can be edited using the **Edit In Place** option of the shortcut menu. **Edit In Place** allows you to change the geometry of the polyline in the workspace without accessing the **Style Manager**. Access **Edit In Place** as shown in Table 3.13.

Command prompt	WALLMODIFIEREDIT
Shortcut menu	Select the wall, right-click, and choose Plan Modifiers >Edit In Place

Table 3.13 *Edit In Place of a Wall Modifier command access*

If the wall modifier was created from a polyline drawn to actual size, you can immediately insert it in the workspace. Wall modifiers not drawn to actual size must first be converted. An AutoCAD message box, as shown in Figure 3.58, will open if the wall modifier was not drawn to actual size. A Yes response to this dialog box will insert the polyline in the workspace as shown in Figure 3.58.

The wall modifier can then be edited by selecting grips or right-clicking and choosing from the shortcut menu. The options of the shortcut menu are described below.

Add Vertex – Executes the **InplaceEditAddVertex** command, and you are prompted to select a point to add in the workspace. The existing polyline will be extended to include the vertex.

Remove Vertex – Executes the **InplaceEditRemoveVertex** command, and you are prompted to select a point to remove in the workspace.

Advanced Wall Features

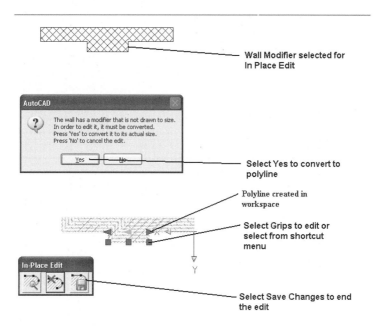

Figure 3.58 *Edit in Place of wall modifier*

Hide Edge – Executes the **InplaceEditHideEdge** command. You are prompted to select an edge to hide in the workspace. The selected polyline segments are turned off.

Show Edge – Executes the **InplaceEditShowEdge** command. You are prompted to select an edge to display of polyline segments that were turned off with the **InplaceEditHideEdge** command.

Replace Modifier – Executes the **InplaceEditReplaceModifier** command, and you are prompted to select an existing polyline in the drawing to be used as the new modifier.

Remove Modifier – Executes the **InplaceEditRemoveModifier** command, removing the modifier from the wall.

Save All Changes – Saves the editing of the modifier with the **InplaceEditSave** command.

Save As New Modifier Style – Saves the edited polyline as a modifier with a different name. The new name is typed in the **New Modifier Style** dialog box.

Discard All Changes – Exits the editing of the polyline without saving any changes.

CREATING WALL MODIFIER STYLES

Wall modifiers can be added to the wall using the Standard style or custom styles that you can create. The **WallModifierStyle** command allows you to create custom wall modifier shapes. Access the **WallModifierStyle** command as shown in Table 3.14.

Menu bar	Format>Style Manager
Command prompt	WALLMODIFIERSTYLE
Shortcut menu	Draw an open polyline, select a wall, right-click, and choose Plan Modifiers>Convert Polyline to Wall Modifier

Table 3.14 *WallModifierStyle command access*

Creating Wall Modifier Styles with the Style Manager

When you select **Format>Style Manager** from the menu bar, the **Style Manager** opens. Expand Architectural Objects in the left pane by selecting the (+) sign, and then select the Wall Modifier Styles folder. Select the **Wall Modifier Style** folder and select the **New Style** button of the **Style Manager** to create a new style. Overtype the name of the new style in the right pane. The content of the wall modifier consists of an open

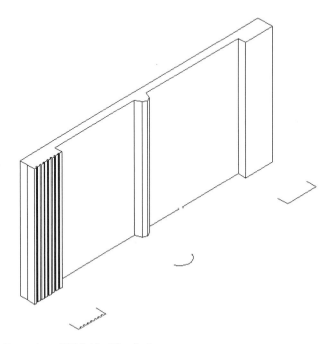

Figure 3.59 *Examples of Wall Modifier Styles*

polyline. Examples of polylines used as wall modifiers are shown in Figure 3.59. To define the polyline for a wall modifier, select the wall modifier name in the **Style Manager**, select **Set From** on the **Style Manager** toolbar and then select the polyline.

The polyline used to define the wall modifier can be drawn the actual size of the desired projection or drawn as a sketch and applied with specified length and depth dimensions in the **Wall Modifiers** dialog box. If a wall modifier style is created from geometry drawn actual size, it can be applied to a wall using the actual dimensions of the original geometry by checking **Use Drawn Size** check box in the **Wall Modifiers** dialog box as shown in Figure 3.53.

Using the Shortcut Menu to Convert a Polyline to a Wall Modifier

If a polyline has been drawn in the workspace, you can convert this geometry to a wall modifier by selecting the wall, right-clicking, and choosing **Plan Modifier>Convert Polyline to Wall Modifier** from the shortcut menu. This process allows you to create a wall modifier for a wall without accessing the **Style Manager**. Access the **Convert Polyline to Wall Modifier** command as shown in Table 3.15.

Command prompt	WALLMODIFIERCONVERT
Shortcut menu	Select a wall, right-click, and choose Plan Modifiers>Convert Polyline to Wall Modifier

Table 3.15 *Wall Modifier Convert command access*

The following steps demonstrate the use of the **Convert Polyline to Wall Modifier** command.

STEPS TO CONVERTING A POLYLINE TO A WALL MODIFIER FOR A SELECTED WALL

1. Open a drawing that consists of an open polyline and a wall.
2. Select the wall, right-click, and choose **Plan Modifiers>Convert Polyline to Wall Modifier**. Respond to the following command line prompts as shown below:

    (Select the wall at p1 as shown in Figure 3.60, right-click, and choose **Plan Modifiers>Convert Polyline to Wall Modifier**.*)*

    Command: WallModifierConvert

    Select a polyline: *(Select the arc at p2 shown in Figure 3.60.)*

    Erase layout geometry? [Yes/No] <No>: ENTER *(Press* ENTER *to retain polyline.)*

Figure 3.60 Creating a Wall Modifier style

3. Type a name in the **New Wall Modifier Style Name** dialog box as shown in Figure 3.60. Click **OK** to dismiss the dialog box and open the **Add Wall Modifier** dialog box.

4. Edit the **Add Wall Modifier** dialog box, specify style and start and end elevations, and click **OK**.

   Wall modifier is placed on the wall as shown below. Note that the wall modifier is placed on the wall near the location where the polyline is drawn. You can specify the location of the wall modifier in the **Wall Modifiers** dialog box by selecting **Plan Modifiers** in the Advanced section of the Properties palette.

 STOP. Do Tutorial 3.3, "Creating Wall Modifiers," at the end of the chapter.

CREATING WALL SWEEPS USING PROFILES

The **Sweep Profile** command is used to generate a wall that has the shape defined from an Aec Profile. The Aec Profile is created from a closed polyline. The closed polyline should be identical to the outline of a typical vertical section of the wall. The shape and size of the profile is then swept along the length of an existing wall. The profile is substituted as a wall component of the wall. Prior to sweeping the wall, the **Profile**

Definition command must be used to create the Aec Profile. Aec Profiles are created in the **Style Manager**. Access the **Profile Definition** command as shown in Table 3.16.

Menu bar	Format>Style Manager
Command prompt	AECPROFILEDEFINE

Table 3.16 *Accessing the Profile Definition command*

When you select **Format>Style Manager** from the menu bar, the **Style Manager** opens as shown in Figure 3.61. Profiles of the drawing are in the Profiles folder of the Multi-Purpose Objects folder. A new style is created by selecting the Profiles folder, and then selecting the **New Style** button on the **Style Manager** toolbar. The new style is named by overtyping the name in the right pane of the **Style Manager** as shown in Figure 3.61.

After a name is created for the style, choose the **Set From** button on the **Style Manager** toolbar, and then respond in the command line as shown below and in the Figure 3.62.

> Command: _AecStyleManager
>
> Select a closed polyline: *(Select a closed polyline at p1 in Figure 3.62.)*
>
> Add another ring? [Yes/No] <No>: ENTER *(Press enter to end selection.)*

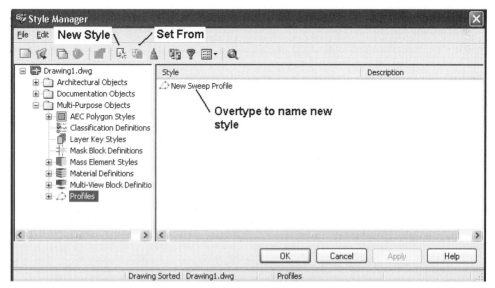

Figure 3.61 *Creating a new Aec Profile in the Style Manager*

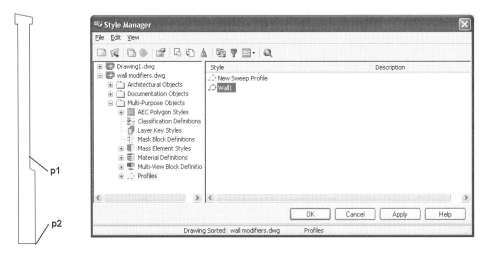

Figure 3.62 *Geometry for the AecProfile*

Insertion Point or <Centroid>: **_endp of** *(Select the lower right corner at p2 with the endpoint object snap.)*

CREATING THE SWEEP

The **WallSweep** command can be used to sweep an existing wall as an extrusion of an Aec Profile. Access the **WallSweep** command as shown in Table 3.17.

Command prompt	WALLSWEEP
Shortcut	Select a wall, right-click, and choose Sweeps>Add

Table 3.17 *Sweep Profile command access*

When you select a wall, right-click, and choose **Sweeps>Add**, the **Add Wall Sweep** dialog box opens. You can specify the wall component and profile for the sweep in the dialog box. The profile definition can be selected from existing profiles or you can use the **Start from scratch** option and edit in place the profile. Described below are the options of the **Add Wall Sweep** dialog box shown in Figure 3.63.

> **Wall Component** – The **Wall Component** list displays the wall components of the selected wall. The sweep will be attached to the specified wall component.
>
> **Profile Definition** – The **Profile Definition** allows you select from **Start from scratch** or the defined profiles of the drawing.

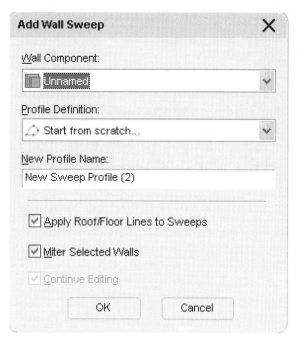

Figure 3.63 Add Wall Sweep dialog box

New Profile Name – The **New Profile Name** is assigned to the wall when a sweep has been performed.

Apply Roof/Floor Lines to Sweeps – The **Apply Roof/Floor Lines to Sweeps** check box allows the sweep to include the roof and the floor line of the wall. Therefore, if the wall has been extended up for the roof line or down for the floor line, the sweep will also be extended accordingly.

Miter Selected Walls – When connecting walls are selected for the sweep, the **Miter Selected Walls** check box, when checked, miters the intersection of the connecting walls.

Continue Editing – Selecting the **Continue Editing** check box allows you to edit in place and grip edit the profile prior to sweeping the wall.

STEPS TO CREATING A SWEEP

1. Open a drawing that consists of an Aec Profile and a wall.
2. Select the wall, right-click, and choose **Sweeps>Add**.
3. Edit the **Add Wall Sweep** dialog box, specify the wall component and profile definition, and check **Apply Roof/Floor Lines to Sweep** and **Miter Selected Walls**.

4. Select **OK** to close the **Add Wall Sweep** dialog box.

Wall is swept as shown in Figure 3.64.

MODIFYING SWEPT WALLS

The intersection of two walls can be swept without miter. The **Miter** options will apply a miter to connecting walls. The **WallSweepMiterAngles** command extends the planes of each wall to a common intersection. The walls shown on the left in Figure 3.65 have not been mitered, whereas the walls on the right have been mitered. Access the **WallSweepMiterAngles** command as shown in Table 3.18.

Command prompt	WALLSWEEPMITERANGLES
Shortcut menu	Select a wall, right-click, and choose Sweeps>Miter

Table 3.18 *WallSweepMiterAngles command access*

When you select walls to apply the sweep, right-click, and choose **Sweeps>Miter**, the walls are swept as shown on the right in Figure 3.65. If you type the command in the command line, you are prompted to select the walls for the miter.

USING EDIT IN PLACE WITH WALL SWEEP

The profile of a swept wall can be edited in place. If you select a wall that has a sweep operation, right-click, and choose **Edit Profile In Place** from the shortcut menu, you

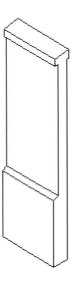

Figure 3.64 *Wall created with wall sweep*

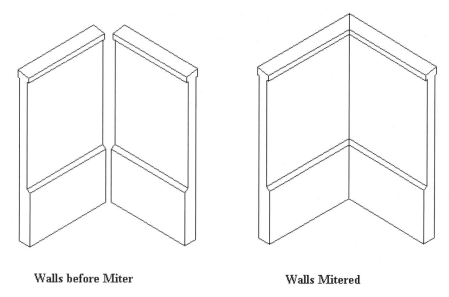

Walls before Miter   Walls Mitered

Figure 3.65 *Application of Sweep Profile Miter Angles command to walls*

can edit the profile. The grips of the profile can be edited as shown in Figure 3.66. The shortcut menu of the selected wall includes additional options for editing. You can select **Vertex Grip** or **Edge Grip** and stretch the grips to a new location to redefine the shape of the profile.

> **Add Vertex** – Executes the **InplaceEditAddVertex** command, and you are prompted to select a point to add in the workspace. The existing polyline will be extended to include the vertex.
>
> **Remove Vertex** – Executes the **InplaceEditRemoveVertex** command, and you are prompted to select a point to remove in the workspace.
>
> **Add Ring** – Executes the **InplaceEditAddRing** command, which allows you to add a ring to the profile. Additional rings can be created as voids.
>
> **Replace Ring** – Executes the **InplaceEditReplaceRing** command. The command line prompts for this command allow you to select a closed polyline, spline, ellipse, or circle.
>
> **Save Changes** – Saves the changes made to the profile.
>
> **Save As New Profile** – Saves the changes in the profile using a new name.
>
> **Discard All Changes** – Ends the in place editing without saving the changes to the profile.

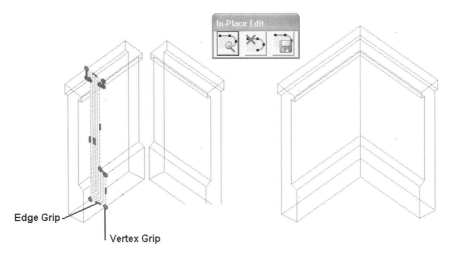

Figure 3.66 Edit In Place of a wall sweep profile

ADDING MASS ELEMENTS USING BODY MODIFIERS

The **WallBody** command is used to combine mass elements with a wall. The mass elements, covered in Chapter 12, can be used to model custom building components. Mass elements such as cylinders, boxes, arch box vault, and right triangle prism can be combined to create building components. The mass element can be combined using Boolean operations of add, subtract, or intersection with the wall. Therefore this command can be used to create projections in the wall or to cut out a portion from the wall. Access the **WallBody** command as shown in Table 3.19.

Command prompt	AECWALLBODY
Shortcut menu	Select a wall, right-click over the drawing area, and choose Body Modifiers>Add

Table 3.19 WallBody command access

When you select a wall, right-click, and choose **Body Modifiers>Add**, you are prompted to select the mass element. When you select the mass element the **Add Body Modifier** dialog opens as shown in Figure 3.67.

The options of the **Add Body Modifier** dialog box are described below.

       Wall Component – The **Wall Component** list allows you to select a wall component to which to attach the mass element.

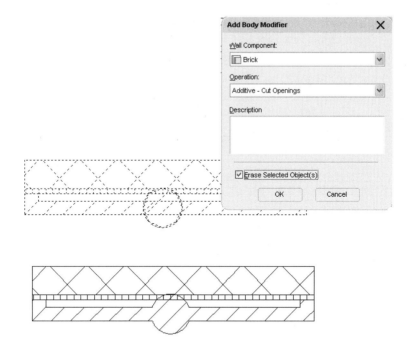

Figure 3.67 *Creating a wall projection with Body Modifier*

> **Operation** – The **Operation** options allow you to attach the mass element using one of the following Boolean operations: Additive - Cut Openings, Additive, Subtractive, and Replace.
>
> **Erase Selected Object** – The **Erase Selected Object** check box, if checked, will erase the mass element after it has been added to the wall.

When you select a wall, right-click, and choose **Body Modifiers>Add** from the shortcut menu, you are prompted to select the body. In Figure 3.67 the cylinder mass element is selected and the **Add Body Modifier** dialog box opens. The brick wall component is specified and the operation is set to **Additive – Cut Openings**. The cylinder is combined with the wall.

After a mass element is added to the wall, select the wall, and right-click—the shortcut menu now includes the **Edit In Place** and **Remove** options for editing the mass element. Choosing **Edit In Place** allows you to edit the size and Boolean operation applied to the mass element. While **Edit In Place** is active, the shortcut menu includes the following additional editing options: **Boolean**, **Trim**, **Split Face**, and **Join Face**. Upon completion of the editing of the wall, save changes by choosing **Save Changes** from the shortcut menu.

Boolean – Allows you to change the Boolean operation applied to Union, Subtract, and Intersect.

Trim – Executes the **MassElementTrim** command, allowing you to trim the mass element by specifying a plane through three points.

Split Face – Executes the **MassElementFaceDivide** command to divide faces of the mass element.

Join Face – Executes the **MassElementFaceJoin** command to combine faces of the mass element.

CREATING ADDITIONAL FLOORS

After you develop the majority of the first floor plan, additional floors can be developed based on its shape. The additional floors can be created by creating a construct for the second floor in the **Project Navigator**. The shortcut menu shown in Figure 3.68 of the **Construct** tab allows you to copy the construct drawing of the floor plan assigned to level 1 to level 2. Therefore, the second floor plan construct consists of exterior walls with the same x,y coordinates, ensuring the vertical alignment of the floors. The second floor plan can be developed by deleting non–load bearing walls and creating the rooms of the second floor.

Floor plans can be developed without the use of a project by copying the files for each floor and assigning different names to each file. When you are ready to create a model

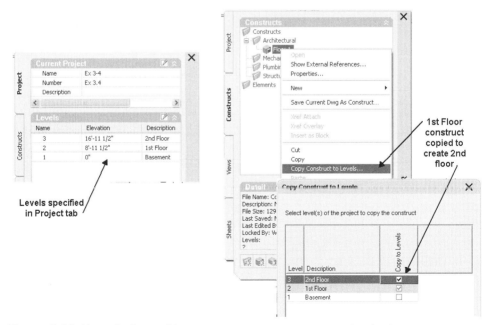

Figure 3.68 *Using the Project Navigator to copy a construct to another level*

to check vertical alignment of features, you can convert each drawing to a construct drawing as part of the project. To convert a file to a construct, open the file and create the project. Select the **Construct** tab of the **Project Navigator**, right-click, and choose **Save Current Dwg As Construct**. The **Add Construct** dialog box opens, allowing you to specify the construct name and level, as shown in Figure 3.69. Each drawing for a floor is created as a construct and assigned to a level defined in the **Project** tab.

After you have created constructs for each level, you can create a model drawing in the **View** tab. The construct drawings for each level are attached as reference files to the model drawing. The model drawing allows you to check the vertical alignment of bearing walls and exterior walls.

If the exterior walls of one floor differ in construction from those of another floor, the wall styles of each floor can be created to permit the vertical alignment of the walls. If exterior walls of each floor are baseline justified, the wall components can be positioned in the wall style to assure vertical alignment. The exterior walls of a basement in residential construction usually differ from those of the first floor. The wall section shown at left in Figure 3.70 illustrates how the baselines are aligned for each level and the components are positioned horizontally to reflect actual construction. The **RoofLine** and **FloorlLne** commands can be applied to project components of the walls to cover the floor system when the floor system is determined.

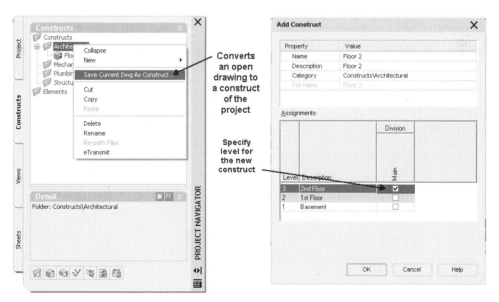

Figure 3.69 *Using the Project Navigator to convert a drawing to a construct*

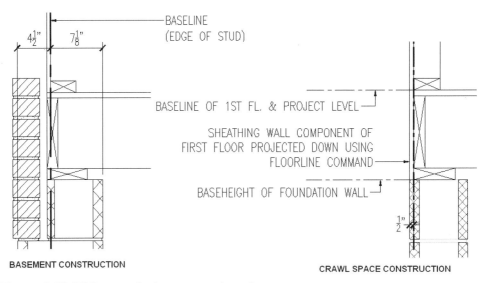

Figure 3.70 *Wall section for basement and crawl space construction*

Figure 3.71 shows the Frame_Brick wall style created in Tutorial 3.1 for the first floor plan and a wall style created for the basement level. The baselines of each wall are shown by centerlines. The wall style for the basement plan includes the brick and block components positioned relative to the baseline.

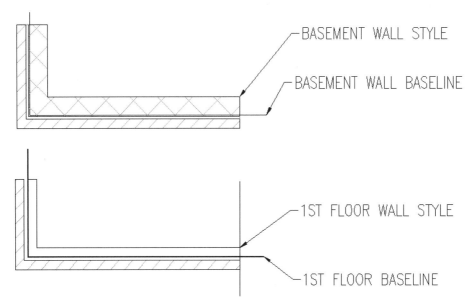

Figure 3.71 *Baseline locations for wall styles*

The components of the basement wall style shown above consist of

Name	Priority	Edge Offset	Width
Block	400	–1/2	7-5/8
Brick	400	–4-1/2	3-5/8

Therefore, if the first floor plan is copied to create the basement level, you can modify the wall style of the basement exterior walls to ensure vertical wall alignment of the baselines.

Note: You can check the vertical alignment of levels by creating a model view in the **Project Navigator** with constructs of floor plan 1 and the basement floor plan attached. A model view of a residence that includes a basement is presented in Chapter 11. The view drawing assists in creating a model for checking vertical alignment of exterior walls and bearing walls.

STOP. Do Tutorial 3.4, "Creating the Basement Floor Plan," at the end of the chapter.

CREATING A FOUNDATION PLAN

A foundation plan for a house that does not have a basement can be developed by copying the first floor construct to a level below the first floor plan. The vertical alignment of the structural components is established through controlling the baseline locations of the walls of each floor. The wall height is reduced to the anticipated height of the foundation wall. The levels defined in the project for a foundation plan are adjusted based upon the foundation wall height. The foundation walls should be baseline justified with the components shifted horizontally as shown in Figure 3.70. Piers can be inserted as walls or columns using layout curves to space the piers along a centerline. Layout curves, presented in Chapter 13, allow you to equally space the piers along a centerline. You will create a wall style for a foundation plan in Tutorial 3.5, which includes a concrete footing wall component in addition to the concrete masonry unit.

Note: To represent double joists and girders, change interior walls to a DOUBLE_JOIST wall style. The DOUBLE_JOIST wall style can be assigned zero width, zero offset, and the linetype of the boundary set to center. Change the wall style of interior walls, which require double joists, to this DOUBLE_JOIST joist wall style.

STOP. Do Tutorial 3.5, "Creating the Foundation Plan," at the end of the chapter.

PLOTTING SHEETS

This chapter has provided you with basic commands for creating floor plans. The construct drawings created can be attached to view drawings to create view drawings for each floor plan. The view drawings are attached to sheet drawings that are created in the **Sheets** tab of the **Project Navigator**. Although you can plot within construct and view drawings, the sheet drawing offers additional flexibility in plotting. The default page setup is for "Arch F (30 × 42) Expand-Dwf 6" layout, and the plot device is DWF6 ePlot. The **Sheets** tab allows you to specify other sheet sizes, layouts, and plot the drawing from the **Project Navigator**. The following steps outline the procedure for setting sheet size and layouts in the **Sheets** tab of the **Project Navigator**.

Published drawings can be viewed and plotted from within the Autodesk DWF Viewer. The Design Web Format file can be viewed but not modified, and is easily transmitted over the Internet. A description of the use of the Autodesk Viewer is included in Appendix A of the CD.

STEPS TO PLOTTING

1. Open the **Project Navigator** and select the **Sheets** tab. Select the **Sheet Set View** toggle, shown in Figure 3.72, to display the Sheet Set View.

2. Select the project sheet set icon as shown at **p1** in Figure 3.72, right-click, and choose **Properties**. The **Properties** dialog box of the sheet set allows you to specify the template and page setup overrides.

3. To specify the Page Setup overrides file, select the **Browse** button shown at **p1** in Figure 3.73 to open the **Select Template** dialog box shown in Figure

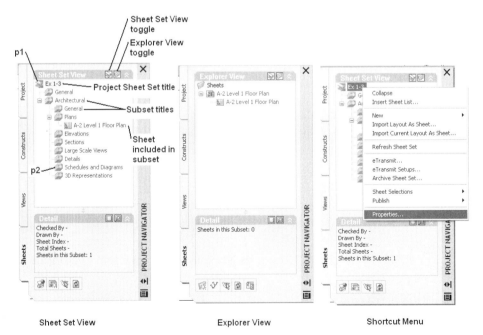

Figure 3.72 *Sheet Set View and Explorer views of a sheet set*

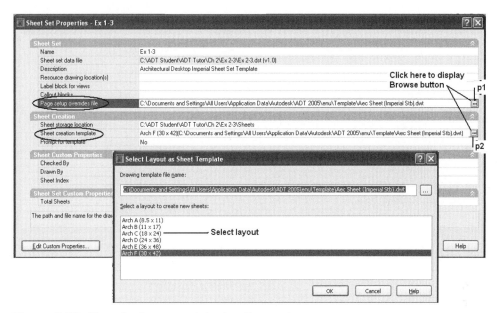

Figure 3.73 *Sheet Set Properties dialog box for a project*

3.74. The template selected from the template list must include layouts for each sheet size. The **Select Template** dialog box, shown in Figure 3.74, lists the Autodesk Architectural Desktop templates displayed for new drawings.

4. To specify the Sheet creation template, select the **Browse** button shown at **p2** in Figure 3.73, to open the **Select Layout as Sheet Template** dialog box. Select a layout as shown in Figure 3.73 and select **OK** to close all dialog boxes.

5. To create a new sheet, select the icon for the Sheet Set (**p1** in Figure 3.72) right-click and choose **New>Sheet**. The **New Sheet** dialog box, shown in Figure 3.75, allows you to specify the sheet number and name.

6. To set the sheet properties of a subset, select a subset (**p2** in Figure 3.72), right-click, and choose **Properties**. The **Subset Properties** dialog box shown in Figure 3.76 includes a **Browse** button. When you select the **Browse** button, the **Select Layout as Sheet Template** dialog box opens as shown in Figure 3.76. The template selected is assigned to all new sheets created within the subset.

7. To create a new sheet, select the icon for the subset, right-click, and choose **New>Sheet**. The **New Sheet** dialog box shown in Figure 3.75 allows you to specify the sheet number and name.

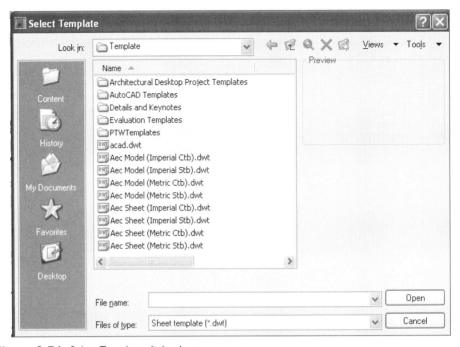

Figure 3.74 Select Template dialog box

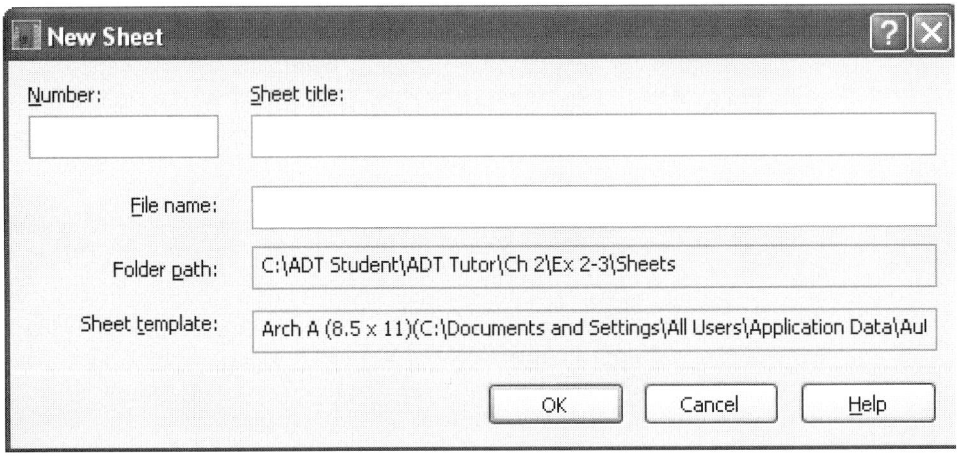

Figure 3.75 New Sheet dialog box

8. Double-click on the name of the new sheet to open the file. The drawing will consist of a Model tab and a layout tab. (The layout tab was named in Step 7.) To configure the layout for printing to a plotter or printer, right-click over the

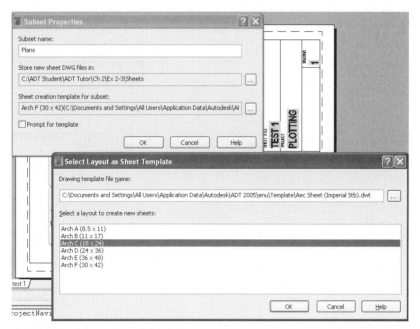

Figure 3.76 *Subset Properties dialog box*

layout tab and choose **Page Setup Manager** to open the **Page Setup Manager**. Choose the **New** button of the **Page Setup Manager** to open the **New Page Setup** dialog box, shown at right in Figure 3.77. Select the **Default output device** option within the **Start with** window and select **OK**. When you select **OK**, the **Page Setup** dialog box opens as shown in Figure 3.78. The **Page Setup** dialog box allows you to specify the plotter, printer, plot style table and other features of the page. To plot from the **Project Navigator**, the **What to plot** field should be set to **Layout** in the **Page Setup** dialog box.

9. To plot an entire subset, select the subset name, right-click, and choose **Publish>Publish to Plotter**. The pages listed in the subset will plot to the plotter defined in the layout.

Tip: You can view and print any size sheet from the Autodesk DWF Viewer. Therefore, when you publish a drawing, you can open the file in the Autodesk Viewer and print the file to fit the printer.

For additional information regarding plotting and the Autodesk DWF Viewer, see Appendix A–"Plotting," located on the CD.

STOP. Do Tutorial 3.6 "Plotting Floor Plans," at the end of the chapter.

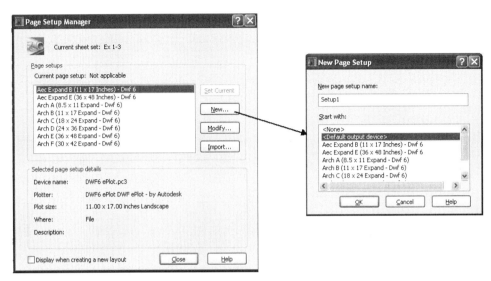

Figure 3.77 *Page Setup Manager and New Page Setup dialog boxes*

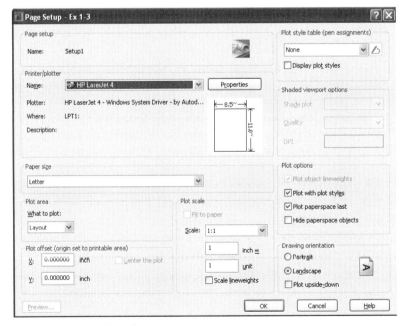

Figure 3.78 *Page Setup dialog box*

SUMMARY

1. Wall styles are created in the **Style Manager** accessed using the **Wall Style** command (**WallStyle**).
2. Wall components are created by the **WallStyle** command; they represent sub-assemblies of a wall and can be defined with unique widths and elevations.
3. The edge offset distance of a wall component positions one edge of the wall component from the baseline.
4. The width of a wall component establishes the distance between the surfaces of a wall component.
5. The wall component priority number controls the clean up of wall components when they intersect with other walls.
6. Wall endcap styles are applied to a wall component to close the end of a wall component at wall openings or wall ends.
7. Wall styles include display properties control settings that control layer, visibility, color, and linetype of wall entities.
8. Display control of a wall style can include the display of materials or hatching of wall components.
9. Wall styles can be exported to resource files and imported from resource files to other files.
10. Endcap styles for wall components are created by the **Endcap Style** command (**WallEndCapStyle**).
11. Wall endcap styles can be edited in place.
12. Applying wall modifier styles to a wall will create vertical projections along a wall.
13. Body modifiers can add a mass element to a wall.
14. Wall modifier styles are created from polylines.
15. The **RoofLine** and **FloorLine** commands allow you to extend the wall above the base height and below the baseline.
16. Copy the first floor construct to other levels to create foundation plans and additional floor plans.
17. Plot sheet drawings from the **Sheets** tab of the **Project Navigator**. View drawings are attached as reference files to sheet drawings.

REVIEW QUESTIONS

1. Create a wall style by selecting the _____ button on the **Style Manager** toolbar.
2. Define specific properties of a wall style by selecting the _____ button on the **Style Manager** toolbar.

3. A wall component with a priority number of 10 will override the display of another wall with a wall component priority number of _____.

4. If the Edge Offset of a wall component is set to –8 and the wall is drawn from left to right, the component will be drawn (above, below) the justification line.

5. If the Width of a wall component is to –8 and the wall is drawn from left to right, the component will be drawn (above, below) the justification line.

6. The top elevation offset of a wall component can be specified relative to _____, _____, _____, and _____,

7. Hatching a wall component is controlled by _____.

8. The geometry used to create an Endcap Style must be a _____.

9. Wall modifier styles are created from _____.

10. Wall modifiers are attached to a wall by editing the _____ of the wall.

11. The Index number of the endcap component determines the _____ of the wall component capped by the Endcap Style.

12. Wall modifiers can be placed along the length of the wall relative to the following locations: _____, _____, and _____.

13. Wall styles are transferred to other drawings with the _____ and _____ feature of the **Style Manager**.

TUTORIAL 3.1 CREATING A WALL STYLE

1. Open Autodesk Architectural Desktop 2005, and select **File>Project Browser** from the menu bar. Use the Project Selector drop-down list to navigate to your *ADT Student\ADT Tutor\Ch 3* student directory. Double-click on **Ex 3-1** to set this project current. Select **Close** to dismiss the **Project Browser**. (If your student folder does not include the project, refer to "Organizing Tutorial Directories" in the Preface.)

2. Select the **Constructs** tab of the **Project Navigator**. Double-click on Floor 1 to open the Floor 1 drawing.

3. Select **Format>Style Manager** from the menu bar.

4. Select **Open Drawing** from the **Style Manager** toolbar to import material definitions into the drawing. Select **Content** in the **Places** panel at right in the **Open Drawing** dialog box. Double-click to open Styles\Imperial\ and select the **Materials Definitions (Imperial).dwg** file. Select Open to load the drawing and close the **Open Drawing** dialog box.

5. Expand the Multi-Purpose Objects folder, and select the Material Definitions folder of the Material Definitions (Imperial).dwg in the left pane of the **Style Manager** to display the list of Material Definitions in the right pane. Scroll down the material list in the right pane and select the Masonry.Unit

Masonry.Brick.Modular-Running.Brown material, right-click, and choose **Copy** from the shortcut menu as shown in Figure 3T.1.

6. Select the Floor 1.dwg in the left pane, right-click, and choose **Paste** from the shortcut menu to paste the material in the current file.

7. To create a new wall style, expand the **Architectural Objects** folder in the left pane for the Floor 1.dwg file.

8. Select the **Wall Styles** folder of the left pane to display the styles in the right pane.

9. Select **New Style** from the **Style Manager** toolbar, and overtype **Frame Brick** as the name of the new wall style as shown in Figure 3T.2.

10. Select the name **Frame Brick** in the right pane of the **Style Manager**, and then select **Edit Style** on the **Style Manager** toolbar.

11. Select the **General** tab of the **Wall Style Properties – Frame Brick** dialog box, and type **Wood frame brick veneer** in the **Description** edit field.

12. In the following steps you will create two components as described below:

Index	Name	Priority	Width	Edge-Offset
1	Wood	200	3.5	0
2	Brick	400	3.625	−5.125

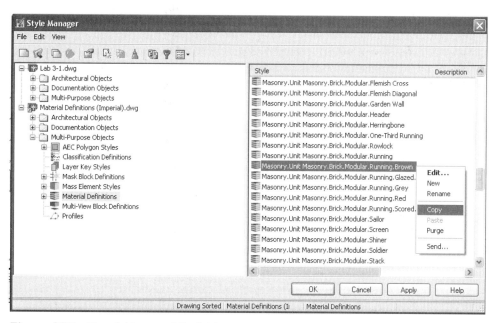

Figure 3T.1 *Material imported for Brick*

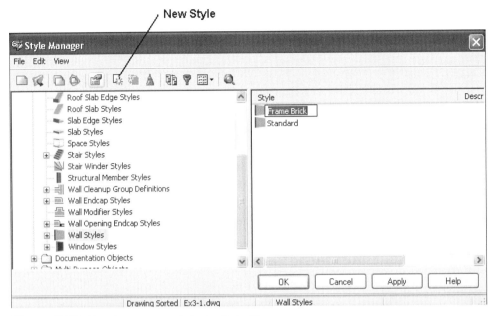

Figure 3T.2 *Creating a new style in the Style Manager*

13. To create the Wood component, select the **Components** tab and select Index 1. Select **Unnamed** in the **Name** edit field and type **Wood**.
14. Select in the **Priority** column and type **200**.
15. Select in the **Width** column and select the arrow to expand the **Width** edit field. Set the fixed value to **3 1/2** and variable value **0** as shown in Figure 3T.3.
16. Select in the **Edge Offset** column and select the arrow to expand the **Edge Offset** field. Set the fixed value to 0 and variable value to **0**.
17. Select the **Add Component** button in the right margin of the dialog box.
18. To create the Brick component, select **2** in the **Index** column, and then click in the **Name** edit field and type **Brick**.
19. Select in the **Priority** column and type **400**.
20. Select in the **Width** column and select the arrow to expand the **Width** edit field. Set the fixed value to **3 5/8** and variable value to **0**.
21. Select in the **Edge Offset** column and select the arrow to expand the **Edge Offset** field. Set the fixed value to **−5 1/8** and variable value to 0.
22. Verify that the **Name**, **Priority**, **Edge Offset**, and **Width** values are as shown in Figure 3T.4.

Advanced Wall Features 219

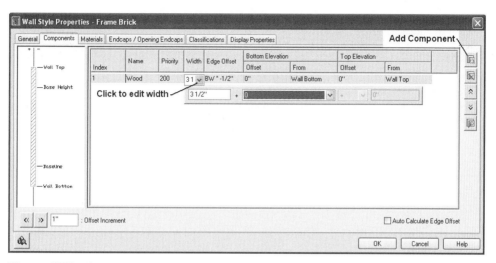

Figure 3T.3 *Creating the Wood component*

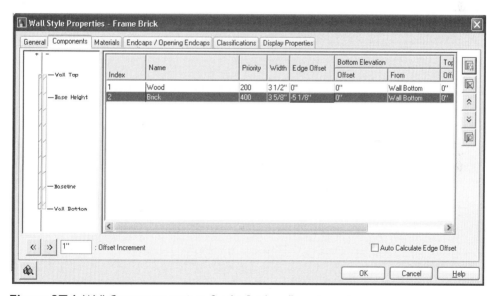

Figure 3T.4 *Wall Component settings for the Brick wall component*

23. To specify materials for wall components, select the **Materials** tab, and click in the **Material Definition** column for the Brick component to expand the drop-down list of materials. Select the Masonry.Unit Masonry.Brick.Modular.Running.Brown material from the drop-down list as shown in Figure 3T.5.

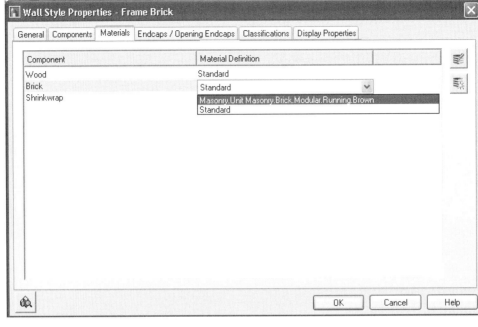

Figure 3T.5 *Material Definition dialog box*

24. Select **OK** to dismiss the **Wall Style Properties** dialog box.
25. Select **OK** to dismiss the **Style Manager**.
26. Select the four exterior walls of the house, right-click, and choose **Properties**.
27. Edit the Style to **Frame Brick** and the Justify to **Baseline** in the Properties palette as shown in Figure 3T.6.
28. Select **Zoom Window** from the **Zoom** flyout of the **Navigation** toolbar and respond to the command line prompts as follows:

    Command:' zoom

    Specify corner of window, enter a scale factor (nX or nXP), or

    [All/Center/Dynamic/Extents/Previous/Scale/Window] <real time>: _w

    Specify first corner: **100',35'** ENTER

    Specify opposite corner: **110',45'** ENTER

    (Notice the interior walls do not merge.)

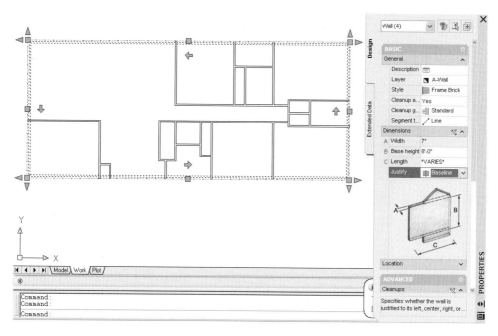

Figure 3T.6 *Editing the style in the Properties palette*

29. Select the interior wall, right-click, and choose **Copy Wall Style and Assign** from the shortcut menu to open the **Wall Style Properties-Standard (2)** dialog box.
30. Select the **General** tab, overtype the name **Interior** in the **Name** edit field to remove the "Standard (2)" name, and type **Interior partitions** in the **Description** edit field.
31. Select the **Components** tab, and edit the Priority of the wall to **200**.
32. Select **OK** to dismiss the **Wall Style Properties** dialog box.
33. Select **Zoom Previous** from the **Zoom** flyout of the **Navigation** toolbar.
34. To change the interior walls to the Interior wall style, select the walls using a crossing selection; select all interior walls from **p1** to **p2** as shown in Figure 3T.7.
35. Edit the Style to **Interior** in the Properties palette, and press ESC twice to end the selection and edit.
36. Select **Zoom Window** from the **Zoom** flyout of the **Navigation** toolbar and respond to the command line prompts as follows:

    Command: '_zoom

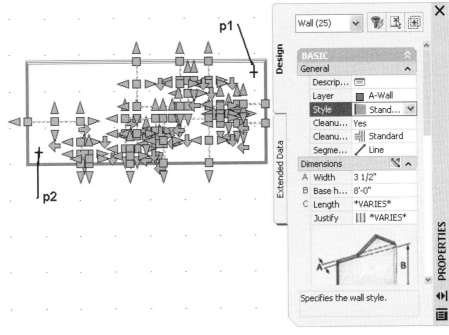

Figure 3T.7 Selection of interior walls

> Specify corner of window, enter a scale factor (nX or nXP), or
>
> [All/Center/Dynamic/Extents/Previous/Scale/Window] <real time>: _w
>
> Specify first corner: **100',35'** ENTER
>
> Specify opposite corner: **110',45'** ENTER
>
> *(Notice the interior walls merge after priority is set for each style.)*

37. Select the **Work** layout tab, double-click in the left viewport, and select **Zoom Extents** from the **Zoom** flyout of the **Navigation** toolbar. Select **Hidden** from the **Shade** flyout of the **Navigation** toolbar to view the house as shown in Figure 3T.8.

38. Save changes and close the drawing.

TUTORIAL 3.2 CREATING AND EDITING AN ENDCAP

1. Open *ADT Student\ADT_Tutor\Ch3\Ex3-2.dwg*.

2. Save the drawing as **Lab3-2** in your student directory.

3. If tool palettes are not displayed, select **Window>Tool Palettes** from the menu bar.

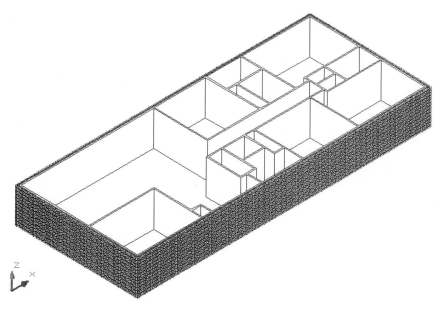

Figure 3T.8 *Pictorial view of house*

4. Double-click in the right viewport, and select **Zoom Extents** from the **Zoom** flyout of the **Navigation** toolbar.
5. To create a new wall style, select **Format>Style Manager** from the menu bar.
6. Expand the Architectural Objects folder of the left pane for Lab 3-2.dwg.
7. Select the **Wall Styles** folder of the left pane.
8. Select **New Style** from the **Style Manager** toolbar, and overtype **10_bull_nose** as the name of the wall style in the right pane.
9. Select the **10_bull_nose** wall style in the right pane right-click, and select **Edit Styl**e.
10. Select the **Components** tab of the **Wall Style Properties** dialog box and create the two components as shown in Figure 3T.9.
11. To edit the wall style to display the footing, select the **Display Properties** tab, and check the **Style Override** check box for the Plan display representation. Verify that the **Layer/Color/Linetype** tab is displayed, click the light bulb ON for the **Below Cut Plane** layer, and click in the Linetype column to open the **Select Linetype** dialog box. Select **Hidden2** from the **Linetype** list, and select **OK** to dismiss the **Select Linetype** dialog box.

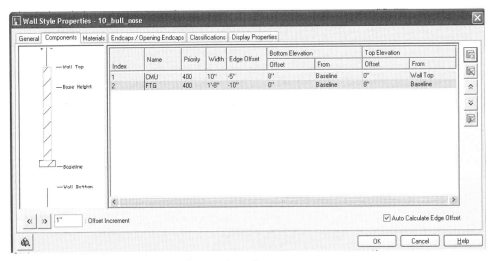

Figure 3T.9 *Components of a wall to apply endcaps*

12. To display the footing with hidden lines, select the **Boundary (2) FTG** component and click in the **Linetype** column. Select the **Hidden2** linetype from the list of the **Select Linetype** dialog box. Select **OK** to dismiss the **Select Linetype** dialog box. Verify the settings of the **Layer/Color/Linetype** tab as shown in Figure 3T.10. Select **OK** to dismiss the **Display Properties** dialog box. Select **OK** to dismiss the **Wall Style Properties** dialog box.

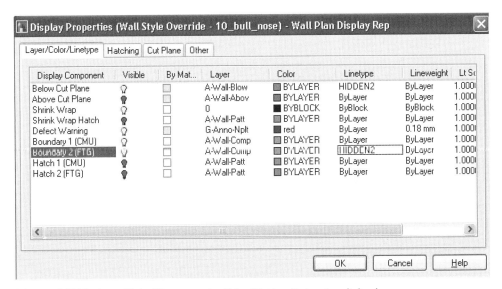

Figure 3T.10 *Layer/Color/Linetype tab of the Display Properties dialog box*

13. To create a new endcap style, select Wall Endcap Styles in the left pane, and select **New Style** on the **Style Manager** toolbar.
14. Overtype the name **Bull_Nose** in the right pane.
15. Verify that **Bull_Nose** style is selected in the right pane, right-click, and select **Set From** from the shortcut menu. Respond to the command sequence as shown in Figure 3T.11.

    Command: _AecStyleManager

    Select a polyline: *(Select the polyline on the right at p1 as shown Figure 3T.11.)*

    Enter component index for this segment <1>: ENTER

    (Pressing ENTER *assigns this polyline to wall component Index 1, the CMU.)*

    Add another component? [Yes/No] <N>: **Y** ENTER

    Select a polyline: *(Select the polyline on the left at p2 as shown in Figure 3T.11.)*

    Enter component index for this segment <2>: ENTER

    (Pressing ENTER *assigns this polyline to the wall component Index 2, the footing.)*

    Add another component? [Yes/No] <N>: ENTER

    Enter return offset <0">: ENTER

    (Pressing ENTER *creates an endcap style without extending the wall by an offset distance.)*

16. Select Wall Styles in the left pane of the **Style Manager**.
17. Double-click on the **10_bull_nose** wall style to open the **Wall Style Properties-10_bull_nose** dialog box.
18. Select the **Endcaps/Openings Endcaps** tab of the **Wall Style Properties-10_bull_nose** dialog box.
19. Select the **Wall Endcap Style** list arrow, and select the **Bull_Nose** wall endcap style as shown in Figure 3T.12.

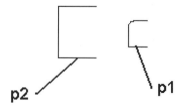

Figure 3T.11 *Polyline geometry for endcaps*

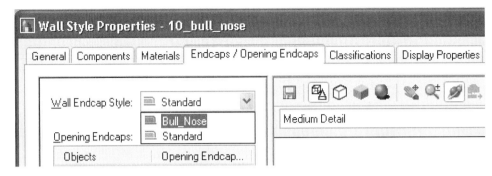

Figure 3T.12 *Selecting Bull_Nose wall endcap style*

20. Select **OK** to dismiss the **Wall Style Properties** dialog box.
21. Select **OK** to dismiss the **Style Manager**.
22. Select **1 1/2"=1'-0"** from the **VP Scale** flyout, and select **Wall** from the Design palette.
23. Edit the Properties palette, and set Style = **10_bull_nose** and Justify = **Baseline**.
24. Draw a wall horizontally 8' long above the two polylines from right to left.
25. Double-click in the left viewport. Select **Zoom Extents** from the **Zoom** flyout of the **Navigation** toolbar. The wall should appear as shown in Figure 3T.13.
26. Verify that POLAR, OTRACK are toggled ON on the status bar. Move the pointer over the OSNAP toggle on the status bar, right-click, and select **Settings**. Verify that **Endpoint** object snap is toggled ON, clear all other object snaps. Toggle OSNAP ON on the status bar. Select **OK** to dismiss the **Drafting Settings** dialog box.
27. Verify that the left viewport is current, select the wall, right-click, and choose **Endcaps>Edit In Place** from the shortcut menu. Respond to the command line prompt as shown below.

    Command: WallEndcapEdit

    Select a point near endcap: *(Select a point near p1 as shown in Figure 3T.13.)*

    (Grips of the endcap are displayed.)

28. Select **Zoom To** from the **In Place Edit** toolbar.
29. Select the endcap to display the grips, right-click, and choose **Remove Vertex** from the shortcut menu. Select a point near p2 in Figure 3T.14. Press ESC to end vertex removal.

Advanced Wall Features 227

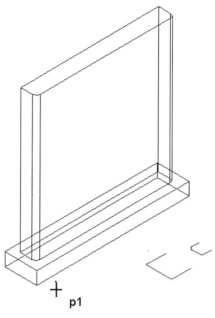

Figure 3T.13 *Endcaps applied to wall*

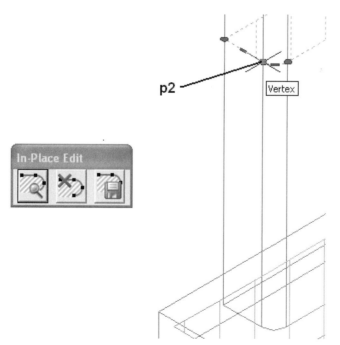

Figure 3T.14 *Grip display using Edit In Place*

30. Select the wall endcap to display the grips. Select the grip shown in Figure 3T.15. Move the pointer to the left to display <0.00° in the Polar tooltip, and type **2.5"** in the command line to stretch the grip.
31. Select **Save All Changes** on the **In Place Edit** toolbar.
32. Select **Zoom All** from the **Zoom** flyout of the **Navigation** toolbar to display the wall as shown in Figure 3T.16.
33. Close the drawing and save changes.

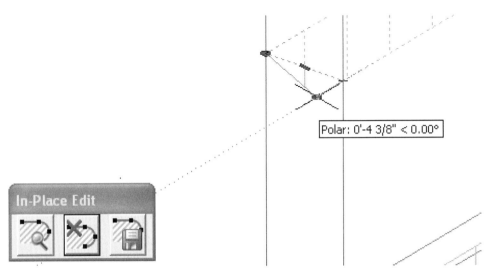

Figure 3T.15 *Stretch wall endcap using Edit In Place*

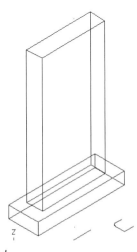

Figure 3T.16 *Wall endcap edited*

TUTORIAL 3.3 CREATING WALL MODIFIERS

1. Open Autodesk Architectural Desktop 2005, and select **File>Project Browser** from the menu bar. Use the Project Selector drop-down list to navigate to your *ADT Student\ADT Tutor\Ch3* directory. Double-click on **Ex 3-3** to set this project current. Select **Close** to dismiss the **Project Browser**. (If your student folder does not include the project, refer to "Organizing Tutorial Directories" in the Preface.)
2. Select the **Constructs** tab of the **Project Navigator**. Double-click on **Floor 1** to open the Floor 1 drawing.
3. Select the Model tab.
4. Move the pointer over OSNAP on the status bar, right-click, and choose **Settings** from the shortcut menu. Clear all object snaps except **Endpoint**, and turn ON Object Snaps. Select the **Polar Tracking** tab, check **Polar Tracking** ON, and select **OK** to dismiss the **Drafting Settings** dialog box.
5. To create a wall modifier style, select **Format>Style Manager** from the menu bar.
6. Expand the **Architectural Objects** folder of the left pane for the Floor 1.dwg drawing.
7. Select the **Wall Modifier Styles** folder of the left pane, right-click, and select **New**.
8. Overtype **Bar_pilaster** in the right pane.
9. Select **Bar_pilaster** in the right pane, right-click, and select **Set From** from the shortcut menu. Select the polyline at p1 as shown in Figure 3T.17. Select **OK** to close the **Style Manager**.
10. Select **Wall** from the Design palette, and set the following properties in the Properties palette: Basic-Style = Interior, Cleanup = Yes, Cleanup Group Definition = Standard, Segment type = Line, Dimensions Width = 3 1/2", Base Height = 42", Justify = Left, Baseline offset = 0, Roof line offset from base height = 0, Floor line offset from baseline = 0 and Advanced-Graphline position = Wall Justification, Cleanup radius = 0, Override start cleanup radius = No, Override end cleanup radius=No, Starting Endcap = ByStyle, Ending Endcap = ByStyle, Priority overrides = (0), and Plan modifier = (0). Respond to the command line prompts to add the wall as shown in the following command sequence and in Figure 3T.18.

    Command: WallAdd

    Start point or
    [STyle/Group/WIdth/Height/OFfset/Justify/Match/Arc]: *(Select the endpoint of existing wall at p1 shown in Figure 3T.18.)*

    (Move the cursor to the right with a polar angle of 0°.)

    End point or
    [STyle/Group/WIdth/Height/OFfset/Justify/Match/Arc]: **9'** ENTER

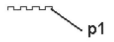

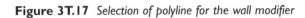

Figure 3T.17 *Selection of polyline for the wall modifier*

Figure 3T.18 *Start point of wall*

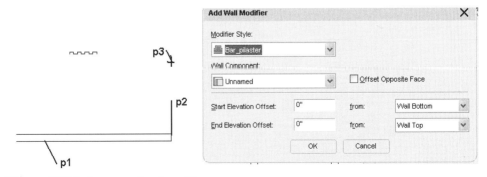

Figure 3T.19 *Location of wall modifier*

End point or
[STyle/Group/WIdth/Height/OFfset/Justify/Match/Arc/Undo]:
ENTER *(Press* ENTER *to end the command.)*

11. Select the wall at **p1** shown in Figure 3T.19, right-click, and choose **Plan Modifiers>Add** from the shortcut menu. Respond to the command line prompts as shown below and refer to Figure 3T.19.

    Command: WallModifierAdd

    Select start point: *(Select the wall at p2 using the running Endpoint object snap as shown in Figure 3T.19.)*

    (Move the pointer left with a polar angle of 180°.)

    Select end point: **8** ENTER

    (Move the pointer up.)

    Select the side to draw the modifier: *(Select a point near p3 as shown in Figure 3T.19.)*

    Enter wall modifier depth <10 1/4">: **1.5** ENTER *(Specify depth of wall modifier.)*

    (**Add Wall Modifier** *dialog box opens; edit the Wall Modifier Style = Bar_Pilaster, Wall Component = Unnamed, Start Elevation = 0, End Elevation = 0 as shown in Figure 3T.19.)*

    (Select **OK** *to close the* **Add Wall Modifier** *dialog box and create the wall modifier.)*

12. Select the wall at **p1** shown in Figure 3T.19, right-click, and choose **Plan Modifiers>Add** from the shortcut menu. Respond to the command line prompts as shown below and refer to Figure 3T.20.

    Command: WallModifierAdd

    Select start point: *(Select the wall at p1 using the running object snap as shown in Figure 3T.20.)*

    (Move the pointer right with a polar angle of 0°.)

    Select end point: **8** ENTER

    (Move the pointer up.)

    Select the side to draw the modifier: *(Select a point near p2 as shown in Figure 3T.20.)*

    Enter wall modifier depth <10 1/4">: **1.5** ENTER *(Specify depth of wall modifier.)*

    (The **Add Wall Modifier** *dialog box opens; edit the Wall Modifier Style = Bar_Pilaster, Wall Component = Unnamed, Start Elevation = 0, End Elevation = 0 as shown in Figure 3T.20.)*

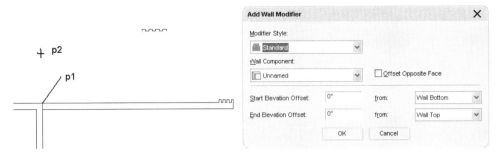

Figure 3T.20 Location of wall modifier at start of wall

(Select **OK** to close the **Add Wall Modifier** dialog box.)

(Wall modifier created.)

13. To create a wall modifier from existing geometry, select **Polyline** from the **Shapes** toolbar, and respond to the command line prompts as shown below to draw the arc shown in Figure 3T.21.

    Command: _pline

    Specify start point: *(Select a point p1 as shown in Figure 3T.21.)*

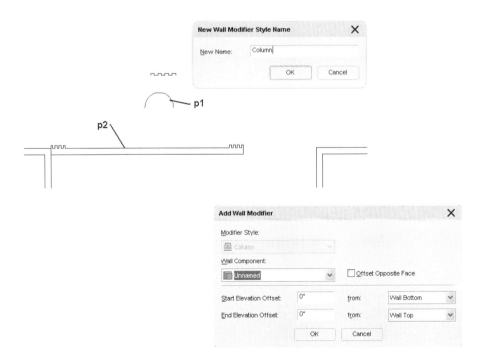

Figure 3T.21 Creating polyline for wall modifier

Current line-width is 0'-0"

Specify next point or [Arc/Halfwidth/Length/Undo/Width]: **a** ENTER

Specify endpoint of arc or

[Angle/CEnter/Direction/Halfwidth/Line/Radius/Second pt/Undo/Width]: **a** ENTER

Specify included angle: 180

Specify endpoint of arc or [CEnter/Radius]: **r** ENTER

Specify radius of arc: **8** ENTER

Specify direction of chord for arc <270.00>: **180** ENTER *(Specify direction for arc.)*

Specify endpoint of arc or

[Angle/CEnter/CLose/Direction/Halfwidth/Line/Radius/Second pt/Undo/Width]: ENTER *(Press ENTER to end the command.)*

14. Select the wall at p2 as shown in Figure 3T.21, right-click, and choose **Plan Modifiers>Convert Polyline to Wall Modifiers**. Respond to the following command line prompts.

    Command: WallModifierConvert

    Select a polyline: *(Select the arc at p1 shown in Figure 3T.21.)*

    Erase layout geometry? [Yes/No] <No> ENTER

15. Type **Column** in the **New Name** edit field of the **New Wall Modifier Style Name** dialog box. Select **OK** to dismiss the **New Wall Modifier Style Name** dialog box.

16. Edit the **Add Wall Modifier** dialog box as shown in Figure 3T.21, and select **OK** to dismiss the dialog box.

17. Select the wall modifier placed in the previous step, scroll down the Properties palette to Worksheets and click the **Plan Modifiers** button to open the **Wall Modifiers** dialog box.

18. Select the **Column** plan modifier and edit the Start Position Offset = **0** from Midpoint of Wall as shown in Figure 3T.22. Select **OK** to dismiss the **Wall Modifiers** dialog box.

19. Select **NE Isometric** from the **View** flyout to view the wall modifiers as shown in Figure 3T.23.

20. Save and close the drawing.

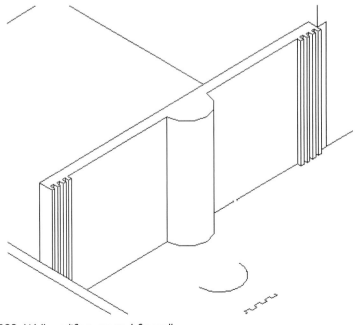

Figure 3T.22 *Wall Modifiers dialog box*

Figure 3T.23 *Wall modifiers created for wall*

TUTORIAL 3.4 CREATING THE BASEMENT FLOOR PLAN

1. Open Autodesk Architectural Desktop 2005, and select **File>Project Browser** from the menu bar. Use the Project Selector drop-down list to navigate to your *ADT Student\ADT Tutor\Ch3* directory. Double-click on **Ex 3-4** to set this project current. Select **Close** to dismiss the **Project Browser**. (If your student folder does not include the project refer to "Organizing Tutorial Directories" in the Preface).

2. Select the **Constructs** tab of the **Project Navigator**. Double-click on **Floor 1** to open the Floor 1 drawing.

3. Select the Model tab, and choose **Top View** from the **View** flyout of the **Navigation** toolbar.

4. In the Project Navigator select Floor 1, right-click, and choose **Copy Construct to Levels** from the shortcut menu to open the **Copy Construct to Levels** dialog box.

5. Check **Level 1** for the new construct. Select **OK** to close the **Copy Construct to Levels** dialog box.

6. Select the Floor 1(1) construct, right-click, and select **Rename** from the shortcut menu. Type **Basement** as the name of the construct. Double-click on Basement to open the file.

7. Select **Format>Style Manager** from the menu bar. Select **File>Close All** from the menu bar of the **Style Manager** to close open files of the **Style Manager**. Verify that **Floor 1** and **Basement** files are open in the left pane.

8. Select **Open Drawing** from the **Style Manager** toolbar, and select **Content** in **Places** at left. Select the \Styles\Imperial folder and choose the Material Definitions (Imperial).dwg file. Select **Open** to close the **Open Drawing** dialog box.

9. Expand the Multi-Purpose Objects folder, and select the Material Definitions folder of the Material Definitions (Imperial).dwg in the left pane of the **Style Manager** to display the list of Material Definitions in the right pane. Hold CTRL down, scroll down the material definition list in the right pane, select the **Concrete.Cast In-Place.Flat Grey** and **Masonry.Unit Masonry.CMU.Stretcher.Running** material definitions, right-click, and choose **Copy**.

10. Select **Basement.dwg** in the left pane, right-click, and choose **Paste** from the shortcut menu to paste the material definitions into the current file.

11. Expand the **Architectural Objects** folder of the **Basement.dwg** in the left pane of the **Style Manager**.

12. Select the **Wall Styles** folder in the left pane to display the styles in the right pane.

13. Select **New Style** on the **Style Manager** toolbar, and overtype **BASEMENT** as the name of the new wall style.
14. Select the name **BASEMENT** in the right pane of the **Style Manager**, and then select **Edit Style** on the **Style Manager** toolbar.
15. Select the **General** tab, and type **Brick veneer basement wall** in the **Description** edit field.
16. Select the **Components** tab and enter the following components as shown in Table 3T.1 and in Figure 3T.24. Select **OK** to dismiss the **Component Offset** dialog box.

Index	Name	Priority	Width	Edge Offset	Bottom Elev.		Top Elev.	
					Offset	From	Offset	From
1	8CMU	400	−7 5/8	7 1/8	4'-0"	Baseline	0	Base Height
2	12CMU	400	−11	7 1/8	−8	Baseline	4'-0"	Baseline
3	Brick	400	−3 5/8	−1 1/2	4'-0"	Baseline	0	Wall Top
4	FTG	400	−2'-0"	1'-1 1/2	−1'-4"	Baseline	−8"	Baseline

Table 3T.1 *Wall components*

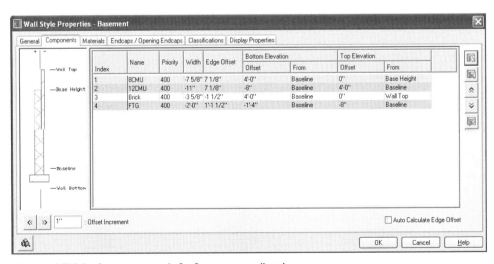

Figure 3T.24 *Components tab for Basement wall style*

17. Select the **Materials** tab, click in the **Material Definition** column, and assign the material definition to the components as shown in Table 3T.2.

Component	Material Definition
8CMU	Masonry, Unit Masonry, CMU, Stretcher, Running
12CMU	Masonry, Unit Masonry, CMU, Stretcher, Running
FTG	Concrete, Cast In Place Flat Grey
Shrinkwrap	Standard

Table 3T.2 *Material Definition assignments*

18. Select the **Display Properties** tab, and check the **Style Override** check box for the current display representation (Plan). Select the **Layer/Color/Linetype** tab and edit the **Below Cut Plane**, click in the **Linetype** column, and select **Hidden2** from the **Select Linetype** dialog box. Select **OK** to dismiss the **Select Linetype** dialog box.

19. Select the **Cut Plane** tab and edit the dialog box as follows: clear **Override Display Configuration Cut Plane**, check **Automatically Choose Above and Below Cut Plane Heights**, and clear the **Manual Above and Below Cut Plane Heights** check box as shown in Figure 3T.25. Select **OK** to dismiss the **Display Properties** dialog box.

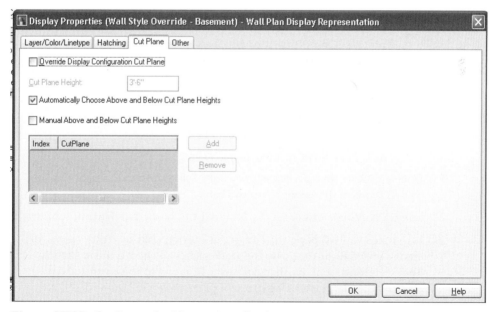

Figure 3T.25 *Cut Plane tab of Basement wall style*

20. Select **OK** to dismiss the **Wall Style Properties** dialog box.
21. Select **OK** to dismiss the **Style Manager**.
22. Select the Work layout tab, and double-click in the left floating viewport. Select **Zoom Extents** from the **View** flyout of the **Navigation** toolbar. Select **2d Wireframe** from the **Shade** flyout of the **Navigation** toolbar. Toggle OFF GRID on the status bar.
23. In this step you will use the **FloorLine** command to project the bottom of the wall down to wall bottom and display the footing. Select the four exterior walls, click in the **Style** edit field of the Properties palette, change the style to **Basement**, and verify that Base height is **8'**. Retain selection of the walls, right-click, and choose **Roof/Floor Line>Modify Floor Line** from the shortcut menu. Respond to the command line prompts as shown below.

    Command: FloorLine

    FloorLine [Offset/Project to polyline/Generate polyline/Auto project/Reset]: **o** ENTER *(Select Offset to extend floors.)*

    Enter offset <0>: **-16** ENTER *(Extends floors down.)*

    FloorLine [Offset/Project to polyline/Generate polyline/Auto project/Reset]: ENTER *(Press ENTER to end FloorLine command.)*

24. In this step you will project the brick component up above the wall base height to cover the floor system. Select the four exterior walls, right-click, and choose **Roof/Floor Line>Modify Roof Line** from the shortcut menu. Respond to the command line prompts as shown below.

    Command: RoofLine

    RoofLine [Offset/Project to polyline/Generate polyline/Auto project/Reset]: **o** ENTER

    Enter offset <-1'-4">: **11.5** ENTER *(Specify distance to extend walls up.)*

    RoofLine [Offset/Project to polyline/Generate polyline/Auto project/Reset]: ENTER *(Press ENTER to end command.)*

    (Walls extended as shown in Figure 3T.26.)

25. Double-click in the right floating viewport. Select **Zoom Extents** from the **Zoom** flyout of the **Navigation** toolbar.
26. Select **2D Wireframe** from the **Shade** flyout of the **Navigation** toolbar.
27. In this step you will begin the edit of the interior walls to create rooms for the basement. Right-click over the workspace, and choose **Basic Modify Tools>Extend** from the shortcut menu. Respond to the command line prompts to extend the wall **p1** to wall **p2** as shown in Figure 3T.27.

    Command: _extend

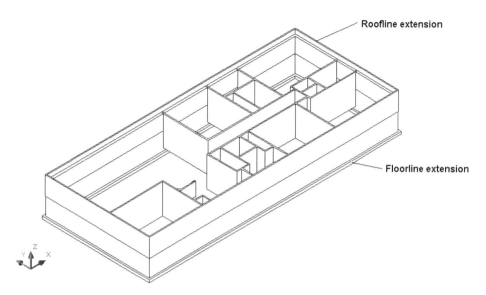

Figure 3T.26 *Walls extended with FloorLine and RoofLine commands*

Current settings: Projection=UCS, Edge=None

Select boundary edges ...

Select objects: 1 found *(Select wall at p1 as shown in Figure 3T.27.)*

Select objects: ENTER *(Press ENTER to end boundary selection.)*

Select object to extend or shift-select to trim or [Project/Edge/Undo]: *(Select wall at p2 as shown in Figure 3T.27.)*

Select object to extend or shift-select to trim or [Project/Edge/Undo] ENTER *(Press ENTER to end selection.)*

28. Right-click over the workspace and choose **Basic Modify Tools>Trim** from the shortcut menu. Select the interior wall shown in Figure 3T.28.

    Command: _trim

    Current settings: Projection=UCS, Edge=None

    Select cutting edges ...

    Select objects: 1 found *(Select wall at p1 as shown in Figure 3T.28.)*

    Select objects: ENTER *(Press ENTER to end cutting edge selection.)*

    Select object to trim or shift-select to extend or [Project/Edge/Undo]: *(Select wall at p2 as shown in Figure 3T.28.)*

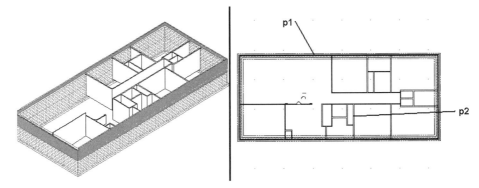

Figure 3T.27 *Walls extended with the Extend command*

Select object to trim or shift-select to extend or
[Project/Edge/Undo]: *(Select wall at p3 as shown in Figure 3T.28.)*

Select object to trim or shift-select to extend or
[Project/Edge/Undo]: ENTER *(Press ENTER to end command.)*

29. Right-click over the workspace and choose **Basic Modify Tools>Erase** from the shortcut menu. Select the interior walls shown in Figure 3T.29 to erase the walls as shown in the following command line prompts.

    Command: _erase

    Select objects: Specify opposite corner: 10 found *(Create crossing selection, click near p1 and then near p2.)*

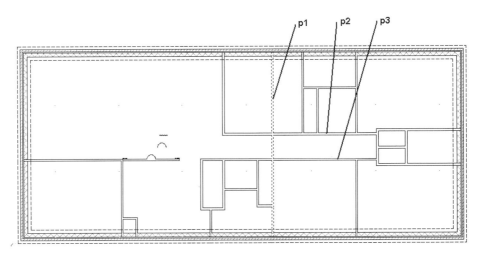

Figure 3T.28 *Selection of walls for the Trim command*

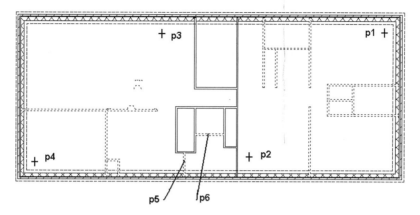

Figure 3T.29 *Walls selected for the Erase command*

   Select objects: Specify opposite corner: 5 found, 15 total *(Create crossing selection, click near p3 and then near p4.)*

   Select objects: 1 found, 16 total *(Select wall at p5.)*

   Select objects: 1 found, 17 total *(Select wall at p6.)*

   Select objects: 1 found, 19 total ENTER *(Press ENTER to end selection.)*

30. Right-click over the workspace, and choose **Basic Modify Tools>Extend** from the shortcut menu. Respond to the command line prompts to extend the wall **p2** to wall **p1** as shown in Figure 3T.30.

    Command: _extend

    Current settings: Projection=UCS, Edge=None

    Select boundary edges ...

    Select objects: 1 found *(Select wall at p1 as shown in Figure 3T.30.)*

    Select objects: ENTER *(Press ENTER to end boundary selection.)*

    Select object to extend or shift-select to trim or [Project/Edge/Undo]: *(Select wall at p2 as shown in Figure 3T.30.)*

    Select object to extend or shift-select to trim or [Project/Edge/Undo] ENTER *(Press ENTER to end selection.)*

31. Select an interior wall, right-click, and choose **Copy Wall Style and Assign**.
32. Select the **General tab of the** Wall Style Properties dialog box and overtype **Basement_Int** as the name of the style.

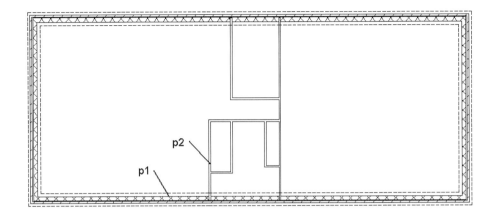

Figure 3T.30 *Walls selected for the Extend command*

33. In this step you will modify the Basement_Int wall style to start the bottom of the wall at the future concrete slab elevation, and set the bottom elevation to 0. In addition, you will edit the priority of wall to display the wall independent of the CMU. Select the **Components** tab of the **Wall Style Properties** dialog box as shown in Figure 3T.31. Edit the component as shown in Table 3T.3.

Index	Name	Priority	Width	Edge Offset	Bottom Elev.		Top Elev.	
					Offset	From	Offset	From
1	Wood	600	BW	BW*-1/2	0	Baseline	0	Base Height

Table 3T.3 *Component definition*

34. Select **OK** to dismiss the **Wall Style Properties -Basement_Int** dialog box.
35. Select all Interior walls, and change the Style to **Basement_Int** in the Properties palette. Select ESC to clear the selection.
36. Click in the left floating viewport of the Work layout tab.
37. Select end Hidden from the end Shade flyout of the end Navigation toolbar to view the interior walls as shown in Figure 3T.32.
38. Select the Model tab and select **Format>Display Manager**. Expand Configurations and select **Medium Detail**. Select the **Cut Plane** tab and edit the Cut Height to **5'-0"**. Select **OK** to dismiss the **Display Manager**.

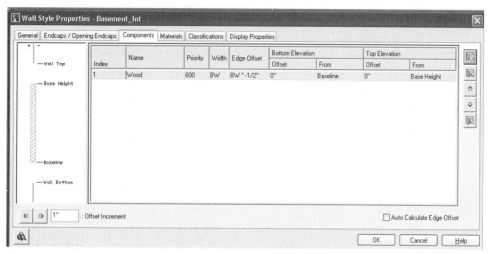

Figure 3T.31 *Components tab for Basement_Int wall style*

39. Save and close your drawing.
40. Select the **Projects** tab of the **Project Navigator** and choose **Launch Project Browser**. Select the current project, right-click, and choose **Close Current Project**.

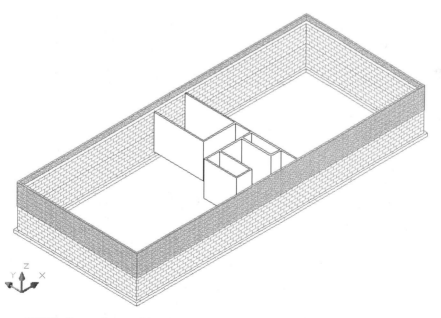

Figure 3T.32 *Pictorial view of basement walls*

TUTORIAL 3.5 CREATING THE FOUNDATION PLAN

1. In this tutorial you will open an existing project and develop the foundation plan for the floor plan. Open Autodesk Architectural Desktop 2005, and select **File>Project Browser** from the menu bar. Use the Project Selector drop-down list to navigate to your *ADT Student\ADT Tutor\Ch3* directory. Double click on **Ex 3-5** to set this project current. Select **Close** to dismiss the **Project Browser**. (If your student folder does not include the project, refer to "Organizing Tutorial Directories" in the Preface.)

2. In this step you will revise the levels for the foundation and first floor plan as shown in Figure 3T.33. Select the **Project** tab, and select the **Edit Levels** button to open the **Levels** dialog box. Edit Level 1 as follows: Level 1 Floor Elevation =0, Floor to Floor Height = 4'-0", Description=Crawl. Edit Level 2 as follows: Floor Elevation = 4'-8", Floor to Floor Height = 8'-0". Select **OK** to close the **Levels** dialog box. Select **Yes** to regenerate all views and dismiss the AutoCAD dialog box.

3. Select the **Constructs** tab of the **Project Navigator**. Select the **Floor 1** construct, right click, and choose **Copy Construct to Levels**. Check **Level 1** for the new construct. Select **OK** to dismiss the dialog box.

4. Select Floor 1(1), right click, and select **Rename**. Overtype **FND** as the name of the new level.

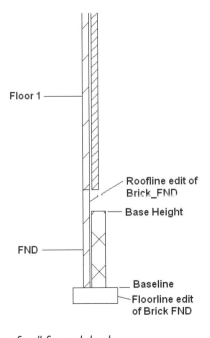

Figure 3T.33 *Wall section of wall for each level*

5. Double-click on **FND** to open the drawing. Select the Work tab, click in the right viewport, and select **Top View** from the **View** flyout of the **Navigation** toolbar.

6. In this step you will remove interior walls. Retain the bearing walls; these walls will be developed into footing wall styles for the support of piers and bearing walls of floor 1. Right-click over the workspace and choose **Basic Modify Tools>Erase**. Erase all the walls except the bearing wall as shown in the following command line sequence.

   Command: _erase

   Select objects: Specify opposite corner: 26 found *(Create a crossing selection, and click at p1 and then at p2 as shown in Figure 3T.34.)*

   Select objects: 1 found, 1 removed, 25 total *(Hold* SHIFT *and select bearing wall p3 to remove it from the selection set.)*

   Select objects: *(Press* ENTER *to end the command.)*

   The interior partition is retained to identify the centerline for the pier footing as shown in Figure 3T.34.

7. Select **Format>Style Manager** from the menu bar. Select **File>Close All** to close unused files from the **Style Manager**.

8. In this step you will modify the wall style for exterior walls to include components designed to cover the floor system. Select **Open Drawing** from the **Style Manager** toolbar, edit the **Look in** directory to your student ADT Student\ADT Tutor\Ch 3\Ex 3-4C\Constructs\Basement.dwg. Choose **Open** to complete the selection.

9. Expand the Architectural Objects folders of the Basement.dwg and the FND.dwg drawings in the left pane. Select the WallStyles type for the

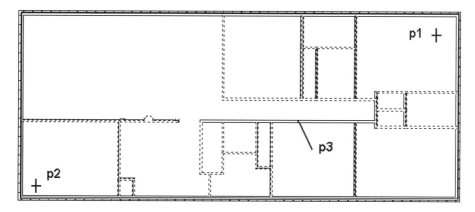

Figure 3T.34 *Walls selected for Erase command*

Basement drawing in the left pane. Select **Toggle View** on the **Style Manager** toolbar as shown in Figure 3T.35.

10. Click on the **Basement** wall style of Basement drawing in the left pane, continue to hold down the left mouse button, and drag the wall style over the FND.dwg and release.

11. Select the Basement wall style of FND in the left pane, right-click, and choose **Rename** from the shortcut menu. Overtype the name **BrickFND**.

12. Double-click **BrickFND** in the left pane to open the **Wall Style Properties BrickFND** dialog box. Select the **Components** tab, delete the **12CMU** wall component, and edit the remaining components as shown in Table 3T.4 and Figure 3T.36. Notice that the brick component extends above the CMU in the left view window. Select **OK** to dismiss all dialog boxes.

Index	Name	Priority	Width	Edge Offset	Bottom Elev.		Top Elev.	
					Offset	From	Offset	From
1	8CMU	400	−7 5/8	7 1/8	0"	Baseline	0	Base Height
2	Brick	400	−3 5/8	−1 1/2	0"	Baseline	0	Wall Top
3	FTG	400	−2'-0"	1'-1 1/2	−8"	Baseline	0	Baseline

Table 3T.4 *Wall components*

13. In this step you will edit the cut plane in the **Display Manager** to pass through the anticipated elevation of the foundation vents. Select the **Format>Display Manager**, and expand the Configurations folder. Select Medium Detail and choose the **Cut Plane** tab. Edit the **Cut Height** to **3'-0"**. Select **OK** to dismiss the dialog box.

14. Select the four exterior walls and edit the Properties palette as follows: Style = BrickFND, Height = 40". Press ESC to clear selection.

15. Select the four exterior walls and the interior wall, right-click, and choose **Roof/Floor Line>Modify Floor Line** from the shortcut menu. Respond to the command line prompts as shown below.

> Command: FloorLine
>
> FloorLine [Offset/Project to polyline/Generate polyline/Auto project/Reset]: **o** ENTER *(Select Offset to extend walls down.)*
>
> Enter offset <0>: **−8** ENTER *(Extends walls down 8" below baseline to display footing.)*

Advanced Wall Features 247

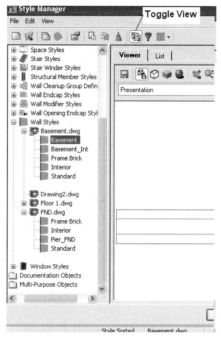

Figure 3T.35 *Style Manager*

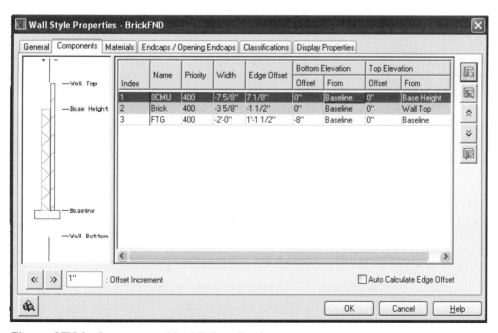

Figure 3T.36 *Components of BrickFND wall style*

FloorLine [Offset/Project to polyline/Generate polyline/Auto project/Reset]: ENTER *(Press enter to end **FloorLine** command.)*

16. Select the four exterior walls, right-click, and select **Roof/FloorLine>Modify Roof Line** from the shortcut menu. Respond to the command line prompts as shown below.

    Command: RoofLine

    RoofLine [Offset/Project to polyline/Generate polyline/Auto project/Reset]: **o** ENTER *(Select Offset to extend walls up.)*

    Enter offset <0>: **11 1/2** ENTER *(Extends walls up.)*

    FloorLine [Offset/Project to polyline/Generate polyline/Auto project/Reset]: ENTER *(Press ENTER to extend the Brick component up to the top of wall above the baseheight.)*

17. Select the interior wall, right-click, and select **Copy Wall Style and Assign** from the shortcut menu.
18. Select the **General** tab, and type **Pier_FND** in the **Name** field and **Pier continuous footing** in the **Description** field.
19. Select the **Components** tab and edit as shown in Table 3T.5. (When you select **Base Width** for the **Width** field, BW will be listed in the **Width** field of the dialog box as shown in Table 3T.5.)

Index	Name	Priority	Width	Edge Offset	Bottom Elev.		Top Elev.	
					Offset	From	Offset	From
1	FTG	400	BW	BW*-1/2	–8"	Baseline	0	Baseline

Table 3T.5 *Wall component settings for the pier footing*

20. Select the **Materials** tab, and edit the FTG component Material Definition to Concrete.Cast-In-Place.Flat.Grey.
21. Select the **Display Properties** tab, and check the **Style Override** check box for the current display representation (Plan). Select the **Layer/Color/Linetype** tab and edit as follows. Edit **Below Cut Plane**, click in the **Linetype** column, and select **Hidden2** from the **Select Linetype** dialog box. Select **OK** to dismiss the **Select Linetype** dialog box. Verify settings in the **Display Properties** dialog box as shown in Figure 3T.37.
22. Select **OK** to dismiss the **Display Properties** dialog box.
23. Select **OK** to dismiss the **Wall Style Properties** dialog box.
24. Select the interior wall at **p3** as shown in Figure 3T.38, edit the Properties palette, and set Width = 24 and Justify = Baseline. Press ESC to end edit and selection.

Advanced Wall Features 249

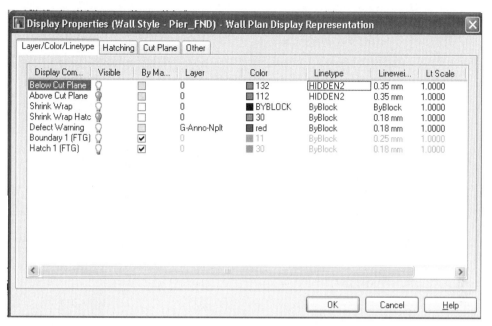

Figure 3T.37 *Display Properties dialog box for Basement wall style*

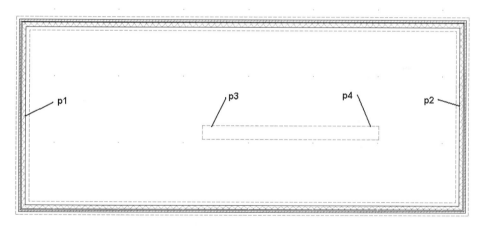

Figure 3T.38 *Wall selection for the Extend command*

25. Right-click over the workspace, and choose **Basic Modify Tools>Extend** from the shortcut menu. Respond to the command line prompts to extend the wall to the exterior walls.

  Command: _extend

Current settings: Projection=UCS, Edge=None

Select boundary edges ...

Select objects: 1 found *(Select wall at p1 as shown in Figure 3T.38.)*

Select objects: 1 found *(Select wall at p2 as shown in Figure 3T.38.)*

Select objects: ENTER *(Press ENTER to end boundary selection.)*

Select object to extend or shift-select to trim or [Project/Edge/Undo]: *(Select wall at p3 as shown in Figure 3T.38.)*

Select object to extend or shift-select to trim or [Project/Edge/Undo]: *(Select wall at p4 as shown in Figure 3T.38.)*

Select object to extend or shift-select to trim or [Project/Edge/Undo] ENTER *(Press ENTER to end selection.)*

26. Select the Work layout tab, and click in the left viewport to view the foundation as shown in Figure 3T.39.
27. Save and close your drawing.

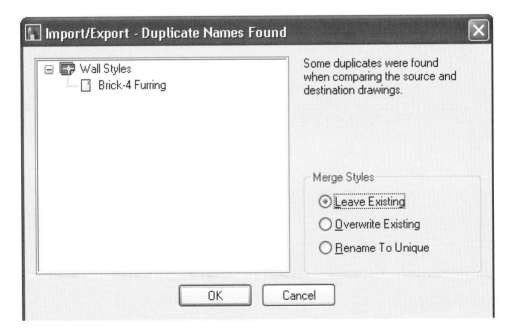

Figure 3T.39 *Foundation walls completed*

TUTORIAL 3.6 PLOTTING FLOOR PLANS

1. Open Autodesk Architectural Desktop 2005, and select **File>Project Browser** from the menu bar. Use the Project Selector drop-down list to navigate to your *ADT Student\ADT Tutor\Ch 3* directory. Double-click on **Ex 3-6** to set this project current. Select **Close** to dismiss the **Project Browser**. (If your student folder does not include the project, refer to "Organizing Tutorial Directories" in the Preface.)

2. Select the **Views** tab of the **Project Navigator**, select the **Views** icon, right-click, and choose **New View Dwg>General**. Type **Floor Plan-Basement** in the name field of the **Add General View** dialog box. Click **Next** to open the **Context** page, and check Level **1** for Division 1. Click **Next** to verify that the Basement construct is checked in the **Content** page. Click **Finish** to close the dialog box.

3. Verify that the **Views** tab of the **Project Navigator** is current, select **Views** icon, right-click, and choose **New View Dwg>General**. Type **Floor Plan - 1** in the name field of the **General** page in the **Add General View** dialog box. Click **Next** to open the **Context** page, check Level **2** for Division 1. Click **Next** to verify the Floor 1 construct on the **Content** page. Select **Finish** to close the dialog box.

4. Select the **Sheets** tab of the **Project Navigator**. Verify that the **Sheet Set View** is current, select the **Plans** subset, right-click, and choose **Properties**. Check the **Prompt for template** check box as shown in Figure 3T.40.

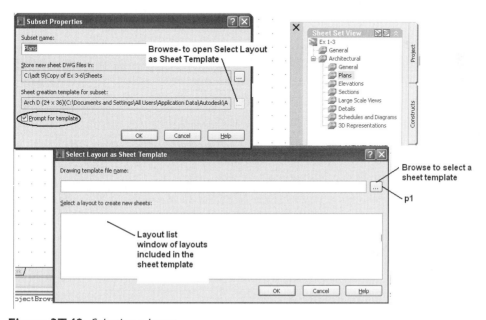

Figure 3T.40 *Selecting a layout*

Select **OK** to dismiss the **Subset Properties** dialog box.

5. Select the **Sheets** tab of the **Project Navigator**. Select the **Plans** subset, right-click, and choose **New> Sheet** to open the **Select Layout as Sheet Template** dialog box. Select the **Browse** button at **p1** shown in Figure 3T.40 to open the **Select Drawing** dialog box. Select the Aec Sheet (Imperial Stb).dwt, and then select **Open** to open the drawing. Select Arch D (24 × 36) layout from the layout list. Select **OK** to dismiss the **Select Layout as Sheet Template** dialog box.

6. Type **A-2** in the **Number** edit field and **Floor Plan Basement Level** in the **Sheet Title** edit field of the **New Sheet** dialog box. Verify that the Sheet template is Arch D (24 × 36). Select **OK** to close the dialog box.

7. Select the **Sheets** tab of the **Project Navigator**. Select the **Plans** subset, right-click, and choose **Properties**. Clear the **Prompt for template** check box. Select the **Browse** button to open the **Select Layout as Sheet Template** dialog box. Select Arch D (24 × 36). Select **OK** to close all dialog boxes.

8. Select the **Plans** subset, right-click, and choose **New >Sheet** to open the **New Sheet** dialog box. Select the **Type A-3** in the **Number** edit field and **Floor Plan Level 2** in the **Sheet Title** edit field. Verify that the Sheet template is Arch D (24 × 36). Select **OK** to close the dialog box.

9. Double click on A-2 Floor Plan Basement Level sheet in the Plans subset to open the sheet.

10. Select the **Views** tab, select the Floor Plan - Basement view, and drag the view into the A-2 sheet. Right-click during insertion and select the **1/4"= 1'-0"** scale.

11. Select the **Sheets** tab and double-click on A-3 sheet in the Plans subset to open the sheet.

12. Select the **Views** tab, select the Floor Plan - 1 view, and drag the view onto the A-3 sheet. Right-click during insertion and select the **1/4"= 1'-0"** scale.

13. Verify sheet A-3 Floor Plan Level 2 is the current drawing, right-click over the A-3 Floor Plan Level 2 layout tab, and select **Page Setup Manager** to open the **Page Setup Manager** dialog box.

14. Choose the **New** button of the **Page Setup Manager** to open the **New Page Setup** dialog box. Type **D Plot** in the **New page setup** name edit field. Select the **<Default output device>** and select **OK** to define the settings for the page in the **Page Setup** dialog box. Specify the plotter available to your computer that will plot D size paper and edit the paper size to **Oversize: ANSI D (landscape)**, **What to plot** to **Layout**, and **Scale 1:1**. Select **OK** to dismiss the dialog box. Select D **Plot**, the new page setup, from the page setups list, and choose the **Set Current** button. Select **Close** to

close the **Page Setup Manager**. Save and close the drawing.

Note: If you do not have a D-size output device, choose the Arch D (24 × 36 Expand-Dwf 6) page setup and select the **Set Current** button in the **Page Setup Manager**. Select the **Close** button to close the **Page Setup Manager**. The Arch D (24 × 36 Expand-Dwf 6) page setup will allow you to create a Design Web Format file.

15. Verify that A-2 Floor Plan Basement Level is the current drawing. Right-click over the A-2 Floor Plan Basement Level and select **Page Setup Manager** to open the **Page Setup Manager**.

16. Choose the **New** button of the **Page Setup Manager** to open the **New Page Setup** dialog box. Type **D Plot** in the **New page setup** name field. Select the **Default output device** and select **OK** to open the **Page Setup** dialog box. Specify the plotter available to your computer that will plot D size paper and edit the paper size to **Oversize: ANSI D (landscape)**, and **What to plot** to **Layout**. Select **OK** to dismiss the **Page Setup** dialog box. Select **D Plot**, the new page setup from the page setups list, and choose the **Set Current** button. Select **Close** to close the dialog box. Save and close the drawing.

17. To plot all sheets of the subset, select the **Sheets** tab. Select the Plans subset, right-click, and choose **Publish>Publish to Plotter**.

PROJECT

Exercise 3.7 Creating a Wall Style

1. Create a new project that includes levels for a crawl space foundation and first floor. Name the project **Ex 3-7**.

2. Create a wall style named Crawl10 to represent the wall as shown in the section of Figure 3T.41. Include a 24" × 8" footing in the wall style.

3. The wall should consist of three components, named brick, CMU, and footing. Import and assign materials to the brick, CMU, and concrete. The completed wall style should appear as shown at right in Figure 3T.41.

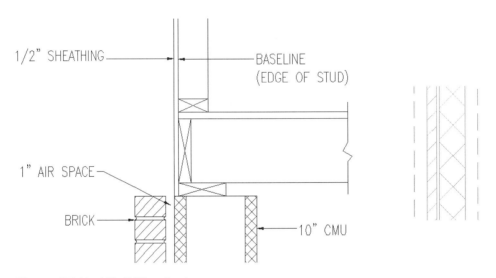

Figure 3T.41 *10" CMU wall style*

CHAPTER 4

Placing Doors and Windows

INTRODUCTION

Doors, windows, and openings are placed in the drawing as objects with three-dimensional properties. When doors, windows, and openings are placed in a wall, the wall is automatically trimmed and edited. The swing, hinge location, and position within and along the wall can be edited in Architectural Desktop. Door and window styles can be created with material assignments. Door and window styles can be imported from resource files, and styles can be exported to resource files to improve design.

OBJECTIVES

After completing this chapter, you will be able to

- Place doors with precision, specifying swing, insertion point, and properties using **DoorAdd**
- Edit swing, opening percent, and hinge location of doors and windows
- Change size, location, and head height, threshold height using the Properties palette
- Create door styles with standard sizes, types, shapes, and display properties
- Use dynamic dimensions to edit the location and size of doors and windows
- Shift doors and windows within and along walls using the **RepositionWithin** and **RepositionAlong** commands
- Place windows in walls with precision using **WindowAdd** command
- Edit the size and style of a window using the Properties palette
- Edit the display representation of windows to include muntins and shutters
- Edit the location, hinge point, and swing of doors and windows using grips

- Create and modify door and window styles using the **Style Manager**
- Place and modify openings in walls using the **OpeningAdd** command
- Import and export door and window styles in the **Style Manager**

The Design palette includes the Door, Window, and Opening tools for inserting doors, windows, and openings. Each tool allows you to specify the properties of the door, window, or opening in the Properties palette. Styles of doors and windows are can be selected from the Doors and Windows palettes as shown in Figure 4.1. Additional styles can be created using the **Style Manager** or i-dropped from the Design Tool Catalog of the **Content Browser**.

PLACING DOORS IN WALLS

Doors are placed in a drawing by selecting the **Door** tool from the Design tool palette of the Design palette set. Additional door styles can be selected from the Doors tool palette. Doors placed from the Design palette are the Standard style. Access the **DoorAdd** command as shown in Table 4.1.

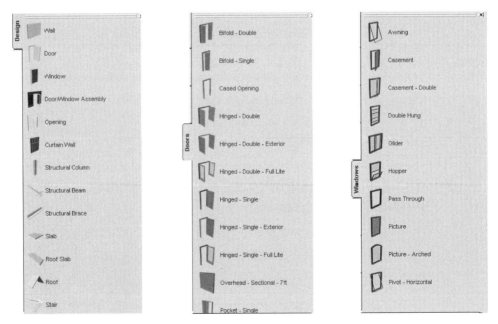

Figure 4.1 *Design, Doors, and Windows tool palettes*

Placing Doors and Windows 257

Command prompt	DOORADD
Palette	Select Door from the Design palette as shown in Figure 4.1
Palette	Select a door style from the Doors tool palette shown in Figure 4.1

Table 4.1 *DoorAdd command access*

To add a door, click **Door** on the Design palette, edit the Properties palette, and then select a wall or space boundary to insert the door. When you select this command, your pointer changes to a pick box to enable you to select the wall or space boundary. After you select the wall, dynamic dimensions are displayed indicating the location of the door as shown in Figure 4.2. Dynamic dimensions can be edited to specify the location of the door. In addition to dynamic dimensions, the door location can be constrained by the **Position along wall** property of the Properties palette. The **Position along wall** feature can be set to **Offset/Center**, and the door will be centered about the midpoint of the wall segment or a specified distance from a wall intersection. The door shown in Figure 4.2 is unconstrained.

PROPERTIES OF A DOOR

The Properties of a door specified in the Properties palette are described below.

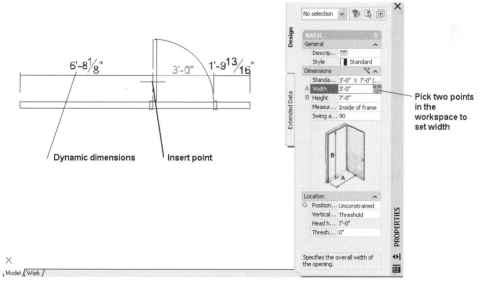

Figure 4.2 *Dynamic dimensions and Properties palette displayed when adding door*

Design–Basic

General

Description – The **Description** button opens a **Description** dialog box to type a description.

Style – The **Style** box displays the current door style, and the drop-down list includes all styles that have been used in the drawing.

Dimensions

Standard sizes – The **Standard sizes** edit field lists the sizes defined in the style; select a predefined size from the drop-down list. Selecting a size from the drop-down list presets the width and height dimensions.

A-Width – The **Width** edit field allows you to type the width of a door. The edit field displays the width of the last door inserted. The door width can also be specified if you click the "Pick two points to set a distance" box to the right of the edit field as shown in Figure 4.2. You can click in this box, and then select two points in the drawing to specify the width.

B- Height – The **Height** edit field allows you to specify the height of the door. The height can be set by selecting the "Pick two points to set a distance" button of the edit field.

Measure to – This field includes the **Inside of frame** and **Outside of frame** selections. The A-Width and B Base Height dimensions can be specified relative to the inside of frame or outside of frame. Doors are usually specified according to the size of door panel; therefore, the **Inside of frame** option would be used. The size of the frame is specified in the definition of the door style.

Swing angle – The **Swing angle** is the angle in degrees that the door is open relative to the frame.

Location

Position along wall – The **Position along wall** includes the **Unconstrained** and **Offset/Center** options. The **Unconstrained** option allows you to position the door at any point along the wall or space boundary. When the **Offset/Center** option is selected, the door will be placed in the center of the wall segment or a specified distance from a wall intersection as shown in Figure 4.3. The **Automatic Offset** distance option is added to the Properties palette when the **Offset/Center** option is selected.

Automatic Offset – The **Automatic Offset** distance edit field allows you to specify the distance between the edge of the door placed and the end of the wall segment or wall intersection. The distance is measured relative to the inside of frame or the outside of frame as specified in the **Measure to** option of the Properties palette.

Vertical Alignment – The **Vertical Alignment** options are **Head** and **Threshold**. The working point for the insertion of the door can be specified from the baseline of the wall relative to the head or threshold of the door as shown in Figure 4.4. The distance is specified to either **Inside of frame** or **Outside of frame** according to the **Measure to** option of the Properties palette.

Head Height – The **Head Height** is the distance from the baseline of the wall to the inside of frame or outside of frame. **Inside of frame** and **Outside of frame** are specified in the **Measure to** edit field of the Properties palette.

Threshold – The **Threshold** height is the distance from the baseline of the wall to the bottom of the door panel.

Rotation – The **Rotation** field displays the angle of the door relative to the 3:00 o'clock position. The door rotation cannot be changed in the Properties palette.

The properties of the door can also be specified by responding to the command line prompts as shown below.

Command: _DoorAdd

Select wall, space boundary or RETURN: *(Select a wall.)*

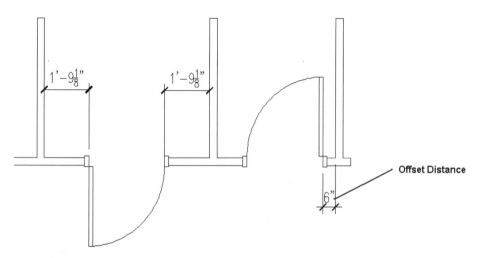

Figure 4.3 *Offset/Center constraint used when placing a door*

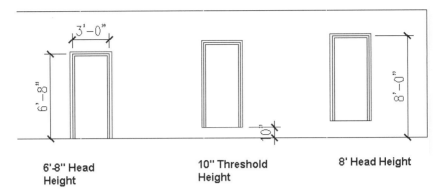

Figure 4.4 Vertical alignment options for doors

> Insert point or [STyle/WIdth/HEight/HEAd height/THreshold height/Auto/Match/CYcle measure to/REference point on]: *(Select an insert point for the door.)*
>
> Insert point or [STyle/WIdth/HEight/HEAd height/THreshold height/Auto/Match/CYcle measure to/REference point on]: ENTER

The first prompt of the command is to select a wall or space boundary or to press ENTER (RETURN). Selecting a wall or space boundary will lock the door anchor to a wall or space boundary without specifying the location along the wall. However, you can place a freestanding door by pressing ENTER rather than selecting a wall or space boundary. If you select a wall or space boundary, the door object will rubber-band along the wall or space boundary when you move your mouse as shown in Figure 4.2. You anchor the door to a location along the wall or space boundary by selecting a point with the mouse when prompted for the insert point in the command line.

Prior to anchoring the door in the wall, you can edit the fields of the Properties palette. When you move the pointer back to the workspace, the **DoorAdd** command is still active, and the door rubber-bands about any wall that you move your pointer over. Each time the Properties palette is edited, the phantom image of the revised door is displayed about the pointer. Toggling between the drawing area and the Properties palette allows you to place doors of various properties in one or more walls while in the **DoorAdd** command.

In addition to setting the sizes and styles in the Properties palette, you can define them by using the options of the command line. Select command line options of the **DoorAdd** command by entering the capitalized letters of the command options as shown below. These options are also available on the shortcut menu if you right-click while the command is active. In the following command sequence, **Width** was chosen from the shortcut menu to change the width of the door prior to insertion.

DOORADD

Select wall, space boundary or RETURN: *(Select a wall.)*

Insert point or [STyle/WIdth/HEight/HEAd height/THreshold height/Auto/OFfset/Match/CYcle measure to/REference point on]: *(Right-click and choose* **Width** *from the shortcut menu.)* WI

Width <2'-0">: **5'** ENTER *(Type new width)*

Insert point or [STyle/WIdth/HEight/HEAd height/THreshold height/Auto/OFfset/Match/CYcle measure to/REference point on]: *(Click a location to place the door.)*

Insert point or [STyle/WIdth/HEight/HEAd height/THreshold height/Auto/OFfset/Match/Undo/CYcle measure to/REference point on]: ENTER *(Press* ENTER *to end the command.)*

Selecting from the shortcut menu during insertion allows you to specify the options of the command listed in the command window. Changing the style of a door using the command window options requires recalling and typing the name of a style. The options of the shortcut menu and command line are described below.

> **STyle** – The **Style** option will prompt you to type the name of the door style in the command line.
>
> **WIdth** – The **Width** option will prompt you to specify a different width for the door panel in the command line.
>
> **HEight** – The **Height** option will prompt you to type a new height of the door panel in the command line.
>
> **HEAd height** – The **Head** height option allows you to type the vertical alignment distance from the head of the door to the baseline.
>
> **THreshold height** – The **Threshold height** option prompts you for the vertical alignment distance from the baseline of the wall to the bottom of the door.
>
> **Auto** – The **Auto** option toggles between Unconstrained and Offset/Center.
>
> **OFfset** – The **Offset** option is displayed if the **Position along wall** option is set to Offset/Center. This option allows you to type the distance for the offset of the door to intersecting walls.
>
> **Match** – The **Match** option prompts you to select an existing door to match the properties of the door for the present insertion.
>
> **Undo** – The **Undo** option, displayed after an insert point is specified, will undo the last insertion of the door.
>
> **CYcle measure to** – The **Cycle measure to** option toggles the handle for the insertion of the door between edge of door and center of door as shown in Figure 4.8.

REference point on – The **Reference point on** option prompts you to select a point to measure from to specify the insertion of the door as shown in Figure 4.9.

SETTING THE SWING

Prior to specifying the insertion point, you can establish the hinge point and swing of the door. A door placed with an Offset/Center constraint allows you to set the hinge location and swing by placing the pointer above, below, left, or right relative to the constrained insert point. The movement of the pointer will cause dynamic display of the hinge point and swing of the door; when the phantom display is correct, left-click, and you have set the hinge point, swing, and insert point. In Figure 4.5, the door is placed with an Offset/Center constraint, and the mouse pointer was moved to the locations shown to set the door swing.

When you place a door without the Offset/Center constraint, you cannot set the hinge point prior to insertion. The swing can be specified by moving the pointer to the other side of the wall prior to specifying the insert point. When the insert point is selected, this finalizes the position for the door swing. The default location of the hinge point of an unconstrained door is toward the start point of the wall. In Figure 4.6 the walls were drawn from right to left. Door placement when unconstrained allows you only to move the pointer location above and below the wall to set the swing as shown in Figure 4.6. The hinge point and swing of doors can be changed by grips, discussed later in this chapter.

LOCATING THE INSERTION POINT OF THE DOOR

Although the insert point of the door can be precisely placed using the Offset/Center constraint, dynamic dimensions allow you to position the door by editing the dynam-

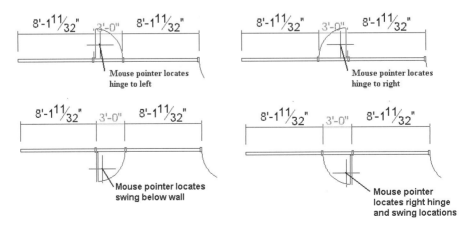

Figure 4.5 *Setting the hinge point and swing of a door placed with Offset/Center constraint*

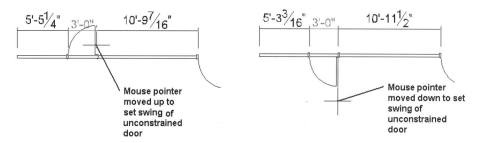

Figure 4.6 *Setting swing of an unconstrained door by moving the mouse pointer above and below the wall*

ic dimension. Options within the **DoorAdd** command allow you to toggle the door insert point to center or edge of the door by cycling through the insert point. In addition, the **Reference point on** option allows you to select points in the drawing to position the door.

Dynamic Dimensions

A door can be placed in a wall with precision using dynamic dimensions. Dynamic dimensions will be displayed when you select a wall or space boundary. The dynamic dimensions shown in Figure 4.7 are displayed from the endpoints of the wall to the door. If you press TAB, a dimension will highlight, allowing you to edit it. You can press TAB to toggle the highlight for each dimension and finally to turn off the highlight. In Figure 4.7, TAB has been pressed to highlight the right dimension. When a dimension is highlighted, you can type a number that will display in the workspace to set the dimension as shown in Figure 4.7.

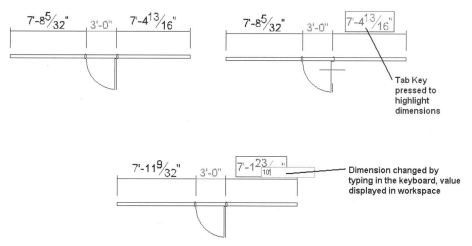

Figure 4.7 *Specifying door placement using dynamic dimensions*

Using the Cycle Measure To Option

The insert point shown on the left in Figure 4.8 is locating the dynamic dimensions to the outside of frame. The dynamic dimension can be changed to the center by right-clicking and choosing **Cycle measure to**. **Cycle measure to** is an option for doors, windows, and openings. This option can be selected from the command line options or the shortcut menu during placement. The **Cycle measure to** option allows you to toggle the dynamic dimension from the edge of the door to the center as shown in Figure 4.8.

Placing a Door Using Reference Point On

The **Reference point on** option of the **DoorAdd** command allows you to measure in the workspace from features other than the selected wall to establish the location of the door. The **Reference point on** option is used in Figure 4.9 to locate the door 5' from the column. The **Reference point on** option requires you to select a point in the drawing to reference the dynamic dimension. After you select the reference point, move the pointer to set the direction for the measurement and type the correct distance in workspace. The door shown in Figure 4.9 was placed in the wall using **Reference point on** as shown in the following command line sequence.

> Command: DOORADD
>
> Select wall, space boundary or RETURN: *(Select the wall at p1 as shown in upper left corner of Figure 4.9.)*
>
> Insert point or [STyle/WIdth/HEight/HEAd height/THreshold height/Auto/Match/CYcle measure to/REference point on]: **re** *(Type RE or select* **Reference point on** *from the shortcut menu to select option.)*
>
> Pick the start point of the dimension: **_endp of** *(SHIFT + right-click, choose* **End** *object snap, and select point p2 as shown in the upper right corner of Figure 4.9.)*
>
> *(Move the pointer to the right.)*

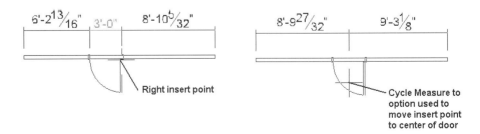

Figure 4.8 *Cycle measure to option used to change dynamic dimension*

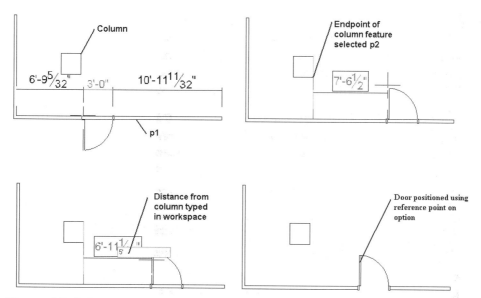

Figure 4.9 Reference point on option used to locate the door

Insert point or [STyle/WIdth/HEight/HEAd height/THreshold height/Auto/Match/CYcle measure to/REference point off]: **5'** ENTER (Type **5'** in the workspace to position the door as shown in lower left corner of Figure 4.9.)

Insert point or [STyle/WIdth/HEight/HEAd height/THreshold height/Auto/Match/Undo/CYcle measure to/REference point off]: ENTER (Press ENTER to end the command.)

Using Grips to Modify a Door

The grips of a door can be used to move the door along the wall, change the hinge location, change the swing, and change the width of the door. Selecting the center grip allows you to stretch the door along the wall to a new location. Editing with this grip is shown in Figure 4.10: the grip is selected and the pointer moved to the right. When you stretch a grip, you can type the distance to move the door in command line. The door can also be shifted along the wall with the **RepositionAlong** command discussed later in this chapter.

In addition to shifting the door along the wall, the center grip can also be toggled using CTRL to shift the door within the wall, or vertically. Selecting the grip when the door is viewed in pictorial will allow you to edit the door vertically with grips as shown in Figure 4.11. After you have selected the center grip, each time you press CTRL the movement of the mouse is restricted to shifting the door vertically, within the wall,

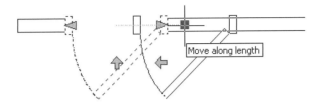

Figure 4.10 Using grips to move the door along the wall

or along the wall. The door shown in Figure 4.11 at left is stretched down using the **Move vertically** option of grips. Shifting the door vertically can be useful when a door from the one level must extend down to the lower level. Garage doors of the first floor often extend down into the foundation wall.

Doors can also be stretched within the wall to position the jamb to the edge of the wall. Pressing CTRL while editing the center grip will allow you to shift the door within the wall. The door in Figure 4.11 is shown stretched out of the wall. This option of grip editing, shown in Figure 4.12, allows you to stretch the doorjamb to the edge of the

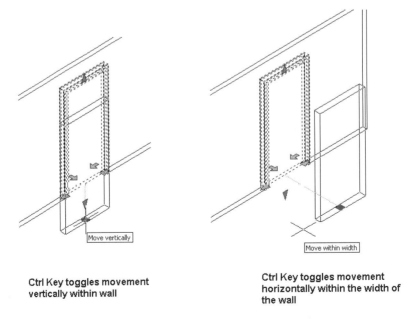

Ctrl Key toggles movement vertically within wall

Ctrl Key toggles movement horizontally within the width of the wall

Figure 4.11 Using grips to move the door within and vertically in the wall

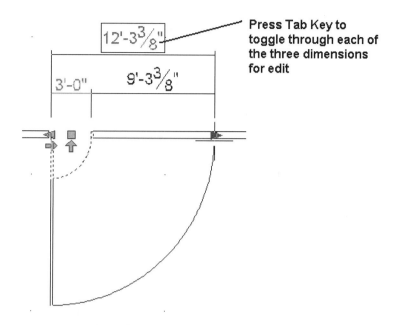

Figure 4.12 *Editing the door frame grip*

wall by tabbing through dynamic dimensions. Doors that have been repositioned within the wall remain anchored to the wall. The shifting of the door within the wall can also be done using the **RepositionWithin** command discussed later in this chapter.

The grips located at the jamb are arrows indicating the direction of edit. When you move the pointer over the frame grip, a dynamic dimension will display the current width of the door. Select the frame grip, and dynamic dimensions are activated. When the dynamic dimension is displayed, movement of the pointer will display two dimensions, the new width of the door and the increase in door width. If you press TAB, you can toggle the highlight to the dimension to increase the door width. The new door size can then be typed in the workspace when the dynamic dimension is highlighted as shown in Figure 4.12. Pressing TAB will toggle the edit of the frame grip, allowing you to type the distance to stretch the grip to the right or left. The left frame location remains anchored when you edit the right frame grip as shown in Figure 4.12.

The **Flip Swing** trigger grip shown in Figure 4.13, if selected, will flip the swing of the door to the opposite side of the door. The hinge can be flipped by selecting the **Flip Hinge** trigger grip shown in Figure 4.13. The hinge and swing of the doors shown in Figure 4.13 were changed using the Flip Hinge and Flip Swing trigger grips. After editing the door using grips, press ESC to clear the grip display.

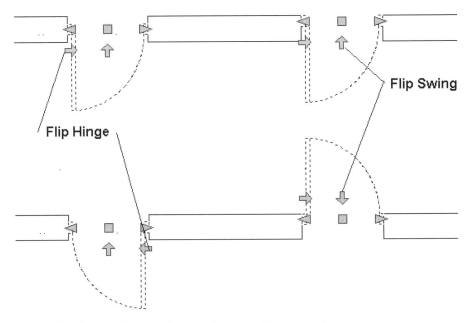

Figure 4.13 *Changing hinge and swing of a door with trigger grips*

The hinge location can also be changed with the **OpeningFlipHinge** command, which flips the location of the hinge.

Access the **OpeningFlipHinge** command as shown in Table 4.2.

Command prompt	OPENINGFLIPHINGE

Table 4.2 *OpeningFlipHinge command access*

When you type this command in the command line, you are prompted to select doors, windows, or openings. The hinge location will flip for all selected doors, windows, or openings. This command allows you to flip the swing of all selected doors without specifying the location of the hinge. The **OpeningFlipHinge** command was applied to the left door to flip the hinge to the right, as shown in the following command sequence and in Figure 4.14.

> Command: OpeningFlipHinge
>
> Select doors, windows, or openings: *(Select the door near p1 as shown in Figure 4.14.)* 1 found
>
> Select doors, windows, or openings: ENTER *(Press ENTER to end the selection.)*

(Hinge of door flipped as shown at right in Figure 4.14.)

The swing location can also be changed by the **OpeningFlipSwing** command, which flips the swing of one or more doors. Access the **OpeningFlipSwing** command as shown in Table 4.3.

Command prompt	OPENINGFLIPSWING

Table 4.3 *OpeningFlipSwing command access*

When you type this command in the command line, you are prompted to select doors or windows. The swing location will flip for all selected doors and windows. The **OpeningFlipSwing** command was applied to the door on the left to flip the swing as shown in the following command sequence and in Figure 4.15.

> Command: OpeningFlipSwing
>
> Select doors or windows: *(Select the door at p1 as shown on the left in Figure 4.15.)* 1 found
>
> Select doors or windows: ENTER *(Select* ENTER *to end door selection.)*
>
> *(Door swing flipped down as shown on right in Figure 4.15.)*

DEFINING DOOR PROPERTIES

The properties of doors can be changed by editing the Properties palette. When you select an existing door, the Properties palette changes to include additional information that is not included when you add a door. Additional edit fields are added to allow you to change features of the inserted door. The Properties palette shown in Figure 4.16 includes the following additional fields. Therefore to change one or more doors, select the doors, and then edit the Properties palette.

Design Tab

General

> **Layer** – The **Layer** edit field displays the name of the layer of the insertion point of the door. The **Layer** drop-down list displays layers of the drawing.

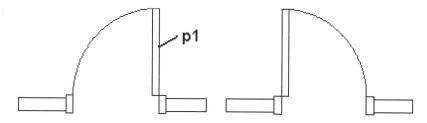

Figure 4.14 *Editing the hinge location with the OpeningFlipHinge command*

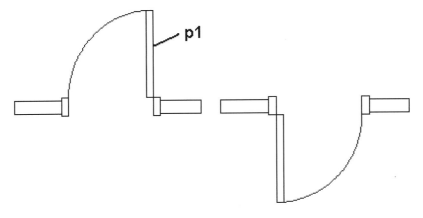

Figure 4.15 Editing door swing with OpeningFlipSwing command

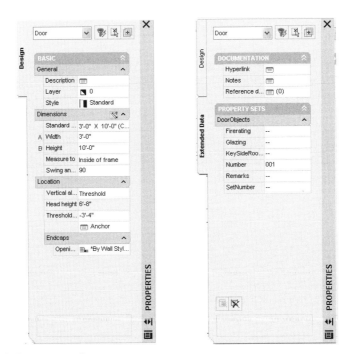

Figure 4.16 Properties palette

Location

    Anchor – The **Anchor** button opens the **Anchor** dialog box shown in Figure 4.17. The **Anchor** dialog box lists the position of the door relative to the justification line. The location of the door can be specified along the X direction, within the Y direction, and vertical to the Z direction. Moving the door by editing the anchor is a precise method of positioning the door relative to the wall anchor line.

Endcaps

Opening Endcaps – The opening endcaps of the wall when the door is placed can be edited by selecting the opening endcaps from the drop-down list.

Extended Data Tab

Documentation

Hyperlink – The **Hyperlink** button opens the Insert Hyperlink dialog box, which allows you to link the door to a file.

Notes – The **Notes** button opens a Notes dialog box for typing notes regarding the door.

Reference Documents – The **Reference Documents** button allows you to list the files and directory that include related information.

Property Sets

The **Property Sets** list will vary according to the object selected and the properties that have been defined for the door. Property data is inserted in the drawing when door tags are placed.

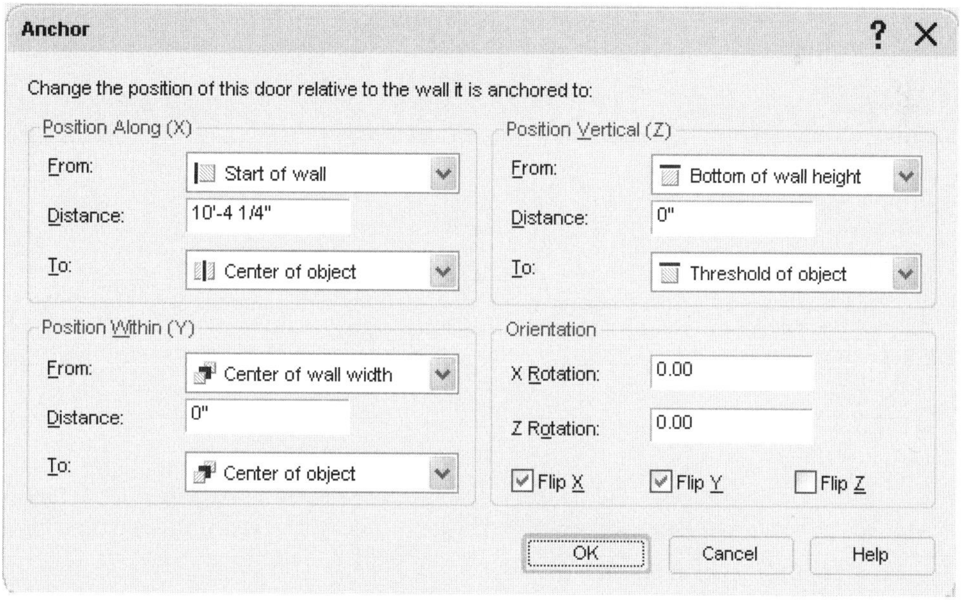

Figure 4.17 *Door Anchor dialog box*

Editing Properties of Multiple Doors

You can edit more than one door using the Properties palette. If you select multiple doors, the Properties palette will display the properties common to all doors. The properties that differ among the doors selected will be described as **Varies**. Therefore you edit **Varies** by typing the desired value in the field for that property.

CREATING AND USING DOOR STYLES

There are 13 door styles included in the Doors palette, which provide customized frame and leaf properties. When a door is selected from the Doors palette, the **DoorAdd** command is selected with style preset according to the tool selected. Additional styles can be imported into the drawing from other drawings or created in the **Style Manager**. Additional door styles are located in the *C:\Documents&Settings\All Users\Application Data\Autodesk\ADT 2005\enu\Styles\Imperial* directory.

Access the **Door Styles** command as shown in Table 4.4.

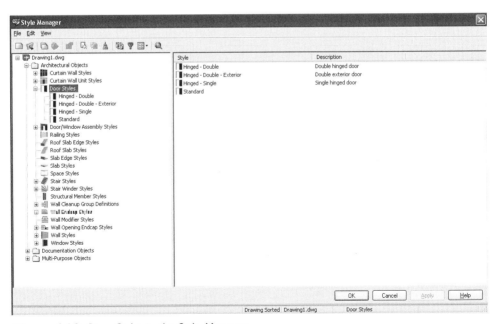

Figure 4.18 *Door Styles in the Style Manager*

Menu bar	Format>Style Manager
Command prompt	DOORSTYLE
Command prompt	AECSTYLEMANAGER
Shortcut menu	Select Door on the Design palette, right-click, and choose Edit Door Styles
Shortcut menu	Select a door style from the Doors palette, right-click, and choose Door Styles

Table 4.4 *Door Styles of the Style Manager command access*

When the **Style Manager** command is executed, the **Style Manager** opens as shown in Figure 4.18. The Door Styles are listed in the Architectural Objects folder of the current drawing. The commands of the **Style Manager** toolbar are described in "Creating and Editing Wall Styles" in Chapter 3.

CREATING A NEW DOOR STYLE

Create a new door style by selecting the Door Styles folder of the left pane of the **Style Manager**, and then selecting **New Style** from the **Style Manager** toolbar. When you select the **New Style** command, the default name New Style will be highlighted in the right pane; you can then overtype a new name. The properties of the door style are defined by selecting the name of the new style in the right pane, and then selecting **Edit Style** from the **Style Manager** toolbar. The **Edit Style** command opens the **Door Style Properties** dialog box, which is used to define the properties of new and existing door styles.

EDITING A DOOR STYLE

Define the properties of a door style by editing the **Door Style Properties** dialog box as shown in Figure 4.19. This dialog box consists of the following tabs: **General**, **Dimensions**, **Design Rules**, **Standard Sizes**, **Materials**, **Classifications**, and **Display Properties**. The purpose and function of each tab follow.

General Tab

> **Name** – The **Name** edit field allows you to edit the name of the door style.
>
> **Description** – The **Description** edit field allows you to add a description for the door.
>
> **Notes** – The **Notes** button will open the **Notes** dialog box, which includes the **Text Docs** and **Reference Docs** tabs. The purpose of these tabs is identical to those of the wall style.

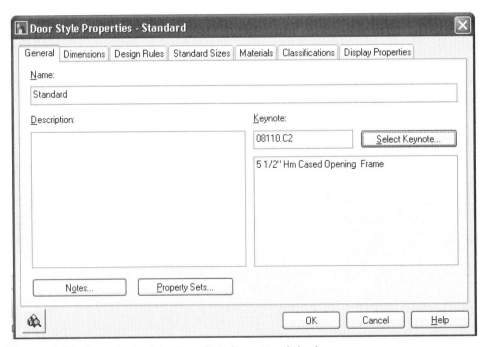

Figure 4.19 *General tab of the Door Style Properties dialog box*

Keynote–Keynotes may be defined for the door style. Choosing the **Keynote** button will open the **Select Keynote** dialog box, which allows you to specify keynotes from the 16 CSI divisions. Keynote commands of the Annotation palette can be used to add the keynote annotation when you select the door.

Dimensions Tab

The **Dimensions** tab of the **Door Style Properties** dialog box allows you to set dimensions of the frame, stop, and door thickness, as shown in Figure 4.20.

Frame

A-Width – The **A-Width** edit field allows you specify the width of the frame. The **Width** dimension for a frame in residential construction should be set equal to the total of the doorjamb and brick molding or exterior trim thickness. Therefore, the **Width** dimension of door styles for residential construction consists of the total unit thickness exclusive of the interior trim. The **Width** dimension of the frame in commercial construction is set according to dimensions of the steel frame.

B-Depth – The **B-Depth** edit field allows you to specify the depth of the frame.

Auto-Adjust to Width of Wall –This check box determines whether the depth is fixed or varies according to the width of the wall. When this check box is selected, the frame depth adjusts to the wall width, as shown at right in Figure 4.21. If **Auto-Adjust to Width of Wall** is not checked, the frame depth is fixed to that specified in the **B-Depth** edit field. The door on the left in Figure 4.21 has a fixed door frame depth.

Stop

C-Width – The **C-Width** of the stop is the dimension the stop projects from the jamb.

D-Depth – The **D-Depth** of the stop is the dimension of stop parallel to the face of the jamb. In residential construction, attaching a wood stop to the frame usually creates the stop. The stop size and style therefore vary. The stop in commercial construction is formed with the steel frame and varies according to the manufacturer of the steel frame.

E-Door Thickness – The **E-Door Thickness** dimension is the thickness of the door. Doors used in residential construction are typically 1 3/8" or 1 3/4". Doors used in commercial construction are usually 1 3/4" or 2" or more in thickness.

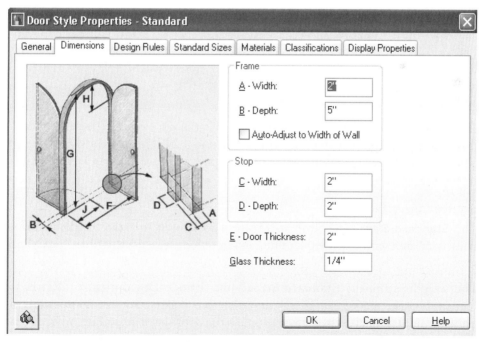

Figure 4.20 *Dimensions tab of the Door Style Properties dialog box*

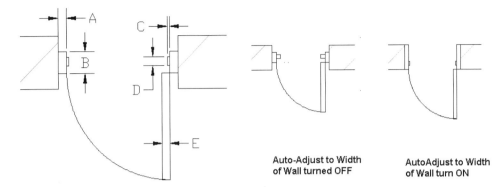

Figure 4.21 *Frame dimensions of a door style*

Design Rules Tab

The **Design Rules** tab defines the shape and type of door used for the style. The **Design Rules** tab shown in Figure 4.22 includes two sections: **Shape** and **Door Type**.

Shape

Predefined – The **Predefined** shapes include rectangular, half-round, quarter-round, arch, gothic, and peak pentagon. This list is extensive and satisfactory for most design applications.

Use Profile – The **Use Profile** option allows you to select an AecProfile from the drop-down list. The AecProfile discussed in Chapter 3 is a created from a closed polyline in the **Style Manager**.

Door Type

The **Door Type** list allows you to select from a list of door types. The types of doors that can be created are shown in Figures 4.23 and 4.24. The name assigned to each type is shown below the figure. Note that the garage door is named Overhead and is shown in Figure 4.24.

Standard Sizes Tab

The **Standard Sizes** tab shown in Figure 4.25 is used to create predefined door sizes associated with a door style. The Standard door style does not have predefined sizes.

The sizes listed in the **Standard Sizes** table include Description, F-Width, G-Height, H-Rise, and J-Leaf. These dimensions are illustrated on the **Dimensions** tab in Figure 4.20. The H-Rise is inactive unless the door shape is Gothic, Arch, or Peak Pentagon. The J-Leaf column is inactive unless the door type is set to Uneven.

Placing Doors and Windows 277

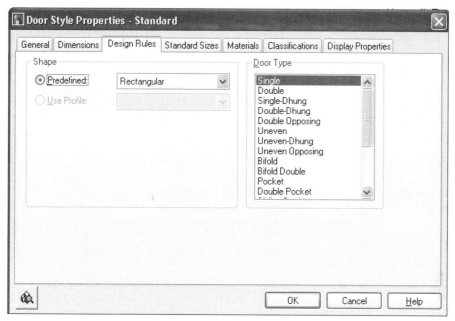

Figure 4.22 *Design Rules tab of the Door Style Properties dialog box*

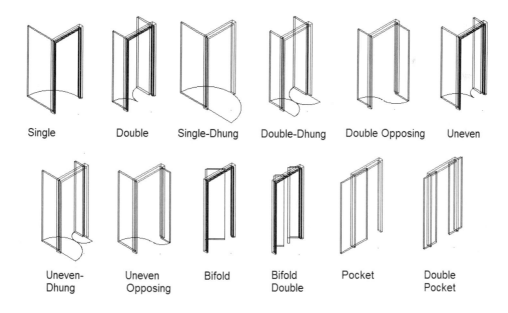

Figure 4.23 *Door types available in Architectural Desktop*

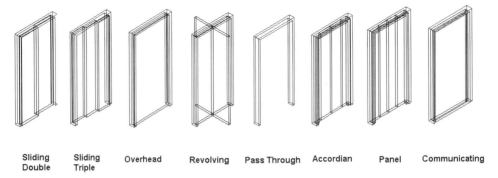

Figure 4.24 Door types including the Overhead garage door

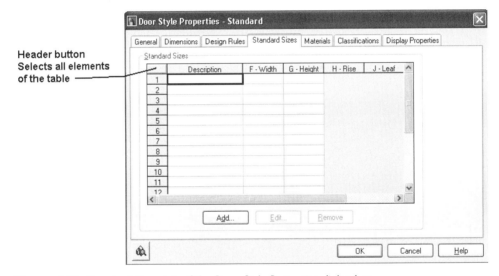

Figure 4.25 Standard Sizes tab of the Door Style Properties dialog box

Add – The **Add** button is used to create a standard size. Selecting the Add button opens the **Add Standard Size** dialog box as shown in Figure 4.26. Create sizes by editing the **Width, Height, Rise,** and **Leaf** edit fields of the **Add Standard Size** dialog box.

Edit – The **Edit** button opens the **Edit Standard Size** dialog box, allowing you to edit the sizes highlighted in the table of standard sizes.

Remove – Selecting the **Remove** button will remove the selected components of the table. Selecting the row number button to the left of the standard size list will select the entire row, allowing you to remove the row. If you select the header button as shown in Figure 4.25 instead, the entire table is selected and can be removed.

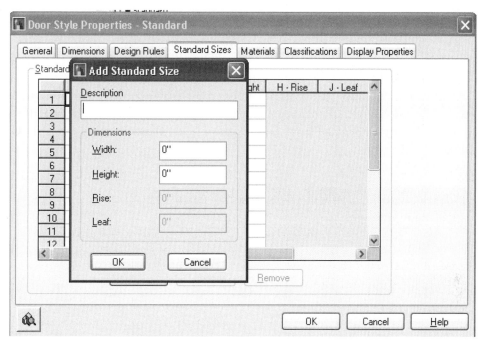

Figure 4.26 *Edit the Add Standard Size dialog box to create a standard size*

Materials Tab

The **Materials** tab allows you to define the material definition for the components of a door. The **Materials** tab shown in Figure 4.27 lists the material definitions for each of the door components. If material definitions have been defined in the drawing, you can click in the **Material Definition** column and select the definitions from the drop-down list. The two buttons in the right margin described below are **Edit Material** and **Add New Material**. The procedure to assign materials and create new material definitions for doors is similar to the procedure outlined in "Assigning Materials to Wall Components" in Chapter 3.

> **Add New Material**– The **Add New Material** button opens the **New Material** dialog box, allowing you to create a name for the new material definition.
>
> **Edit Material** – The **Edit Material** button opens the **Material Definitions Properties** dialog box; the **Display Properties** tab is shown in Figure 4.28. This tab lists the display representations and display property sources for the material definition. When you select the **Edit Display Properties** button of the **Material Definition Properties** dialog box, the **Display Properties – Materials Display** dialog box opens as shown in Figure 4.29, which allows you to specify layer, hatching, and rendering properties of the material display

for the door. The **Layer/Color/Linetype** tab of the **Display Properties – Material Display** dialog box is shown in Figure 4.29. The **Hatching** and **Other** tabs are shown in Figure 4.30.

Classifications Tab

If classification definitions have been defined in the drawing, a classification can be specified for a door style. Classifications can be defined to specify the construction status of an object style such as "new" and "existing." The **Classifications** tab is shown in Figure 4.31.

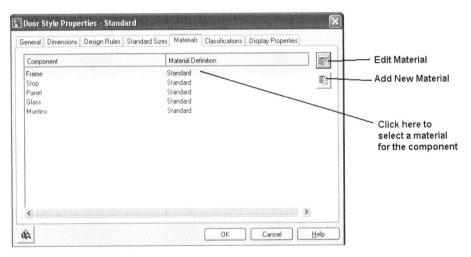

Figure 4.27 *Materials tab of the Door Style Properties dialog box*

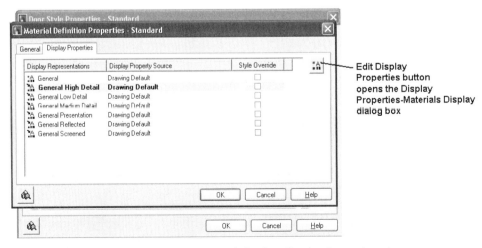

Figure 4.28 *Material Definitions Properties dialog box, Display Properties tab*

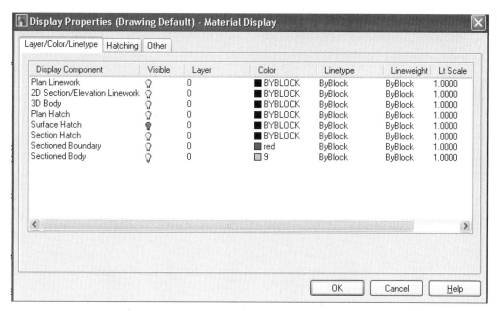

Figure 4.29 *Layer/Color/Linetype tab of Display Properties – Material Display dialog box*

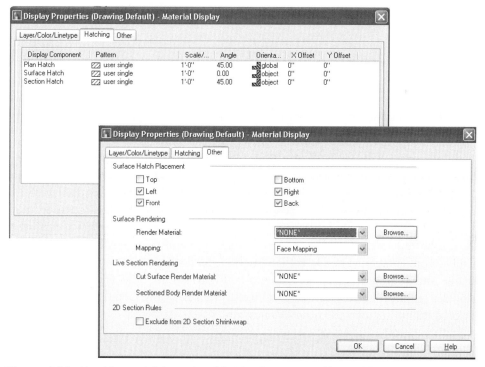

Figure 4.30 *Hatching and Other tabs of Display Properties – Material Display dialog box*

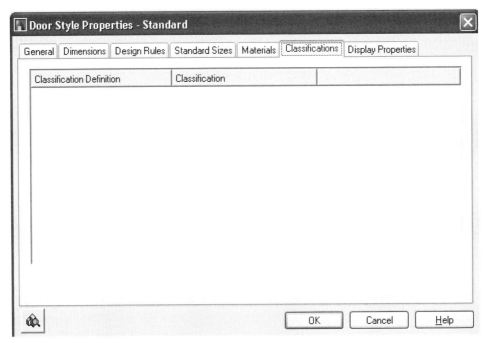

Figure 4.31 *Classifications tab of the Door Style Properties dialog box*

Display Properties Tab

The **Display Properties** tab allows you to control the display of a door. The display representations for the door style are listed in the display representations list. The display representations for a door style are shown at left in Figure 4.32.

The display representations shown bold are the current display representations used in the viewport. If you check the **Style Override** box, the **Display Properties – Door Display Representation** dialog box opens. The **Display Properties – Door Display Representation** dialog boxes allow you to edit how the door is displayed. The **Layer/Color/Linetype** tab of the **Display Properties – Door Display Representation** dialog box is shown in Figure 4.32. The content of the dialog box varies according to the display representation. Figure 4.32 is an example of the Door Plan High Detail display representation. The Door Plan High Detail display representation includes the **Other** and **Frame Display** tabs shown in Figures 4.33. The **Frame Display** tab provides "L" and "U" shaped frame options. The Plan Low Detail display representation does not include a **Frame Display** tab.

You can add thresholds to the display of a door by editing the Threshold display representation. The Threshold and Plan display representations include **Threshold A** and **Threshold B** components. Turning on the display of the threshold component will dis-

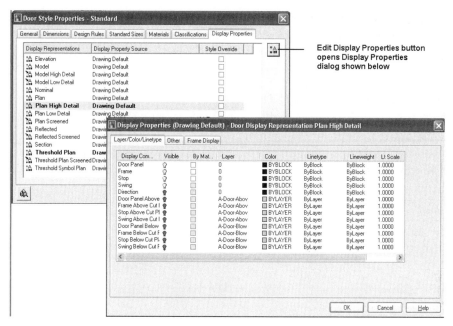

Figure 4.32 *Display representations and Display Properties dialog box of a door style*

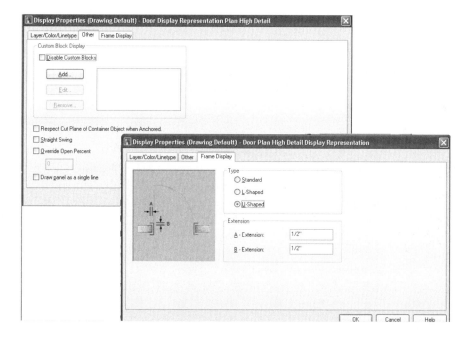

Figure 4.33 *Other and Frame Display tabs of the Display Properties – Door High Display Representation dialog box*

play a threshold on the door. The **Layer/Color Linetype** tab of the **Display Properties – Door Threshold Plan Display Representation** dialog box is shown in Figure 4.34. The **Other** tab, also shown in Figure 4.34, allows you to specify the extension of the threshold along and beyond the wall.

SHIFTING A DOOR WITHIN THE WALL

A door frame can be shifted within a wall using grips or the **RepositionWithin** command. The prompts of this command allow you to measure the distance from a point on the frame to a point on the wall. The distance between the two points can then be edited with dynamic dimension to adjust the offset. Access **RepositionWithin** as shown in Table 4.5.

Command prompt	REPOSITIONWITHIN
Shortcut menu	Select a door, right-click, and choose Reposition Within Wall

Table 4.5 *RepositionWithin command access*

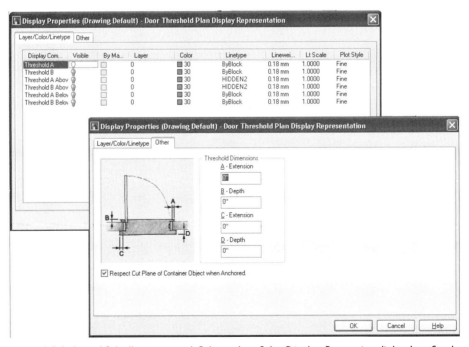

Figure 4.34 *Layer/Color/Linetype and Other tabs of the Display Properties dialog box for the Threshold display representation*

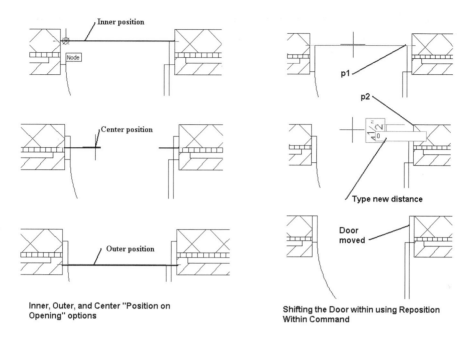

Figure 4.35 *Position on the opening options and editing frame location*

When the **Reposition Within Wall** command is selected from the shortcut menu of a selected door, you are prompted to specify a position on the opening. When you move the pointer over the frame, the center, inner, and outer edges of the frame are highlighted to help you select the position on the opening as shown at right in Figure 4.35.

The next prompt is to select a reference point. The reference point should be the point within the wall where you want to move the "position on the opening." After you select these two points, a dynamic dimension is displayed that allows you to edit the distance between the points. The following command line sequence moved the edge of the frame (**p1**) to the face of wall (**p2**) as shown in Figure 4.35.

>Command: RepositionWithin
>
>Select position on the opening: *(Select a point on the frame p1 shown at right in Figure 4.35.)*
>
>Select a reference point *(Select a point on a feature of the wall p2 shown at right in Figure 4.35.)*
>
>Enter the new distance between the selected points <0">: **0** *(Overtype a distance in the dynamic dimension.)*

SHIFTING THE DOOR ALONG A WALL

The **RepositionAlong** command will shift a door along a wall a specified distance. Access **RepositionAlong** as shown in Table 4.6.

Command prompt	REPOSITIONALONG
Shortcut menu	Select door, right-click, and choose Reposition Along Wall

Table 4.6 *RepositionAlong command access*

When you select a door, right-click, and choose **Reposition Along Wall** from the shortcut menu, you are prompted to specify a position on the opening, which will serve as a handle to position the door. The position on the opening can be the center or either edge of the frame as shown in Figure 4.36. The next command line prompt is to select a reference point, which can be any point on the wall. After the two points are selected, a dynamic dimension is displayed, allowing you to edit the distance between the two points.

The **RepositionAlong** command was used in the following command line sequence to move the door 4' to the end of the wall.

> Command: RepositionAlong
>
> Select position on the opening: *(Select center of frame at p1 in Figure 4.37.)*
>
> Select a reference point *(Select node at end of wall near p2 in Figure 4.37.)*
>
> Enter the new distance between the selected points <5'-0">: **4'**
> *(Type **4'** in the dynamic dimension.)*

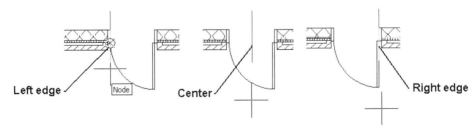

Figure 4.36 *Center and Edge options to position frame*

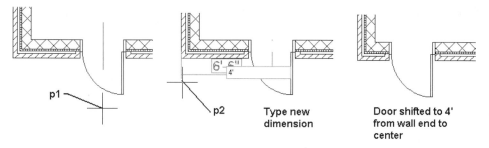

Figure 4.37 Door adjusted with RepositionAlong command

The **RepositionAlong** command can be used to edit doors inserted with the incorrect Automatic Offset/Center value. The offset value of the **RepositionAlong** command positions an existing door in a manner similar to placing a new door with the Automatic Offset/Center option of the **DoorAdd** command.

 Note: The **RepositionWithin** and **RepositionAlong** commands can also be used to shift windows and openings in a wall.

 STOP. Do Tutorial 4.1, "Inserting Doors," at the end of the chapter.

PLACING WINDOWS IN WALLS

The **WindowAdd** command is used to place windows in a drawing. Access the **WindowAdd** command as shown in Table 4.7.

Command prompt	WINDOWADD
Palette	Select Window from the Design palette as shown in Figure 4.1
Palette	Select a window style from the Windows tool palette

Table 4.7 WindowAdd command access

When you select the **WindowAdd** command, you are prompted to select a wall or space boundary. Windows are inserted in the drawing with precision using the dynamic dimensions, **Reference point on**, and **Cycle measure to** techniques as presented in the **DoorAdd** command. The content of the Properties palette displayed when you insert windows is described below.

Design–Basic

General

Description – Selecting the **Description** button opens the **Description** dialog box, allowing you to type a description.

Style – The **Style** box lists the current window style; the drop-down list includes all styles used in the drawing.

Dimensions

Standard sizes – The **Standard sizes** edit field displays the defined sizes for the window style.

A-Width – The **Width** allows you to type the width of the window in the edit field. The edit field displays the width of the last window inserted.

B-Height – The **Height** edit field allows you to specify the height of the window.

Measure to – The **Measure to** edit field includes the **Inside of frame** and **Outside of frame** options for specifying the width and height. Use **Measure to Outside of frame** to specify the Rough Opening or Masonry Opening. Use **Measure to Inside of frame** to specify sash dimensions.

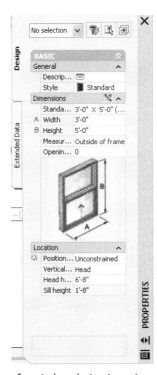

Figure 4.38 *Properties palette of a window during insertion*

Swing angle – The **Swing angle** is the angle in degrees that the window is open relative to the frame. This property is displayed for hinged windows such as casement.

Opening Percent – The **Opening Percent** is the percentage of sash opening. This property is displayed when inserting window components that slide open such as double hung windows.

Rise – The **Rise** edit field will be active on window styles created from the arch, gothic, peak pentagon, and trapezoid shapes. The rise is the distance from the top of the window in which the arch begins. A one-foot rise was used to create the arch, gothic, peak pentagon and trapezoid windows shown in Figure 4.39.

Location

Position along wall – The **Position along wall** includes the **Unconstrained** and **Offset/Center** options. The position along wall options are identical to those discussed for doors.

Automatic Offset – The **Automatic Offset** edit field allows you to specify the distance from a wall intersection for window placement. The **Automatic Offset** option is displayed when the **Position along wall** option is set to **Offset/Center** in the Properties palette.

Vertical Alignment – The **Vertical Alignment** options include Head or Threshold.

Head Height – The **Head Height** is the distance from the baseline of the wall to the inside of frame or outside of frame at the top of the window.

Sill Height – The **Sill Height** is the distance from the baseline of the wall to the bottom of the window frame

Rotation – The **Rotation** display indicates the angle of the window relative to the 3:00 o'clock position. The rotation option is active for free standing windows. The **Rotation** field is not displayed in Figure 4.38 because a wall was selected for insertion of the window.

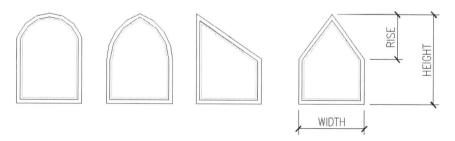

Figure 4.39 *Rise of windows*

After you select the wall during the insertion of the window, the window will dynamically slide along the wall until an insertion point is defined by a left-click of the mouse. The Properties palette can be edited before or after an insertion point for the window is specified. After selecting the wall, right-click, and the shortcut menu includes the options of the command. These options are also displayed in the command line. The options of the shortcut menu are described below:

STyle – The **Style** option will prompt you to type the name of the window style in the command line.

WIdth – The **Width** option will prompt you to specify a different width in the command line.

HEight – The **Height** option will prompt you to type a new height in the command line.

HEAd height – The **Head height** option allows you to type the vertical alignment distance from the head of the window to the baseline.

Sill Height – The **Sill Height** option prompts you for the vertical alignment distance from the baseline of the wall to the bottom of the window.

Auto – The **Auto** option toggles between **Unconstrained** and **Offset/Center**.

OFfset – The **Offset** option is displayed if the **Position along wall** option is set to Offset/Center. This option allows you to type the distance of offset for the window from intersecting walls.

Match – The **Match** option prompts you to select an existing window to match the properties of the window for the present insertion.

Undo – The **Undo** option, displayed after an insertion, will undo the last insertion of the window.

CYcle measure to – The **Cycle measure to** option toggles the handle for the insertion of the window between edge of window and center of window.

REference point on – The **Reference point on** option prompts you to select a point in the drawing to measure from to locate the insertion point of the window.

Tip: The Array command can be used to array a series of windows or doors within a wall.

DEFINING WINDOW PROPERTIES

When you select existing windows, the Properties palette includes additional information. The Properties palette can be used to edit the properties of one or more windows. The additional information is described below.

Design Tab

General

> **Layer** – The **Layer** edit field displays the name of the layer of the insertion point of the window.

Location

> **Anchor** – The **Anchor** button opens the **Anchor** dialog box, which lists the position of the window relative to justification line.
>
> **Endcaps** – The **Opening Endcaps** of the wall when the door is placed can be edited from the drop-down list.

Extended Data Tab

Documentation

> **Hyperlink** – The **Hyperlink** button opens the **Insert Hyperlink** dialog box, which allows you to link the window to a file.
>
> **Notes** – The **Notes** button opens a **Notes** dialog box for typing notes regarding the window.
>
> **Reference Documents** – The **Reference Documents** button opens the **Reference Documents** dialog box, which allows you to list the files and directory information related to the window style.

Property Sets

The **Property Sets** list will vary according to the object selected and the properties that have been defined for the window. Properties are inserted in the drawing when window tags are placed. The **A-Width** and **B-Height** values are properties extracted from the drawing and used in the Window Schedule table. Therefore if you enter the rough or masonry opening values in the **Width** and **Height** fields of the Properties palette, these values will automatically be displayed in the Window Schedule. Additional properties can be defined to customize the window schedule to include sash size and model numbers. (The development of schedules is presented in Chapter 10.) The **Property Sets** list for a window is shown at right in Figure 4.40. The **RoughHeight** and **RoughWidth** properties included in the WindowObjects properties shown in Figure 4.40 add the frame width defined in the window style to the **A-Width** and **B-Height** values. The **RoughHeight** and **RoughWidth** values can be included in a window schedule.

When the **Measure to Outside of frame** is selected, the width and height measurements will be applied to the window frame rather than the sash opening size. The **Measure to Outside of frame** dimension would be appropriate if you want to specify the masonry opening or rough opening. **Measure to Inside of frame** will apply the width and height dimensions to specify the window sash opening size. The width of

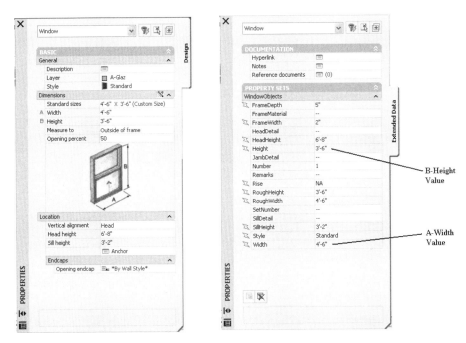

Figure 4.40 *Properties palette of an existing window*

frame dimension specified in the window style is added to the width and height of the sash opening size to determine the overall **RoughHeight** and **RoughWidth** properties.

 Note: When you use the **Measure to Outside of frame**, the **A-Width** and **B-Height** values are equal to the **RoughWidth** and **RoughHeight** property values. The default window schedule table automatically extracts the **A-Width** and **B-Height** property values regardless of "Measure to" options.

When you use **Measure to Inside of frame**, the frame width is added to the **A-Width** and **B-Height** values to define the **RoughWidth** and **RoughHeight** property values. The default window schedule table can be modified to include the **RoughWidth** and **RoughHeight** property values.

EDITING A WINDOW WITH GRIPS

The grips of a window can be used to edit the swing, hinge point, size, head height, sill height, and location within, and location along a wall as described for doors earlier in this chapter. The grip locations of a window are shown in Figure 4.41.

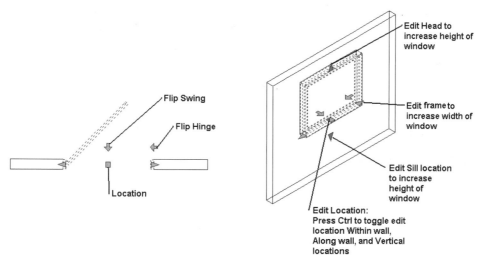

Figure 4.41 *Grip locations of a window*

THE WINDOWS PALETTE

Windows can also be inserted by selecting a window style from the Windows palette as shown in Figure 4.1. There are 10 styles listed on the Windows palette. When you select a tool from the palette, the **WindowAdd** command is executed with the style preset. Additional styles can be imported from other drawings or created in the **Style Manager**. Additional styles are located in the *C:\Documents&Settings\All Users\Application Data\Autodesk\ADT 2005\enu\Styles\Imperial* directory. There are 72 styles included in the Window Style (Imperial) drawing in the Imperial directory. Bay and Bow Window styles are included in the Window Style (Imperial) drawing.

CREATING A NEW WINDOW STYLE

Window styles can be created that include a saved list of sizes, frame, and sash parameters. Using styles of windows improves uniformity in placing windows. You can create and edit window styles using the **Style Manager** or the **WindowStyle** command. Access **WindowStyle** as shown in Table 4.8.

Command prompt	WINDOWSTYLE
Shortcut menu	Select Window in the Design palette or Windows palette, right-click, and choose Window Styles
Shortcut menu	Select a window, right-click, and choose Edit Window Style

Table 4.8 *WindowStyle command access*

Selecting the **WindowStyle** command opens the **Style Manager** to the Window Styles folder.

Window styles are created in the **Style Manager**. You can open the **Style Manager** directly to the Architectural Objects\Window Styles folder by selecting the **Window** tool from the Design palette, right-clicking, and choosing **Window Styles**. This shortcut menu selection will display the **Style Manager** with only the Window Styles folder displayed in the left pane. Because the Window Styles folder is preselected when the **Style Manager** opens, to create a new style, select **New Style** from the **Style Manager** toolbar. The default name for the new style is displayed in the right pane. Overtype a name for the new style in the right pane. Double-click on the name of the new style to open the **Window Style Properties** dialog box. You then edit the tabs of the **Window Style Properties** dialog box to define the style. The following tabs are described below: **General**, **Dimensions**, **Design Rules**, **Standard Sizes**, **Materials**, **Classifications**, and **Display Properties**.

General

The General tab is shown in Figure 4.42.

> **Name** – The **Name** edit field allows you to edit the name of the window style.

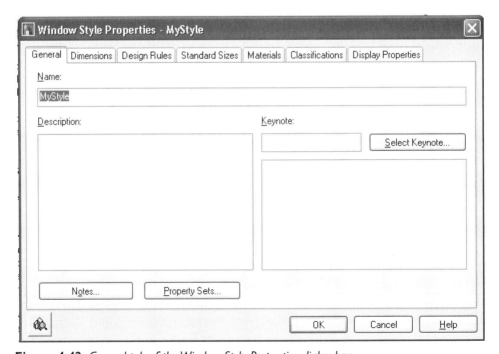

Figure 4.42 *General tab of the Window Style Properties dialog box*

Description – The **Description** edit field allows you to add a description for the window.

Notes – The **Notes** button will open the **Notes** dialog box. This dialog box allows you to add text and file descriptions related to the window.

Property Sets – The **Property Sets** button will open **Edit Property Sets Data** dialog box. This dialog box allows you to edit the style-based property sets of the window. Property set data is extracted for schedules.

Keynote – The **Keynote** button opens the **Select Keynote** dialog box, which allows you to select keynotes from the 16 CSI divisions.

Dimensions Tab

The **Dimensions** tab allows you to define the size of the frame, sash, and glass components. The components are shown in Figure 4.43 and described below.

Frame

A-Width – The **A-Width** dimension is the total width of the window jamb and brick mold or interior trim. Refer to Dimension "A" in Figure 4.43.

B-Depth – The **B-Depth** dimension is the total depth of the window jamb and trim which is specified based upon the width of wall. **Auto-Adjust to Width of Wall** can be toggled ON to adjust the frame depth value to equal the wall width.

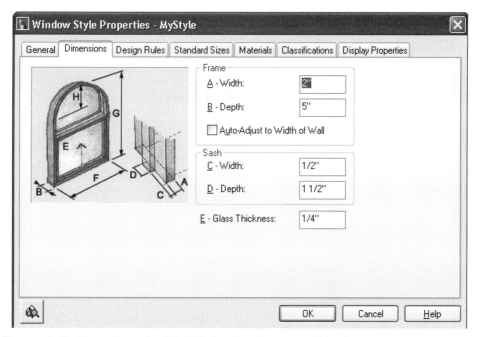

Figure 4.43 *Dimensions tab of the Window Style Properties dialog box*

> **Auto-Adjust to Width of Wall** – This check box determines whether the depth is fixed or varies to match the width of the wall.

Sash

> **C-Width** – The **C-Width** dimension is the distance the sash projects from the frame.
>
> **D-Depth** – The **D-Depth** dimension is the thickness of the sash.
>
> **E-Glass Thickness** – The **E-Glass Thickness** is the thickness of the glass.

Design Rules Tab

The **Design Rules** tab allows you to specify the shape and window type of the style. The **Design Rules** tab is shown in Figure 4.44 and described below.

> **Predefined** – The **Predefined** radio button toggles ON the predefined shapes, which include thirteen shapes, including rectangle, round, half round, quarter round, oval, arch, gothic, Isosceles triangle, right triangle, peak pentagon, octagon, hexagon, and trapezoid. Each of these shapes can be applied to a window type to create a window style.
>
> **Use Profile** – The **Use Profile** option allows you to select an AecProfile from the drop-down list. Profiles presented in Chapter 3 are created from a closed polyline in the **Style Manager**. This option allows you to create an irregular shaped polyline that when converted to a profile becomes the shape of the window.

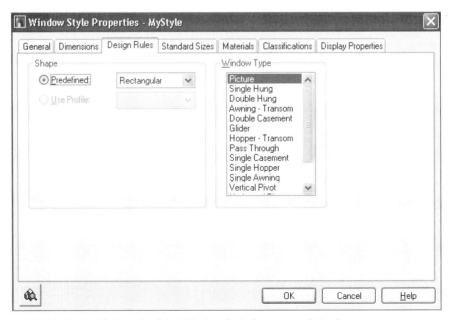

Figure 4.44 *Design Rules tab of the Window Style Properties dialog box*

Window Type – The **Window Type** list box includes various types of windows that are manufactured, such as awning, casement, double hung, and glider.

Standard Sizes Tab

The **Standard Sizes** tab provides a table of sizes for a window style. The **Standard Sizes** tab is shown in Figure 4.45.

Standard Sizes – The sizes are listed in a size table that includes F-Width, G-Height, H-Rise, and J-Leaf. These dimensions are illustrated on the **Dimensions** tab of the **Window Style Properties** dialog box. The H-Rise is inactive unless the window shape is Arch, Gothic, Peak, or Trapezoid. The J-Leaf column is inactive unless you specify an "Uneven" window type.

Add – The **Add** button located below the size table is used to create a standard size. Selecting the **Add** button opens the **Add Standard Size** dialog box as shown in Figure 4.46.

Edit – The **Edit** button located below the size table opens the **Edit Standard Size** dialog box, which is used to edit the sizes that are highlighted in the table of standard sizes.

Remove – The **Remove** button allows you to remove part or all of the standard sizes. If you select a size from the table, the **Remove** button is activated and you can select the **Remove** button to delete the selected size. If you select the header button at the top of the column, as shown in Figure 4.45, the entire schedule will be selected, and then if you select the **Remove** button, the entire table is deleted.

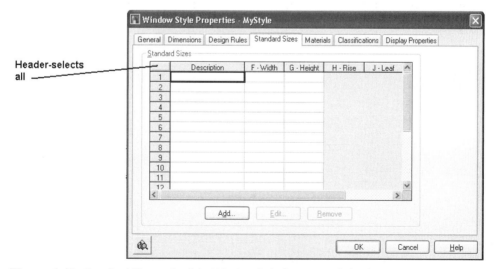

Figure 4.45 *Standard Sizes tab of the Window Style Properties dialog box*

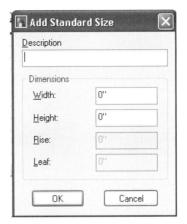

Figure 4.46 *Creating standard sizes by editing the Add Standard Size dialog box*

Materials Tab

The **Materials** tab allows you to define the material definition for the components of the window. The **Materials** tab shown in Figure 4.47 includes the four components of the window. You can click in the **Material Definition** column and select a material definition for each component. Material definitions can be imported to the drawing using the **Style Manager**. The two buttons in the right margin are described below. The procedure to assign materials and create new material definitions for windows is similar to the procedure outlined in "Assigning Materials to Wall Components" in Chapter 3.

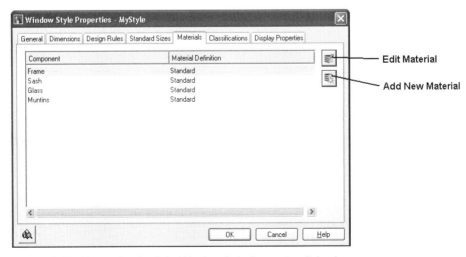

Figure 4.47 *Materials tab of the Window Style Properties dialog box*

Add New Material – The **Add New Material** button opens the **New Material** dialog box, which allows you to create a name for the material definition.

Edit Material – The **Edit Material** button opens the **Material Definition Properties** dialog box as shown in Figure 4.48. Select the **Edit Display Properties** button to open the **Display Properties – Material Definition** dialog box for the display representation as shown in Figure 4.48. The **Display Properties – Material Definition** dialog box consists of three tabs, which allow you to specify color, layer, linetype, hatching, and rendering properties in this dialog box. The **Other** tab of the **Display Properties – Material Definition** dialog box is shown in Figure 4.48.

Classifications Tab

The **Classifications** tab includes **Classification Definition** and **Classification** columns, as shown in Figure 4.49. If classifications are defined in the drawing, classifications can be assigned to the window style.

Display Properties Tab

The **Display Properties** tab allows you to control the display of a window. The display representations for current viewport of the window style are displayed bold in the

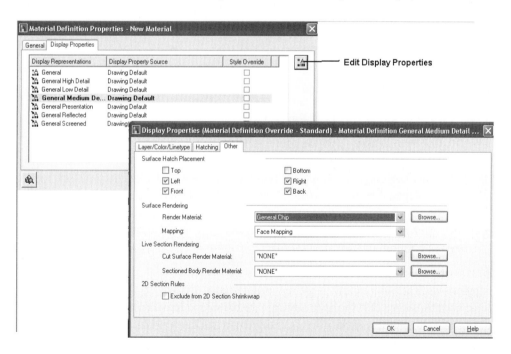

Figure 4.48 *Material Definition Properties and Display Properties – Material Definition dialog boxes*

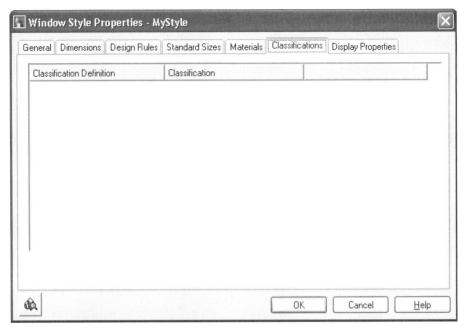

Figure 4.49 *Classifications tab of the Window Style Properties dialog box*

list. If you check the **Style Override** box, the **Display Properties – Window Plan Display Representation** dialog box opens as shown in Figure 4.50. This dialog box allows you specify the layer, color, linetype, and set the display of custom blocks.

CREATING MUNTINS FOR WINDOWS

The **Display Properties** dialog box for the Elevation and Model display representations includes a **Muntins** tab, which allows you to define muntins for the window style. Muntins can also be defined for individual windows as unique exceptions to the style in the **Display Properties** dialog box of the **Edit Object Display** command. The muntins divide the glass area of the window to make the window attractive. The muntin patterns can be Rectangular, Diamond, Prairie 9 Lights, Prairie 12 Lights, Starburst, Sunburst, or Gothic as shown in Figure 4.51.

The starburst, sunburst, and gothic patterns are restricted to certain types and shapes of windows. Table 4.9 below lists the window types and window shapes in which the Starburst, Sunburst, and Gothic muntins can be used. The window shape and type are defined in the window style. The Rectangular and Diamond muntin patterns can be used on any window type or shape.

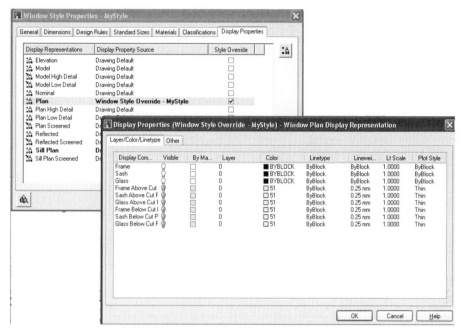

Figure 4.50 *Window Style Properties and Display Properties dialog boxes*

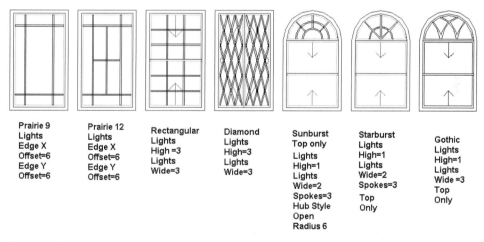

Figure 4.51 *Window muntins available using Display Properties*

Window Types for Starburst-Sunburst-Gothic Muntins	Window Shapes for Starburst	Window Shapes for Sunburst	Window Shapes for Gothic
Awning	Round	Round	Round
Single Hopper	Half Round	Half Round	Half Round
Single Transom	Quarter Round Top	Quarter Round Top	Gothic
Vertical Pivot			Peak Pentagon
Horizontal Pivot			Arch
Double Hung			
Glider			
Single Hung			
Single Casement			
Picture			

Table 4.9 *Muntin options for window types*

Muntins are defined by creating a **Style Override** for the Elevation or Model display representations. When you check the **Style Override** check box, the **Display Properties** dialog box opens as shown in Figure 4.52. The **Muntins tab** of the **Display Properties** dialog box is shown at right in Figure 4.52. To create a muntin, select the **Add** button. When you select the **Add** button, the **Muntins Block** dialog box opens as shown in Figure 4.53.

The definition of the muntin block is created by editing the fields shown on the left and the resulting muntin pattern is displayed on the right. The number of divisions or spokes can be specified for each pattern. The rectangular pattern allows you to specify the number of lights high and wide. Muntins can be applied to the top, bottom, or both sashes.

        Name – The **Name** edit field is defined by AutoCAD, or you can type a specific name of the muntin pattern.

Window Pane

        Top – Select the **Top** radio button to turn on the display of muntins only in the top sash area.

        Other – Select the **Other** radio button to specify muntins in all sash areas or a single sash specified by its index number. If **Other** and **Single** are select-

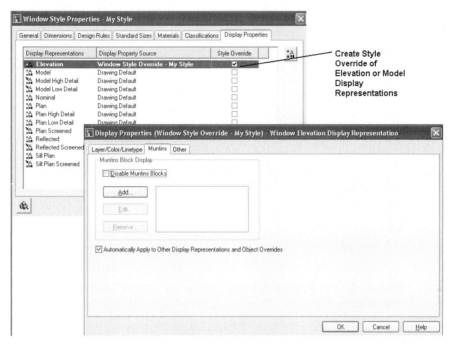

Figure 4.52 Window Style Properties and Display Properties – Window Elevation display representation

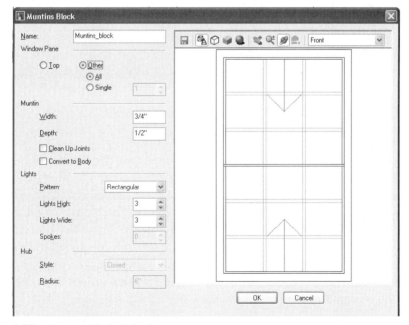

Figure 4.53 Muntins Block dialog box

ed, you specify the sash number to the right of the **Single** radio button. The index number is a number identifying the sash location within the window. The number system starts numbering the sash in the lower left corner and increases the number assigned as you move in a counterclockwise direction from the lower left sash location. The index numbers of the two sashes of a double hung are 1 for the lower window and 2 for the upper sash.

Muntin

Width – The **Width** edit field specifies the width dimension of the muntins.

Depth – The **Depth** edit field specifies the depth dimension of the muntins.

Clean Up Joints – Check ON **Clean Up Joints** to trim the intersection of muntins lines as shown in Figure 4.54.

Convert to Body – The **Convert to Body** check box creates a 3D body to represent the muntins.

Lights

Patterns – The **Patterns** drop-down list consists of Rectangular, Diamond, Prairie 9 Lights, Prairie 12 Lights, Starburst, Sunburst, and Gothic patterns.

Lights High – Specify in the **Lights High** edit field the number of divisions in each sash vertically by clicking on the arrow to the right.

Lights Wide – Specify in the **Lights Wide** edit field the number of divisions to divide the sash in a horizontal direction.

Spokes – Specify the number of spokes to create in the sash if the starburst, or sunburst patterns are selected.

Hub

Style – The **Hub Style** can be set to closed or open style can be selected for the sunburst style.

Radius – The **Radius** of the hub can be specified.

ADDING SHUTTERS TO A WINDOW STYLE

Shutters can be created in the window style by importing the Shutters-Dynamic or Shutters-12 style into the drawing using the **Style Manager**. The files are located in the *Window Styles (Imperial).dwg* of the *C:\Documents & Settings\All Users\Application Data\Autodesk\ADT 2005\enu\Styles\Imperial* directory. The Shutters-Dynamic style adjusts the shutter width according to the width of the window. The Shutters-12 style includes a window with a fixed 12" wide shutter. Each style has added the shutters to a window with rectangular shape using the picture window type. You can copy the window style in the **Style Manager** and change the window type to Double Hung or other types in the **Design Rules** tab of the **Window Style Properties** dialog box. Therefore, if you use the Shutters-Dynamic or Shutters-12 style as the base window style, you can create any type and shape window with shutters.

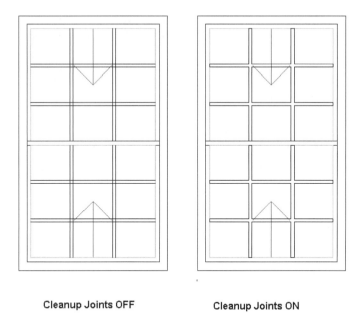

Cleanup Joints OFF   Cleanup Joints ON

Figure 4.54 *Display of muntins using Clean Up Joints toggle*

The shutters created in the Shutters-Dynamic and Shutters-12 styles are custom blocks applied to the Elevation and Model display representations. Therefore, if the shutter styles have been used in a drawing, you can add the custom blocks to other windows in the drawing. The custom blocks are added by selecting the Add button on the **Other** tab of the **Display Properties – Window Elevation Display Representation** or the **Display Properties – Window Model Display Representation** dialog box. Selecting the **Add** button opens the **Custom Block** dialog box as shown in Figure 4.55. Select the **Select Block** button to open the **Select A Block** dialog box. The shutters are then selected as custom blocks from the block list and attached to the left and right sides of the window as shown in Figure 4.55.

 STOP. Do Tutorial 4.2, "Inserting Windows and Creating Window Styles," at the end of the chapter.

CREATING OPENINGS

You can create openings in walls without inserting doors or windows by using the **OpeningAdd** command. Access the **OpeningAdd** command as shown in Table 4.10.

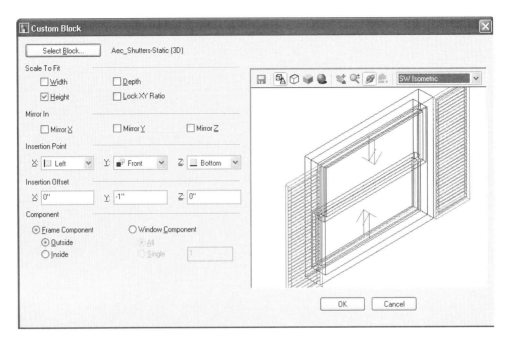

Figure 4.55 Adding custom blocks to the window

Command prompt	OPENINGADD
Palette	Select Opening from the Design palette

Table 4.10 OpeningAdd command access

Openings are placed in a wall or space boundary by setting the shape, width, and height in the Properties palette in a manner similar to placing doors or windows. The opening does not include a frame. The features of the opening are specified in the Properties palette as shown in Figure 4.56. The features are described below.

Basic

General

> **Description** – The **Description** button opens the **Description** dialog box, allowing you to type a description.
>
> **Shape** – The **Shape** of the opening can be selected from the drop-down list in the Properties palette. The shape list includes the following: rectangular, round, half round, quarter round, oval, arched, trapezoid, gothic, isosceles triangle, right triangle, peak pentagon, octagon, hexagon, or Custom. Selecting the Custom option allows you to specify an AecProfile to define the shape.

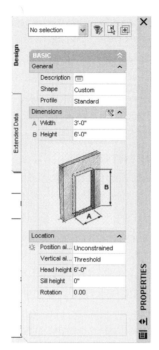

Figure 4.56 *Properties palette of an opening*

Profile – The **Profile** feature is added to the Properties palette if **Custom** is selected from the **Shape** drop-down list. The **Profile** lists the AEC Profiles in the drawing.

Dimensions

A–Width – The **A-Width** edit field allows you to type the width of the opening. Note that openings do not include frames.

B–Height – The **B-Height** edit field allows you to type the height of the opening.

Location

Position along wall – The **Position along wall** includes the Unconstrained and Offset/Center options.

Offset Distance – If the **Offset/Center** option is selected, in the **Position along wall** option, the offset distance is typed in this field.

Vertical Alignment – Vertical alignment options include **Sill** and **Head**. If the **Sill** option is selected, the opening is placed by controlling distance from the wall baseline to bottom of the opening. If **Head is** selected, the opening is positioned in the wall by controlling the distance from the opening head to the wall baseline.

Head height – The **Head height** allows you to specify distance from the wall baseline to the top of the opening.

Sill height – The **Sill height** edit field allows you to specify the distance from the wall baseline to the bottom of the opening.

Rotation – The **Rotation** edit field displays the angle of the openings relative to the 3:00 o'clock position for opening placed freestanding and not anchored to a wall.

DEFINING OPENING PROPERTIES

Editing the Properties palette can change the properties of openings. When you select an existing opening, the Properties palette is expanded to include the following additional features.

Layer – The **Layer** edit field displays the name of the layer of the insertion point of the opening

Anchor – The **Anchor** button opens the **Anchor** dialog box, which lists the position of the opening relative to the justification line.

Endcaps – The opening endcaps of the wall for the opening can be edited by selecting the opening endcaps from the drop-down list.

ADDING AN OPENING WITH PRECISION

The opening can be added by editing the dynamic dimensions, or by selecting **Reference point on** or **Cycle measure to** to insert the opening with precision. After you select the wall or space boundary, the shortcut menu includes the following options:

Shape – The **Shape** option will prompt you to type the number that identifies the predefined list of shapes. The list can be seen if you press F2 to view the text window. The **Shape** drop-down list of the Properties palette includes the list of polygons and the Custom option, which expands the Properties palette to include the **Profile** edit field. The **Profile** drop-down list includes profiles defined in the drawing.

WIdth – The **Width** option will prompt you to specify a different width for the opening in the command line.

HEight – The **Height** option will prompt you to type a new height of the opening in the command line.

HEAd height – The **Head height** option allows you to type the vertical distance from the head to the baseline.

SIll height – The **Sill** option prompts you for the vertical distance from the baseline of the wall to the bottom of the opening.

Auto – The **Auto** option toggles between Unconstrained and Offset/Center.

OFfset – The **Offset** option is displayed if Offset/Center has been selected for the **Position along wall** option. Type the distance for the offset of the opening from intersecting walls in the command line or in the **Automatic offset** field of the Properties palette.

Match – The **Match** option prompts you to select an existing opening to match its properties for the current insertion.

Undo – The **Undo** option, displayed after an insertion, will undo the last insertion of the opening.

CYcle measure to – The **Cycle measure to** option toggles the handle for the insertion of the opening between edge of opening and center of opening.

REference point on – The **Reference point on** option prompts you to select a point to measure from to specify the insertion of the opening.

MODIFYING OPENINGS

Openings can be modified by selecting the opening and editing its features in the Properties palette. The grips of the opening allow you to change its location, width, height, sill height, and head height as shown in Figure 4.57. Openings can be edited by selecting the grip and editing the dynamic dimensions in the workspace.

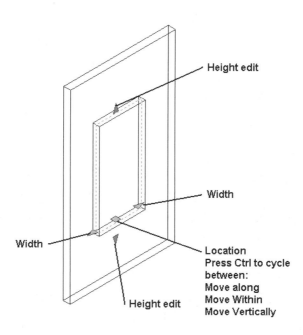

Figure 4.57 *Grips of the opening*

ADDING AND EDITING THE PROFILE OF DOORS, WINDOWS, AND OPENINGS

When you place doors and windows, the shape is defined in the style definition. When you place openings, the shape is defined in the Properties palette. The shape can be a predefined shape or Custom. The Custom option allows you to select a profile defined in the drawing, which defines the opening shape. The **OpeningAddProfile** command allows you to create or assign a profile to an existing door, window, or openings created without the use of a profile. Access the **OpeningAddProfile** command as shown in Table 4.11.

Command	OPENINGADDPROFILE
Shortcut menu	Select a door, window, opening, or door/window assembly, right-click, and choose Add Profile

Table 4.11 *OpeningAddProfile command access*

If you select a door, window, opening, or door/window assembly (created without a shape profile), right-click, and choose **Add Profile**, the **Add Opening Profile** dialog box opens as shown in Figure 4.58. The **Add Opening Profile** dialog box includes a **Profile Definition** edit field that allows you to add the profile from scratch or select an existing profile from the drop-down list. The **New Profile Name** edit field allows you to type a new name for the profile. When you select **OK**, the **Add Opening Profile** dialog box closes, and the profile is hatched. You can select the grips as shown in Figure 4.58 to edit the profile. You can select the negative grip symbol at the midpoint of the segment and stretch the segment to a new location, or select a vertex grip to edit the location of a vertex of the profile.

The **In-Place Edit** toolbar is displayed to assist you in editing the profile. You can select the grip to modify the shape, or you can right-click and choose from the shortcut menu to modify the shape. The options of the shortcut menu are described below. After editing the profile, select **Save All Changes** from the **In Place Edit** toolbar. When the changes are saved, the **In Place Edit** toolbar closes and the grips are cleared.

> **Add Vertex** – Executes the **InplaceEditAddVertex** command, and you are prompted to select a point to add in the workspace. The existing polyline will be extended to include the vertex.
>
> **Remove Vertex**– Executes the **InplaceEditRemoveVertex** command, and you are prompted to select a point to remove in the workspace.
>
> **Add Ring** – Executes the **InplaceEditAddRing** command, which allows you to add a ring to the profile. Additional rings can be created as voids.

Replace Ring – Executes the **InplaceEditReplaceRing** command. The command line prompts allow you to select a closed polyline, spline, ellipse, or circle.

Save Changes – Saves the changes made to the profile.

Save As New Profile – Saves the changes in the profile using a new name.

Discard All Changes – Ends the in-place editing without saving the changes to the profile.

The **OpeningAddProfile** command added a profile to the door, window, or opening. The **OpeningProfileEdit** command can be used to modify the profile. Access the **OpeningProfileEdit** command as shown in Table 4.12. The **OpeningProfileEdit** command displays the grips of the profile and displays the **In-Place Edit** toolbar.

Command	OPENINGPROFILEEDIT
Shortcut menu	Select an opening (defined by a profile), right-click, and choose Edit Profile In Place

Table 4.12 *OpeningProfileEdit command access*

APPLYING TOOL PROPERTIES

Existing doors, windows, and openings can be changed to the settings of the current tool with the **Apply Tool Properties to** command. When a tool of the tool palette is

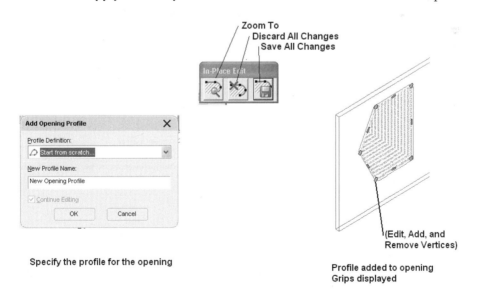

Figure 4.58 *Changing the profile with Edit In Place*

highlighted, right-click, and choose **Apply Tool Properties to** as shown in Figure 4.59. The **Apply Tool Properties to** command will prompt you to select objects for edit. The objects selected are changed to current tool, you can edit the Properties palette for additional size editing. An application of the **Apply Tool Properties to** command is shown in the following steps.

STEPS TO USING APPLY TOOL PROPERTIES FOR DOORS

1. Select a tool, such as **Bifold-Double** tool, from the Doors tool palette, right-click, and choose **Apply Tool Properties to Door** from the shortcut menu as shown in Figure 4.59.

2. Select doors in the drawings for edit as shown in the following command line sequence.

   Command: ApplyToolToObjects

   Select Door(s): 1 found *(Select a door at p1 as shown in Figure 4.59.)*

   Select Door(s): 1 found, 2 total *(Select a door at p2 as shown in Figure 4.59.)*

   Select Door(s): ENTER

3. Doors remain selected; edit the Properties palette to complete the edit.

4. Press ESC to end the **Apply Tool Properties to** command.

   Doors are changed to Bi-fold type.

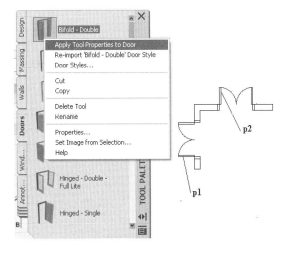
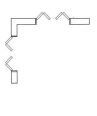

Doors changed to the properties of the current tool-bifold doors

Figure 4.59 *Apply Tool Properties to command*

IMPORTING AND EXPORTING DOOR AND WINDOW STYLES

Door and window styles can be imported into a drawing from the Design Tool Catalog - Imperial and Design Tool Catalog-Metric of the **Content Browser**. Once styles are i-dropped into the drawing, they are listed in the **Style Manager**. Styles can be imported and exported within the **Style Manager**. You can access all styles included in Architectural Desktop from within the **Style Manager** by opening the files in the *C:\Documents&Settings\All Users\Application Data\Autodesk\ADT 2005\enu\Styles* directory. The files included in the style directory are shown in Table 4.13.

Imperial	Description
Door Styles (Imperial)	50 styles including panel, arched, half round and overhead garage doors with windows
Window Styles (Imperial)	73 styles including jalousie, bay, bow, shutters and arched windows
Metric	
Door Styles (Metric)	50 styles including 6 panel, arched, glazed, and louvered
Window Styles (Metric)	64 styles including louvre and shapes such as arched, Gothic, half round, oval, trapezoid

Table 4.13 *Door and Window Styles of Autodesk Architectural Desktop*

Styles can be exported to a styles directory. The styles can then be imported into new drawings from the styles directory as you develop them. The resource files then become a depository of frequently used styles that can be imported into future drawings. The steps for importing door styles are shown below.

STEPS TO IMPORTING DOOR STYLES

1. Select **Door** from the Design tool palette, right-click, and choose **Door Styles**.
2. Select **Open Drawing** from the **Style Manager** toolbar.
3. Select **Content** from the **Places** panel at left, and then select the Styles to access Imperial or Metric folders and the Door Styles (Imperial).dwg and Door Styles (Metric).dwg files.
4. Select **Door Styles (Imperial)**, and select the **Open** button to select the drawing and close the **Open Drawing** dialog box.

5. Select the Door Styles (Imperial).dwg file in the left pane of the Style Manager. Expand the Architectural Objects\Door Styles folder of the Door Styles (Imperial) file in the left pane.

6. Select a style from the Door Styles (Imperial) file, right-click, and choose **Copy** from the shortcut menu.

7. Select the current drawing file name in the left pane, right-click, and choose **Paste**.

8. The imported door style is displayed in the Architectural Objects\Door Styles folder of the current file.

Tip: Additional door and window styles can be i-dropped into the drawing from the Architectural Desktop Design Tool Catalog-Imperial>Doors and Windows>Doors tool catalog from the Content Browser.

STOP. Do Tutorial 4.3, "Creating an AEC Profile and Using the Profile in a Door Style," at the end of the chapter.

SUMMARY

1. Doors are inserted in a drawing with the **DoorAdd** command.
2. Door styles are created in the **Style Manager**.
3. The insertion handle for doors and windows is located on the end of the unit near the start of the wall.
4. The insertion point for doors, windows, and openings can be toggled from each edge of the frame and center by selecting the CYcle measure to option.
5. Change the hinge and swing of doors and windows with **Flip Hinge** and **Flip Swing**.
6. Use grips to edit the location along a wall, within a wall, and the vertical position of doors and windows.
7. Dynamic dimensions can be toggled to edit the size and location of doors and windows.
8. Shutters can be added as custom blocks to windows, and shutter windows styles can be imported into a drawing.
9. Openings can be created with a specified width and height.
10. Doors, windows, and openings can be created in the shape of an AEC Profile.
11. Door and window styles can be imported from and exported to resource files.

REVIEW QUESTIONS:

1. The style of a door is defined with the _____ _____ command of the **Style Manager** toolbar.

2. To shift a frame to the outer surface of a wall, the _____ _____ command should be used.

3. Door styles are imported into a drawing by the _____ _____ command.

4. The overhead garage door style is located on the _____ palette.

5. The Opening Endcaps of a wall can be changed for a specific door unit by selecting the _____ _____ edit field of the Properties palette.

6. Openings do not have defined _____, unlike doors and windows.

7. Describe the procedure to automatically center a door or window in a wall.

8. Door sizes can be preset in the _____ tab of the **Door Style Properties** dialog box.

9. Toggle the edit of dynamic dimensions by selecting the _____ key.

10. The _____ of an Arch window defines its radius.

11. You can create custom shapes of doors and windows by specifying a _____ for the shape.

TUTORIAL 4.1 INSERTING DOORS

1. Open Autodesk Architectural Desktop 2005, and select **File>Project Browser** from the menu bar. Use the Project Selector drop-down list to navigate to your *ADT Student\ADT Tutor\Ch 4* student directory. Double-click on **Ex 4-1** to set this project current. Select **Close** to dismiss the **Project Browser**. (If your student folder does not include the project, refer to "Organizing Tutorial Directories" in the Preface.)

2. Select the **Constructs** tab of the **Project Navigator**. Double-click on **Floor 1** to open the Floor 1 drawing.

3. If the tool palettes are not displayed, select **Window>Tool Palettes** from the menu bar.

4. Verify that Running Object Snap is turned OFF on the status bar.

5. Select **Zoom Window** from the **Navigation** toolbar and view the area shown in the following command line sequence.

    Command: '_zoom

    Specify corner of window, enter a scale factor (nX or nXP), or

    [All/Center/Dynamic/Extents/Previous/Scale/Window] <real time>: _w

Specify first corner: **100',58'** ENTER

Specify opposite corner: **119',45'** ENTER

6. Select **Door** from the Design tool palette.

7. Edit the Properties palette as follows: General-Style = Standard, Dimensions-A-Width = 2'-0", B-Height = 6'-8", Measure to = Inside of Frame, Swing angle = 45, Location-Position along wall = Offset/Center, Automatic Offset = 6", Vertical Alignment = Head, and Head height = 6'-8".

8. Place the door by responding to the AutoCAD prompts as follows:

   Command: DoorAdd

   Select wall, space boundary or RETURN: *(Select the wall at p1 as shown in Figure 4T.1 to place the 2'-0" door.)*

   Insert point or [STyle/WIdth/HEight/HEAd height/THreshold height/Auto/OFfset/Match/CYcle measure to/REference point on]: *(Select a point below the selected wall near p2 to specify the swing as shown in Figure 4T.1.)*

   *(Click in the Properties palette, edit the A-Width to **2'-6"**, and click in the workspace.)*

   Insert point or [STyle/WIdth/HEight/HEAd height/THreshold height/Auto/OFfset/Match/Undo/CYcle measure to/REference point on]: *(Select a point near p1 as shown in Figure 4T.2.)*

   Insert point or [STyle/WIdth/HEight/HEAd height/THreshold height/Auto/OFfset/Match/Undo/CYcle measure to/REference point on]: *(Select a point near p1 as shown in Figure 4T.3.)*

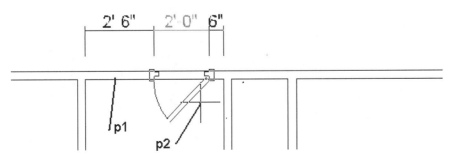

Figure 4T.1 *Selecting wall for placing door*

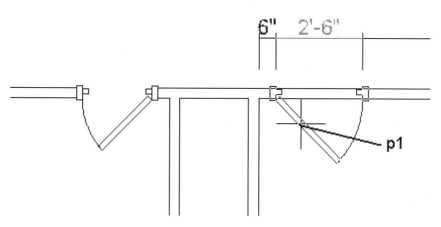

Figure 4T.2 *Determining door swing during door placement*

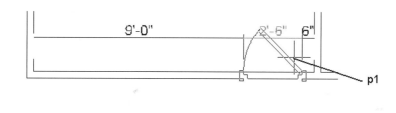

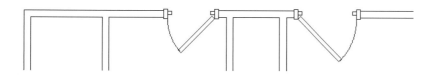

Figure 4T.3 *Placing additional door*

Insert point or [STyle/WIdth/HEight/HEAd height/Threshold height/Auto/OFfset/Match/Undo/CYcle measure to/REference point on]: ENTER *(Press* ENTER *to end the command.)*

9. Select **Zoom Window** from the Navigation toolbar and respond to the command line prompts as shown below.

Command: '_zoom

Specify corner of window, enter a scale factor (nX or nXP), or

[All/Center/Dynamic/Extents/Previous/Scale/Window] <real time>: _w

Specify first corner: **72',72'** ENTER

Specify opposite corner: **105',60'** ENTER

10. Select the Doors tool palette, and select **Hinged-Single-Exterior** from the tool palette.

11. Edit the Properties palette as follows: A-Width = 3'-0" and Position along wall to **Unconstrained**. Verify the remaining settings: General-Style = **Hinged-Single-Exterior**, Dimensions-B-Height = **6'-8"**, Measure to = Inside of Frame, Swing angle = **45**, Vertical Alignment = **Head**, and Head height = **6'-8"**.

12. Select the wall at **p1**, move the pointer down to set the door swing inside as shown in Figure 4T.4, press TAB until the left dimension is highlighted as shown, and then type **17'** ENTER in the edit field of the workspace. Press ESC to end the command.

13. Select the door at **p2** as shown in Figure 4T.4, right-click, and choose **Edit Door Style** from the shortcut menu.

14. Select the **Dimensions** tab of the **Door Style Properties** dialog box. Edit the Frame: A-Width= **2 1/4"**, B-Depth = **5-9/16"**, Stop: C-Width = **3/8"**, D-Depth = **2"**, E-Door Thickness = **1 3/4"**, and Glass Thickness = **3/16"** as shown in Figure 4T.5. Select the OK button to accept the changes and close the **Door Style Properties** dialog box.

15. Select **Format>Style Manager** from the menu bar.

16. Select **Open** from the **Style Manager** toolbar, select **Content** from the **Places** panel at left, and select Styles\Imperial. Select **Material Definitions (Imperial).dwg**, and select **Open** to complete the selection and close the **Open Drawing** dialog box.

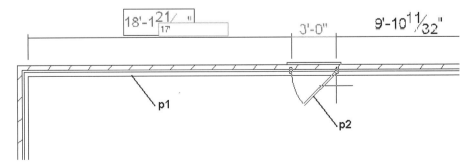

Figure 4T.4 *Editing dynamic dimension to place window*

Placing Doors and Windows 319

17. Expand the Multi-Purpose Objects\Material Definitions folders of the Material Definitions (Imperial).dwg file. Select **Doors & Windows.Wood Doors.Ash**, right-click, and choose **Copy**. Select **Floor 1** in the left pane, right-click, and choose Paste from the shortcut menu. Select **OK** to close the **Style Manager**.

18. Select the door at **p2** as shown in Figure 4T.4, right-click, and choose **Edit Door Style** from the shortcut menu. Select the **Materials** tab of the **Door Style Properties-Hinged Single Exterior** dialog box. Select in the **Material Definition** column of the **Frame**, and select **Doors & Windows.Metal Doors & Frames.Steel.Doors.Painted.White** from the **Material Definition** list to change the material of the frame.

19. To change the material of the stop, select in the **Material Definition** column of the **Stop**, and select **Doors & Windows.Metal Doors & Frames.Steel Doors.Painted.White** from the **Material Definition** list.

20. To change the material of the panel, select in the **Material Definition** of the **Panel**, and select **Doors & Windows.Wood.Doors.Ash** from the **Material Definition** list.

21. Verify that **Panel** is selected in the **Materials** tab, select the **Edit Material** button shown in Figure 4T.6, and then select the **Display Properties** tab. Verify **General Medium Detail** is the current display representation, and

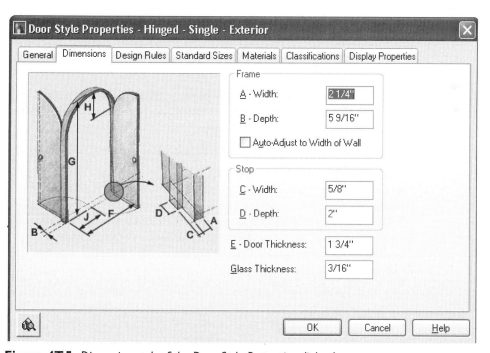

Figure 4T.5 *Dimensions tab of the Door Style Properties dialog box*

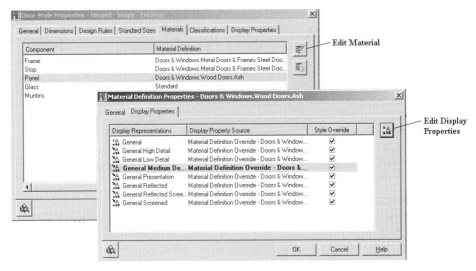

Figure 4T.6 *Material Definition Properties dialog box*

select the **Edit Display Properties** button, shown in Figure 4T.6, to open the **Display Properties (Material Definition Override-Doors& Windows.WoodDoors.Ash) – Material Definition General Medium Detail Display Representation** dialog box.

22. Select the **Other** tab as shown in Figure 4T.7, click in the **Render Material** list, and verify that **Doors & Windows.Wood.Doors.Ash** is the render material. Verify that the remaining settings are as shown in Figure 4T.7.

23. Select **OK** to dismiss the **Display Properties (Material Definition Override-Doors&Windows.WoodDoors.Ash) – Material Definition General Medium Detail Display Representation** dialog box.

24. Select **OK** to dismiss the **Material Definition Properties-Doors & Windows.Wood Doors.Ash** dialog box.

25. Select the **Display Properties** tab of the **Door Style Properties – Hinged-Single-Exterior** dialog box. Select **Threshold Plan** display representation, verify that **Style Override** is checked, and select the **Display Properties** button to open the **Display Properties (Door Style Override-Hinged-Single) – Door Threshold Plan Display Representation** dialog box.

26. Select the **Layer/Color/Linetype** tab, verify that **Threshold B** is set visible. Select the **Other** tab, and edit the threshold extension as follows: C-Extension = **2"** and D-Depth = **1"** as shown in Figure 4T.8. Select **OK** to dismiss the **Display Properties (Door Style**

Placing Doors and Windows 321

Figure 4T.7 Other tab of the Display Properties – Material Definition dialog box

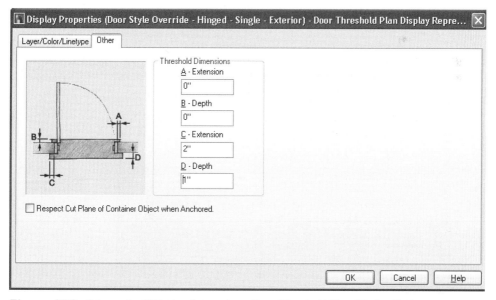

Figure 4T.8 Other tab of Display Properties – Door Threshold Plan Display Representation dialog box

Override-Hinged-Single-Exterior) – Door Threshold Plan Display Representation dialog box.

27. Select **OK** to close the **Door Style Properties – Hinged-Single-Exterior** dialog box.

28. Select the exterior door at **p2,** as shown in Figure 4T.4, right-click, and choose **Object Viewer** from the shortcut menu. Select **View>SE Isometric**, edit the display representation to **Medium Detail**, and select **Flat Shaded** to view the door with materials applied as shown in Figure 4T.9. Press ESC to close the Object Viewer.

29. Select **Zoom Window** from the **Navigation** toolbar and respond to the command line prompts as shown below.

    Command: '_zoom

    Specify corner of window, enter a scale factor (nX or nXP), or

    [All/Center/Dynamic/Extents/Previous/Scale/Window] <real time>: _w

    Specify first corner: **90',70'** ENTER

    Specify opposite corner: **100',65'** ENTER

30. Select the exterior door, right-click, and choose **Reposition Within Wall** from the shortcut menu. Respond to the command line prompts as shown below to move the jamb flush with the interior wall.

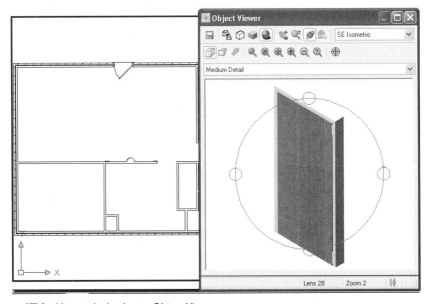

Figure 4T.9 *Materials display in Object Viewer*

Placing Doors and Windows   323

Command: RepositionWithin

Select position on the opening: *(Select a point near p1 as shown in Figure 4T.10.)*

Select a reference point_endp of *(SHIFT + right-click, choose Endpoint from the shortcut menu, and select the endpoint of the wall at p2 shown in Figure 4T.10.)*

Enter the new distance between the selected points <0">: **0** ENTER *(Type the desired distance between the two points to reposition the door as shown in Figure 4T.11.)*

31. Save and close the drawing.

TUTORIAL 4.2   INSERTING WINDOWS AND CREATING WINDOW STYLES

1. Open Autodesk Architectural Desktop 2005, and select **File>Project Browser** from the menu bar. Use the Project Selector drop-down list to navigate to your student directory. Double-click on **Ex 4-2** to set this project current. Select **Close** to dismiss the **Project Browser**. (If your student folder does not include the project, refer to "Organizing Tutorial Directories" in the Preface.)

2. Select the **Constructs** tab of the **Project Navigator**. Double-click on **Floor 1** to open the Floor 1 drawing.

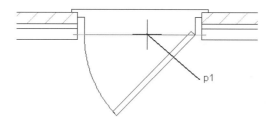

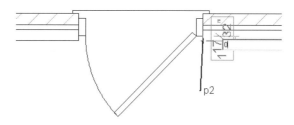

Figure 4T.10 *Moving jamb with Reposition Within Wall*

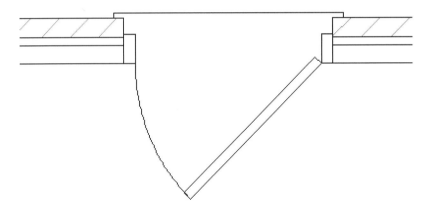

Figure 4T.11 Door repositioned in wall

3. Select **Zoom Window** from the **Navigation** toolbar and respond to the following command line prompts.

   Command: '_zoom

   Specify corner of window, enter a scale factor (nX or nXP), or

   [All/Center/Dynamic/Extents/Previous/Scale/Window] <real time>: _w

   Specify first corner: **72',72'** ENTER

   Specify opposite corner: **110',64'** ENTER

4. Verify that Object Snap is OFF.
5. Choose the Windows tool palette, and choose **Casement-Double** tool. Select the wall at p1 as shown in Figure 4T.12.

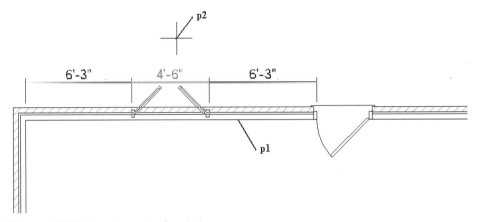

Figure 4T.12 Insertion point for window

6. Edit the Properties palette as follows: General-Style = **Casement-Double**, Dimensions-A-Width = **4'-6"**, B-Height = **3'-6"**, Measure to = **Outside of Frame**, Swing angle = **45**, Location-Position along wall = **Offset/Center**, Automatic Offset = 6", Vertical Alignment = **Head**, and Head height = **6'-8"**. Respond to the command line prompts as shown below.

> Command: WindowAdd
>
> Select wall, space boundary or RETURN: *(Select the wall at p1 as shown in Figure 4T.12.)*
>
> Insert point or [STyle/WIdth/HEight/HEAd height/SIll Height/Auto/OFfset/Match/CYcle measure to/REference point on]: *(Select a point near p2 as shown in Figure 4T.12.)*
>
> *(Move the pointer to the right beyond the door.)*
>
> Height/Auto/OFfset/Match/Undo/CYcle measure to/REference point on]: **CY** ENTER *(Turn cycle ON to move the reference for the insertion to the center of the window.)*
>
> *(Press* TAB *to highlight the dimension on the right as shown in Figure 4T.13.)*
>
> Insert point or [STyle/WIdth/HEight/HEAd height/SIll Height/Auto/OFfset/Match/Undo/CYcle measure to/REference point on]: **5'** ENTER *(Type the desired distance in the dynamic dimension.)*
>
> Insert point or [STyle/WIdth/HEight/HEAd height/SIll Height/Auto/OFfset/Match/Undo/CYcle measure to/REference point on]: ENTER *(Press* ENTER *to end the command.)*

7. Select **Content Browser** from the **Navigation** toolbar, double-click on the Design Tool Catalog-Imperial catalog, and open the Doors and Windows category of the left pane. Select page 2 of the Windows category.

8. Select the i-drop for the **Bay** window and drag the i-drop into the drawing. Press ESC to end the **AddWindow** command before selecting a wall. Close the **Content Browser**.

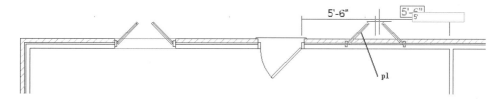

Figure 4T.13 *Editing dynamic dimension of the window*

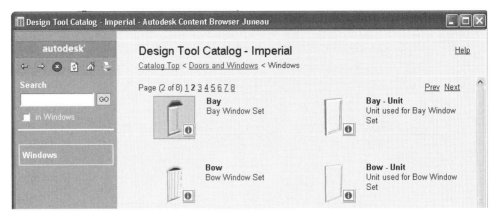

Figure 4T.14 *Bay window in the Content Browser*

9. Select the casement window at right, **p1** as shown in Figure 4T.13, and edit the style to **Bay** in the Properties palette.
10. Select the left grip of the Bay window, and type **5'** in the dynamic dimension as shown in Figure 4T.15. Press ESC to clear the grips selection.
11. Select **Format>Style Manager** from the menu bar.
12. Expand the Architectural Objects of Floor 1.dwg in the left pane and expand the Window Styles of the left pane. Select the Window Style folder in the left pane, right-click, and choose New from the shortcut menu.
13. Overtype **Andersen_Casement** in the right pane as the name of the new style.
14. Select the **Andersen_Casement** name in the right pane, right-click, and choose **Edit** from the shortcut menu.
15. Select the **General** tab, and type **Double Casement Windows** in the **Description** field.

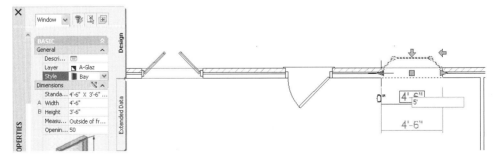

Figure 4T.15 *Editing window width using dynamic dimensions*

16. Select the **Dimensions** tab, and edit the Frame section A-Width = **2**, B-Depth = **5 9/16**, Sash C-Width = **2**, D-Depth = **2**, and E-Glass Thickness = **3/16** as shown in Figure 4T.16.

17. Select the **Design Rules** tab, and edit the Shape to **Predefined Rectangular** and Window Type to **Double Casement**.

18. Select the **Standard Sizes** tab, select the **Add** button, and edit the **Add Standard Size** dialog box shown in Figure 4T.17 as follows: Description = **CW235**, Width = **2'-4"**, Height = **3'-4 13/16"**. Select **OK** to close the **Add Standard Size** dialog box.

19. Select the **Add** button, and edit the **Add Standard Size** dialog box as follows: Description = **CW135**, Width = **4'-8 1/2"**, Height = **3'-4 13/16"**. Select **OK** to close the **Add Standard Size** dialog box.

20. Select the **Materials** tab, and edit the Frame, Sash, and Muntins material definition to **Doors & Windows.Metal Doors & Frames.Aluminum Windows.Painted.White** and Glass to **Doors & Windows.Glazing.Glass.Clear**. The material definitions were imported with the Bay window from the **Content Browser**.

21. Select **OK** to dismiss the **Window Style Properties** dialog box. Select **OK** to dismiss the **Style Manager**.

22. Select the casement window at **p1** shown in Figure 4T.18. Click in the **Style** field of the Properties palette, select **Andersen_Casement** and select the **4'-8 1/2" x 3'-4 13/16"** standard size.

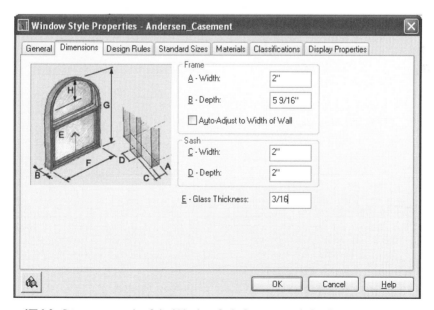

Figure 4T.16 *Dimensions tab of the Window Style Properties dialog box*

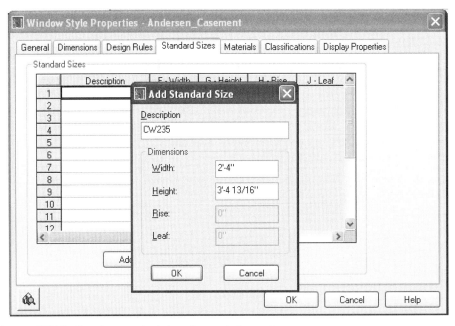

Figure 4T.17 *Creating standard sizes for a window style*

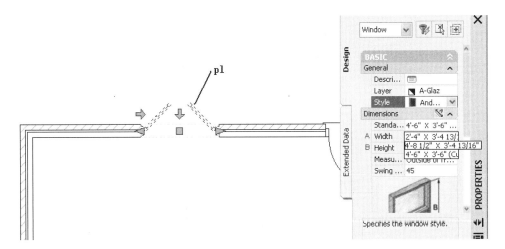

Figure 4T.18 *Editing existing windows*

23. Press ESC to clear the grips selection.
24. Select **NE Isometric** from the **Navigation** toolbar.

25. To verify that the model drawing display is up to date, select **Regenerate Model** from the **Standard** toolbar and press ENTER to select all the objects in the drawing.

26. Save and close the drawing.

29. Select **File>Project Browser**, select the current project, right-click, and choose **Close current project**.

TUTORIAL 4.3 CREATING AN AEC PROFILE AND USING THE PROFILE IN A DOOR STYLE

1. Open **Ex 4-3** in your ADT Student\ADT Tutor\Ch4 directory.
2. Choose **File>SaveAs** from the menu bar and save the drawing as **Lab 4-3** in your student directory.
3. Choose **Format>Style Manager** from the menu bar.
4. Expand Multi-Purpose Objects, and select Profiles in the left pane.
5. Select the Profiles folder, right-click, and choose **New**.
6. Overtype the name **Chamfer** in the right pane as the name for the new style.
7. Select the **Chamfer** profile, right-click, and choose **Set From** from the shortcut menu. Select the polyline as shown in the following command line sequence and Figure 4T.19.

    Command: _AecStyleManager

    Select a closed polyline: *(Select the polyline line at p1 shown in Figure 4T.19.)*

    Add another ring? [Yes/No] <No>: **y** ENTER

    Select a closed polyline: *(Select the polyline line at p2 shown in Figure 4T.19.)*

    Ring is a void area? [Yes/No] <Yes>: **y** ENTER

    Add another ring? [Yes/No] <No>: **n**

    Insertion Point or <Centroid>: _endp of *(Press* SHIFT, *right-click, and choose the Endpoint object snap; then select the end of the line at p3 as shown in Figure 4T.19.)*

8. Select **OK** to dismiss the **Style Manager**.
9. Select the Doors tool palette, select a door style, right-click, and choose **Door Styles** from the shortcut menu.
10. Select the Door Styles folder, right-click, and choose **New** from the shortcut menu. Overtype the name **Chamfer** in the right pane to name the door style.

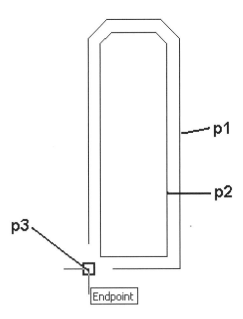

Figure 4T.19 *Selected polyline for profile*

11. Select the **Chamfer** style, right-click, and choose **Edit** from the shortcut menu.
12. To assign the profile to the new style, select the **Design Rules** tab, select the **Use Profile** radio button, and select the **Chamfer** Profile. Verify that Door Type is set to **Single**.
13. Select **OK** to close the **Door Style Properties** dialog box.
14. Select the Chamfer door style in the left pane. Select the **Viewer** tab in the right pane, select **SW Isometric** view, **Hidden** shade, and **High Detail** display as shown in Figure 4T.20.
15. Select **OK** to dismiss the **Style Manager**.
16. Select Door from the Design tool palette. Edit the Properties palette as follows: Style = **Chamfer**, Width = **3'**, Height = **6'-8"**, Measure to = **Inside** of frame, Swing angle = **45**, Position along wall = **Offset/Center**, Automatic offset = **6**. Place the door in the center of the wall as shown in the following command line sequence.

    Command: DoorAdd

    Select wall, space boundary or RETURN: *(Select the wall at p1 as shown in the drawing).*

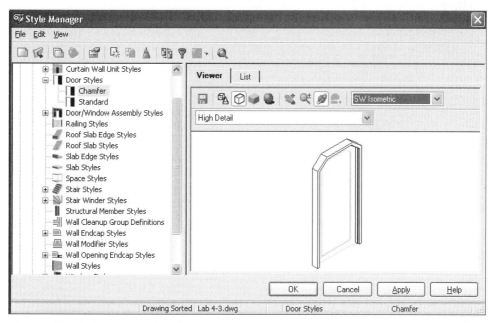

Figure 4T.20 *Door created in the shape of the profile*

> Insert point or [STyle/WIdth/HEight/HEAd height/THreshold height/Auto/OFfset/Match/CYcle measure to/REference point on]: *(Select the point at p2 shown in the drawing.)*
>
> Insert point or [STyle/WIdth/HEight/HEAd height/THreshold height/Auto/OFfset/Match/CYcle measure to/REference point on]: *(Press ESC to end the command.)*

17. Select **SW Isometric** from the **View** flyout of the **Navigation** toolbar.
18. Toggle ON POLAR on the status bar.
19. Select the door, right-click, and choose **Edit Profile In Place** from the shortcut menu. Select **Yes** to convert the profile to its actual size and dismiss the AutoCAD warning dialog box. Select the middle grip at the bottom edge of the inner panel at **p1** shown in Figure 4T.21, move the pointer up (polar angle=90), and type **6** in the command line to stretch the segment up. Select **Save All Changes** on the **In Place Edit** toolbar. Profile of door is modified as at right in Figure 4T.21.
20. Close and save the drawing.

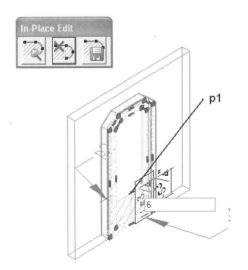 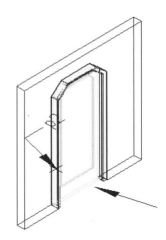

Figure 4T.21 *Profile of door modified in place*

PROJECTS

Ex 4.4 Inserting Doors in a Floor Plan

Open project *Ex 4-4* from your *ADT Student\Adt Tutor\Ch4* folder. Open the Floor 1 construct drawing and insert additional doors as shown in Figure 4T.22 and Table 4T.1. Use Offset/Center set to 3" to place the doors. View the building from SW Isometric when complete.

DOOR MARK	SIZE	STYLE	Note
A	2'-6" x 6'-8"	Hinged Single	
B	2'-0" x 6'-8"	Hinged Single	
C	3'-0" x 6'-8"	Hinged Single Exterior	
D	6'-0" x 6'-8"	Bifold Double	
E	4'-0" x 6'-8"	Bifold Double	
F	4'-0" x 6'-8"	Hinged Double-6 Panel Half Lite	I-drop from the Design Tool Catalog>Doors and Windows>Doors

Table 4T.1 *Door sizes for project*

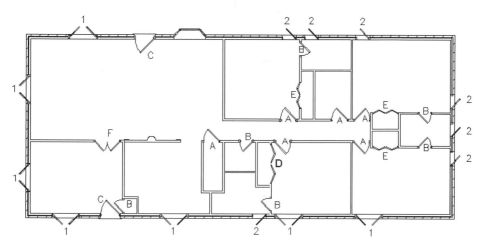

Figure 4T.22 Door and window placement in the floor plan

Ex 4.5 Inserting Windows in a Floor Plan

Open project **Ex 4-5** from your *ADT Student\ADT Tutor\Ch4* folder. Open the Floor 1 construct drawing and insert the following windows, as shown in Figure 4T.22 and Table 4T.2. Create a window style named Pella_Case_Sgl, and set window type to Single Casement and frame A-Width = 2, B-Depth = 5 9/16", Sash C-Width = 3/8", D-Depth = 2", and E-Glass Thickness = 3/16". Define the following material definitions: Doors & Windows.MetalDoors&Frames.AluminumWindows. Painted.White material definition to the Frame, Sash, and Muntins and Doors&Windows.Glazing.Glass.Clear material definition to the Glass component.

Create one standard size for the style that is 2'-4" x 3'-4 13/16" and enter a description **PW2434**. Use Offset/Center with 6" offset when placing windows. Reposition all exterior windows except the bay window to the interior surface of the wall. Reposition the exterior door frames to the interior surface of the wall.

WINDOW MARK	SIZE	STYLE
1	4'-8 1/2" x 3'-4 13/16"	Anderson_Casement
2	2'-4" x 3'-4 13/16"	Pella_Case_Sgl

Table 4T.2 *Window sizes*

Ex 4.6 Inserting Doors and Windows in a Basement Plan

Open the **Project Browser**, and set project Ex 4-6 current from your *ADT Student\Adt*

Tutor\Ch4 folder. Open the Basement construct drawing and insert the following windows and doors, as shown in Figure 4T.23 and Table 4T.3. Use **Cycle Measure to** and **Reference point on** to place windows with precision. Select **Auto Adjust to Width of Wall** in the **Window Style Properties** dialog box for the awning window style. Edit the Awning window style, select Display Representation – Sill Plan, Style Override, select the **Other** tab, and increase the B-Depth 2".

Note: Edit the wall style by selecting **Display Properties**, selecting the **Other** tab, and then clearing the **Hide Lines Below Openings at Cut Plane** check box.

WINDOW MARK	SIZE	STYLE
A	3'-6" x 2'-4"	AWNING
DOOR MARK	**SIZE**	**STYLE**
1	16' x 7'	Import Overhead 4 window door from Door Style (Imperial)
2	3'-0" x 6'-8"	HINGED-SINGLE
3	4'-0" 3 6'-8"	BIFOLD-DOUBLE
4	2'-6" 3 6'-8"	HINGED-SINGLE

Table 4.16 *Sizes and styles for Exercise 4.6*

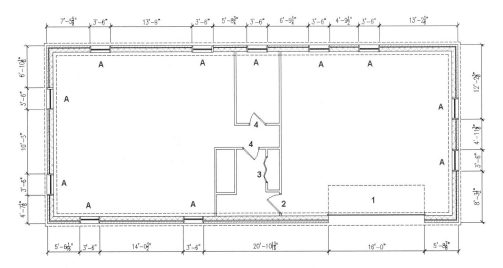

Figure 4T.23 *Door and window placement in the basement plan*

CHAPTER 5

Door and Window Assemblies

INTRODUCTION

Door/window assemblies allow you to create a unit with multiple window, door, and panel components. The door/window assembly is created based upon a grid. The grid allows you to divide the horizontal and vertical dimensions of the unit into a number of equal divisions or manually specify grid locations.

OBJECTIVES

After completing this chapter, you will be able to

- Insert door/window assemblies using the **DoorWinAssemblyAdd** command
- Edit the properties of the door/window assembly
- Create and edit door/window assemblies, including the specification of divisions, infills, frames, and mullions
- Define fixed cell dimension, fixed number of cells, and manual location of grid
- Assign door and window styles as infills
- Add or modify the profile used for frame and mullion components
- Edit the miter angle of door/window assemblies
- Apply interference to a door/window assembly to trim the Aec object

ADDING DOOR/WINDOW ASSEMBLIES

The style of a door/window assembly defines the content of the grid, which includes the divisions, frames, cell infills, and mullions. There are 125 window assembly styles included in the *C:\Documents and Settings\All Users\Application Data\Autodesk\ADT 2005\enu\Styles\Imperial\Door-Window Assembly Style (Imperial)* file. Included are styles with arched, peaked, and trapezoid shapes that can include transoms and skylights. Additional styles can be i-dropped from the Design Tool Catalog - Imperial>Doors and Windows>Door and Window Assemblies of the **Content Browser** as shown in Figure 5.1.

The Standard style is provided in the software; however, additional styles can be imported into the drawing. Door/window assemblies are placed in the drawing by selecting Door/Window Assembly from the Design tool palette. Access the **DoorWinAssemblyAdd** command as shown in Table 5.1.

Command prompt	DOORWINASSEMBLYADD
Palette	Select Door/Window Assembly from the Design palette shown in Figure 5.2.

Table 5.1 *Door/Window Assembly (DoorWinAssemblyAdd) command access*

When you select **Door/Window Assembly** from the Design palette, you are prompted to select a wall or space boundary. Unless styles have been imported, the Standard style is the only option and is inserted. The door/window assembly is anchored to the wall or space boundary. The width, height and other features of the Door/window assembly can be specified in the Properties palette as shown in Figure 5.2.

The features specified in the Properties palette for a door/window assembly are described below.

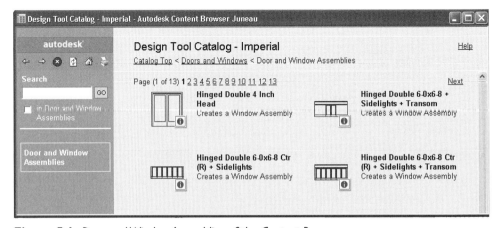

Figure 5.1 *Door and Window Assemblies of the Content Browser*

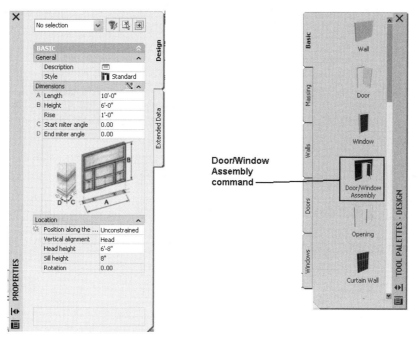

Figure 5.2 Door and Window Assembly command and Properties palette

Design

General

Description – The **Description** button opens a **Description** dialog box to add a description.

Style – The **Style** displays the current window and door assembly; the drop-down list includes all styles loaded in the drawing.

Dimensions

A-Length – The **Length** edit field allows you to specify the length of the assembly.

B-Height – The **Height** edit field allows you to type the distance from the baseline to the top of the unit. The **Height** value includes the rise.

Rise – The **Rise** edit field allows you to type the vertical dimension of the assembly from the base height to the top of the window assembly. This dimension applies to arched, trapezoid, and peaked shapes.

C-Start Miter – The **Start Miter** is the angle of the frame, mullion, and infill at the start of the assembly.

D-End Miter – The **End Miter** is the angle of the frame, mullion, and infill at the end of the assembly.

Location

Position along wall – The options of **Position along wall** are **Offset/Center** or **Unconstrained**. The **Offset/Center** option will restrict placement to the center of the wall segment or a specified distance from an intersecting wall. The **Unconstrained** option allows placement at any location along the wall or space boundary.

Offset Distance – The **Offset Distance** option is available when the Offset/Center constraint is selected. The offset distance is the distance between the unit and the intersecting wall.

Vertical Alignment – **Vertical alignment** can be specified to the head or sill.

Head Height – The **Head Height** is the distance from head of the assembly to the baseline of the wall.

Sill Height – The **Sill Height** is the height of the sill from the baseline.

Rotation – The **Rotation** angle is the angle of the door and window assembly from the 3:00 o'clock position.

When you select the **Door/Window Assembly** tool from the Design palette, you are prompted to select a wall or space boundary as shown in the following command line

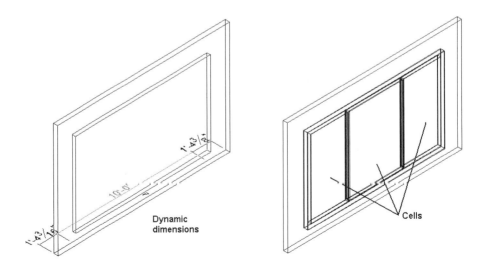

Figure 5.3 *Specifying the insert point of the Door/Window Assembly*

sequence. The handle for the insert point when inserting the door/window assembly without the Offset/Center constraint is located near the start point of the wall. In Figure 5.3 the handle is located on the right; however, the **Cycle measure to** option of the shortcut menu allows you to change the insert point. The dynamic dimensions can be edited to insert the assembly with precision in the same manner as discussed for doors and windows.

> Command: DoorWinAssemblyAdd
>
> Select wall or RETURN: *(Select a wall.)*
>
> Insert point or [STyle/LEngth/HEight/HEAd height/SIll Height/Auto/Match/Cycle measure to/REference point on]: *(Click to specify the insert point of the assembly.)*

The options of the command line can be selected by typing the letters capitalized in the command line or, if you right-click, you can choose them from the shortcut menu. A description of the options follows:

> **STyle** – The **Style** option allows you to type the name of another assembly style in the command line.
>
> **LEngth** – The **Length** option allows you to type a different length in the command line.
>
> **HEight** – The **Height** option allows you to type a different height in the command line.
>
> **HEAd height** – The **Head height** option allows you to type a different height in the command line.
>
> **SIll Height** – The **Sill Height** allows you to type a different sill height in the command line.
>
> **Auto** – The **Auto** option toggles between Unconstrained and Offset/Center.
>
> **Match** – The **Match** option prompts you to select an existing assembly to match its properties for the current insertion.
>
> **Cycle measure to** – The **Cycle measure to** option toggles the handle for the insertion of the assembly between the center and edges of the assembly.
>
> **REference point on** – The **Reference point on** option prompts you to select a point to measure from to specify the insertion point of the assembly.

The Standard style for a door/window assembly is shown in Figure 5.3. This assembly consists of three cells. The cells could include doors or windows. The style of the assembly controls what is placed in the cells and the layout of the assembly. The style of the door/window assembly is defined in the **Door/Window Assembly Style Properties** dialog box. You can access this dialog box when you edit the style of the door/window assembly. Access the **DoorWinAssemblyStyle** command as shown in Table 5.2.

Menu bar	Format>Style Manager
Command prompt	DOORWINASSEMBLYSTYLE
Shortcut menu	Select the Door/Window Assembly tool from the Design palette, right-click, and choose Door/Window Assembly Styles
Shortcut menu	Select a door/window assembly, right-click, and choose Edit Door/Window Assembly Style

Table 5.2 *Edit Door/Window Assembly Style (DoorWinAssemblyStyle) command access*

When you select the Door/Window Assembly tool from the Design palette, right-click, and choose **Door/Window Assembly Styles**, the **Style Manager** opens to the Door/Window Assembly Styles folder. Select a style, right-click, and choose **Edit** to open the **Door/Window Assembly Style Properties** dialog box. The **Door/Window Assembly Style Properties** dialog box shown in Figure 5.4 includes tabs, which allow you to specify the shape and contents of the assembly.

COMPONENTS OF A DOOR/WINDOW ASSEMBLY STYLE

The style of the door/window assembly is based upon a grid. The length and height dimensions specified when adding the assembly become the dimensions of horizontal and vertical dimensions of the grid. The style defines the size and number of divisions in the grid. The intersection of the divisions creates a cell, which can be filled with a window, door, or panel unit. Each grid consists of the following elements as defined in the **Design Rules** tab of the **Door/Window Assembly Style Properties** dialog box.

Divisions – The Divisions of the grid specify the horizontal or vertical divisions of the grid to create the number of cells. The default dimension of a division is 3'.

Infills – The Cell Infills define the components for each cell. The cell infill can be a panel, door, or window. The default panel size is 2" thick.

Frames – The Frame is the component that surrounds the outer edge of the primary grid. The default shape is rectangular with default sizes of 3" wide and 3" deep.

Mullions – The Mullion is the component that separates cells. The default shape of the mullion is rectangular with default sizes of 1" wide and 3" deep.

The definition of the door/window assembly style is specified in the **Door/Window Assembly Style Properties** dialog box, which consists of **General**, **Shape**, **Design Rules**, **Overrides**, **Materials**, **Classification**, and **Display Properties** tabs.

General Tab

The **General** tab shown in Figure 5.4 is described below.

> **Name** – The **Name** edit field allows you to edit the name of the style.
>
> **Description** – The **Description** edit field allows you to type a description for the assembly.
>
> **Keynote**–Choose the **Select Keynote** button to open the **Select Keynote** dialog box, which allows you to specify keynotes from the 16 CSI divisions. Keynote commands of the Annotation palette can be used to extract the keynote annotation. Additional information regarding keynotes is presented in Chapter 11.

Shape Tab

The **Shape** tab shown in Figure 5.4 allows you to define the shape of the window assembly. There are two radio buttons, **Predefined** and **Use Profile**, which include drop-down lists of shapes. The options of the **Shape** tab are described below.

> **Predefined** – The **Predefined** shapes are specific shapes listed in the drop-down list such as rectangular, round, half round, quarter round, oval, arch,

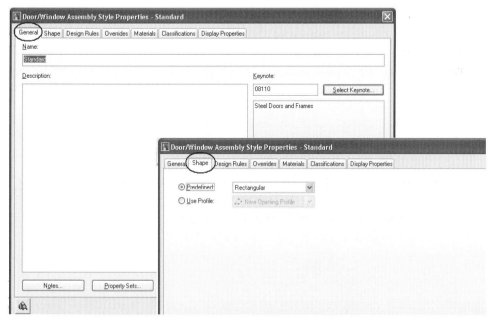

Figure 5.4 *General and Shape tabs of the Door/Window Assembly Style Properties dialog box*

gothic, isosceles triangle, right triangle, peak pentagon, octagon, hexagon, or trapezoid.

Use Profile – The **Use Profile** option allows you to select from a list of profiles. Therefore, you can create a window assembly based on any customized shape that has been defined as a profile. If there are no profiles defined in the drawing, this option is inactive.

Design Rules Tab

The majority of the components defined in the door/window assembly are specified in the **Design Rules** tab as shown in Figure 5.5. This tab includes a toolbar for creating new cell assignments, frames, mullions, or deleting components. The commands of the toolbar are shown to the left in Figure 5.5.

Overrides Tab

The **Overrides** tab of the **Door/Window Assembly Style Properties** dialog box lists any overrides that have been defined as exceptions to the style as shown in Figure 5.6. This tab includes information only if a specific style includes exceptions defined in the style.

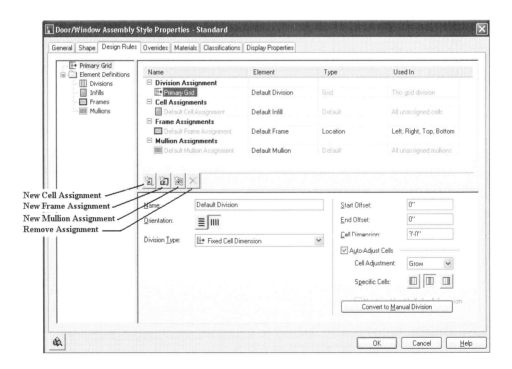

Figure 5.5 *Design Rules tab of the Door/Window Assembly Style Properties dialog box*

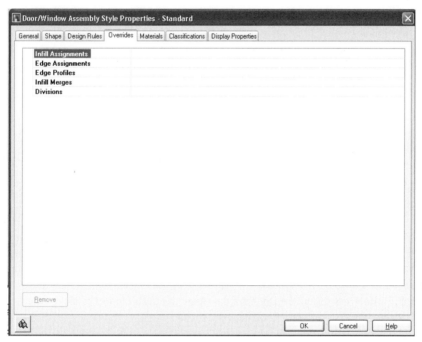

Figure 5.6 Overrides tab of the Door/Window Assembly Style Properties dialog box

Materials Tab

The **Materials** tab allows you to define the material definition for the components of a door/window assembly. The **Materials** tab shown in Figure 5.7 lists the components of the assembly in the left column and the material definition assigned to the components in the right column. The components are edited by clicking in the **Material Definition** column for the component and selecting a material definition from the drop-down list. Additional material definitions can be created by selecting the **Add New Material** button shown in Figure 5.7. Additional material definitions can be imported into the drawing from the *C:\Documents and Settings\All Users\Application Data\Autodesk\ADT 2005\enu\Styles\Imperial\Material Definitions (Imperial).dwg* in the **Style Manager**.

If you select the **Edit Material** button located in the right margin, the **Material Definition Properties** dialog box opens, which allows you to edit how the material definition is applied according to the display representation. This dialog box, shown in Figure 5.7, lists the display representations and the display property sources for the door/window assembly. An override to the material definition can be defined by selecting the **Style Override** check box. Selecting the **Edit Display Properties** button opens the **Display Properties – Material Definition** dialog box as shown in

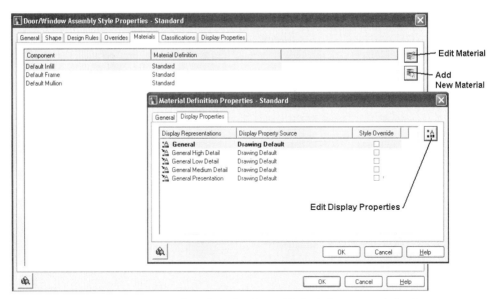

Figure 5.7 *Materials tab of the Door/Window Assembly Style Properties dialog box*

Figure 5.8, which allows you to set the properties of the components in a similar manner as discussed in "Assigning Materials to Wall Components" in Chapter 3.

Classifications Tab

The **Classifications** tab shown in Figure 5.9 allows you to assign classifications that are defined in the drawing to the style. Classifications can be defined for the status of the building such as new and existing, as discussed in the wall style definitions described in Chapter 3.

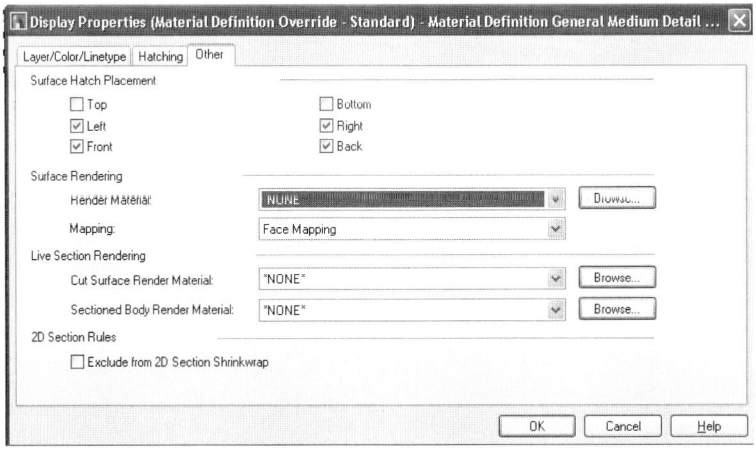

Figure 5.8 *Display Properties – Material Definition dialog box*

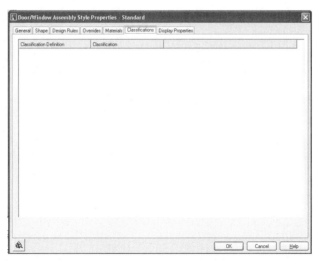

Figure 5.9 *Classifications tab of the Door/Window Assembly Style Properties dialog box*

Display Properties Tab

The **Display Properties** tab includes a list of display representations and display property sources. The check box shown in Figure 5.10 indicates if a **Style Override** has been defined for the display representation. Selecting the **Edit Display Properties** button opens the **Display Properties** dialog box for a display representation. The **Display Properties** dialog box includes the following tabs: **Layer/Color/Linetype**, **Hatch**, **Custom Plan Components**, and **Cut Plane**. The **Layer/Color/Linetype** tab allows you to specify display control of each component of the assembly. Display of each component can be controlled by material or by its layer or color as shown in Figure 5.10.

DEFINING THE COMPONENTS USING DESIGN RULES

The **Design Rules** tab consists of a tree on the left, which consists of the **Primary Grid** and its **Element Definitions**. If you select an element in the tree of the left pane, its properties will be displayed in the upper right pane and can be edited in the lower right pane. As shown in Figure 5.11, the primary grid has a vertical orientation. Selecting the **Orientation** buttons in the lower right pane will change the direction in which the grid is subdivided.

The **Design Rules** tab allows you to control the size and quantity of divisions within the primary grid. The door/window assembly unit length can vary; therefore, the vertical divisions divide the unit length and a horizontal division will divide the height. The division can be set to **Fixed Cell Dimension**, **Fixed Number of Cells**, or **Manual**. Therefore the first step in creating a unit is to select Primary Grid in the left

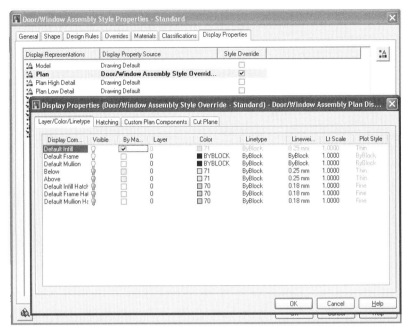

Figure 5.10 *Display Properties tab and Display Properties dialog box for selected display representation*

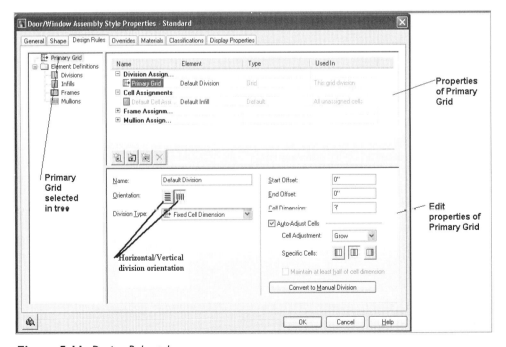

Figure 5.11 *Design Rules tab*

pane and define the orientation and division type. After specifying how the primary grid is divided you can specify the cell assignments. **Cell Assignments** can be an infill or nested grid. An infill can be a Simple Panel or a Style. The style option allows you to specify an **Aec Polygon**, **Curtain Wall Unit**, **Door Style**, or **Window Style**. If the nested grid is specified, the grid can subdivide the primary grid division. Finally, you can specify the frame and mullion dimensions.

Defining Element Definitions

Division

The orientation and division type can be specified for a division. The division can be applied to the Primary Division or nested grids.

> **Name** – A **Name** can be assigned to a division definition.
>
> **Orientation** – The **Orientation** can be set to horizontal or vertical.
>
> **Division Types** – The method of dividing the division is specified in the lower portion of the dialog box. As you select the division type, the dialog box changes allowing you to specify the parameters for one of the following division types: **Fixed Cell Dimension**, **Fixed Number of Cells**, or **Manual**.
>
> **Fixed Cell Dimension** – Allows you to specify the size of a cell, as shown in Figure 5.12. If fixed cell dimension is selected and the orientation is vertical, the total length will be divided by the fixed cell dimension. The additional unit length can be defined to automatically adjust to the length. This adjustment can be assigned to a cell and set to grow or shrink.
>
> **Fixed Number of Cells** – Divides the unit length or height into an equal number of spaces to create a cell as shown in Figure 5.13.
>
> **Manual** – Allows you to specify the location of the cell relative to the start, middle, or end of the grid as shown in Figure 5.13. Grid locations can be specified from the Start, Middle, or End for vertical divisions and from the Bottom, Middle, and Top for horizontal divisions.

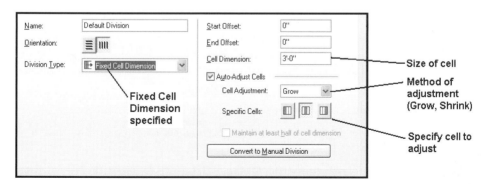

Figure 5.12 *Specifying Fixed Cell Dimension*

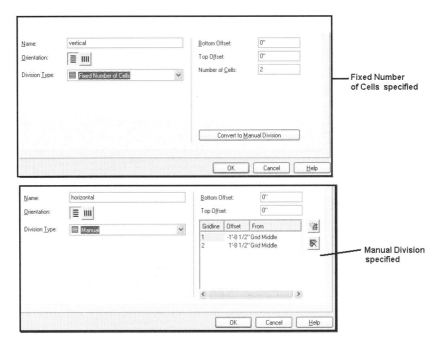

Figure 5.13 *Specifying Fixed Number of Cells and Manual division*

> **Rise** – The **Rise** option is only available for horizontal orientation. The offset dimension can be specified for the rise.

Specifying Infill
After specifying the division options, you can specify the infill. Infills are assigned to cell assignments. Cell assignments can be an infill or a nested grid. The infill can be a simple panel, Aec Polygon, curtain wall style, door style, or window style. The options for the infill are described below.

> **Name** – The **Name** edit field allows you create a name for the infill definition.
>
> **Infill Type** – The **Infill Type** options are simple panel or style. The panel and style allow you to specify the thickness of the panel and the name of the style in the lower right corner as shown in Figure 5.14.
>
> **Alignment** – The **Alignment** options are front, center, and back.
>
> **Offset** – The **Offset** option allows you to specify a distance towards the front or back to place the panel.
>
> **Default Orientation** – The **Default Orientation** allows you to flip the infill about the x and y axes.

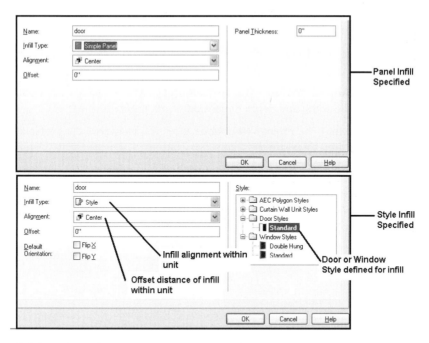

Figure 5.14 *Defining the simple panel and style infills*

A nested grid will create a grid within the primary cell divisions and can have a vertical or horizontal orientation. Cell assignments can be defined for the nested grid. Cell assignments can be specified for the start, end, middle, or default location.

Creating Custom Frames

The Frame of the window assembly style encloses the cell assignments to provide an edge around the outside dimensions of the primary grid. When you select the Default Frame assignment in the top portion as shown in Figure 5.15, the size and profile used for the frame is displayed in the lower portion of the dialog box. Finally, the size of the frame and mullion can be specified in the lower right section of the Design Rules tab.

New Frame – The New Frame button allows you to create a new frame definition.

Remove Frame – Select a frame definition in the top window and select the Remove Frame button, and the frame definition will be removed.

Name – A Name can be specified for the frame settings.

Width – The Width of the default frame is specified.

Depth – The **Depth** of the default frame is specified.

Use Profile – The **Use Profile** field is active if the drawing includes a profile. The profile can be used as an extrusion to create the frame.

Profile – The **Profile** drop-down list allows you to select the name of the profile.

Auto-Adjust Profile – The **Auto-Adjust Profile** options allow you to adjust the width and depth dimension of the profile to the frame.

Mirror In – The **Mirror In** options allow you to mirror the profile about the x and y axes to create the frame.

Rotation – The **Rotation** option will rotate the frame a specified angle.

Offsets – The **Offsets** edit field allows you to specify a distance to offset the frame in the x and y directions at the start and end of the unit.

Creating Custom Mullions

The size and shape of the mullion of the door/window assembly can be defined by selecting **Mullions** in the left pane and editing its properties in the right pane. The components of the mullion are identical to the frame as shown in Figure 5.16.

After defining the element definitions, you can assign them to components of the **Primary Grid** of the door/window assembly in the upper right pane of the dialog box.

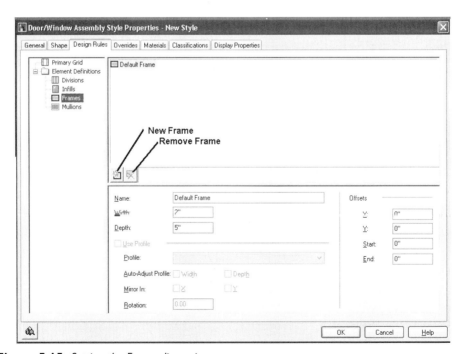

Figure 5.15 *Setting the Frame dimensions*

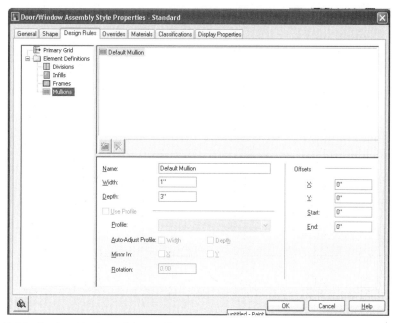

Figure 5.16 *Setting the size of the mullion*

After you select the **Primary Grid** in the left pane, selecting a component field in the upper right pane of the dialog box allows you to edit its components in the lower right pane. You can click in the **Element**, **Type**, and **Used In** columns shown in Figure 5.17 to select from the drop-down list.

In Figure 5.17, the **New Cell Assignment** field was selected to expand the lower portion of the dialog box and display the settings for the cell. A **Double Door** is assigned to the **New Cell Assignment** as shown in Figure 5.17. The **Double Door** is applied to the cell at the start of the unit. The remaining units are assigned the **sidelight**, since it is the default. The **Type** and **Used In** columns allow you to specify where the element will be used. Additional cell assignments, frame assignments, and mullion assignments can be created and deleted when you select the buttons shown in Figure 5.17 as follows:

> **New Cell Assignment** – Creates a new cell assignment definition.
>
> **New Frame Assignment** – Defines a new frame assignment.
>
> **New Mullion Assignment** – Defines a new mullion assignment.
>
> **Remove** – Select cell, frame, or mullion assignment and then choose **Remove** to delete the assignment definition.

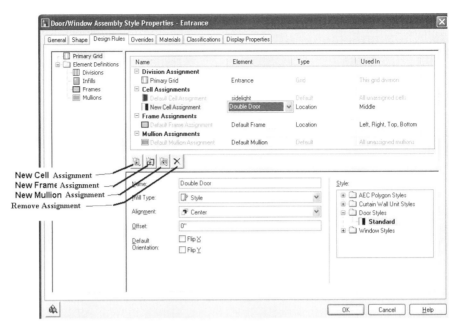

Figure 5.17 *Editing the Design Rules tab*

MODIFYING DOOR/WINDOW ASSEMBLIES

Door/window assemblies can be edited by selecting the door/window assembly and then editing the features of the Properties palette. The layer, style, length, height, rise, start miter angle, end miter angle, vertical alignment, head height, sill height, and rotation can be changed on the **Design tab** of the Properties palette. In addition, you can select and edit each component of the assembly. Therefore, if the assembly consists of two double hung windows and a picture window, you can select the picture window and edit it in the Properties palette. Additional details regarding the door/window assembly style can be changed by selecting options from the shortcut menu.

CHANGING THE DOOR/WINDOW ASSEMBLY STYLE IN THE DRAWING

The components of the door/window assemblies that have been placed in the drawing can be changed in the workspace, allowing you to see the changes, rather than editing the **Design Rules** tab and returning to the drawing to view the changes. These changes can be saved to the style or remain as an override to the style. A door/window assembly is changed by selecting the grips and editing the grips in the workspace. Commands located on the shortcut menu allow additional editing.

Editing the Grid in the Workspace

The grid definition of a door/window style can be changed in the workspace with or without editing the style. If you select a door/window assembly, the grips of the unit are displayed as shown at left in Figure 5.18. The grips shown at left allow you to change the location, height, and width of the assembly. The Edit Grid trigger grip (gray circle grip) toggles ON In Place Editing. When you select the Edit Grid trigger grip, you are prompted to select an edge to edit. Selecting the edge starts In Place Editing and the remainder of the drawing is inactive. When In Place Editing begins, additional grips are displayed on the assembly to help you edit the grid as shown at right in Figure 5.18. The **In-Place Edit** toolbar is displayed, which includes the **Zoom To**, **Discard All Changes**, and **Save All Changes** commands. The grips for editing a door/window assembly vary based upon the design rules defining the unit. The door/window assembly shown in Figure 5.18 is defined with the Fixed Cell Dimension division type.

The grips shown at right allow you to change the fixed cell dimension and to change the spacing and the offsets at the start and end. The grid can be edited by selecting the grips or selecting the Set Fixed Cell Dimension Rules grip. Selecting the Set Fixed Cell Dimension Rules grip will open the **Set Fixed Cell Dimension Rules** dialog box as shown in Figure 5.19, allowing you to specify a new set of design rules. When you select **OK** to close the **Set Fixed Cell Dimension Rules** dialog box, the In Place Edit is still active. Selecting **Save All Changes** on the **In-Place Edit** toolbar will open the **Save Changes** dialog box as shown in Figure 5.20. The changes can be saved to current division name, or you can specify a new division name. If you save the change to the current division name, all door/window assemblies that use that division name will change. If you select the **New** button and specify a new division name, the change will be saved only to the current door/window assembly, and other assemblies of the same style will not change. Selecting the **Discard** button of the **Save Changes** dia-

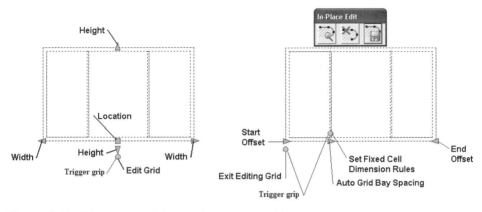

Figure 5.18 *Editing grips of the door/window assembly*

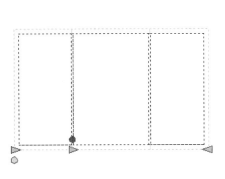
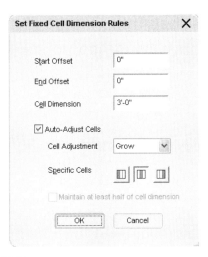

Figure 5.19 *In Place Edit of the Fixed Cell Dimension Division*

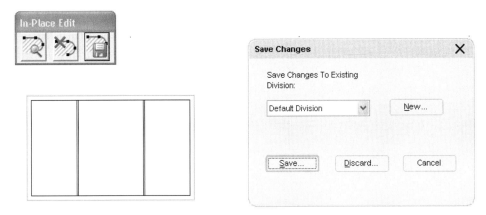

Figure 5.20 *Save Changes dialog box and ending In Place Edit*

log box ends In Place Editing and discards all changes. Selecting the **Save** button shown in Figure 5.20 will save the changes to the door/window assembly as exceptions to the style. The Exit Editing Grid trigger grip can also be selected during In Place Edit to open the **Save Changes** dialog box.

The dynamic dimensions and grips displayed for In Place Edit of assemblies created with fixed number of cells, manual division, and fixed cell dimensions are shown in Figure 5.21.

When the division definition is set to a **Fixed Number of Cells**, positive and negative signs are displayed on the grid. When you move your pointer over the positive sign,

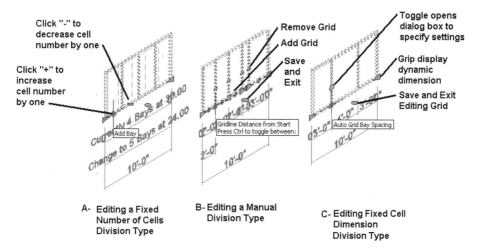

Figure 5.21 *Dynamic dimensions of division types*

dynamic dimensions will be displayed to allow you to add a bay as shown at A in Figure 5.21. If you move the pointer over the negative sign, you can click to decrease the number of bays.

The editing of a **Manual** division type will allow you to click on the arrow grip to display dynamic dimensions. You can click on a grid division to shift its position as shown at B in Figure 5.21. A plus grip is displayed at the midpoint of the cell; when this grip is selected you can add a manual location at the midpoint of the cell. The Remove manual grid grip can be selected to remove a manual location for the grid as shown at B of Figure 5.21.

When you edit a **Fixed Cell Dimension** division type, a Set Fixed Cell Dimension Rules toggle is displayed as shown at C of Figure 5.21. This toggle opens a **Set Fixed Cell Dimension Rules** dialog box, which allows you to edit the cell dimension and auto adjustment of the division.

SAVING IN PLACE EDIT CHANGES TO THE STYLE

After you save the changes with In Place Edit to a **New** division name, the changes can be saved to the style by transferring control to the object. The **Transfer to Object** (**GridAssemblyCopyFromStyle**) option can be selected before or after an In Place Edit. This option toggles control to the workspace for either editing the style definition of the selected door/window assembly or reverting a door/window assembly to its original definition. Access **GridAssemblyCopyFromStyle** as shown in Table 5.3.

Command prompt	GRIDASSEMBLYCOPYFROMSTYLE
Shortcut menu	Select a door/window assembly, right-click, and choose Design Rules>Transfer to Object

Table 5.3 *Transfer to Object (GridAssemblyCopyFromStyle) command access*

When you select a door/window assembly and choose the **Design Rules>Transfer to Object** option of the shortcut menu, you can continue to edit the grid of the door/window assembly in the workspace. When editing is complete, you can save the changes reflected in the door/window assembly back to the style by selecting **Design Rules>Save to Style** from the shortcut menu (see Table 5.4 for command access). When you select this command, the **Save Changes** dialog box opens, allowing you to save the changes to the current style or to select the **New** button to create a new style name to save the changes.

Command prompt	GRIDASSEMBLYSAVECHANGES
Shortcut menu	Select the door/window assembly, right-click, and choose Design Rules>Save to Style

Table 5.4 *Save to Style (GridAssemblySaveChanges) command access*

The changes can be discarded and the door/window assembly returned to its original definition if you select a door/window assembly and then select the Design Rules>Revert to Style Design Rules command (see Table 5.5 for command access). This command removes all edits and redisplays the door/window assembly based upon the last saved definition of the style.

Command prompt	GRIDASSEMBLYMAKESTYLEBASED
Shortcut menu	Select the door/window assembly, right-click, and choose Design Rules>Revert to Style

Table 5.5 *Revert to Style (GridAssemblyMakeStyleBased) command access*

Modifying the Infill

The infill can be edited by merging the cells or overriding the infill assignment to a cell. Prior to editing the infill, the cell markers must be turned on. Turning ON the cell markers allows you to select a cell for editing. Cell Markers are displayed in the Elevation or Model view in the center of the cell as shown in Figure 5.22. Access the

Infill>Show Markers (**GridAssemblySetEditDepthAll**) command to turn on cell markers as shown in Table 5.6.

Command prompt	GRIDASSEMBLYSETEDITDEPTHALL
Shortcut menu	Select the door/window assembly, right-click, and choose Infill>Show Markers

Table 5.6 *Show Markers (GridAssemblySetEditDepthAll) command access*

The infill of a cell can be merged with another cell; this allows you to convert a 4-division door/window assembly to a 3-division assembly as shown at right in Figure 5.22. Access the **Infill>Merge** (**GridAssemblyMergeCells**) command to merge cells as shown in Table 5.7.

Command prompt	GRIDASSEMBLYMERGECELLS
Shortcut menu	Select the door/window assembly, right-click and choose Infill>Merge

Table 5.7 *Merge (GridAssemblyMergeCells) command access*

The command requires you select the cell marker for each cell as shown in the following command sequence.

> Command: GridAssemblyMergeCells
> Select cell A: *(Select cell marker at p1 as shown at left in Figure 5.22.)*
> Select cell B: *(Select cell marker at p2 as shown at left in Figure 5.22.)*
> *(Cells merged as shown at right in Figure 5.22.)*

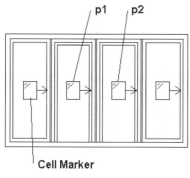
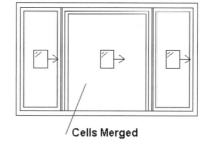

Figure 5.22 *Infill cells merged*

Overriding the Infill

Overriding Infill Assignments allows you to define a different infill for a cell. This allows you to change one cell without changing all cells defined in the style. The options for the infill override must first be defined as an infill within the style definition before you can apply it as an override. Access the **Override Assignment** (**GridAssemblyAddCellOverride**) command as shown in Table 5.8.

Command prompt	GRIDASSEMBLYADDCELLOVERRIDE
Shortcut menu	Select the door/window assembly, right-click, and choose Infill>Override Assignment

Table 5.8 *Override Assignment (GridAssemblyAddCellOverride) command access*

When the cell markers have been turned ON, you can select the door/window assembly, right-click, and choose **Infill>Override Assignment** from the shortcut menu. You are prompted to select a cell to override; select the cell marker of the cell to edit. The **Infill Assignment Override** dialog box opens, allowing you to select a different infill element. The following steps outline the procedures for creating an infill and applying the new infill as an Override Assignment.

STEPS TO CREATING AN OVERRIDE ASSIGNMENT

1. View the door/window assembly in Elevation or Pictorial view.
2. Select the door/window assembly, right-click, and choose **Infill>Show Markers**.
3. To create a new infill, select the door/window assembly, right-click, and choose **Edit Door/Window Assembly Style** from the shortcut menu.
4. Select the **Design Rules** tab of the **Door/Window Assembly Style Properties** dialog box.
5. Select **Infills** in the left pane. Move the pointer to the upper right pane, right-click, and choose **New**.
6. Verify that the new infill is selected in the upper right pane, and edit the properties of the new infill in the lower right pane. Select **OK** to close the dialog box.
7. To create the Infill Override, select the door/window assembly, right-click, and choose **Infill>Override Assignment**. Respond to the command line prompts as shown below.

    Command: GridAssemblyAddCellOverride

    Select infill to override: *(Select a cell marker as shown at p1 at left in Figure 5.23.)*

Door and Window Assemblies 359

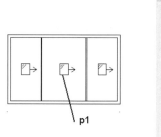

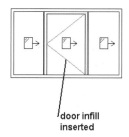

Figure 5.23 *Overriding the Infill Assignment*

Select infill to override: ENTER *(Press* ENTER *to end selection and select* **door**, *the new infill, in the drop-down list of the* **Infill Assignment Override** *dialog box as shown in Figure 5.23.)*

(Select **OK** *to dismiss the* **Infill Assignment Override** *dialog box; door infill is assigned to center cell as shown at right in Figure 5.23.)*

EDITING THE PROFILE OF THE FRAME OR MULLION

The profile of the frame or mullion can be edited in the workspace. The profile can be edited with the following three commands: **Add Profile**, **Edit Profile In Place**, and **Override Assignment**. The profile can be saved as an override to the style or an additional profile created and saved in the style. Access the **Frame/Mullion>Add Profile** (**GridAssemblyAdd ProfileOverride**) command as shown in Table 5.9.

Command prompt	GRIDASSEMBLYADDPROFILEOVERRIDE
Shortcut menu	Select the door/window assembly, right-click, and choose Frame/Mullion>Add Profile

Table 5.9 *Frame/Mullion>Add Profile (GridAssemblyAdd ProfileOverride) command access*

When you select the **GridAssemblyAdd ProfileOverride** command from the shortcut menu, you are prompted to select a frame or mullion. The **Add Frame Profile** dialog box opens if you are editing a frame, which allows you to select a profile from the drop-down list. The drop-down list includes **Start from Scratch** and other profiles that have been created in the drawing. The **Start from Scratch** option creates a new profile, numbered as shown in Figure 5.24. The profile can be applied as a shared frame element definition or as a frame profile override. The **Add Profile** command was used to change the frame as shown in the following command line sequence.

STEPS TO ADDING A PROFILE OVERRIDE

1. Select a door/window assembly, right-click, and choose **Frame/Mullion>Add Profile**.

2. Select a frame or mullion, select frame at **p1** as shown in Figure 5.24. The **Add Frame Profile** dialog box opens as shown in Figure 5.24. Specify the name of the new profile or use the **Start from scratch** default name. Select **OK** to dismiss the **Add Frame Profile** dialog box and begin in place editing.

3. Select **Zoom To** from the **In-Place Edit** toolbar. An enlarged view of the frame or mullion is displayed hatched.

4. Retain the hatched display, right-click, and choose from the following shortcut menu options: **Add Vertex**, **Remove Vertex**, **Add Ring**, **Replace Ring**, **Save Changes**, **Save As New Profile**, **Discard All Changes**, or select the grips and stretch or move the edge as shown at right in Figure 5.24.

5. Select the **Save All Changes** command from the **In-Place Edit** toolbar to end the edit.

Editing a Profile In Place

The **Edit Profile In Place** command is a shortcut menu option (see Table 5.10) that allows you to edit a profile that has been added or edited to define a frame or mullion. This command applies the tools of **Edit In Place** to edit the vertices of the modified profile.

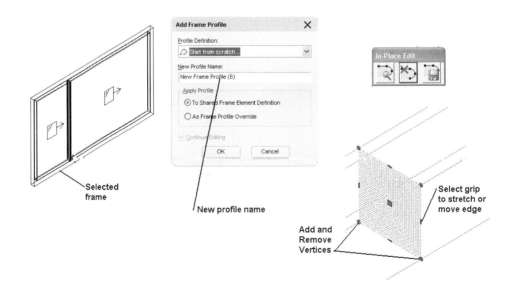

Figure 5.24 *Editing a frame profile*

Command prompt	GRIDASSEMBLYPROFILEEDIT
Shortcut menu	Select the door/window assembly, right-click, and choose Frame/Mullion>Edit Profile In Place

Table 5.10 *Frame/Mullion>Edit Profile In Place (GridAssemblyProfileEdit) command access*

Start this command by selecting a door/window assembly that includes a profile that has been previously edited. Right-click and choose **Frame/Mullion>Edit Profile In Place** from the shortcut menu. When you select a frame that has a profile, the **In-Place Edit** toolbar will be displayed as shown in Figure 5.25 at left. The **Zoom To** option will enlarge the view of the frame or mullion, allowing you to easily edit the profile as shown on right in Figure 5.25. The profile is shown hatched; select the profile to display its grips. The profile can then be modified by grip editing, or you can select **Add Vertex** or **Delete Vertex** from the shortcut menu. The procedure to edit a profile is shown in the following steps:

STEPS TO EDITING A MODIFIED PROFILE USING IN PLACE EDIT

1. Select a previously modified door/window assembly, right-click, and choose **Frame/Mullion>Edit Profile In Place**.
2. Select a frame or mullion of the grid assembly to edit. The **In-Place Edit** toolbar is displayed and grips displayed on the profile as shown at left in Figure 5.25.

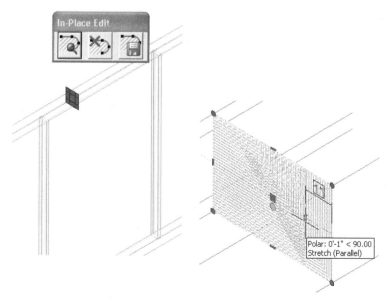

Figure 5.25 *Editing a modified frame profile*

3. Select **Zoom To** from the **In-Place Edit** toolbar.
4. Click in the workspace to end the **Zoom** command, and select the profile to display its grips as shown at right in Figure 5.25.
5. Retain the hatched display of the profile, right-click, and choose one of the following: **Add Vertex, Remove Vertex, Add Ring, Replace Ring, Save Changes, Save As New Profile, Discard All Changes**. You can select the grips and stretch the profile to a new location.
6. Select **Save All Changes** from the **In-Place Edit** toolbar.

Overriding the Frame or Mullion Assignment

The **Override Assignment** (**GridAssemblyAddEdgeOverride**) command allows you to define a different frame for the top, bottom, or sides of the door/window assembly. Access the **Override Assignment** command as shown in Table 5.11.

Command prompt	GRIDASSEMBLYADDEDGEOVERRIDE
Shortcut menu	Select a door/window assembly, right-click and choose Frame/Mullion>Override Assignment

Table 5.11 *Frame/Mullion>Override Assignment (GridAssemblyAddEdgeOverride) command access*

When you execute this command, you are prompted to select a frame, and the **Frame Assignment Override** dialog box opens, which allows you to change the frame. The **Frame Assignment Override** dialog box is shown in Figure 5.26. The drop-down list of frames is limited to frame definitions defined in the style. The **Frame Assignment Override** dialog box also allows you to delete the frame at the selected location. This option would allow you to set the frame at the top or header position wider than the other frame components. The following steps list the procedure to create a frame assignment override.

STEPS TO CREATING A FRAME OVERRIDE TO THE EDGE ASSIGNMENT

1. Select the door/window assembly, right-click, and choose **Frame/mullion>Override Assignment**.
2. Select the header frame or an edge to open the **Frame Assignment Override** dialog box.
3. Select a frame definition from the drop-down list of the **Frame Assignment Override** dialog box.
4. Select **OK** to close the **Frame Assignment Override** dialog box. Frame header is increased as shown in Figure 5.27.

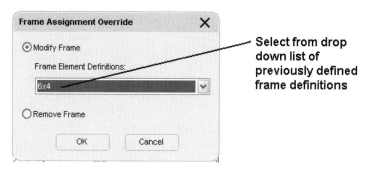

Figure 5.26 *Selecting frame element definition of the Frame Assignment Override*

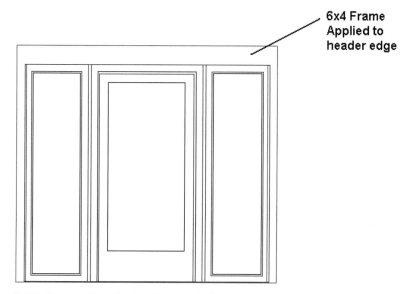

Figure 5.27 *Frame adjusted using Override Assignment*

Creating an Override of Division Assignment

The division definition of the door/window assembly can be overridden. The division definition can be changed when you select the **Division>Override Assignment** command from the shortcut menu as shown in Table 5.12.

Command prompt	GRIDASSEMBLYADDDIVISIONOVERRIDE
Shortcut menu	Select the door/window assembly, right-click, and choose Division>Override Assignment

Table 5.12 *Division>Override Assignment command access*

This allows you to change how the unit is divided without editing the style definition. In Figure 5.29 the door/window assembly on the left was edited to create four equal divisions. The division definition must exist in the style prior to selecting the Override Assignment. The division definitions are included in the drop-down list shown in Figure 5.28.

STEPS TO CREATING AN OVERRIDE OF DIVISION ASSIGNMENT

1. Select the door/window assembly, right-click, and choose **Division>Override Assignment**.
2. Select an edge of the door/window assembly.
3. Select a division definition from the drop-down list of the **Division Assignment Override** dialog box as shown in Figure 5.28.
4. Select **OK** to close the **Division Assignment Override** dialog box. Divisions of the unit are changed to four equal divisions as shown in Figure 5.29.

 STOP. Do Tutorial 5.1, "Creating and Editing Door/Window Assemblies for a Front Entrance Unit," at the end of the chapter.

Setting the Miter Angle of Door/Window Assemblies

When ends of door/window assembly units intersect, the **Set Miter Angle** command will modify the ends to create a miter angle at the start or end of the door/window

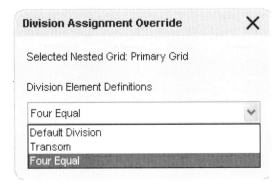

Figure 5.28 *Editing the division of the door/window assembly*

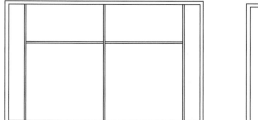
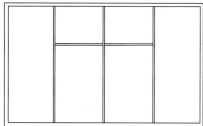

Figure 5.29 *Overriding door/window assembly division definition*

assembly. The angle of miter is created according to the angle of the intersecting door/window units. Access the **Set Miter Angle** command as shown in Table 5.13.

Command prompt	GRIDASSEMBLYSETMITERANGLES
Shortcut menu	Select the door/window assembly, right-click, and choose Set Miter Angle

Table 5.13 *Set Miter Angle (GridAssemblySetMiterAngles) command access*

STEPS TO CREATING A MITER

1. Select the door/window assembly at **p1** as shown in Figure 5.30, right-click, and choose **Set Miter Angle**.
2. Select the second grid assembly at **p2** in Figure 5.30. Units are mitered as shown in Figure 5.30.

Applying Interference Add to the Assembly

If a door/window assembly intersects a column, the assembly can be trimmed to the intersection of the Aec object with the **Interference>Add** command. Access the **Interference>Add** command as shown in the Table 5.14.

Command prompt	GRIDASSEMBLYINTERFERENCEADD
Shortcut menu	Select door/window assembly, right-click and choose Interference>Add

Table 5.14 *Interference>Add (GridAssemblyInterferenceAdd) command access*

The **Interference>Add** command allows you to trim the assembly by an Aec object. The procedure for trimming an object by the door/window assembly is shown below.

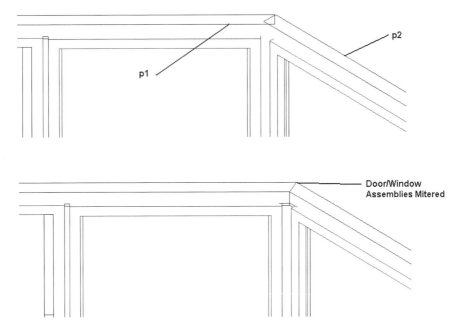

Figure 5.30 *Miter start and end of door/window assembly*

STEPS TO USING INTERFERENCE ADD TO TRIM THE ASSEMBLY

1. Select the door/window assembly at **p1** as shown in Figure 5.31, right-click, and choose **Interference>Add**.
2. Respond to the command line prompts as shown below.

    Command: GridAssemblyInterferenceAdd

    Select AEC objects to add: 1 found *(Select a column at p2.)*

    Select AEC objects to add: ENTER *(Press ENTER to end selection.)*

    Apply to infill? [Yes/No] <Yes>: ENTER *(Press enter to apply intersection to the infill.)*

    Apply to frames? [Yes/No] <Yes>: ENTER *(Press ENTER to apply intersection to the frames.)*

    Apply to mullions? [Yes/No] <Yes>: ENTER *(Press ENTER to apply intersection to the mullions.)*

    1 object(s) added as interference [1193]

    (Interference applied as shown in Figure 5.31.)

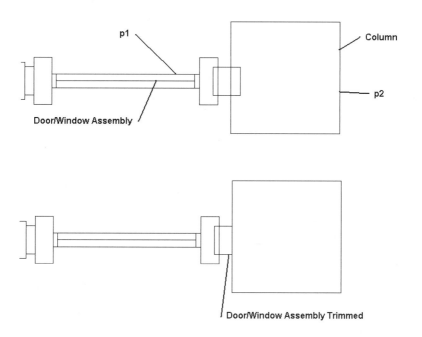

Figure 5.31 *Applying interference to the door/window assembly unit*

Removing Interference with Aec Objects

The **Interference>Remove** command on the shortcut menu will remove the interference between Aec objects and the door/window assembly. When the interference is removed, the assembly will overlap the column as shown in Figure 5.31. Access the **Interference>Remove** command as shown in Table 5.15.

Command prompt	GRIDASSEMBLYINTERFERENCEREMOVE
Shortcut menu	Select the door/window assembly, right-click, and choose Interference>Remove

Table 5.15 *Interference>Remove (GridAssemblyInterferenceRemove) command access*

SUMMARY

1. Door/window assemblies are created based upon a grid.
2. The primary grid can be horizontal or vertical and defined as a fixed cell dimension, fixed number of cells, or manual.
3. Components of each grid intersection can be a simple panel, style, or nested grid.
4. A nested grid defined for a cell can create additional divisions of a cell.
5. Frame and mullions can be created from profiles.
6. Display of cell markers allows you to select cells using the **GridAssemblyMergeCells** command.
7. Profiles used for frames and mullions can be edited in place.
8. The divisions, cells, frames, and mullions assignments can be overridden for a style.
9. The **GridAssemblyAddInterference** command allows you to trim the assembly by an Aec object such as a column.

REVIEW QUESTIONS

1. The _____ command is used to insert door/window assemblies in the drawing.
2. Door/window assembly cell assignment can be _____, _____, or _____ _____.
3. The orientation of the primary grid can be _____ or _____.
4. A grid that has a _____ _____ _____ allows you to specify the size of a cell.
5. The _____ _____ _____ will divide the unit length or height into a number of equal divisions.
6. When a grid is defined to a fixed cell dimension, the Auto-Adjust toggle can be used specify a cell to _____ or _____.
7. The Cell _____ must be turned ON to select the cell for an override.
8. Profiles are created from _____ and can be used to develop frames and mullions.
9. Changes created during In Place Edit of a profile must be _____ or _____ to exit the In Place Edit.
10. If a Design Rule is transferred to the object, the _____ can be edited in the workspace and saved to the style.

TUTORIAL 5.1 CREATING AND EDITING DOOR/WINDOW ASSEMBLIES FOR A FRONT ENTRANCE UNIT

1. Open **Ex 5-1** from the your *ADT Student\ADT Tutor\Ch5* directory.
2. Choose **File>SaveAs** from the menu bar and save the drawing as **Lab 5-1** in your student directory.
3. Choose **Format>Style Manager** from the menu bar.
4. Expand the Architectural Objects folder and then expand the Door/Window Assembly Styles folder.
5. Select Door/Window Assembly Styles in the left pane, right-click, and choose **New** from the shortcut menu.
6. Overtype **Entrance** as the name for the new style in the lower right pane.
7. Select **Entrance**, right-click, and choose **Edit** from the shortcut menu.
8. Select the **General** tab, and type **Entrance door with side lights** in the **Description** field.
9. Select the **Select Keynote** button, and verify that the keynote database is *C:\Documents and Settings\All Users\Application Data\Autodesk\ADT 2005\enu\Details\Details (US)\AecKeynotes (US).mdb*.
10. Expand Division 08-Doors and Windows to 08400 Entrances and Storefronts\08410 Metal Framed Storefronts.
11. Select the **Design Rules** tab, select **Divisions** in the left pane, edit the lower right pane as follows: Name = **Entrance**, Orientation = **Vertical**, Division Type = **Manual**, Start Offset = **0**, End Offset = **0**, select **Add Gridline** and create the following grid lines as shown in Table 5T.1 and in Figure 5T.1.

Gridline	Offset	From
1	−1'-8 1/2"	Grid Middle
2	1'-8 1/2"	Grid Middle

Table 5T.1 *New gridline settings*

12. Select **Infills** in the left pane, and type **Sidelight** in the name edit field of the lower right pane.
13. Edit the **Infill Type** in the lower right pane to **Style**, expand the Window Style folder in the style window, and select **Standard** as shown in Figure 5T.2.

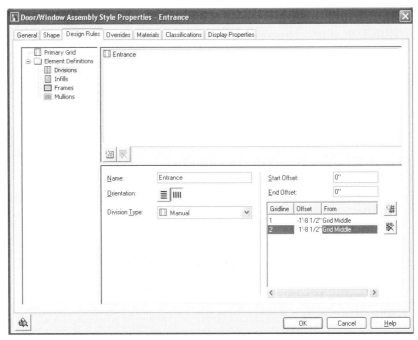

Figure 5T.1 *Specifying the Divisions in the Design Rules*

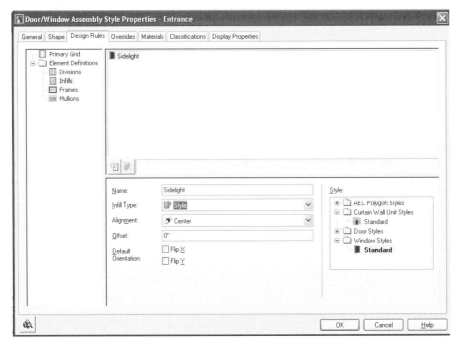

Figure 5T.2 *Specifying the window style for the infill*

14. Edit the Alignment to **Center**.
15. Right-click over the upper right pane and choose **New**. Select the **New Style** in the upper right pane and type **Door** in the name edit field of the lower right pane.
16. Edit the **Infill Type** in the lower right pane to **Style**, expand the Door Styles in the Style window, and select **Standard**.
17. Edit the Alignment to **Center**.
18. Select **Frames** in the left pane, and edit the frame dimension in the lower right pane as follows: Name = **Default Frame**, Width = **2"**, Depth = **4 9/16"**, Offset X = **0**, Y = **0**, Start = **0**, and End = **0** as shown in Figure 5T.3.
19. Select **Mullions** in the left pane, and edit the mullion dimension in the lower right pane as follows: Name = **Default Mullion**, Width = **1"**, Depth = **3"**, Offset X = **0**, Y = **0**, Start = **0**, and End = **0**.
20. Select **Primary Grid** in the left pane, select **Cell Assignments** in the upper right pane, and choose the **New Cell Assignment** button in the right pane.
21. Edit the New Cell Assignment, click in the **Element** column and select **Door** from the drop-down list. Click in the **Used In** column, edit the Cell Location Assignment to **Middle**, and clear the check boxes for **start** and **end** as shown in Figure 5T.4. Select **OK** to close the **Cell Location Assignment** dialog box.
22. Select **OK** to close the **Door/Window Assembly Style Properties** dialog box. Select **OK** to close the **Style Manager**.
23. Choose **Door/Window Assembly** from the Design palette.

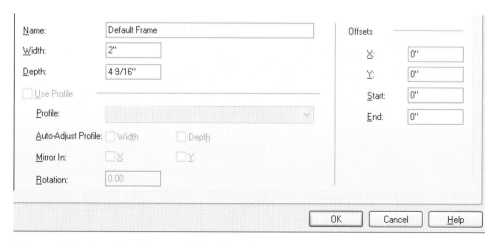

Figure 5T.3 *Specifying the size of the default frame*

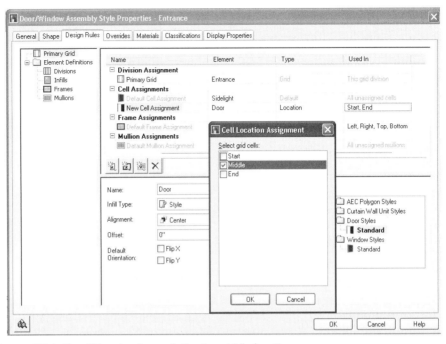

Figure 5T.4 *Specifying the door unit for the middle location*

24. Edit the Properties palette as follows: Style = Entrance, A-Length = 7'-6", B-Height = 6'-8", Rise = 1'-0", Start Miter = 0, End Miter = 0, Position along wall = Offset/Center, Automatic offset = 6", Vertical Alignment = Head, and Head Height = 6'-8".

25. Specify the Insert point as shown in the following command line sequence.

    Command: DoorWinAssemblyAdd

    Select wall or RETURN:

    Insert point or [STyle/LEngth/HEight/HEAd height/SIll Height/Auto/Match/Cycle

    measure to/REference point on]: *(Select a point near the middle of the wall.)*

    Insert point or [STyle/LEngth/HEight/HEAd height/SIll

    Height/Auto/Match/Undo/Cycle measure to/REference point on]: ENTER *(Press ENTER to end the command.)*

26. Select **SW Isometric** from the **View** flyout of the **Navigation** toolbar.

27. Select the left window unit as shown in Figure 5T.5, right-click, and choose **Edit Window Style**.

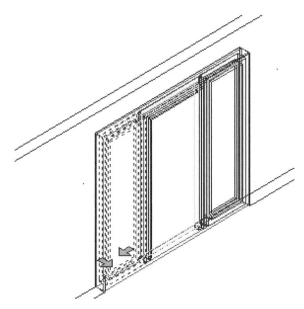

Figure 5T.5 *Selecting the window unit of the door/window assembly*

28. Select the **Display Properties** tab, and select the **Edit Display Properties** button. Select the **Muntins** tab, and select the **Add** button to open the **Muntins Block** dialog box as shown in Figure 5T.6. In the **Window Pane** section click the Other button, and click **All**. In the **Muntin** section, set Width = **3/4**, Depth = **1/2**. In the **Lights** section, set the Pattern = **Rectangular**, Lights High = **6**, and Lights Wide = **2**.

29. Select **OK** to dismiss the **Muntins Block** dialog box.

30. Check **Automatically Apply to Other Display Representations**. Select **OK** to dismiss the **Display Properties – Window Model Display Representation** dialog box. Select **OK** to dismiss the **Window Style Properties** dialog box.

31. Select the door of the door/window assembly, right-click, and choose **Edit Door Style**.

32. Select the **Dimensions** tab, and set the Frame- A-Width = **2"**, B-Depth = **4 9/16"**, Stop-C-Stop Width = **2"**, D-Depth = **1 1/2"**, E-Door Thickness = **1 3/4"**, and Glass Thickness = **1/4"**. Select **OK** to dismiss the **Door Style Properties** dialog box.

33. Select **Front View** from the **View** flyout of the **Navigation** toolbar.

34. Select the frame of the door/window assembly, right-click, and choose **Edit Door/Window Assembly Style**.

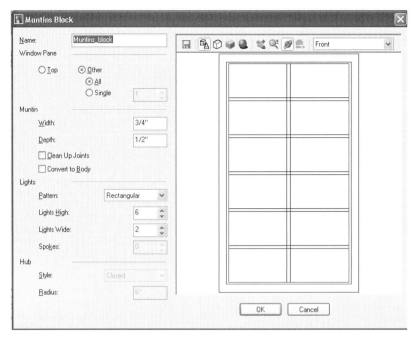

Figure 5T.6 *Editing the Muntins Block dialog box*

35. Select the **Design Rules** tab, select **Frames** in the left pane, move the pointer over the upper right pane, right-click, and choose **New** from the shortcut menu. Edit the lower right pane as follows: Name = **Threshold**, Width = **5/8**, Depth = **6**, check **Use Profile**, select the **Threshold** profile from the drop-down list (this profile was created in the current file from the geometry at right prior to your starting this tutorial), and for Auto-Adjust Profile, check **Width** and **Depth**. Select **OK** to dismiss the **Door/Window Assembly Style Properties** dialog box.

36. Select the frame of the door/window assembly, right-click, and choose **Frame/Mullion>Override Assignment** from the shortcut menu. Respond to the command line prompts as shown below.

    Command: GridAssemblyAddEdgeOverride

    Select an edge: *(Select the sill of the frame.)*

37. Edit the **Frame Assignment Override** dialog box: select the **Modify Frame** radio button and select **Threshold** from the drop-down list as shown in Figure 5T.7. Select **OK** to dismiss the **Frame Assignment Override** dialog box.

38. Select **SW Isometric** from the **View** flyout of the **Navigation** toolbar (see Figure 5T.8).

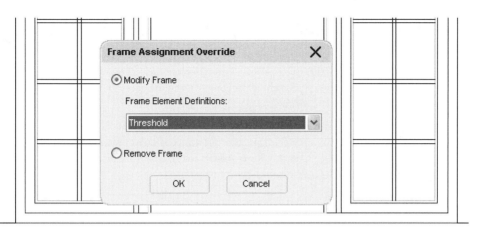

Figure 5T.7 *Overriding the frame assignment*

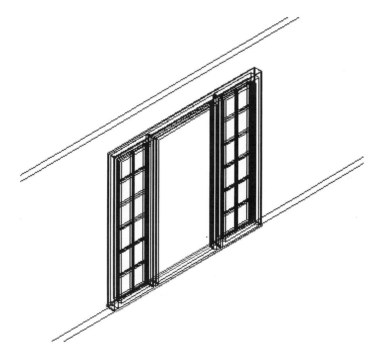

Figure 5T.8 *Front door entrance created as a door/window assembly*

39. Save and close the drawing.

PROJECT

Ex 5-2 Creating a Door/Window Assembly

1. Open **Ex 5-2** from the *ADT Student\ADT Tutor\Ch5* directory. Save the drawing as **Lab 5-2** in your student directory. Insert a double hung window without attaching it to the wall.

2. Create a new door/window assembly style named **DH-Fixed** assembly that has a vertical primary grid. Set the cell to fixed cell dimension and the cell dimension to 24". Use the Auto-Adjust feature in the Design Rules to grow the middle cell. Create two new Infills:

Name	Infill Type	Alignment	Style
DH	Style	Center	Double Hung window style
Fixed	Style	Center	Standard window style

Table 5.16 *Infill type and style*

3. Set the default cell assignment to the **DH** infill. The **Fixed** infill should not be used in any location.

4. Insert a door/window assembly using the **DH-Fixed** style, A-Length=8', B-Height=5' with the Position along wall set to Offset/Center constrained. Place the assembly in the middle of the wall. View the wall in the SW Isometric view.

5. Turn on cell markers and merge the inner two cells. Create an override of the cell infill for the inner cells, and set the inner cell to the **Fixed** cell infill as shown in Figure 5T.9.

6. Close and save the drawing.

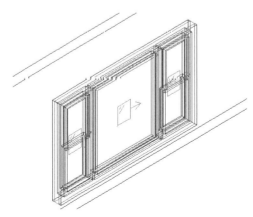

Figure 5T.9 *DH-Fixed door/window assembly style*

CHAPTER 6

Creating Roofs and Roof Slabs

INTRODUCTION

Simple or complex roofs can be placed on a building through the **RoofAdd** command. The roof can be applied to the walls of the building or created from a polyline. This chapter includes tutorials that develop the following roofs: gable, gambrel, hip, and dormer. Additional detail and flexibility in editing the roof is obtained by converting the roof to roof slabs. The use of roof slabs to develop overhang features will be presented. A roof slab can be extended, trimmed, or mitered to create a complex roof. The commands for creating roofs and roof slabs are included on the Design tool palette.

OBJECTIVES

After completing this chapter, you will be able to

- Use **RoofAdd** to create hip, gable, shed, and gambrel roofs
- Edit roof plate height, roof slope, thickness, overhang, and fascia angle using the Properties palette
- Use grips to change a hip roof to a gable roof
- Use a double sloped roof to create a gambrel roof
- Convert walls and linework to a roof
- Convert roofs, walls, and linework to roof slabs
- Create roof slabs and dormer roofs that intersect the main roof
- Edit roof slabs to create complex roofs
- Extend walls to the roof line

CREATING A ROOF WITH ROOFADD

The commands for creating and editing roofs are located on the Design tool palette as shown in Figure 6.1. The **Roof tool** (RoofAdd) is used to place a roof. New and existing roofs are changed by editing the Properties palette. Existing roofs are edited by their grips and by choosing commands from the shortcut menu of the selected roof. Existing walls, lines, and arcs can be converted to a roof by the **Apply Tool Properties to** shortcut option of the **RoofAdd** command located on the Design tool palette.

Use the **Roof** tool (**RoofAdd** command) to create roofs by selecting the walls at the corners of the roof. See Table 6.1 for command access.

Command prompt	ROOFADD
Palette	Select Roof from the Design palette

Table 6.1 *RoofAdd command access*

Selecting the **Roof** tool from the Design tool palette will open the Properties palette as shown in Figure 6.2. The roof is created by editing the values in the Properties palette and selecting points **p1, p2, p3, p4,** and **p5**. The completed roof is shown in the lower left corner of Figure 6.2.

The features specified in the Properties palette are described below.

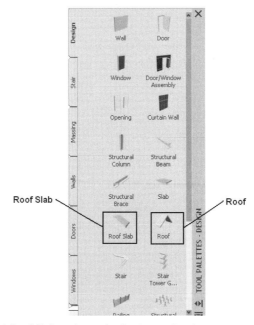

Figure 6.1 *Roof and Roof Slab tools on the Design tool palette*

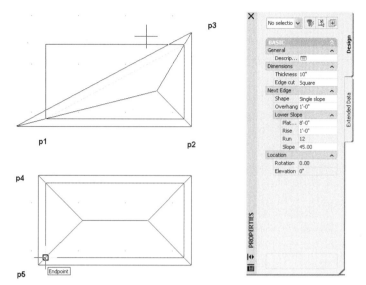

Figure 6.2 *Properties palette for creating a roof*

Design–Basic

General

    Description – Select the **Description** button to open the **Description** dialog box to type a description.

Dimensions

    Thickness – The **Thickness** edit field is the distance between the faces of the roof. The default value is 10". Roof thickness should be set according to the dimensions of the rafters.

    Edge cut – The options for the **Edge cut** are **Square** and **Plumb**. The **Square** option creates a fascia perpendicular to the face of the roof. The **Plumb** option creates a vertical fascia.

Next Edge

    Shape – There are three **Shape** options: **Single slope**, **Double slope**, and **Gable** (see Figure 6.3). A single slope roof extends a roof plane at an angle from the **Plate Height**. A double slope roof includes a single slope and adds another slope, which begins at the intersection of the first slope and the height specified for the second slope. The *gable* roof is a vertical roof component. Select the Shape option in the command line by typing **S**.

    Overhang – The **Overhang** is the horizontal distance the roof plane is extended down from the plate height at the angle of the roof to create the overhang. Type **V** to select the overhangValue option in the command line.

Lower Slope – The **Lower Slope** is the angle of the roof plane from horizontal. Slope angles can be entered directly or derived from the rise value. If slope angles are entered, the equivalent rise value is displayed in the **Rise** edit field. Type **PS** to select the PSlope option in the command line.

> **Plate height** – The **Plate Height** edit field allows you to specify the height of the top plate from which the roof plane is projected. The height is relative to the XY plane, which has a Z coordinate of zero. The plate height can vary for one or more planes of a roof. Type **PH** in the command line to select the PHeight option and set the plate height.
>
> **Rise** – The **Rise** is the slope value entered to obtain the slope angle of a single slope roof. The rise to run ratio value specifies the angle of the roof based upon a run value of 12. As shown in Figure 6.2, a rise of 12 creates a 12/12 roof, which forms a slope angle of 45 degrees. The angle formed is the arc tangent of the rise value divided by 12. Type **PR** to select the PRise option in the command line.
>
> **Run** – The **Run** is the base value of 12.

Upper Slope – The upper slope fields are displayed only for the double slope shaped roofs.

> **Upper height** – The **Upper Height** edit field is the height defined to begin the second slope of a double slope roof. If **Shape** is toggled to **Double slope**, the **Upper Height** and upper **Rise** edit fields are displayed in the Properties palette. The second slope begins at the **Upper Height** distance defined from the XY plane. Type **UH** to select the UHeight option in the command line.
>
> **Rise** – The **Rise** value determines the slope angle of the second slope. The **Rise** value for the upper slope is based upon a run of 12. Type **UR** to select the URise option in the command line.
>
> **Run** – The **Run** has a base value of 12.
>
> **Slope** – The **Slope** is the angle of the roof plane from horizontal. Slope angles can be entered directly or derived from the rise value. If slope angles are entered, the equivalent rise value is displayed in the **Rise** edit field. Type **US** to select the USlope option in the command line.

Location

> **Rotation** – The **Rotation** field displays the rotation angle of the roof.
>
> **Elevation** – The **Elevation** field displays the elevation of the roof relative to insert points.

The Properties palette consists of edit fields to define the shape, overhang, and slope of the roof. The creation of the roof is based upon a specified slope for the roof and the height of the plate. Some of the options of the Properties palette are display only.

The options of the command can be selected from the Properties palette or the command line. The options listed in the command line are as follows:

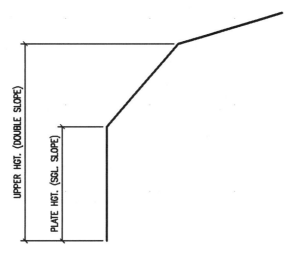

Figure 6.3 Plate heights for single and double slopes

    Command: RoofAdd

    Roof point or

    [Shape/Gable/Overhang/overhangValue/PHeight/PRise/PSlope/ UHeight/URise/USlope/Ma

    tch]: m ENTER *(Type **M** to select match option.)*

    Select a roof to match: *(Select a roof to match the properties.)*

    Match [Shape/Overhang/PHeight/PRise/PSlope/UHeight/URise/ USlope] <All>: ENTER *(Press ENTER to match all properties.)*

    Roof point or

    [Shape/Gable/Overhang/overhangValue/PHeight/PRise/PSlope/ UHeight/URise/USlope/Match]: *(Select points for the new roof with matching properties.)*

To select the options, type in the command line the letters capitalized in each of the options. The Match option was selected in the command line sequence shown above. Editing the Properties palette to select an option is a faster method of defining roof properties.

When the **RoofAdd** command is executed, the Properties palette opens, and you are prompted to select roof points for each corner of the roof. The points selected for the roof edge are located at the bottom of the wall with a Z coordinate of zero. Selecting the points using the **Intersection** object snap will locate the edge of the roof plane on either the inner or outer wall surface of a corner. The **Node** object snap locates the roof plane on the justification line of the wall. Selecting the outer wall surface for the roof

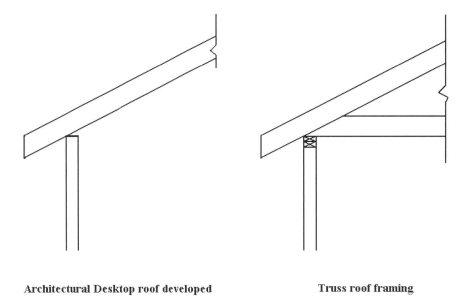

Architectural Desktop roof developed   **Truss roof framing**

Figure 6.4 *Roof developed by the selection of outer lines of the wall to define the roof*

points will generate a roof construction similar to the truss roof construction as shown in Figure 6.4. Note that the lower plane of the roof intersects the outer wall surface.

If the inner wall surface is selected to determine the roof plane, the construction is similar to that of conventional framed rafters. Using the **Intersection** object snap mode to select the inner roof plane will result in the lower plane of the roof being generated from the top of the inner wall, as shown in Figure 6.5.

As points are selected, the roof is dynamically constructed. If errors are made in selecting points, right-click and choose **Undo** from the shortcut menu or type **U** in the command line. Roof planes will be created at the slope angle specified in the Properties palette from the edge specified by each pair of points. Because a sloped roof plane is generated from each edge, a hip roof is created by default when the **Single slope** shape is specified.

The plate height for a roof should be set equal to the wall height. The plate height is measured relative to the XY plane with a zero Z coordinate. The plate height and slope can be changed as you select vertices of the roof.

 Note: You can select **Zoom Window** from the **Navigation** toolbar to zoom a window around desired roof points while creating a roof. **Zoom** on the **Navigation** toolbar is a transparent command and does not cancel the active command. Selecting transparent **Zoom** and **Pan** commands from the **Navigation** toolbar while creating a roof can increase precision in selecting roof points.

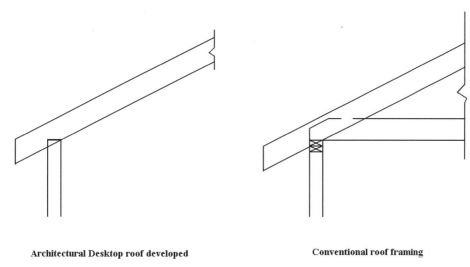

Figure 6.5 *Roof developed by the selection of inner lines of the walls to define the roof*

CREATING A HIP ROOF

A hip roof is created by default when the **Single slope Shape** option is selected. To create a hip roof, set the **Shape** to **Single slope** and set the **Rise** for the roof planes. Selecting the corners of the building identifies each edge of the roof. The slope of each edge can be set independently of other roof edges. The slope of an edge should be defined prior to selecting the second point that defines that edge of the roof. The roof is created dynamically as the corners of the building are selected; therefore the **Undo** button can be selected to deselect a corner of the roof as it is developed.

 STOP. Do Tutorial 6.1, "Creating a Hip Roof," at the end of the chapter.

DEFINING ROOF PROPERTIES

The properties of a roof can be set prior to the creation of the roof or after the roof has been developed. Selecting a roof opens the Properties palette, which allows you to edit the roof properties. Changes made in the Properties palette are reflected immediately in the roof. Additional edit fields are displayed in the Properties palette as shown in Figure 6.6. The additional fields of the **Design** tab are **Edges/Faces** and **Additional Information**, as described below. The **Extended Data** tab includes the **Hyperlink**, **Notes**, and **Reference Documents** buttons.

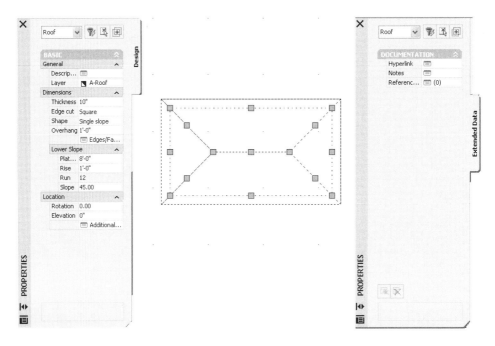

Figure 6.6 *Properties palette of an existing roof*

Design Tab

    Edges/Faces – When you select the **Edges/Faces** button in the Properties palette, the **Roof Edges and Faces** dialog box opens, listing the edges and faces of the roof. The **Roof Edges and Faces** dialog box shown in Figure 6.7 consists of **Roof Edges** and **Roof Faces (by Edge)** sections.

The **Roof Edges** list box includes the roof planes that have been created by edge. The properties of each roof edge are specified according to edge number. Edge number 0 is the first edge, specified by selecting the first two points to create the first roof segment. The remaining edges are listed in ascending order in the same sequence as they were created. The properties of each edge are described in columns of the list box as follows:

    Edge – The number listed in this column identifies the edge number of the roof plane.

    (A) Height – The **Height** is the plate height or the distance from the XY plane that the roof plane begins. The roof is generated at the specified slope from this point.

    (B) Overhang – The **Overhang** is the horizontal distance the edge projects from the wall line. The roof extends at the specified slope angle to create the

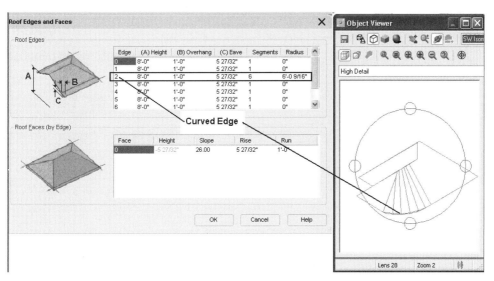

Figure 6.7 The Roof Edges and Faces dialog box for a roof with a curved edge

overhang equal to the horizontal overhang distance. The overhang creates the soffit component of the roof.

(C) Eave – The **Eave** is the vertical distance that the overhang projects down from the top plate over the distance of the overhang. Increasing the overhang and roof slope will cause the Eave dimension to increase.

Segments – The **Segment** number is the number of divisions created to cover a curved roof edge.

Radius – The **Radius** is the curve of the roof generated to fit a curved wall.

In Figure 6.7, Edge 2 covers the curved portion of the wall and consists of six segments.

The **Roof Faces (by Edge)** section of the dialog box includes a list of faces created for each edge. If you select an edge from the top portion of the dialog box, the properties of the face for that edge will be displayed in the **Roof Faces (by Edge)** list box. The height, slope, rise, and run of each face are described. The slope of the face can be edited; however, the remaining dimensions of a face cannot be edited.

Face – The **Face** column identifies the number of the face for the roof edge. Double slope roofs include two faces.

Height – The **Height** is the vertical distance the overhang projects down from the top plate.

Slope – The **Slope** is the angle of the roof measured in degrees. The slope value can be edited to specify the angle of the roof.

Rise – The **Rise** angle of the roof as the vertical change in the roof over a horizontal distance of one foot.

Run – The **Run** is the horizontal distance of the roof used to define the slope based upon the specified rise.

The elevation view of the roof face and roof edge are shown in Figure 6.8. The **(C) Eave** equals 5 27/32, which is the vertical distance the roof projects down due to the overhang distance. Because the roof is single slope, there is only one face in the **Roof Faces (by Edge)** list.

Additional Information – The **Additional Information** button on the **Design** tab opens the **Location** dialog box. The **Location** dialog box displays the **Insertion Point**, **Normal**, and **Rotation** of the roof relative to the World Coordinate System.

Extended Data Tab

The **Extended Data** tab includes the following buttons when an existing roof is selected.

Hyperlink – The **Hyperlink** button opens the **Insert Hyperlink** dialog box, which allows you to link the roof to a file.

Notes – The **Notes** button opens a **Notes** dialog box for typing notes regarding the roof.

Reference Documents – The **Reference Documents** button allows you to list the files and directory that may contain related information.

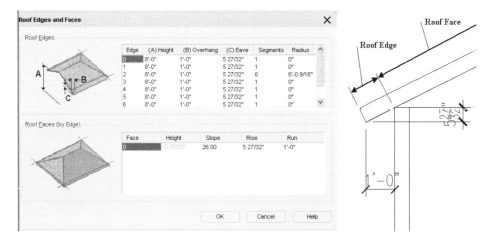

Figure 6.8 *Roof elevation created by editing the Roof Properties palette*

CREATING A GABLE ROOF

A gable roof is created in a procedure similar to that for creating the hip roof. Define the gable ends of the roof by selecting **Gable** in the Properties palette when the vertical edges of the roof plane are defined. When a gable roof edge is created, the overhang on that edge can be changed in the Properties palette. The steps to creating a gable roof are described below and in Figure 6.9.

STEPS TO CREATING A GABLE ROOF

1. Select the **Roof** tool from the Design palette.
2. Edit the Properties palette: set Shape = **Single slope**, Overhang = **16"**, and select point **p1** as shown at left in Figure 6.9.
3. Retain settings in the Properties palette, and select point p2.
4. Edit the Properties palette: set Shape = **Gable**, Overhang = **3"**, and select point **p3**.
5. Edit the Properties palette: set Shape = **Single slope**, Overhang = **16"**, and select point **p4**.
6. Edit the Properties palette: set Shape = **Gable**, Overhang = **3"**, and select point **p5**.

 STOP. Do Tutorial 6.2, "Creating a Gable Roof," at the end of the chapter.

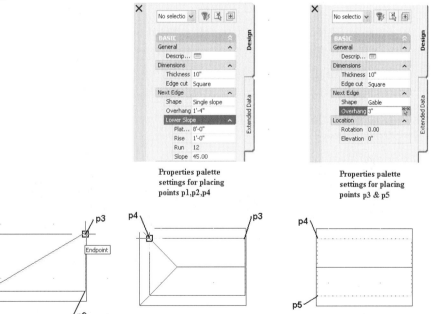

Figure 6.9 *Selection of points to create a gable roof*

EDITING AN EXISTING ROOF TO CREATE GABLES

It is often easier to create roof gables by developing a roof for the entire structure and then editing selected roof planes to create the gable. A roof for an entire structure can be developed with a single slope. The single slope roof creates a hip roof. The **Roof Edges and Faces** dialog box can be used to edit the roof planes. If you select a roof and then select the **Edges/Faces** button of the Properties palette, all roof edges and faces will be displayed in the **Roof Edges and Faces** dialog box. You could select the edge for edit in this dialog box; however, it is difficult to identify which edge of the roof you are editing. The roof edges are listed in the **Roof Edges and Faces** dialog box in the same order as the edges were created when the roof was developed. This approach is appropriate if you need to edit a feature for all edges.

USING EDIT EDGES/FACES TO EDIT ROOF PLANES

Each plane of a roof can be edited with the **Edit Edges/Faces** command (**RoofEditEdges**). This command will allow you to select a plane of the roof and edit its properties. See Table 6.2 for command access.

Command prompt	ROOFEDITEDGES
Shortcut menu	Select the roof, right-click, and choose Edit Edges/Faces

Table 6.2 *Edit Edges/Faces command access*

When the **Edit Edges/Faces** command (**RoofEditEdges**) is selected from the shortcut menu, as shown in Figure 6.10, you are prompted to select an edge as shown below.

    Command: RoofEditEdges

    Select a roofedge:*(Select a roof edge and press* ENTER*)*

Selecting an edge opens the **Roof Edges and Faces** dialog box. In contrast to selecting the **Edges/Faces** button in the Properties palette, this command limits the display to only information regarding the plane selected. The **Edit Edges/Faces** command allows you to select one or more roof edges for edit, and change the edge in the **Roof Edges and Faces** dialog box. When the slope is changed, the adjoining roof planes extend or adjust to the different slope. The steps to editing an edge to create a gable are described below.

STEPS TO EDITING ROOF EDGES AND FACES

1. Select the roof, right-click, and choose **Edit Edges/Faces** from the shortcut menu as shown in Figure 6.10. Respond to the command line prompts as shown below.

    Command: RoofEditEdges

Select a roofedge: *(Select roof edge at p1 in Figure 6.10.)*

Select a roofedge: ENTER *(Press* ENTER *to end selection.)*

2. Edit the **Roof Edges and Faces** dialog box: click in the **Overhang** column of the top **Roof Edges** section, and overtype the overhang value to **3**. Then click in the **Roof Faces (by Edge)** section, click in the **Slope** column, and overtype the slope value to **90** as shown in Figure 6.11. Click **OK** to close the **Roof Edges and Faces** dialog box. Slope of roof has changed as shown at right in Figure 6.11.

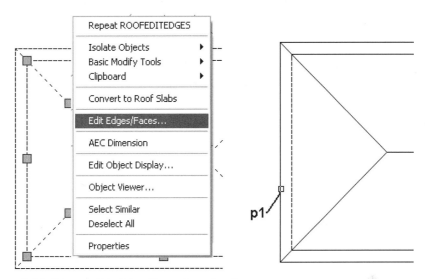

Figure 6.10 *Selecting Edit Edges/Faces from the shortcut menu*

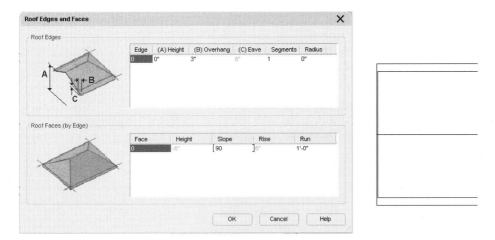

Figure 6.11 *Editing the edge in the Roof Edges and Faces dialog box*

If more than one edge is selected, the edges will be listed in the **Roof Edges** section of the **Roof Edges and Faces** dialog box. To select more than one edge, hold down CTRL to add to the selection as shown in Figure 6.12. After selecting multiple edges, click in the column for overhang or slope, and the change will be applied to all selected edges.

USING GRIPS TO EDIT A ROOF

The roof planes that are created have grips on each edge of the roof plane. Selecting a roof will display the grips. Grips are located on the midpoint and endpoints of the roof edge at the point of intersection between the roof and the wall at the plate height. Grips are also located at the top plane of the roof where each plane intersects with the adjoining roof plane. Each grip can be selected and stretched to a new position. The grips of the ridge can be stretched to create a gable roof. Editing the grips on the ridge of a roof is a quick method of changing a plane from hip to a gable.

STEPS TO CREATING A GABLE USING GRIPS

1. Select a roof to display its grips.
2. Select the ridge grip at **p1** as shown in A of Figure 6.13.
3. Stretch the hot grip to the left; specify the new location by selecting a point to the left near **p2** as shown in B of Figure 6.13.

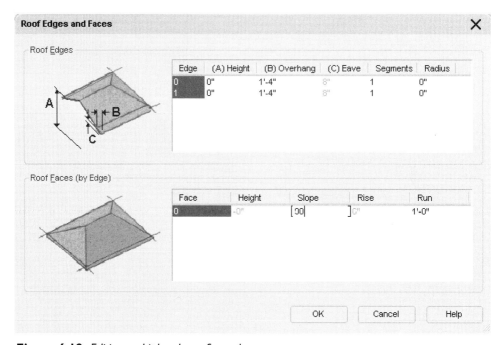

Figure 6.12 *Editing multiple edges of an edge*

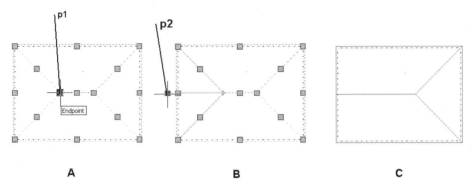

Figure 6.13 Editing the grip on the ridge of the roof

4. Press ESC to remove the display of grips. The hip roof plane is converted to a gable as shown in C of Figure 6.13.

Grips allow each plane to be stretched to a new position, and the adjoining planes adjust to the new position. Each ridge grip can be stretched to any point near or outside the building to convert the hip roof to a gable.

CREATING A SHED ROOF

Create a shed roof by selecting the **Gable** shape in the Properties palette for three of the four sides of a roof. The slope of the non-gable side of the roof creates the shed. The **Gable** shape cannot be selected for more than three sides. Each time it is used, the roof for that edge of the building has a vertical slope. Shed roofs are often used in combination with other roof types.

CREATING A GAMBREL ROOF USING DOUBLE SLOPED ROOF

A double sloped roof allows you to create a gambrel roof, which consists of two roof planes, with the first beginning at the top plate height. This first roof plane shown in Figure 6.14 is projected from the top plate at the lower slope angle of 30/12 until it intersects with the height of the upper slope. The second roof plane shown in Figure 6.14 begins at 14' from the intersection with the lower slope and continues to the ridge at a slope of 6/12. Walls are not necessary for specifying the position of the upper slope. The gambrel roof is created when you use the **Double slope** shape for two sloped edges and the **Gable** shape to create the gable ends of the building.

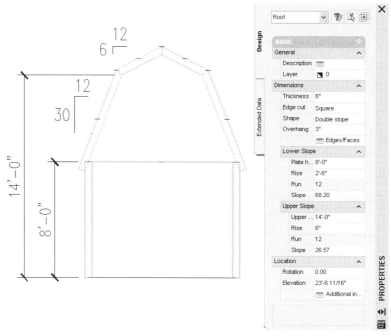

Figure 6.14 Creating a Gambrel roof with double slope

 STOP. Do Tutorial 6.3, "Creating and Editing a Gambrel Roof," at the end of the chapter.

CREATING A FLAT ROOF

A flat roof is often used in commercial construction. Create a flat roof using a slab, which is presented in Chapter 7. The slab can include the cant strip edge style around the perimeter of the slab.

CONVERTING POLYLINES OR WALLS TO ROOFS

The **Apply Tool Properties to** (**RoofToolToLinework**) command will create a roof from a closed polyline or a closed series of walls without specifying each corner of the roof. The walls or polyline can be selected using a crossing or window selection method or they can be selected individually. See Table 6.3 for **Apply Tool Properties to>Linework and Walls** command access.

| Shortcut menu | Select the Roof tool from the Design tool palette, right-click, and choose Apply Tool Properties to>Linework and Walls |

Table 6.3 *Apply Tool Properties to>Linework and Walls to create a Roof command access*

When the **Apply Tool Properties to>Linework and Walls** command is executed, you are prompted to select walls or polylines as shown below. In response to this prompt, select all walls or a closed polyline to create a roof over the walls or polyline. After you select the walls or polyline, you can specify the properties in the Properties palette. The **RoofToolToLinework** command is a quick method of creating a roof over curved walls.

(Select the **Roof** *tool on the Design palette, right-click, and choose* **Apply Tool Properties to>Linework and Walls** *as shown in Figure 6.15.)*

Command: RoofToolToLinework

Choose walls or polylines to create roof profile:

Select objects: *(Select polyline at p1 as shown in Figure 6.15 to convert to roof.)*

Select objects: ENTER *(Press* ENTER *to end selection.)*

Erase layout geometry? [Yes/No] <No>: ENTER *(Press* ENTER *to retain the original geometry.)*

CREATING ROOF SLABS

Roof slabs include features such as fascia, soffit, and frieze components. Unlike the roof discussed earlier, each roof slab is a single object that can be edited independent of adjoining roof slabs. If you edit the slope of an edge of a roof, the adjoining edges will adjust to continue the roof. Therefore, the initial roof should be developed using

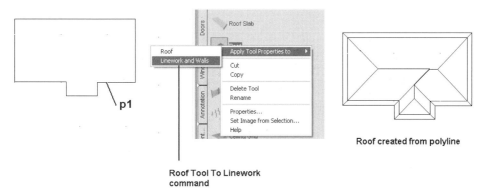

Figure 6.15 *Creating a roof from the polyline geometry*

the **Roof** command and then the roof object converted to roof slabs to customize the edges and individual planes of the roof. Roof slabs created from the roof object are not linked; therefore any changes to the roof object are not reflected in the roof slab. Roof slabs can be extended, trimmed, and mitered to create custom roof shapes. Mass elements and polygons can be used as a cutting edge to cut holes in a roof slab. Roof slabs are also used to trim a roof for a dormer. The treatment of the edge of a roof slab can include fascia, soffit, and frieze components, which are saved in roof edge style. There are 14 roof slab styles and 11 roof slab edge styles included in the *C:\Documents and Settings\All Users\Application Data\Autodesk\ADT 2005\enu\Styles\Imperial\Roof Slab & Roof Slab Edge Styles (Imperial)* drawing. Figure 6.16 includes examples of applying various roof slab edges to a slab.

The slab is created by specifying the two points, and the slope of the slab is hinged about the baseline, which connects the first two points. The overhang, edge style, and cut can be specified for each edge of the slab. The first point specified becomes the pivot point. A pivot point marker can be turned on in the object display. The pivot point is shown in Figure 6.17.

Roof slabs are created by selecting **Roof Slab** (**RoofSlabAdd** command) from the Design palette as shown in Table 6.4.

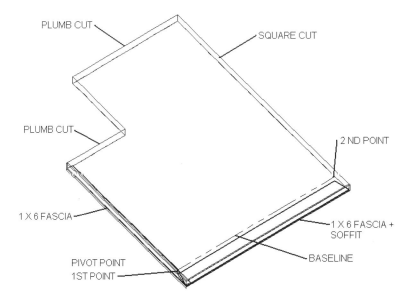

Figure 6.16 *Roof slab edge style applied to a roof slab*

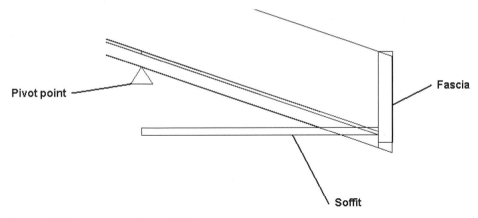

Figure 6.17 *Pivot point of a roof slab*

Command prompt	ROOFSLABADD
Palette	Select Roof Slab from the Design palette as shown in Figure 6.1

Table 6.4 *RoofSlabAdd command access*

When you select the **RoofSlabAdd** command, the Properties palette displays the default values of the roof slab as shown in Figure 6.18. You are prompted to specify the start point. The start point becomes the default slab insertion point and the pivot point. The next point specified creates the baseline, which is the hinge line for the slope of the slab. The roof slab must consist of at least three vertices and a slope angle.

The Properties palette allows you to set the style, mode, thickness, base height, overhang, justify, slope, and direction. The options of the Properties palette are described below

Basic

General

    Description – The **Description** button opens a **Description** dialog box to type a description of the roof slab.

    Style – The **Style** drop-down list consists of the Standard style. Styles can include fascia, frieze, and soffit components. Type **ST** in the command line to edit this option.

    Mode – The **Mode** options include **Direct** or **Projected**. The **Direct** mode creates the roof plane from the vertices specified. The Projected mode pro-

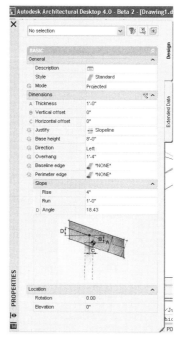

Figure 6.18 *Properties palette of a Roof Slab*

jects from the specified vertices up the distance as specified in the base height from the selected points. Type **MO** in the command line to edit this option.

Dimensions

A-Thickness – The **Thickness** specifies the thickness or depth of the roof slab object. Type **T** in the command line to edit this option.

B-Vertical offset – The **Vertical offset** specifies the vertical distance from the top of the wall to the roof slab baseline. The top of the wall is established by the base height property for a projected slab. A direct slab is developed with a 0 base height.

C-Horizontal offset The **Horizontal offset** distance shifts the slab horizontally relative to the baseline. A positive horizontal offset will shift the slab to the right of the baseline as shown in Figure 6.19. A negative horizontal offset will shift the roof slab to the left of the baseline.

Justify – The **Justify** option specifies the position of the handle on the roof slab relative to the first selected point to create the roof. Options include **Top**, **Center**, **Bottom**, and **Slopeline** as shown in Figure 6.20. Type **J** in the command line to edit the **Justify** option.

Top – Generates the roof slab with the pivot point located at the top of the roof.

Center – Generates the roof slab about the pivot point located at the midpoint of the thickness value.

Bottom – Generates the roof slab located at the bottom of the slab.

Slopeline – Generates the roof slab with the pivot point located at the slopeline of the roof slab. The thickness offset value specified in the Design Rules tab of the roof slab style will position the slopeline offset from the slab bottom.

Base height – The **Base height** is the distance the slab is created from the Z=0 plane. The base height is applied to the slab created in the projected mode. Type **H** in the command line to edit the base height option.

Direction – The **Direction** toggle allows you to create a roof slab to the left or right of first two selected points when you use **Ortho close**. Type **SL** in the command line to edit the direction option.

Overhang – The **Overhang** edit field specifies the horizontal distance from the pivot point to the fascia line. Type **OV** in the command line to edit the overhang option.

Baseline edge – The **Baseline edge** edit field allows you to specify roof slab edge style for the edge of the roof along the baseline edge.

Perimeter edge – The **Perimeter edge** edit field allows you to specify a roof slab edge style for the perimeter edges of the roof slab, excluding the baseline edge.

SLOPE:

Rise – The **Rise** specifies the vertical displacement per horizontal run unit. You can edit the Rise edit field to change the angle of existing roof slabs.

Run – The **Run** specifies the horizontal component of the slope. This value is usually 12 as base value. You can edit the **Run** value to change the angle of existing roof slabs.

Slope – The **Slope** displays the corresponding angle as a result of the rise and run values. The angle will be displayed resulting from the rise and run, or you can type a value for the angle. Type **SL** in the command line to edit the angle option. You can edit the **Slope** value to edit the angle of existing roof slabs.

The options of the command can be selected from the Properties palette or the command line. The options listed in the command line are shown below.

Command: RoofSlabAdd

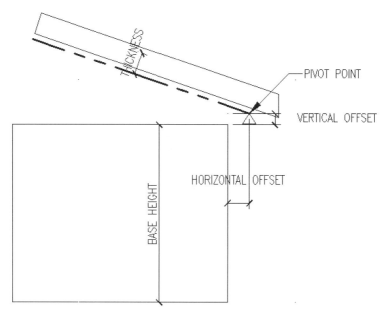

Figure 6.19 *Horizontal and Vertical Offset of a roof slab*

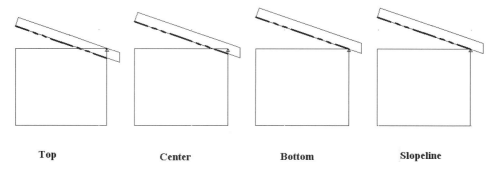

Figure 6.20 *Justify options of the roof slab*

Specify start point or

[STyle/MOde/Height/Thickness/SLope/OVerhang/Justify/MAtch]: **MA** ENTER

Select a roof slab to match: *(Select an existing roof slab.)*

Match [STyle/Thickness/SLope/Overhang] <All>: ENTER *(Press ENTER to match all properties.)*

The command line options are also included in the Properties palette. The Match option has been selected from the command line as shown above by typing **MA** and then selecting a roof slab to match. To select an option, type in the command line the letters capitalized in each of the options. Editing the Properties palette to select an option is a faster method of defining roof properties because it requires limited typing.

When you select a roof slab, the properties of the slab are displayed in the Properties palette. The Properties palette allows you to change the style, thickness, slope, vertical offset, horizontal offset, and edges.

USING THE DIRECTION PROPERTY TO CREATE ROOF SLABS

Roof slabs can be created by selecting a series of points to define the vertices of the roof slab. The **Direction** property specified in the Properties palette allows you to create a roof slab to the left or right of the first roof edge line. The roof slab is created by selecting two points to specify the pivot point and baseline, and then selecting **Ortho close** from the shortcut menu. Selecting **Ortho close** from the shortcut menu will generate the roof to the left or right of the baseline as shown at A in Figure 6.21. The roof is generated to the left or right of the first edge segment according to the **Direction** specified in the Properties palette.

The **Ortho close** option can also be used if additional points are specified as shown at B in Figure 6.21. The slab is specified by selecting points **p1**, **p2**, **p3**, **p4**, and **p5**, and then right-clicking and choosing **Ortho close** from the shortcut to close the roof to point **p1**. The roof shape is completed as necessary to form a right angle between the last roof edge segment and the first roof edge segment.

The **Close** option of the shortcut menu can be used to close the roof shape from the last specified point to the first specified point as shown at C in Figure 6.21. The **Close** option of the shortcut menu will close the last segment to the first edge segment without regard to the angle formed.

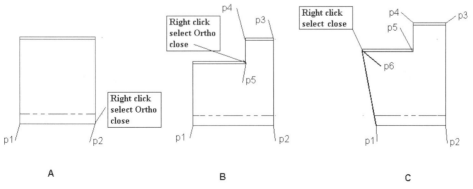

Figure 6.21 *Creating a slab using Close and Ortho close options*

CREATING THE ROOF SLAB FROM A WALL, ROOF, OR POLYLINE

A roof slab can be created from existing walls, roof, or a closed polyline. The shortcut menu of the **Roof Slab** tool of the Design tool palette includes the option **Apply Tool Properties to>Linework, Walls and Roof** (**RoofSlabToolToLinework** command) as shown in Table 6.5. When you select this command you can convert walls, roofs, or linework to a roof slab.

Palette	Select Roof Slab from the Design palette, right-click, and choose Apply Tool Properties to>Linework, Walls and Roof

Table 6.5 Apply Roof Slab Properties to>Linework, Walls and Roof command access

The roof slab created in Figure 6.22 was created from the closed polyline shown at **p1**. The first point selected to create the polyline becomes the pivot point, and the baseline starts at the first point. As shown in the following command line sequence, you can choose to erase the original geometry, and specify the mode, projection distance, and slab justification.

Command: RoofSlabToolToLinework

Select roof, walls or polylines: 1 found *(Select polyline at p1 shown in Figure 6.22.)*

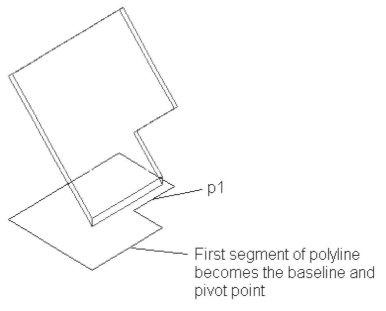

Figure 6.22 Converting the closed polyline to a roof slab

Select roof, walls or polylines: ENTER *(Press* ENTER *to end selection.)*

Erase layout geometry? [Yes/No] <No>: ENTER *(Press* ENTER *to retain the polyline.)*

Creation mode [Direct/Projected] <Projected>: ENTER *(Press* ENTER *to project the roof above the polyline.)*

Specify base height <0">: **8'** ENTER *(Type the distance value to project the roof.)*

Specify slab justification [Top/Center/Bottom/Slopeline] <Bottom>: ENTER *(Press* ENTER *to create a bottom justified slab.)*

Roof slabs that represent the planes in a two-dimensional plan view can be created by converting the two-dimensional plan to polylines and then converting the polylines to a roof slab.

Creating the Roof Slab from a Wall

The **RoofSlabToolToLinework** command can also be used to generate the roof slab from a wall. The wall base height is used as the projection distance, and the wall length establishes the length and width of the roof slab. The roof slab created in Figure 6.23 was created based upon the wall shown at **p1**. The roof slab is created with equal length and width. The default slope of 45° and thickness of the roof can be edited in the Properties palette.

Command: RoofSlabToolToLinework

Select roof, walls or polylines: 1 found *(Select wall at p1 shown in Figure 6.23.)*

Select roof, walls or polylines: ENTER *(Press* ENTER *to end selection.)*

Erase layout geometry? [Yes/No] <No>: ENTER *(Press* ENTER *to retain the wall.)*

Specify slab justification [Top/Center/Bottom/Slopeline] <Bottom>: ENTER *(Press* ENTER *to specify a bottom justification.)*

Specify wall justification for edge alignment [Left/Center/Right/Baseline]<Baseline>: **r** *(Specify the right justification.)*

Specify slope Direction [Left/Right] <Left>: ENTER *(Press* ENTER *to select the default left justification.)*

Converting an Existing Roof to a Roof Slab

The **RoofSlabToolToLinework** command can be used to convert a roof to a roof slab. The roof slab will remain separate from the original roof; however, the slope and thickness of the roof slab will be equal to the roof as shown at A in Figure 6.24. If you retain the roof when creating the roof slab, you can turn off the **A-Roof** layer to turn off the display of the roof. The roof slab is created on layer **A-Roof-Slab**.

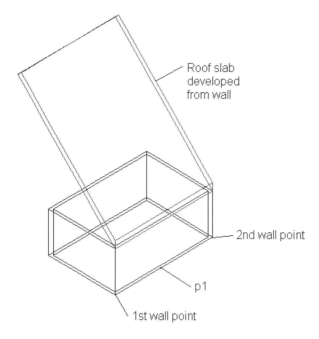

Figure 6.23 *Creating a roof slab based upon a wall*

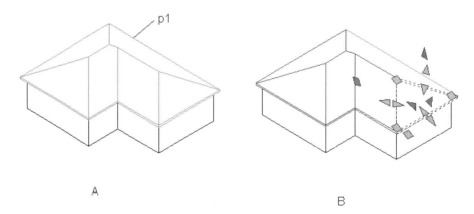

Figure 6.24 *Creating a roof slab from a roof*

In addition, you can select a roof, right-click, and select **Convert to Roof Slabs** from the shortcut menu to convert a roof to a roof slab. Access **RoofSlabConvert** as shown in Table 6.6.

Command prompt	ROOFSLABCONVERT
Shortcut menu	Select a roof, right-click, and select Convert to Roof Slabs

Table 6.6 *Converting a roof to a roof slab*

(Select a roof, right-click, and select **Convert to Roof Slabs** from the shortcut menu.)

Command: RoofSlabConvert

Erase layout geometry? [Yes/No] <No>: *(Press* ENTER *to retain geometry.)*

Specify roof slab style name or [?] <Standard>: *(Press* ENTER *to accept the Standard style.)*

Since the dimensions of roof and roof slab are identical, it is difficult to distinguish between the two objects. However, you can cycle through the selection if you hold down CTRL before you click on the object, and then continue clicking until just the roof slab is shown selected. One of the roof slabs created from the hip roof is shown selected at B in Figure 6.24.

Command: RoofSlabToolToLinework

Select roof, walls or polylines: 1 found *(Select the roof object at p1 shown at A in Figure 6.24.)*

Select roof, walls or polylines: ENTER *(Press* ENTER *to end the selection.)*

Erase layout geometry? [Yes/No] <No>: ENTER *(Press* ENTER *to retain the roof object.)*

MODIFYING ROOF SLABS

A roof slab can be edited by editing its properties in the Properties palette or editing the grips. In addition, roof slabs can be trimmed, extended, or mitered to obtain a roof shape. If you select an existing roof slab, additional fields are added to the roof slab Properties palette as shown in Figure 6.25. The additional fields of the Properties palette shown in Figure 6.25 include the **Edges** button, which opens the **Roof Slab Edges** dialog box. The **Roof Slab Edges** dialog box allows you to assign roof edge styles to the edges of the roof slab. The following **additional** edit fields in the Properties palette for existing roof slabs can be modified to change a roof slab.

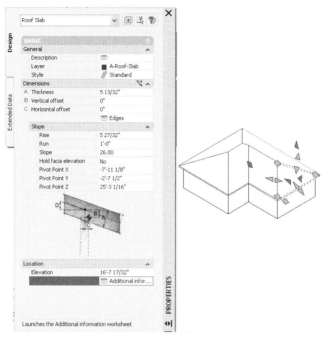

Figure 6.25 *Using the Properties palette to modify a roof slab*

Design–Basic

General

> **Layer** – The **Layer** edit field displays the name of the layer of the insertion point of the roof slab. Roof slabs are placed on the **A-Roof-Slab** layer when the AIA (256 Color) Layer Key Style is used in the drawing.

Dimensions

> **Edges** – The **Edges** button opens the **Roof Slab Edges** dialog box. The **Roof Slab Edges** dialog box shown in Figure 6.26 describes the hip plane shown at right in the object viewer. The options of the dialog box are as follows:
>
>> **A-Overhang** – The **Overhang** is the horizontal distance the roof extends beyond the wall.
>>
>> **Edge Style** – The **Edge Style** drop-down list shows the roof edge styles of the drawing. Roof edge styles can be imported in the **Style Manager** from *C:\Documents and Settings\All Users\ApplicationData\ Autodesk\ADT 2005\enu\Styles\Imperial\Roof Slab & Roof Slab Edge Styles (Imperial).dwg*.

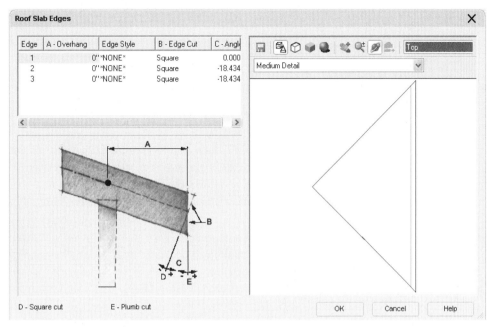

Figure 6.26 *Roof Slab Edges dialog box*

B-Edge Cut – The **Edge Cut** options include the plumb or square cut.

C-Angle – The **Angle** is the slope angle of the roof edge.

SLOPE:

Hold facia elevation – The **Hold facia elevation** edit field includes the **No** option, which will allow the fascia elevation to change as the slope of the roof is changed. The **By adjusting overhang** option will increase or decrease the overhang to retain the same fascia height. The **By adjusting baseline height** option increases or decreases the baseline height to retain the overhang elevation.

Pivot Point X – The **Pivot Point X** value is the world coordinate X value of the pivot point. You can type a different value in the **Pivot Point X** field to move the pivot point.

Pivot Point Y – The **Pivot Point Y** value is the world coordinate Y value of the pivot point. You can type a different value in the **Pivot Point Y** field to move the pivot point.

Pivot Point Z – The **Pivot Point Z** value is the world coordinate Z value of the pivot point. You can type a different value in the **Pivot Point Z** field to move the pivot point.

Location

Elevation – The **Elevation** edit field displays the elevation of the insertion point of the roof slab.

Additional information – The **Additional information** button opens the **Location** dialog box, which lists the insertion point, normal, and rotation values of the roof slab.

Using Grips to Edit a Roof Slab

The grips of a roof slab shown in Figure 6.27 allow you to change the properties and vertices of the roof slab. The *circle* grip markers allow you to change the **vertices** of the roof slab. The cyan colored *rhombus* grip allows the edit of the **pivot point**. The *wedge* shaped grip markers on each edge allow you to change the **edge overhang**. The *negative* symbol grip allows you to edit the location of the edge. If you hold down CTRL, you can toggle through the options to add an edge, move the edge maintaining the slope, or move the edge changing the slope. The remaining wedge shaped grips located on the face allow you to edit the **thickness, horizontal offset**, and **vertical offset**. The slope of the plane can be changed by selecting the grey colored *rhombus* shaped grip. When you select a grip, the dynamic dimensions are displayed; you can then press TAB to change the dimension that is highlighted. When a dynamic dimension is highlighted, you can overtype a new dimension in the workspace to change the size of the feature.

The *edge* grip at the top edge of the roof has been selected for the roof slab shown at left in Figure 6.28. The location of this edge has been stretched using the perpendicular object snap to a point on the wall while retaining the slope of the roof slab. The *edge offset* grip at the right side of the roof has been selected for the roof shown at right in Figure 6.28 to change the overhang of the roof slab.

TOOLS FOR EDITING ROOF SLABS

The commands to edit roof slabs are on the shortcut menu of a selected roof slab. Included on the shortcut menu are the commands to trim, extend, miter, cut, edit the vertices, edges, dormer, and modify using Boolean operations.

TRIMMING A ROOF SLAB

The **RoofSlabTrim** command can be used to trim a roof slab that intersects with a polyline, wall, or slab. The cutting object is projected to intersection in the current user coordinate system. Therefore, if the cutting object does not intersect the roof slab, it is projected up based upon the current UCS to identify the cutting plane. Access the **RoofSlabTrim** command as shown in Table 6.7.

Creating Roofs and Roof Slabs 407

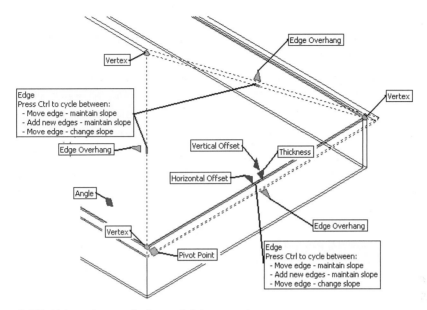

Figure 6.27 *Using grips to edit the roof slab properties*

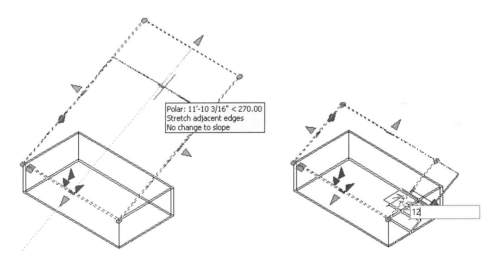

Figure 6.28 *Editing roof slab edges*

Command prompt	ROOFSLABTRIM
Shortcut menu	Select the roof slab, right-click, and choose Trim

Table 6.7 *Roof SlabTrim command access*

When you select a roof slab, right-click, and choose **Trim** from the shortcut menu shown in Figure 6.29, you are prompted to select the trimming object and to specify the side to remove when trimmed. In the following command line sequence, the roof slab is trimmed by the wall. The wall height is below the roof slab; however, because the slab and wall are viewed from the top view, the wall is projected based upon the current UCS to identify the cutting edge.

(Select the roof slab at p1 at left in Figure 6.30, right-click, and choose **Trim** *from the shortcut menu. Respond to the following command line prompts.)*

Command: RoofSlabTrim

Select trimming object (a slab, wall, or polyline): *(Select the wall at p2 at left as shown in Figure 6.30.)*

Specify side to be trimmed: *(Select a point near p3 as shown at left in Figure 6.30. Roof slab is trimmed as shown in the right side view at right in Figure 6.30.)*

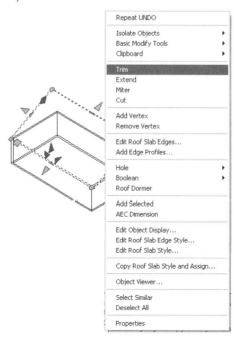

Figure 6.29 *Commands of the roof slab shortcut menu*

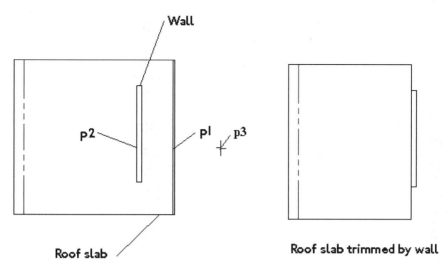

Figure 6.30 *Trim Roof Slab command used to trim the roof slab*

EXTENDING A ROOF SLAB

The **RoofSlabExtend** command can be used to extend a roof slab to a polyline, wall, or slab. The cutting object is projected based upon the current user coordinate system. Therefore, if the object does not intersect the roof slab, it is projected up based upon the current UCS to identify the extension. Access the **RoofSlabExtend** command as shown in Table 6.8.

Command prompt	ROOFSLABEXTEND
Shortcut menu	Select a roof slab, right-click, and choose Extend

Table 6.8 *Extend Roof Slab command access*

When you select a slab, right-click, and choose **Extend** from the shortcut menu, you are prompted to select the object to extend to and the edges to lengthen. In the following command line sequence, the roof slab was extended to the wall as shown in Figure 6.31.

> (*Select a roof slab at p1 as shown in Figure 6.31, right-click, and choose* **Extend** *from the shortcut menu. Respond to the following command line prompts.*)
>
> Command: RoofSlabExtend

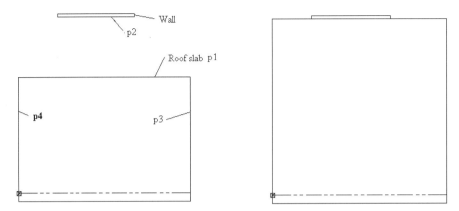

Figure 6.31 *Extend Roof Slab used to extend the roof slab*

Select an object to extend to (a slab or wall): *(Select the wall at p2 at left in Figure 6.31.)*

(Slab will be extended by lengthening two edges.)

Select first edge to lengthen: *(Select the roof slab at p3 at left in Figure 6.31.)*

Select second edge to lengthen: *(Select the slab at p4 at left in Figure 6.31. Wall is extended as shown at right in Figure 6.31.)*

MITERING ROOF SLABS

The **RoofSlabMiter** command can be used to extend or trim the roof slabs that are of the same type. The **RoofSlabMiter** command edits only one edge of each roof slab. Access the Miter command as follows shown in Table 6.9.

Command prompt	ROOFSLABMITER
Shortcut menu	Select the roof slab, right-click, and choose Miter

Table 6.9 *Miter Roof Slab command access*

When you select the **RoofSlabMiter** command, you are prompted to miter either the Intersection or Edges of the roof slabs. If you miter by intersection, the slabs will be trimmed according to the intersection of the roof slabs. The miter by edges requires you to select the edges of the roof slabs. The two roof slabs shown in Figure 6.32 were mitered by intersection as described in the following command line sequence.

*(Select a roof slab, right-click, and choose **Miter** from the shortcut menu.)*

Command: RoofSlabMiter

Miter by [Intersection/Edges] <Intersection>: ENTER *(Press* ENTER *to select the Intersection option.)*

Select first roof slab at the side to keep: *(Select the roof slab at p1 as shown at left in Figure 6.32.)*

Select second roof slab at the side to keep: *(Select the roof slab at p2 as shown at left in Figure 6.32.)*

(Roof slabs are mitered as shown at right in Figure 6.32.)

The Edges miter method can be used to extend two roof slabs to create the miter. Although the roof slabs shown in Figure 6.33 do not intersect, edges **p1** and **p2**, if extended, would create a gable roof. The two roof slabs were mitered using the Edges option as shown in the following command line sequence.

(Select a roof slab, right-click, and choose **Miter** *from the shortcut menu.)*

RoofSlabMiter

Miter by [Intersection/Edges] <Intersection>: **e** ENTER *(Type* **e** *to select the Edges option.)*

Select edge on first slab: *(Select the edge at p1, shown at left in Figure 6.33.)*

Select edge on second slab: *(Select the edge at p2, shown at left in Figure 6.33.)*

(Edges extended to form miter as shown at right in Figure 6.33.)

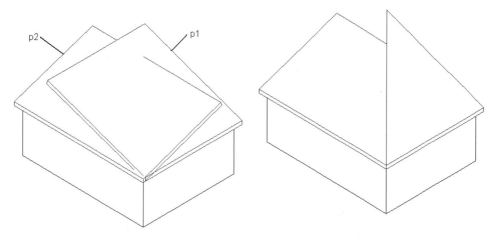

Figure 6.32 *Roof slabs mitered*

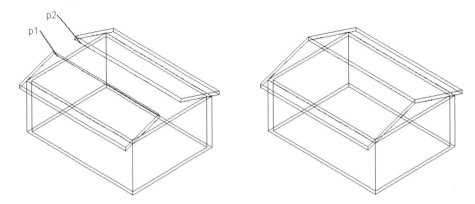

Figure 6.33 *Roof slabs mitered by Edges method*

CUTTING A ROOF SLAB

The **RoofSlabCut** command can be used to cut a roof slab by any 3D object or polyline. Therefore this command could be used to cut a roof slab by a wall or chimney. The intersecting object must intersect the perimeter line of the roof slab. Access the **RoofSlabCut** command as shown in Table 6.10.

Command prompt	ROOFSLABCUT
Shortcut menu	Select the roof slab, right-click, and choose Cut

Table 6.10 *Cut Roof Slab command access*

When you select a slab, right-click, and choose **Cut** from the shortcut menu, you are prompted to select the objects that intersect the roof slab. The **RoofSlabCut** command was used to cut the roof slab by the intersecting wall shown in Figure 6.34 and the following command line sequence.

> *(Select the roof slab at p1 shown at left in Figure 6.34, right-click, and choose **Cut** from the shortcut menu.)*
>
> Command: RoofSlabCut
>
> Select cutting objects (a polyline or connected solid objects): 1 found *(Select the wall at p2 shown in Figure 6.34.)*
>
> Select cutting objects (a polyline or connected solid objects): 1 found, 2 total *(Select the wall at p3 shown in Figure 6.34.)*
>
> Select cutting objects (a polyline or connected solid objects): 1 found, 3 total *(Select the wall at p4 shown in Figure 6.34.)*

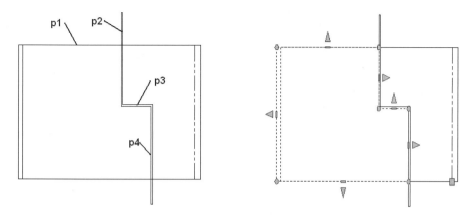

Figure 6.34 *Roof slab cut using the RoofSlabCut command*

      Select cutting objects (a polyline or connected solid objects) ENTER
      (Press ENTER *to end object selection.)*

      (Verify the roof slab is cut by selecting the slab as shown at right in Figure 6.34.)

ADDING A ROOF SLAB VERTEX

The **RoofSlabAddVertex** command can be used to add a vertex along the edge of a roof slab. The new vertex will have a grip that can be stretched to a location for editing the roof. The **RoofSlabAddVertex** command is used to modify the perimeter of the slab. Each edge of the perimeter consists of grips at the midpoint and ends of the roof edge. Access the **RoofSlabAddVertex** command as shown in Table 6.11.

Command prompt	ROOFSLABADDVERTEX
Shortcut menu	Select the roof slab, right-click, and choose Add Vertex

Table 6.11 *Add Roof Slab Vertex command access*

When you select the **RoofSlabAddVertex** command, you are prompted to select the roof slab and specify the location for the vertex. The **RoofSlabAddVertex** command was used to add a vertex to the roof slab at the intersection as shown in Figure 6.35 and the following command line sequence.

      (Select a roof slab, right-click, and choose **Add Vertex** *from the shortcut menu.)*

      Command: RoofSlabAddVertex

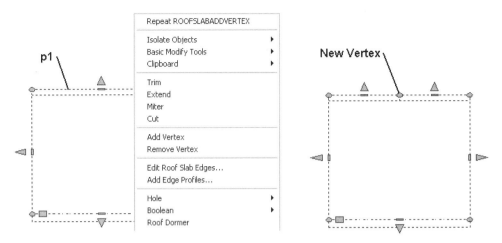

Figure 6.35 Vertex added to roof slab

> Specify point for new vertex: _mid of (Hold SHIFT down, right-click, and choose **Mid** from the shortcut menu; then select the edge near p1 shown at left in Figure 6.35.)
>
> (Select the slab to verify the new grip point as shown at right in Figure 6.35.)

Grips can be used to edit the location of the new vertex and change the perimeter of the roof as shown at left in Figure 6.36. The SW Isometric view of the roof slab is shown at right in Figure 6.36.

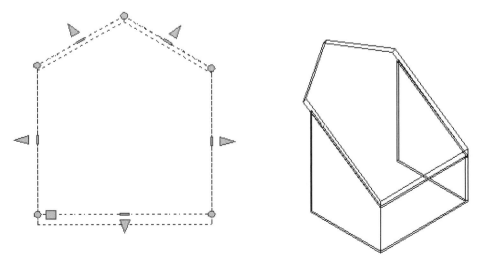

Figure 6.36 Grips used to edit the roof at the new vertex

REMOVING THE VERTEX OF THE ROOF SLAB

The **RoofSlabRemoveVertex** command can be used to remove a vertex from the perimeter of the roof. The **RoofSlabRemoveVertex** command is used to modify the perimeter of the slab. Access the **RoofSlabRemoveVertex** command as shown in Table 6.12.

Command prompt	ROOFSLABREMOVEVERTEX
Shortcut menu	Select the roof slab, right-click, and choose Remove Vertex

Table 6.12 *Remove Roof Slab Vertex command access*

When you select the **RoofSlabRemoveVertex** command, you are prompted to select a roof slab vertex. The **RoofSlabRemoveVertex** command can be used to remove a vertex of a roof slab at the midpoint as shown at left in Figure 6.37 and the following command line sequence.

>   RoofSlabRemoveVertex
>   Select a vertex: *(Select the vertex at p1 as shown at left in Figure 6.37.)*
>   *(Roof slab vertex removed as shown at right in Figure 6.37.)*

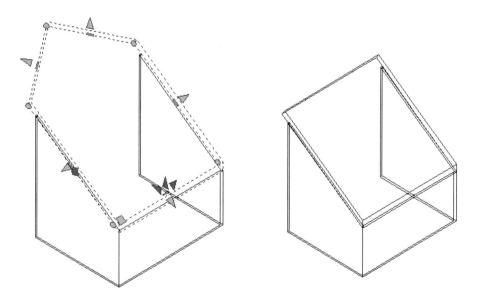

Figure 6.37 *Remove Vertex command used to remove midpoint vertex*

CREATING HOLES IN A ROOF

The **RoofSlabAddHole** command can be used to add a hole in a roof slab. The hole is projected from closed polylines or a 3D object. The **Hole** command on the shortcut menu can be used to create a hole or remove a hole from the roof slab. The projection of the polyline is based upon the current UCS. When 3D objects that consist of voids are used to cut the hole, you are prompted to select the inside or outside to cut the hole. Access the **RoofSlabAddHole** command as shown in Table 6.13.

Command prompt	ROOFSLABADDHOLE
Shortcut menu	Select the roof slab, right-click, and choose Hole>Add

Table 6.13 RoofSlabAddHole command access

When you select a slab, right-click, and choose **Hole>Add** from the shortcut menu, you are prompted to select a closed polyline or other connected solid objects. The objects will be projected to the roof plane to cut the hole. The **RoofSlabAddHole** command was used to cut a hole in the roof slab as shown in Figure 6.38 and the following command line sequence.

>(Select a roof slab, right-click and choose **Hole>Add** from the shortcut menu.)
>
>Command: RoofSlabAddHole
>
>Select a closed polyline or connected solid objects to define a hole: 1 found (Select the polyline at p1 as shown in Figure 6.38.)
>
>Select a closed polyline or connected solid objects to define a hole: ENTER (Press ENTER to end selection.)

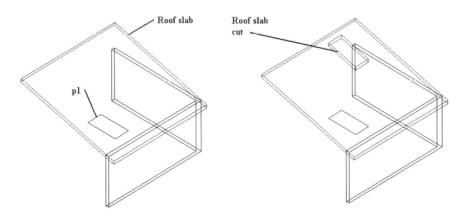

Figure 6.38 Hole created in the roof slab with the RoofSlabAddHole command

Erase layout geometry? [Yes/No] <No>: ENTER *(Press* ENTER *to retain the polyline.)*

(Polyline is projected to the roof plane to cut the hole as shown at right in Figure 6.38.)

(Roof slab cut by the projection of the closed polyline as shown in Figure 6.38.)

Removing Holes from a Roof

The **RoofSlabRemoveHole** command is used to remove holes that exist in slabs. Access the **RoofSlabRemoveHole** command as shown in Table 6.14.

Command prompt	ROOFSLABREMOVEHOLE
Shortcut menu	Select the roof slab, right-click, and choose Hole>Remove

Table 6.14 *Roof SlabRemoveHole command access*

When you select a slab, right-click, and choose **Hole>Remove** from the shortcut menu, you are prompted to select a hole to remove. The hole is removed when you select it, as shown in the following command line sequence and Figure 6.39.

(Select a roof slab, right-click, and choose **Hole>Remove** *from the shortcut menu.)*

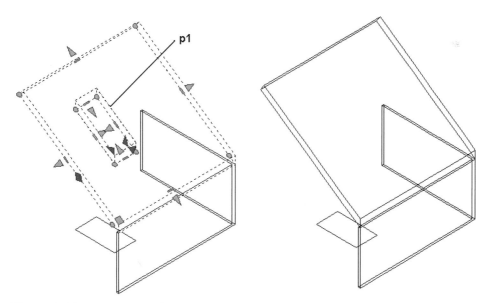

Figure 6.39 *Hole removed from slab using the RoofSlabRemoveHole command*

Command: RoofSlabRemoveHole

Select an edge of hole to remove: *(Select the hole at p1 as shown at left in Figure 6.39.)*

(Hole is removed as shown at right in Figure 6.39.)

USING BOOLEAN ADD/SUBTRACT/DETACH TO COMBINE MASS ELEMENTS

The **RoofSlabBoolean** command includes the **Add**, **Subtract**, and **Detach** options. The **Add** option can be used to add a mass element to the roof slab. The command can also be used to subtract the mass element from the roof slab. The **Detach** option removes the Boolean operation of add or subtract from the roof object. Mass elements are presented in Chapter 12. Mass elements are 3D objects in such shapes as the following: arch, barrel vault, box, cylinder, or triangular prism. The shapes can be added together using Boolean operations to form a group of elements. These elements can be used to add or subtract their shape to or from a roof.

Access the **Boolean>Add** command as shown in Table 6.15.

Command prompt	ROOFSLABBOOLEANADD
Shortcut menu	Select the roof slab, right-click, and choose Boolean>Add

Table 6.15 *Boolean>Add command access*

When you select a roof slab, right-click, and choose **Boolean>Add**, you are prompted to select a mass element. The mass element is then added to or subtracted from the roof slab. The mass element in Figure 6.40 was added to the roof slab as shown in the following command line sequence.

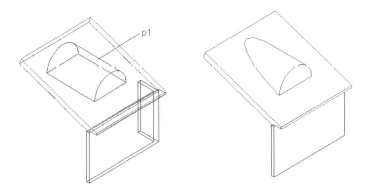

Figure 6.40 *Mass element added to roof using Boolean>Add*

(Select a roof slab, right-click, and choose **Boolean>Add** *from the shortcut menu.)*

Command: RoofSlabBooleanAdd

Select objects: 1 found *(Select the mass element at p1 as shown at left in Figure 6.40.)*

Select objects: ENTER *(Press* ENTER *to end selection.)*

1 object(s) attached.

(Select **Hidden** *from the* **Shade** *flyout of the* **Navigation** *toolbar to view the union as shown at right in Figure 6.40.)*

The mass element can also be used to subtract the shape from the roof slab. If you move the mass element after an addition or subtraction, the roof slab will be modified based upon the new location. Access the **Boolean>Subtract** command as shown in Table 6.16.

Command prompt	ROOFSLABBOOLEANSUBTRACT
Shortcut menu	Select the roof slab, right-click, and choose Boolean>Subtract

Table 6.16 *Boolean>Subtract command access*

The cylinder mass element in Figure 6.41 is subtracted from the roof slab in the following command sequence.

(Select the roof slab, right-click, and choose **Boolean>Subtract**.*)*

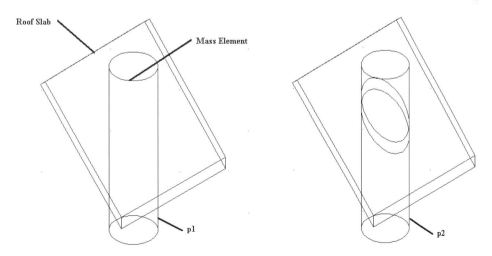

Figure 6.41 *Mass element subtracted from a roof slab*

Command: RoofSlabBooleanSubtract

Select objects: 1 found *(Select the mass element at p1 as shown at left in Figure 6.41.)*

Select objects: ENTER *(Press ENTER to end selection.)*

1 object(s) attached.

(Mass element subtracted from roof slab as shown at right in Figure 6.41.)

The mass element can also be detached from the roof slab. Access the **Boolean>Detach** command as shown in Table 6.17.

Command prompt	ROOFSLABBOOLEANDETACH
Shortcut menu	Select the roof slab, right-click, and choose Boolean>Detach

Table 6.17 *Boolean>Detach command access*

Command: RoofSlabBooleanDetach

Select objects: 1 found *(Select mass element at p2 as shown at right in Figure 6.41.)*

Select objects: ENTER

(Mass element detached as shown at left in Figure 6.41.)

CREATING A ROOF DORMER

The **Roof Dormer** (**RoofSlabDormer**) command allows you to create a dormer from walls that penetrate the roof. The walls may or may not be trimmed by the main roof. The dormer walls must include four walls, which penetrate the roof slabs. Access the **Roof Dormer** command as shown in Table 6.18:

Command prompt	ROOFSLABDORMER
Shortcut menu	Select the roof slab, right click, and choose Roof Dormer

Table 6.18 *Roof Dormer command access*

When you select the **Roof Dormer** command, you are prompted to select the roof slab and the other objects that form the dormer. The roof slab is sliced by the other elements to form the dormer as shown in the following command line sequence.

(Select the roof slab at p1 shown at left in Figure 6.42, right-click, and choose **Roof Dormer** *from the shortcut menu.)*

Command:

Command: RoofSlabDormer

Select objects that form the dormer: 1 found *(Select wall at p2 as shown at left in Figure 6.42.)*

Select objects that form the dormer: 1 found, 2 total *(Select wall at p3 as shown at left in Figure 6.42.)*

Select objects that form the dormer: 1 found, 3 total *(Select wall at p4 as shown at left in Figure 6.42.)*

Select objects that form the dormer: 1 found, 4 total *(Select wall at p5 as shown at left in Figure 6.42.)*

Select objects that form the dormer: 1 found, 5 total *(Select dormer roof slab at p6 as shown at left in Figure 6.42.)*

Select objects that form the dormer: 1 found, 6 total *(Select dormer roof slab at p7 as shown at left in Figure 6.42.)*

Select objects that form the dormer: ENTER *(Press* ENTER *to end selection.)*

Slice wall with roof slab [Yes/No] <Yes>: ENTER *(Press* ENTER *to slice the dormer wall to remove the lower portion of dormer wall.)*

(Dormer created as shown at right in Figure 6.42. To complete the operation, use the Erase command to erase the wall at p8.)

 STOP. Do Tutorial 6.4, "Creating Dormers and Roof Slabs," at the end of the chapter.

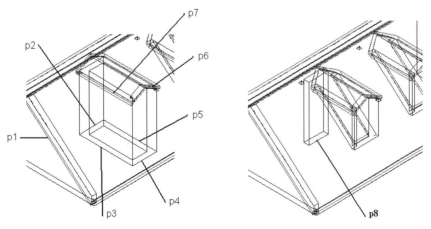

Figure 6.42 *Dormer geometry to create a roof dormer*

DETERMINING ROOF INTERSECTIONS

The **Roof Dormer** command determines the intersection of the dormer roof with the main roof. However, other roofs can intersect with the main roof without forming a dormer. When a roof slab intersects with another roof slab with different slope and plate height, the intersection is not automatically determined. To determine the valley intersections of multiple roof slabs:

1. Create a mass group and attach the intersecting roof slabs to the group to determine the intersection.
2. Draw a polyline that traces the intersection, and use this polyline as a cutting object with the RoofSlabCut command to cut each roof slab.
3. Remove the unnecessary extensions of each roof slab.

CREATING A MASS GROUP

Create a mass group by selecting the **Mass Group** command from the Massing tool palette. The purpose of this command is to group together two or more mass elements to form a mass model of a building. The development of the mass models is discussed in detail in Chapter 12. However, this command can also be used to unite roof objects of the drawing. Access the **Mass Group** command as shown in Table 6.19.

Command prompt	MASSGROUPADD
Palette	Select Mass Group from the Massing palette

Table 6.19 *MassGroupAdd command access*

When the **MassGroupAdd** command is selected from the Massing palette, you are prompted to select elements to attach to the group as shown in the command sequence below:

Command: MassGroupAdd

Select elements to attach: 1 found *(Select p1 as shown in Figure 6.43.)*

Select elements to attach: 1 found *(Select p2 as shown in Figure 6.43.)*

Select elements to attach: 1 found *(Select p3 as shown in Figure 6.43.)*

Select elements to attach: ENTER *(Press ENTER to end selection.)*

Location: *(Select a point in the drawing as shown at p4 in Figure 6.43.)*

The mass group marker is inserted at the Location point. However, it is not displayed until you select an element that belongs to the group. When an element of the group is selected, the grips of the elements and the mass group marker are displayed, as shown in Figure 6.44. The grips of the marker allow you to add or remove

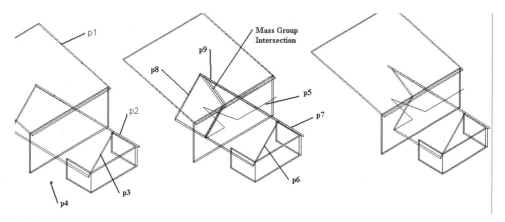

Figure 6.43 Cutting a roof slab at the intersection of roof planes

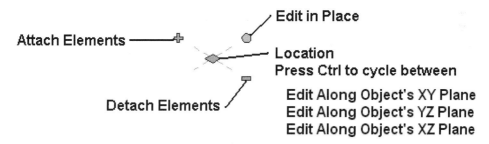

Figure 6.44 Grips of the mass group marker

elements from the group, and the Edit In Place toggle grip allows direct editing of the geometry of the elements in the mass group. Elements can be added to the group by selecting the plus (+) grip symbol or detached by selecting the (-) grip symbol. Access the **MassGroupAttach** command as shown in Table 6.20.

Command prompt	MASSGROUPATTACH
Shortcut menu	Select a mass group member, right-click, and choose Attach Elements

Table 6.20 MassGroupAttach command access

If you select the mass group member, right-click, and choose **Attach Elements**, you are prompted to select elements as shown in the following command line sequence:

Command: MassGroupAttach

Select elements to attach: *(Select additional elements.)*

Elements can be removed by selecting a member of the group, right-clicking, and selecting **Detach Elements**. The elements you select will be removed from the group. Access the **MassGroupDetach** command as shown in Table 6.21.

Command prompt	MASSGROUPDETACH
Shortcut menu	Select a mass group member, right-click, and choose Detach Elements

Table 6.21 *MassGroupDetach command access*

The elements of the mass group are linked together; therefore, if you change the slope of the roof, the mass group will automatically update to the new slope. The mass group is placed on layer **A-Area-Mass-Grps** with color 13. After creating the mass group, perform the following steps to edit the roof slabs.

STEPS TO CUTTING AND EDITING ROOF SLABS

1. Convert each roof to roof slabs.
2. Select **Mass Group** from the **Massing** toolbar and attach each roof slab (**p1**, **p2**, and **p3** shown in Figure 6.43) to the group. Select any coordinate for the location of the mass group marker.
3. View the roofs from Top, select **Polyline** from the **Shapes** toolbar, and trace the intersection of the roof using the **Intersection** object snap to create a polyline. The isometric view of the polyline is shown at **p5** in Figure 6.43.
4. Select **SW Isometric** from the **View** flyout.
5. Select the group to display the group marker, select the negative sign of the group marker, and then select the roof slabs to detach them from the group.
6. Select the roof slab at **p6** as shown in Figure 6.43, right-click, and choose **Cut** from the shortcut menu. Respond to the command line prompt as shown below.

    Command: RoofSlabCut

    Select cutting objects (a polyline or connected solid objects): *(Select the polyline at p5 as shown in Figure 6.43.)*

    Select cutting objects (a polyline or connected solid objects): ENTER

    Erase layout geometry [Yes/No] <No>:N ENTER

7. Repeat step 6 to cut the plane **p7**.

8. Select the roof planes at **p8** and **p9** that penetrate into the building, right-click, and choose **Basic Modify Tools>Erase** from the shortcut menu. View the slab intersection as shown at right in Figure 6.43.

CREATING ROOF SLAB STYLES

Roof slab styles include preset values for thickness and predefined roof slab edge styles. Additional details such as fascia, soffit, and frieze components are added by assigning roof slab edge styles to the edge of the roof slab. Materials can be assigned to the roof slab to represent the roofing materials. The Standard roof slab style is the default when a roof slab is inserted. The Standard roof slab style does not include soffit, fascia, and frieze components. Fourteen roof slab styles are included in the Roof Slab & Roof Slab Edge Styles (Imperial) drawing in the *C:\Documents and Settings\All Users\Application Data\Autodesk\ADT 2005\enu\Styles\Imperial* directory. The styles can be copied from this drawing and pasted into the current drawing in the **Style Manager**. When you paste a roof slab style into your drawing, any associated roof slab edge styles are also included. The styles can also be i-dropped from the Design Tool Catalog-Imperial>Roof Slabs catalog of the **Content Browser**. Access the **Roof Slab Styles** command as shown in Table 6.22.

Menu bar	Format>Style Manager, expand Architectural Objects, select Roof Slab Styles folder
Command prompt	RoofSlabStyleEdit
Shortcut menu	Select Roof Slab tool on the Design tool palette, right-click, and select Roof Slab Styles

Table 6.22 *Roof Slab Styles command access*

When you access this command by selecting the **Roof Slab** tool, right-clicking, and choosing **Roof Slab Styles** from the shortcut menu, the **Style Manager** opens to the Roof Slab Styles folder. Select the **Open Drawing** command from the **Style Manager** toolbar, select **Content** from the **Places** panel and navigate to *Styles\Imperial* folder to select the **Roof Slab and Roof Slab Edge Styles (Imperial).dwg**. Roof slab styles can be copied from the **Roof Slab & Roof Slab Edge Styles** drawing and pasted into the current drawing in the **Style Manager**. Roof slab styles are included in the **Architectural Objects\Roof Slab Styles** folder. After you have pasted a roof slab style into the current drawing, select the style in the **Style Manager**, right-click, and choose **Edit** to open the **Roof Slab Styles** dialog box for the selected style. The **General** tab for the 04 – 1×4 Fascia + Soffit roof slab style imported from the *C:\Documents and Settings\All Users\Application Data\Autodesk\ADT 2005\enu\Styles\Imperial\Roof Slab & Roof Slab Edge Styles (Imperial)* drawing is shown in Figure 6.45.

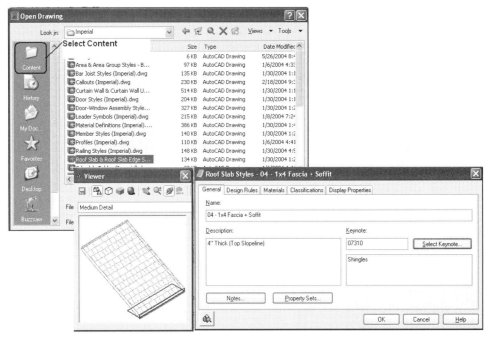

Figure 6.45 *General tab of the Roof Slab Styles dialog box*

The tabs of the **Roof Slab Styles** dialog box for the 04 – 1×4 Fascia + Soffit roof slab style are described below.

General Tab

Name – The **Name** edit field displays the name of the roof style.

Description – The **Description** edit field allows you to type text to describe the roof slab style.

Select Keynote – Choose the **Select Keynote** button to open the **Select Keynote** dialog box, which allows you to specify keynotes from the 16 CSI divisions. Keynote commands of the Annotation palette can be used to extract the keynote annotation

Notes – The **Notes** button opens the **Notes** dialog box, allowing you to type text to describe the roof slab.

Property Sets – The **Property Sets** button opens the **Edit Property Set Data** dialog box to add or delete property sets to the roof slab style.

Design Rules Tab

The Design Rules tab shown in Figure 6.46 allows you to define the thickness property of the roof slab in the style. The options of this dialog box are described below.

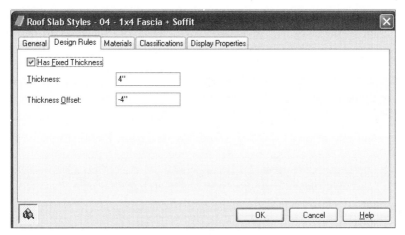

Figure 6.46 *Design Rules tab of the Roof Slab Styles dialog box*

Has Fixed Thickness – If the **Has Fixed Thickness** check box is checked, the thickness is defined in the style and specified on the **Design Rules** tab. If this check box is cleared, the thickness is defined in the **A-Thickness** edit field accessed from the Properties palette.

Thickness – The **Thickness** specifies the thickness of the roof slab when defined in the style.

Thickness Offset – The **Thickness Offset** specifies the distance to offset the roof slab from the insertion point. A negative thickness offset value will shift the slab below the baseline.

Materials Tab

The **Materials** tab allows you to specify the material definition for the components of the slab. The options of the **Materials** tab are shown in Figure 6.47 and described below.

Component – The **Component** column lists the name of the roof slab components to apply materials.

Material Definition – The **Material Definition** column lists the material definitions defined for the component. If you click in this column you can select from the drop-down list the materials used in the drawing. The Thermal & Moisture.Shingles.Asphalt Shingles.Deep Shadow.Beige is the material definition assigned in the style for the Slab component as shown in Figure 6.47.

Edit Material – The **Edit Material** button at right opens **Material Definition Properties** dialog box to edit the material definition. A material definition is defined in the **Other** tab of the **Display Properties** dialog box for the display representation.

Add New Material – The **Add New Material** button opens the **New Material** dialog box, which allows you to specify a name for the new material.

Classifications Tab

The **Classifications** tab allows you to specify predefined classifications of building status or other categories. The components of the **Classifications** tab are shown in Figure 6.47.

> **Classification Definition** – The **Classification Definition** column includes a list of the classifications defined in the drawing. The Uniformat II classification listed in Figure 6.47 was imported from the *C:\Documents and Settings\All Users\Application Data\Autodesk\Autodesk\ADT 2005\enu\Styles\Imperial\ Uniformat II Classifications (1977 edition)*.
>
> **Classification** – The **Classifications** included in the Uniformat II Classifications are included in the drop-down list.

Display Properties Tab

The **Display Properties** tab shown in Figure 6.48 allows you to set the display property source for a display representation. The options of the tab are as follows:

> **Display Representations** – The **Display Representations** column lists the display representations available for the roof slab. The **Plan** display representation is shown bold because that is the current view of the object.

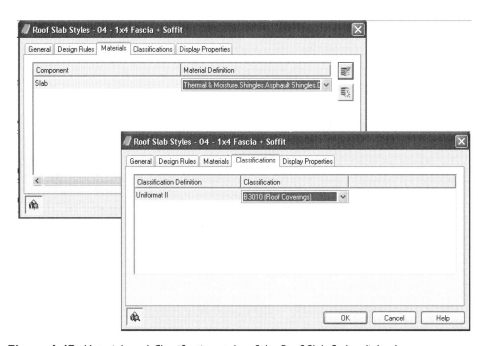

Figure 6.47 *Materials and Classifications tabs of the Roof Slab Styles dialog box*

Display Property Source – The **Display Property Source** column lists object display categories for the current viewport. The **Display Property Source** is either controlled by the drawing default global settings or by an override to the style.

Style Override – The **Style Override** check box indicates if an override has been defined in the display properties. If this box is clear, **Drawing Default** is controlling the display of the roof style. When you check this box, the **Display Properties** dialog box opens.

Edit Display Properties – The **Edit Display Properties** button opens the **Display Properties** dialog box. The **Display Properties** dialog box for the Plan display representation of a roof slab is shown in Figure 6.48. This dialog box consists of three tabs: **Layer/Color/Linetype**, **Hatching**, and **Other**.

Note: The High Detail display configuration displays the roofing material pattern, and the Low Detail turns off the roofing material pattern and displays the outline of the roof. Select the display configuration from the Drawing Window status bar.

Layer/Color/Linetype – The **Layer/Color/Linetype** tab lists the display components and the settings for visibility, by material, layer, color, linetype, lineweight, and linetype scale. The list of components varies according to the display representation.

Hatching – The **Hatching** tab lists the display components and the pattern name, scale, angle orientation, x offset, and y offset. This tab controls the display of the hatch pattern of the roof slab. The **Hatch** component is turned Off as shown in Figure 6.48; therefore the pattern settings will not be displayed active in Figure 6.49.

Other – The **Other** tab allows you to set the height of the cut plane and provides a toggle to override the cut plane of the display configuration.

CREATING A NEW ROOF EDGE STYLE

A new roof edge style can be created to define the shape of the fascia, frieze, and soffit. The roof edge style is created in the **Style Manager**. Roof edge styles are included in the Architectural Objects\Roof Edge Style folder as shown in Figure 6.50. Access the **RoofSlabEdgeStyleEdit** command as shown in Table 6.23.

Command prompt	ROOFSLABEDGESTYLEEDIT
Menu bar	Select Format>Style Manager, expand the Architectural Objects\Roof Slab Edge Styles folder, then edit style
Shortcut menu	Select a roof slab that includes a roof slab edge style, right-click, and choose Edit Roof Slab Edge

Table 6.23 *RoofSlabEdgeStyleEdit command access*

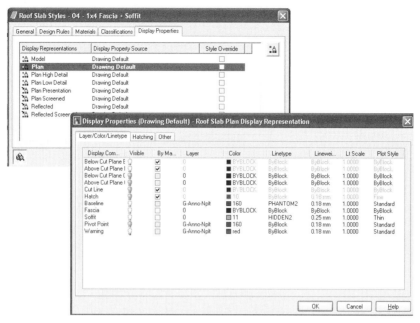

Figure 6.48 *Display Properties tab and Display Properties (Drawing Default) – Roof Slab Plan Display Representation dialog box*

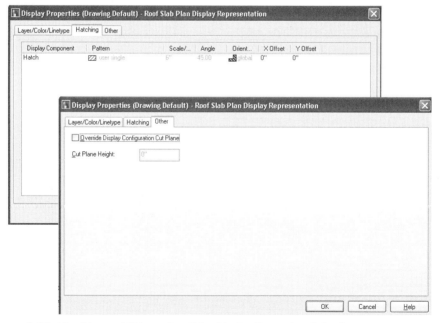

Figure 6.49 *Hatching and Other tabs of the Display Properties dialog box*

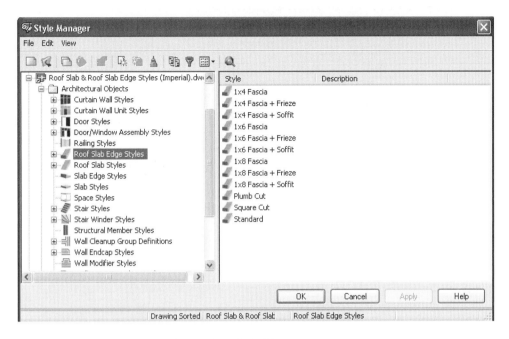

Figure 6.50 *Roof Slab Edge Styles of the Style Manager*

When you select the **Style Manager** and expand the Architectural Objects>Roof Slab Edge Styles, the roof slab edge styles are listed in the right pane. The styles can be copied for export or edited in the **Style Manager**. The right pane of the **Style Manager** shown in Figure 6.50 lists the Roof Slab Edge Styles included in the Roof Slab & Roof Slab Edge Styles (Imperial) drawing of the *C:\Documents and Settings\All Users\Application Data\Autodesk\ADT 2005\enu\Styles\Imperial* directory.

You can edit the roof slab edge style by double-clicking on the style name in the **Style Manager** to open the **Roof Slab Edge Styles** dialog box. The **Roof Slab Edge Styles** dialog box consists of four tabs: **General**, **Defaults**, **Design Rules**, and **Materials**. A description of each of the tabs follows.

General Tab

The **General** tab shown in Figure 6.51 allows you to create a name and description for the new roof slab edge style.

Defaults Tab

The **Defaults** tab shown in Figure 6.52 allows you to define the properties of the new style. A description of the options in the tab follows.

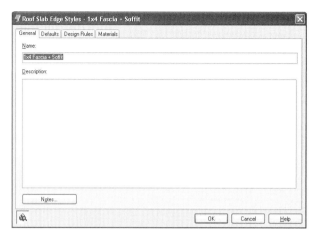

Figure 6.51 *General tab of the Roof Slab Edge Styles dialog box*

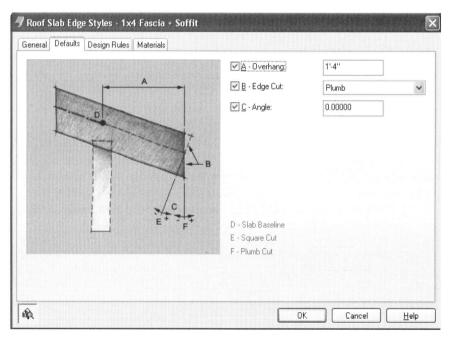

Figure 6.52 *Defaults tab of the Roof Slab Edge Styles dialog box*

A-Overhang – The **Overhang** specifies the horizontal distance to extend the roof edge from the insertion point.

B-Edge Cut – The **Edge Cut** specifies the roof edge vertical plane as either plumb or square. A square orientation places the fascia plane perpendicular to the roof-decking plane. The plumb option orients the fascia plane perpendicular to the floor or a horizontal plane.

C-Angle – The **Angle** edit field specifies the angle of the fascia relative to the plane specified by the orientation. Negative angles are measured clockwise from the specified orientation plane, whereas positive angles are measured counterclockwise from the orientation plane.

Design Rules

The **Design Rules** tab shown in Figure 6.53 consists of edit fields for defining the profiles and location of fascia and soffit components. The shapes of the fascia and soffit components are developed from profiles. The profiles of the current drawing are listed in the drop-down list for the components. Custom profiles can be created from closed polylines in the **Style Manager**. A description of the options of the **Design Rules** tab follows.

Fascia

Fascia – The **Fascia** check box turns on the display of a fascia. If it is toggled ON, the **Profile** drop-down list is active, and you can select a profile for the fascia.

A-Auto-Adjust to Edge Height – The **Auto-Adjust to Edge Height** check box adjusts the fascia automatically according to the roof slab height changes. If this is toggled OFF, the fascia dimensions are fixed according to the size defined in the profile.

Soffit

Soffit – The **Soffit** check box turns on the display of the soffit. If it is toggled ON, you can select a profile for the soffit from the drop-down list.

B-Auto-Adjust to Overhang Depth – The **Auto-Adjust to Overhang Depth** check box allows the soffit to automatically scale according to changes in the roof slab. If this is toggled OFF, the soffit is fixed according to the dimensions defined in the profile.

C-Angle – The **Angle** specifies the soffit angle from horizontal. A zero angle places the soffit in a horizontal position, whereas a positive angle will tilt the soffit up from the insertion point at J as shown in Figure 6.53.

D-Horizontal Offset from Roof Slab Baseline – The **Horizontal Offset from Roof Slab Baseline** edit field specifies the horizontal distance from the soffit to the roof slab baseline.

E-"Y"Direction – The **"Y" Direction** edit field specifies the distance from the fascia insertion point to the soffit insertion point.

F-"X"Direction – The **"X" Direction** edit field specifies the distance in x direction from the fascia insertion point to the soffit insertion point.

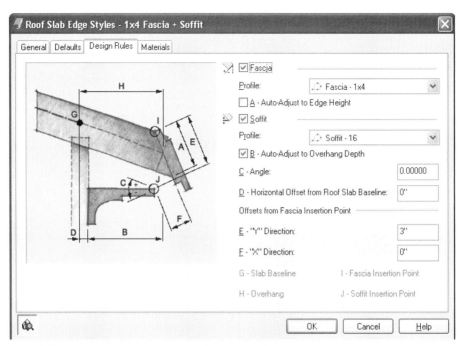

Figure 6.53 *Design Rules tab of the Roof Slab Edge Styles dialog box*

Materials Tab

The **Materials** tab allows you to specify the name and definition of materials applied to the roof slab edge. The options of the **Materials** tab are shown in Figure 6.54 and described below.

> **Component** – The **Component** column lists the name of the roof slab edge components to apply materials.
>
> **Material Definition** – The **Material Definition** column lists the materials defined for the component. If you click in this column, you can select from the drop-down list the materials used in the drawing. Materials are defined on the **Display Properties** tab of the **Material Definition Properties** dialog box.
>
> **Edit Material** – The **Edit Material** button opens the **Material Definition Properties** dialog box to change material definitions. The **Material Definition Properties** dialog box includes a **General** tab and a **Display Properties** tab. The **General** tab allows you to edit the name, description, notes and property sets. The **Display Properties** tab lists the **Display Representations**, **Display Property Sources**, **Style Override** check box, and the Edit Display Properties button. Selecting the **Edit Display Properties** button opens the **Display Properties – Material Definition** dialog box as shown in Figure 6.55. The **Layer/Color/Linetype** tab of the

Creating Roofs and Roof Slabs 435

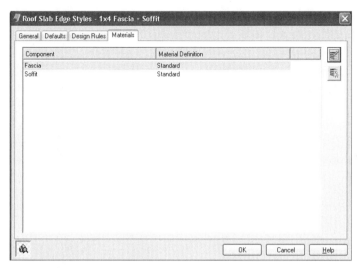

Figure 6.54 *Materials tab*

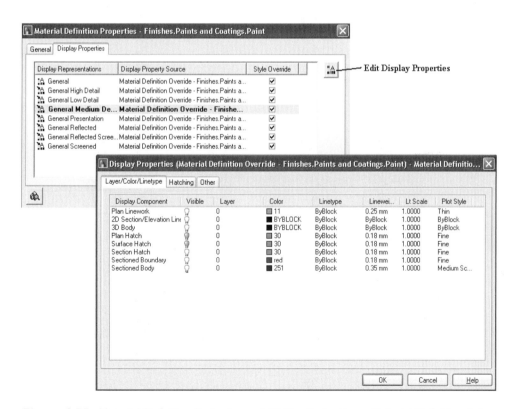

Figure 6.55 *Material Definition Properties and Display Properties – Material Definition dialog boxes*

Display Properties – Material Definition dialog box lists the components of the roof edge style, and you can specify the visibility, color, layer, linetype, lineweight, linetype scale, and plot style of each component.

The **Hatching** tab allows you to specify the hatch pattern for plan, surface, and section hatch. The **Other** tab includes options for defining **Surface Hatch Placement**, **Surface Rendering**, **Live Section Rendering**, and **2D Section Rules** as shown in Figure 6.56.

Add New Material – The **Add New Material** button opens the **New Material** dialog box, which allows you to specify a name for the new material. The material mapped to this name is specified in the **Edit Material** dialog box.

ASSIGNING ROOF SLAB EDGE STYLES TO SLAB EDGES

Roof slab edge styles can be assigned to a roof slab edge by selecting the **Edges** button in the Properties palette of an existing roof slab to open the **Roof Slab Edges** dialog box. The **Roof Slab Edges** dialog box lists all edges of the roof slab. Although the edges are numbered according to the sequence of development, it may be difficult to determine which edge of the slab you are editing in the dialog box as shown in Figure 6.57.

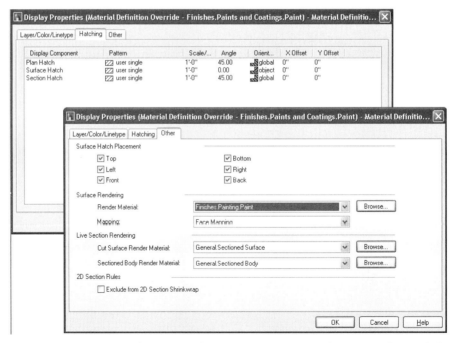

Figure 6.56 *Hatching and Other tabs of the Display Properties – Material Definition dialog box*

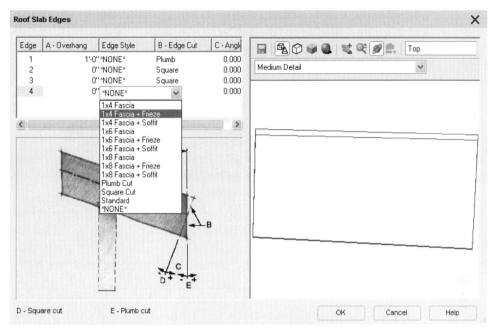

Figure 6.57 *Editing roof slab edge styles in the Roof Slab Edges dialog box*

However, the **Edit Roof Slab Edges** command allows you to select an edge for edit, and the **Edit Roof Slab Edges** dialog box will open with information regarding only the selected edge as shown in Figure 6.58. Access the **Edit Roof Slab Edges** command as shown in Table 6.24.

Command prompt	ROOFSLABEDGEEDIT
Shortcut menu	Select a roof slab, right-click, and choose Edit Roof Slab Edges

Table 6.24 *Edit Roof Slab Edges command access*

When you select a roof slab, right-click, and choose **Edit Roof Slab Edges** from the shortcut menu, you are prompted to select an edge as shown in the following command line sequence.

> (Select a roof slab, right-click, and choose **Edit Roof Slab Edges** from the shortcut menu.)
>
> RoofSlabEdgeEdit
>
> Select edges of one roof slab: *(Select an edge of a roof slab.)*

Select edges of one roof slab: ENTER *(Press* ENTER *to end selection and open the* **Edit Roof Slab Edges** *dialog box as shown in Figure 6.58.)*

The top left section of the **Edit Roof Slab Edges** dialog box lists the edge number, and you can click in the columns for each edge and edit the **A-Overhang**, **Edge Style**, **B-Edge Cut**, and **C-Angle**. The roof edge styles included in the drawing are listed in the **Edge Style** drop-down list as shown in Figure 6.58.

USING ADD EDGE PROFILE IN THE WORKSPACE

A profile can be added to the drawing for the fascia without opening the **Style Manager** and creating a new profile. The **Add Edge Profiles** command is included in the shortcut menu of a selected roof slab. This command allows you to create a profile while remaining in the workspace. The edge selected for the command must have a defined edge style. Access the **Add Edge Profiles** command as shown in Table 6.25.

Command prompt	ROOFSLABADDEDGEPROFILES
Shortcut menu	Select a roof slab, right-click, and choose Add Edge Profiles

Table 6.25 *Add Edge Profiles command access*

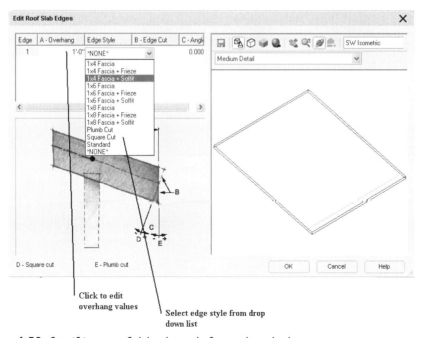

Figure 6.58 *Specifying a roof slab edge style for a selected edge*

When you select a roof slab, right-click, and choose **Add Edge Profiles** from the shortcut menu, you are prompted to select a roof slab edge, which should include a defined roof slab edge style. The command allows you to specify a profile in the drawing for the selected edge and opens the profile for edit using Edit In Place. The **In-Place Edit** toolbar includes a **Zoom To** option that creates an enlarged view of the profile. The profile can then be edited by adding or deleting vertices that define the profile.

STEPS TO ADDING AN EDGE PROFILE

1. Select a roof slab, right-click, and choose **Add Edge Profiles**.

2. Select a roof slab edge. If no roof slab edge style and overhang are defined for the edge, you are prompted in an AutoCAD dialog box shown in Figure 6.59. Select **Yes** to specify these parameters in the **Edit Roof Slab Edges** dialog box. If you select **No**, the command will end.

3. Click **Yes** in response to the AutoCAD dialog box shown in Figure 6.59 to open the **Edit Roof Slab Edges** dialog box. Edit the **Edit Roof Slab Edges** dialog box, click in the style drop-down list, select **Standard**, and then type a distance for the overhang.

4. Click **OK** to close the **Edit Roof Slab Edges** dialog box and open the **Add Fascia/Soffit Profiles** dialog box. Select a profile from the drop-down lists of profile definitions for the fascia and soffit as shown in Figure 6.60.

5. Click **OK** to close the **Add Fascia/Soffit Profiles** dialog box, and the **In-Place Edit** toolbar opens. This toolbar, as discussed in Chapter 5 and as shown in Figure 6.61, includes the **Zoom to** command.

6. Select the **Zoom to** option of the **In-Place-Edit** toolbar. The profile is shown in cross section.

7. Select a grip on the profile and stretch the grip to a new location, or right-click and choose one of the following options of the shortcut menu: **Add Vertex**, **Remove Vertex**, **Replace Ring**, **Save All Changes**, **Save As New Profile**, or **Discard All Changes** as shown in Figure 6.61.

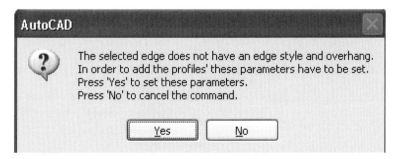

Figure 6.59 *AutoCAD message box*

Figure 6.60 Add Fascia/Soffit Profiles dialog box

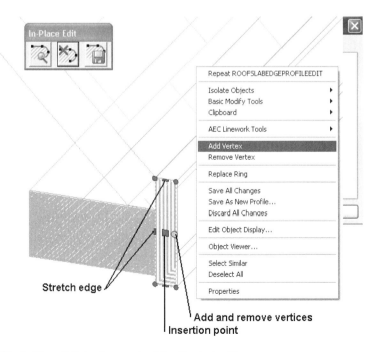

Figure 6.61 In-Place Edit toolbar and editing fascia profile

8. Select the **Save All Changes** from the **In-Place Edit** toolbar to save the edit.

CHANGING THE PROFILE WITH EDIT EDGE PROFILE IN PLACE

After a profile has been changed using the **Add Edge Profiles** command, additional editing can be performed with the **Edit Edge Profile In Place** command. This command allows you to perform additional editing of the profile. It allows you to add or remove vertices from the profile in the workspace. Access the **Edit Edge Profile In Place** command as shown in Table 6.26.

Command prompt	ROOFSLABEDGEPROFILEEDIT
Shortcut menu	Select a roof slab, right-click, and choose Edit Edge Profile In Place

Table 6.26 *Edit Edge Profile In Place command access*

When you select a roof slab, right-click, and choose the **Edit Edge Profile In Place** command from the shortcut menu, the **In-Place Edit** toolbar is displayed. The **Zoom To** command of the **In-Place Edit** toolbar will create an enlarged view of the profile. Edit the profile by selecting a grip, or right-click and choose one of the following from the shortcut menu: **Add Vertex**, **Remove Vertex**, **Replace Ring**, **Save All Changes**, **Save As New Profile**, or **Discard All Changes**.

EXTENDING WALLS TO THE ROOF

The walls below roofs and roof slabs do not automatically extend to the roof when a roof or roof slab is created. The **RoofLine** command allows you to extend the walls to the roof or roof slab. See Table 3.2 of Chapter 3 for command access.

SUMMARY

1. Roofs are created with the **RoofAdd** command located on the Design tool palette.
2. Using the **RoofAdd** command with a single slope shape will create a hip roof.
3. The Properties palette allows you to define the thickness of the roof plane, slope, and overhang.
4. Existing roofs can be edited by changing the slope, plate height, shape, and overhang in the Properties palette.
5. The angle of the roof can be defined by the ratio of the rise to run or by a slope angle.

6. Create a gable end of a gable roof by setting the roof edge slope to 90 degrees.
7. Create a shed roof by selecting the **Gable** shape for three of the four roof edges.
8. Create the gambrel roof by applying the double slope shape to two edges of the roof.
9. The **Apply Tool Properties to>Roof** command is a shortcut menu option of the **Roof** command located on the Design palette. The **Apply Tool Properties to** command can be used to create a roof based upon linework or walls.
10. A roof object can be converted to roof slabs by selecting the roof, right-clicking, and choosing **Convert to Roof Slabs**.
11. Roof slabs can be trimmed, extended, and mitered to create complex roofs.
12. Fascia, soffit, and frieze millwork components can be added to a roof slab by editing the roof slab edge style.
13. A profile is used to create components for the roof slab edge style.
14. A profile used in a roof slab edge style can be created or modified by the **Add Edge Profiles** or **Edit Edge Profile in Place** command of the roof slab shortcut menu.

REVIEW QUESTIONS

1. The Gable slope option of the Properties palette sets the slope angle to _____.
2. The slope of a roof is defined as _____ / _____.
3. Create a gambrel roof by setting the shape to _____.
4. The Properties palette displays the _____, _____, and _____ properties of each edge of a roof.
5. The plate height of the roof should be set equal to _____.
6. The **Apply Tool Properties to** command will develop a roof from a _____ or _____.
7. The **Edit Edges/Faces** command allows you to _____.
8. The plate height of a roof is relative to _____.
9. Mass groups markers are located on the _____ tool palette.
10. Two disconnected roof slabs can be joined by the _____ option of the _____ command.
11. Walls can be extended to the gable roof plane with the _____ command.
12. Represent rafter size and the decking thickness by adjusting the _____ in the _____ palette.
13. Describe the procedure for creating a shed roof.

TUTORIAL 6.1 CREATING A HIP ROOF

1. Open Autodesk Architectural Desktop 2005, and select **File>Project Browser** from the menu bar. Use the Project Selector drop-down list to navigate to your *ADT Student\ADT Tutor\Ch6* student directory. Double-click on **Ex 6-1** to set this project current. Select **Close** to dismiss the **Project Browser**. (If your student folder does not include the project, refer to "Organizing Tutorial Directories" in the Preface.)

2. Select the **Constructs** tab of the **Project Navigator**, and double-click on **Floor 1** of the **Constructs** tab.

3. Select **Top View** from the **View** flyout of the **Navigation** toolbar.

4. Move the mouse pointer to the OSNAP toggle on the status bar, right-click, and choose **Settings** from the shortcut menu. Edit the **Object Snap** tab of the **Drafting Settings** dialog box: clear all object snap modes, and check the **Intersection** object snap. Check the **Object Snap On (F3)** check box and clear the **Object Snap Tracking On (F11)** check box. Select **OK** to dismiss the **Drafting Settings** dialog box.

5. Turn off the display of brick and brick hatching by selecting the exterior wall; right-click and choose **Edit Wall Style** from the shortcut menu.

6. Select the **Display Properties** tab of the **Wall Style Properties** dialog box.

7. Check the **Style Override** check box for the **Plan** display representation.

8. Turn OFF the visibility of the following components: Shrink Wrap, Boundary 2 (Brick) and Hatch 2 (Brick) components as shown in Figure 6T.1.

9. Select **OK** to dismiss the **Display Properties (Wall Style Override-Frame Brick) – Wall Plan Display Representation** dialog box. Select **OK** to dismiss the **Wall Style Properties – Frame Brick** dialog box.

10. Select the **Roof** command from the Design palette and edit the Properties palette as follows: Basic–Dimensions: Thickness = 6", Edge cut = Plumb, Next Edge: Shape = Single slope, Overhang = 20", Lower Slope: Plate height = 8', Rise = 6 as shown in Figure 6T.2. Respond to the following command line prompts by selecting roof points as specified below.

    Command: RoofAdd

    Roof point or

    [Shape/Gable/Overhang/overhangValue/PHeight/PRise/PSlope/UHeight/URise/USlope/Match]: *(Select the intersection of lines representing the outside of stud, point p1 in Figure 6T.2.)*

    Roof point or

    [Shape/Gable/Overhang/overhangValue/PHeight/PRise/PSlope/UHeight/URise/USlope/Match/Undo]: *(Select the intersection of lines representing the outside of stud, point p2 in Figure 6T.2.)*

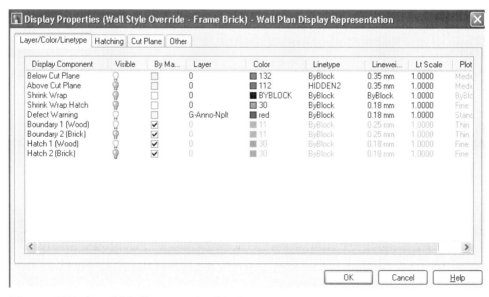

Figure 6T.1 *Layer/Color/Linetype tab of the Display Properties (Wall Style Overide – Frame Brick) – Wall Plan Display Representation dialog box*

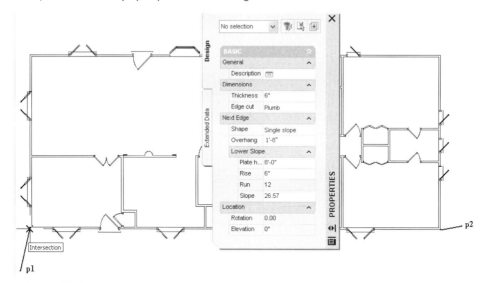

Figure 6T.2 *Settings in the Properties palette*

Roof point or

[Shape/Gable/Overhang/overhangValue/PHeight/PRise/PSlope/UHeight/URise/USlope/Match/Undo]: *(Select the intersection of lines representing the outside of stud, point p3 in Figure 6T.3.)*

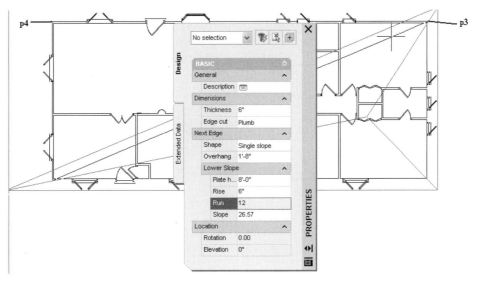

Figure 6T.3 *Location of points for hip roof*

Roof point or

[Shape/Gable/Overhang/overhangValue/PHeight/PRise/PSlope/ UHeight/URise/USlope/Match/Undo]: *(Select the intersection of lines representing the outside of stud, point p4 in Figure 6T.3.)*

Roof point or

[Shape/Gable/Overhang/overhangValue/PHeight/PRise/PSlope/ UHeight/URise/USlope/Match/Undo]: ENTER *(Press* ENTER *to end the selection of points.)*

11. Turn ON the display of brick and brick hatching by selecting an exterior wall, right-clicking, and choosing **Edit Wall Style** from the shortcut menu.

12. Select the **Display Properties** tab of the **Wall Style Properties** dialog box.

13. Clear the check for **Style Override** of the **Plan** display representation. Select **OK** to close the **Wall Style Properties – Frame Brick** dialog box.

14. Select **SW Isometric** from the **View** flyout on the **Navigation** toolbar.

15. Select **Hidden** from the **Shade** flyout of the **Navigation** toolbar to view the roof as shown Figure 6T.4.

16. Save and close your drawing.

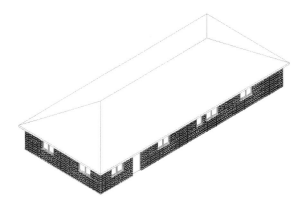

Figure 6T.4 *Isometric view of hip roof*

TUTORIAL 6.2 CREATING A GABLE ROOF

1. Open your *ADT Student\ADT Tutor\Ch6\Ex6-2.dwg* and save the drawing as **Lab 6-2** in your ADT Student directory.

2. Move the mouse pointer to the OSNAP toggle of the status bar and right-click; then choose **Settings** from the shortcut menu. Check ON the **Intersection** mode of object snap and clear all other modes of object snap. Check ON **Object Snap On(F3)** at the top of the **Object Snap** tab and select **OK** to dismiss the **Drafting Settings** dialog box.

3. Select the **Roof** command from the Design palette. Edit the Properties palette as follows: Basic-Dimensions: Thickness = 6", Edge cut = Plumb, Next Edge: Shape = Gable, Overhang = 6", as shown in Figure 6T.5. Respond to the following command line prompts by selecting roof points as specified below.

    Command: RoofAdd

    Roof point or

    [Shape/Gable/Overhang/overhangValue/PHeight/PRise/PSlope/ UHeight/URise/USlope/Match]: *(Select the wall using the intersection of the outer lines of the corner at p1 as shown in Figure 6T.5.)*

    Roof point or

    [Shape/Gable/Overhang/overhangValue/PHeight/PRise/PSlope/ UHeight/URise/USlope/Match/Undo]: *(Select the wall using the intersection of the outer lines of the corner at p2 as shown in Figure 6T.5.)*

    (Edit the Properties palette as follows: Shape = **Single slope**, *Overhang =* **20**, *and Rise =* **8".**)

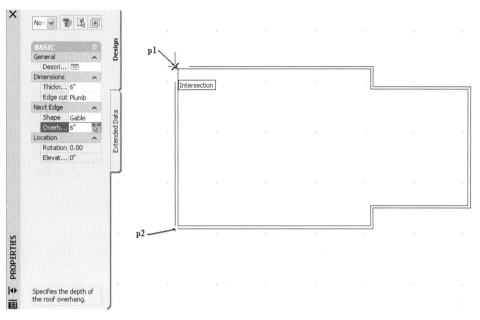

Figure 6T.5 *Editing the Properties palette for roof edge 1*

Roof point or

[Shape/Gable/Overhang/overhangValue/PHeight/PRise/PSlope/UHeight/URise/USlope/Match/Undo]: *(Select the wall using the intersection of the **outer** lines of the corner at p3 as shown in Figure 6T.6.)*

*(Edit the Properties palette as follows: Shape = **Gable**, Overhang = **6**.)*

Roof point or

[Shape/Gable/Overhang/overhangValue/PHeight/PRise/PSlope/UHeight/URise/USlope/Match/Undo]: *(Select the wall using the intersection of the outer lines of the corner at p4 as shown in Figure 6T.6.)*

*(Edit the Properties palette as follows: Shape = **Single slope**, Overhang = **20**, and Rise = **8"**.)*

Roof point or

[Shape/Gable/Overhang/overhangValue/PHeight/PRise/PSlope/UHeight/URise/USlope/Match/Undo]: *(Select the wall using the intersection of the outer lines of the corner at p5 as shown in Figure 6T.6.)*

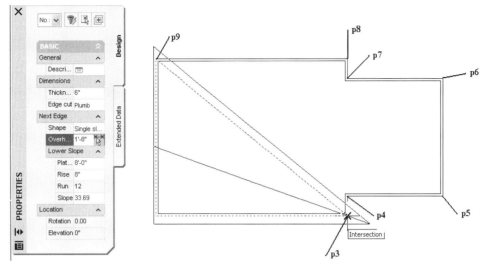

Figure 6T.6 *Editing the Properties palette for point 3*

*(Edit the Properties palette as follows: Shape = **Gable**, Overhang = **6**.)*

Roof point or

[Shape/Gable/Overhang/overhangValue/PHeight/PRise/PSlope/UHeight/URise/USlope/Match/Undo]: *(Select the wall using the intersection of the outer lines of the corner at p6 as shown in Figure 6T.6.)*

*(Edit the Properties palette as follows: Shape = **Single slope**, Overhang = **20**, and Rise = **8"**.)*

Roof point or

[Shape/Gable/Overhang/overhangValue/PHeight/PRise/PSlope/UHeight/URise/USlope/Match/Undo]: *(Select the wall using the intersection of the outer lines of the corner at p7 as shown in Figure 6T.6.)*

*(Edit the Properties palette as follows: Shape = **Gable**, Overhang = **6**.)*

Roof point or

[Shape/Gable/Overhang/overhangValue/PHeight/PRise/PSlope/UHeight/URise/USlope/Match/Undo]: *(Select the wall using the intersection of the outer lines of the corner at p8 as shown in Figure 6T.6.)*

*(Edit the Properties palette as follows: Shape = **Single slope**, Overhang = **20**, and Rise = **8"**.)*

Roof point or

[Shape/Gable/Overhang/overhangValue/PHeight/PRise/PSlope/UHeight/URise/USlope/Match/Undo]: *(Select the wall using the intersection of the outer lines of the corner at p9 as shown in Figure 6T.6.)*

Roof point or

[Shape/Gable/Overhang/overhangValue/PHeight/PRise/PSlope/UHeight/URise/USlope/Match/Undo]: ENTER

4. Select **SE Isometric** from the **View** flyout on the **Navigation** toolbar.
5. Select **Hidden** from the **Shade** flyout of the **Navigation** toolbar; the final roof should appear as shown in Figure 6T.7.
6. Save and close the drawing.

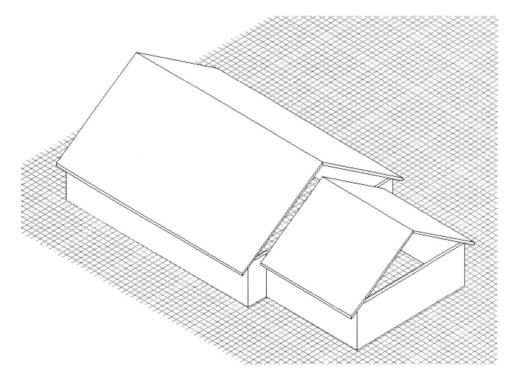

Figure 6T.7 *Completed gable roof*

TUTORIAL 6.3 CREATING AND EDITING A GAMBREL ROOF

1. Open your *ADT Student\ADT Tutor\Ch6\Ex 6-3.dwg* and save the drawing as **Lab 6-3** in your student directory.

2. Set the object snap by moving the mouse pointer to the OSNAP toggle of the status bar and right-clicking; choose **Settings** from the shortcut menu. Select the **Intersection** mode and clear all other object snap modes. Verify that **Object Snap On(F3)** is checked. Select **OK** to dismiss the **Drafting Settings** dialog box.

3. Clear all toggles except OSNAP on the status bar.

4. Right-click over the **Roof** tool of the Design palette and choose **Apply Tool Properties to>Linework and Walls**. Respond to the command line prompts as shown in the following command line sequence:

    Command: RoofToolToLinework

    Choose walls or polylines to create roof profile: *(Create a crossing selection by selecting a point near p1 and moving the pointer to the left and selecting a point near p2 as shown at left in Figure 6T.8.)*

    Select objects: Specify opposite corner: 7 found

    Select objects: ENTER *(Press ENTER to end selection.)*

    Erase layout geometry? [Yes/No] <No>: ENTER *(Press ENTER to retain the walls in the drawing.)*

    (Roof is created as shown at right in Figure 6T.8.)

5. Convert the Hip roof to the Gambrel shape by editing the Properties palette: select the roof, and edit the Properties palette as follows: Basic–Layer = A-Roof, Dimensions: Thickness = **6"**, Edge cut = Plumb, Shape = **Double slope**, Overhang = **6"**, Lower Slope: Plate height = **8'**, Rise = **24"**, Upper Slope: Upper height = **16'**, Rise = **5"** as shown in Figure 6T.9.

6. Press ESC, and toggle ON ORTHO on the status bar.

7. Create a gable using grips, verify that grips are displayed, and then select the ridge grip of the upper slope at **p1** as shown in Figure 6T.10. Stretch the grip to a point near **p2** as shown at right in Figure 6T.10.

8. Create a gable for the remaining slope using grips by selecting the grip of the lower slope at **p1** as shown in Figure 6T.11. Stretch the grip to a point near **p2** as shown at right in Figure 6T.11.

9. Edit grips at **p3** and **p4** as shown at left in Figure 6T.11 by repeating steps 7 and 8 above.

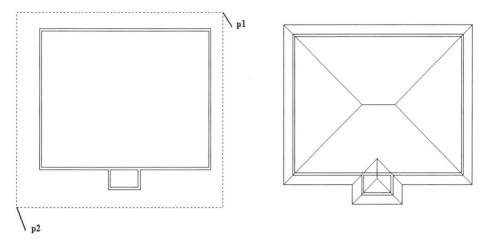

Figure 6T.8 *Roof created from walls*

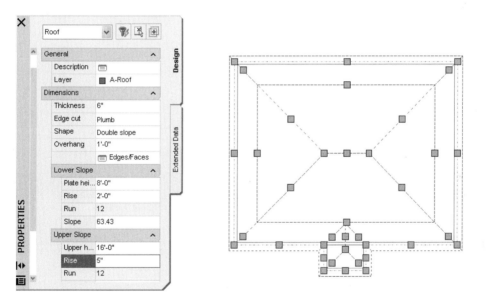

Figure 6T.9 *Properties of roof defined*

10. Press ESC, and select the **SE Isometric** from the **View** flyout of the **Navigation** toolbar.
11. Select **Top View** from the **View** flyout of the **Navigation** toolbar.
12. Select **2D Wireframe** from the **Shade** flyout of the **Navigation** toolbar.

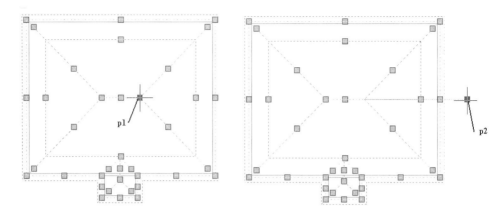

Figure 6T.10 *Upper slope ridge grip edited*

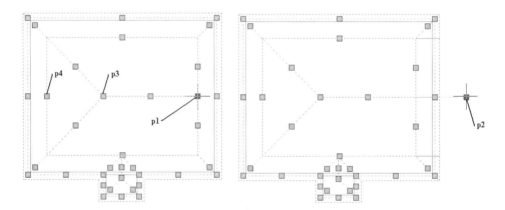

Figure 6T.11 *Lower slope grip edited*

13. Select the roof, right-click, and choose **Edit Edges/Faces** from the shortcut menu. Respond to command prompts as shown below.

    Command: RoofEditEdges

    Select a roofedge: *(Select edge at p1 as shown in Figure 6T.12.)*

    Select a roof edge: ENTER *(Press ENTER to end selection and open the* **Roof Edges and Faces** *dialog box.)*

14. Select **Edge 0** in the **Roof Edges** section of the **Roof Edges and Faces** dialog box. Click in the **Slope** column of Face 0 and change the slope to **90**. Click in the **Slope** column of Face 1 and change the slope to **90**. Select **OK** to dismiss the **Roof Edges and Faces** dialog box.

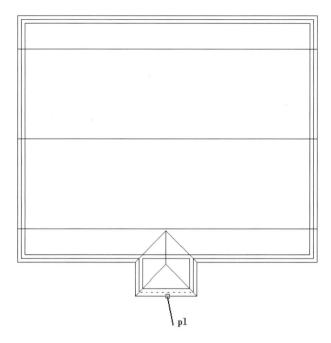

Figure 6T.12 *Creating gable using RoofEditEdges command*

15. Select the roof, right-click, and choose **Edit Edges/Faces** from the shortcut menu. Select edges at **p1** and **p2** as shown in Figure 6T.13.
16. Select Edge 0 in the **Roof Edges** section of the **Roof Edges and Faces** dialog box. Click in the **Slope** column of Face 0 and change the slope to **22.62**.

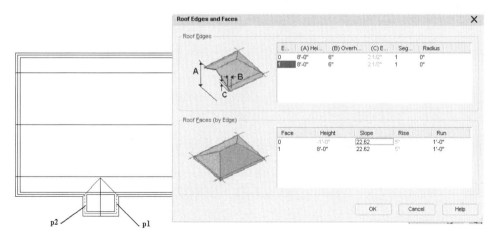

Figure 6T.13 *Creating equal slope for upper and lower faces*

Select Edge 1 in the **Roof Edges** section and change the slope of Face 0 to have a slope of **22.62** as shown in Figure 6T.13. Select **OK** to close the **Roof Edges and Faces** dialog box.

17. Select the **SE Isometric** from the **View** flyout of the **Navigation** toolbar.
18. Select **2D Wireframe** from **Shade** flyout of the **Navigation** toolbar.
19. Select the wall at **p1** and **p2** as shown at left in Figure 6T.14, right-click, and choose **Roof/Floor Line>Modify Roof Line** from the shortcut menu. Respond to the command line prompts as shown below.

    Command: RoofLine

    RoofLine [Offset/Project to polyline/Generate polyline/Auto project/Reset]: **a** ENTER

    Select objects: 1 found *(Select the gambrel roof.)*

    Select objects: ENTER *(Press ENTER to end the selection.)*

20. Select **Hidden** from the **Shade** flyout of the **Navigation** toolbar to view the walls extended as shown at right in Figure 6T.15.
21. Close and save the drawing.

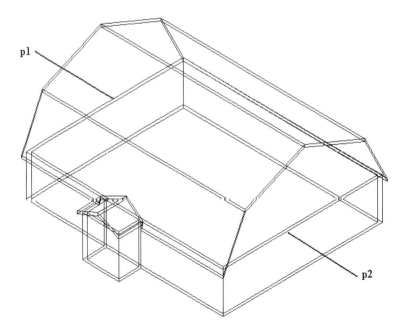

Figure 6T.14 *Extending walls to roof*

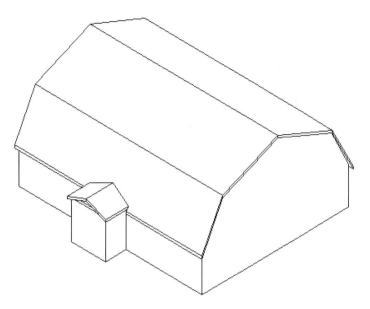

Figure 6T.15 *Gambrel roof completed*

TUTORIAL 6.4 CREATING DORMERS AND ROOF SLABS

1. Open *ADT Tutor\Ch6\Ex 6-4.dwg* and save the drawing as **Lab 6-4** in your student directory. This file includes a polyline developed by tracing the outline of the first floor plan.

2. Set the object snap by moving the mouse pointer to the OSNAP toggle of the status bar and right-clicking; choose **Settings** from the shortcut menu. Select the **Intersection** mode and clear all other object snap modes. Verify that **Object Snap On(F3)** is checked. Select **OK** to dismiss the **Drafting Settings** dialog box.

3. Clear all toggles except OSNAP on the status bar.

4. Select the **Roof** command from the Design palette. Edit the Properties palette as follows: Basic–Dimensions: Thickness = 6", Edge cut = Plumb, Next Edge: Shape = Gable, Overhang = 12", as shown in Figure 6T.16. Respond to the following command line prompts by selecting roof points as specified below.

   Command: RoofAdd

   Roof point or

   [Shape/Gable/Overhang/overhangValue/PHeight/PRise/PSlope/ UHeight/URise/USlope/Match]: *(Select the wall using the Intersection object snap at the outer lines of the corner at p1 as shown in Figure 6T.16.)*

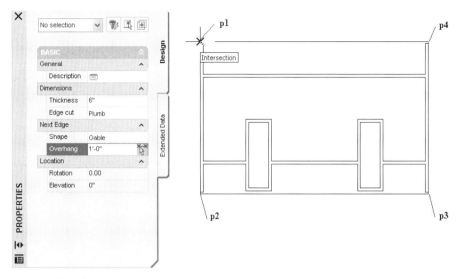

Figure 6T.16 *Vertex location for main roof*

Roof point or

[Shape/Gable/Overhang/overhangValue/PHeight/PRise/PSlope/ UHeight/URise/USlope/Match/Undo]: *(Select the wall using the intersection of the **outer** lines of the corner at p2 as shown in Figure 6T.16.)*

*(Edit the Properties palette as follows: Shape = **Single slope**, Overhang = **12**, Plate height = **0** and Rise = **12"**.)*

Roof point or

[Shape/Gable/Overhang/overhangValue/PHeight/PRise/PSlope/ UHeight/URise/USlope/Match/Undo]: *(Select the wall using the intersection of the outer lines of the corner at p3 as shown in Figure 6T.16.)*

*(Edit the Properties palette as follows: Shape = **Gable**, Overhang = **12**.)*

Roof point or

[Shape/Gable/Overhang/overhangValue/PHeight/PRise/PSlope/ UHeight/URise/USlope/Match/Undo]: *(Select the wall using the intersection of the outer lines of the corner at p4 as shown in Figure 6T.16.)*

*(Edit the Properties palette as follows: Shape = **Single slope**, Overhang = **12**, Plate height = **0** and Rise = **12"**.)*

Roof point or

[Shape/Gable/Overhang/overhangValue/PHeight/PRise/PSlope/ UHeight/URise/USlope/Match/Undo]: *(Select the wall using the intersection of the outer lines of the corner at p1 as shown in Figure 6T.16.)*

Roof point or

[Shape/Gable/Overhang/overhangValue/PHeight/PRise/PSlope/ UHeight/URise/USlope/Match/Undo]: ENTER *(Press* ENTER *to end Roof command.)*

5. Add the dormer roofs by selecting the **Roof** command from the Design palette. Edit the Properties palette as follows: Basic–Dimensions: Thickness = 6", Edge cut = Plumb, Next Edge: Shape = Single slope, Overhang = 0", Lower Slope-Plate Height = 8', Rise = 4. Respond to the following command line prompts by selecting roof points as specified in Figure 6T.17.

   Command: RoofAdd

   Roof point or

   [Shape/Gable/Overhang/overhangValue/PHeight/PRise/PSlope/ UHeight/URise/USlope/Match]: *(Select the wall using the intersection of the outer lines of the corner at p1 as shown in Figure 6T.17.)*

   Roof point or

   [Shape/Gable/Overhang/overhangValue/PHeight/PRise/PSlope/ UHeight/URise/USlope/Match/Undo]: *(Select the wall using the intersection of the outer lines of the corner at p2 as shown in Figure 6T.17.)*

   (Edit the Properties palette as follows: Shape = **Gable***, Overhang =* **0***.)*

   Roof point or

   [Shape/Gable/Overhang/overhangValue/PHeight/PRise/PSlope/ UHeight/URise/USlope/Match/Undo]: *(Select the wall using the intersection of the outer lines of the corner at p3 as shown in Figure 6T.17.)*

   (Edit the Properties palette as follows: Shape = **Single slope***, Overhang =* **0***, Lower Slope-Plate Height =* **8'***, Rise =* **4***.)*

   Roof point or

   [Shape/Gable/Overhang/overhangValue/PHeight/PRise/PSlope/ UHeight/URise/USlope/Match/Undo]: *(Select the wall using the intersection of the outer lines of the corner at p4 as shown in Figure 6T.17.)*

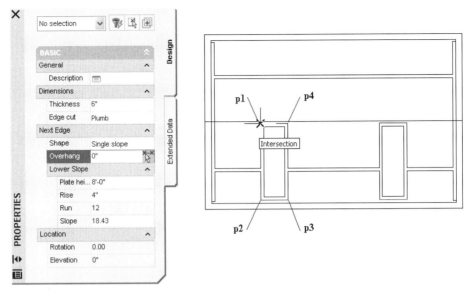

Figure 6T.17 *Vertex locations for dormer roof*

(*Edit the Properties palette as follows: Shape =* **Gable**, *Overhang =* **0**.)

Roof point or

[Shape/Gable/Overhang/overhangValue/PHeight/PRise/PSlope/ UHeight/URise/USlope/Match/Undo]: *(Select the wall using the intersection of the outer lines of the corner at p1 as shown in Figure 6T.17.)*

[Shape/Gable/Overhang/overhangValue/PHeight/PRise/PSlope/ UHeight/URise/USlope/Match/Undo]: ENTER *(Press* ENTER *to end the command.)*

6. Repeat step 5 to add a roof for the dormer at right.
7. Select **SW Isometric** from the **Navigation** toolbar.
8. Select the **Roof Slab** command on the Design palette, right-click, and choose **Apply Tool Properties to>Linework, Walls and Roof**. Respond to the command line prompts as shown below:

Command: RoofSlabToolToLinework

Select roof, walls or polylines: 1 found *(Select main roof at p1 as shown in Figure 6T.18.)*

Select roof, walls or polylines: 1 found, 2 total *(Select dormer roof at p2 as shown in Figure 6T.18.)*

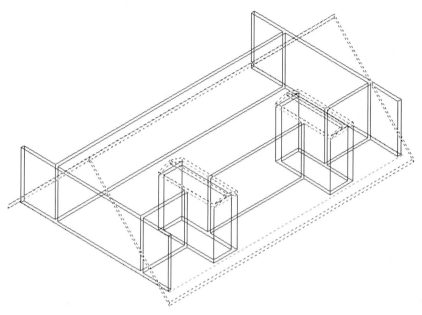

Figure 6T.18 *Selection of existing roofs to convert to roof slabs*

Select roof, walls or polylines: 1 found, 3 total *(Select dormer roof at p3 as shown in Figure 6T.18.)*

Select roof, walls or polylines: ENTER *(Press ENTER to end selection.)*

Erase layout geometry? [Yes/No] <No>: ENTER *(Press ENTER to retain the roof.)*

9. Select **Top View** from the **View** flyout of the **Navigation** toolbar. Select the **Layer Manager** from the **Properties** toolbar. Freeze the A-Roof layer.

10. To remove the cut plane lines from the main roof slab, select the front roof slab of the main roof at p1 as shown in Figure 6T.19, right-click, and choose **Edit Object Display**. Select the **Display Properties** tab of the **Object Display** dialog box. Select the **Edit Display Properties** button to open the **Display Properties (Drawing Default) Roof Slab Plan Display Representation** dialog box. Select the **Other** tab, check the **Override Display Configuration Cut Plane** check box, and set the Cut Plane Height to **0**. Select **OK** to close the **Other** tab and select **OK** to close the **Object Display** dialog box.

11. Select **SW Isometric** from the **View** flyout of the **Navigation** toolbar.

12. Select the front roof slab near **p1** as shown in Figure 6T.20, right-click, and choose **Roof Dormer** from the shortcut menu. Respond to the command line prompts as shown below.

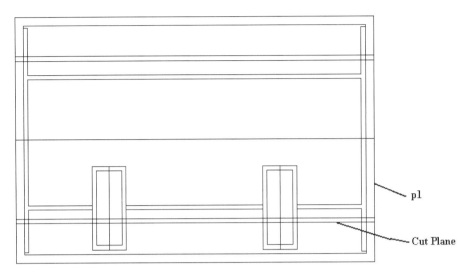

Figure 6T.19 *Selection of roof to edit display of cut plane*

Command: RoofSlabDormer

Select objects that form the dormer: *(Select p2 as shown at left in Figure 6T.20.)* 1 found

Select objects that form the dormer: *(Select p3 as shown at left in Figure 6T.20.)* 1 found, 2 total

Select objects that form the dormer: *(Select p4 as shown at left in Figure 6T.20.)* 1 found, 3 total

Select objects that form the dormer: *(Select p5 as shown at left in Figure 6T.20.)* 1 found, 4 total

Select objects that form the dormer: *(Select p6 as shown at left in Figure 6T.20.)* 1 found, 5 total

Select objects that form the dormer: *(Select p7 as shown at left in Figure 6T.20.)* 1 found, 6 total

Select objects that form the dormer: ENTER *(Press ENTER to end selection.)*

Slice wall with roof slab [Yes/No] <Yes>: ENTER

13. Repeat step 8 to create the dormer at right.
14. Select the walls **p1** through **p8** shown in Figure 6T.21, right-click, and choose **Roof/Floor Line>Modify Roof Line** from the shortcut menu. Respond to the command line prompts as shown below.

    Command: RoofLine

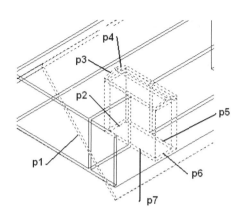 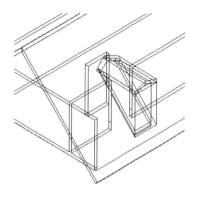

Figure 6T.20 *Selection of components to create dormer*

RoofLine [Offset/Project to polyline/Generate polyline/Auto project/Reset]: **a** ENTER

Select objects: 1 found *(Select roof slab at p9 as shown in Figure 6T.21.)*

Select objects: 1 found, 2 total *(Select roof slab at p10 as shown in Figure 6T.21.)*

Select objects: 1 found, 3 total *(Select roof slab at p11 as shown in Figure 6T.21.)*

Select objects: 1 found, 4 total *(Select roof slab at p12 as shown in Figure 6T.21.)*

Select objects: 1 found, 5 total *(Select roof slab at p13 as shown in Figure 6T.21.)*

Select objects: 1 found, 6 total *(Select roof slab at p14 as shown in Figure 6T.21.)*

Select objects: ENTER *(Press ENTER to end selection.)*

[8] Wall cut line(s) converted.

RoofLine [Offset/Project to polyline/Generate polyline/Auto project/Reset]: ENTER *(Press ENTER to end the command.)*

15. Select **Erase** from the **Modify** toolbar and respond to the command line prompts as shown below.

    Command: _erase

    Select objects: 1 found *(Select wall A as shown in Figure 6T.21.)*

    Select objects: 1 found, 2 total *(Select wall B as shown in Figure 6T.21.)*

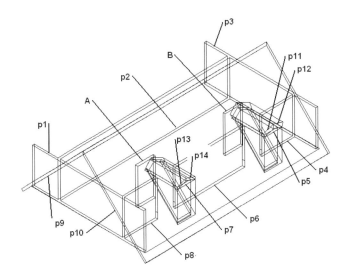

Figure 6T.21 *Selection of walls for edit to roofline*

Select objects: ENTER *(Press* ENTER *to end the selection.)*

16. Select **SW Isometric** view from the **View** flyout of the **Navigation** toolbar.
17. Select **Hidden** from the **Shade** flyout of the **Navigation** toolbar to view the house as shown in Figure 6T.22.
18. Close and save the drawing.

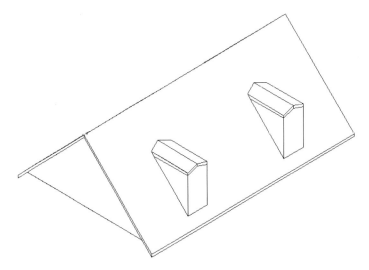

Figure 6T.22 *Completed roof slab edit*

PROJECT

Exercise 6-5 Creating a Shed Roof

1. Open *ADT Student\ADT Tutor\Ch6\Ex6-5.dwg*.
2. Save the drawing to your student directory as **Lab 6-5**.
3. Referring to Figure 6T.23, create a gable roof with a slope of 6/12 through points **p1, p2, p3**, and **p4**. Set the thickness = 6, edge cut = plumb, plate height = 8', overhang to 12 on the sloped edges. Set the overhang of the gables to 0.
4. Convert the roof to roof slabs.
5. Create a roof slab with the pivot point at p5 with the baseline endpoint at **p6**. Set the Mode = Projected, Thickness = 6, Vertical Offset = 0, Horizontal Offset = 0, Justify = Bottom, Base Height = 8', Direction = Left, Overhang = 12, Baseline edge = none, Perimeter edge = none, and Rise = 3/12.
6. Use the **Content Browser** to import the 04 – 1x6 Fascia + Soffit roof slab style from the Design Tool Catalog - Imperial>Roof Slabs & Slabs>Roof Slabs category.
7. Edit edges **p1–p2, p3–p4**, and **p5–p6** to include the **1x6 Fascia + Soffit**.
8. Trim or extend the roof slab beginning at **p5–p6** to the wall that includes **p2–p3**.
9. Extend all walls to roof as shown at right in Figure 6T.23.
10. Select **SE Isometric** from the **Navigation** toolbar.

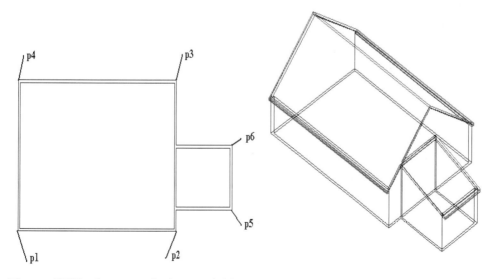

Figure 6T.23 *Creating and editing roof slabs*

CHAPTER 7

Creating Slabs for Floors and Ceilings

INTRODUCTION

Slabs are used to represent the floor or roof system of a building. The slab used in wood frame construction represents the thickness of the floor joist and flooring system. If a concrete slab is used, the slab thickness is set to the thickness of the concrete slab. Slabs can also be used to represent a flat or low-sloped roof in commercial construction.

OBJECTIVES

After completing this chapter, you will be able to

- Use the **SlabAdd** command to add a slab
- Modify a slab by editing the Properties palette and using grips
- Create a slab that includes a slab edge style
- Convert polylines and walls to a slab
- Edit the slab edge style of a slab

ADDING AND MODIFYING A FLOOR SLAB

A floor slab is added to the drawing by selecting the **Slab** tool (**SlabAdd** command) located on the Design tool palette as shown in Figure 7.1. Access the **Slab** tool as shown in Table 7.1. When you select the Slab tool, the Properties palette opens, displaying the properties of the floor slab. The floor slab has properties similar to the roof slab. The slab has thickness and can be sloped to accommodate the concrete slab requirements of a garage. The Properties palette is shown in Figure 7.1 and described below.

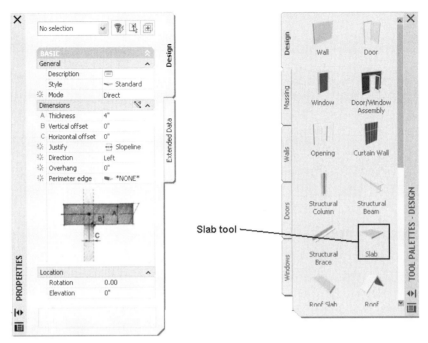

Figure 7.1 *Slab command of the Design tool palette and Properties palette*

Command prompt	SLABADD
Palette	Select the Slab tool from the Design palette as shown in Figure 7.1

Table 7.1 *SlabAdd command access*

Basic

General

Description – The **Description** button opens the **Description** dialog box, allowing you to type a description.

Style – The **Style** drop-down list displays the styles that have been created or imported into the drawing. Additional styles can be imported from the *C:\Documents and Settings\All Users\Application Data\Autodesk\ADT 2005\enu\Styles\Imperial\Slab and Slab Edge Styles (Imperial).dwg* drawing. Type **ST** in the command line to edit this option.

Mode – The **Mode** options include **Direct** and **Projected**. The **Direct** mode creates the slab at the z coordinate of the vertices specified. The **Projected** mode projects from the specified vertices up from the distance specified in the **Base height**. Type **MO** in the command line to edit this option.

Dimensions

A-Thickness – The **Thickness** specifies the thickness or depth of the slab. Type **T** in the command line to edit this option.

B-Vertical offset – The **Vertical offset** specifies the vertical distance from the insert points to the slab baseline.

C-Horizontal offset – The **Horizontal offset** distance shifts the slab horizontally relative to the baseline.

Justify – The **Justify** options include **Top**, **Center**, **Bottom**, and **Slopeline**. Justification is a property used when placing slabs and is not retained in the Properties palette for existing slabs. Type **J** in the command line to edit the Justify option.

> **Top** – Positions the top of the slab with the first point or pivot point.
>
> **Center** – Positions the center of the slab with the first point or pivot point.
>
> **Bottom** – Positions the bottom of the slab with the first point or pivot point.
>
> **Slopeline** – Generates the slab with the pivot point located at the slopeline of the slab. The slopeline is the position on the slab baseline relative to the bottom face.

Base height – The **Base height** is the distance the slab is created from the Z=0 plane. The base height is applied to a slab created in the **Projected** mode. Type **H** in the command line to edit the base height option. The **Base height** is not an option with a Direct Mode slab.

Direction – The **Direction** toggle allows you to create a slab to the left or right of first two selected points when you use **Ortho close**. Type **SL** in the command line to edit the direction option.

Overhang – The **Overhang** value specifies the horizontal distance from the pivot point to extend the slab. Type **OV** in the command line to edit the overhang option.

Perimeter edge – The **Perimeter edge** edit field allows you to specify a slab edge style for the perimeter.

SLOPE:

Rise – The **Rise** specifies the vertical displacement per horizontal run unit.

Run – The **Run** specifies the horizontal component of the slope. The base value is usually 12.

Slope – The **Slope** displays the corresponding angle as a result of the rise and run values. Type **SL** in the command line to edit the angle option.

Location

Rotation – The **Rotation** field displays the rotation angle of the slab.

Elevation – The **Elevation** field displays the current elevation of the slab. The elevation of existing slabs can be edited.

When you select the **Slab** tool (**SlabAdd** command) from the Design palette, you are prompted to specify points for the vertices of the slab. The location of the slab in the Z direction is controlled by the **Mode** property of the slab. The **Direct** mode will place the slab vertices with the same z component as the feature selected. The **Projected** mode will place the slab with a z component as specified in the **Base height** edit field of the Properties palette. The slab thickness is specified according to joist size; the section shown in Figure 7.2 includes the sill, band board, and subfloor for 2 × 10 joists. The base height is set equal to the distance from the **Z=0** elevation or the **Baseline** location of the wall style. The slab in the following command sequence is placed using the **Projected** mode as required for the section shown in Figure 7.2 to place the slab on top of the foundation wall.

*(Select the **Slab** command from the Design palette, and edit the Properties palette as follows: General: Style = Standard, Mode = Projected, Dimensions: Thickness = 11.5", B-Vertical = 0, C-Horizontal = 0, Justify = Bottom, Base height = 3'-4", Direction = Left, Overhang = 0, Perimeter edge = None, Slope: Rise = 0, Run = 12.)*

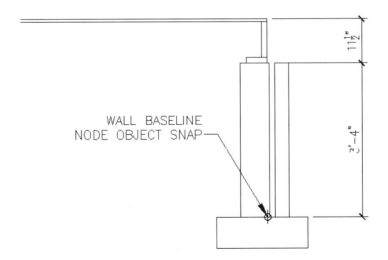

Figure 7.2 *Determining slab thickness and base height from a section*

Command: SlabAdd

Specify next point or

[STyle/MOde/Height/Thickness/SLope/OVerhang/Justify/MAtch]: *(Use the Node object snap to select point p1 as shown in Figure 7.3.)*

Specify next point or

[STyle/MOde/Height/Thickness/SLope/OVerhang/Justify/MAtch/Undo/Ortho]: *(Use the Node object snap to select point p2 as shown in Figure 7.3.)*

Specify next point or

[STyle/MOde/Height/Thickness/SLope/OVerhang/Justify/MAtch/Undo/Ortho/Close]: *(Use the Node object snap to select point p3 as shown in Figure 7.3.)*

Specify next point or

[STyle/MOde/Height/Thickness/SLope/OVerhang/Justify/MAtch/Undo/Ortho/Close]: *(Use the Node object snap to select point p4 as shown in Figure 7.3.)*

Specify next point or

[STyle/MOde/Height/Thickness/SLope/OVerhang/Justify/MAtch/Undo/Ortho/Close]: *(Use the Node object snap to select point p1 as shown in Figure 7.3.)*

[STyle/MOde/Height/Thickness/SLope/OVerhang/Justify/MAtch/Undo/Ortho/Close]: *(Press ENTER to end the command.)*

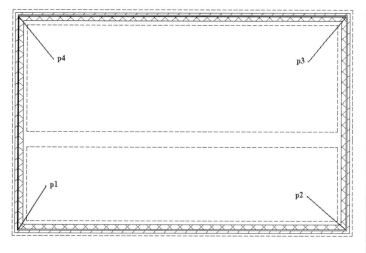

Figure 7.3 *Location of vertex points for the slab*

The slab is created on layer A-Slab; therefore, you can isolate the view of the slab by turning off all layers except **A-Slab** to view the slab. A slab can be created in the Foundation plan and exported to the first floor plan using the **Wblock** command. When the slab is inserted in the first floor plan, the pictorial view is realistic because a floor system is present. In the example shown above, the wall style of the foundation wall positioned the baseline exactly on the edge of the slab.

 Tip: When you select the slab location with the Node object snap of a wall, there are nodes at the top and bottom of the baseline. Therefore view the wall in isometric to verify location of points.

However, if the baseline of the wall and slab edge do not line up, a horizontal offset or overhang can be specified in the Properties palette. In addition, you can draw a polyline using object snaps that represents the outline of the slab and convert this polyline shape to a slab with the **Apply Tool Properties to** command. The **Apply Tool Properties to** command of the **Slab** shortcut menu can be used to create a slab in the shape of the polyline. Access the **Apply Tool Properties to** command as shown in the Table 7.2.

Shortcut menu	Select the Slab tool on the Design palette, right-click, and choose Apply Tool Properties to

Table 7.2 *Apply Tool Properties to command access*

When you select the Slab tool on the Design palette, right-click, and choose **Apply Tool Properties to**, the menu cascades with the option **Linework and Walls**. The **Linework and Walls** option allows you to select a polyline and create the slab as shown in the following command line sequence.

> *(Select the **Slab** tool from the Design palette, right-click, and choose* **Apply Tool Properties to>Linework and Walls** *from the shortcut menu.)*
>
> Command: SlabToolToLinework
>
> Select walls or polylines: 1 found *(Select the polyline at p1 as shown in Figure 7.4.)*
>
> Select walls or polylines: ENTER *(Press ENTER to end selection.)*
>
> Erase layout geometry? [Yes/No] <No>: ENTER *(Press ENTER to retain the polyline.)*
>
> Creation mode [Direct/Projected] <Projected>: ENTER *(Press ENTER to project the slab in the Z direction.)*
>
> Specify base height <0">: **36** ENTER *(Specify distance to project slab in Z direction.)*

Specify slab justification [Top/Center/Bottom/Slopeline] <Bottom>: ENTER *(Specify the slab justification.)*

(The slab is created as shown in Figure 7.4; edit the slab properties in the Properties palette if necessary.)

The **Apply Tool Properties to** command can also be applied to a straight wall. The wall geometry defines the baseline of the slab. The wall **base height** is used as the slab base height. As shown in the following command line sequence, you are prompted to specify the wall justification to align the slab. The **slab baseline** is then located at the specified **wall justification**. The slope direction is specified as left or right. The left or right justification is from the orientation of the start point of the wall.

*(Select the **Slab** tool on the Design tool palette, right-click, and choose* **Apply Tool Properties to>Linework and Walls** *from the shortcut menu.)*

Command: SlabToolToLinework

Select walls or polylines: 1 found *(Select the wall at p1 in Figure 7.5.)*

Select walls or polylines: ENTER *(Press ENTER to end selection.)*

Erase layout geometry? [Yes/No] <No>: ENTER *(Press ENTER to retain the geometry.)*

Specify slab justification [Top/Center/Bottom/Slopeline] <Bottom>: ENTER *(Press ENTER to justify the slab to the bottom.)*

Specify wall justification for edge alignment [Left/Center/Right/Baseline]

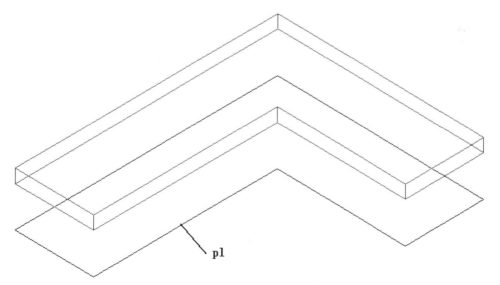

Figure 7.4 *Slab created from polyline shape*

<Baseline>: **R** *(Select right justification to position the slab to the right of the wall.)*

Specify slope Direction [Left/Right] <Left>: **R** *(Specify Right as the direction of slope.)*

Command:

Tip: You can select the wall and display the grips to identify the start point and justification of the wall prior to applying slab properties to the wall.

MODIFYING THE SLAB USING PROPERTIES AND GRIPS

A slab can be edited by changing its properties in the Properties palette or editing the grips. Slabs can be trimmed, extended, or mitered in a manner similar to roof slabs. If you select an existing slab, additional fields are added to the slab Properties palette as shown in Figure 7.6. The additional fields of the Properties palette include the **Layer edit** field and the **Edges** button, which allows you to assign slab edge styles to the edges of the slab. The additional edit fields of the Properties palette to modify a slab are described below.

Basic

General

Layer – The **Layer** edit field displays the name of the layer of the slab. Slabs are placed on the **A-Slab** layer when the AIA (256 Color) Layer Key Style is used in the drawing.

Dimensions

Edges – The **Edges** button opens the **Slab Edges** dialog box, which is identical to the **Slab Edges** dialog box of roof slabs discussed in Chapter 6. This dialog box allows you to specify an overhang, edge style, edge cut, and angle.

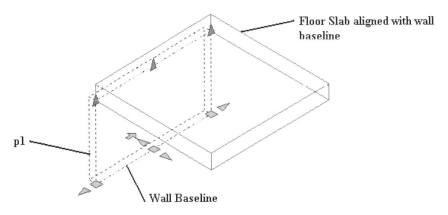

Figure 7.5 *Slab created based upon wall*

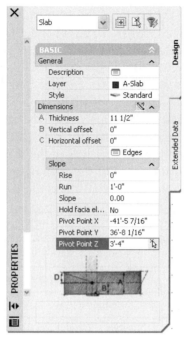

Figure 7.6 *Properties dialog box of an existing slab*

SLOPE:

Hold facia elevation – The **Hold facia elevation** edit field includes the following options: **No**, **By adjusting overhang**, and **By adjusting baseline height**, as discussed in Chapter 6.

Pivot Point X – The **Pivot Point X** value is the world coordinate X value of the pivot point.

Pivot Point Y – The **Pivot Point Y** value is the world coordinate Y value of the pivot point.

Pivot Point Z – The **Pivot Point Z** value is the world coordinate Z value of the pivot point.

Location

Elevation – The **Elevation** edit field displays the elevation of the insertion point of the roof slab.

Additional information – The **Additional information** button opens the Location dialog box, which lists the insertion point, normal, and rotation descriptions of the slab.

USING GRIPS

The **grips** of a slab allow you to change the pivot point, thickness, edge location, angle, and vertices. The grips of a slab are identical to those of the roof slab; therefore refer to "Using Grips to Edit a Roof Slab" in Chapter 6. The shortcut menu of a selected slab includes the commands shown in Table 7.3. The commands are listed in the order in which they appear on the shortcut menu. The commands are similar to the commands used to edit roof slabs. Holes can be added to a slab using the **SlabAddHole** command to represent framing openings for stairs, skylight shafts, and masonry fireplaces. Holes are created by cutting the hole in the shape of a closed polyline. The closed polyline is drawn in the shape of the framing opening. Refer to "Tools for Editing Roof Slabs" in Chapter 6 to review the commands used to edit slabs.

Command prompt	Shortcut menu
SLABTRIM	Select a slab, right-click, and choose Trim
SLABEXTEND	Select a slab, right-click, and choose Extend
SLABMITER	Select a slab, right-click, and choose Miter
SLABCUT	Select a slab, right-click, and choose Cut
SLABADDVERTEX	Select a slab, right-click, and choose Add Vertex
SLABREMOVEVERTEX	Select a slab, right-click, and choose Remove Vertex
SLABADDHOLE	Select a slab, right-click, and choose Hole>Add
SLABREMOVEHOLE	Select a slab, right-click, and choose Hole>Remove
SLABBOOLEANADD	Select a slab, right-click, and choose Boolean>Add
SLABBOOLEANSUBTRACT	Select a slab, right-click, and choose Boolean>Subtract
SLABBOOLEANDETACH	Select a slab, right-click, and choose Boolean>Detach

Table 7.3 *Slab Edit commands of the shortcut menu*

 STOP. Do Tutorial 7.1, "Creating a Cathedral Ceiling," at the end of the chapter.

ACCESSING SLAB STYLES

The Standard slab style is used when you create a slab. Additional slab styles can be i-dropped from the Design Tool Catalog>Roof Slabs and Slabs>Slabs of the **Content Browser** shown in Figure 7.7. The styles can be accessed in the **Style Manager** when you open the Slab & Slab Edge Styles (Imperial) drawing located in the *C:\Documents and Settings\All Users\Application Data\Autodesk\ADT 2005\enu\Styles\ Imperial* directory. Slab styles can be copied from the Slab & Slab Edge Style (Imperial) drawing and pasted into the current drawing in the **Style Manager**. The slab styles include preset thickness, roof slab edge styles, and material definitions. The features of a slab style represent ceilings with surface finishes, floors with surface finishes, grass, and landscaping surfaces.

The slab styles can be edited in the **Slab Styles** dialog box, accessed by the **SlabStyleEdit** command. Access the **SlabStyleEdit** command as shown in Table 7.4.

Menu bar	Format>Style Manager, expand Architectural Objects, select Slab Styles folder
Command prompt	SLABSTYLEEDIT
Shortcut menu	Select a slab, right-click, and select Edit Slab Style

Table 7.4 *Edit Slab Style command access*

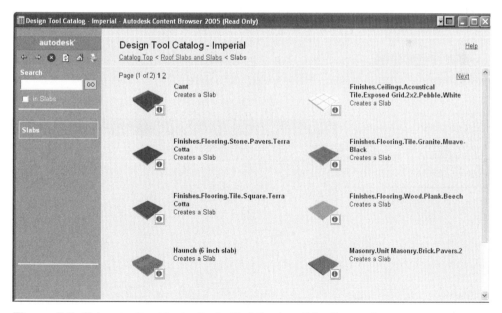

Figure 7.7 *Slab styles listed in the Design Tool Catalog of the Content Browser*

When you select a slab, right-click, and choose **Edit Slab Style** from the shortcut menu, the **Slab Styles** dialog box opens as shown in Figure 7.8. The **Design Rules** tab of the **Slab Styles** dialog box allows you to specify the slab thickness and thickness offset. The **Materials** tab lists the material definitions used in the style. The **Slab Styles** dialog box shown includes the same tabs and content as the **Roof Slab Styles** dialog box; therefore refer to "Roof Slab Styles" in Chapter 6 for a review of the tab contents.

ATTACHING A SLAB EDGE STYLE TO A SLAB

Slab edge styles are used in a slab style to create edges to represent curbs, recesses, and projections typically used in concrete slab construction. Slab edge styles are created from profiles. You can edit a slab edge to include a slab edge style by selecting **Edit Slab Edges** from the shortcut menu of a selected style. Access the **Edit Slab Edges** command as shown in Table 7.5.

Command prompt	SLABEDGEEDIT
Shortcut menu	Select a slab, right-click. and choose Edit Slab Edges

Table 7.5 *SlabEdgeEdit command access*

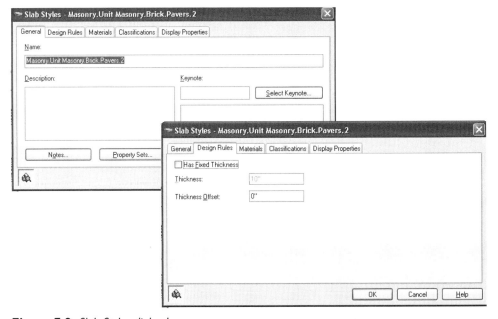

Figure 7.8 *Slab Styles dialog box*

Creating Slabs for Floors and Ceilings   477

When you select **Edit Slab Edges** from the shortcut menu, you are prompted in the command line to select an edge as shown in the following command line sequence. When you select one or more slabs, the **Edit Slab Edges** dialog box opens as shown in Figure 7.9. The upper left window of the dialog box allows you to select a slab edge style from the **Edge Style** drop-down list, which displays the slab edge styles included in the drawing. Many of the slab styles shown in Figure 7.7 include slab edge styles within their slab style. Therefore, if you import a slab style from the Slab & Slab Edge Style (Imperial) drawing, the associated slab edge style is imported. The slab edge styles included in the Slab & Slab Edge Style (Imperial) drawing are shown in Figure 7.10.

> Command: SlabEdgeEdit
>
> Select edges of one slab: *(Select a slab.)*
>
> Select edges of one slab: ENTER *(Press* ENTER *when finished selecting edges.)*
>
> (**Edit Slab Edges** *dialog box opens as shown in Figure 7.9.)*

The **Edit Slab Edges** dialog box is similar to the **Edit Roof Slab Edges** dialog box. The **A-Overhang**, **B-Edge Cut**, and **C-Angle** can be edited to modify the slab edge. A negative value can be entered in the overhang field to shrink the edge from the baseline edge. Slab edge styles can be assigned to the edge in the **Edge Style** column. The slab edge styles shown in Figure 7.10 are included in the Slab & Slab Edge Styles (Imperial) drawing located in the *C:\Documents and Settings\All Users\Application Data\ADT 2005\enu\Styles\Imperial* directory.

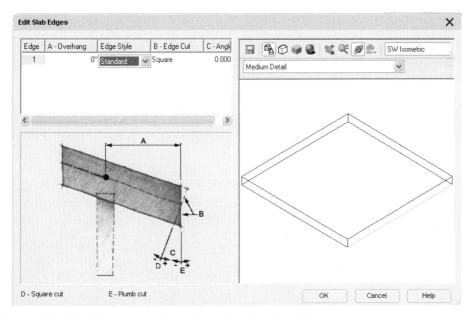

Figure 7.9 *Specifying a slab edge style for the edge of a slab*

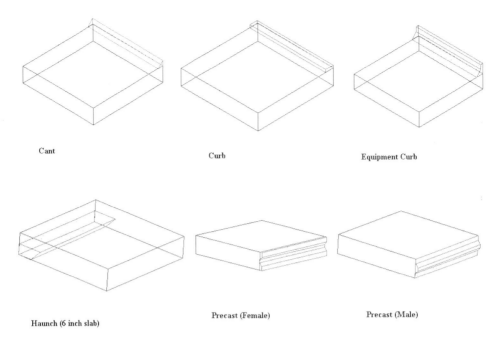

Figure 7.10 Edit Slab Edge Styles

Slab edge styles can also be assigned to a slab edge by selecting the **Edges** button in the Properties palette. Selecting the **Edges** button opens the **Slab Edges** dialog box, which lists all edges of the slab as shown in Figure 7.11. The edges are numbered according to the sequence of development; however, it may be difficult to determine which edge of the slab you are editing in the dialog box. Therefore, use the **Edit Slab Edge** command to select an edge for edit and edit the **Edit Slab Edges** dialog box.

 STOP. Do Tutorial 7.2, "Creating a Flat Roof with a Slab," at the end of the chapter.

EDITING THE SLAB EDGE STYLES OF A SLAB

The style of the slab edge can be edited by choosing **Edit Slab Edge Style** from the shortcut menu of a selected slab. Refer to Table 7.6 for SlabEdgeStyleEdit command access.

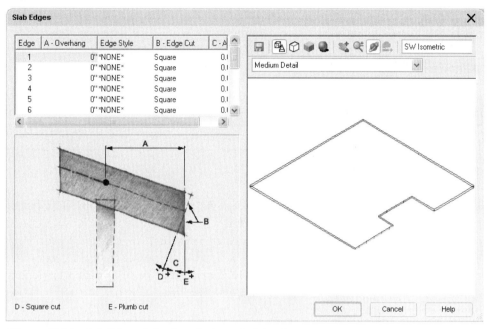

Figure 7.11 *Slab Edges dialog box displays all slab edges*

Command prompt	SLABEDGESTYLEEDIT
Menu bar	Format>Style Manager, expand the Architectural Objects\Slab Edge Styles, then edit the style
Shortcut menu	Select a slab that includes a slab edge style, right-click, and choose Edit Slab Edge Style

Table 7.6 *SlabEdgeStyleEdit command access*

When you select a slab, right-click, and choose **Edit Slab Edge Style** from the shortcut menu, you are prompted to select a slab edge. If the selected edge does not include an edge style, the command ends. If the slab edge includes a slab edge style, the **Slab Edge Styles** dialog box opens, displaying the properties of the slab edge style. The **Slab Edge Styles** dialog box is similar to the **Roof Slab Edge Styles** dialog box. The **Design Rules** tab of this dialog box, shown in Figure 7.12, allows you to specify the profile for the edges of the slab.

ADDING AN EDGE PROFILE IN THE WORKSPACE
A profile can be added to the drawing for the edge of the slab while in the workspace without opening the **Style Manager** and creating a new profile. The **Add Edge**

Profiles command is included in the shortcut menu of a selected slab. This command allows you to create a profile while remaining in the workspace. The edge selected for the command must have a defined edge style. Access the **Add Edge Profiles** command as shown in Table 7.7.

Command prompt	SLABADDEDGEPROFILES
Shortcut menu	Select a slab, right-click, and choose Add Edge Profiles

Table 7.7 *Add Edge Profiles command access*

When you select a slab, right-click, and choose **Add Edge Profiles** from the shortcut menu, you are prompted to select a slab edge. The slab edge must include a defined roof slab edge style. If the slab edge includes an overhang, you can add a profile for the soffit and fascia. If no overhang is included, only the profile for the fascia can be changed. After you select the edge, the **Add Fascia/Soffit Profiles** dialog box opens as shown in Figure 7.13. The **Add Fascia/Soffit Profiles** dialog box includes a drop-down list of profiles defined in the drawing. Selecting from the list will change the profile used in the edge style. The **Start from scratch** profile option will allow you to edit the current profile. After you make a selection of the profile, the **In-Place Edit** tool-

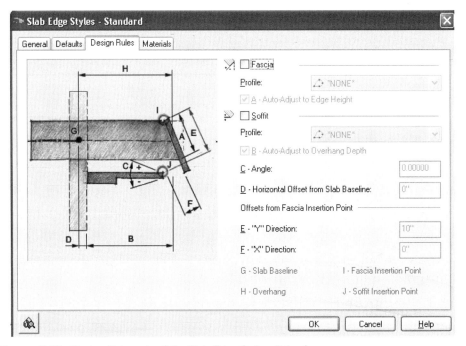

Figure 7.12 *Design Rules tab of the Slab Edge Styles dialog box*

Creating Slabs for Floors and Ceilings

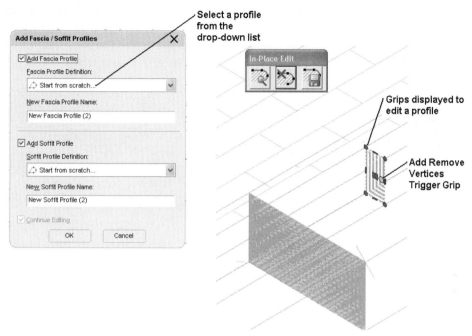

Figure 7.13 *Adding and editing a profile with Edit In Place*

bar opens, allowing you to zoom to the profile. The grips of the profile shown at right in Figure 7.13 can be selected to edit the profile. The Edit In Place procedures are similar to editing the roof slab.

CHANGING THE PROFILE WITH EDIT EDGE PROFILE IN PLACE

After a profile has been changed using the **Add Edge Profiles** command, additional editing can be performed with the **Edit Edge Profile In Place** (**SlabEdgeProfileEdit**) command. This command allows you to perform additional editing of the profile. The **Edit Edge Profile In Place** command allows you to add or remove vertices from the profile in the workspace. Access the **Edit Edge Profile In Place** command as shown in Table 7.8.

Command prompt	SLABEDGEPROFILEEDIT
Shortcut menu	Select a slab, right-click, and choose Edit Edge Profile In Place

Table 7.8 *Edit Edge Profile In Place command access*

When you select a roof slab, right-click, and choose the **Edit Edge Profile In Place** command from the shortcut menu, the **In-Place Edit** toolbar is displayed. The **Zoom To** command of the **In-Place Edit** toolbar will create an enlarged view of the profile. The profiles used for slabs are edited in a manner similar to profiles for roof slab edges.

SUMMARY

1. Slabs are added with the **Slab** command located on the Design tool palette.
2. A closed polyline can be converted to a slab.
3. The Direct Mode slab is created at the same elevation as the points selected.
4. A Projected Mode slab is created at a base height elevation from the selected points.
5. The shortcut menu of a selected slab allows you to trim, extend, miter, cut, and add holes to the slab.
6. Slab edge styles attached to a slab edge customize the edge of the slab.
7. Slab edge styles and slab styles are defined in the **Style Manager**.
8. Profiles of the slab edge can be edited in place with the **Add Edge Profiles** and **Edit Edge Profile in Place** commands.

REVIEW QUESTIONS

1. The _____ command is used to create a slab.
2. The _____ mode is used to create a slab at Z=8'.
3. The _____ command is used to create a slab from a polyline or wall.
4. The slope of a slab is hinged about the _____ of the slab.
5. A slab is created on the _____ layer.
6. A slab is edited to create a stairwell opening with the _____ command.
7. Two slabs can be extended to intersection with the _____ command.
8. The _____ slab edge style is used to create a representation of the cant strip of a flat roof.
9. Slab edge styles are created in the _____ _____.
10. Profiles for slab edge styles are created from a _____.

TUTORIAL 7.1 CREATING A CATHEDRAL CEILING

1. Open your *ADT Student/ADT Tutor\Ch7\EX 7-1.dwg*.

2. Choose **File>SaveAs** from the menu bar and save the drawing as **Lab 7-1** in your student directory.

3. Move the pointer to the OSNAP toggle of the status bar, right-click, and choose **Settings** from the shortcut menu. Edit the **Object Snap** tab of the **Drafting Settings** dialog box: clear all object snap modes, check **Node** object snap. Check the **Object Snap ON (F3)** check box and check **Object Snap Tracking ON (F11)**. Select the **Polar Tracking** tab; check ON **Polar Tracking (F10)**. Select **OK** to dismiss the **Drafting Settings** dialog box.

4. Toggle OFF SNAP, GRID, and ORTHO on the status bar.

5. Select **Slab** from the Design palette, and edit the Properties palette as follows: Basic–General: Style = **Standard**, Mode = **Projected**, Dimensions: Thickness = **6"**, Vertical Offset = **0**, Horizontal Offset = **0**, Justify = **Bottom**, Base height = **8'**, Direction = **Left**, Overhang = **0**, Perimeter = **None**, Slope: Rise = **3**, Run = **12**, and Angle = **14.04**. Then respond to the following command line prompts to place the slab.

   Command: SlabAdd

   Specify start point or

   [STyle/MOde/Height/Thickness/SLope/OVerhang/Justify/MAtch]: *(Select a point near p1 using the Node object snap as shown in Figure 7T.1.)*

   Specify next point or

   [STyle/MOde/Height/Thickness/SLope/OVerhang/Justify/MAtch/Undo/Ortho]: *(Select a point near p2 using the Node object snap as shown in Figure 7T.1.)*

   (Move the pointer left to set the polar tracking angle vector to 90.)

   Specify next point or

   [STyle/MOde/Height/Thickness/SLope/OVerhang/Justify/MAtch/Undo/Ortho]: **6'** ENTER *(Type **6'** distance to create a vertical edge of the slab 6' from the front wall.)*

   *(Move the pointer to the left to display a polar angle of 180°, right-click, and choose **Ortho close** from the shortcut menu.)*

   Specify next point or

   [STyle/MOde/Height/Thickness/SLope/OVerhang/Justify/MAtch/Undo/Ortho/Close]: **o** *(**Ortho close** selected from the shortcut menu to close the slab as shown in Figure 7T.2.)*

   (Press ESC to end the command.)

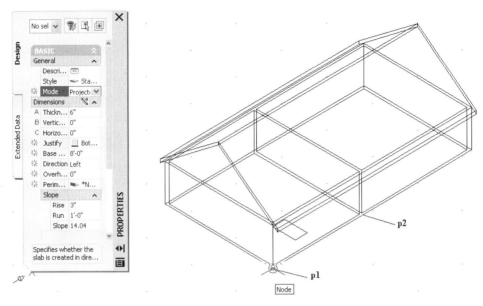

Figure 7T.1 Insert point for slab

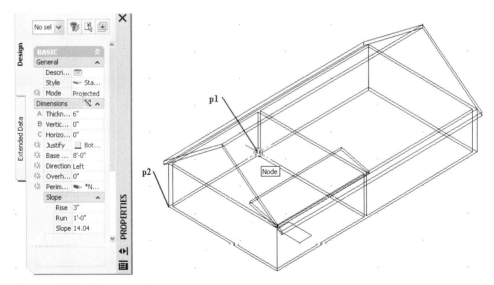

Figure 7T.2 Slab inserted

6. Select **Slab** from the Design palette, and edit the Properties palette as follows: Basic–General: Style = **Standard**, Mode = **Projected**; Dimensions: Thickness = **6"**, Vertical Offset = **0**, Horizontal Offset = **0**, Justify = **Bottom**,

Base height = **8'**, Direction = **Right**, Overhang = **0**, Perimeter = **None**; Slope: Rise = **3**, Run = **12**, and Angle = **14.04**. Then respond to the following command line prompts to place the slab.

> Specify start point or
>
> [STyle/MOde/Height/Thickness/SLope/OVerhang/Justify/MAtch]: *(Select a point near p1 using the Node object snap as shown in Figure 7T.2.)*
>
> Specify next point or
>
> [STyle/MOde/Height/Thickness/SLope/OVerhang/Justify/MAtch]: *(Select a point near p2 using the Node object snap as shown in Figure 7T.2.)*
>
> *(Move the pointer **right** to set the polar tracking angle vector to 270.)*
>
> Specify next point or
>
> [STyle/MOde/Height/Thickness/SLope/OVerhang/Justify/MAtch/Undo/Ortho]: **6'** ENTER *(Type 6' distance to create a vertical edge of the slab 6' below the wall.)*
>
> *(Move the pointer to the right, set the polar angle = **0**, right-click, and choose **Ortho close** from the shortcut menu.)*
>
> Specify next point or
>
> [STyle/MOde/Height/Thickness/SLope/OVerhang/Justify/MAtch/Undo/Ortho/Close]: **O** (**Ortho close** *selected from the shortcut menu to close the slab.)*
>
> *(Press ESC to end the command.)*

7. Select **Top View** from the **View** flyout of the **Navigation** toolbar.
8. In this step you will miter the two slabs: respond to the prompts of the command line as shown below.

   > *(Select slab at p1 as shown in Figure 7T.3, right-click, and choose **Miter** from the shortcut menu.)*
   >
   > Command: SlabMiter
   >
   > Miter by [Intersection/Edges] <Intersection>: **e** ENTER *(Select the Edges method of miter.)*
   >
   > Select edge on first slab: *(Select slab at p1 as shown in Figure 7T.3.)*
   >
   > Select edge on second slab: *(Select slab at p2 as shown in Figure 7T.3.)*

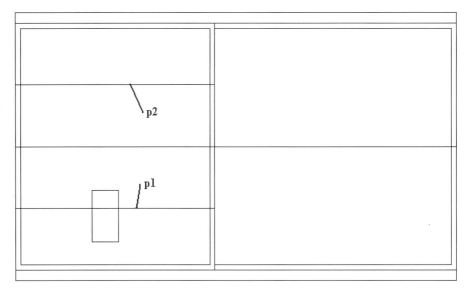

Figure 7T.3 *Selected of slabs for Miter*

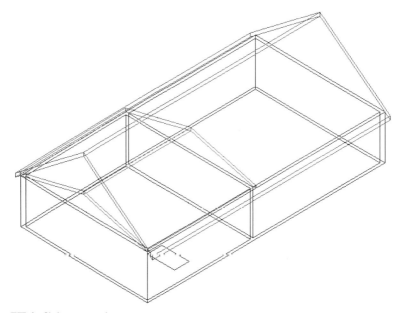

Figure 7T.4 *Slabs mitered*

9. Select **SW Isometric** from the **Navigation** toolbar to view the slabs mitered as shown in Figure 7T.4.

10. In this step you cut a hole in the ceiling: respond to the following command line prompts.

    *(Select slab at p1 as shown in Figure 7T.5, right-click, and choose **Hole>Add** from the shortcut menu.)*

    Command: SlabAddHole

    Select a closed polyline or connected solid objects to define a hole: 1 found *(Select the polyline shown at p2 in Figure 7T.5.)*

    Select a closed polyline or connected solid objects to define a hole: ENTER *(Press ENTER to end selection.)*

    Erase layout geometry? [Yes/No] <No>: ENTER *(Press ENTER to retain the polyline.)*

11. Select the roof slab at **p1** shown in Figure 7T.6, right-click, and choose **Hole>Add** from the shortcut menu. Create a hole in the roof slab by selecting the polyline at **p2** shown in Figure 7T.6.

12. Select **Slab** from the Design palette and add a slab with vertices at **p1, p2, p3**, and **p4** as shown in Figure 7T.7 with the following properties: Basic–General: Style = **Standard**, Mode = **Direct**; Dimensions: Thickness = **6"**, Vertical Offset = **0**, Horizontal Offset = **0**, Justify = **Bottom**, Direction = **Left**, Overhang = **0**, and Perimeter edge = **None**.

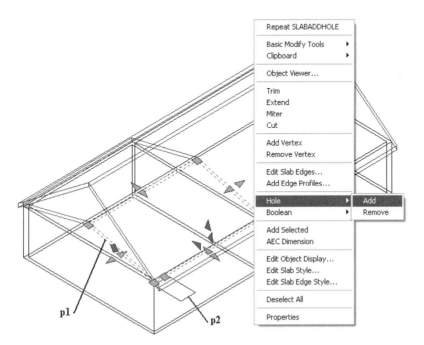

Figure 7T.5 *Selecting Hole>Add from the shortcut menu*

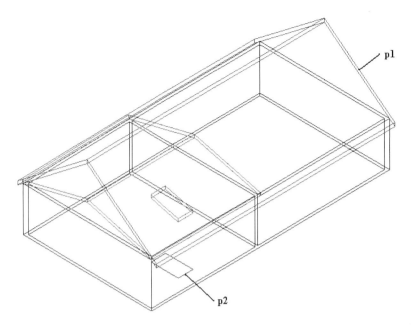

Figure 7T.6 Hole added to slab

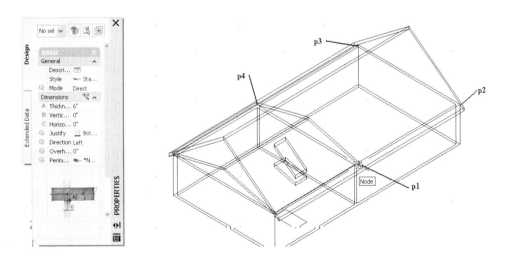

Figure 7T.7 Slab vertex points

13. Select **SE Isometric** from the **Navigation** toolbar to view the house as shown in Figure 7T.8.
14. Close and save the drawing.

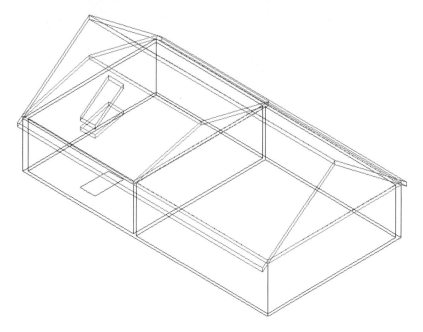

Figure 7T.8 *Slabs inserted in the drawing*

TUTORIAL 7.2 CREATING A FLAT ROOF WITH A SLAB

1. Open your *ADT Student/ADT Tutor\Ch7***EX 7-2.dwg**.

2. Choose **File>SaveAs** from the menu bar and save the drawing as **Lab 7-2** in your student directory.

3. Move the mouse pointer to the OSNAP toggle of the status bar, right-click, and choose **Settings** from the shortcut menu. Edit the **Object Snap** tab of the **Drafting Settings** dialog box: clear all object snap modes, check **Node** object snap. Check the **Object Snap On (F3)** check box and check **Object Snap Tracking On (F11)**. Select the **Polar Tracking** tab; check **Polar Tracking (F10)**. Select **OK** to dismiss the **Drafting Settings** dialog box.

4. Turn OFF ORTHO and SNAP on the status bar.

5. Select **Format>Style Manager** from the menu bar.

6. Select **Open Drawing** from the toolbar of the **Style Manager**. Select **Content** from the left **Places** panel and then open the Styles\Imperial directory of the **Open Drawing** dialog box. Select the **Slab and Slab Edge Styles (Imperial)** drawing as shown in Figure 7T.9, and select the **Open** button to close the **Open Drawing** dialog box.

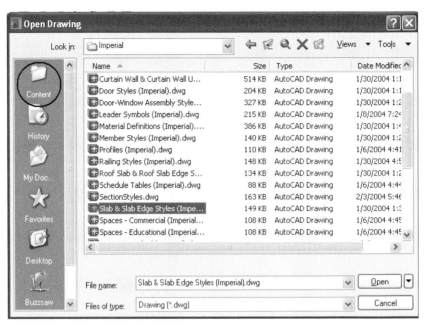

Figure 7T.9 *Open Drawing dialog box*

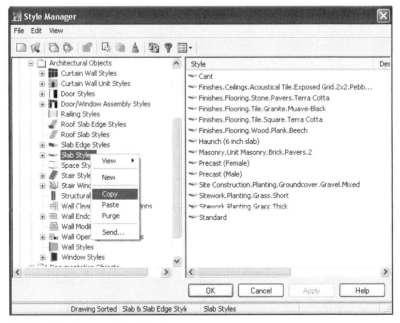

Figure 7T.10 *Selecting the Slab Edge Styles*

Creating Slabs for Floors and Ceilings   491

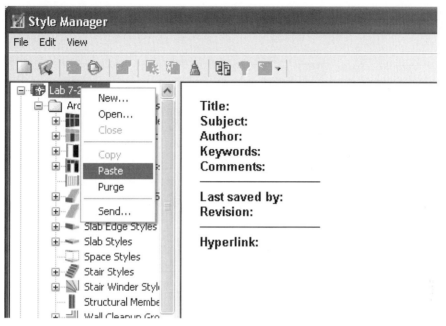

Figure 7T.11 *Pasting edge slab style in the current drawing*

7. Select the **Slab and Slab Edge Styles (Imperial)** drawing in the left pane and then select the (+) sign to expand the Architectural Objects>Slab Styles folder in the left pane. Select the Slab Styles folder in the left pane, right-click, and choose **Copy** from the shortcut menu as shown in Figure 7T.10.

8. Scroll up in the left pane and select **Lab 7-2**, the current drawing, in the left pane, right-click, and choose **Paste** from the shortcut menu as shown in Figure 7T.11.

9. Select **OK** to dismiss the **Style Manager** dialog box.

10. To place a slab with a Cant strip edge, select **Slab** from the Design palette, and edit the Properties dialog box as follows: Basic-General: Style = **Cant**, Mode = **Projected**; Dimensions: Thickness = **6"**, Vertical Offset = **0**, Horizontal Offset = **0**, Justify = **Bottom**, Base height = **8'**, Direction = **Left,** Overhang = **0**, Perimeter = **Cant**; Slope: Rise = **–.02**, Run = **12**, and Angle = **359.90**. Then respond to the following command line prompts to place the slab.

   Command: SlabAdd

   Specify start point or

   [STyle/MOde/Height/Thickness/SLope/OVerhang/Justify/MAtch]:
   (Select a point near p1 using the Node object snap as shown in Figure 7T.12.)

Specify next point or

[STyle/MOde/Height/Thickness/SLope/OVerhang/Justify/MAtch]: *(Select a point near p2 using the Node object snap as shown in Figure 7T.12.)*

Specify next point or

[STyle/MOde/Height/Thickness/SLope/OVerhang/Justify/MAtch/Undo/Ortho]: *(Select a point near p3 using the Node object snap as shown in Figure 7T.12.)*

Specify next point or

[STyle/MOde/Height/Thickness/SLope/OVerhang/Justify/MAtch/Undo/Ortho/Close]: *(Select a point near p4 using the Node object snap as shown in Figure 7T.12.)*

Specify next point or

[STyle/MOde/Height/Thickness/SLope/OVerhang/Justify/MAtch/Undo/Ortho/Close]: *(Select a point near p1 using the Node object snap as shown in Figure 7T.12.)*

Specify next point or

[STyle/MOde/Height/Thickness/SLope/OVerhang/Justify/MAtch/Undo/Ortho/Close]: ENTER *(Press ENTER to end the command.)*

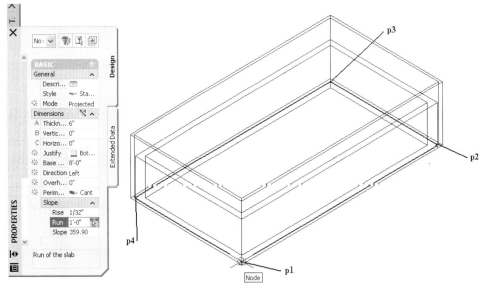

Figure 7T.12 *Insert points of the slab*

11. Select **Front View** from the **View** flyout of the **Navigation** toolbar.

12. Select **Zoom Window** from the **Zoom** flyout of the **Navigation** toolbar, and respond to the following command line prompts.

   Command: z ZOOM

   Specify corner of window, enter a scale factor (nX or nXP), or

   [All/Center/Dynamic/Extents/Previous/Scale/Window] <real time>: **128',12'** ENTER *(Specify the corner of the zoom window.)*

   Specify opposite corner: **133',6'** ENTER *(Specify the second corner of the zoom window. Cant strip added to slab as shown in Figure 7T.13.)*

13. Select **SW Isometric** from the **View** flyout of the **Navigation** toolbar to view the flat roof as shown in Figure 7T.14.

14. Select **Flat Shaded** from the **Shading** toolbar to view the roof materials.

15. Close and save the drawing.

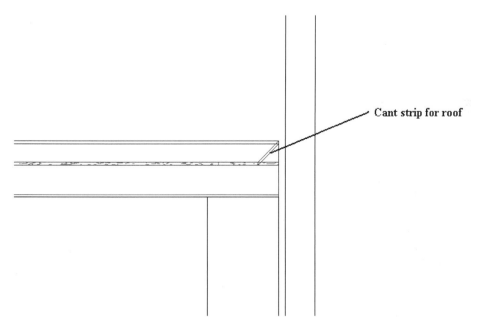

Figure 7T.13 *Front view of Cant strip*

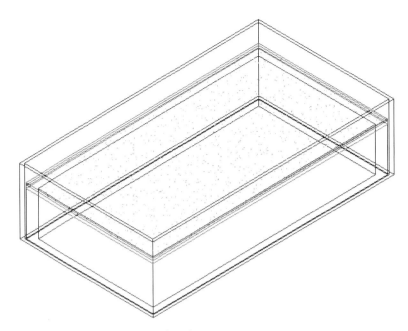

Figure 7T.14 *SW Isometric view of roof*

PROJECT

EX 7-3 Creating a Slab for a Basement Plan

1. Open Autodesk Architectural Desktop 2005, and select **File>Project Browser** from the menu bar. Use the Project Selector drop-down list to navigate to your *ADT Student\ADT Tutor\Ch7* directory. Double-click on **Ex 7-3** to set this project current. Select **Close** to dismiss the **Project Browser**. (If your student folder does not include the project, refer to "Organizing Tutorial Directories" in the Preface).

2. Select the **Constructs** tab of the **Project Navigator** and open the Basement file.

3. Create a slab using the **Node** object snap for the first floor. Create the slab **11.5"** thick, **Bottom** justified, and projected **8'-0'** to represent the wood floor system.

4. Cut a hole in the slab for a proposed stairwell represented by the polyline at **p6** shown in the drawing file.

5. Select **Hidden** from the **Shade** flyout of the **Navigation** toolbar.

6. Select the **SW Isometric** from the **View** flyout of the **Navigation** toolbar to view the hole in the slab.

7. Select **2D Wireframe** from the **Shade** flyout and **Top View** from the **View** flyout of the **Navigation** toolbar.

8. Create a slab for the garage area that passes through point p1 to **p2** as the baseline. Convert to a slab the polyline drawn on layer A-Garage_Slab shown at **p5**. The garage slab should be Mode = **Direct**, Thick = **4"**, Justify = **Top**, and Slope: Rise = **–.125** and Run = **12** slope down toward the right wall.

9. Create a slab for the living space of the basement area that passes through **p1, p2, p3,** and **p4**. The basement slab should be Mode = **Direct**, Thick = **4"**, and Justify = **Top**.

10. Create a slab using the polyline drawn for the stairwell to represent the ceiling above the proposed stair. The stairwell slab should be Mode = **Projected**, Base Height = **8'**, Justify = **Bottom**, Thickness = **6"**, and Slope = **9/12**. After creating the slab, use grips to retain the slope of the slab and stretch the slab **5'** toward the inside of the house.

11. Select **Hidden** from the **Shade** flyout of the **Navigation** toolbar.

12. Select **SW Isometric** from the **View** flyout of the **Navigation** toolbar.

13. Turn off all layers except the **A-Slab** layer to view the slabs as shown in Figure 7T.15. Save the file and close the project upon completion.

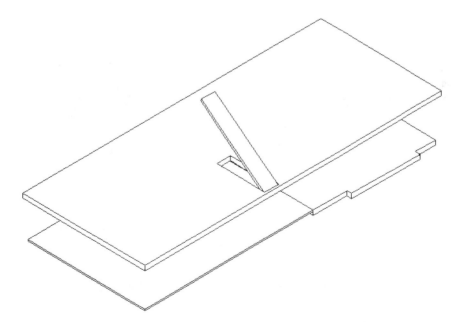

Figure 7T.15 *Slab added to the basement floor plan*

CHAPTER 8

Stairs and Railings

INTRODUCTION

During the development of a floor plan, stairs must be inserted to determine the required space for the stairwell. This chapter will explain how to insert straight, U-shaped, multi-landing, and spiral stairs. When you place a stair, the design rules of a stair style automatically calculate the riser dimension, number of risers, size of tread, and the straight length. The style of the stair can control the landing extensions and landing size. The application of railings to the stairs and platforms will be presented in this chapter. Railing styles allow you to create guardrails appropriate for decks or railings, with custom shapes for interior stair applications.

OBJECTIVES

After completing this chapter, you will be able to

- Use the **Stair** command (**StairAdd**) to create Straight, Multi-landing, U-shaped, U-shaped Winder, and Spiral stairs
- Create a stair tower in a project
- Constrain the number of risers, tread size, riser size, and run of a stair using **Calculation Rules** of the Properties palette
- Edit the style, shape, turn type, width, height, and justification in Properties palette to create stairs
- Identify the purpose of grips located on various stair shapes
- Create, edit, import, and export stair and railing styles using the **Style Manager** or **Content Browser**
- Control the visibility of the lower or upper portions of stairs using the display properties

- Modify the stair edge with the following **Customize Edge** options: Offset, Project, Remove Customization, and Generate Polyline
- Anchor an isolated stair to a stair landing and anchor a railing to an AEC object
- Use the **Railing** command (**RailingAdd**) to create railings that are freestanding, attached to stairs, and attached to stair flights
- Create railing styles

CREATING THE STAIR CONSTRUCT

Stairs can be placed in a drawing without the use of a project; however, the tools of the **Project Navigator** assist in the development of the stair. A stair is created as a construct that spans multiple levels. The construct is assigned to each level the stair serves. Therefore the stair would be assigned to levels 1 and 2 of a two-story house. When a view drawing is developed for each floor plan, the floor plan construct and stair construct for that level come together in the view drawing.

The stair construct can also be included in the model view drawing to visualize how the stair fits throughout the various levels. To create a stair construct, create a new construct drawing and attach the floor plan for level 1 as a reference file. This allows you to determine an appropriate start and end point for the flight based upon existing walls of the floor plan. After creating the stair construct, you can detach the floor plan construct to remove the duplication of walls. When the stair construct is inserted in a level as part of a view drawing, its upper and lower flight components are correctly displayed as shown at right in Figure 8.1.

STEPS TO CREATING A STAIR CONSTRUCT

1. Open the **Project Navigator** and select the **Constructs** tab of the project.
2. Select the Construct category, right-click, and choose **New Construct**. Type **Stair**, the name of the construct, and assign levels 1 and 2 as shown at left in Figure 8.1.
3. Double-click on the **Stair** construct to open the new construct.
4. Open the **Project Navigator**, select the **Construct** tab. Select the Floor 1 construct, right-click, and choose **Xref Overlay** from the shortcut menu to view the walls of the first floor.
5. Create the stair. Verify that the stair is displayed based upon display configuration in the **Display Properties** of the stair style.
6. Select **Manage Xrefs** in the Drawing status bar, select Floor 1 and choose **Detach** to remove the display of level 1 walls.

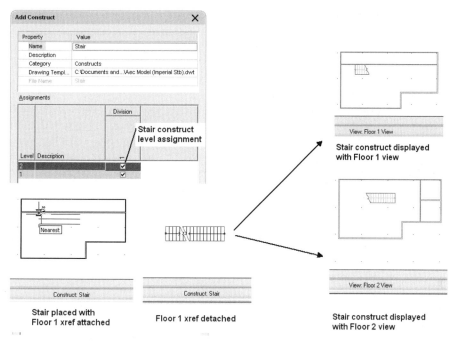

Figure 8.1 *Creating a stair construct drawing*

7. Select the **View** tab, and open the View drawing for each level in the **Project Navigator**, to view the stair as shown at right in Figure 8.1.

USING THE STAIRADD COMMAND

Stairs are placed in a drawing with the **Stair** (**StairAdd**) command selected from the Design palette as shown in Figure 8.2. Access the **Stair** command as shown in Table 8.1. The available shapes of stairs include U-shaped, Multi-landing, Spiral, and Straight. The available space for the stair is a major consideration for the selection of the shape. The Properties palette for stairs shown in Figure 8.2 allows you to set the height and run of the stair based upon the **Calculation Rules**. The **Calculation Rules** button opens the **Calculation Rules** dialog box shown in Figure 8.5, which allows you to define the relationship of tread size, riser size, straight length, and riser count. Stair styles can be imported into the drawing that include Cantilever, Concrete, Half Wall Rail, Ramp Concrete, Ramp Conc Curb, Ramp Steel, Standard, Steel, Wood-Housed, and Wood Saddled. Each of these styles changes the appearance of the stair and the means of support for the stair. Prior to placing the stair, you should first review the content of the Properties palette for stairs.

Command prompt	STAIRADD
Palette	Select Stair from the Design palett

Table 8.1 *StairAdd command access*

Selecting the **Stair** command opens the Properties palette, and you are prompted in the command line to specify the start point of the flight. The start flight and end flight points of a multi-landing stair are shown in Figure 8.2. The command line sequence for the stair created in Figure 8.2 is shown below. The polar tracking tip indicates the distance between flight start and flight end. The number of risers is dynamically displayed in fraction format as you move the mouse to the flight end. The numerator of the fraction is the riser count currently displayed, and the denominator is the total risers required.

>Command: STAIRADD
>
>Flight start point or [SHape/STyle/Tread/Height/Width/Justify/ Next/Flight length/Match/Undo]: *(Select flight start at p1 as shown in Figure 8.2.)*
>
>Flight end point or [SHape/STyle/Tread/Height/Width/Justify/ Next/Flight length/Match/Undo]: *(Select flight end at p2 as shown in Figure 8.2.)*

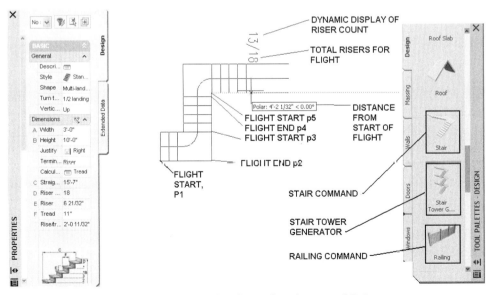

Figure 8.2 *Accessing the Stair command and specifying location of flight*

Flight start point or [SHape/STyle/Tread/Height/Width/Justify/Next/Flight length/Match/Undo]: *(Select flight start point at p3 as shown in Figure 8.2.)*

Flight end point or [SHape/STyle/Tread/Height/Width/Justify/Next/Flight length/Match/Undo]: *(Select flight start point at p4 as shown in Figure 8.2.)*

Flight start point or [SHape/STyle/Tread/Height/Width/Justify/Next/Flight length/Match/Undo]: *(Select flight start at p5 as shown in Figure 8.2.)*

Flight end point or [SHape/STyle/Tread/Height/Width/Justify/Next/Flight length/Match/Undo]: *(Dynamic display of current pointer location as shown in Figure 8.2.)*

PROPERTIES OF A STAIR

The options of the **Stair** command are set in the Properties palette shown at left in Figure 8.2. The options can also be selected from the command line prompt. The options of a stair displayed in the Properties palette are described below.

Basic

General

Description – The **Description** button opens a **Description** dialog box, allowing you to type a description of the stair.

Style – The **Style** drop-down list displays the styles included in the drawing.

Shape – The **Shape** drop-down list displays the following shapes, shown in Figure 8.3:

U-shaped – Includes two runs of equal length with a half landing.

Multi-landing – Includes stair runs of different lengths, and the turns can include any of the following: quarter landing, quarter turn, half landing, and half turn.

Spiral – Develops a curved stair.

Straight – Creates a stair without turns.

Turn type – The **Turn type** drop-down list includes the following turn options, as shown in Figure 8.3: 1/2 landing, 1/2 turn, 1/4 landing, and 1/4 turn.

Horizontal Orientation – The **Horizontal Orientation** edit field is displayed if the U-shaped or Spiral stairs are selected. A U-shaped stair can be created clockwise or counterclockwise from the start point.

Vertical Orientation – The **Vertical Orientation** drop-down list includes Up and Down options, which allows you to create the stair that goes up or down from the start point. The start point originates at the z=0 vertical position.

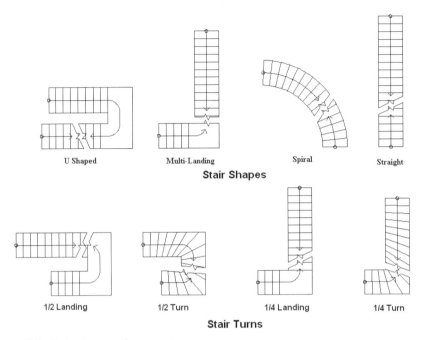

Figure 8.3 *Stair shape and turn options*

Dimensions

A-Width – The **Width** edit field allows you to enter the width of the stair. The width dimension is the total width of the stair unit. Regardless of the construction style, the 3'-0" dimension is inclusive of all stair elements.

B-Height – The **Height** edit field allows you to enter the finish floor to finish floor design dimension for the stair.

Justify – The **Justify** drop-down list provides the following justifications **Left**, **Center**, and **Right**. The justify option specifies the location of the handles for placing the stair. If you choose **Left**, the start and end points will be located on the left edge of the stair.

Terminate with – The **Terminate with** option specifies the stair flight to end with a riser, tread, or landing. The end of flight was specified beyond the stair flight in Figure 8.4; however, a tread riser or landing was appended to each stair.

Calculation Rules – The **Calculation Rules** button opens the **Calculation Rules** dialog box shown in Figure 8.5. This dialog box allows you to select among the following control options: **A-Straight Length**, **B-Riser Count**, **C-Riser**, and **D-Tread**. If **Automatic** is toggled ON for all fields, the stair dimensions are calculated based upon the **Design Rules** tab of the **Stair Styles** dialog box. Selecting a control option such as **Riser Count** will

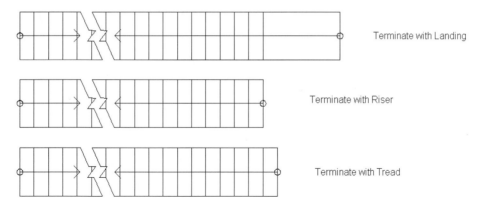

Figure 8.4 *"Terminate with" options*

calculate the stair dimensions based upon the **Design Rules** tab of the **Stair Styles** dialog box, and the results will be displayed in the **Calculation Rules** dialog box. The tread dimension can be overridden in the **Calculation Rules** dialog box as shown in Figure 8.5. The name of the control option selected is then displayed in the **Calculation Rules** edit field. If you insert values in the **Calculation Rules** dialog box that violate the parameters specified in the **Design Rules** tab for the stair style, a warning dialog box is displayed, and a warning symbol is displayed on the stair, as shown at right in Figure 8.5.

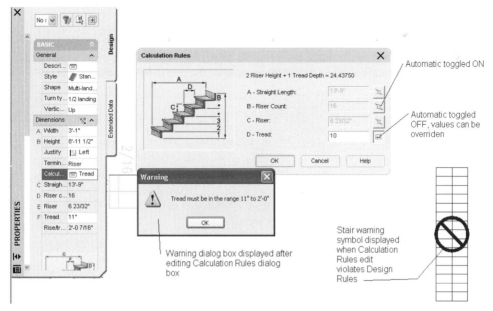

Figure 8.5 *Calculation Rules dialog box*

Straight Length – The resulting straight length based upon the **B-Height** and the stair calculation is displayed. If the **Straight Length** is set to **Automatic** in the **Calculation Rules**, it is calculated in the **Design Rules** tab of the style, based on the tread size and riser count. The **Straight Length** edit field can be edited when it has been selected as an option in the **Calculation Rules** dialog box, and the riser, riser count, and tread are adjusted. The **Riser Count** and **Straight Length** can be unlocked and specific values entered; the **Tread** dimension is then adjusted.

Riser Count – The **Riser Count** is the number of risers of the stair. This field can be edited when it has been selected as an option in the **Calculation Rules** dialog box. Values entered must comply with rules of the **Design Rules** tab of the style; otherwise an error message box is displayed. If the riser count is **Automatic** in the **Calculation Rules**, the number of risers is calculated based upon the **Design Rules** tab according to the **height** and **overall length**.

Riser – The **Riser** dimension is displayed in this field. If control is specified **Automatic** in the **Calculation Rules**, the value is calculated according to the rules specified in the **Design Rules** tab of the **Stair Styles** dialog box. The **Riser** dimension can be changed by editing the **number** of risers.

Tread – The calculated or default **Tread** dimension is displayed. This field can be edited when it has been selected as an option in the **Calculation Rules** dialog box. If the **Tread** is specified **Automatic** in the **Calculation Rules** dialog box, the tread dimension is calculated based upon the straight length and number of risers. The **Tread** dimension can be adjusted by unlocking and editing the **Riser Count** and the **Straight Length**.

Riser/Tread Calculation – The **Riser/Tread Calculation** edit field displays the results of the calculation of 2 Riser+Tread formula. This rule is toggled ON or OFF in the **Design Rules** tab of the **Stair Styles** dialog box.

Advanced

Floor Settings

Top Offset – The **Top Offset** value allows you to enter a value that represents the thickness of flooring material at the upper floor of the stair. A **Top Offset** value of zero creates a stair calculated using the tread thickness as the thickness of the finished floor as shown in Figure 8.6. A top offset of 4" is shown in Figure 8.6 to illustrate the change in the vertical dimension of the stair to accommodate the thickness.

Top Depth – The **Top Depth** allows you to specify the depth of the floor system at the top of the stair.

Bottom Offset – The **Bottom Offset** value represents the thickness of the flooring material at the bottom of the stair. A Bottom Offset value of zero creates a stair riser that begins at Z=0.

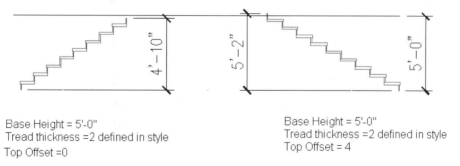

Base Height = 5'-0"
Tread thickness =2 defined in style
Top Offset =0

Base Height = 5'-0"
Tread thickness =2 defined in style
Top Offset = 4

Figure 8.6 *Stair adjustment for Top Offset*

Flight Height

Minimum Limit Type – The **Minimum Limit Type** options allow you to set **no limits** or **limit** the number of risers or height per flight. If **Risers** or **Height** is selected, the Properties palette is expanded to include **Minimum Height** and **Minimum Risers** edit fields.

Minimum Height – The **Minimum Height** edit field allows you to type a distance as the minimum height.

Minimum Risers – The **Minimum Risers** edit field allows you to type a value for the minimum number of risers.

Maximum Limit Type – The **Maximum Limit Type** includes **no limit** or **limit**. The **limit** option sets a maximum number of risers or height for a flight. If a Risers or Height limits is selected, the Properties palette is expanded to include the **Maximum Risers** and **Maximum Height** options. Setting the maximum height limit allows the stair to be developed with a landing automatically placed in compliance with building code landing requirements.

Maximum Height – The **Maximum Height** edit field allows you to type a distance as the maximum height.

Maximum Risers – The **Maximum Risers** edit field allows you to type a value for the maximum number of risers.

Interference

Headroom – The **Headroom** edit field allows you to specify a distance for the headroom of the stair. The headroom value is used with the Interference Condition of spaces or walls to cut vertically from the AEC object to create the headroom as shown in Figure 8.7.

Side Clearance – The **Side Clearance** edit field is the value of the interference condition on all four sides exclusive of the starting edge of the first and last step.

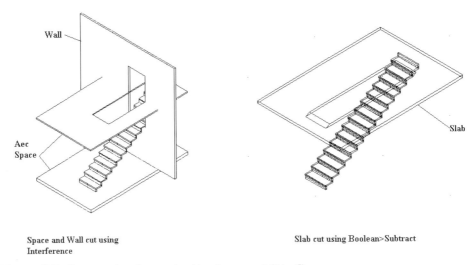

Figure 8.7 *Space and wall cut using Headroom and Side Clearance*

 Note: The slab shown at right in Figure 8.7 was cut using the **Boolean>Subtract** command.

Worksheets

Components – The **Components** button opens the **Stair Components** dialog box as shown in Figure 8.8. This dialog box allows you to set the flight and landing dimensions. This dialog box is inactive, as shown in Figure 8.8, because control is set in the style. If the **Allow Each Stair to Vary** is toggled OFF in the **Components** tab of the **Stair Styles** dialog box, the contents of this dialog box are not editable.

Landing Extensions – The **Landing Extensions** button opens the **Landing Extensions** dialog box as shown in Figure 8.9. This dialog box allows you to set the distance to extend features of the landings. This dialog box as shown in Figure 8.9 is inactive because the **Allow Each Stair to Vary** option is toggled OFF in the **Landing Extensions** tab of the **Stair Styles** dialog box.

Using Properties to Modify an Existing Stair

A stair is modified by editing the Properties palette. Additional properties can be assigned to the stair in the stair style. After the stair is placed, handrails, balusters, and guardrails can be added through the **Railing** command. Existing railings are modified in the Properties palette. The following additional fields are displayed to edit an existing stair in the Properties palette.

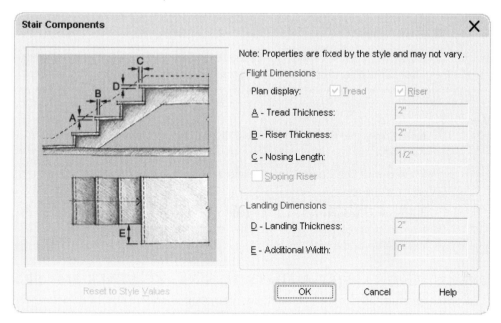

Figure 8.8 *Stair Components dialog box*

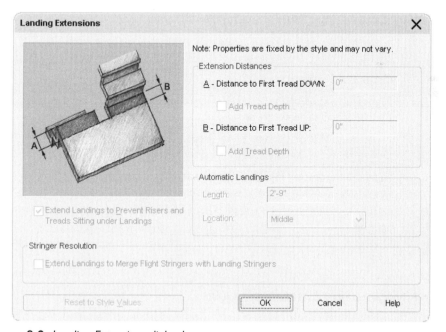

Figure 8.9 *Landing Extensions dialog box*

Basic

General

> **Layer** – The name of the layer of the stair is displayed in the **Layer** edit field. The stair is placed on the **A-Flor-Strs** layer when the AIA (256 Color) Layer Key Style is used in the drawing.

Location

> **Rotation** – The **Rotation** edit field displays the rotation of the stair in the XY plane.
>
> **Elevation** – The **Elevation** edit field display the elevation of the insertion point of the stair.
>
> **Additional Information** – The **Additional Information** button opens the **Location** dialog box, which lists the insertion point, normal, and rotation description of the stair.

SPECIFYING FLIGHT POINTS TO CREATE STAIR SHAPES

The straight, multi-landing, U-shaped, and spiral stairs are created by selecting the **Shape** edit field in the Properties palette. The shape presets the sequence of prompts to create the stair. It is helpful to turn on polar tracking when placing straight, multi-landing, and U-shaped stairs. The polar tracking tip displays the distance you have moved the pointer from the last specified point. The straight stair can be created by selecting the start point of the stair at **p1** as shown in Figure 8.10, and the end of flight can be selected along the stair or beyond the dynamic display of the stair. The command line prompts for the straight stair created in Figure 8.10 are shown below.

> Command: StairAdd
>
> Flight start point or [SHape/STyle/Tread/Height/Width/Justify/Match/Undo]: *(Select start of flight at p1 as shown in Figure 8.10.)*
>
> Flight end point or [SHape/STyle/ Tread/Height/Width/Justify/Match/Undo]: *(Select end of flight at p2 as shown in Figure 8.10.)*
>
> Flight start point or [SHape/STyle/ Tread/Height/Width/Justify/Match/Undo]: ENTER *(Press ENTER to end the command.)*

 STOP. Do Tutorial 8.1, "Creating Straight Stairs with Vertical Orientations" at the end of the chapter.

CREATING A MULTI-LANDING STAIR

The multi-landing stair can include one or more landings and stair flights. The second stair flight can be straight or at an angle to the first flight. If the multi-landing stair turns at an angle, you can specify the type of turn in the **Turn type** edit field of the Properties palette. The turn types are shown in Figure 8.3. The **1/2 Landing** turn type was selected to place the stair shown in Figure 8.11 for the command line sequence shown below. The **1/2 Landing** is created by starting the second flight at the end of the landing. The display of the riser count is dynamically displayed as you place the endpoint of the flight.

> Command: StairAdd
>
> Flight start point or [SHape/STyle/Tread/Height/Width/Justify/Next/Flight length/Match/Undo]: *(Using Endpoint object snap, select point p1 as shown at left in Figure 8.11.)*
>
> Flight end point or [SHape/STyle/Tread/Height/Width/Justify/Next/Flight length/Match/Undo]: *(Using Endpoint object snap, select point p2 as shown at left in Figure 8.11.)*
>
> Flight start point or [SHape/STyle/Tread/Height/Width/Justify/Next/Flight length/Match/Undo]: *(Move the pointer to the **left** until polar track tip distance equals at least 3', and then left-click to specify point p3 as shown at left in Figure 8.11.)*
>
> Flight end point or [SHape/STyle/Tread/Height/Width/Justify/Next/Flight length/Match/Undo]: *(Move the pointer to the left, and select a point beyond the dynamic display of the stair near p4 as shown at right in Figure 8.11.)*
>
> Flight start point or [SHape/STyle/Tread/Height/Width/Justify/Next/Flight length/Match/Undo]: ENTER *(Press ENTER to end the command.)*

CREATING A U-SHAPED STAIR

U-shaped stairs divide the total run into two equal flights. The stair can include a flat landing or a winder landing between the two runs. The turn of the landing is created by specifying the **1/2 turn** or the **1/2 landing**. The stair placed in Figure 8.12 includes 1/2 landing and requires the justification of the stair be set to the right to place the stair edge at **p1**. The stair can be generated with clockwise or counterclockwise

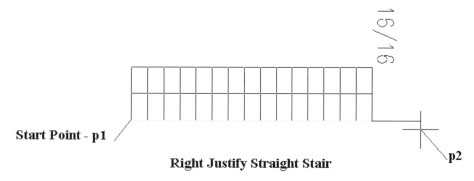

Figure 8.10 *Creating a straight stair*

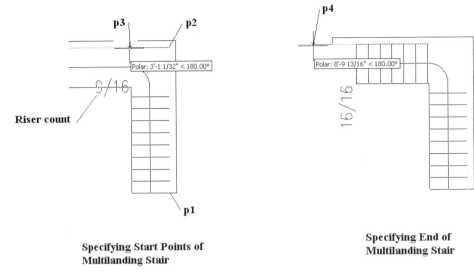

Figure 8.11 *Start and end flight points of a multi-landing stair*

Horizontal Orientation from the start point as specified in the Properties palette. The stair shown in Figure 8.12 was created with counterclockwise horizontal orientation. When placing a U-shaped stair with counterclockwise orientation, move the pointer to the left of the start point and specify a point greater than the width of the stair flights at **p2** as shown in Figure 8.12. The polar tracking tooltip displays the distance the pointer is moved from the start point of the flight.

Command: STAIRADD

Flight start point or [SHape/STyle/Tread/Height/Width/Justify/Next/Match/Undo]: *(Select the end of the wall at p1 as shown in Figure 8.12.)*

Flight end point or [SHape/STyle/Tread/Height/Width/Justify/Next/Match/Undo]: *(Select a point near p2 as shown in Figure 8.12.)*

Flight start point or [SHape/STyle/Tread/Height/Width/Justify/Next/Match/Undo]: ENTER *(Press ENTER to end the command.)*

After the stair is created, you can use the **Move** command to move it to fit existing walls. A U-shaped winder stair can be created if you set the **Turn type = 1/2 turn** as shown in Figure 8.13. When you specify a 1/2 turn, the **Winder** property is added to the Properties palette. The **Balanced** winder is the default winder; however, additional winders can be defined in the **Style Manager**. Winder styles are located in the Architectural Objects>Stair Winder Styles folder of the **Style Manager**. The U-shaped winder stair is placed in a similar manner as the 1/2 landing U-shaped stair.

Stairs and Railings 511

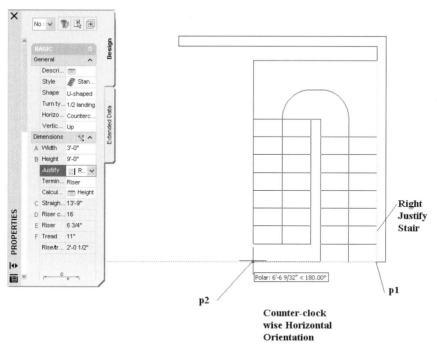

Figure 8.12 U-shaped stair with landing

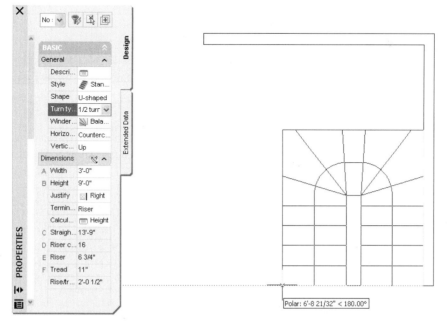

Figure 8.13 Winder U-shaped stairs

CREATING SPIRAL STAIRS

The **Spiral** shape is used to create curved or spiral stairs. A spiral stair is created by specifying the center and radius of the spiral stair; the stair **Calculation Rules** determine the number of risers and angle of the sweep. The angle of the arc can be specified as constrained or free in the **Arc Constraint** edit field of the Properties palette. The justification of the stair places the handle for insertion at the left, center, or right.

To place a spiral stair to fit an existing curved wall, select the Center object snap when specifying the center of the stair. The center of the spiral stair can be stretched by selecting the grip of the spiral stair. Editing the grips allows you to change the radius and location of the start point. In Figure 8.19, the **Radius** grip of the spiral stair can be stretched to change the radius; therefore, the start point for the flight remains stationary. When you select the **Spiral** shape, the Properties palette expands to include the following additional edit fields to place the stair.

> **Horizontal Orientation** – The **Horizontal Orientation** field allows the stair to be developed from the start point clockwise or counterclockwise.
>
> **Specify on Screen** – The **Specify on Screen** drop-down list includes **Yes** or **No** options. Select the **Yes** option and select two locations in the workspace to specify the radius in the workspace.
>
> **Radius** – If **Specify on Screen** is set to **No**, the **Radius** can be typed in this edit field.
>
> **Arc Constraint** – The **Arc Constraint** includes options to constrain the angle of the arc of the stair. The drop-down list is described below.
>
>> **Free** – The **Free** arc is not constrained.
>>
>> **Total Degrees** – The **Total Degrees** field specifies the degrees the arc can sweep for the flight. Total degrees of the flight can exceed 360°.
>>
>> **Degrees per Tread** – The **Degrees per Tread** specifies the degree for each tread.
>
> **Arc Angle** – The **Arc Angle** edit field displays the total angle the stair will sweep with free constraint. The **Arc Angle** edit field displays the angle for the **Total Degrees** and **Degrees per Tread** constraints.

The **Specify on Screen** option allows you to select the center of a wall using the Center object snap and then specify the start point to specify the radius of the stair. The stair shown in Figure 8.14 was placed by setting the following options in the Properties palette and responding to the command line sequence: Style = **Standard**, Shape = **Spiral**, Horizontal = **Counterclockwise**, Vertical = **Up**, Width = **3'-8"**, Height = **10'**, Justify = **Right**, Terminate with = **Riser**, Specify on Screen = **Yes**, Arc Constraint = **Free**.

> Command: StairAdd

Center of spiral stair or [SHape/STyle/Tread/Height/Width/Justify/
Radius/Match/Undo] _cen of *(Press* SHIFT *+ right-click with mouse,
select Center from the Object Snap option, move the pointer over the
wall, left-click when the Center object snap marker is displayed, and left-
click to specify point p1 as shown in Figure 8.14.)*

Start point or [SHape/STyle/Tread/Height/Width/Justify/Radius/
Match/Undo]: *(Select the start point p2 as shown in Figure 8.14.)*

Start point or [SHape/STyle/Tread/Height/Width/Justify/Match/
Undo]: ENTER *(Press* ENTER *to end the command.)*

In addition to the riser and tread code requirements, most building codes require that the tread depth **12" from the narrow edge be at least 7 1/2"**, to provide an adequate tread surface. The Properties palette does not include a code check for this tread dimension. To create a spiral stair that complies, set the **A-Width** to **24**, and then use the **StairOffset** command discussed in "Customizing the Edge of a Stair" in this chapter to offset the edge of the stair to the total desired stair width. You can check the minimum tread distance by drawing an arc with a radius equal to the spiral stair, offsetting the arc 12" and then measuring the tread at this location. Measuring the tread along the arc that is 12" from the narrow edge will allow you to check the spiral stair for code compliance, as shown in Figure 8.15.

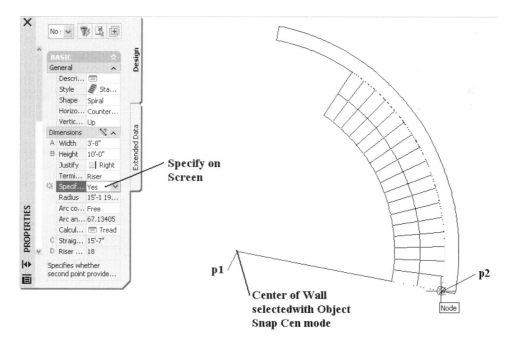

Figure 8.14 *Creating a spiral stair with Free Constraint*

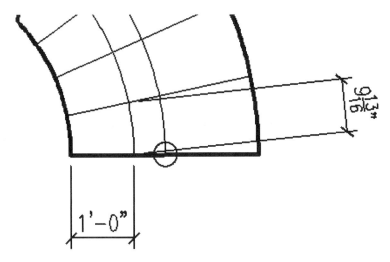

Figure 8.15 *Checking the minimum tread 12" from the narrow edge of the spiral stair*

EDITING A STAIR USING GRIPS

Each stair shape can be edited by its grips. Edge grips can be toggled off by selecting the trigger grip, and the grips can be used to edit the location, flight start, and flight end as shown at top in Figure 8.16. When **Edit Edge grips** is toggled ON, you can adjust the width of the stair and the taper of the stair as shown at bottom in Figure 8.16.

Stretching the grips of the stair allows you to quickly adjust the **start** and **end** points of the stair flight. If a multi-landing stair is edited by grips, it is **shortened** or **lengthened** a distance equal to the tread size. The **Flight Taper** and **Flight Width** grips allow you to stretch the edge of the stair. The **Location** grip will move the stair without stretching its components. The **Lengthen Flight** grip will stretch the straight length of a Straight stair if **Straight Length** is not set to **Automatic** in the **Calculation Rules**. **Automatic** can be toggled OFF in **Calculation Rules** for an existing stair. **Flight Start** and **Flight End** grips adjust the location of the insertion point of the start and end of the flight. The **Construction Line** and **Graphic Path** grips edit the location of the path arrows. Movement of grip points to locations that violate the **Design Rules** will cause the display of the stair warning symbol shown in Figure 8.5.

GRIPS OF THE MULTI-LANDING STAIR

The grips of the multi-landing stair consist of grips for each of the flights and the landing as shown in Figure 8.17. The grips of **Edit Edges** toggled ON are shown at left, and the grips of **Edit Edges** toggled OFF are shown at right in Figure 8.17. The **Flight**

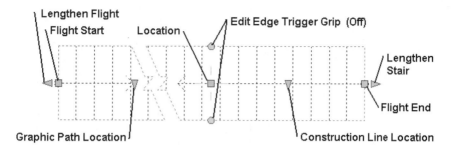
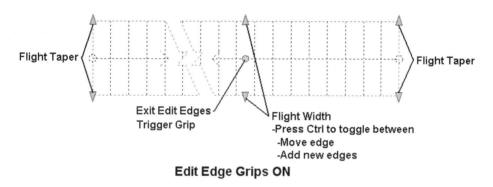

Figure 8.16 *Grips of the straight stair*

Taper grip allows you to taper each flight of the mult-landing flight. The **Move Flight** and **Turn Points** grips will move the landing and the flight. The **Edge** and **Landing Width** grips will change the shape of the landing without stretching the flight.

GRIPS OF THE U-SHAPED STAIR

The grips of the U-shaped stair, shown in Figure 8.18, are similar to those of the multi-landing stair. The U-shaped stair consists of two flights that are fixed in length; therefore, selecting the **Flight Start** and **Flight End** grips will **not** stretch the flight. Selecting the **Flight Start** or the **Flight End** will **rotate** the unit rather than stretch the flight. The **Flip** grips shown in Figure 8.18 can be used to mirror flip the stair horizontally or vertically.

GRIPS OF THE SPIRAL STAIR

The grips of the spiral stair are shown in Figure 8.19. The **Flight Taper** and **Flight Width** grips are accessed when **Edit Edges** is toggled ON. The entire stair can be

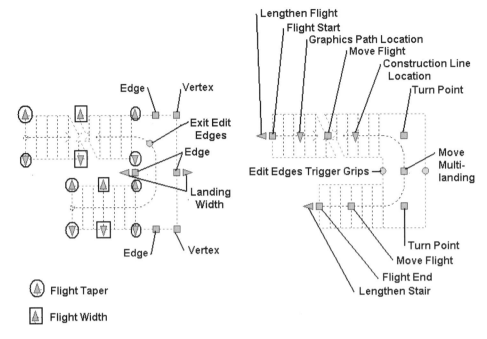

Figure 8.17 Grips of the multi-landing stair

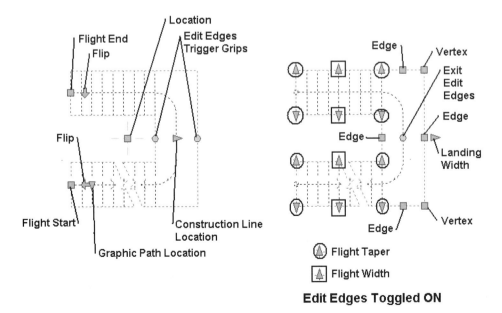

Figure 8.18 Grips of the U-shaped stair

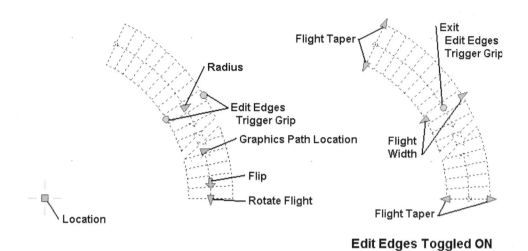

Figure 8.19 *Grips of the spiral stair*

moved with the **Location** grip. The **Radius** grip allows you to drag the grip relative to the **Location** grip to change the radius.

CUSTOMIZING THE EDGE OF A STAIR

The edge of the stair can be customized to increase or decrease its width by a specific value or extend the stair to a wall or polyline. Previous edits of the stair can be removed with the customize edge commands. The **Customize Edge** shortcut menu option includes the following commands: **Offset, Project, Remove Customization**, and **Generate Polyline**. The **Offset** option (**StairOffset** command) allows you to extend the width of the stair by a specified distance. Access **Offset** (**StairOffset** command) as shown in Table 8.2.

Command prompt	STAIROFFSET
Shortcut menu	Select the stair, right-click, and choose Customize Edge>Offset

Table 8.2 *StairOffset command access*

When you select the **Customize Edge>Offset** command from the shortcut menu for a stair, you are prompted in the command line to type a positive distance to move the stair edge out or a negative distance to move the edge toward the inside of the stair. This command is used to adjust the width of the stair to fit the structure. The command allows you to control which edge of the stair to increase or decrease in width.

The following command line sequence was used to increase the width of the stair as shown in Figure 8.20.

(Select a stair, right-click, and choose **Customize Edge>Offset** *from the shortcut menu.)*

Command: StairOffset

Select an edge of a stair: *(Select the stair at p1 as shown in Figure 8.20.)*

Enter distance to offset (positive = out, negative = in): **12"** ENTER
(Type positive distance for offset toward the outside of the stair.)

(Edge extended as shown in Figure 8.20.)

PROJECTING STAIRS TO WALLS

The **Project** option of the **Customize Edge** shortcut menu (**StairProject** command) allows you to extend the stair to a wall or polyline. This option can be used to trim or extend a stair to a wall or polyline. Access the **StairProject** command as shown in Table 8.3.

Command prompt	StairProject
Shortcut menu	Select the stair, right-click, and choose Customize Edge>Project

Table 8.3 *StairProject command access*

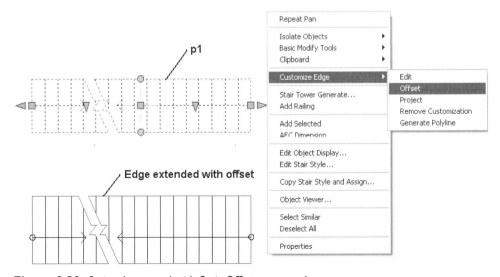

Figure 8.20 *Stair edge moved with Stair Offset command*

The following command line sequence was used to extend the stair shown in Figure 8.21 to the wall.

> (Select a stair, right-click, and choose **Customize Edge>Project** from the shortcut menu.)
>
> Command: StairProject
>
> Select an edge of a stair: (Select the stair at p1 as shown in Figure 8.21.)
>
> Select a polyline or connected AEC objects to project to: 1 found (Select the wall at p2 as shown in Figure 8.21.)
>
> Select a polyline or connected AEC objects to project to: ENTER (Press ENTER to end selection.)

REMOVING STAIR CUSTOMIZATION

Previous customizations of the stair using the **Offset** and **Project** options can be removed with the **Remove Customization** option of the **Customize Edge** shortcut menu. The **Remove Customization** option (**StairRemoveCustomization**) will remove all previous edits to the edge of the stair. Access the **StairRemoveCustomization** command as shown in Table 8.4.

Command prompt	STAIRREMOVECUSTOMIZATION
Shortcut menu	Select the stair, right-click, and choose Customize Edge>Remove Customization

Table 8.4 *StairRemoveCustomization command access*

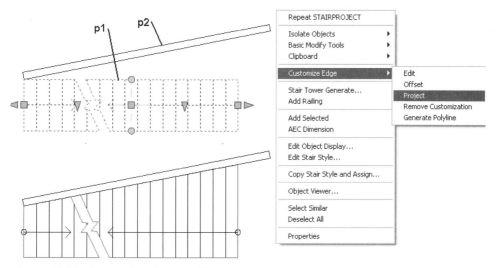

Figure 8.21 *Stair projected to a wall*

The customization of the stair edge was removed in the following command line sequence as shown in Figure 8.22.

> *(Select a stair, right-click, and choose* **Customize Edge>Remove Customization** *from the shortcut menu.)*
>
> Command: StairRemoveCustomization
>
> Select an edge of a stair: *(Select the edge of the stair at p1 to remove changes to the edge.)*
>
> *(Stair changed to original state as shown in Figure 8.22.)*

CREATING A POLYLINE FROM THE EDGE OF THE STAIR

The **Generate Polyline** option of the **Customize Edge** shortcut menu is used to create a polyline from the edge of the stair. The polyline can be edited to develop walls or other stair edges in the drawing. Access the **Generate Polyline** command as shown in Table 8.5.

Command prompt	STAIRGENERATEPOLYLINE
Shortcut menu	Select the stair, right-click, and choose CustomizeEdge>Generate Polyline

Table 8.5 *StairGeneratePolyline command access*

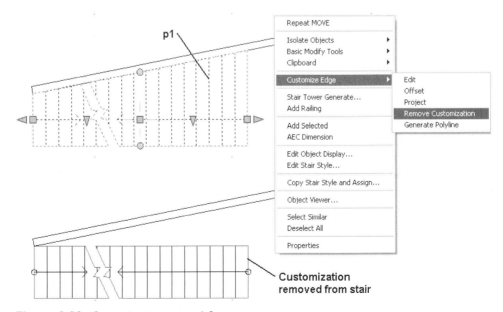

Figure 8.22 *Customization removed from a stair*

A polyline was created from the edge of the stair as shown in Figure 8.23 and the following command line sequence. If the stair is created with a down vertical orientation, the polyline is generated at the top of the stair; whereas the polyline is created at the base of a stair with an up vertical orientation.

>(Select a stair, right-click, and choose **Customize Edge>Generate Polyline** from the shortcut menu.)
>
>Command: StairGeneratePolyline
>
>Select an edge of a stair: (Select the edge of the stair at p1 in Figure 8.23.)
>
>(Select the polyline to display the grips of the new polyline as shown in Figure 8.23.)

CREATING STAIR STYLES

The Standard stair style is included in drawings created from the Aec Model templates; therefore the Standard stair style is inserted in a new drawing. If other stair styles have been imported or created in the drawing, the style and other settings of the last

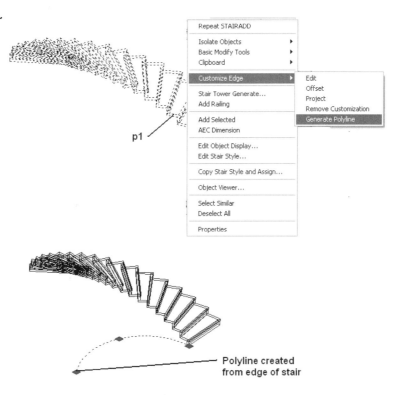

Figure 8.23 *Polyline created from edge of stair with up vertical orientation*

stair inserted become the defaults for current insertion. The Standard style is the simplest because it includes only risers and treads, with no stringer. The Standard stair style provides a simple representation of a stair. However, other stair styles add more detail regarding the stringer position and better represent the complexity of the stair. You can i-drop stair styles from the Design Tool Catalog -Imperial>Stairs and Railing>Stairs category of the **Content Browser**. Stair styles can be imported in the **Style Manager** from the *Stairs Styles (Imperial)* file located in the *C:\Documents and Settings\All Users\Application Data\Autodesk\ADT 2005\enu\Styles\Imperial* directory. The Stair Styles (Imperial) file includes the following styles: Cantilever, Concrete, Half Wall Rail, Ramp Concrete, Ramp Concrete Curb, Ramp Steel, Standard, Steel, Wood Housed, and Wood Saddle as shown in Figures 8.24 and 8.25.

USING THE STYLE MANAGER TO CREATE STAIR STYLES

Stair styles can be created, edited, imported, and exported in the **Style Manager**. The style of an existing stair can be edited by selecting the **Edit Stair Style** command. Access the **Edit Stair Style** (**StairStyleEdit**) command as shown in Table 8.6. The Stair Style definition includes control over the riser and tread dimensions and the insertion of stringers. The specific location and size of the stringer can be established through the **Stair Styles** command. In addition, the visibility of the upper or lower half of the stairs can be controlled in the display properties of the stair style.

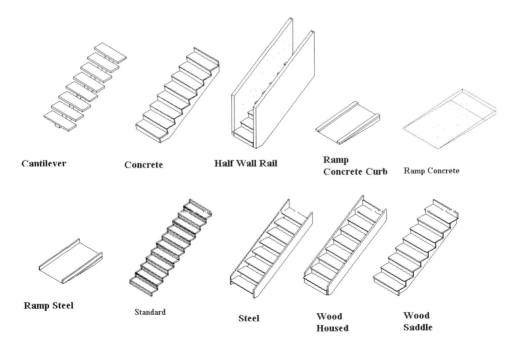

Figure 8.24 *Stair styles of the Stair Styles drawing*

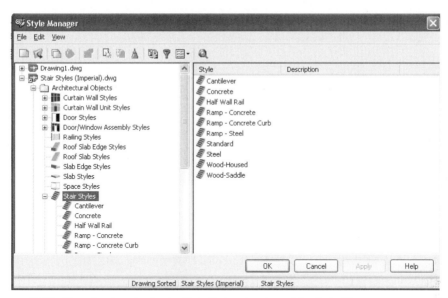

Figure 8.25 List of stair styles in the Stair Styles drawing of the Style Manager

Command prompt	STAIRSTYLEEDIT
Menu bar	Format>Style Manager, expand the Architectural Objects>Stair Styles in the left pane
Shortcut menu	Select the stair, right-click, and choose Edit Stair Style
Tool palette	Right-click over the Stair tool and choose Stair Styles

Table 8.6 Edit Stair Style command access

To create a stair style, select the **Format>Style Manager**, and then expand the Architectural Objects folder in the left pane. Select the **Stair Styles** folder in the left pane, and then select **New Style** from the **Style Manager** toolbar. A new style named **New Style** is created; overtype a name for the new style.

The properties of a new style are defined by selecting the name of the new style in the right pane and selecting **Edit Style** from the **Style Manager** toolbar. The style is then defined by editing the **Stair Styles** dialog box, which consists of eight tabs. The content of the **Stair Styles** dialog box for the Wood-Housed stair style is presented below. The Wood-Housed stair style is included in the Stair Styles (Imperial) drawing.

General Tab

The **General** tab of the Wood-Housed stair style is shown in Figure 8.26.

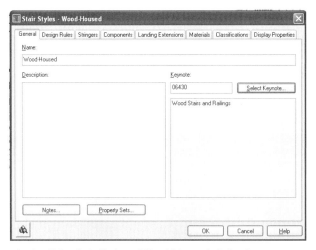

Figure 8.26 *General tab of the Stair-Styles – Wood-Housed dialog box*

Name – The **Name** edit field lists the name of the stair style.

Description – The **Description** field allows you to add text describing the style.

Notes – The **Notes** button opens the **Notes** dialog box, which allows you to add text regarding the style.

Property Sets – The **Property Sets** button opens the **Edit Property Set Data** dialog box, which allows property sets to be added to the stair style.

Keynote – The **Keynote** button opens the **Select Keynote** dialog box, which allows you to select keynotes from the 16 CSI divisions.

Design Rules Tab

The **Design Rules** tab includes **Code Limits** and **Calculator Rule** sections as shown in Figure 8.27. The **Code Limits** section allows you to set maximum, minimum, and optimum dimensions for the riser and tread. An **Optimum Slope** sets the target for riser and tread dimensions. The **Calculator Rule** section of the tab includes a formula check that can be turned on or off using the **Use Rule Based Calculator** check box.

Code Limits

A-Maximum Slope – The **Maximum Slope** edit fields allow you to type the tallest riser and shortest tread that is acceptable in the building code.

B-Optimum Slope – The **Optimum Slope** represents the target values for the riser and tread when applied to the stair height and number of risers.

C-Minimum Slope – The **Minimum Slope** edit fields allow you to type the shortest riser and longest tread that is acceptable.

Calculator Rule

Use Rule Based Calculator – Selecting the **Use Rule Based Calculator** check box toggles ON/OFF the test of the riser and tread dimensions to comply with the formula.

Minimum Maximum Limits – The **Calculator Rule** specifies the maximum and minimum value of the sum of two risers plus a tread. The formula rule shown in Figure 8.27 requires that two riser heights plus one tread depth must be no more than 2'-1". The **Calculator Rule** defines a range of dimensions acceptable for the riser and tread.

Stringers Tab

The **Stringers** tab is shown in Figure 8.28.

Stringer – Click in the **Stringer** column to edit the name of a stringer.

Type – Click in the **Type** column to display the drop-down list, which includes the following types of stringers: Housed, Saddled, Slab, and Ramp.

Alignment – Click in the **Alignment** column to display the Alignment drop-down list that includes Align Left, Align Right, Center, or Full Width.

A-Width – The **Width** is the actual width of the stringer. Click in this column and overtype the width distance.

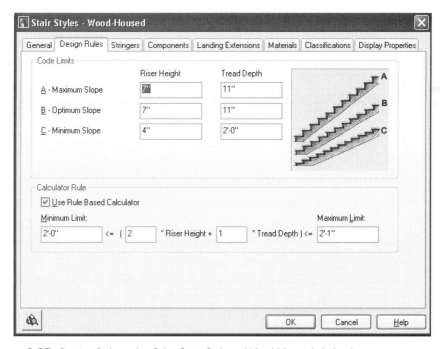

Figure 8.27 *Design Rules tab of the Stair Styles – Wood-Housed dialog box*

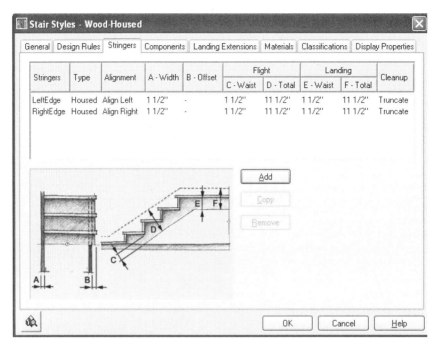

Figure 8.28 *Stringers tab of the Stair Styles dialog box*

B-Offset – The **Offset** locates the distance from the stringer to the edge of the stair.

FLIGHT

C-Waist – The **Waist** is the minimum width between the bottom surface of the stringer and the tread location on the stringer.

D-Total – The **Total** represents the dimension of the stringer inclusive of the stringer material.

LANDING

E-Waist – The **Waist** represents the thickness of the flooring system of the landing.

F-Total – The **Total** represents the total thickness of the slab and stringer of the landing.

Cleanup – The **Cleanup** drop-down list allows you to choose **Cleanup** or **Truncate** the end of the stringer. **Truncate** will stop the stringer at the landing. **Cleanup** will allow the stringer to continue under the landing.

Add – Select the **Add** button to add a new stringer.

Copy – Select a stringer in the list window and then select the Copy button to create a copy of a stringer.

Remove – Select a stringer in the list window, and then select the Remove button to delete the stringer.

Components Tab

The Components tab allows you to specify the dimensions of the various components of the stair. This tab, shown in Figure 8.29, consists of **Flight Dimensions** and **Landing Dimensions** sections. The riser, tread, and nosing dimensions are set in the **Flight Dimensions** section. The **Landing Dimensions** section allows you to specify the thickness and width of the landing. A description of the options of this tab follows:

Allow Each Stair to Vary – The **Allow Each Stair to Vary** check box, if checked, allows the dimensions in the **Components** tab to be edited in the Properties palette when the stair is placed. If this box is cleared, component sizes are controlled in the stair style.

Flight Dimensions

Display Tread – The **Display Tread** check box will turn ON/OFF the display of the tread.

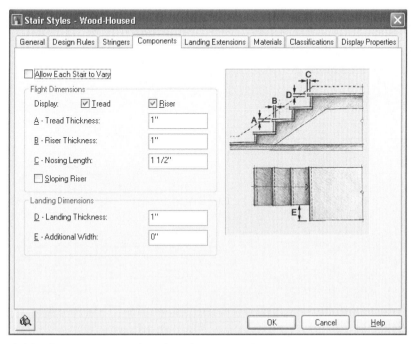

Figure 8.29 *Components tab of the Stair Styles dialog box*

Display Riser – The **Display Riser** check box will turn ON/OFF the display of the riser. Turn OFF the riser to create an open riser stair.

A-Tread Thickness – The **Tread Thickness** represents the thickness of the material used for the tread.

B-Riser Thickness – The **Riser Thickness** represents the thickness of the material used for the riser.

C-Nosing Length – The **Nosing Length** represents the distance the nosing overhangs the riser.

Sloping Riser – The **Sloping Riser** check box, if checked, creates a riser tilted from vertical.

Landing Dimensions

D-Landing Thickness – The **Landing Thickness** represents the material thickness of the landing.

E-Additional Width – The **Additional Width** represents the additional width of the landing beyond the stair justification line.

Landing Extensions Tab

The landing extensions can be adjusted in the **Landing Extensions** tab as shown in Figure 8.30

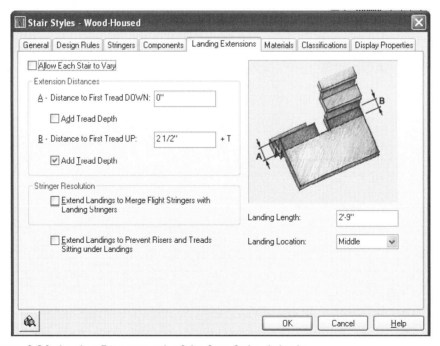

Figure 8.30 *Landing Extensions tab of the Stair Styles dialog box*

Allow Each Stair to Vary – The **Allow Each Stair to Vary** check box, if checked, allows the dimensions of the **Landing Extensions** tab to be edited in the Properties palette by selecting the **Landing Extensions** button located in the **Advanced–Worksheets** section. If this box is clear, the landing extension components are controlled in the stair style definition.

Extension Distances

A-Distance to First Tread DOWN – The **Distance to First Tread DOWN** is the horizontal distance the landing is extended before the first step down.

Add Tread Depth – The **Add Tread Depth** check box, if checked, extends the landing before the down step by one tread depth.

B-Distance to First Tread UP – The **B-Distance to First Tread UP** is the horizontal distance the landing is extended before the first step up.

Add Tread Depth – The **Add Tread Depth** check box, if checked, extends the landing before the up step by one tread depth.

Stringer Resolution

Extend Landings to Merge Flight Stringers with Landing Stringers – This check box, if checked, extends the landing stringer to merge with flight stringers.

Extend Landing to Prevent Risers and Treads Sitting Under Landings – This check box, if checked, adjusts the landing position to create flush rectangular landings.

Landing Length – The **Landing Length** is the length of the landings placed with automatic landing placement. Automatic landings are placed when you set a minimum and maximum **Flight Height** in the **Advanced** section of the Properties palette.

Landing Location – The **Landing Location** drop-down list includes top, middle, and bottom locations. When **Automatic** landings are placed based upon the minimum and maximum **Flight Height**, the landing can be placed at the top, middle, and bottom locations to comply with the flight height restrictions.

Materials Tab

The **Materials** tab allows you to specify the name and definition of materials applied to the stair. The options of the **Materials** tab shown in Figure 8.31 are similar to other styles. Refer to "Editing a Door Style" in Chapter 4 for the procedure to assign a material to an object. The components of the **Materials** tab are described below.

Component – The **Component** column lists the names of the stair components to apply the materials. Materials can be applied to the tread, riser, and landing.

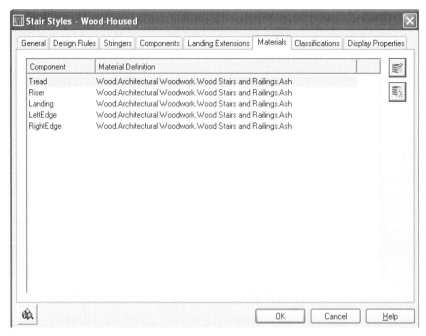

Figure 8.31 *Materials tab of the Stair Styles dialog box*

Material Definition – The **Material Definition** column lists the materials defined for the component.

Edit Material – The **Edit Material** button opens the **Material Definition Properties** dialog box to change material definitions.

Add New Material – The **Add New Material** button opens the **New Material** dialog box, which allows you to specify a name for the new material.

Classifications Tab

The **Classifications** tab allows you to specify classifications within the style as shown in Figure 8.32. Classifications can be imported from the *C:\Documents and Settings\All Users\Application Data\Autodesk\ADT 2005\enu\Styles\Imperial\Uniformat II Classifications (1997 ed).dwg*. The classifications of this drawing are located in the Multi-Purpose\Classifications folder within the **Style Manager**. If classifications are imported into the drawing, they can be assigned in this tab.

Classification Definition – The **Classification Definition** includes a list of classifications defined in the drawing.

Classifications – The classifications included in the drawing are listed in the **Classification**s drop-down list.

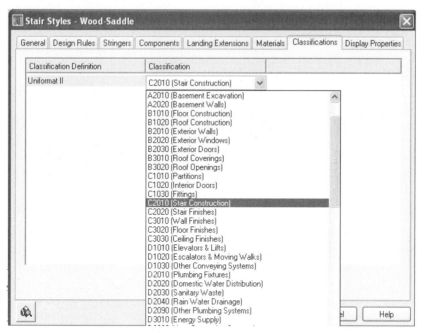

Figure 8.32 *Classifications tab of the Stair Styles dialog box*

Display Properties Tab

The **Display Properties** tab shown in Figure 8.33 allows you to set the display of components of the stair in the stair style. The options of the tab are as follows:

> **Display Representations** – The **Display Representations** column lists the display representations available for the stair.
>
> **Display Property Source** – The **Display Property Source** column lists object display categories for the current floating viewport. The **Display Property Source** is either controlled by the drawing default global settings or by an override to the style.
>
> **Style Override** – The **Style Override** check box indicates if an override has been defined in the display properties. If this box is **clear**, the **Drawing Default** display property source is controlling the display of the stair style. When the box is checked, additional display properties are defined when the **Display Properties** dialog box opens as shown in Figure 8.33.
>
> **Edit Display Properties** – The **Edit Display Properties** button opens the **Display Properties** dialog box. The **Display Properties** dialog box for the **Plan** display representation of a stair style is shown in Figure 8.33. This dialog box consists of three tabs: **Layer/Color/Linetype**, **Other**, and **Riser Numbering**.

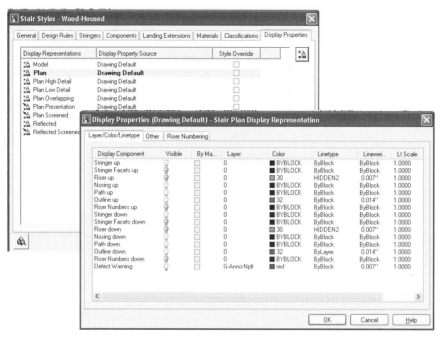

Figure 8.33 *Display Properties tab and the Display Properties Stair Plan Display Representation*

Layer/Color/Linetype – The **Layer/Color/Linetype** tab lists the display components and the settings for visibility, by material, layer, color, linetype, lineweight, and linetype scale. The list of components varies according to the display representation.

Other – The **Other** tab allows you to set the height of the cut plane, directional arrow, stair line, and break mark as shown in Figure 8.34.

CUT PLANE

Override Display Configuration Cut Plane – If the **Override Display Configuration Cut Plane** check box is checked, the stair is displayed independent of the display configuration cut plane.

Elevation – The **Elevation** is the vertical distance from the Z=0 coordinate of the cutting plane for creating the location of the lower half and the upper half of the stairs. The elevation of the cutting plane determines the location of the break line for the stair.

Distance – The **Distance** is the horizontal distance separating the break lines of the upper and lower stair sections.

Angle – The **Angle** is the angle formed relative to the treads for the break line symbol.

ARROW

Size – The **Size** edit field allows you to define the dimensions of the directional arrow for the stair.

Offset – The **Offset** is the distance from the break line to the head of the directional arrow.

Dim Style – The **Dim Style** edit field lists the dimensioning style used to define the style of the directional arrow.

STAIR LINE

Shape – The **Shape** drop-down list includes curved or straight options.

Apply to – The **Apply to** drop-down list includes Entire Stair, Cut Plane Parallel, and Cut Plane Opposite.

Display Path – The **Display Path** drop-down list includes Graphics path and Construction line options for the direction of stair.

BREAK MARK

Type – The **Type** lists the break mark symbol options to represent the break, including None, Curved, Zigzag, Custom shapes. If **Custom shape** is selected, the blocks of the drawing can be selected from the **Block** list. The break mark selected is displayed in the window below the **Block** edit field.

Block – The **Block** list displays blocks of the drawing that can be used for a custom break mark.

Riser Numbering – The **Riser Numbering** tab allows you to set the style and location of the text used to number the risers. You can edit the digit used as the first riser number as shown in Figure 8.34.

Style – The **Style** section allows you to specify the style, alignment, orientation, and height of text used for the riser numbers.

Location – The **Location** section allows you to specify the justification, x and y offsets of the text used for the riser numbers.

First Riser Number – The **First Riser Number** edit field allows you to specify the number for beginning the riser numbering.

Number Final Riser – The **Number Final Riser** check box turns ON/OFF the display of the final riser number.

ANCHORING A STAIR TO A LANDING

The **AnchorToStairLanding** command allows you to create two or more flights of stairs that share a single landing or link an AEC object to a landing. Objects can be anchored to landings created using the multi-landing or U-shaped stairs that include a landing. The anchor links the flights such that changes in one flight will cause the AEC object to change accordingly. Access the **AnchorToStairLanding** command as shown in Table 8.7.

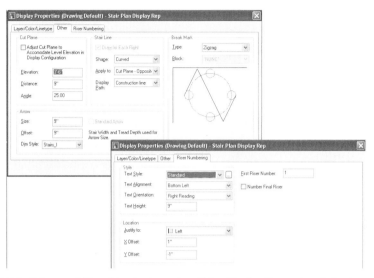

Figure 8.34 *Other and Riser Numbering tabs of the Display Properties dialog box*

Command prompt	ANCHORTOSTAIRLANDING
Shortcut menu	Select the stair with a landing, right-click, and choose Stair Landing Anchor>Anchor Object

Table 8.7 *AnchorToStairLanding command access*

When you select a multi-landing stair, right-click, and choose **Stair Landing Anchor>Anchor Object**, you are prompted to select an AEC object to anchor to stair and to select the stair landing. The selected landing and the AEC object are then anchored together. The **AnchorRelease** command will remove the anchor between the AEC object and the stair. Access the **AnchorRelease** command as shown in Table 8.8.

Command prompt	ANCHORRELEASE
Shortcut menu	Select the stair, right-click, and choose Stair Landing Anchor>Release

Table 8.8 *AnchorRelease command access*

CREATING A STAIR TOWER

The **StairTowerGenerate** command is used to generate multi-landing or U-shaped stairs within a project. See Table 8.9 and Figure 8.2 for **StairTowerGenerate** command access.

Command prompt	STAIRTOWERGENERATE
Tool palette	Select Stair Tower from the Design tool palette
Shortcut menu	Select a stair, right-click, and choose Generate Stair Tower

Table 8.9 *StairTowerGenerate*

The stair should be generated from a first floor stair within a construct drawing assigned to each level of the project. The first floor stair should have its display controlled by the display configuration. (The stair display control is set by clearing the **Override Display Configuration Cut Plane** check box of the **Other** tab in the **Display Properties** dialog box. The **Display Properties** dialog box is accessed from the **Edit Display Properties** button in the **Display Properties** tab of the **Stair Style** dialog box.) The **StairTowerGenerate** command creates additional stairs, slabs, and railing above the selected stair the distance specified in the floor to floor height for the levels of the project. The interference subtraction is also applied to each new stair and slab. Each stair can be edited independently of the stair for the first level. The stair shown at left in Figure 8.35 was generated to the next level as shown in the following command line sequence.

(*Select the stair at p1, right-click, and choose* **Generate Stair Tower**.)

Command: StairTowerGenerate

Select railings and slabs: 1 found (*Select slab at p2 as shown in Figure 8.35.*)

Select railings and slabs: ENTER (*Press* ENTER *to end selection.*)

(*The* **Select Levels** *dialog box opens, which specifies the floor to floor height, select* **OK** *to accept the settings and dismiss the* **Select Levels** *dialog box. The stair tower is generated as shown in the middle of Figure 8.35.*)

If the floor to floor height is changed after the stair tower is generated, the stair does not update to the changes. When the stair construct drawing is attached to the model view and to floor plan views, you can create the appropriate plan view of the stair when you edit the cutting plane elevation to an elevation appropriate to floor plan level in the view drawing, as shown at right in Figure 8.35.

 STOP: Do Tutorial 8.2 "Creating a U-Shaped Stair and a Stair Tower" at the end of the chapter.

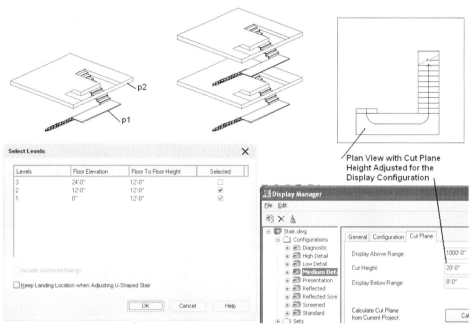

Figure 8.35 *Creating a stair tower*

CREATING RAILINGS

The **RailingAdd** command is used to add a railing to each side or the center of the stair. Railings can consist of posts, balusters, guardrails, bottom rails, and handrails. Railings can be added to a stair or stair flight or be freestanding. See Table 8.10 for **RailingAdd** command access.

Command prompt	RAILINGADD
Tool palette	Select Railing from the Design tool palette
Shortcut menu	Select a stair, right-click, and choose Add Railing

Table 8.10 *RailingAdd command access*

When you select the **Railing** tool (**RailingAdd** command) from the Design palette, the Properties palette opens, which displays the properties of the railing as shown in Figure 8.36. The Properties palette for the **RailingAdd** command is described below.

Stairs and Railings 537

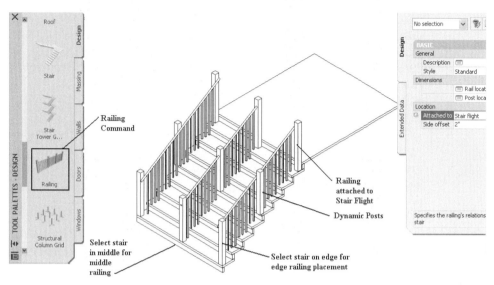

Figure 8.36 *Railing command of the Design palette and the Properties palette*

Basic

General

> **Description** – The **Description** button opens the **Description** dialog box, which includes edit fields for typing a description of the railing.
>
> **Laye**r – The **Layer** edit field is displayed when you select existing stairs; it lists the name of the layer for the railing. The railing is placed on the A-Flor-Hral layer if the AIA (256) Color layer key style is used.
>
> **Style** – The **Style** drop-down list displays the railing styles of the drawing. The Standard style is the default style; additional styles must be created or imported.

Dimensions

> **Rail locations** – The **Rail locations** button opens the **Rail Locations** dialog box. The **Rail Locations** dialog box is inactive for the Standard railing style because the features are defined in the railing style. The **Rail Locations** dialog box can be edited if **Allow Each Railing to Vary** is toggled ON in the **Rail Locations** tab of the **Railing Styles** dialog box.
>
> **Post locations** – The **Post locations** button opens the **Post Locations** dialog box. The **Post Locations** dialog box is inactive for the Standard railing style. The **Post Locations** dialog box can be edited if **Allow Each Railing to Vary** is toggled ON in the **Post Locations** tab of the **Railing Styles** dialog box.

Location

> **Attach to** – The **Attached to** drop-down list includes the Stair, Stair Flight, and None options. If **Attached to** is set to **Stair,** the railing is created as an

attachment to the entire stair. The **Stair** option expands the Properties palette to include the **Automatic Placement** option discussed below. The **Stair Flight** option will allow the railing to be attached to a flight segment of the multi-landing stair. The **None** option allows you to place a freestanding railing or guardrail.

Automatic Placement – The **Automatic Placement** options include Yes and No. If **Yes** is selected, the railing will be attached to the stair at the selected stair edge or the center of the stair. If **Automatic Placement** is set to **No**, the railing will be placed on the edge or center selected, and you will be prompted to select the start and end points of the railing.

Side offset – The **Side offset** is the distance from the centerline of the railing to the edge of the stair.

ADDING A RAILING

When you select **Railing** from the Design palette, the **Attach to** option in the Properties palette governs the prompts of the command line. When **Attach to** is set to **Stair Flight**, the pointer changes to a pick box, and you are prompted to select the stair flight for the railing. When you select the stair, you are also specifying the location for the railing. Therefore, if you select the stair in the middle of the tread width, the railing is applied in the center as shown at A in Figure 8.37.

If the **Railing** command is selected and **Attach to** is set to **Stair**, then **Automatic Placement** is added to the Properties palette. When **Automatic Placement** is set to **Yes**, selecting the stair will add the railing to the entire stair edge as shown at B in Figure 8.37. If **Automatic Placement** is set to **No**, you are prompted to select the start and end points of the railing as shown in the following command line sequence.

> Command: RailingAdd
>
> Select a stair or [Style/Attach/Match/Undo] *(Select the stair at p1, determines stair edge for railing as shown at C in Figure 8.37.)*
>
> Railing start point or [Style/Attach/Match/Undo] *(Select a point p2 to begin the railing as shown at C in Figure 8.37.)*
>
> Railing end point or [Style/Attach/Match/Undo] *(Select a point p3 to end the railing as shown at C in Figure 8.37.)*

Using the Attached to: None option

When the **Railing** command is selected and **Attach to** is set to **None**, you are prompted to specify the start and end points of the railing. The railing is not anchored to an AEC object; therefore, when you move an adjoining object, the railing does not move. This option can be used for placing railings on decks and balconies.

DEFINING RAILING STYLES

The design of the Standard railing is simple; therefore railing styles can be created to

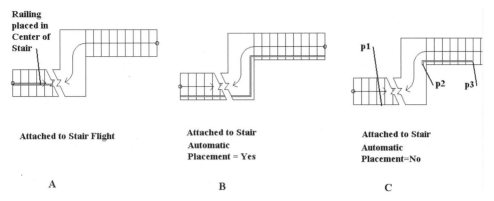

Figure 8.37 Attach to options of the Railing command

represent typical interior and exterior railings. The style can specify the size, shape, and placement of handrails, guardrails, post, and balusters as shown in Figure 8.38. Additional styles can be created or pasted into the drawing using the **Style Manager**. There are 22 styles in the Railing Styles (Imperial) drawing located in the *C:\Documents and Settings\All Users\Application Data\Autodesk\ADT 2005\enu\Styles\Imperial* directory. To access this drawing, select the **Open Drawing** icon of the **Style Manager** toolbar; then select **Content** from the **Places** panel, and navigate to the Styles\Imperial folder.

Railing styles are created in the **Style Manager**. The railing style can be edited in the **Style Manager** or with the **Edit Railing Style** command; see Table 8.11 for command access.

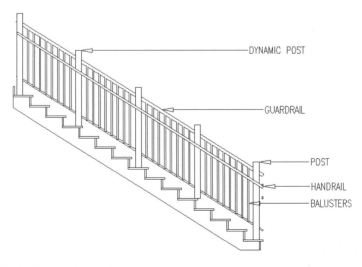

Figure 8.38 Defining railing styles

Command prompt	RAILINGSTYLEEDIT
Menu bar	Format>Style Manager, expand the Architectural Objects>Railing Styles, select a railing style, and then select Edit Style from the Style Manager toolbar
Shortcut menu	Select the railing, right-click, and choose Edit Railing Style
Tool palette	Right-click over the Railing tool and choose Railing Styles

Table 8.11 *Style Manager and Edit Railing Style command access*

STEPS TO IMPORTING AND EDITING A RAILING STYLE IN THE STYLE MANAGER

1. Select **Format>Style Manager** from the menu bar.
2. Select **Open** from the **Style Manager** toolbar, select **Content** from the **Places** panel at left of the **Open Drawing** dialog box, and navigate to the Styles\Imperial folder. Select the **Railing Styles (Imperial)** drawing.
3. Select **Open** to complete selection and close the **Open Drawing** dialog box.
4. Select **Railing Styles (Imperial).dwg**, and expand the Architectural Objects>Railing Styles in the left pane.
5. Select a railing style listed in the Railing Styles folder of the left pane, right-click, and choose **Copy** from the shortcut menu.
6. Scroll up or down the left pane to select the name of the current drawing, right-click, and choose **Paste**. The railing style is pasted into the current drawing.
7. After pasting the style in the current drawing, select the railing style in the right pane, right-click, and choose **Edit** from the shortcut menu. The **Railing Styles** dialog box opens, allowing you to edit the properties of the style.

COMPONENTS OF A RAILING STYLE

A railing style can be edited in the **Style Manager**. If a railing has been inserted, select the railing, right-click, and choose **Edit Railing Style** from the shortcut menu to edit a railing style in the drawing. The **Edit Railing Style** command opens the **Railing Styles** dialog box as shown in Figure 8.39. The **Railing Styles** dialog box consists of the following tabs: **General**, **Rail Locations**, **Post Locations**, **Components**, **Extensions**, **Materials**, **Classifications**, and **Display Properties**. A description of the tabs for the Guardrail – Wood Balusters 02 railing style follows. The Guardrail – Wood Balusters 02 railing style includes some features not included in the Standard rail style.

General Tab

Name – The **Name** edit field lists the name of the railing style.

Description – The **Description** edit field allows you to add text to describe the railing.

Notes – The **Notes** button opens the **Notes** dialog box, which allows you to type text to describe the railing.

Property Sets – The **Property Sets** button opens the **Property Sets** dialog box to add property sets to the style.

Keynote– Choose the **Select Keynote** button to open the **Select Keynote** dialog box, which allows you to specify keynotes from the 16 CSI divisions.

Rail Locations Tab

The **Rail Locations** tab shown in Figure 8.40 allows you to define the horizontal and vertical components of the railing.

Allow Each Railing to Vary – When this check box is toggled ON, you can set values for the rail locations of the style in the Properties palette for each insertion in the drawing. If this check box is clear, the settings of this tab are specified only in the style and cannot be edited in the Properties palette.

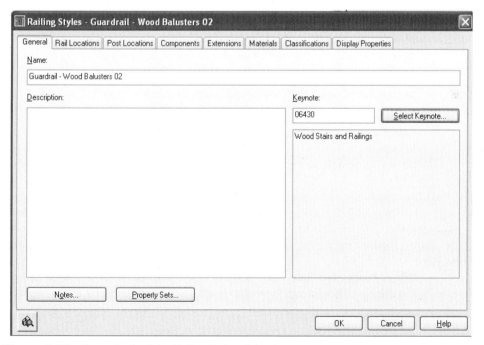

Figure 8.39 *General tab of the Railing Styles dialog box*

Upper Rails

Guardrail – Check the **Guardrail** check box to display a guardrail for the style. If the **Guardrail** check box is clear, the dimension edit fields are inactive.

Handrail – Check the **Handrail** check box to display a handrail for the style. If the **Handrail** check box is clear, the dimension edit fields are inactive.

Horizontal Height – The **Horizontal Height** is the height of the guardrail, handrail, or bottomrail in the Z direction.

Sloping Height – The **Sloping Height** is the height of the guardrail, handrail, and bottomrail as it is placed on the stair parallel to the stair.

Offset from Post – The **Offset from Post** is the distance from the guardrail, handrail, or bottomrail positioned from the centerline of the vertical posts.

Side for Offset – The **Side for Offset** specifies the direction for placing the handrail, guardrail, and bottomrail relative to the posts. A positive offset distance will place the railing the specified distance toward the center of the stair. Side of offset options in the drop-down list include right, left, center, and auto.

Bottomrail

Bottomrail – The **Bottomrail** check box, if checked, will display a bottomrail.

Horizontal Height – The **Horizontal Height** sets the vertical distance from Z=0 to the bottomrail when placed in a horizontal position.

Sloping Height – The **Sloping Height** sets the distance between stair and bottomrail as the railing is developed along the stair.

Number of Rails – The **Number of Rails** specifies the number of bottomrails. If more than one bottomrail is specified, you can edit the Spacing of Rails in inches.

Post Locations Tab

The **Post Locations** tab allows you set the frequency and position of posts placed along the rail. The following locations are specified in the **Post Locations** tab shown in Figure 8.41.

Fixed Posts

Fixed Posts – The **Fixed Posts** check box allows you to turn ON or OFF the display of fixed posts located at the beginning and end of the stair flight.

A-Extension of ALL Posts from Top Railing – The **A-Extension of ALL Posts from Top Railing** specifies the distance to extend the post above the top railing.

B-Extension of ALL Posts from Floor Level – The **B-Extension of ALL Posts from Floor Level** will extend the bottom of the post up the specified distance from the floor level.

Fixed posts at Railing Corners – The **Fixed posts at Railing Corners** check box when checked forces the development of a post at the corners or change of direction of the railing.

Dynamic Posts

Dynamic Posts – The **Dynamic Posts** check box allows you to turn on or off the display of posts between the fixed posts located at the beginning and end of the stair.

C-Maximum Center to Center Spacing – The **Maximum Center to Center Spacing** edit field allows you to specify the frequency of dynamic posts. If this distance is set to 4', a dynamic post will be placed along the rail no more than 4' apart.

Balusters

Balusters – The Balusters check box, if checked, specifies the display of balusters.

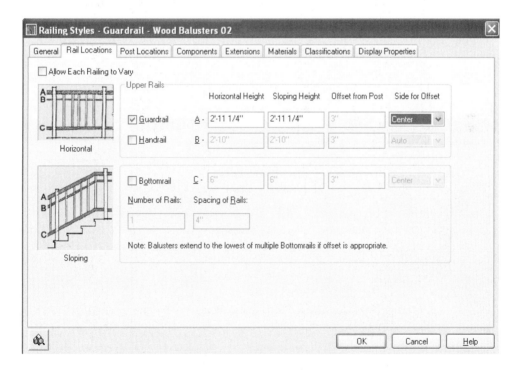

Figure 8.40 *Rail Locations tab of the Railing Styles dialog box*

D-Extension of Balusters from Floor Level – The **Extension of Balusters from Floor Level** offsets the bottom of the baluster from the floor level a specified distance.

E-Maximum Center to Center Spacing – The **Maximum Center to Center Spacing** edit field specifies the distance between the centerline of the balusters.

Stair Tread Length Override – The **Stair Tread Length Override** check box, if checked, will override the distance between balusters to place a specified number of balusters on a stair tread.

F-Number per Tread – The **Number per Tread** edit field specifies the number of balusters per tread.

Note: Decks can be created by specifying guardrail and bottom rails. Porch posts can be created by editing the A-Extension of ALL Posts from Top Railing to extend the posts above the guardrail as shown at right in Figure 8.41.

Components Tab

The **Components** tab allows you to specify the profile or shape used in extruding railing components. You can specify an Aec Profile for a component by clicking in the **Profile Name** column shown in Figure 8.42. The **Profile Name** field becomes an active list, which displays circular, rectangular, or other available Aec Profiles of the drawing.

Component – The Components listed in Figure 8.43 are components of the railing developed from Profiles. Guardrail, handrail, posts, and balusters can be

Figure 8.41 *Post Locations tab of the Railing Styles dialog box*

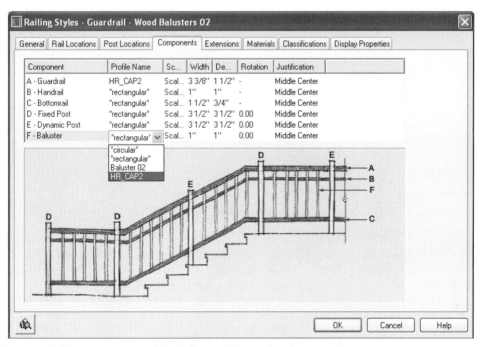

Figure 8.42 *Components tab including profiles used as baluster shapes*

created by extruding profiles to create custom shapes. Components can be created that are circular or rectangular.

Profile Name – The **Profile Name** drop-down list shows the Aec Profiles loaded in the drawing. A profile can be defined for the component by selecting in the **Profile Name** column of the component and selecting from the drop-down list. Figure 8.43 includes examples of polylines used to form the profile of components.

Scale – The **Scale** drop-down list allows you to scale the profile to Scale to Fit, Scale to Fit Width, Scale to Fit Depth, or No Scale.

Width – The **Width** dimension of the railing component.

Depth – The **Dept**h dimension of the railing component.

Rotation – The **Rotation** dimension of the railing component in a counter-clockwise direction.

Justification – The **Justification** option allows the profile to be positioned relative to the center line of the railing using the following justifications: Top Left, Top Center, Top Right, Middle Left, Middle Center, Middle Right, Bottom Left, Bottom Center, Bottom Right, and Insertion point.

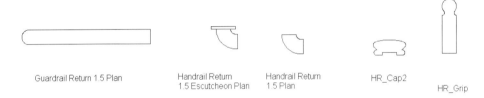

Figure 8.43 *Polylines of profiles used as handrail shapes*

Extensions Tab

The amount of railing extension beyond the end of the flight can be specified in the Extensions tab as shown in Figure 8.44.

> **Allow Each Railing to Vary** – When the **Allow Each Railing to Vary** check box is checked, the contents of the tab can be edited in the Properties palette for each railing inserted. If the **Allow Each Railing to Vary** check box is clear, the settings in this tab are controlled by the style definition.

At Floor Levels
> **Use Stair Landing Extension** – If **Use Stair Landing Extension** check box is ON at floor levels, the extension of the railing will match the stair

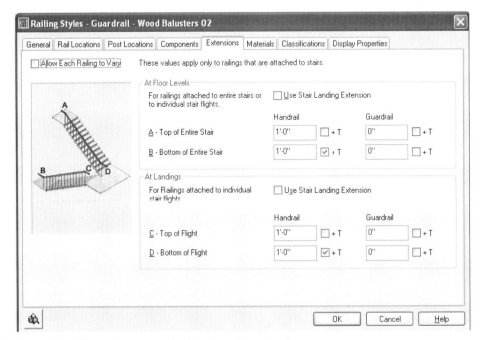

Figure 8.44 *Extensions tab of the Railing Styles dialog box*

extension for the stairs and stair flights. When this check box is clear, the extension of the railing can be defined in this tab.

A-Top of Entire Stair – The **Top of Entire Stair** edit fields define the distance to extend the railing beyond the stair for the handrail and guardrail. The distance can be defined as a distance plus the depth of the tread to obtain a total distance.

B-Bottom of Entire Stair – The **Bottom of Entire Stair** edit fields define the distance to extend the railing beyond the stair at the bottom of the stair for the guardrail and handrail. The distance can be defined as a distance plus the depth of the tread to obtain a total distance.

At Landings

Use Stair Landing Extension – If the **Use Stair Landing Extension** check box is ON, the extension of the railing will match the stair flights at the landing. When this toggle is clear the extension of the railing can be defined in this tab.

Use Stair Landing Extension – If **Use Stair Landing Extension** is toggled ON at floor levels, the extension of the railing will match the stair extension for the stairs and stair flights. When this toggle is clear, the extension of the railing can be defined in this tab.

C-Top of Flight – The **Top of Flight** edit field defines the distance to extend the railing beyond the stair for the handrail and guardrail at the landing for the top flight. The distance can be defined as a distance plus the depth of the tread to obtain a total distance.

D-Bottom of Flight – The **Bottom of Flight** defines the distance to extend the railing beyond the stair for the handrail and guardrail at the landing for the bottom flight. The distance can be defined as a distance plus the depth of the tread to obtain a total distance.

Materials Tab

The **Materials** tab shown in Figure 8.45 is similar to the **Materials** tabs of other object styles. Refer to "Editing a Door Style" in Chapter 4 for the procedure used to define materials to an object.

Component – The **Component** field lists the name of the railing components to apply the materials.

Material Definition – The **Material Definition** column lists the material definitions in the drop-down list for the component.

Edit Material – The **Edit Material** button opens the **Material Definition Properties** dialog box to change material definitions.

Add New Material – The **Add New Material** button opens the **New Material** dialog box, which allows you to specify a name for the new material.

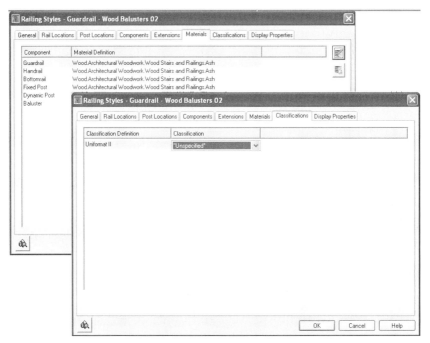

Figure 8.45 *Materials and Classifications tab of the Railing Styles dialog box*

Classifications Tab

The **Classifications** tab allows you to assign classifications defined in the drawing to the style. The classifications of the drawing can be based upon the status of the construction. The **Classifications** tab is shown in Figure 8.45.

> **Classification Definition** – The **Classification Definition** includes a list of classifications defined in the drawing.
>
> **Classification** – The classifications included in the drawing are listed in the **Classification** drop-down list.

Display Properties Tab

The **Display Properties** tab shown in Figure 8.46 allows you to define the railing components to display in the drawing. Refer to the **Display Properties** tab description of the stair style in this chapter. The display up and down components of the railing can be turned off in the **Layer/Color/Linetype** tab in the **Display Properties** dialog box as shown in Figure 8.46.

The **Display Properties** tab shown in Figure 8.46 allows you to set the display of components of the railing in the railing style. The options of the **Display Properties** tab are as follows:

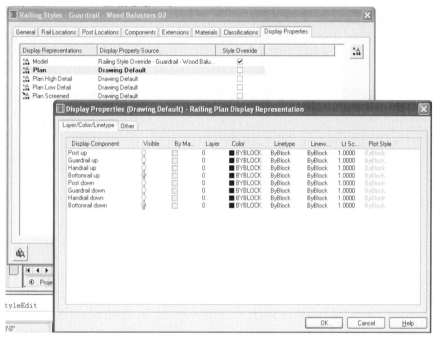

Figure 8.46 *Display Properties tab of the Railing Styles dialog box*

Display Representations – The **Display Representations** column lists the display representations available for the railing.

Display Property Source – The **Display Property Source** column lists object display categories for the current viewport. The **Display Property Source** is controlled either by the drawing default global settings or a style override.

Style Override – The **Style Override** check box indicates if an override has been defined in the display properties. If this box is clear, the **Drawing Default** display property source is controlling the display of the railing style. When you check this box, the **Display Properties** dialog box opens.

Edit Display Properties – The **Edit Display Properties** button opens the **Display Properties** dialog box. The **Display Properties** dialog box for the Railing Plan Display Representation is shown in Figure 8.47. This dialog box consists of two tabs, **Layer/Color/Linetype** and **Other**, described below.

Layer/Color/Linetype – The **Layer/Color/Linetype** tab lists the display components and the settings for visibility, by material, layer, color, linetype, lineweight, and linetype scale. The list of components varies according to the display representation. This tab can be edited to turn OFF the display of the upper or lower half of the stair in plan view by turning OFF the up or down

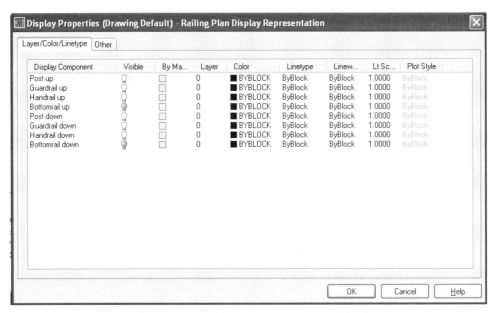

Figure 8.47 *Layer/Color/Linetype tab of the Display Properties dialog box*

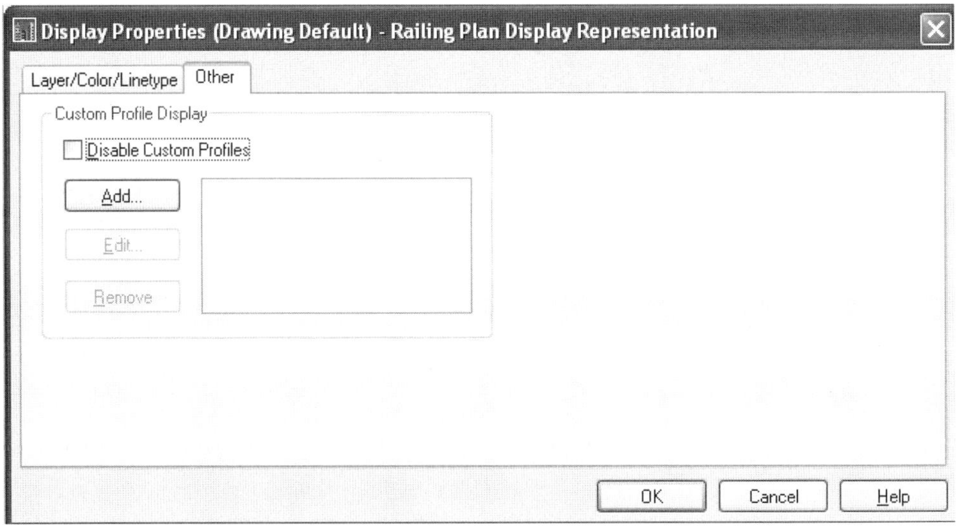

Figure 8.48 *Other tab of the Display Properties dialog box*

components of the railing. This display control technique should be used when the railing display is not controlled by the Display Configuration.

Other – The **Other** tab allows you to assign custom blocks for display as shown in Figure 8.48.

MODIFYING THE RAILING

Existing railings can be modified in the Properties palette; however, the grips of the railing can be used to change the location of the ends and vertices of the railing. The **Start** and **End** grips as shown in Figure 8.49 can be selected to move the end of the railing segment. The **Edge** grip will stretch the railing segment. The **Fixed Post Position** grip allows you shift the position of the end post. Additional post editing can be performed with the **Post Location** option of the shortcut menu.

EDITING POST LOCATIONS

The shortcut menu of a selected railing includes **Post Placement**, which cascades to include options to **Add**, **Remove**, **Hide**, **Show**, and **Redistribute** the posts. The **Post Placement>Add** option selects the **RailingPostAdd** command as shown in Table 8.12.

Command prompt	RAILINGPOSTADD
Shortcut menu	Select a railing, right-click, and choose Post Placement>Add

Table 8.12 *RailingPostAdd command access*

When you select this option, you are prompted in the command line to select a location for the additional posts, as shown in the following command line sequence. Object

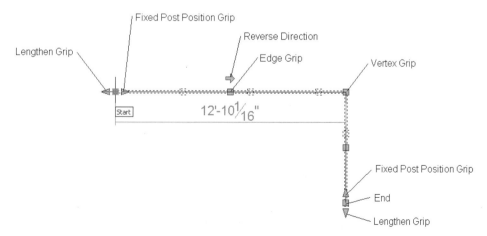

Figure 8.49 *Grip editing of railings not attached to stairs or stair flights*

snap and object tracking can be used to precisely place posts. The remaining posts of the railing are redistributed evenly along the railing when a post is added.

> Command: RailingPostAdd
>
> Specify position for Post: *(Select a location on the railing at p1 as shown at left in Figure 8.50.)*
>
> Specify position for Post: ENTER *(Press ENTER to end the command.)*
>
> *(Post is added as shown at right in Figure 8.50.)*

Removing Posts

Posts that have been added with the **RailingPostAdd** command can be removed from the railing using the **RailingPostRemove** command. Removing an added post will redistribute the posts to be evenly placed along the railing. Access the **RailingPostRemove** command as shown in Table 8.13.

Command prompt	RAILINGPOSTREMOVE
Shortcut menu	Select a railing, right-click, and choose Post Placement>Remove

Table 8.13 *RailingPostRemove command access*

The following command line sequence was used to remove the post placed on the corner as shown at left in Figure 8.51. The remaining posts were redistributed along the railing after the post was removed.

> *(Select the railing, right-click, and choose* **Post Placement>Remove.***)*

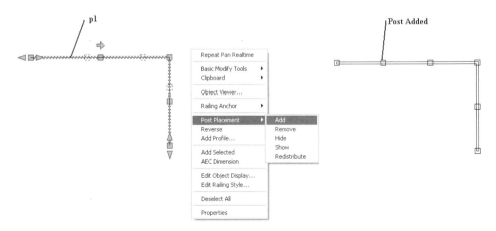

Figure 8.50 *Post added with Post Placement>Add*

Command: RailingPostRemove

Select a Railing Post to Remove: *(Select the manually added post at p1 in Figure 8.51.)*

Select a Railing Post to Remove: ENTER *(Press ENTER to end selection.)*

Hiding Posts

Posts that have been added to a railing or located on the end of the railing can be hidden. *Posts placed dynamically between the fixed post position grips cannot be hidden.* The **RailingPostHide** command is used to hide posts. Access the **RailingPostHide** command as shown in Table 8.14.

Command prompt	RAILINGPOSTHIDE
Shortcut menu	Select a railing, right-click, and choose Post Placement>Hide

Table 8.14 *RailingPostHide command access*

The RailingPostHide command was used in the following command sequence to hide the posts as shown at right.

(Select a railing, right-click, and choose **Post Placement>Hide***.)*

Command: RailingPostHide

Select a Railing Post to Hide: *(Select posts at p1 and p2 as shown at left in Figure 8.52.)*

(Posts are hidden as shown at right in Figure 8.52.)

Showing Posts

Posts that have been hidden can be displayed for the railing when you select **Show** from the **Post Placement** shortcut menu. The **RailingPostShow** command will display all hidden posts of the railing. Access the **RailingPostShow** command as shown in Table 8.15.

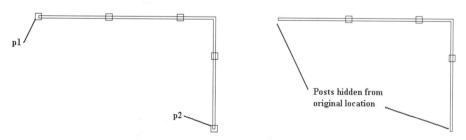

Figure 8.51 *Post removed with RailingPostRemove command*

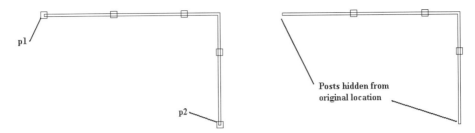

Figure 8.52 *Post hidden with Post Placement>Hide*

Command prompt	RAILINGPOSTSHOW
Shortcut menu	Select a railing, right-click, and choose Post Placement>Show

Table 8.15 *RailingPostShow command access*

Redistributing Posts

Posts that have been added to or removed from a railing create post locations not uniformly placed along the railing. The start and end posts can be stretched to a different location using grips. The **Redistribute** option resets the post distribution along the railing. Access **Redistribute** as shown in Table 8.16.

Command prompt	RAILINGREDISTRIBUTEPOSTS
Shortcut menu	Select a railing, right-click, and choose Post Placement>Redistribute

Table 8.16 *RailingRedistributePosts command access*

REVERSING THE RAILING

The **Reverse** option of the shortcut menu of a railing allows you to reverse the start and end of a railing. Access the **Reverse** command as shown in Table 8.17. A railing that has been edited to include additional posts can be reversed. The railing shown at right in Figure 8.53 was reversed from its orientation shown on the left. The baluster design of the railing shown in Figure 8.53 is justified to **Back-Right**; therefore, when the railing is reversed, the balusters shift to the opposite side of the railing centerline. If the baluster position is center justified, the **Reverse** option does not change the appearance of the railing.

Command prompt	RAILINGREVERSE
Shortcut menu	Select a railing, right-click, and choose Reverse

Table 8.17 *RailingReverse command access*

ANCHORING A RAILING

The **Anchor Railing** (**RailingAnchorToObjects**) command can be used to link a railing to an AEC object. Access the **Anchor Railing** command as shown in Table 8.18. The railing shown at left has been anchored to the ramp shown in Figure 8.54.

Command prompt	RAILINGANCHORTOOBJECTS
Shortcut menu	Select a railing, right-click, and choose Railing Anchor>Anchor to Object

Table 8.18 *RailingAnchorToObjects command access*

(*Select the Railing, right-click, and choose* **Railing Anchor>Anchor to Object**.)

Command: RailingAnchorToObjects

Select AEC Objects: 1 found (*Select the ramp at p1 as shown at left in Figure 8.54.*)

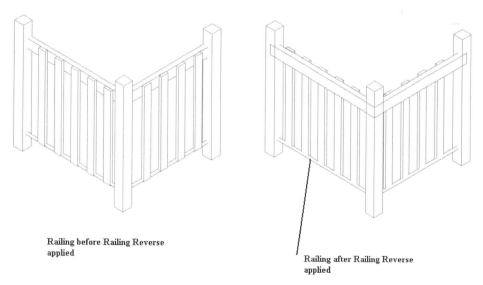

Railing before Railing Reverse applied

Railing after Railing Reverse applied

Figure 8.53 *Changing the start and end of the railing with RailingReverse command*

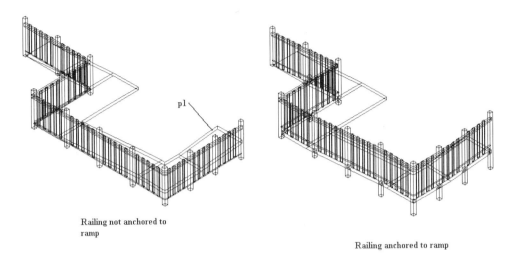

Figure 8.54 *Railing anchored to ramp*

Select AEC Objects: ENTER *(Press* ENTER *to end selection.)*

Automatic cleanup [Yes/No] <Yes>: **y** ENTER

Calculate height [Follow surface/At post locations only] <Follow surface>: ENTER

(Railing anchored to ramp as shown at right in Figure 8.54.)

The railing can be released from the AEC object by selecting the **AnchorRelease** command as shown in Table 8.19. When you select a railing, right-click, and choose **Railing Anchor> Release**, the railing remains in its position; however, future changes to the AEC object are independent of the railing.

Command prompt	ANCHORRELEASE
Shortcut menu	Select a railing, right-click, and choose Railing Anchor>Release

Table 8.19 *Railing Anchor>Release command access*

ADDING A PROFILE FOR RAILING COMPONENTS

A profile can be created in the workspace for the components of the railing. The **Add Profile** (**RailingAddComponentProfile**) command can be used to create a custom profile in the workspace for a selected railing component. Access the **Add Profile** command as shown in Table 8.20.

Command prompt	RAILINGADDCOMPONENTPROFILE
Shortcut menu	Select a railing, right-click, and choose Add Profile

Table 8.20 *RailingAddComponentProfile command access*

When you select the railing, right-click, and choose **Add Profile**, you are prompted to select a railing component. After you select the component, the **Add Post Profile** dialog box opens, allowing you to define the name for the new profile. The grips of a profile on the selected component are displayed and the **In-Place Edit** toolbar opens as shown in Figure 8.55. The profile can then be edited by modifying the grips or selecting options from the shortcut menu as discussed in the "Adding an Edge Profile in the Workspace" topic of Chapter 7.

CHANGING THE PROFILE WITH EDIT IN PLACE

After a profile has been changed using the **RailingAddComponentProfile** command, additional editing of the profile can be performed with the **RailingProfileEdit** command. Access the **Edit Profile In Place** (**RailingProfileEdit**) command as shown in Table 8.21.

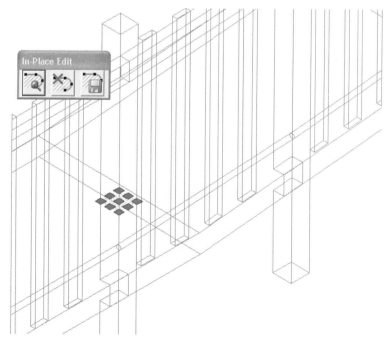

Figure 8.55 *Editing the profile of a railing*

Command prompt	RAILINGPROFILEEDIT
Shortctut menu	Select a railing, right-click, and choose Edit Profile In Place

Table 8.21 *RailingProfileEdit command access*

When you select a railing, right-click, and choose the **Edit Profile In Place** command from the shortcut menu, the **In-Place Edit** toolbar is displayed. The profile can then be edited by selecting the grips or selecting **Add Vertex** or **Remove Vertex** from the shortcut menu.

DISPLAYING THE STAIR IN MULTIPLE LEVELS

Stairs are displayed on each level of a floor plan. The lower half of the stair is displayed on the plans of the lower levels, while the upper half above the stair break line is displayed on the top level. Therefore the same stair can be used in each of two floor plans. Display of the upper or lower half of the stair can be controlled by the settings of the object's display representation or by the display configuration.

You may choose to not display stairs as a construct or part of a project and therefore need to control stair display. The display representation settings of the Standard style

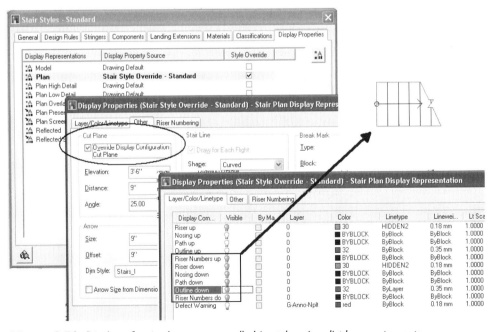

Figure 8.56 *Display of stair elements controlled in style using display components*

stair have the **Override Display Configuration Cut Plane** checked ON as seen in the **Other** tab for **Display Properties(Stair Style Override)** of the **Plan** display representation as shown in Figure 8.56. This check box allows you to specify within the stair style the vertical elevation for the break in the stair. The up or down components can then be turned on or off in the **Layer/Color/Linetype** tab of the **Display Properties (Stair Plan Display Representation)** dialog box shown in Figure 8.56.

When stairs are created as part of a project, you may choose to control stair display in the definition of the display configuration. Display of the upper and lower sections of the stair can be controlled using the display configuration settings, which require the **Override Display Configuration Cut Plane** to be toggled OFF. The upper half of the stair will not be displayed in plan view according to the settings of cut plane defined in the display configuration. The cut plane for the Medium Detail configuration is defined for 3'-6". Therefore stair components above 3'-6" are not displayed. This technique can also be applied to the model view drawing to create each floor plan from the model. The display configuration settings for 2 display configuration include a cut plane at 13'-6" as shown in Figure 8.57. The display of the lower half of the stair for the 2nd floor plan is controlled by the cut plane cut height. A cut height can be defined for each display configuration to display each level and slice through the model.

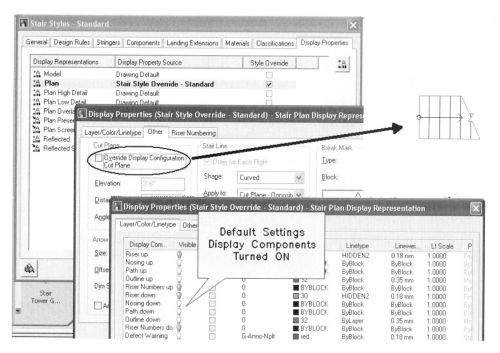

Figure 8.57 *Display of stair elements controlled in style using display configuration*

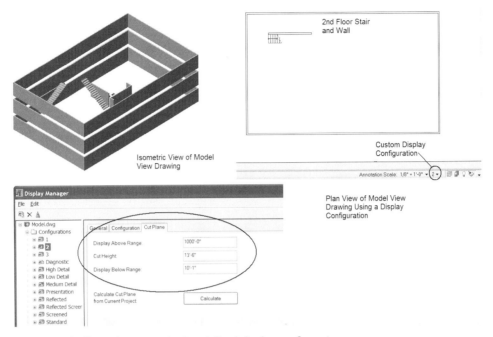

Figure 8.58 *Floor plan created using defined display configuration*

If there is no project associated with the drawing, you can display the stair in an upper level plan by creating a WBLOCK of the stair in the lower level plan drawing and insert it in the upper level floor plan. When you create the WBLOCK of the stair, specify the base point equal to 0,0,0, which will ensure the vertical alignment of the stair. This procedure requires that one floor level be created from the geometry of the other, so that the absolute x,y coordinates of walls remain identical. When the stair block is then inserted in the upper level floor plan, it is inserted at the same X,Y location but at an elevation below the upper level equal to the stair rise. If the stair rise is 9'-2", the stair block would be inserted at 0,0,–9'-2". You can then explode the stair and edit the entity display properties of the stair to freeze the lower level of the stair in the upper floor plan.

 STOP. Do Tutorial 8.3, "Creating Stair and Railing Styles" at the end of the chapter.

SUMMARY

1. Stairs are created in a drawing with the **StairAdd** command.
2. Design of the number of risers, tread dimension, and run of the stair is set in the Properties palette.
3. Stairs are generated in the drawing from start of the flight up or down from the Z=0 coordinate.
4. The shapes of a stair can be U-shaped, U-shaped winder, Multi-landing, Spiral, and Straight.
5. The start point and center of a spiral stair can be edited with grips.
6. The Properties palette is used to change the size, components, and style of a stair.
7. The **Style Manager** is used to create, edit, import, export, and purge stair styles from a drawing.
8. The **Calculation Rules** dialog box allows you to constrain the stair by the following properties: A-Straight Length, B-Riser Count, C-Riser, and D-Tread dimensions.
9. The display of the lower or upper half of the stair and railing is controlled in the **Display Properties** dialog box for a stair style.
10. The **RailingAdd** command is used to place a railing attached to a stair or free-standing railing on a balcony.
11. A stair is placed on the A-Flor-Strs layer, and railings are placed on the A-Flor-Hral layer.
12. Railing heights and posts extensions can be edited to create deck post and guardrails.
13. Railing styles allows you to edit the size, location, and profile used for the railing posts, handrail, guardrail, and balusters.

REVIEW QUESTIONS

1. A stair shape with two equal runs and a landing is the _____.
2. A stair shape with different length runs and a landing is the _____.
3. The shape used to create a stair which does not change directions is the _____.
4. The number of risers can be specified in the _____ dialog box.
5. The display of the lower half of the railing is controlled by the settings in the _____ in the _____ tab of the style.
6. Create Housed stairs by editing this field of the _____ Properties palette.
7. Aec Profiles can be specified for the handrail in the _____ tab of the

_____ dialog box.

8. Stair default settings are established in the _____ palette.
9. The default values for the calculation of stair dimensions are set in the _____ tab of the _____ dialog box.
10. The grips of the spiral stair are located at the _____ of the arc formed by the spiral stair.
11. The size and position of the stringer are defined in the _____ command.
12. To turn off the upper half of the stair, the _____ of components are turned off in the **Display Properties** dialog box.
13. Describe the procedure for inserting a freestanding railing.
14. A handrail and guardrail can be placed in center of a stair by _____.
15. Stair styles created in a drawing can be used in other drawings by _____.
16. The stair is created as a _____ drawing of a project.

TUTORIAL 8.1 CREATING STRAIGHT STAIRS WITH VERTICAL ORIENTATIONS

1. Open Autodesk Architectural Desktop 2005, and select **File>Project Browser** from the menu bar. Use the Project Selector drop-down list to navigate to your *ADT Student\ADT Tutor\Ch8* directory. Double-click on **Ex 8-1** to set this project current. Select **Close** to dismiss the **Project Browser**.
2. Select the **Constructs** tab of the **Project Navigator**, click on the Constructs category, right-click, and choose **New>Construct**. Type **Stair** in the **Name** edit field of the stair in the **Add Construct** dialog box. Check **Level 1** and **2** for **Division 1** of the construct. Select **OK** to close the **Add Construct** dialog box.
3. Double-click on the **Stair** construct to open the new drawing. Select the **Construct** tab of the **Project Navigator**, select **Floor 1** construct, right-click and choose **Xref Overlay** from the shortcut menu.
4. Select zoom extents to view the existing walls of **Floor 1** reference file.
5. Right-click over the OSNAP button on the status bar, and choose **Settings** from the shortcut menu. Edit the **Object Snap** tab of the **Drafting Settings** dialog box as follows: clear all object snap modes, check the **Intersection** object snap. Check the **Object Snap On (F3)** check box and clear **Object Snap Tracking On (F11)**. Select **OK** to dismiss the **Drafting Settings** dialog box.
6. Toggle ON POLAR and toggle OFF SNAP and GRID on the status bar.

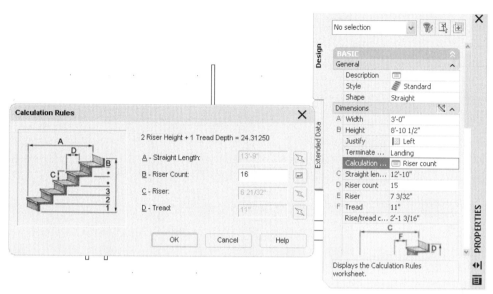

Figure 8T.1 *Calculation Rules dialog box*

7. Right-click over the command window, choose **Options**, and select the **AEC Object Settings** tab of the **Options** dialog box. Clear the **Presentation Format (No Cut Lines or Path)** check box, set Node Osnap = **Flight & Landing Corners**, Measure Stair Height = **Finished Floor to Floor**, and Calculator Limits = **Relaxed**. Select **OK** to dismiss the **Options** dialog box.

8. To create a straight stair for a finish floor to finish floor rise equal to 8'–10 1/2", select the **Stair** command from the Design palette. Edit the Properties palette for the stair as follows: Basic–General: Style = Standard, Shape = Straight, Dimensions: A-Width = 36, B-Height = 8'–10 1/2", Justify = Left, Terminate with = Landing; select **Calculation Rules**, toggle ON **Riser Count**, set Riser Count = **16**, and verify all edit fields are set to automatic as shown in Figure 8T.1. Select **OK** to dismiss the **Calculation Rules** dialog box. Verify Advanced–Floor Settings are A-Top Offset = **0**, Top Depth = **10**, Bottom Offset = **0**, and Bottom Depth = **0**. Flight Height: Minimum = **None**, Flight Maximum = **None**. Respond to the following command line prompts.

   Command: StairAdd

   Flight start point or [SHape/STyle/Tread/Height/Width/Justify/Match/Undo]: *(Select the wall near p1 in Figure 8T.2 using the Intersection object snap.)*

   Flight end point or [SHape/STyle/Tread/Height/Width/Justify/ Match/Undo]: *(Select beyond the wall near p2 using Polar tracking 90-degree angle beyond the stair bounding box as shown in Figure 8T.2.)*

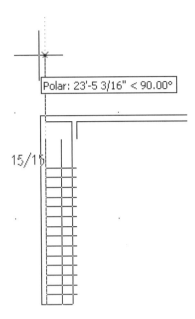

Figure 8T.2 *Start flight and end flight locations for stair placement*

Flight start point or [SHape/STyle/Tread/Height/Width/Justify/Match/Undo]: ENTER *(Press ENTER to end the command.)*

9. To create a stair that goes down, select the **Stair** command from the Design palette. Edit the Properties palette: Terminate with = Riser, to display the **Vertical Orientation** edit field. Edit the Properties palette for the stair as follows: Basic–General: Style = **Standard**, Shape = **Straight**, Vertical Orientation = **Down**; Dimensions: A-Width = **36**, B-Height = **14'-0"**, Justify = **Left**, select **Calculation Rules**, toggle on **Automatic** for all options. Select **OK** to close the **Calculation Rules** dialog box. Verify Calculation rules = **Height**. Verify that Advanced–Floor Settings are A-Top Offset = **0**, Top Depth = **10**, Bottom Offset = **0**, and Bottom Depth = **0**. Set Flight Height: Minimum = **Height**, Minimum Height = **3'-0"**, Flight Maximum = **Height**, Maximum Height = **12'**. Respond to the following command line prompts.

Command: StairAdd

Flight start point or [SHape/STyle/Tread/Height/Width/Justify/Match/Undo]: *(Select the wall near p1 in Figure 8T.3 using the Intersection object snap.)*

Flight end point or [SHape/STyle/Tread/Height/Width/Justify/Match/Undo]: *(Select beyond the stair bounding box near p2 in Figure 8T.3.)*

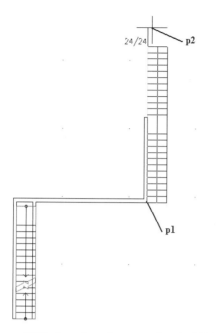
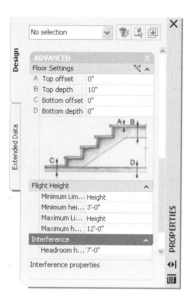

Figure 8T.3 *Location of start and end of flight down*

        Flight start point or [SHape/STyle/Tread/Height/Width/Justify/Match/Undo]: ENTER *(Press* ENTER *to end the command.)*

10. Select **SE Isometric** from the **Navigation** toolbar to view the stairs as shown in Figure 8T.5.

11. Select **Zoom Window** from the **Zoom** flyout of the **Navigation** toolbar. Respond to command line as shown below:

        Command: '_zoom

        Specify corner of window, enter a scale factor (nX or nXP), or

        [All/Center/Dynamic/Extents/Previous/Scale/Window] <real time>:_w

        Specify first corner: **24',6'** ENTER *(First corner specified.)*

        Specify opposite corner: **26',18'** ENTER *(Opposite corner specified.)*

12. Press F3 to turn OFF object snaps.

13. Select a wall, right-click, and choose **Edit Xref in place**. Verify that **Automatically select all nested objects** is selected. Select **OK** to dismiss the **Reference Edit** dialog box.

14. Select the wall at **p1** in Figure 8T.4 to display the grips of the wall. Select the wall grip as shown at right in Figure 8T.4. Move the pointer to the **right**, and press TAB to highlight the dimension shown at right in Figure 8T.4. Type 3" in the workspace, and press ENTER to move the wall end under the stair.

15. Select the wall at **p1** as shown in Figure 8T.4, right-click, and choose **Roof/Floor Line>Modify Roof Line** from the shortcut menu. Respond to the command line as shown below.

    Command: RoofLine

    RoofLine [Offset/Project to polyline/Generate polyline/Auto project/Reset]: **A** ENTER

    Select objects: 1 found *(Select the stair at p2 as shown in Figure 8T.4.)*

    Select objects: ENTER

    [1] Wall cut line(s) converted.

    RoofLine [Offset/Project to polyline/Generate polyline/Auto project/Reset]: ENTER *(Press* ENTER *to end the command; wall is trimmed by the stair as shown in Figure 8T.5.)*

16. Right-click and choose **Close Refedit Session>Save Reference Edits**. Select **OK** to save all changes and dismiss the AutoCAD dialog box.

17. Select **Manage Xrefs** from the Drawing Window status bar to open the **Xref Manager**. Select **Floor 1** and then select the **Detach** button. Select **OK** to dismiss the **Xref Manager**. Save and close the Stair drawing.

18. To view the stairs in the view drawing, select the **View** tab of the **Project Navigator**, and double-click on the Floor 2 view drawing. Save and close the drawing.

19. Select **File>Project Browser**. Select the **Ex 8-1** project, right-click, and choose **Close Current Project**.

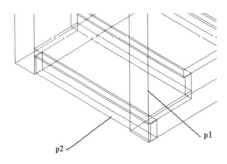 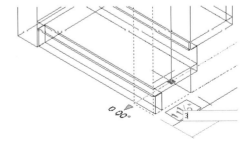

Figure 8T.4 *Grip of wall selected*

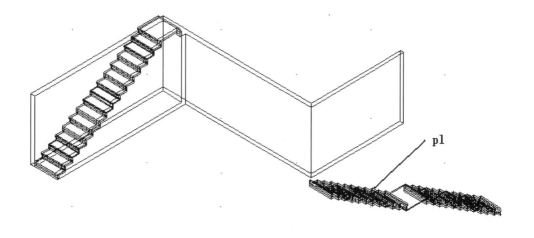

Figure 8T.5 *RoofLine command applied to stair and wall*

TUTORIAL 8.2 CREATING A U-SHAPED STAIR AND A STAIR TOWER

1. Open Autodesk Architectural Desktop 2005 and select **File>Project Browser** from the menu bar. Use the Project Selector drop-down list to navigate to your *ADT Student\ADT Tutor\Ch8* directory. Double-click on **Ex 8-2** to set this project current. Select **Close** to dismiss the **Project Browser**.

2. Select the **Constructs** tab of the **Project Navigator**, click on the Constructs category, right-click and choose **New>Construct**. Type **Stair** in the name edit field of the stair in the **Add Construct** dialog box. Check **Level 1, 2, 3,** and **4** for **Division 1** of the construct. Select **OK** to close the **Add Construct** dialog box.

3. Double-click on the **Stair** construct to open the new drawing. Select the **Constructs** tab of the **Project Navigator**, select **Floor 1** construct, right-click, and choose **Xref Overlay** from the shortcut menu.

4. Select **Zoom Window** from the **Zoom** flyout of the **Navigation** toolbar. Respond to the command line as follows to view the stair well.

    Command: '_ ZOOM

    Specify corner of window, enter a scale factor (nX or nXP), or

[All/Center/Dynamic/Extents/Previous/Scale/Window/Object] <real time>: _w

Specify first corner: 142',40' ENTER *(Coordinates of zoom window specified.)*

Specify opposite corner: **155',60'** ENTER *(Coordinates of zoom window specified.)*

5. Right-click over the OSNAP button of the status bar, and choose **Settings** from the shortcut menu. Edit the **Object Snap** tab of the **Drafting Settings** dialog box as follows: clear all object snap modes, and check **Endpoint** object snap mode. Check the **Object Snap On (F3)** check box and clear **Object Snap Tracking On (F11)**. Select **OK** to dismiss the **Drafting Settings** dialog box.

6. Toggle POLAR ON, and toggle OFF SNAP and GRID on the status bar.

7. To create a U-shaped stair for a finish floor to finished rise equal to 12', select the **Stair** command from the Design palette. Edit the Properties palette for the stair as follows: Basic–General: Style = Standard, Shape = U-shaped, Turn type = 1/2 landing, Horizontal Orientation = Clockwise, Vertical Orientation = Up, Dimensions: A-Width = 44, B-Height = 12', Justify = Outside, Terminate with = Landing; select **Calculation Rules** and verify all options are set to automatic. Select **OK** to dismiss the **Calculation Rules** dialog box. Verify that Advanced–Alignment type = Tread to tread, Alignment offset = 0, Extend alignment = Lower flight, and Uneven tread on = Lower flight; Floor Settings are A-Top Offset = 0, Top Depth = 6, Bottom Offset = 0, and Bottom Depth = 6. Flight Height: Minimum Limit Type = None, Maximum Limit Type = Height, and Maximum Height = 12'; Interference: Headroom height = 7'-0", Side clearance = .5. Respond to the following command line prompts:

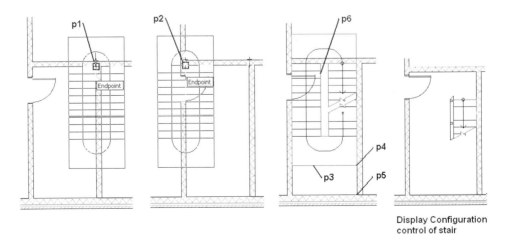

Display Configuration control of stair

Figure 8T.6 *Location of flight points of U-shaped stair*

Command: StairAdd

Flight start point or [SHape/STyle/Tread/Height/Width/Justify/Match/Undo]: *(Select the wall near p1 using the Endpoint object snap as shown in Figure 8T.6.)*

Flight end point or [SHape/STyle/Tread/Height/Width/Justify/Match/Undo]: *(Select the wall near p2 using the Endpoint object snap as shown in Figure 8T.6.)*

Flight start point or [SHape/STyle/Tread/Height/Width/Justify/Match/Undo]: ENTER *(Press ENTER to end the command.)*

8. Select **Move** from the **Modify** toolbar, and respond to the command line as follows:

    Command: _move

    Select objects: *(Select the stair at p3 as shown in Figure 8T.6)* 1 found

    Select objects: *(Press ENTER to end selection.)*

    Specify base point or displacement: *(Select a point near p4 as shown in Figure 8T.6)*

    Specify second point of displacement or <use first point as displacement>: *(Select a point near p5 as shown in Figure 8T.6)*

9. Select the **Content Browser** from **Navigation** toolbar and open the Design Tool Catalog-Imperial>Stairs and Railings>Railings. Select the i-drop for **Guardrail-Rect Balusters** from page 2 of the Railings category and drag this style into the workspace. Edit the Properties palette for the railing as follows: Attached to = **Stair**, Side offset = **2"** and Automatic placement = **yes**. Select the stair near p6 as shown in command sequence below and Figure 8T.6.

    Select a stair or [Style/Attach/Match/Undo]: *(Select the stair near p6 as shown in Figure 8T.6.)*

    Select a stair or [Style/Attach/Match/Undo]: ENTER *(Press ENTER to end selection.)*

10. Select the Stair, right-click, and choose **Edit Stair Style**. Select the **Display Properties** tab of the **Stair Styles – Standard** dialog box. Choose the **Edit Display Properties** button of the **Display Properties** tab. Select the **Other** tab of the **Display Properties (Drawing Default) – Stair Plan Display Representation**. Clear the **Override Display Configuration Cut Plane** check box. Select **OK** to dismiss all dialog boxes and view the stair as shown at far right in Figure 8T.6.

11. Select **SW Isometric** from the **View** flyout of the **Navigation** toolbar.

12. Select the Stair, right-click, and choose **Stair Tower Generate** from the shortcut menu. Respond to the command line as follows:

    Command: StairTowerGenerate

    Select railings and slabs: *(Select the railing at p1 as shown in Figure 8T.7.)* 1 found

    Select railings and slabs: ENTER *(Press ENTER to end selection and open the **Select Levels** dialog box; select **OK** to dismiss the **Select Levels** dialog box.)*

    (Stair tower generated as shown at right in Figure 8T.7.)

13. Select **Manage Xrefs** from the Drawing Window status bar. Select **Floor 1** and select the **Detach** button of the **Xref Manager** dialog box. Select **OK** to dismiss the **Xref Manager**. Save and close the Stair drawing.

14. Select the **Views** tab. Select the **Floor Level 3** drawing on the **Views** tab, right-click, and choose **Properties**. Review the **General**, **Context**, and **Content** pages. Select OK to dismiss the dialog box. Double-click on the **Floor Level 3** drawing to open the drawing and view the stair and floor plan for level 3.

15. Save the Floor Level 3 drawing.

16. Select **File>Project Browser**, select the Ex 8-2 project, right-click, and choose **Close Current Project**.

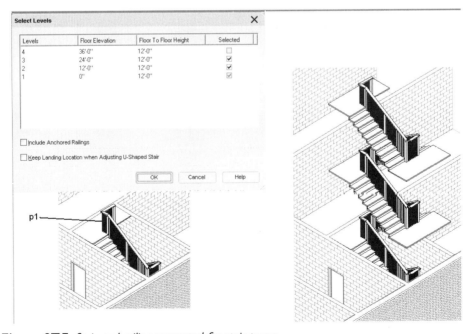

Figure 8T.7 *Stair and railing generated for stair tower*

TUTORIAL 8.3 CREATING STAIR AND RAILING STYLES

1. Open Autodesk Architectural Desktop 2005 and select **File>Project Browser** from the menu bar. Use the Project Selector drop-down list to navigate to your *ADT Student\ADT Tutor\Ch8* directory. Double-click on **Ex 8-3** to set this project current. Select **Close** to dismiss the **Project Browser**.

2. Select the **Constructs** tab of the **Project Navigator**, click on the Constructs category, right-click, and choose **New>Construct**. Type **Stair** in the name edit field of the stair in the **Add Construct** dialog box. Check **Level 1** and **2** for **Division 1** of the construct. Select **OK** to close the **Add Construct** dialog box.

3. Double-click on the **Stair** construct to open the new drawing. Select the **Constructs** tab of the **Project Navigator**, select the Basement construct, right-click, and choose **Xref Overlay** from the shortcut menu.

4. Right-click over the OSNAP button of the status bar and choose **Settings** from the shortcut menu. Edit the **Object Snap** tab of the **Drafting Settings** dialog box as follows: clear all object snap modes, and check the **Endpoint** object snap mode. Check the **Object Snap On (F3)** check box and clear **Object Snap Tracking On (F11)**. Select **OK** to dismiss the **Drafting Settings** dialog box.

5. Toggle POLAR ON on the status bar. Toggle OFF SNAP and GRID on the status bar.

6. Verify that the Model tab is selected. Select **Zoom Extents** from the **Zoom** flyout of the **Navigation** toolbar.

7. Select the **Content Browser** from the **Navigation** toolbar. Open the Design Tool Catalog-Imperial>Stairs and Railings>Stairs. To insert a style from the **Content Browser**, select the i-drop of the **Stair** and drag it into the Floor 1 drawing. Press ESC to end the command. This procedure inserts the **Wood-Saddle** stair style into the drawing when **Stair** content tool is i-dropped from the **Content Browser**.

8. To edit the new stair style, select **Format>Style Manager** from the menu bar.

9. Expand the Architectural Objects folder of the Stair.dwg in the left pane. Expand the Stair Styles folder, select the **Wood-Saddle** stair style, right-click, and choose **Rename**. Overtype the name **Residential**.

10. Double-click on the **Residential** stair style name in the left pane to open the **Stair Styles – Residential** dialog box. Refer to Figure 8T.8, select the **Design Rules** tab, and then edit as follows: A-Maximum Slope: Riser Height = 7 3/4" and Tread Depth = 10"; B-Optimum Slope: Riser Height = 7 3/4" and Tread Depth = 10 1/2"; C-Minimum Slope: Riser Height = 6" and Tread Depth = 11"; clear the **Use Rule Based Calculator** check box.

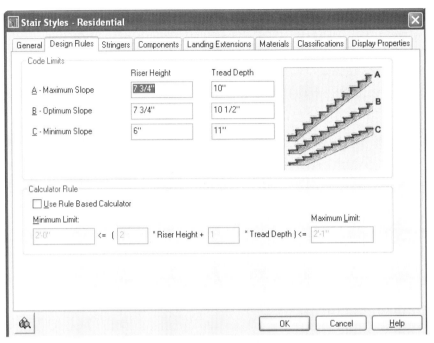

Figure 8T.8 *Design Rules tab of the Stair Styles dialog box*

11. Refer to Figure 8T.9, select the **Stringers** tab, and then edit as follows: select the **Add** button, select the **Unnamed** stringer in the **Stringers** column, and overtype **Middle** to name the new stringer. Click in the **Alignment** column of the new stringer and select **Center** from the drop-down list.

12. Refer to Figure 8T.10, select the **Components** tab, and then edit as follows: verify **Allow Each Stair to Vary** is clear and **Display Tread** and **Riser** check boxes are checked. Set A-Tread Thickness = 1", B-Riser Thickness = 3/4", C-Nosing Length = 1", clear the **Sloping Riser** check box, and set D-Landing Thickness = 1" and E-Additional Width = 0.

13. Refer to Figure 8T.11, select the **Landing Extensions** tab, and edit as follows: clear the **Allow Each Stair to Vary** check box, set A-Distance to First Tread Down = **0**, clear the **Add Tread Depth** check box, set B-Distance to First Tread UP = **0**, clear the **Add Tread Depth** check box, clear the **Extend Landings to Merge Flight Stringers with Landing Stringers** check box, clear the **Extend Landings to Prevent Risers & Treads Sitting under Landings** check box, and set Landing Length = **2'-9"** and Landing Location = **Middle**.

14. Select **OK** to close the **Stair Styles – Residential** dialog box. Select **OK** to close the **Style Manager**.

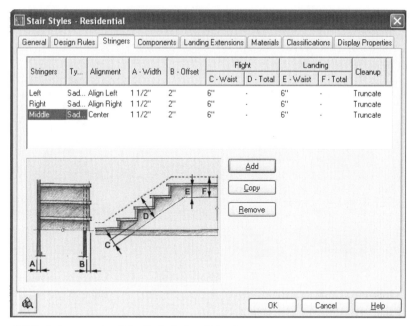

Figure 8T.9 *Stringers tab of the Stair Styles dialog box*

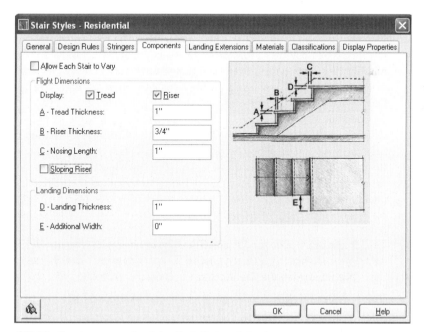

Figure 8T.10 *Components tab of the Stair Styles dialog box*

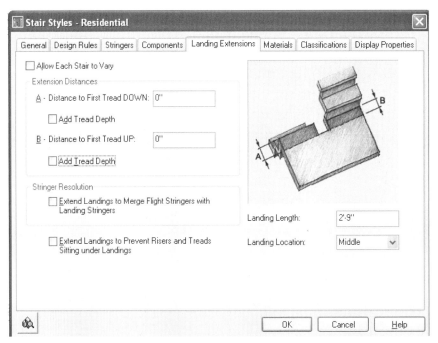

Figure 8T.11 Landing Extensions tab of the Stair Styles dialog box

15. To create a Multi-landing stair, select **Stair** from the Design palette and edit the Properties palette as follows: Basic–General: Style = Residential, Shape = Multi-landing, Turn type = 1/2 Landing, Terminate with = Riser , Vertical Orientation = Up; Dimensions: A-Width = 36, B-Height = 8'–11 1/2", Justify = Left, select **Calculation Rules**, and verify all edit fields are set to **Automatic** except B-Riser Count and D-Tread. Set B-Riser Count = 14 and D-Tread = 10", and then select **OK** to close the **Calculation Rules** dialog box. Verify Advanced–Floor Settings are A-Top Offset = 0, Top Depth = 10, Bottom Offset = 0, and Bottom Depth = 0; Flight Height: Minimum Limit Type= None, Maximum Limit Type = None. Respond to the following command line prompts:

    Command: StairAdd

    Flight start point or [SHape/STyle/Tread/Height/Width/Justify/Next/Flight length/Match/Undo]: (SHIFT + *right-click and choose* **Nearest** *from the* **Osnap** *shortcut menu.*)

    _nea to (*Select a point near p1 as shown in Figure 8T.12.*)

    Flight end point or [SHape/STyle/Tread/Height/Width/Justify/Next/Flight length/Match/Undo]: (*Select a point near p2 using the Endpoint object snap as shown in Figure 8T.13.*)

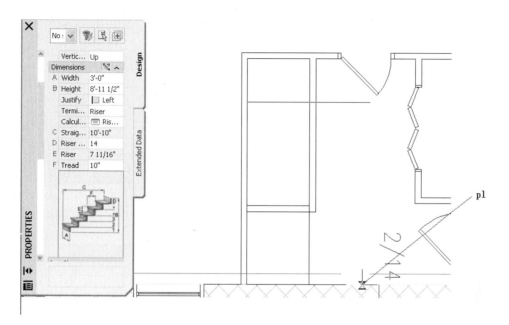

Figure 8T.12 *Start of stair flight location*

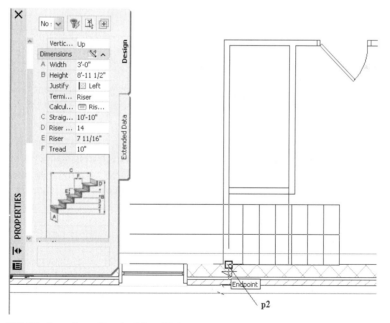

Figure 8T.13 *Location of end of first flight*

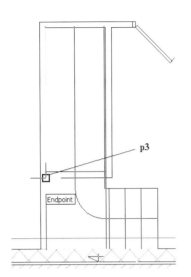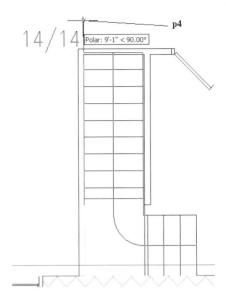

Figure 8T.14 Location of the start and end of the second flight

Flight start point or [SHape/STyle/Tread/Height/Width/Justify/Next/Flight Length/Match/Undo]: *(Select a point near p3 using the Endpoint object snap as shown at left in Figure 8T.14.)*

Flight end point or [SHape/STyle/Tread/Height/Width/Justify/Next/Flight length/Match/Undo]: *(Select a point near p4 beyond the stair bounding box as shown at right in Figure 8T.14.)*

Flight start point or [SHape/STyle/Tread/Height/Width/Justify/Next/Flight length/Match/Undo]: ENTER *(Press ENTER to end the command.)*

16. This step imports a railing style into the drawing from the **Content Browser.** Select the **Content Browser** from the **Navigation** toolbar. Open the Design Tool Catalog-Imperial>Stairs and Railings>Railings. Go to page 2 of the catalog, select the i-drop of the **Guardrail-Wood Balluster 01** style, drag it to the workspace, and press ESC to terminate the command without selecting a stair. The Guardrail-Wood Balluster 01 style is imported into the drawing.

17. Select **Format>Style Manager** from the menu bar and expand the Architectural Objects folder and Railings Styles folder in the left pane for the Stair.dwg drawing. Choose the Guardrail-Wood Balusters 01 in the right pane, right-click, and choose **Rename**. Overtype the name **Handrail**.

18. Double-click on the **Handrail** railing style to open the **Railing Styles – Handrail** dialog box. To create a simple handrail, refer to Figure 8T.15, select

Stairs and Railings 577

the **Rail Locations** tab, and edit as follows: clear the **Allow Each Railing to Vary** check box, in the **Upper Rails** section check **Guardrail**; edit the A-Horizontal Height = **38**, Sloping Height = **38**, and Side for Offset = **Center**. Clear the **Handrail** and **Bottomrail** check boxes.

19. Refer to Figure 8T.16, select the **Post Locations** tab, clear **Allow Each Railing to Vary**, check **Fixed Posts**, set A-Extension of All Post from Top Railing = 0, B-Extension of All Posts from Floor Level = 0, check **Fixed Posts at Railing Corners**; clear **Dynamic Posts**, and clear **Balusters**.

20. Select the **Components** tab and edit the D-Fixed Post: edit Width = **2.5** and Depth = **2.5**.

21. Refer to Figure 8T.17, select the **Extensions** tab, and clear the **Allow Each Railing to Vary** check box, check the **Use Stair Landing Extension** check box for **At Floor Levels**, and check the **Use Stair Landing Extension** check box for **At Landings**. Select **OK** to dismiss the **Railing Styles – Handrail** dialog box. Select **OK** to dismiss the **Style Manager**.

22. Select **Railing** from the Design palette. Edit the Properties palette as follows: Style = Handrail, Attach to = Stair, Side offset = 2, and Automatic Placement = Yes. Respond to the command prompts as shown below to place the railing.

    Command: RailingAdd

    Select a stair or [Style/Attach/Match/Undo] *(Select the stair at p1 as shown in Figure 8T.18.)*

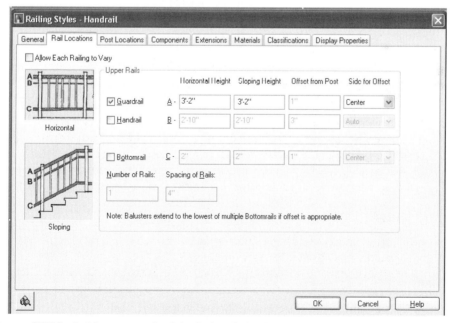

Figure 8T.15 *Rail Locations tab of the Railing Styles–Handrail dialog box*

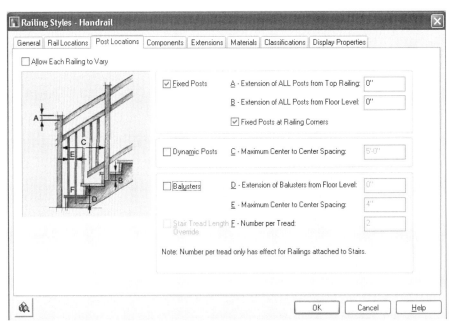

Figure 8T.16 *Post Locations tab of the Railing Styles–Handrail dialog box*

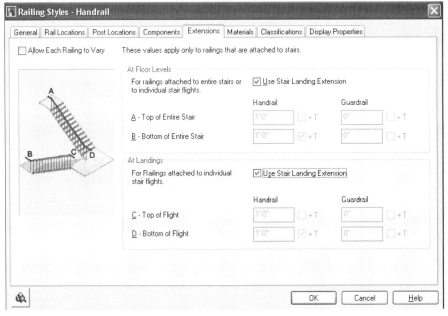

Figure 8T.17 *Extensions tab of the Railing Styles–Handrail dialog box*

Stairs and Railings 579

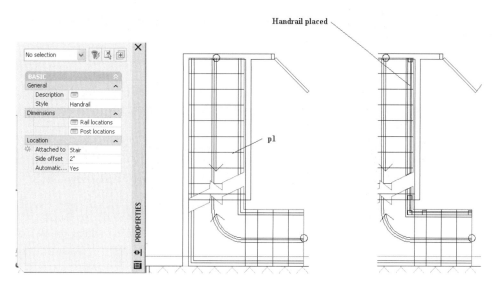

Figure 8T.18 *Selection of stair for automatic placement of railing*

Select a stair or [Style/Attach/Match/Undo] ENTER *(Press* ENTER *to end the command.)*

23. To create a guardrail railing style for the steps leading to the landing, select **Format>Style Manager** from the menu bar. Double-click on Architectural Objects of Stair.dwg drawing. Double-click on the Railing Styles folder of the Stair.dwg drawing. Select the Handrail style, right-click, and choose **Copy** from the shortcut menu. Select the Railing Styles folder in the left pane, right-click, and choose Paste. Select Handrail (2), right-click, and choose **Rename**. Overtype **Guardrail** to rename the new style. Double-click on the Guardrail style to open the **Railing Styles – Guardrail** dialog box.

24. Select the **Post Locations** tab and edit as follows: clear **Allow Each Railing to Vary** check box, check **Fixed Posts**, set A-Extension of All Posts from Top Railing = **0**, B-Extension of All Posts from Floor Level = **0**, check **Fixed Posts at Railing Corners**, clear **Dynamic Posts**, check **Balusters** ON, and set Extension of Balusters from Floor Level = **0**, and E-Maximum Center to Center Spacing = **4"**.

25. Select **OK** to close **Railing Styles – Guardrail** dialog box. Select **OK** to close the **Style Manager**.

26. Select **Railing** from the Design palette, and edit the Properties palette as follows: Style = Guardrail, Location: Attach to = Stair Flight, and Side Offset = 2.

    Command: RailingAdd

Select a stair or [Style/Attach/Match/Undo] *(Select the stair at p1 as shown in Figure 8T.19.)*

Select a stair or [Style/Attach/Match/Undo]: ENTER *(Press ENTER to end the command.)*

27. Select the handrail at **p1** as shown in Figure 8T.20, right-click, and choose **Post Placement>Hide** from the shortcut menu. Select the posts shown at **p1**, **p2**, and **p3**.

28. To edit the reference file, select a wall, right-click, and choose **Edit Xref in-place**. Select **Prompt** to select nested objects radio button, and select **OK** to close the **Reference Edit** dialog box. Select the wall at **p1** shown in Figure 8T.21 to open the **Refedit** toolbar.

29. Select the wall shown at p1 in Figure 8T.21, right-click, and choose **Roof/Floor Line>Modify Roof Line**. Respond to the command line as follows to modify the wall.

    Command: RoofLine

    RoofLine [Offset/Project to polyline/Generate polyline/Auto project/Reset]: **a** ENTER *(Select Auto project to project the wall to the stair.)*

    Select objects: 1 found *(Select the stair at p2 as shown in Figure 8T.21.)*

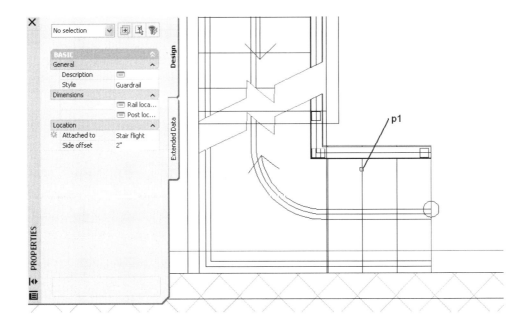

Figure 8T.19 *Selection of stair for location of railing*

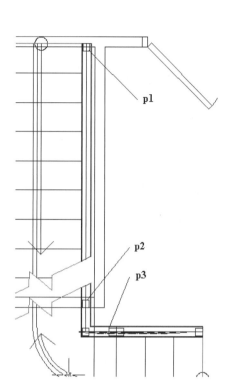

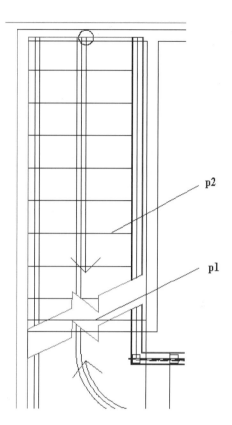

Figure 8T.20 *Selection of posts to hide*

Figure 8T.21 *Selection of walls for roof line edit*

Select objects: ENTER *(Press* ENTER *to end selection.)*

[1] Wall cut line(s) converted.

RoofLine [Offset/Project to polyline/Generate polyline/Auto project/Reset]: ENTER *(Press* ENTER *to end the command.)*

30. Select **Save back changes to reference** of the **Refedit** toolbar. Select **OK** to dismiss the AutoCAD dialog box.

31. Select the stair, right-click, and choose **Edit Stair Style**. Choose **Display Properties** in the **Stair Styles - Residential** dialog box. Verify that the **Plan** display representation is current, and select the **Edit Display Properties** button. Choose the **Other** tab and clear the **Override Display Configuration Cut Plan** check box. Select **OK** to dismiss all dialog boxes and view the lower half of the stair, which is now controlled by the display configuration cutting plane of the display configuration.

32. Select **Manage Xref** in the Drawing Window status bar to open the **Xref Manager** dialog box. Select **Basement** and choose **Detach in the Xref Manager**. Select **OK** to dismiss the **Xref Manager** dialog box.

33. Select **NE Isometric View** from the **View** flyout of the **Navigation** toolbar. Select **Hidden** from the **Shade** flyout of the **Navigation** toolbar to view the stair.

34. Save and close the drawing.

35. Select the **Views** tab of the **Project Navigator**. Double-click on the **Floor Plan-Basement** view. Select the Floor Plan-Basement in the **Project Navigator**, right-click, and choose **Properties**. Select the check boxes for the **General**, **Context,** and **Content** pages of the **Properties** dialog box. Verify that the Stair construct drawing is listed in the **Content** page. Select **OK** to close the **Modify General View** dialog box.

36. Select the **Views** tab of the **Project Navigator**. Double-click on the **Floor Plan-1** view. Select the Floor Plan-1 in the **Project Navigator**, right-click, and choose **Properties**. Select the check boxes for the **General**, **Context**, and **Content** pages of the **Properties** dialog box. Verify that the Stair construct drawing is listed in the **Content** page. Select **OK** to close the **Modify General View** dialog box.

37. To edit the stair display as shown at right in Figure 8T.22, select **Format>Display Manager** from the menu bar. Expand Configurations and select **Medium Detail**. Select the **Cut Plane** tab. Edit the **Display Below Range** to -7'.

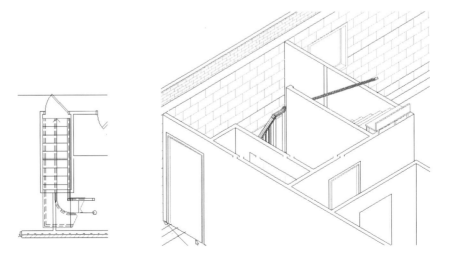

Plan view of stair Floor 1   NE Isometric view of stair

Figure 8T.22 *Stair created for basement*

38. Select **NE Isometric View** from the **View** flyout of the **Navigation** toolbar. Select **Hidden** from the **Shade** flyout of the **Navigation** toolbar to view the stair as shown at right in Figure 8T.22.

39. Close and save all drawings.

PROJECT

EX 8.4 Creating a Multi-landing Stair with Railings

1. Open Autodesk Architectural Desktop 2005, and select **File>Project Browser** from the menu bar. Use the Project Selector drop-down list to navigate to your *ADT Student\ADT Tutor\Ch8* directory. Double-click on **Ex 8-4** to set this project current. Select **Close** to dismiss the **Project Browser**.

2. Select the **Constructs** tab of the **Project Navigator**, click on the Constructs category, right-click and choose **New>Construct**. Type **Stair** in the name edit field of the stair in the **Add Construct** dialog box. Check **Level 1** and **2** for **Division 1** of the construct. Double-click on the Stair construct to open the drawing and develop the stair.

3. Create a stair style named **Residential2** in the Stair drawing. Set the **Design Rules** as follows: Maximum Slope Riser Height = **7-3/4"**, Maximum Slope Tread Depth = **10**, Optimum Slope Riser Height = **7 1/2**, Optimum Slope Tread Depth = **10**, Minimum Slope Riser Height = **7** and Minimum Slope Tread Depth = **11**. Toggle OFF **Use Rule Based Calculator** in the **Design Rules** tab of the **Stair Styles** dialog box.

4. Overlay Xref Floor 1 onto the Stair drawing. Using the Residential2 stair style, create a multi-landing stair that begins at **p1** with the second flight beginning at **p2** as shown in Figure 8T.23. Use the 1/2 Landing turn type.

5. The height of the stair is **9'-6"** and the width is **44"** Edit the **Calculation Rules** to **16** risers and record below the riser height, tread depth, and straight length.

6. Using Xref edit select the stair, right-click, choose **Customize Edge>Project**, and project the stair to the wall at **p4**.

7. Attach the Handrail Round (page 2 of the **Content Browser**) to the stair at **p1**. The extensions of the handrail should be based on **Use Stair Landing Extension**. Attach the handrail to the stair with a side offset equal to 2" and use **Automatic** placement.

8. Import Guardrail Wood Balusters 01 from the Design Tool Catalog-Imperial of the **Content Browser**.

9. Insert the Wood Balusters 01 from **p5** to **p6** on the landing. Select the **RailingAnchor>Anchor to Objects** command to attach the railing to the landing object. Use **Automatic Cleanup** and **Follow Surface** to anchor the railing to the landing.

10. View the stair from **SW Isometric** and select **Hidden** from the **Navigation** toolbar. The completed stair and handrail should appear as shown in Figure 8T.23.

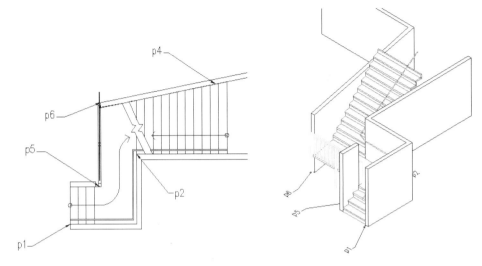

Figure 8T.23 *Multi-landing stair and railing*

CHAPTER 9

Using and Creating Symbols

INTRODUCTION

Architectural Desktop includes hundreds of symbols created for two-dimensional and three-dimensional representations of building components. The two-dimensional symbols include the plan and elevation views. The symbols are selected from the Design Tool Catalog or the AutoCAD DesignCenter and include blocks, multi-view blocks, and mask blocks. The contents of the AutoCAD DesignCenter can also be customized to include new blocks through the **AEC Content Wizard**.

OBJECTIVES

After completing this chapter, you will be able to

- Set the scale of a drawing for inserting symbols
- Specify the default symbol menu
- Use the DesignCenter and the **Content Browser** to insert symbols in a drawing
- Insert and modify multi-view blocks
- Define the display representation for a multi-view block
- Import and export multi-view blocks
- Create and insert masking blocks
- Attach masking blocks to Architectural Desktop objects to control display
- Create a masking or multi-view block using the **AEC Content Wizard**

SETTING THE SCALE AND LAYER FOR SYMBOLS AND ANNOTATION

Prior to symbols being inserted in the drawing, the scale of the plotted drawing should be established. Some of the symbols included in the DesignCenter are created at a specific size, while others are sized according to the scale of the drawing. For example, an electrical receptacle is sized according to the scale of the drawing, and a cabinet will be inserted at its actual size. If a receptacle symbol is placed in a drawing with the scale set to 1/4"=1'–0", it is drawn with a 6" diameter circle. This symbol, if placed in a drawing with the 1/8"=1'–0" scale, would be drawn with a 12" diameter circle because the scale has changed. The scale of a drawing sets the scale factor, which is used as a multiplier for selected symbols and annotation. The scale factor is the ratio between the size of the AutoCAD entity and its display printed on paper. The typical architectural scales and the associated scale factors are listed in the **Scale** tab of the **Drawing Setup** dialog box when the **AecDwgSetup** command is selected. Refer to Table 1.11 of Chapter 1 for **AecDwgSetup** command access.

Architectural Desktop elements placed in a drawing are automatically placed on the appropriate layer according to the layer key style. Therefore, if the AIA 256 Color layer key style is used, a refrigerator is inserted on A-Flor-Appl layer. The Current layer key style places the object on the layer that is current at the time of insertion. Table 9.1 shows the layer names for an appliance when the respective layer key style options are selected.

Layer for an Appliance	Layer Key Style
Layer 0	Current
A-Flor-Appl	AIA (256 Color)
A-Flor-Appl	AIA (256 Color) ADT 3
A700G4	BS 1192 Aug Version 2 (256 Color)
A-Appliance-G	BS 1192 Description (256 Color)
610-Ausstattung	DIN 276
A-Appliances	Generic Architectural Desktop
027-Ausstatung	STLB

Table 9.1 *Layer names for an appliance using layer key styles*

Using a layer standard enhances the drawing because the display of similar objects can be controlled as a group in the drawing. Architectural Desktop places the objects on the correct layer automatically when the object is inserted in the drawing. If the layer key style is changed in the middle of a drawing, all previous objects remain on the layer of their initial insertion. Therefore, you should determine the layer key style prior to beginning the drawing.

USING THE CONTENT BROWSER

Symbols can be selected from the Design Tool Catalogs of the **Content Browser**. Access the **Content Browser** from the **Navigation** toolbar shown in Figure 9.1. (Symbols are organized in catalogs as shown in Table 9.3.) Therefore to place plumbing fixtures, select from the Mechanical catalog, whereas cabinets are placed in the drawing from the Furnishings folder. The Design Tool Catalog shown in Figure 9.2 includes a preview of the symbol. You can select the i-drop of the symbol and drag the symbol into the drawing, or drag the i-drop of the symbol onto a custom tool palette.

The content of the tool palettes is stored in the directory as specified for the Workspace Catalog defined by the login user name. The **Tool Path File Locations** is

Figure 9.1 *DesignCenter and Content Browser commands of the Navigation toolbar*

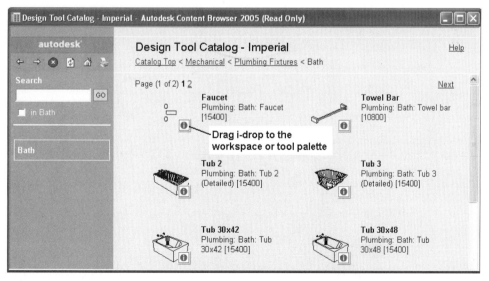

Figure 9.2 *Design Tool Catalog of the Content Browser*

specified in the **Files** tab of the **Options** dialog box as shown in Figure 9.3. Therefore, each time you log in and open Autodesk Architectural Desktop, the content of tool palettes is read from the login directory. The path shown in Figure 9.3 is as follows: *C:\Documents and Settings\Login Name\Application Data\Autodesk\ADT 2005\enu\Support\WorkspaceCatalog(Imperial)*. The default path to the DesignCenter and tool catalogs are specified in the **AEC Content** tab of the **Options** dialog box.

Custom tool palettes can be created and content from the catalogs of the **Content Browser** or the folders of the DesignCenter can be added to a custom palette.

STEPS TO CREATING A TOOL PALETTE AND ADDING CONTENT

1. Right-click over the Tool Palettes title bar, and choose **New Palette** as shown at left in Figure 9.4.
2. Overtype a new name to name the new palette as shown in Figure 9.4.
3. To add content from the catalogs of the **Content Browser**, select **Content Browser** from the **Navigation** toolbar.
4. Double-click on Design Tool Catalog as shown at right in Figure 9.4 to open the catalog.
5. Click on a category to open the tool catalog.
6. Click on the i-drop of the tool in the tool catalog and drag the tool to the tool palette as shown in Figure 9.5.

   (The Ceiling Fan shown in Figure 9.5 at bottom was added to the new tool palette with content from the Architectural Desktop Design Tool Catalog-Imperial catalog.)

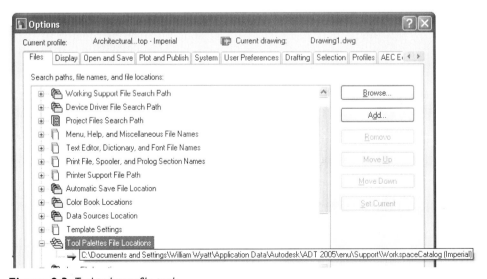

Figure 9.3 *Tool palettes file path*

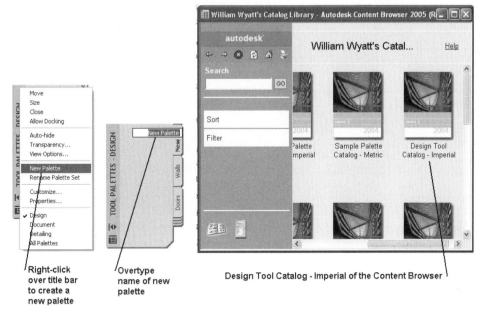

Figure 9.4 Adding content to new tool palettes

7. To add content from the DesignCenter, choose the **DesignCenter** button on the **Navigation** toolbar.

8. Left-click on a symbol in the DesignCenter, and drag the symbol to a tool palette. When the pointer is placed over the tool palette, a plus sign will be added to the pointer as shown at top of Figure 9.6. When you release the pointer over the tool palette, the content is added as shown at the bottom of Figure 9.6.

After you place symbols on your custom tool palette, the tools can be copied from a tool palette to a tool catalog, package, or palette in the **Content Browser**. A tool catalog consists of tools placed in categories. A package can be created to store a group of tools for distribution. The My Tool Catalog of the **Content Browser** shown in Figure 9.7 is designed to receive tools from the user. Tool palettes, categories, and packages can be created in this catalog by selecting the buttons in the lower left corner of Figure 9.7. If you place your tools in a tool palette in the My Tool Catalog, the entire palette can be copied to the tool palettes of a drawing. You can open additional windows of the **Content Browser** by pressing CTRL + N when **Content Browser** is opened. If **different catalogs** are opened when you drag tools from one **Content Browser** window to another, you copy the tool. Pressing CTRL when you drag across different catalogs will move the tool. If the **same catalog** is opened in each of the **Content Browser** windows, dragging the tools will move them to a different category,

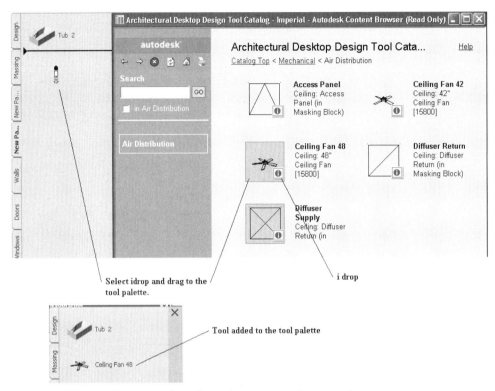

Figure 9.5 *Dragging content from the tool category to the new palette*

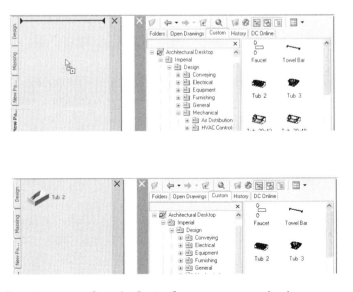

Figure 9.6 *Dragging content from the DesignCenter to a new tool palette*

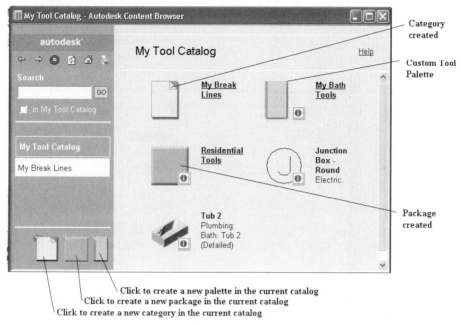

Figure 9.7 *Placing tools in the "My Tool Catalog"*

thereby removing the original. Pressing CTRL when you drag across the same catalog will copy the tool.

USING THE DESIGNCENTER

Architectural symbols can also be inserted in a drawing from the DesignCenter. Access the AutoCAD DesignCenter as shown in Table 9.2. The AutoCAD DesignCenter is modified to include the display of a **Custom** view. Figure 9.8 shows the DesignCenter window and the commands of the DesignCenter.

Command prompt	ADCENTER
Toolbar	Select DesignCenter from the Navigation toolbar shown in Figure 9.1
Keyboard Shortcut	CTRL +2

Table 9.2 *DesignCenter command access*

The commands of the **DesignCenter** toolbar are described below:

    Load – Opens the **Load** dialog box to load symbols from Projects.

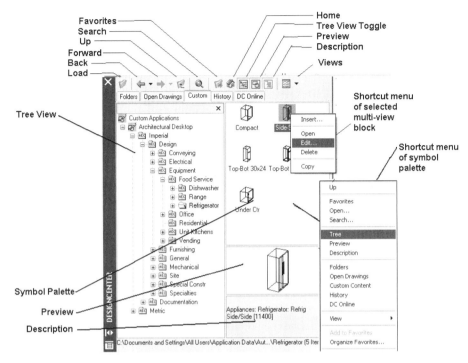

Figure 9.8 *AutoCAD DesignCenter*

Back/Forward – Toggles the history of selections in the Tree View of the DesignCenter. The flyout lists the path of the Back/Forward options.

Up – Opens the folder one level above the current container.

Search – Opens the **Search** window to search the hard drive for files.

Favorites – Toggles to the **Favorites** folder of Architectural Desktop.

Home – Opens the DesignCenter located in the *Program Files\ Architectural Desktop 2005\Sample\DesignCenter* directory.

Tree View – Toggles ON/OFF the Tree View.

Preview – Toggles open/closed a preview of a selected symbol.

Description – Opens/closes a text description window at the bottom of the palette.

Views – Allows you to select from a flyout Large Icon, Small Icon, List, or Detail View of the block.

The tabs located below the **DesignCenter** toolbar allow you to quickly edit the Tree View to access directories and the Internet for symbols. The tabs are described below.

Folders – The **Folders** tab edits the Tree View to display the folder level of ADT 2005.

Open Drawings – The **Open Drawings** tab displays the Tree View of the current open drawing. Content for the current drawing includes Blocks, Dimstyles, Layers, Layouts, Linetypes, Textstyles, and Xrefs.

Custom – The **Custom** tab opens the Tree View to the Custom Applications folder.

History – The **History** tab lists the drawing previously accessed in the DesignCenter.

DC Online – The **DC Online** tab opens the DesignCenter Online, allowing access to additional symbols available via the Internet.

The **Custom** view includes all the symbols and documentation of Architectural Desktop. When symbols are selected from the menu bar or from a toolbar, the AutoCAD DesignCenter opens to the Architectural Desktop node in the Tree View as shown in Figure 9.8. The symbols are located in the *C:\Documents and Settings\All Users\Application Data\Autodesk\ADT 2005\enu\AEC Content* directory. The Imperial and Metric directories include Design and Documentation subdirectories. You can set the path of the default symbol menu for the DesignCenter on the **AEC Content** tab of the **Options** dialog box. The **AEC Content** tab includes a **Browse** button, which allows you to edit the path to the **Content Browser** and DesignCenter.

Table 9.3 shows the metric and imperial content of the Design Tool Catalog categories and the DesignCenter folders.

INSERTING SYMBOLS

When you select symbols from the **Content Browser**, click and drag the i-drop onto the workspace of the drawing. Release the left mouse button while over the workspace and then left-click in the workspace to specify the insert point. The symbol rotation is preset; however, you can type r in the command line to select the rotation option and type a value for the rotation prior to specifying the insert point.

To select a symbol from the DesignCenter, you **double-click** on the preview of the symbol displayed in the right pane of the DesignCenter and specify the insert point in the workspace. The dynamic preview of the symbol is displayed at the mouse pointer as you are prompted to specify the insert point of the symbol in the command line.

However, if you click and drag the symbol from the DesignCenter, the insert point location is specified at the location where you release the left mouse button. Therefore, if you drag the symbol into the drawing, you should set Running Object Snap modes prior to beginning the drag operation. If you drag the symbol, as you move the mouse over the DesignCenter, a slash circle symbol appears to warn you not to

Design Tool Catalog - Imperial	Imperial\Design Folder of the DesignCenter	Design Tool Catalog-Metric	Metric\Design Folder of the DesignCenter
Conveying	Conveying	Bathroom	Bathroom
Curtain Walls			Curtain Walls
Doors and Windows			Doors and Windows
Electrical	Electrical	Domestic Furniture	Domestic Furniture
Equipment	Equipment	Electrical Services	Electrical Services
Furnishings	Furnishing	Kitchen Fittings	Kitchen Fittings
General	General	Office Furniture	Office Furniture
Mechanical	Mechanical	Piped & Ducted Services	Piped and Ducted Services
Roof Slabs and Slabs			Roof Slabs and Slabs
Site	Site	Site	Site
Spaces			Spaces
Special Construction	Special Construction		
Specialties	Specialties		
Stairs and Railings			Stairs and Railings
Structural			Structural
Walls			Walls

Table 9.3 *Content folders of the DesignCenter and Design Tool Catalog*

release the symbol over the DesignCenter, which will void the drag operation. The symbol will be displayed attached to the pointer as the mouse is moved over the drawing area. The default rotation of the symbol is displayed prior to your releasing the left mouse button. The rotation can be set in the command line after the left mouse button is released.

Symbols inserted from the DesignCenter or **Content Browser** are multi-view or masking block styles that will be listed in the Multi Purpose Objects\Mask Block Definitions and Multi Purpose Objects\Multi-View Block Definitions folders of the **Style Manager**.

STEPS TO INSERTING SYMBOLS INTO A DRAWING FROM THE CONTENT BROWSER

1. Right-click over the OSNAP button on the status bar, and choose **Settings** from the shortcut menu.
2. Select desired Running Object Snap modes from the **Drafting Settings** dialog box.
3. Select **OK** to close the **Drafting Settings** dialog box.
4. Select **Drawing Menu** from the Drawing Window status bar, and then select **Drawing Setup**. Select the **Scale** tab in the **Drawing Setup** dialog box.
5. Select the scale of the drawing, and then select **OK** to dismiss the **Drawing Setup** dialog box.
6. Choose the **Content Browser** from the **Navigatio**n toolbar.
7. Open the **Design Tool Catalog** and expand the tool palettes or categories displayed in the left pane.
8. Click on the i-drop of the desired tool and drag the i-drop into the workspace. Prior to specifying the insert point, edit the Properties palette if necessary, and then left-click in the drawing to specify the insertion point. Press ENTER to end the command.

PROPERTIES OF MULTI-VIEW BLOCKS

Many of the symbols inserted from the Design Tool Catalog or DesignCenter are multi-view blocks. Multi-view blocks have special display control properties, which allow the block to be displayed or not displayed when viewed from specified directions. A multi-view block of a duplex outlet can be defined to display only in the top viewing position, and not in other viewing positions. Controlling display according to viewing direction is defined for each display representation.

When you insert symbols from the **Content Browser** or DesignCenter, the command line displays the name of the multi-view block; however, the process usually consists of more than three lines of text then you are prompted to specify the rotation and insert point as shown below.

> Rotation <0.00>: *(Rotation angle specified.)*
>
> Insert point or [NAme/X scale/Y scale/Z scale/Rotation/Match/Base point]: *(Specify insertion point.)*

Insert point or [NAme/X scale/Y scale/Z scale/Rotation/Match/
Undo/Base point]: *(Press* ENTER *to end the command.)*

Additional features of the multi-view block can be specified in the Properties palette prior to specifying the insertion point. The symbol is attached to the pointer scaled and rotated as specified in the Properties palette. Prior to specifying the insertion point, you can return to the Properties palette and edit the scale, rotation, and other features of the multi-view block. The options of the Properties palette of a multi-view block are shown below.

Basic

General

Description – The **Description** button opens the **Description** dialog box, which allows you to add additional text.

Definition – The **Definition** edit field lists the name of the multi-view block. The drop-down list displays all blocks inserted in the drawing. A multi-view block can be selected from the drop-down list and inserted into the drawing.

Scale

Specify on Screen – The **Specify on Screen** edit field includes **Yes** and **No** options. If the **No** option is selected, you can type a scale factor in this field or in the command line. The **Yes** option prevents you from entering values in the Properties palette. Scale factors can also be specified by selecting points in the drawing area.

X – The **X** edit field displays the current scale along the X axis.

Y – The **Y** edit field displays the current scale along the Y axis.

Z – The **Z** edit field displays the current scale along the Z axis.

Location

Specify rotation on screen – The **Specify Rotation on Screen** edit field includes **Yes** and **No** options. If the **No** option is selected, you type an angle in this field or in the command line. Rotation angles can also be specified by selecting points in the workspace with the pointer.

Elevation – The multi-view block is inserted at Z=0 coordinate elevation. Editing the elevation value can change the elevation of existing multi-view blocks.

Advanced

Insertion offsets – The **Insertion offsets** button opens the **Multi-view Blocks Offsets** dialog box as shown in Figure 9.9.

Note: The insertion point offsets for all views defined for the multi-view block should be edited. If both insertion point offsets are not edited, the block will be positioned differently when viewed in the plan versus the model view of the block.

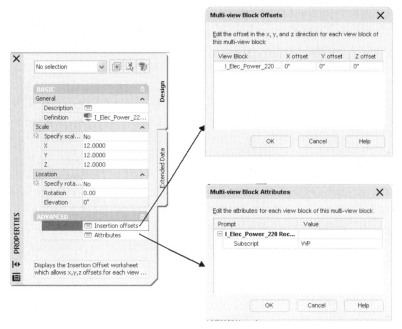

Figure 9.9 *Multi-view Block Offsets and Multi-view Block Attributes dialog boxes*

Attributes –The **Attributes** button opens the **Multi-view Blocks Attributes** dialog box, as shown in Figure 9.9. You can add the values for attributes for a block by entering text in the **Value** column. The **Prompt** column cannot be edited. This dialog box allows you to add text for the attributes of the block.

EDITING MULTI-VIEW BLOCKS USING PROPERTIES

Selecting a multi-view block, right-clicking, and choosing **Properties** displays the features of an existing multi-view block. The Properties palette can be used to edit the features of the multi-view block. Additional edit fields described below are added to the Properties palette for existing blocks.

Basic

General

Layer – The **Layer** edit field displays the name of the layer of the multi-view block. The layer drop-down list displays the layers of the drawing.

Location

Additional information – The **Additional information** button opens the **Location** dialog box shown in Figure 9.10, which lists the x, y, z coordinates of the multi-view block, normal extrusion, and rotation.

Options of the **Location** dialog box are as follows:

World Coordinate System/Current Coordinate System – Select the **World Coordinate System or Current Coordinate System** radio button to indicate which system is used to display the coordinate location of the multi-view block.

Insertion Point – The **Insertion Point** section includes **X**, **Y**, and **Z** fields, which specify the coordinates of the insertion point of the multi-view block. You can change the coordinates of the multi-view block to move the multi-view block.

Normal – The **Normal X**, **Y**, and **Z** edit fields indicate the direction a three-dimensional block is projected. A 1.000 in the **Z** edit field indicates that the symbol is developed perpendicular to the XY plane.

Rotation – The **Rotation** section indicates the angular rotation of the multi-view block. You can change the rotation angle to rotate the multi-view block.

Insertion offsets – The **Insertion Offsets** button opens the **Multi-view Block Offsets** dialog box, which displays each of the existing view blocks included in the multi-view block. The **Multi-view Block Offsets** dialog box shown in Figure 9.10 includes view blocks designed for model, plan, front, left, and right views. The **X offset**, **Y offset**, and **Z offset** values allow you to modify the location of the insertion point for placing the multi-view block. The insertion point offset for each view of the multi-view block should be edited to maintain the objects insertion point in each view direction.

Note: If the insertion point offsets for plan and model view blocks are not edited with the same values, the view block when viewed in the plan versus the model view will be positioned differently based upon the respective insertion points.

Extended Data Tab

Hyperlinks – Opens the **Insert Hyperlinks** dialog box, which allows you to define hyperlinks for the multi-view blocks.

Notes – The **Notes** button opens the **Notes** dialog box, which allows you to type additional notes.

Reference Documents – The **Reference Documents** button opens the **Reference Docs** dialog box, which allows you to display the path and file name of other files related to the multi-view block.

STOP. Do Tutorial 9.1, "Inserting Multi-View Blocks with Precision," at the end of the chapter.

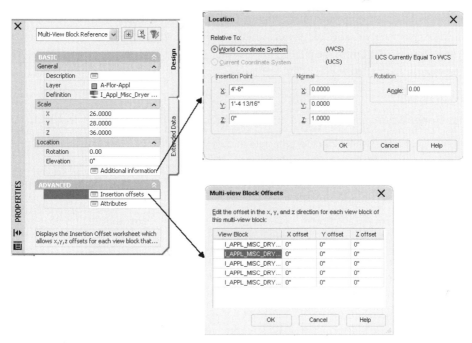

Figure 9.10 *Location and Multi-view Block Offsets dialog boxes*

EDITING THE MULTI-VIEW BLOCK DEFINITION PROPERTIES

The Properties palette can be used to edit attributes, dimensions, location, and rotation of a multi-view block. However, the definition of multi-view block can be edited by selecting a multi-view block, right-clicking, and choosing **Edit Multi-View Block Definition** (**MvBlockDefEdit** command) from the shortcut menu. Access the **MvBlockDefEdit** command as shown in Table 9.4.

Command prompt	MVBLOCKDEFEDIT
Shortcut menu	Select a multi-view block, right-click, and choose Edit Multi-View Block Definition

Table 9.4 *Edit Multi-View Block Definition command access*

When you select a multi-view block, right-click, and choose **Edit Multi-View Block Definition**, the **Multi-View Block Definition Properties** dialog box opens, which allows you to edit the definition of view blocks used to create the multi-view block. The **Multi-View Block Definition Properties** dialog box can also be accessed from the Multi Purpose Objects>Multi-view Blocks folder of the **Style Manager**. The display representations and views assigned to the view blocks can be edited in the **Multi-View Block Definition Properties** dialog box. The options of the **Multi-View Block Definition Properties** dialog box are described below.

General Tab

The **General** tab includes **Name** and **Description** edit fields, as shown in Figure 9.11.

> **Name** – The **Name** edit field allows you to change the name of the multi-view block.
>
> **Description** – The **Description** edit field allows you to enter a description for the multi-view block.
>
> **Notes** – Selecting the **Notes** button opens the **Notes** dialog box for adding text notes.
>
> **Keynotes** – Keynotes may be defined for the symbol. Choosing the **Select Keynote** button will open the **Select Keynote** dialog box, which allows you to specify keynotes from the 16 CSI divisions or other keynote databases.
>
> **Property Sets** – The **Property Sets** button opens the **Edit Property Set Data** dialog box. The **Edit Property Set Data** dialog box includes the **Add**

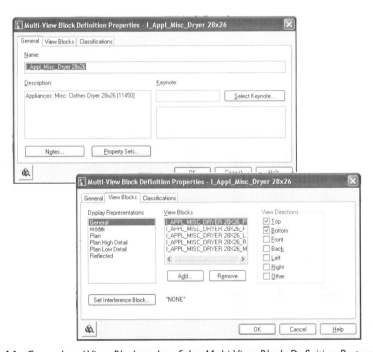

Figure 9.11 *General and View Blocks tabs of the Multi-View Block Definition Properties dialog box*

property sets and the **Remove property sets** buttons, which allow you to add or remove property sets. Property sets can be imported in the **Style Manager** from the Schedule Tables (Imperial).dwg of the *C:\Documents and Settings\All Users\Application Data\Autodesk\ADT 2005\enu\Styles\Imperial* directory. The **Add property sets** button is active if property sets have been imported into the drawing. Property sets are presented in Chapter 10.

View Blocks Tab

The **View Blocks** tab shown in Figure 9.11 includes a list of the display representations of the view blocks assigned to the multi-view block.

> **Display Representations** – The **Display Representations** list includes each of the display representations used by the multi-view block. Multi-view blocks can be defined for the General, Model, Plan, Plan High Detail, Plan Low Detail, and Reflected display representations.
>
> **View Blocks** – The **View Blocks** list is used to define the multi-view block's appearance for each of the display representations. Each of the names in the **Display Representations** list box represents a display method for controlling the view of a single block. Some of the view blocks defined in the Architectural Desktop folders have a suffix to the block name to indicate the viewing direction that will apply the block. In Figure 9.11, suffixes were used: F (front), L (left), M (model), and P (plan). However, the same view block can be defined for each of the display representations and view. The view blocks list varies according to the display representation selected.
>
> **Add** – The **Add** button opens the **Select a Block** dialog box, which lists the blocks of the drawing. If you select a block from this list, it is assigned to the selected display representation and view directions for the multi-view block.
>
> **Remove** – To remove a view block, select it from the view block list and then select the **Remove** button. View blocks are added and removed for each of the display representations.
>
> **View Directions** – The **View Directions** check box allows you to turn ON or OFF the visibility of the block when the drawing is viewed from the direction listed.
>
> **Set Interference Block** – The **Set Interference Block** button opens the **Select a Block** dialog box.

Classifications Tab

The **Classifications** tab lists the classification definition and classifications associated with the multi-view block.

> **Classification Definition** – The **Classification Definition** lists the classification definition that has been attached to the drawing in the **Style Manager**.

Classification – The **Classification** can be selected from the drop-down list of the classifications.

CREATING A MULTI-VIEW BLOCK

Most multi-view blocks are developed from a three-dimensional object. The multi-view block is defined from one or more view blocks. A view block is an AutoCAD block created from an elevation or plan view drawing of a three-dimensional object. Therefore, view blocks can be created for the top, front, right side, left side, bottom, and model views of the three-dimensional object. The tools used to create elevations, discussed in Chapter 11, can be used to develop each of the view blocks. You can include additional insertion points on each of the view blocks. The insertion points are placed on the Defpoints layer using the AutoCAD **POINT** command. The view blocks shown in Figure 9.12 are the components of the Dryer 28x26 multi-view block. View blocks are created as a projection drawing from a three-dimensional object; however, lines can be added or deleted based upon the representation conventions for that view. The left view of the dryer shown in Figure 9.12 does not include hidden lines to represent the top plane because the conventional representation of a dryer does not include the hidden lines.

Multi-view blocks are created in the Multi-Purpose Objects folder of the **Style Manager**. The name and definition of the multi-view block is specified in the **Style Manager**. The definition requires reference to other multi-view blocks or view blocks in a drawing. The multi-view block definition includes control of the visibility of the block according to the viewing direction. See Table 9.5 for **Style Manager** command access.

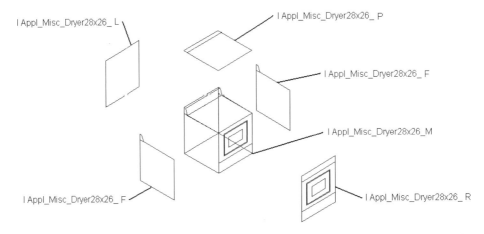

Figure 9.12 *View blocks used to create a multi-view block*

Command prompt	AECSTYLEMANAGER
Menu bar	Select Format>Style Manager, select Multi-View Block Definitions folder in the Multi-Purpose Objects folder

Table 9.5 *Style Manager access to create a multi-view block*

Before you can create a new multi-view block, you must create or have used other blocks or multi-view blocks in the drawing to assign to the new multi-view block. The view blocks or multi-view blocks are used as a base for creating the new multi-view block. When you create a multi-view block, a view block must be defined for each possible orthographic view orientation of the multi-view block. View blocks are created from blocks or other multi-view blocks and are defined for one or more view directions. The view blocks of the I Appl_Misc_Dryer 28x26 multiview block are shown in Figure 9.13. Each of these blocks is assigned to a display representation and view direction in the Multi-View Block Definition Properties dialog box.

After you create a name for the multi-view block and view blocks as necessary to describe the object, the **View Blocks** tab of the **Multi-View Block Definition Properties** dialog box is used to identify which view blocks and view directions to connect to one or more display representations. As shown in Figure 9.13, the I Appl_Misc_Dryer 28x26 multi-view block is defined only for the General and

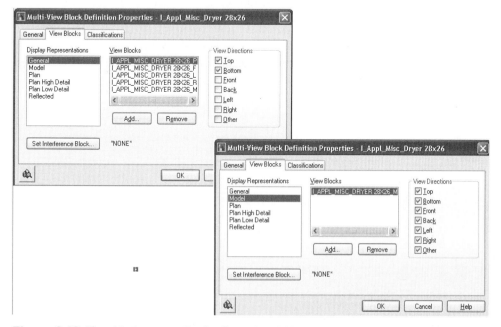

Figure 9.13 *View block settings for the General and Model display representations*

Model display representations. The General display representation is used in the Plan display set. The Plan display set is used for the Medium Detail display configuration when the drawing is viewed from the top. Therefore the **Multi-View Block Definition Properties** dialog box links the display representation to a view block for describing the dryer.

The following steps outline the procedure to create a new multi-view block in the **Style Manager**.

STEPS TO CREATING AND DEFINING A NEW MULTI-VIEW BLOCK

1. Select **Format>Style Manager** from the menu bar.
2. Double-click on the **Multi-Purpose Objects** folder of the current drawing in the left pane.
3. Double-click on the **Multi-View Block Definitions** folder in the Multi-Purpose Objects folder.
4. Select **New Style** from the **Style Manager** toolbar and overtype a name for the new multi-view block.
5. Double-click on the new multi-view block name to open the **Multi-View Block Definition Properties** dialog box.
6. Edit the **Multi-View Block Definition Properties** dialog box to define the view blocks used per display representation and view direction.

DEFINING THE PROPERTIES OF A NEW MULTI-VIEW BLOCK

After you create a name for the new multi-view block, selecting the **Edit Style** button on the **Style Manager** toolbar allows you to define the specific properties. Selecting the **Edit Style** button or double-clicking on the name of the multi-view block

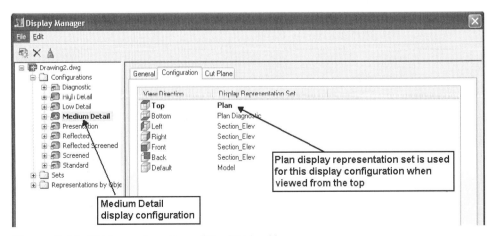

Figure 9.14 *Display configurations of the Display Manager*

will open the **Multi-View Block Definition Properties** dialog box, which includes the three tabs described above.

The **Multi-View Block Definition Properties** dialog box for the I Appl_Misc_Dryer 28x26 multi-view block is shown in Figure 9.13. The **View Blocks** tab shown at left lists five view blocks defined for the General display representation. When this multi-view block was defined, the General display representation was selected and the **Add** button was used to specify each of the four view blocks. A view direction was specified for each of the view blocks selected for the General display representation. The results of this edit will display the I Appl_Misc_Dryer 28x26_P view block when the General display representation is used and the drawing is viewed from the Top or Bottom. In contrast, the Model display representation shown at right includes only the Appl_Misc_Dryer 28x26_M view block, and it is used for all view directions when the Model display representation is used.

There are no view blocks defined for the Reflected display representation of the I Appl_Misc_Dryer 28x26 multi-view block. If the Reflected display configuration is assigned to a viewport and the view direction is top, the dryer will not be displayed.

The view blocks and display representations of a multi-view block should be defined based upon the intended use of the multi-view block. The Configurations folder of the **Display Manager** allows you to identify the display representation set that is used for the associated view direction, as shown in Figure 9.14. The Sets folder allows you to identify the display representation used per object type within a display set as shown in Figure 9.15. The Medium Detail display configuration uses the Plan display representation set when viewed from the top or plan view. The Plan display representation set includes General and Plan display representations for multi-view blocks.

Symbols created for a floor plan should be assigned the General display representation. If you are developing an electrical plan to include receptacles, switches, and incandescent lights, all of these symbols must share a common display representation or set. The multi-view block definition for duplex receptacles and switches is General. However, the incandescent light symbol has the Reflected display representation. Therefore, the switch and receptacle symbols are displayed when the Medium Detail display configuration is applied to a viewport, and the incandescent light is hidden, except when the Reflected display configuration is applied. If you edit the incandescent light symbol to the General display representation, it will be displayed with the switch and receptacle symbols when the Medium Detail display configuration is used.

Tutorial 9.2, "Creating and Modifying Multi-View Blocks," will step you through the procedure to create a new symbol and to modify the display representation of an incandescent light symbol in the **Multi-View Block Definition Properties** dialog box.

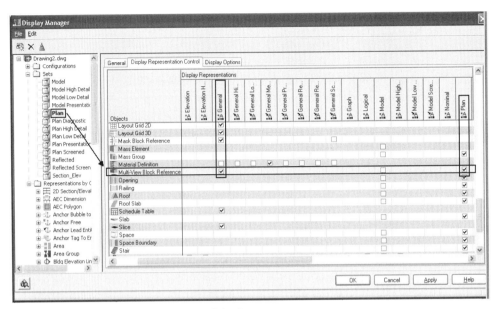

Figure 9.15 *Display representations of display sets*

IMPORTING AND EXPORTING MULTI-VIEW BLOCKS

Multi-view blocks allow you to insert symbols in a drawing with more display control than with blocks. You can transfer multi-view blocks from one drawing to another drawing by exporting the multi-view block definition. Multi-view blocks can be transferred across drawings in the **Style Manager**.

STEPS TO EXPORTING MULTI-VIEW BLOCKS

1. Open a drawing that has the desired multi-view blocks.
2. Choose **Format>Style Manager** from the menu bar.
3. Double-click on **Multi-Purpose Objects** folder of the current drawing in the left pane.
4. Select the **Multi-View Block Definitions** folder, right-click, and choose **Copy** as shown in Figure 9.16.
5. Select **Open** from the **Style Manager** toolbar to display the **Open Drawing** dialog box. Edit the directory and select the target drawing for the multi-view blocks. Select **Open** to close the **Open Drawing** dialog box.
6. Select the target drawing in the left pane, right-click, and choose **Paste** to paste the multi-view blocks into the target drawing.
7. If the **Import/Export Duplicate Names Found** dialog box is displayed when you paste, verify that **Leave Existing** is toggled ON, and then select **OK** to dismiss the **Import/Export Duplicate Names Found** dialog box.

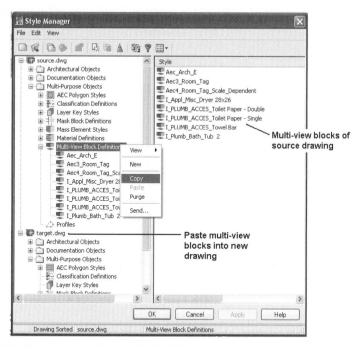

Figure 9.16 *Copying the multi-view block definitions of drawing in the Style Manager*

INSERTING NEW MULTI-VIEW BLOCKS

Multi-view blocks are usually inserted from the Design Tool Catalog or DesignCenter; however, new multi-view blocks or imported multi-view blocks will **not** be included in these resources. The **Multi-View Block Reference** tool can be accessed from the Stock Tool Catalog>Helper Tools as described in Table 9.6. The AutoCAD **INSERT** command cannot be used to insert multi-view blocks; however, the **INSERT** command can be used to insert the view blocks of a multi-view block.

Command prompt	MVBLOCKADD
Content Browser	Open the Stock Tool Catalog>Helper Tools category and select Multi-View Block Reference

Table 9.6 *Multi-View Block Reference command access*

You can drag the Multi-View Block Reference i-drop from the Helper Tools catalog onto a new tool palette or into the workspace. When you select the **Multi-View Block Reference** tool, select the name of new multi-view blocks from the Definition drop-down list of the Properties palette.

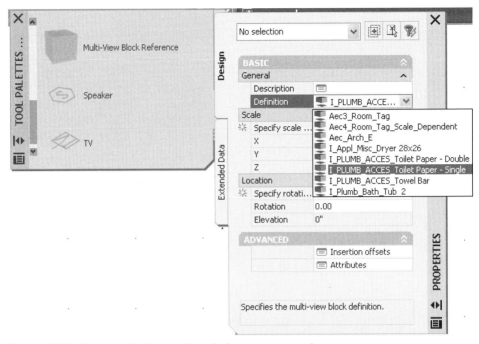

Figure 9.17 *Using the Multi-view Block Reference command*

IN-PLACE EDITING OF VIEW BLOCK OFFSETS

You can edit the location of insertion offsets and attribute locations of view block within the multi-view block in the workspace without accessing the Properties palette. The **Edit View Block Offset** command is located on the shortcut menu of a multi-view block as shown in Table 9.7.

Command prompt	MVBLOCKEDITVIEWBLOCKOFFSETS
Shortcut menu	Select a multi-veiw block, right-click, and choose Edit View Block Offsets

Table 9.7 *Edit View Block Offsets command access*

STEPS TO IN-PLACE EDIT VIEW BLOCK INSERTION OFFSETS

1. Select a multi-view block, right-click, and choose **Edit View Block Offsets** to display the grip of the insertion point.
2. Select the grip and move the cursor to display the dynamic dimensions as shown in Figure 9.18.

3. Press TAB to specify a horizontal or vertical dimension for edit as shown in Figure 9.18, and overtype the desired distance for edit.

4. Right-click and choose **Exit Edit View Block Offsets** from the shortcut menu.

5. Press ESC to remove block selection.

You can change the location of the attributes if you select **Edit Attribute Locations** from the shortcut menu of a multi-view block as shown in Table 9.8.

Command prompt	MVBLOCKEDITATTRIBUTELOCATIONS
Shortcut	Select a multi-view block, right-click, and select Edit Attribute Locations

Table 9.8 *Edit Attribute Locations command access*

The steps for editing the attribute location are as follows:

STEPS TO IN-PLACE EDIT OF ATTRIBUTE LOCATION

1. Select a multi-view block, right-click, and choose **Edit Attribute Locations** to display the grip of the attribute.

2. Select the grip and move the cursor to display the dynamic dimensions as shown in Figure 9.19.

3. Press TAB to specify a horizontal or vertical dimension for edit as shown in Figure 9.19, and overtype the desired distance for edit.

4. Right-click and choose **Exit Edit Attribute Locations** from the shortcut menu.

5. Press ESC to remove selection of the multi-view block.

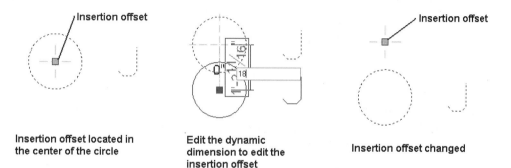

Figure 9.18 *Editing Insertion Offsets in the workspace*

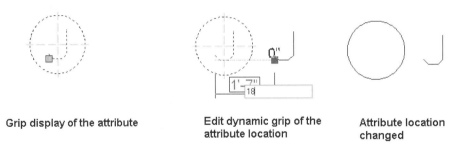

| Grip display of the attribute | Edit dynamic grip of the attribute location | Attribute location changed |

Figure 9.19 *Editing the location of attributes in the workspace*

INSERTING LAYOUTS WITH MULTIPLE FIXTURES

The **Content Browser** and the DesignCenter include layouts that consist of several fixtures and symbols appropriate for restroom layouts. The layouts are located in the Design Tool Catalog- Imperial\Mechanical\Plumbing Fixtures\Layouts folder and the Design Tool Catalog-Metric\ Bathroom\Layouts of the **Content Browser**. Multiple lavatories are spaced evenly in a counter as shown in Figure 9.20. The symbols are three-dimensional AutoCAD blocks, and they can be used in the development of interior elevations. When you drag the i-drop from the **Content Browser** catalog, the **Insert** dialog box opens, allowing you to set the insertion point, scale, and rotation and toggle **Explode**. When the layouts are inserted in the drawing, you cannot adjust the length of the counter position of other fixtures unless you check **Explode** or select **Edit Block in-Place** from the shortcut menu to edit the block. If you check **Explode**, the changes apply only to the selected block. If you apply **Edit Block in-Place** to the layout, when you save your changes, you are editing the original block definition, and future insertions of the layout will be changed to comply to the changes made in the current drawing.

When you explode the layout upon insertion or use **Edit Block in-Place**, you can edit the components of the layout. The layout shown in Figure 9.21 has been exploded upon insertion. The layout consists of a custom wall style that serves as the counter. The counter has wall grips, which are displayed in Figure 9.21. The grips can be stretched to fit the design of a restroom according to the existing room shape, as shown in Figure 9.21. The lavatories are anchored to a layout curve to evenly space the fixtures. Each component of the original layout can then be manipulated to create customized layouts of fixtures. The grips of the anchor nodes of the layout curve are shown in Figure 9.22. The anchor nodes are located on the A-Grid-Layo layer, a no plot layer. When you select a node, the properties of the layout curve are shown in the Properties palette shown at right in Figure 9.22. The shortcut menu of the selected

Using and Creating Symbols 611

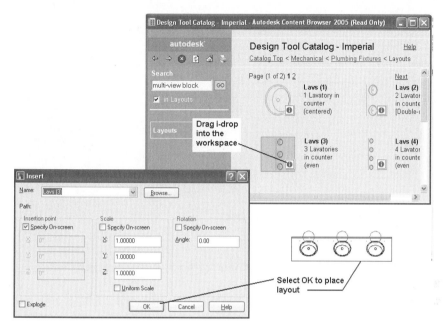

Figure 9.20 *Selecting layout tool from the Content Browser*

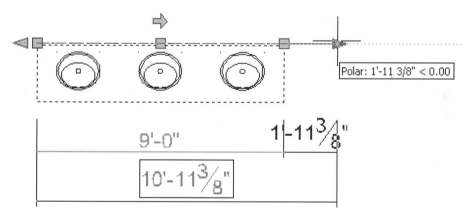

Figure 9.21 *Placing a layout multi-view block from the Design Tool Catalog-Imperial*

node also allows you to add or remove a node and change the layout to manual, repeat, or space evenly. You can select the (+) and (-) grips to edit the layout curve in the workspace as shown in Figure 9.22. (Layout curves are discussed in detail in Chapter 13, "Drawing Commercial Structures.")

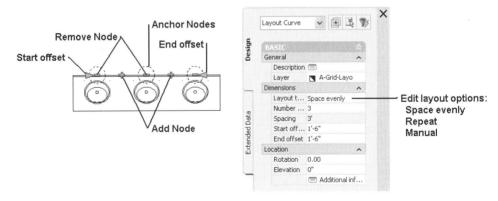

Figure 9.22 Editing a layout multi-view block

VIEWING MULTI-VIEW BLOCKS WITH REFLECTED DISPLAY REPRESENTATION

The majority of the multi-view blocks in Design Tool Catalog include a General display representation, which displays in plan view. Some multi-view blocks are designed as symbols for reflected ceiling plans. The multi-view blocks, listed in Table 9.9, are visible only with the Reflected display configuration and are defined with view blocks that are displayed only when the Reflected display configuration is current. Therefore if the multi-view block is inserted with a Medium Detail display configuration, the symbol will not be displayed.

 Tip: Prior to inserting multi-view blocks with Reflected only display representations, select Reflected from the **Display Configuration** flyout of the Drawing Window status bar.

Design Tool Catalog-Imperial and DesignCenter\Design Locations	Description	Display Representation
Design\Electrical\Lighting	Incandescent and track lights	Reflected
Design\Electrical\Communications	Bells and Speakers	Reflected
Design\Special Construction\Detection and Alarm	Exit Signs, Heat Detector, and Smoke Detector	Reflected

Table 9.9 Multi-view blocks with Reflected display configuration

STOP. Do Tutorial 9.2, "Creating and Modifying Multi-View Blocks," at the end of the chapter.

CREATING MASKING BLOCKS

Some symbols in the DesignCenter and the Design Tool Catalog-Imperial are masking blocks. These symbols can be used to mask or block the display of other Architectural Desktop objects that are located within the masking block boundary. Masking blocks are often used in reflected ceiling plans or other plan views to block the display of ceiling grids or other objects. Masking blocks are in the folders as shown in Table 9.10, and include symbols for ceiling mounted devices. The masking blocks are visible in the Reflected display configuration.

Design Tool Catalog-Imperial and DesignCenter\Imperial\ Design Locations	Description	Display Configuration Required
Mechanical\Air Distribution	Diffusers	Reflected
Electrical\Lighting\Fluorescent	Fluorescent lights	Reflected

Table 9.10 *Folders with masking block content using Reflected display representation*

The masking blocks of the Fluorescent and Air Distribution folders include fluorescent lights and Access Panels, Diffuser Returns, and Diffuser Supplies. If you select the symbol by double-clicking on the symbol in the DesignCenter or dragging the i-drop from the tool catalog, you must use Node object snap to snap to a ceiling grid. These symbols are designed to fit in a suspended ceiling grid. If you drag the symbols to the workspace from the DesignCenter, you are prompted to select a layout node. The Architectural Desktop ceiling grid includes layout nodes; therefore when you select the grid to insert the symbol, it will snap in the grid without the use of object snaps. (Ceiling grids and other Architectural Desktop grids will be discussed in Chapter 13.).

Masking blocks do not automatically block all entities that are within their boundaries. The masking block must be attached to the objects you want to block before it performs the mask function. The fluorescent fixture shown at **p1** in Figure 9.23 was attached to the ceiling grid at **p2**. The resulting mask of the ceiling grid is shown at **p3** on the right in Figure 9.23. The polyline representing the raceway that crosses the fluorescent fixture is not masked by the fixture **p4** because it is not an AEC object.

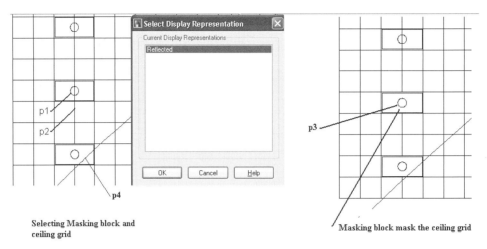

Figure 9.23 *Applying masking blocks*

PROPERTIES OF THE MASKING BLOCK

When you insert a masking block from the DesignCenter, the Properties palette displays the name of the masking block and other properties. The Properties palette can be edited to change the description, scale, and rotation.

Basic

General

    Description – The **Description** button opens the **Description** dialog box, which allows you to add additional text to describe the masking block.

    Definition – The **Definition** edit field lists the name of the masking block. The drop-down list displays all masking blocks previously inserted in the drawing.

Scale

    Specify scale on screen – **Specify scale on screen** edit field includes a yes and no option. If the **No** option is selected, you can type a scale factor in this field or the command line. Scale factors can also be specified by selecting points in the drawing area.

    X – The **X** edit field displays the current scale along the X axis.

    Y – The **Y** edit field displays the current scale along the Y axis.

    Z – The **Z** edit field displays the current scale along the Z axis.

Location

    Specify rotation on screen – The **Specify rotation on screen** edit field includes **Yes** and **No** options. If the **No** option is selected, you type an angle in this field or the command line. Selecting points with the pointer can also specify rotation angles.

Elevation – The multi-view block is inserted at Z=0 coordinate elevation. The Node object snap allows the mask block to be inserted in ceiling grid. The elevation value of existing mask blocks can be changed.

USING PROPERTIES TO EDIT EXISTING MASKING BLOCKS

The Properties palette is used to set the initial settings during insertion of a masking block. Additional edit fields are displayed in the Properties palette when you select an existing masking block such as a ceiling diffuser. The additional edit fields to the Properties palette of an existing mask block reference are described below.

Basic

General

Layer – The **Layer** edit field displays the name of the layer of the masking block.

Location

Additional information – The **Additional information** button opens the **Location** dialog box, which includes the x, y, z coordinates of the masking block, normal extrusion, and rotation. The **Location** dialog box of the masking block is identical to the multi-view block; therefore refer to Figure 9.10 and the description presented in the multi-view blocks section of this chapter.

Extended Data Tab

Hyperlinks – Opens the **Insert Hyperlinks** dialog box, which allows you to define hyperlinks for the masking blocks.

Notes – The **Notes** button opens the **Notes** dialog box, which allows you to type additional notes.

Reference Documents – The **Reference Documents** button opens the **Reference Documents** dialog box, which allows you to display the path and file name of related text or drawing files.

ATTACHING OBJECTS TO MASKING BLOCKS

If the masking blocks have been inserted in the drawing, they perform their masking function if attached to selected Architectural Desktop objects. *Masking blocks will only block the display of selected Architectural Desktop objects when attached to the objects.* Attached objects can also be detached to return the visibility of the object. The **Attach Objects to Mask** command (**MaskAttach**) is used to associate the masking block with the selected Architectural Desktop object. See Table 9.11 for **MaskAttach** command access.

Command prompt	MASKATTACH
Shortcut menu	Select a masking block, right-click, and choose Attach Objects

Table 9.11 *MaskAttach command access*

The command sequence for attaching the masking block to the ceiling grid using the **MaskAttach** command is shown below.

> *(Select a masking block at p1 shown in Figure 9.23, right-click, and choose* **Attach Objects** *from the shortcut menu.)*
>
> Command: MaskAttachSelect AEC entity to be masked: *(Select the ceiling grid at p2 to attach the ceiling grid to the masking block.)*
>
> [1] Masks attached.
>
> *(The* **Select Display Representation** *dialog box opens, listing the current display representation as shown in Figure 9.23; select* **OK** *to dismiss the dialog box and mask the ceiling grid as shown at p3.)*

The result of attaching the ceiling grid to the masking block is shown at right in Figure 9.23. If the fixture is moved to another location, it will continue to block the display of the ceiling grid. The Architectural Desktop objects that are attached to a masking block can be identified by applying the **LIST** command to the masking block.

DETACHING MASKING BLOCKS

Masking blocks can be detached from an Architectural Desktop object by the **Detach Objects** option of the shortcut menu of the masking block. See Table 9.12 for the **Detach Objects** command access.

Command prompt	MASKDETACH
Shortcut menu	Select a masking block, right-click, and choose Detach Objects

Table 9.12 *MaskDetach command access*

The command sequence for detaching a masking block from a wall using the MaskDetach command is shown below.

> *(Select a masking block at p1 in Figure 9.24, with masking applied, right-click, and choose* **Detach Objects***.)*
>
> Command: MaskDetach
>
> Select AEC object to be detached: *(Select the ceiling grid at p2.)*
>
> 1 mask(s) detached.

Detaching the block returns the visibility of the ceiling grid through the symbol, as shown at right in Figure 9.24.

CREATING MASKING BLOCKS

Masking blocks are included in the Design Tool Catalog and DesignCenter; however, any closed polyline can be used to create a masking block. Custom masking blocks are created in the **Style Manager**.

STEPS TO CREATING A MASKING BLOCK

1. Select **Format>Style Manager**.
2. Double-click on the **Multi-Purpose Objects** folder in the left pane.
3. Select the **Mask Block Definitions** folder in the left pane, right-click, and choose **New**.
4. Overtype **MyMaskBlock** as the name of the masking block.
5. Select the name of the new masking block, and select **Set From** on the **Style Manager** toolbar. Respond to the command line prompts as shown below.

   Command: _AecStyleManager

   Select a closed polyline: *(Select the polyline at p1 as shown in Figure 9.25.)*

   Add another ring? [Yes/No] <No>: ENTER *(Press* ENTER *to end selection.)*

   Insertion base point: *(Select a point near p2 as shown in Figure 9.25.)*

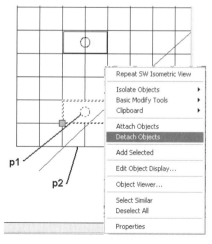
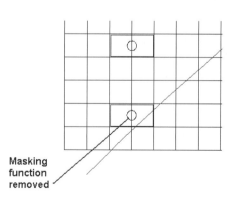

Figure 9.24 *Detaching the masking block from the ceiling grid*

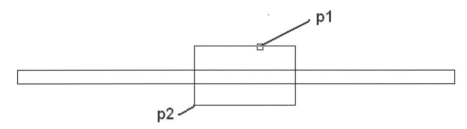

Figure 9.25 *Selecting a closed polyline for the mask block*

Select additional graphics *(Select additional entities or press* ENTER *to end selection.)*

6. Double-click on the new name of the masking block to open the **Mask Block Definition Properties** dialog box. Select tabs to edit the properties of the mask block.

In the command sequence above, you are prompted to select a closed polyline, an insertion point, and finally additional graphics to define the masking block. After selecting the first polyline you are prompted: **Add another ring?** If you respond Yes, additional polylines can be added to the mask block. The additional polylines can be defined as a void. Polylines defined as a void will be subtracted or form a hole, and that boundary will not block the display of Architectural Desktop objects. The next prompt is to select an insertion point. The insertion point can be located on or off the polylines. The final prompt is to select additional graphics, which can include any AutoCAD entity. The additional graphics prompt allows you to include text as part of the mask block.

SETTING DEFINITION PROPERTIES OF A NEW MASK BLOCK

After a new mask block has been created in the **Style Manager**, double-click on the name of the new mask block to open the **Mask Block Definition Properties** dialog box. The **Mask Block Definition Properties** dialog box consists of three tabs: **General**, **Classifications**, and **Display Properties**. The tabs of the **Mask Block Definition Properties** dialog box are described below.

General Tab

The **General** tab includes **Name** and **Description** edit fields, as shown in Figure 9.26.

Name – The **Name** edit field allows you to change the name of the masking block.

Description – The **Description** field allows you to enter a description.

Notes – Selecting the **Notes** button opens the **Notes** dialog box for adding text.

Property Sets – The **Property Sets** button opens the **Edit Property Set Data** dialog box. The **Edit Property Set Data** dialog box includes the **Add property sets** and **Remove property sets** buttons to add or remove related property sets included in the drawing. The **Add property sets** button is active only after the masking block has been inserted and related property sets are included in the drawing.

Classifications Tab

The **Classifications** tab allows you to specify classifications of building components such as the status of construction as shown in Figure 9.26.

Classfication Definition – The **Classification Definition** includes a list of classifications defined in the drawing. Classifcation definitions can be imported in the **Style Manager** from the Multi-Purpose Objects\Classification Definitions folder. The Uniformat II Classifications (1997 ed) .dwg is included in the *C:\Documents and Settings\All Users\Application Data\Autodesk\ADT 2005\enu\Styles\Imperial* directory.

Classification – The classifications included in the drawing are listed in the drop-down list.

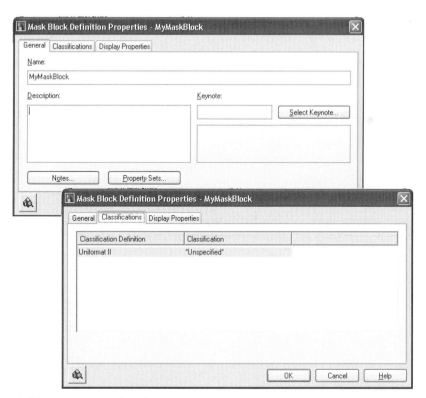

Figure 9.26 *General and Classifications tabs of the Mask Block Definition Properties*

Display Properties Tab

The **Display Properties** tab shown in Figure 9.27 allows you to set the display of components of the masking block. The options of the tab are as follows:

> **Display Representations** – The **Display Representations** column lists the display representations available for the masking block.
>
> **Display Property Source** – The **Display Property Source** column lists drawing default global control or a style override.
>
> **Style Override** – The **Style Override** check box indicates if an override has been defined in the display properties. If this box is clear, the **Drawing Default** is controlling the display of the stair style.
>
> **Edit Display Properties** – The **Edit Display Properties** button opens the **Display Properties** dialog box. The **Display Properties** dialog box consists of the **Layer/Color/Linetype** tab, which allows you to edit the visibility, color, linetype, lineweight and linetype scale for the Boundary Profile or Additional Graphics components for the display representation.

INSERTING NEW MULTI-VIEW AND MASKING BLOCKS

Masking blocks of Architectural Desktop will be inserted from the Design Tool Catalog; however, new masking blocks or imported masking blocks will not be included in the tool catalog. You can open the **Style Manager** and drag a new masking block from the **Style Manager** to a tool palette. The masking block can then be inserted into the drawing from the tool palette. Masking blocks can also be inserted

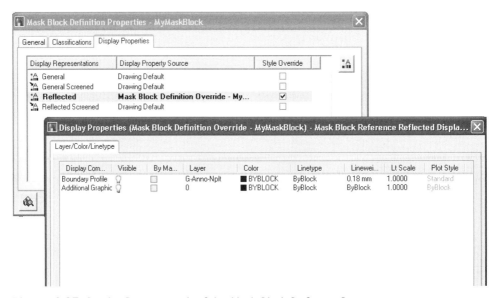

Figure 9.27 *Display Properties tab of the Mask Block Definition Properties*

in the drawing by choosing the **Mask Block** tool from the Document tool group, Drafting palette as described in Table 9.13. When you select **Mask Block** from the drafting tool palette, specify the new Definition in the Properties palette.

Command prompt	MASKADD
Tool Palette	Select the Document tool palette group, and choose Mask Block from the Drafting tool palette

Table 9.13 *MaskAdd command access*

 STOP. Do Tutorial 9.3, "Accessing Mask Blocks from the DesignCenter and Creating Mask Blocks," at the end of the chapter.

CREATING SYMBOLS FOR THE DESIGNCENTER

New blocks, multi-view blocks, and masking blocks can be created in a drawing file; however, the layer and scale of the symbols are not preset. The **Create AEC Content Wizard** can be used for adding symbols, drawings, or custom commands to the DesignCenter. The **Create AEC Content Wizard** includes options to add each block of the current drawing to the contents of the AutoCAD Architectural Desktop Content. Options of the **Create AEC Content Wizard** allow you to predefine the scale, rotation, attribute text angle, and layer. Placing the symbol on a predefined layer improves a drawing's compliance with office layer standards. See Table 9.14 for **Create AEC Content** command access.

Menu bar	Format>AEC Content Wizard
Command prompt	CREATECONTENT

Table 9.14 *Create AEC Content command access*

Selecting the **CreateContent** command opens the **Create AEC Content Wizard**, as shown in Figure 9.28.

The **Create AEC Content Wizard** consists of three pages: **Content Type**, **Insert Options**, and **Display Options**. You can display each page by selecting the **Next** or **Back** button located at the bottom of each page. Each page is identified in the upper left corner.

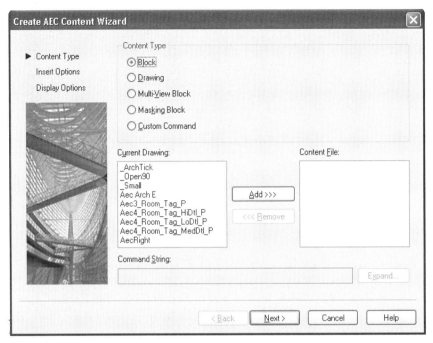

Figure 9.28 *Content Type page of the Create AEC Content Wizard*

Content Type

The **Content Type** page consists of the following options:

> **Content Type** – The **Content Type** section allows you to specify the type of content you want to add to the DesignCenter: **Block**, **Drawing**, **Multi-View Block**, **Masking Block**, and **Custom Command**. When you select a content type, all named objects of that type will be displayed in the **Current Drawing** list box. In Figure 9.28, the **Block** type is selected; therefore the blocks of the drawing are displayed in the **Current Drawing** list.
>
> **Current Drawing** – The **Current Drawing** list includes objects in the drawing classified according to the content type specified in the **Content Type** section.
>
> **Content File** – The **Content File** list includes objects of the specified content type that have been exported to the AutoCAD Architectural Desktop Content.
>
> **Add/Remove** – The **Add** and **Remove** buttons allow you to add or delete the content from the **Current Drawing** and **Content File** lists. Content must be added to the **Current Drawing** list to complete the **Create AEC Content Wizard** operations.

Command String – The **Command String** edit field allows you to enter a command string and command responses to execute a command in AutoCAD. Selecting the command string icon created for the DesignCenter will execute the command. The **Expand** button allows you to enter up to 255 characters in the command string.

Insert Options

The **Insert Options** page shown in Figure 9.29 allows you to predefine the scale, elevation, layer, and attribute properties of the content.

Explode on Insert – The **Explode on Insert** check box, if selected, will apply the **EXPLODE** command to the block upon insertion.

Preset Elevation – The value entered in the **Preset Elevation** field presets the Z coordinate value of the content.

Anchor Type – The **Anchor Type** drop-down list allows you to attach the object to one of the following: cell, curve, leader, node, space boundary, tag, volume, or wall.

Scale

X, Y, Z – The X, Y, and Z **Scale** edit fields allow you to preset the scale value in the X, Y, Z directions of the content upon insertion.

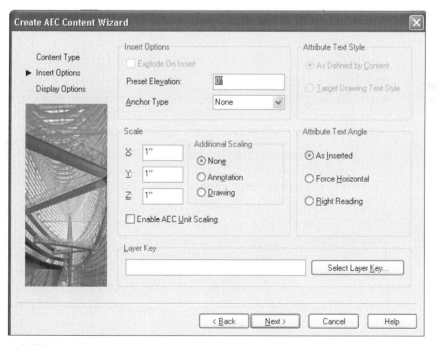

Figure 9.29 *Insert Options page of the Create AEC Content Wizard*

ADDITIONAL SCALING

None – The **None** option scales a multi-view block at the scale defined in the **X, Y, Z** edit fields. Select the **None** option if the block was drawn actual size and the units of the multi-view block match the units of target drawings.

Annotation – The **Annotation** option applies the **Annotation Plot Size** factor to multi-view blocks. If a multi-view block was created 1 unit long, when the annotation plot size is set at 1/8" and the scale factor set at 48, the multi-view block would be inserted 6" long. The actual multi-view block dimension is multiplied by the annotation factor and the drawing scale factor.

Drawing – The **Drawing** scale factor adjusts the scale of the multi-view block by multiplying the drawing scale factor of the drawing by the multi-view block size. Table 9.15 indicates the scaling effects of each option given the annotation plot size and the drawing scale factor.

Additional Scaling	Multi-View Block Size	Annotation Plot Size	Drawing Scale Factor	Size when Inserted
None	1	1/8"	48	1
Annotation	1	1/8"	48	6 (1 x 1/8 x 48)
Drawing	1	1/8"	48	48 (1 x 48)

Table 9.15 *Effects of additional scaling options*

Enable AEC Unit Scaling – The **Enable AEC Unit Scaling** check box applies scaling factors to blocks.

Attribute Text Style – The **Attribute Text Style** toggles allow you to select the text style for the attribute text to be either the text style as defined in the block or the default text style of the target drawing.

Attribute Text Angle – The **Attribute Text Angle** toggles allow you to choose the text orientation of attribute text. The attribute text angle options are **As Inserted**, **Force Horizontal**, or **Right Reading**.

Select Layer Key – The **Select Layer Key** button allows you to specify the layer for the block to be inserted on. Selecting the **Select Layer Key** button will open the **Select Layer Key** dialog box shown in Figure 9.30. It includes a list of the layer keys, names, and descriptions for the layer standard used in the drawing. A layer key is a name assigned to the layer in Architectural Desktop. The name of the layer displayed in the Properties palette will be the name displayed in the **Layer Name** column.

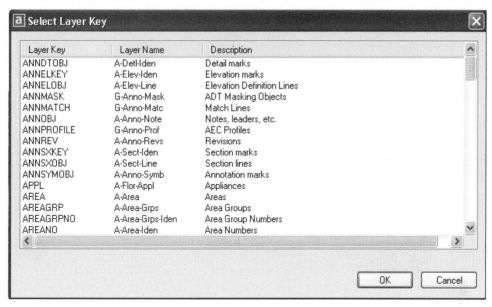

Figure 9.30 *Layer Key list in the Select Layer Key dialog box*

Display Options

The Display Options page, shown in Figure 9.31, includes options to define the location of the symbol within the DesignCenter and the icon used to represent the symbol.

File Name – The **File Name** edit field allows you to enter the path and file name for the symbol. Selecting the **Browse** button will open the **Save Content File** dialog box, which allows you to easily specify the name and path for the symbol to be placed. The **File Name** is required and must be specified to create the AEC Content.

Icon – The **Icon** for the symbol can be specified as a new icon or default icon. The default icon is the preview graphics for the file of the symbol. The **New Icon** button allows you to select a bitmap file to be used as the symbol.

Detailed Description – The **Detailed Description** is an edit field for entering text that describes the block. The text entered will be displayed in the description field of the DesignCenter when the symbol is selected.

Save Preview Graphics – The **Save Preview Graphics** check box saves the last view of the symbol drawing as the preview of the symbol.

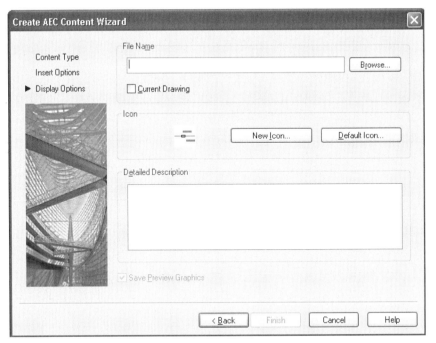

Figure 9.31 *Display Options page of the Create AEC Content Wizard*

SUMMARY

1. The scale for symbols and annotation is set with the **AecDwgSetup** command.
2. Define a layer standard for the drawing by selecting the **Layering** tab of the **Drawing Setup** dialog box.
3. Symbols are inserted in the drawing from the **Custom** view of the DesignCenter.
4. The content path is set in the **AEC Content** tab of the **Options** dialog box.
5. Multi-view blocks include display control based upon the viewing direction and the display representation.
6. View blocks can be created from other multi-view blocks or blocks.
7. The **Multi-View Block Reference Properties** dialog box allows you to edit the location of the insertion point, insert attribute data, scale, rotate, and edit the style of the multi-view block.
8. Display representations for multi-view blocks can be set to General, Model, or Reflected.
9. Multi-view blocks can be imported from other drawings or exported to other drawings.

10. Mask blocks are created to block the display of other Architectural Desktop objects.
11. Mask blocks are created from closed polylines.
12. Mask blocks must be attached to other Architectural Desktop objects to perform their blocking function.
13. The **AEC Content Wizard** is used to add blocks, drawings, multi-view blocks, masking blocks, and custom commands to the DesignCenter.
14. The **AEC Content Wizard** allows you to define the scale, rotation, layer, icon, and description of the object added to the DesignCenter.

REVIEW QUESTIONS

1. The scale and layering standard of a drawing can be defined with the _____ command.
2. Setting the scale of a drawing establishes the _____ of dimensions.
3. The default layer key style included Architectural Building Model and View (Imperial-ctb) template is _____.
4. Multi-view blocks can be created from AutoCAD _____.
5. The name of a multi-view block is displayed in the _____ edit field of the Properties palette.
6. The attributes of a multi-view block can be entered in the _____ section of the Properties palette.
7. If the Insertion offset values are edited for a multi-view block, the values should be changed for _____ _____ _____.
8. The _____ folder of the DesignCenter allows you to insert multiple lavatories in a counter spaced evenly.
9. The majority of multi-view blocks are displayed with the _____ display representation.
10. Multi-view and mask blocks designed for ceiling grids are displayed with the _____ display representation.
11. Blocks created to block the display of other Architectural Desktop objects are _____ blocks.
12. The **AEC Content Wizard** allows you to preset the _____, _____, and _____ of the block.

TUTORIAL 9.1 INSERTING MULTI-VIEW BLOCKS WITH PRECISION

1. Open Autodesk Architectural Desktop 2005, and select **File>Project Browser** from the menu bar. Use the Project Selector drop-down to navigate to your student directory. Double click on **Ex 9-1** to set this project current. Select **Close** to dismiss the **Project Browser**. (If your student folder does not include the project, refer to "Organizing Tutorial Directories" in the Preface.)

2. Select the **Constructs** tab of the **Project Navigator**. Double-click on **Floor 1** to open the Floor 1 drawing.

3. Select **Zoom Window** from the **Navigation** toolbar and respond to the following command line prompts.

   Command: '_zoom

   Specify corner of window, enter a scale factor (nX or nXP), or

   [All/Center/Dynamic/Extents/Previous/Scale/Window] <real time>: _w

   Specify first corner: **135',60'** ENTER

   Specify opposite corner: **145',50'** ENTER

4. Right-click over the OSNAP button on the status bar; then choose **Settings** from the shortcut menu. Select the **Clear All** button, and then select the **Endpoint** and **Intersection** object snap modes. Select **OK** to close the **Drafting Settings** dialog box.

5. Toggle POLAR and OTRACK ON in the status bar. Turn OFF SNAP and GRID.

6. Right-click over the title bar of the Design tool palette, and choose **New Palette**. Overtype the name **Accessing** to name the new palette.

7. Select **Content Browser** from the **Navigation** toolbar. Open the Design Tool Catalog -Imperial and expand Mechanical>Plumbing Fixtures\Toilet folder as shown in Figure 9T.1.

8. Drag the i-drop of **Tank 1** onto the new Accessing tool palette. Select **Tank 1** from the tool palette.

9. Move the mouse to the drawing area to determine the default symbol rotation, as shown in Figure 9T.1.

10. Move the pointer over the Properties palette, click in the **Specify rotation on screen** edit field, and set it to **No**, click in the **Rotation** edit field, and type **180**. Verify that **Specify Scale on Screen** is set to **No**.

11. Move the pointer over the workspace, select point **p1** at the edge of the window shown in Figure 9T.1 using the Endpoint object snap mode to locate the toilet. Press ESC to end insertion.

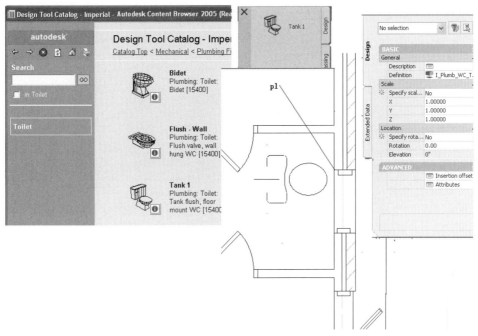

Figure 9T.1 *Default rotation of the symbol when placing Tank 1*

12. Select the **Content Browser** from the Drawing Window status bar, and open the Mechanical>Plumbing Fixtures>Lavatory category of the Design Tool Catalog. Navigate to page 2 of the catalog, and drag the i-drop for the Oval-3D symbol onto the Accessing tool palette.

13. Select the Oval-3D tool from the tool palette, and move the pointer to the drawing area to identify the default rotation.

14. Move the point to the Properties palette, click in the **Specify rotation on screen** edit field of the Properties palette, and select **No** from the drop-down list. Click in the **Rotation** edit field, and type **180**.

15. Move the pointer over the workspace, pause over **p2** as shown in Figure 9T.2 until the Endpoint osnap marker is displayed, and then click to select the location of the lavatory. Press ESC to end insertion.

16. Select the **Content Browser** from the Drawing Window status bar, and open the Mechanical>Plumbing Fixtures>Bath category of the Design Tool Catalog. Drag the i-drop for the Tub 30x60 symbol onto the tool palette created in this tutorial. Select the Tub 30x60 tool from the tool palette.

17. Move the pointer over the workspace to determine default rotation. In the Properties palette, click in the Specify rotation on screen edit field and select No. Click in the Rotation edit field and type –90.

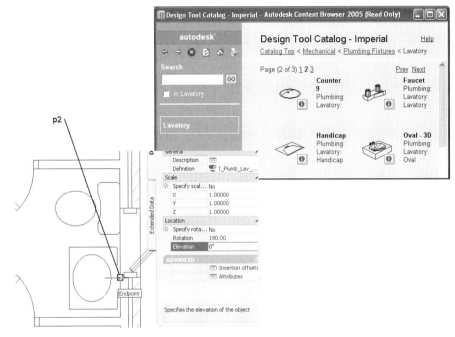

Figure 9T.2 *Selecting Endpoint for placement of lavatory*

18. Select the Insertion offsets button of the Advanced section of the Properties palette to open the Multi-view Block Offsets dialog box. To edit the insertion point, edit the Y Offset field for each of the view blocks to 15 as shown in Figure 9T.3. Select OK to close the Multi-view Block Offsets dialog box.

19. Insert the tub using the Endpoint object snap at p1 as shown in Figure 9T.4. Press ESC to end insertion.

20. Select **Zoom Window** from the **Zoom** flyout of the **Navigation** toolbar. Respond to the command line prompts as shown below.

    Command: z ZOOM

    Specify corner of window, enter a scale factor (nX or nXP), or

    [All/Center/Dynamic/Extents/Previous/Scale/Window] <real time>: w

    Specify first corner: **86',60'** ENTER *(Type coordinates for the first corner of the zoom window.)*

    Specify opposite corner: **108',36'** ENTER *(Type coordinates for the oppopsite corner of the zoom window.)*

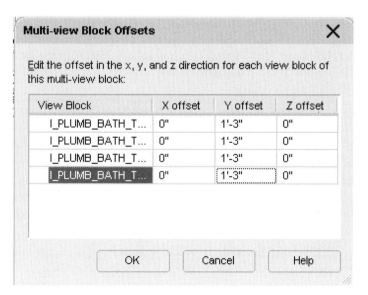

Figure 9T.3 *Edit of insertion Y offset values*

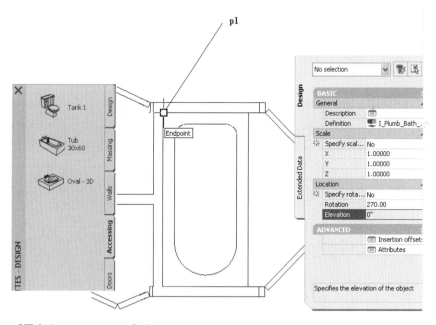

Figure 9T.4 *Insertion point of tub*

21. Select the Design Tool Catalog-Imperial from the Windows task bar, and open the Equipment\Food Service\Refrigerator category. Drag the **Top-Bottom Refrigerator 31x28** i-drop onto the palette created in this tutorial. Select the Top-Bot 31x28 tool from the tool palette and insert the refrigerator as shown in Figure 9T.5

22. Select **Content Browser** from the **Navigation** toolbar, and open the Furnishing\Casework\Base Cabinet category. Drag the **15 in Wide** i-drop onto the palette created in this tutorial. Select the **15 in Wide** tool from the tool palette and insert the cabinet as shown in Figure 9T.5.

23. Select **Content Browser** from the **Navigation** toolbar, and open the Equipment\Food Service\Range category. Drag the **Range 30x26** i-drop onto the palette created in this tutorial. Select the **Range 30x26** tool from the tool palette and insert the range as shown in Figure 9T.5.

24. Select **Content Browser** from the **Navigation** Content Browser from Navigation toolbar, and open the Furnishing\Casework\Base Cabinet category. Drag the **21 in Wide** i-drop onto the palette created in this tutorial. Select the **21 in Wide** tool from the tool palette and insert the cabinet (rotate –90) as shown in Figure 9T.5.

25. Select **Content Browser** from the **Navigation** toolbar, and open the Furnishing\Casework\Base Cabinet category. Drag the **36 in Wide** i-drop

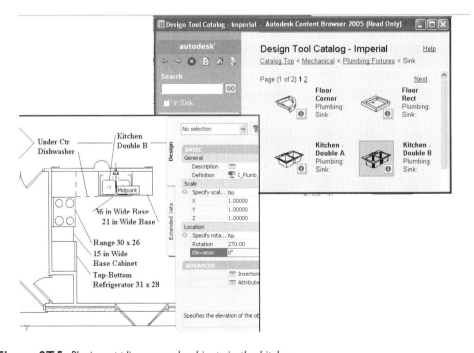

Figure 9T.5 *Placing appliances and cabinets in the kitchen*

onto the palette created in this tutorial. Select the **36 in Wide** tool from the tool palette and insert the cabinet (rotate –90) as shown in Figure 9T.5

26. Select **Content Browser** from the **Navigation** toolbar, and open the Equipment\Food Service\Dishwasher category. Drag the **Under Ctr** dishwasher i-drop onto the palette created in this tutorial. Select the **Under Ctr** dishwasher tool from the tool palette and insert it, rotate –90, as shown in Figure 9T.5

27. Select **Content Browser** from the **Navigation** toolbar, and open the Mechanical\Plumbing Fixtures\Sink category. Drag the **Kitchen Double B** i-drop onto the palette created in this tutorial. Select the **Kitchen Double B** tool from the tool palette, edit its rotation to –90, SHIFT+ right–click, and choose the Midpoint object snap mode and insert the sink at the midpoint of the 36" wide base as shown in Figure 9T5.

28. Open the Design Tool Catalog-Imperial>Walls>Casework category. Drag **Casework - 36 (Counter)** onto the palette created in this tutorial. To import the style into the drawing, select the **Casework - 36 (Counter)**, and click in the drawing area to begin the wall. However, press ESC to end the command.

29. Select **Format>Style Manager** from the menu bar. Expand Architectural Objects>Wall Styles folder and double-click on the **Casework 36 (Counter)** wall style.

30. Select the **Endcaps/Opening Endcaps** tab, change the **Wall Endcap Style** to **Standard**. To adjust the height of the counter, select the **Components** tab and edit **Case-Counter** Bottom Elevation Offset = **2' 10 1/2"** from **Baseline**.

31. Select **OK** to close the **Wall Style Properties** dialog box. Select **OK** to close the **Style Manager**.

32. Select **Casework - 36 (Counter)** tool from the Accessing palette. Edit the Properties palette to Style = **Casework 36 (Counter)**, Base Height = **8'-0"**, Justify = **Left**. Set the Cleanup Group definition = **Standard**. Draw the counter top wall segments as shown below.

    Command: WallAdd

    (Edit the Style = Casework 36 Counter, Cleanup group definition = Standard.)

    Start point or [STyle/Group/WIdth/Height/OFfset/Justify/Match/Arc]: *(Select p1 shown in Figure 9T.6.)*

    End point or [STyle/Group/WIdth/Height/OFfset/Justify/Match/Arc]:]: *(Select p2 shown in Figure 9T.6.)*

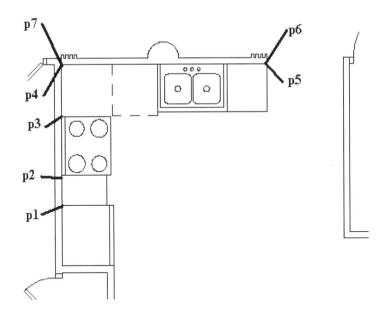

Start and end of counter tops

Figure 9T.6 *Selection points for counter top wall style*

End point or [STyle/Group/WIdth/Height/OFfset/Justify/Match/Arc/Undo]: ENTER *(Press* ENTER *will end the command; then the next* ENTER *will repeat the last command.)*

Command: WALLADD

Start point or [STyle/Group/WIdth/Height/OFfset/Justify/Match/Arc]: *(Select p3 shown in Figure 9T.6.)*

End point or [STyle/Group/WIdth/Height/OFfset/Justify/Match/Arc]: *(Select p4 shown in Figure 9T.6.)*

End point or [STyle/Group/WIdth/Height/OFfset/Justify/Match/Arc/Undo]:]: *(Select p5 shown in Figure 9T.6.)*

End point or [STyle/Group/WIdth/Height/OFfset/Justify/Match/Arc/Undo/Close/ORtho] ENTER ENTER *(Press* ENTER *will end the command; then the next* ENTER *will repeat the last command.)*

33. Open the Design Tool Catalog>Walls>Casework category. Drag **Casework - 42 (Bar)** into the drawing to start the **WallAdd** command. Respond to the command line prompts to create the bar counter top. Edit the Properties palette: set Style = **Casework 42 (Bar)** and Justify = **Left**.

Command: WALLADD

Start point or [STyle/Group/WIdth/Height/OFfset/Justify/Match/Arc]: *(Select p6 shown in Figure 9T.6.)*

End point or [STyle/Group/WIdth/Height/OFfset/Justify/Match/Arc]: *(Select p7 shown in Figure 9T.6.)*

34. Select **Layer Manager** from the **Properties** toolbar, and freeze the **A-Roof** layer. Select **OK** to close the **Layer Manager**.

35. Since the construct will be viewed using the High Detail and 1/4" = 1'-0", select the **High Detail** display configuration from the Drawing window display configuration flyout. The Casework 36 (Counter) and Casework 42 (Bar) wall styles are hatched.

36. To turn off the hatch of the Casework 36 (Counter) wall style, select the Casework 36 (Counter), right-click, and choose **Edit Wall Style**. Select the **Display Properties** tab, select the **Edit Display Properties** button, and then select the **Layer/Color/Linetype** tab. Toggle off the **Hatch 1 (Case-Counter)** and **Hatch 2 (Case-Backsplash)**. Select **OK** to close all dialog boxes.

37. To turn off the hatch of the Casework 42 (Bar) wall style, select the Casework 42 (Bar), right-click and choose **Edit Wall Style**. Select the **Display Properties** tab, click the **Edit Display Properties** button, and then select the **Layer/Color/Linetype** tab. Toggle off the **Hatch 1 (Stud)** and **Hatch 2 (Case-Counter)**. Select **OK** to close all dialog boxes.

38. Select **SE Isometric** from the **View** flyout of the **Navigation** toolbar.

39. Select the **Kitchen Double B** sink, right-click, and choose **Properties**. Edit the **Elevation** edit field to 37".

40. Select **Hidden** from the **Shade** flyout of the **Navigation** toolbar to view the complete kitchen as shown in Figure 9T.7.

41. Save and close the drawing.

42. Select **File>Project Browser** from the menu bar. Select **Ex 9-1** from the project list of the **Project Browser**, right-click, and choose **Close Current Project**.

TUTORIAL 9.2 CREATING AND MODIFYING MULTI-VIEW BLOCKS

1. Open the *ADT Student\ADT Tutor\Ch9***EX 9-2.dwg**.

2. Choose **File>SaveAs** from the menu bar and save the drawing as **Lab 9-2** in your student directory.

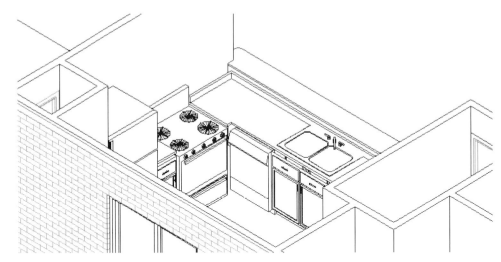

Figure 9T.7 SE Isometric view of cabinets

3. Select **1/4"=1'-0"** from the **Drawing Scale** flyout of the Drawing Window status bar.

4. Right-click over the OSNAP button on the status bar; choose **Settings**, check **Nearest** object snap mode, and clear all other object snap modes from the **Drafting Settings** dialog box. Select **OK** to dismiss the **Drafting Settings** dialog box.

5. Right-click over the title bar of the Design tool palette, and choose **New Palette**. Overtype the name **Electrical** to name the new palette.

6. Select **Content Browser** from the **Navigation** toolbar. Open the Design Tool Catalog-Imperial and open the Electrical\Power folder. Open the Switch palette as shown in Figure 9T.8.

7. Drag the **Single Pole switch** symbol onto the new custom palette named Electrical. Select the **Single Pole switch** tool from the palette and move the pointer over the drawing area to determine the default rotation angle.

8. Edit the Properties palette: select **No** for **Specify rotation on screen** in the Properties palette. Type **90** in the rotation edit field of the Properties palette.

9. Move the pointer over the workspace, and select a location along the bottom wall using the Nearest object snap near p1 as shown in Figure 9T.8. Press ESC to end the command.

10. Open the Design Tool Catalog-Imperial and open the Electrical\Power folder. Open the Receptacle palette and drag the 220 Recpt symbol onto the Electrical tool palette. Select the **220 Recpt** symbol from the tool palette to open the Properties palette. Edit the Rotation to **90** in the Properties palette.

Using and Creating Symbols 637

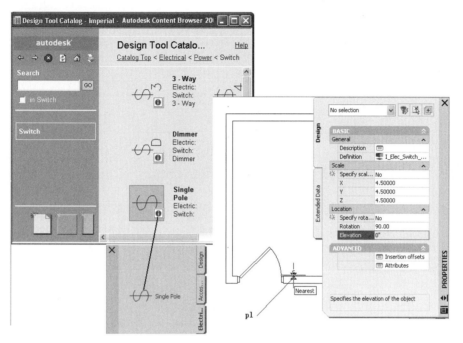

Figure 9T.8 *Selecting Single Pole Switch from the Design Tool Catalog-ImperialCenter*

11. Select a location for the **220 Recpt** near **p1** as shown in Figure 9T.9 along the bottom wall using the Nearest object snap. Press ESC to end the command.

12. Select **Reflected** from the **Display Configuration** flyout of the Drawing Window status bar to change the display of the building. (The receptacle and switch are not displayed in the Reflected display configuration.)

13. Open the Electrical>Lighting>Incandescent category of the Design Tool Catalog. Drag the **Ceiling** symbol onto the Electrical tool palette of the drawing. Select the **Ceiling** symbol from the tool palette and place the symbol near the middle of the room as shown in Figure 9T.10.

14. To edit the Ceiling light symbol display representation, select the **Ceiling** symbol, right-click, and choose **Edit Multi-view Block Definition** from the shortcut menu to open the **Multi-view Block Definition Properties** dialog box.

15. Select the **View Blocks** tab, and select the **Reflected** display configuration to verify that the name of the view block used in plan view is **I_ELEC_INC_CEILING_P**.

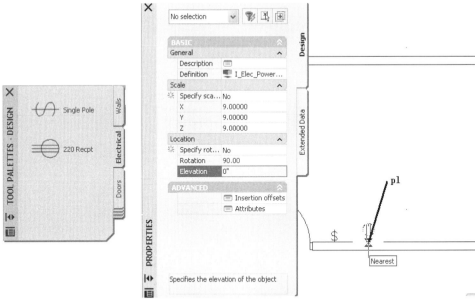

Figure 9T.9 *Location of 220 Recpt symbol*

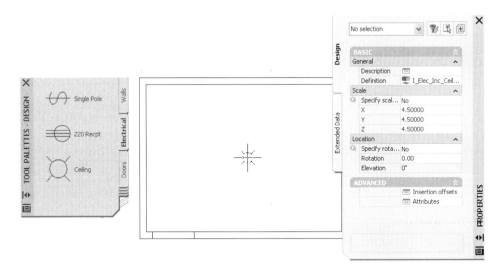

Figure 9T.10 *Incandescent light symbol placed near center of room*

16. Verify that the **View Blocks** tab is current as shown in Figure 9T.11. Select the **General** display representation at left, and select the **Add** button to

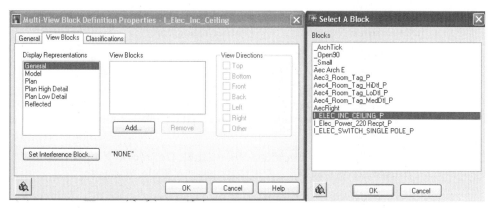

Figure 9T.11 *View block added to the General display representation*

open the **Select A Block** dialog box. Select the **I_ELEC_INC_CEILING_P** block from the **Select A Block** dialog box as shown in Figure 9T.11. (Note, the blocks listed in the **Select A Block** dialog box of your drawing may include additional blocks not shown in Figure 9T.11.) Select **OK** to close the **Select a Block** dialog box. Verify that all **View** Directions are checked ON for the **General** display configuration. Select **OK** to close the **Multi-view Block Definition Properties** dialog box.

17. Select **Medium Detail** from the **Display Configuration** flyout of the Drawing Window status bar to change the display in the viewport. (The receptacle and switch are now displayed with the ceiling light.)

18. To create a circle for a view block, select **Circle**, **Center**, **Radius** from the **Shapes** toolbar. Respond to the command line prompts as shown below.

    Command: _circle

    Specify center point for circle or [3P/2P/Ttr (tan tan radius)]: **33',33'** ENTER

    Specify radius of circle or [Diameter] <0'-6">: **6** ENTER

19. To create a view block, select **Format>Blocks>Block Definition** from the menu bar. Type **12Diffuser** in the **Name** edit field, in **X** insertion point type **33'** and in the **Y** insertion point type **33'**. Select the **Select Objects** button, and then select the circle drawn in Step 18. Press ENTER to end object selection, and then select **OK** to dismiss the **Block Definition** dialog box.

20. Select **Format>Multi-View Block>Multi-View Block Definitions** from the menu bar.

21. Select the **Multi-view Block Definitions** folder in the left pane for Lab 9-2, right-click, and choose **New** from the shortcut menu. Overtype the name **12_Diffuser** as the name of the new multi-view block.

22. Double-click on the **12_Diffuser** name to open the **Multi-View Block Definition Properties – 12 Diffuser** dialog box.

23. Verify that the **View Blocks** tab is current as shown in Figure 9T.12, select the **General** display configuration, and select the **Add** button to open the **Select a Block** dialog box. Select **12diffuser** block from the block list. Select **OK** to close the **Select a Block** dialog box. Verify that all view directions are checked for the **General** display representation and **12diffuser** view block as shown in Figure 9T.12.

24. Select **OK** to dismiss the **Multi-View Block Definition Properties – 12 Diffuser** dialog box. Select **OK** to dismiss the **Style Manager**.

25. Select the **Ceiling** multi-view block, right-click, and choose **Add Selected** from the shortcut menu. Select the **Definition** edit field of the Properties palette, and choose **12 Diffuser** from the drop-down list of multi-view blocks. Select the **Specify scale on screen** edit field and choose **Yes**. Respond to command line prompts as follows:

    Command: MvBlockAddSelected

    Insert point or [Name/X scale/Y scale/Z scale/Rotation/Match]: *(Definition of multi-view block selected in the Properties palette.)*

    Insert point or [Name/X scale/Y scale/Z scale/Rotation/Match]: *(Edit the Properties palette: Specify scale on Screen = Yes)*

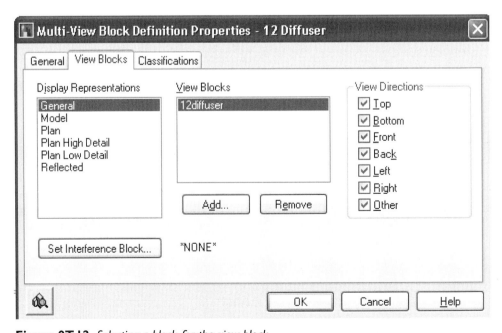

Figure 9T.12 *Selecting a block for the view block*

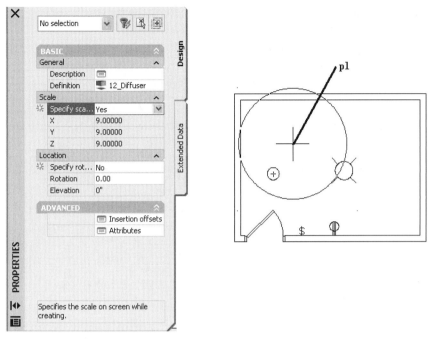

Figure 9T.13 *Location of 12 Diffuser multi-view block*

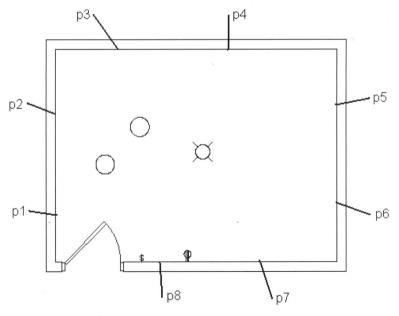

Figure 9T.14 *Locations of receptacles*

Insert point or [Name/Rotation/Match]: *(Select the location for the diffuser near p1 as shown in Figure 9T.13.)*

Scale or [Name/Rotation/Match/Undo]: **1** ENTER *(Scale specified in the command line.)*

Insert point or [Name/Rotation/Match]: ENTER *(Press* ENTER *to end the command.)*

26. To complete the tutorial, expand the Electrical\Power\Receptacle folder and drag the **Duplex Recept** i-drop symbol onto the Electrical tool palette. Select the **Duplex Recept** symbol from the Electrical tool palette and add the symbol at points **p1** to **p8** as shown in Figure 9T.14.

27. Save and close the drawing.

TUTORIAL 9.3 ACCESSING MASK BLOCKS FROM THE DESIGNCENTER AND CREATING MASK BLOCKS

1. Open your *ADT Student\ADT Tutor\Ch9\EX9-3.dwg*.

2. Choose **File>SaveAs** from the menu bar and save the drawing as **Lab 9-3** in your student directory.

3. Choose **1/4"=1'-0"** from the Drawing Window status bar **Scale** flyout.

4. Move the pointer over the OSNAP button on the status bar, right-click, and choose **Settings**. Select the **Intersection** and **Node** object snap modes, and clear all other object snap modes. Select **OK** to dismiss the **Drafting Settings** dialog box. Verify the OSNAP button on the status bar is pushed in.

5. Select **Format>Style Manager** from the menu bar.

6. Double-click to expand the **Multi-Purpose Objects** folder in the left pane.

7. Select **Mask Block Definitions** from the **Multi-Purpose Objects** folder, right-click, and choose **New** from the shortcut menu. Overtype **Speaker** as the name of the new mask block.

8. Select **Speaker**, the new mask block name, in the right pane, right-click, and choose **Set From** from the shortcut menu. Respond to the command line prompts as shown below to select the geometry for the mask block.

    Command: _AecStyleManager

    Select a closed polyline: *(Select the rectangle at p1 shown in Figure 9T.15.)*

    Add another ring? [Yes/No] <No>: ENTER *(Press* ENTER *to respond No.)*

Insertion base point: *(Select the lower left corner of the rectangle at p2 shown in Figure 9T.15 using the Intersection object snap mode.)*

Select additional graphics: 1 found *(Select the rectangle at p3, shown in Figure 9T.15.)*

Select additional graphics: 1 found, 2 total *(Select the hexagon at p4, shown in Figure 9T.15.)*

Select additional graphics: 1 found, 3 total *(Select the letter S at p5, shown in Figure 9T.15.)*

Select additional graphics: ENTER *(Press ENTER to end selection.)*

9. Select **OK** to close the **Style Manager**.
10. Select **Zoom Extents** from the **Zoom** flyout of the **Navigation** toolbar.
11. Select **DesignCenter** from the **Navigation** toolbar.
12. Expand Imperial\Design\Electrical\Lighting\Fluorescent of the DesignCenter.
13. Drag the **2x4** fluorscent light symbol from the DesignCenter to the workspace. Release the left mouse button over the workspace. Respond to the command line prompt as shown below.

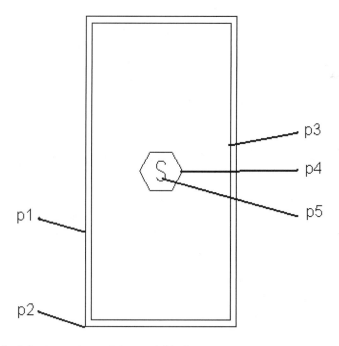

Figure 9T.15 *Selecting entities of the mask block*

Command:

Select Layout Node: *(Select the ceiling grid at p1 shown in Figure 9T.16 to place the masking block.)*

14. Select the **2x4** fluorescent light symbol in the drawing area, right-click, and choose **Add Selected** from the shortcut menu. Edit the **Definition** edit field in the Properties palette to **Speaker**. Select the location for the speaker at **p1** as shown in Figure 9T.17. Press ESC to end the command.

15. Select the **2x4** light symbol and the **Speaker** symbol, right-click, and choose **Attach Objects** from the shortcut menu. Respond to the command line prompts as shown below.

    Command: MaskAttach

    Select AEC object to be masked: *(Select the ceiling grid.)*

    *(Select **OK** to dismiss the **Select Display Representation** dialog box.)*

    (Mask blocks hide the ceiling grid as shown in Figure 9T.18.)

16. Insert additional **2x4** light symbols as shown in Figure 9T.19.

17. Expand Imperial\Design\Electrical\Lighting\Incandescent in the DesignCenter. Double-click on **Emergency Square 1** symbol in the DesignCenter, and select the **Attributes** button of the Properties palette. Type **23** in the **Value** field of the **Multi-view Block Attributes** dialog box as shown in Figure

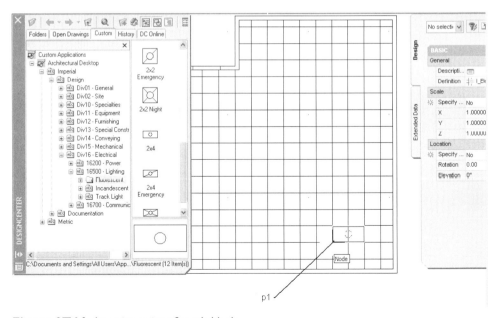

Figure 9T.16 *Insertion point of mask block*

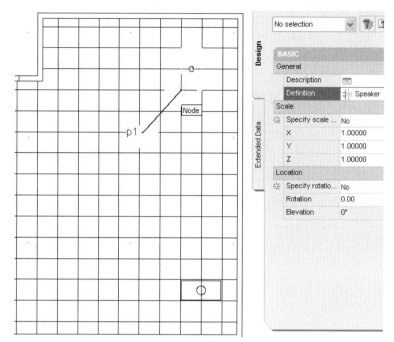

Figure 9T.17 *Insertion point of Speaker mask block*

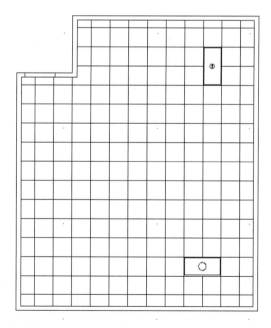

Figure 9T.18 *Mask block hides ceiling grid*

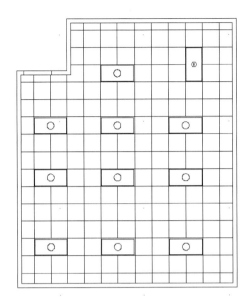

Figure 9T.19 Insertion locations for additional lights

9T.20. Select **OK** to dismiss the **Multi-view Block Attributes** dialog box. Insert the **Emergency Square 1** symbol at **p1** using the Node object snap as shown in Figure 9T.20.

18. Save and close the drawing.

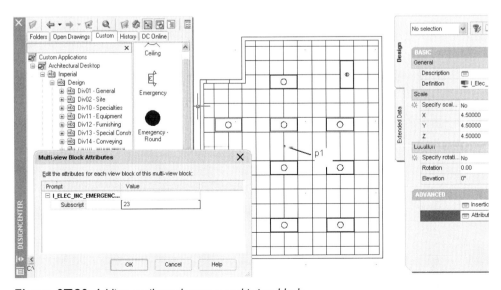

Figure 9T.20 Adding attribute data to a multi-view block

PROJECT

Ex 9.4 Inserting Symbols

Open the **Project Browser** and use the Project Selector drop-down list to navigate to your *ADT Student\ADT Tutor\Ch9\Ex9-4* project. Select the **Constructs** tab of the **Project Navigator**, double-click to open the **Floor 1** drawing and then insert the following symbols from the Design Tool Catalog- Imperial, as shown in Figure 9T.21.

BATH 1

1. Insert a **Tub 30 x 60** from the Mechanical\Plumbing Fixtures\Bath folder.
2. Insert the **Tank 1** water closet from the Mechanical\Plumbing Fixtures\Toilet folder.
3. Insert the **Wall** lavatory from the Mechanical\Plumbing Fixtures\Lavatory folder.

BATH 2

1. Insert the **Vanity** symbol at the midpoint of the wall from the Mechanical\Plumbing Fixtures\Lavatory folder.
2. Insert a **Tub 30 x 60** from the Mechanical\Plumbing Fixtures\Bath folder.
3. Insert the **Tank 2** water closet from the Mechanical\Plumbing Fixtures\Toilet folder.

BATH 3

1. Insert the **Tank 1** water closet from the Imperial\Design\Mechanical\Plumbing Fixtures\Toilet folder.

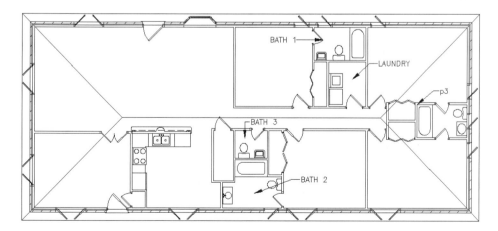

Figure 9T.21 *Symbol insertion locations*

2. Insert the **Wall** lavatory from the Mechanical\Plumbing Fixtures\Lavatory folder.

LAUNDRY

1. Insert the **Washer 26 x 26** and **Dryer 28 x 26** from the Equipment\Residential folder.

CHAPTER 10

Annotating and Documenting the Drawing

INTRODUCTION

The annotation and documentation of a drawing includes the dimensions, notes, schedules, and associated symbols for specifying the size and location of architectural features. The symbols, dimensions, and leaders for annotating the drawing are selected from the tool palettes of the Document palette set. The Document palette set includes Annotation, Callouts, Scheduling, and Drafting tool palettes. The commands included on these tool palettes represent a sample of the commands located in the **Content Browser** and the DesignCenter. The commands of the Callouts palette and keynoting commands will be presented in Chapter 11.

The Scheduling palette includes commands for creating schedules and areas of the building. The content of tags and schedules is defined in the schedule table style. Each schedule table style utilizes one or more property sets, which are defined for objects listed in the schedule. Schedules can be created that automatically update when changes are made in the drawing. The content of a schedule can be exported to external database files to enhance the use of the schedule data.

OBJECTIVES

After completing this chapter, you will be able to

- Place Aec dimensions, text, and leaders from the Annotation palette
- Place fire rating lines on walls
- Create match lines, north arrows, and datum elevations
- Create revision clouds using various styles of revision clouds
- Create tags for doors, windows, rooms, finish, objects, and walls
- Create Aec Space objects for room and place finish tags
- Insert schedule data in tags using the **Edit Property Set Data** command

- Create door and window schedules using the **ScheduleAdd** command
- Edit and update the content of existing schedules and tags

PLACING ANNOTATION ON A DRAWING

The content of annotation is unique to a drawing; therefore it is usually placed on the View drawings in a project. The scale of View drawings is specified in the properties of the view when placed on a sheet. Prior to the placing of annotation in any project or non-project drawing, the scale and layer key style of the drawing must be set with the **Drawing Setup** (**AecDwgSetup**) command. The scale can be selected from the **Scale** flyout located on the Drawing Window status bar. The commands of the Annotation palette shown in Figure 10.1 represent a sample of the commands available in the Documentation Tool Catalog-Imperial and the Documentation Tool Catalog-Metric of the **Content Browser**. The commands within the **Content Browser** also reside in the Documentation folder in the DesignCenter. The commands included in the Annotation palette provide control over layer and scale, whereas text and dimensioning tools of AutoCAD do not provide such control.

PLACING TEXT

The **Text Tool** tool, accessed as shown in Table 10.1, applies the **AecDtlAannoLeaderAdd** command to create text scaled similar to dimensions and leaders. The text style, font, height, and layer are preset by the command. Text is placed

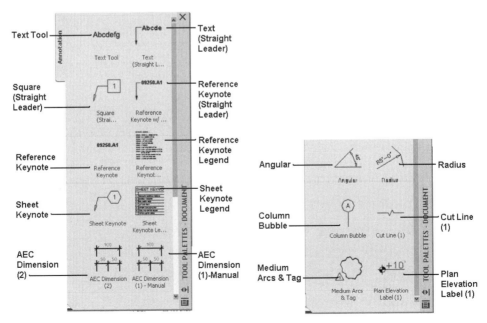

Figure 10.1 *Tools of the Annotation tool palette*

on the A-Anno-Note layer. The text height is determined from the scale factor specified in **Drawing Setup**, which is multiplied by the annotation plot size of 3/32". Therefore, if you are placing text with the drawing scale set to 1/2"=1'-0", the scale factor (24) is multiplied by 3/32 to obtain a text height of 2 1/4".

Command prompt	AECDTLANNOLEADERADD
Palette	Select Text Tool from the Annotation tool palette of the Document palette set

Table 10.1 *Text Tool command access*

When you select **Text Tool**, you are prompted to specify the location for the anticipated text as shown in the following command line prompts:

> Command: AecDtlAnnoLeaderAdd
>
> Select MText insertion point: *(Select a point to specify the location of the text.)*
>
> Select text width<0">: *(Press ENTER to accept text width.)*
>
> Enter first line of text <Mtext>: *(Type text in the command line or press ENTER to open the Mtext window.)*

The Mtext formatting window shown in Figure 10.2 allows you to create paragraph-style text and modify the style, font, and text height. The shortcut menu within the text window allows you to change the justification, case, and use AutoCAPS. The **AutoCAPS** option can be used to set all text to uppercase. After text is placed, you can double-click on the text to open the **Text Formatting** toolbar and editor.

PLACING LEADERS IN THE DRAWING

Leaders are placed in a drawing to identify building components and sizes. Leaders are part of the tools used to dimension a drawing. The **Text** (**Straight Leader**) tool can be selected from the Annotation tool palette as shown in Table 10.2.

Tool Palette	Select Straight Text from the Annotation palette as shown in Figure 10.1

Table 10.2 *AecDtlAnnoLeaderAdd command access*

When you select the **Text** (**Straight Leader**) tool, you are prompted in the command line to select the start point of the leader and additional vertices of the leader. You are prompted to specify the text width, and then insert the text after a null response to the vertices prompt, as shown in Figure 10.3 and the following command sequence.

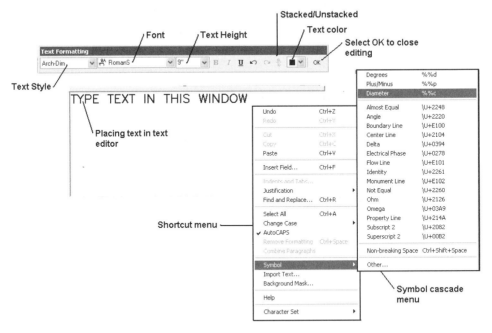

Figure 10.2 *Text editor for placing text*

Command: AecDtlAnnoLeaderAdd

Specify first point of leader line: *(Specify the start point of the leader near p1 as shown in Figure 10.3.)*

Specify next point of leader line: *(Specify leader vertex near p2 as shown in Figure 10.3.)*

Specify next point of leader line: ENTER

Select text width<0">: ENTER

Enter first line of text <Mtext>: *(Type text in the command line or press* ENTER *to enter text in the Multiline Text Editor.)*

(Select **OK** *to dismiss the Multiline Text Editor.)*

The geometry and text of the leader are scaled based on the scale specified in the **Drawing Setup** dialog box. The text is placed on the **A-Anno-Note** layer when the AIA (256) Color layer key style is used. Additional leaders can be placed in the drawing from the Documentation Tool Catalog-Imperial or Documentation Tool Catalog-Metric of the **Content Browser** shown in Figure 10.4. The leader tools included in the Documentation Tool Catalog-Imperial and Documentation Tool Catalog-Metric create an AutoCAD block. The text for the tag symbols is inserted in the **Edit**

Annotating and Documenting the Drawing 653

Figure 10.3 *Placing a leader on a drawing*

Attributes dialog box shown in Figure 10.4, which is displayed during placement of the leader.

CREATING DIMENSIONS

When you work in a project, dimensions are usually placed in a View drawing because this drawing may consist of one or more Construct drawings to create a floor plan. The View drawing can be developed into a floor plan by attaching deck or porch plans to the main floor. Working within the View drawing, you can use Object Snap modes to locate features in the Construct drawings for the dimensions. The primary source of dimensioning tools resides on the Annotation palette.

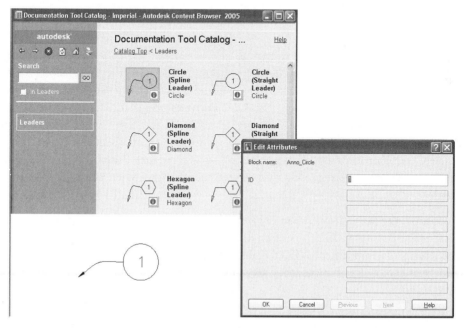

Figure 10.4 *Leaders category of the Documentation Tool Catalog-Imperial*

The AEC Dimension (2) tool (see Table 10.3) and the AEC Dimension (1) – Manual tool are used to place linear dimensions on the A-Anno-Dims layer and scale the dimensioning geometry based upon the display configuration. Additional dimension commands shown in Table 10.4 can be accessed from the DesignCenter and the **Content Browser**, which insert the dimension on the A-Anno-Dims layer and are scaled according to the viewport scale specified in **Drawing Setup**.

Command prompt	DIMADD
Tool Palette	Select AEC Dimension (2) from the Annotation palette as shown in Figure 10.1

Table 10.3 *Automatic AEC Dimension command access*

The AutoCAD Dimensioning toolbar is available; however, avoid using these commands because layer and scale settings are not controlled upon execution of the command. AutoCAD dimensioning commands insert the dimension on the current layer using the current dimensioning style settings without regard to the scale settings of the **AecDwgSetup** command.

AEC dimensions can be placed by selecting **AEC Dimension (2)** or **AEC Dimension (1) – Manual** from the Annotation palette. The **AEC Dimension (2)** command of the Annotation palette (refer to Table 10.3 for command access) is used to place dimensions by selecting objects. Automatic AEC dimensions can be applied to Aec objects or linework. The advantage of the AEC dimension is that, if the location or size of wall components is changed, the dimension is modified to reflect the change. Therefore if you add, delete, or move a window in the wall, the dimensions for the object will update to reflect the change. Automatic AEC dimensions are tied to the logical points of the AEC object. The **AEC Dimension (2)** tool allows you to select a wall from within an external reference drawing. Therefore, when you apply an **AEC Dimension (2)** dimension within a View drawing, you select the walls of a reference Construct drawing. If the walls of the Construct drawing are edited, the dimension in the View drawing will reflect the changes when the reference file is reloaded.

In contrast, the **AEC Dimension (1) – Manual** tool places the dimension by selecting specific points within an external reference drawing or the current drawing. Dimensions placed with the **AEC Dimension (1) – Manual** tool are not automatically updated as the object changes; they are updated only as the specific points are moved in the drawing.

The scale (DimScale) of AEC dimensions is controlled by the display configuration assigned to the viewport. Aec dimensions placed when the Medium Detail display con-

Source	Tool	Scale Control	Layer
Annotation tool palette	DimAdd – applying AEC Dimension (2) & AEC Dimension 1 – Manual	Display Representation	A-Anno-Dims
Shortcut	DimAdd – applying AEC Dimension with Standard style	Display Representation	A-Anno-Dims
Content Browser			
(Documentation Tool Catalog-Imperial Catalog> Miscellaneous> Dimensions)	Aligned, Angular, Baseline, Continue, Linear, and Radius	Drawing Setup (AecDwgSetup)	A-AnnoDims
DesignCenter			
(Architectural Desktop\Imperial\ Documentation\ Miscellaneous\ Dimensions)	Aligned, Angular, Baseline, Continue, Linear, and Radius	Drawing Setup (AecDwgSetup)	A-AnnoDims
AutoCAD Dimensioning Toolbar	Aligned, Angular, Baseline, Continue, Linear, and Radius	Current default, no control	Current default, no control

Table 10.4 *Dimensioning sources*

figuration is utilized apply the Aec Arch I-96 dimension style, which specifies a 1:96 DimScale. The dimension scale factors of each display configuration is shown in Table 10.8.

INSERTING AEC DIMENSION (2) DIMENSIONS

When you select **AEC Dimension (2)** from the Annotation tool palette, you are prompted to select the geometry to dimension, as shown in the following command line sequence. The objects can be selected with a crossing or window selection. The default dimension style applied during the dimension operation is the 2 Chain dimension style. This style also provides control of the scale, units, and the number dimension strings.

Command: DimAdd

Select geometry to dimension or [Style]: *(Select a point near p1 as shown in Figure 10.5.)*

Specify opposite corner: 20 found *(Select a point near p2 as shown in Figure 10.5.)*

13 were filtered out

Select geometry to dimension or [Style]: ENTER *(Press ENTER to end selection.)*

Specify insert point or [Rotation/Align]: *(Select a point near p3 as shown in Figure 10.5 to specify the location of the dimension string.)*

7 added

(AEC Dimension created as shown at right in Figure 10.5.)

CREATING AEC MANUAL DIMENSIONS

AEC manual dimensions are placed by selecting **AEC Dimension (1) – Manual** from the Annotation palette (refer to Table 10.5). **AEC Dimension (1) – Manual** dimensions can be used to dimension the location of both AEC objects and non-AEC objects. Changes in AEC objects dimensioned are not automatically updated in the **AEC Dimension (1) – Manual** dimension. The AEC manual dimension requires you to select each extension line point to create a single dimension string as shown in the following command line sequence.

Command: DimManAdd

Pick dimension points on screen to add: *(Select point p1 as shown at left in Figure 10.6.)*

Pick dimension points on screen to add: *(Select point p2 as shown at left in Figure 10.6.)*

Pick dimension points on screen to add: ENTER *(Press ENTER to end selection of points.)*

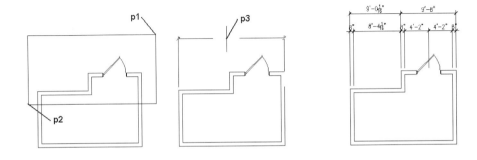

Select walls for dimensions   Location of dimension string specified   2 Chain Dimension string created

Figure 10.5 *Placing an AEC Dimension (2)*

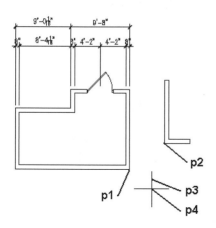
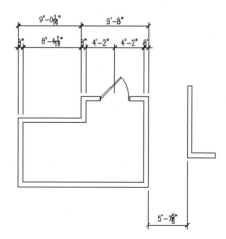

Figure 10.6 *Placing an AEC Manual dimension*

Pick side to dimension or [Style]: *(Select a point near p3 as shown at left in Figure 10.6 to specify dimension line location.)*

Second point: *(Select a point near p4 as shown at left in Figure 10.6 to specify dimension alignment.)*

2 added *(Dimensions added as shown at right in Figure 10.6.)*

Command prompt	DIMMANADD
Palette	Select AEC Dimension (1) – Manual from the Annotation palette as shown in Figure 10.1

Table 10.5 *AEC Manual Dimension command access*

AEC Dimension (1) - Manual and **AEC Dimension (2)** dimensions are placed on the A-Anno-Dims layer, which has color 211 when the AIA (256) Color layer key style is used. The text of **AEC Dimension (1) – Manual** and **AEC Dimension (2)** can be superscripted to add notes. In addition, the dimensions can be grouped and the extension line length edited.

USING THE STYLE MANAGER TO CREATE AN AEC DIMENSION STYLE

When you insert AEC dimensions, the style is preset to **2 Chain** for AEC Dimension (2) or **1 Chain** for AEC Dimension (1) – Manual. The 1 Chain, 2 Chain, and 3 Chain dimensioning styles are included in drawings developed from the Imperial and Metric Aec Model templates. You can select the tool on the tool palette, right-click, and choose **Properties** to review style properties of the tool. Additional styles can be created, or you can edit any AEC dimension style in the **Style Manager**. The dimension style determines what is included in the dimension and the number of dimension strings. The dimension string can be defined to include only selected components such as outer boundaries of the wall, wall lengths, width of wall components, wall intersections, and openings in the wall. The style of an AEC dimension can include up to 10 dimension strings. Therefore, before placing manual or automatic AEC dimensions, you should verify the style in the Properties palette and refine that style to include the necessary components.

Access the **AecDimStyle** command from **Format>AEC Dimension Styles** on the menu bar to edit Aec dimension styles. Refer to Table 10.6 for the **AecDimStyle** command access. When you access the **AecDimStyle** command, the **Style Manager** opens to the **Documentation Objects\AEC Dimension Styles** folder as shown in Figure 10.7. The dimension style is displayed in the Viewer of the **Style Manager** to illustrate how the objects are dimensioned in the style. A style is edited by double-clicking on the style name in the left pane of the **Style Manager** to open the **AEC Dimension Style Properties** dialog box shown in Figure 10.8.

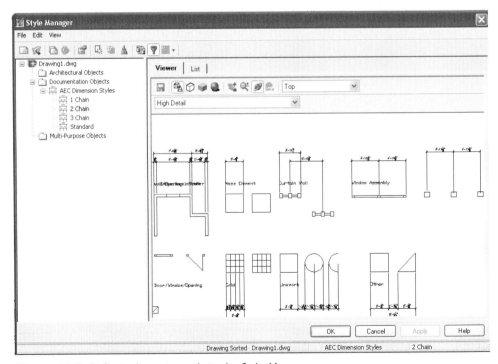

Figure 10.7 *2 Chain dimension style in the Style Manager*

Menu bar	Select Format>AEC Dimension Styles
	Select Format>Style Manager, expand the Documentation Objects\AEC Dimension Styles
Command prompt	AECDIMSTYLE

Table 10.6 *Edit AEC Dimension Style command access*

The **AEC Dimension Style Properties** dialog box can also be accessed directly for a selected dimension when you select the **AecDimStyleEdit** command. Access the **AecDimStyleEdit** command as shown in Table 10.7.

Shortcut menu	Select an AEC Dimension, right-click, and choose Edit Aec Dimension Style
Command prompt	AECDIMSTYLEEDIT

Table 10.7 *Edit Aec Dimension Style command access*

The tabs of the **AEC Dimension Style Properties** dialog box for the 2 Chain AEC dimension style are described below.

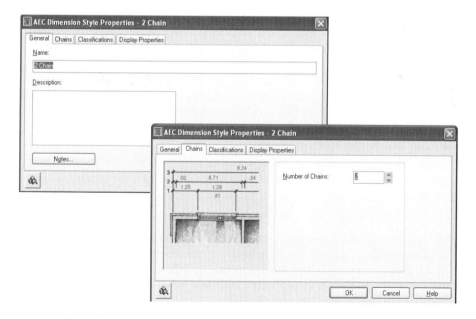

Figure 10.8 *General and Chains tabs of the AEC Dimension Style Properties - 2 Chain dialog box*

General Tab

The **General** tab shown in Figure 10.8 allows you to edit the **Name** and **Description** of the AEC dimension style.

Chains Tab

The **Chains** tab shown in Figure 10.8 includes an edit field to edit the number of chains.

Classifications Tab

The **Classifications** tab allows you to specify a classification to the dimension style. Classifications can be used to identify the style with a particular status of the construction.

Display Properties Tab

The **Display Properties** tab, shown in Figure 10.9, includes a list of the **Display Representations**, **Display Property Source**, and **Style Override** fields. Each display representation is assigned a dimension style, shown in Table 10.8; therefore the scale and other properties vary with the display configuration.

AEC Dimemsioning Style	DimScale	Display Configuration
Aec-Arch-I-48	48	High Detail
Aec-Arch-I-96	96	Medium Detail, Presentation, Reflected, Reflected Screened, Screened, Standard
Aec-Arch-I-192	192	Low Detail

Table 10.8 *Scale factors of AEC dimensions*

Selecting the **Edit Display Properties** button for a display representation opens the **Display Properties (AEC Dimension Style Override – 2 Chain)** dialog box, which includes three tabs as described below.

> **Layer/Color/Linetype** – **Layer/Color/Linetype** tab allows you to turn off or on the display of the components of the dimension. The **AEC Dimension Group** consists of the dimension chains of the dimension. The AEC Dimension Group component is displayed for all display representations. The **AEC Dimension Group Marker** is shown in Figure 10.10 at left. The marker can be selected to edit the grips of the dimension. The **Removed Points Marker** component is usually not visible, since it is used to represent points that have been removed from the chain. The **Override Text & Lines**

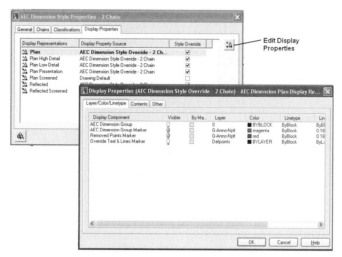

Figure 10.9 *Display Properties tab, and the Layer/Color/Linetype tab of the Display Properties (AEC Dimension Style Override – 2 Chain) dialog box*

Marker component consists of an overline placed over text when it has been overridden.

Contents – The **Contents** tab is the most significant tab, because it lists the contents to be included in each of the chains. The **Contents** tab shown in Figure 10.11 consists of the following components:

Apply to – The **Apply to** window lists the objects included in each of the selected chains. This tab is edited for a selected chain by selecting an object in the **Apply to** window, and then editing the aspects of the object to be included in the dimension at right.

Chain – The **Chain** list located in the lower left corner specifies the chains of the style. The chain that is highlighted can be edited by selecting objects in the **Apply to** window.

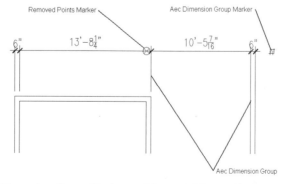

Figure 10.10 *AEC Dimension Group Marker and Removed Points Marker displayed*

Building Object Dimension Points – The check boxes located in the right panel refer to the building object dimension points for the object selected in the **Apply to** window at left. The building object dimension points listed in Figure 10.11 are for the selected Wall object.

 Note: Changes made for a display representation must be repeated for all other display representations in order for the dimension to be displayed identically for all display configurations.

Therefore, to create a dimension chain as shown in Figure 10.12, select the **1 Chain** style. Edit the **Contents** tab as follows: select the chain number in the lower left list, select the Wall object in the **Apply to** window, and check **Overall** and **Length of Wall (Outer Boundaries)** check boxes as shown in Figure 10.11. Select the **Opening in Wall** building object in the **Apply to** window and check **Center of Opening**. The **Contents** tab allows you to customize which features of the building objects are included in the dimension.

Other – The **Other** tab, shown in Figure 10.13, includes the **AutoCAD Dimension Settings** and the **AEC Dimension Settings** sections. The **AutoCAD Dimension Settings** allows you to specify the dimension style used by the AEC dimension. The **Edit** button shown to the right of the dimension style name opens the **Dimension Style Manager**. The AutoCAD **Dimension Style Manager** can be used to change the format of the dimensions.

The **AEC Dimension Settings** section allows you to edit the distance between the chains, toggle ON fixed length extension lines, and toggle ON the display of the opening height below the width dimension. The dimension of the window shown in Figure 10.14 includes the display of the window height below the dimension line for its width.

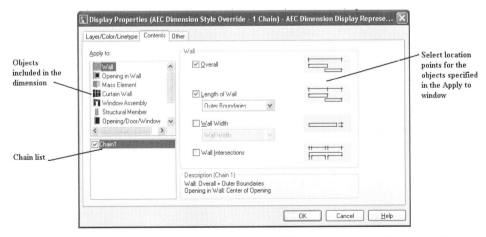

Figure 10.11 *Contents tab of the Display Properties (AEC Dimension Style Override – 2 Chain) dialog box*

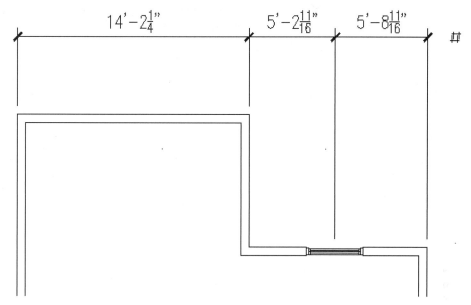

Figure 10.12 *AEC Dimension style to display dimension to the center of opening*

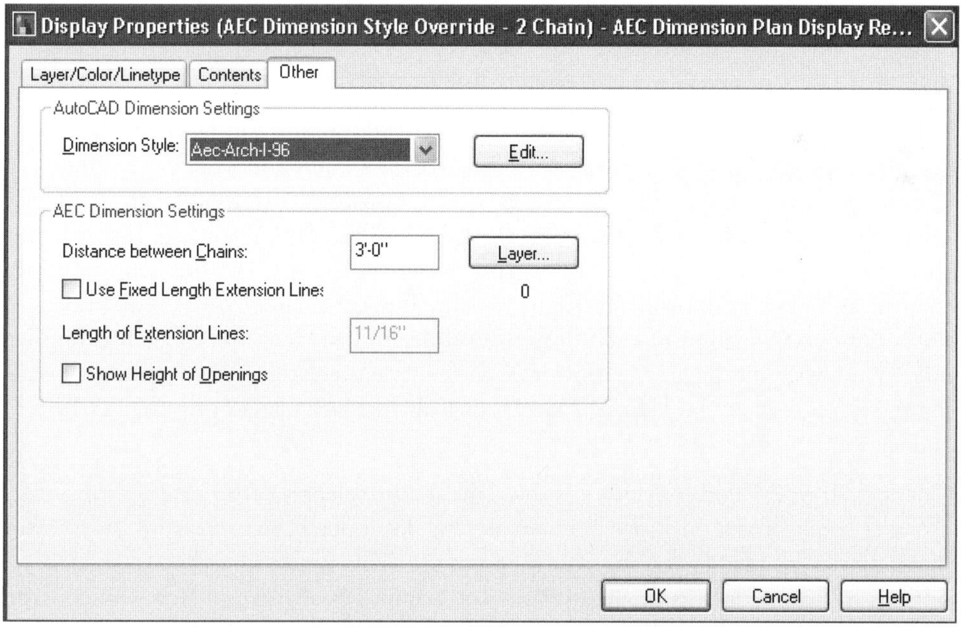

Figure 10.13 *Other tab of the Display Properties (AEC Dimensions Style Override - 2 Chain) dialog box*

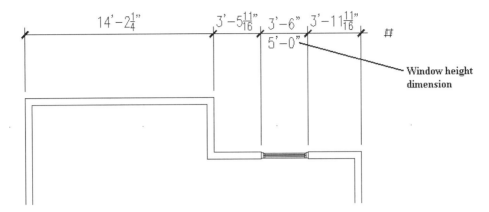

Figure 10.14 *Window height dimension placed below window width dimension*

EDITING AEC DIMENSIONS

The features included in an AEC dimension can be defined in the **AEC Dimension Style** dialog box; however, grips of the dimension provide additional editing. The grip locations and functions of AEC dimensions differ from the grips of AutoCAD dimensions. In addition to the grips, the shortcut menu of a selected AEC dimension allows you to add or remove dimension points, attach or detach objects, and override the text.

Grip Editing AEC Dimensions

When you select an AEC dimension, the **Move All Chains** and **Edit In Place** trigger grips are displayed as shown in Figure 10.15. When you select the **Move All Chains** grip, you can stretch the dimension group to a different location. If you select the **Edit In Place** trigger grip, the grips of the dimension are displayed as shown in the middle in Figure 10.15. You can select the grips of the text and move the text as shown at the right in Figure 10.15. The grips allow you to edit the location of the text, location of each dimension line, and the location of each extension line. When you edit the grips, an override to the style is created. When you finish editing the grips, reselect the **Edit In Place/Exit Edit In Place** trigger grip to end the edit and save changes.

Changes to an AEC dimension can be removed by selecting from the options of its shortcut menu. If text positions or extension line locations have been edited, the **Reset** menu option is added to the shortcut menu of the AEC dimension. The cascade menu options of **Reset** include the **Reset Text to Original Position** and **Reset Extension Lines**. If you select **Reset Text to Original Position**, all text will return to its original location. Refer to Table 10.9 for command access for **Reset Text to Original**

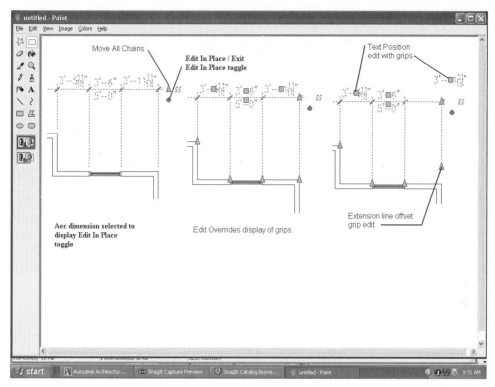

Figure 10.15 *Grip editing of AEC dimensions*

Position (**DimRemoveOverrideTextOffsets**); this command cannot be typed in the command line. The **Reset Extension Lines** command will remove the changes made to the extension lines. Refer to Table 10.10 for command access for **Reset Extension Lines** (**DimRemoveOverrideExtLines**); this command cannot be typed in the command line.

Shortcut menu	Select a dimension that includes text position edits, right-click, and choose Reset>Reset Text to Original Position

Table 10.9 *Reset Text to Original Position of an AEC Dimension command access*

Shortcut menu	Select a dimension that includes extension line edits, right-click, and choose Reset>Reset Extension Lines

Table 10.10 *Reset Extension Lines command access*

EDITING AEC DIMENSIONS WITH THE SHORTCUT MENU

The shortcut menu of an AEC dimension includes the following options: **Attach Objects**, **Detach Objects**, **Add Dimension Points**, **Remove Dimension Points**, **Override Text & Lines**, **Edit In-Place**, and **Reset**. The shortcut menu of an AEC dimension also includes **Edit AEC Dimension Style**, which can be used to edit the style of the dimension.

 Caution: Prior to editing AEC dimensions, verify that the display configuration and scale of the drawing are correct.

Attaching Building Objects to AEC Dimensions

The **Attach Objects** option allows you to add building objects to an Aec dimension. Access the **DimAttach** command as shown in Table 10.11.

Command prompt	DIMATTACH
Shortcut menu	Select an AEC Dimension, right-click, and choose Attach Objects

Table 10.11 *DimAttach command access*

The wall at **p1** was attached to the AEC dimension as shown in the following command line sequence.

> Command: DimAttach
>
> Select Building Elements: 1 found *(Select the wall at p1 as shown at left in Figure 10.16.)*
>
> Select Building Elements: ENTER *(Press ENTER to end selection.)*
>
> 1 added
>
> *(Dimensions added as shown at right in Figure 10.16.)*

Detaching Building Objects

Building objects attached to an AEC dimension can be removed. See Table 10.12 for **DimDetach** command access. Select the dimension, right-click, and choose **Detach Objects** from the shortcut menu. The command line prompts shown below removed the building objects from the dimension as shown at right in Figure 10.17.

> Command: DimDetach
>
> Select Building Elements: *(Select a point near p1 as shown at left in Figure 10.17.)*

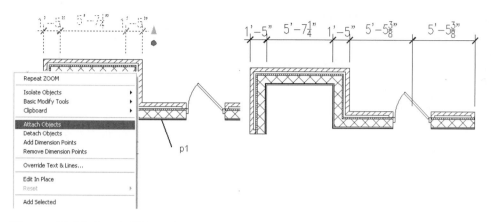

Figure 10.16 Attach Objects shortcut menu option

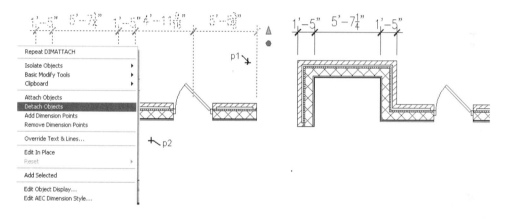

Figure 10.17 Points detached from AEC dimension

Specify opposite corner: 3 found *(Select a point near p2 as shown at left in Figure 10.17.)*

1 was filtered out

Select Building Elements: ENTER *(Press ENTER to end selection.)*

1 removed, 1 were filtered out.

(Dimension removed as shown at right in Figure 10.17.)

Command prompt	DIMDETACH
Shortcut menu	Select an AEC Dimension, right-click, and choose Detach Objects

Table 10.12 *DimDetach command access*

If you type **DIMDETACH** in the command line, you must select the dimension group prior to selecting the building elements. However, if you choose **Detach Objects** from the shortcut menu of the dimension, the selection of the dimension group is not required.

Adding Dimension Points

Additional points on building objects can be added to a dimension group without adding all the points associated with that object. Access the **DimPointsAdd** command as shown in Table 10.13.

Command prompt	DIMPOINTSADD
Shortcut menu	Select an AEC Dimension, right-click, and choose Add Dimension Points

Table 10.13 *DimPointsAdd command access*

The point at the right end of the wall shown at left in Figure 10.18 was added to the dimension group, as shown in the following command line sequence.

(Select the dimension, right-click, and choose **Add Dimension Points**.*)*

Command: DimPointsAdd

Pick dimension points on screen to add: _endp of *(*SHIFT *+ right-click, choose the Endpoint object snap, and select p1 shown at left in Figure 10.18.)*

Pick dimension points on screen to add: ENTER *(Press* ENTER *to end selection.)*

Select Dimension chain: *(Select the dimension chain at p2 as shown at left in Figure 10.18.)*

1 added

(Dimension point added to the chain as shown at right.)

Removing AEC Dimension Points

Points of an AEC dimension can be removed to revise the dimension to include only

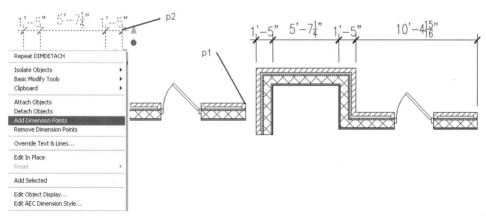

Figure 10.18 *Dimension point added to AEC*

selected points on the building objects. Access the **DimPointsRemove** command as shown in Table 10.14.

Command prompt	DIMPOINTSREMOVE
Shortcut menu	Select an AEC Dimension, right-click, and choose Remove Dimension Points

Table 10.14 *DimPointsRemove command access*

The points representing the width of the wall shown at left in Figure 10.19 were deleted from the dimension group as shown in the following command line sequence.

> (Select the dimension, right-click, and choose **Remove Dimension Points**.)
>
> Command: DimPointsRemove
>
> Select extension lines to remove dimension points: 1 found *(Select the extension line at p1 as shown in Figure 10.19.)*
>
> Select extension lines to remove dimension points: *Cancel*
>
> Select extension lines to remove dimension points: 1 found *(Select the extension line at p2 as shown in Figure 10.19.)*
>
> Select extension lines to remove dimension points: *Cancel*
>
> Select extension lines to remove dimension points: ENTER *(Press* ENTER *to end selection.)*

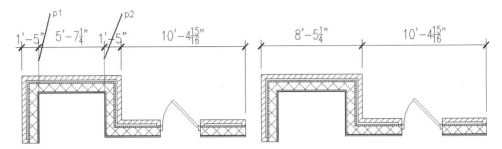

Figure 10.19 Revised AEC Dimensions

Override set for 1 Dimension Group(s).

(Dimension revised as shown at right in Figure 10.19.)

Overriding Text of AEC Dimensions

The display and contents of the text used in an AEC dimension can be overriden. Access the **DimTextOverride** command as shown in Table 10.15.

Command prompt	DIMTEXTOVERRIDE
Shortcut menu	Select an AEC Dimension, right-click, and choose Override Text & Lines

Table 10.15 *Override Text & Lines command access*

The **DimTextOverride** command opens the **Override Text & Lines** dialog box, which allows you to add notes or change the dimension text. The "FIELD VERIFY" note was added in the following command line sequence to change the dimension shown in Figure 10.20 at right. The overline can be turned off by turning off the **Override Text & Lines Marker** display component in the **Display Properties** dialog box of the dimension style.

Command: DimTextOverride

Select dimension text to change: *(Select a dimension to edit.)*

(The **Override Text & Lines** *dialog box opens as shown in Figure 10.20.)*

[hide Text/hide text and Lines/Override text/Underline/Prefix/Suffix/Reset all]: DBOX

(Type **Field Verify** *in the* **Text Override** *edit field of the* **Override Text & Lines** *dialog box. Select* **OK** *to dismiss the* **Override Text &**

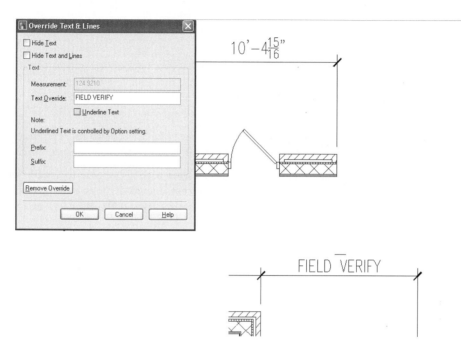

Figure 10.20 *Overriding text of the AEC dimension*

Lines *dialog box. To indicate the text is overridden a line is displayed above the edited text.)*

Using the AEC Dimension Style Wizard

The **AEC Dimension Style Wizard** allows you to change display features for each AEC Dimension Style of the drawing. Refer to Table 10.16 to access the **AEC Dimension Style Wizard** command.

Command prompt	AECDIMWIZARD
Menu bar	Select Format>AEC Dimension Style Wizard

Table 10.16 *AEC Dimension Wizard command access*

The **AEC Dimension Display Wizard** dialog box consists of four pages, shown in Figures 10.21 and 10.22. The **AEC Dimension Display Wizard** dialog box allows you to change the display representation, layer, size, and color of the geometry used in the AEC dimension. The pages of the **AEC Dimension Display Wizard** dialog box are described below.

Select Style – The **Select Style** page includes a drop-down list of AEC dimension styles of the drawing.

Lines and Arrows – The **Lines and Arrows** page allows you to specify the display representation, arrowhead block, dimension, and extension lines. The distance between the dimension chains and the distance the extension lines extend beyond the dimension line can be specified as shown in Figure 10.21.

Text – The **Text** page allows you to specify for the display representation the text style, text height, and text round off values as shown in Figure 10.22.

Color and Layer – The **Color and Layer** page allows you to specify for the display representation the color of the text, dimension lines, extension lines, and the layer assignment. If the layer is set to Layer 0, the layer key assigns the layer.

ANGULAR DIMENSIONS

Angular dimensions are added to a drawing by selecting **Angular** from the Annotation tool palette. When you select the **Angular** tool, you are prompted in the command line to select objects to dimension. The dimension is placed with components scaled as specified in **Drawing Setup** and placed on the **A-Anno-Dims** layer. You can edit the arrowhead and other features in the Properties palette of the dimension. Access the **Angular** command as shown in Table 10.17. Access from the command line does not apply the automatic dimension scaling and layering from **Drawing Setup**.

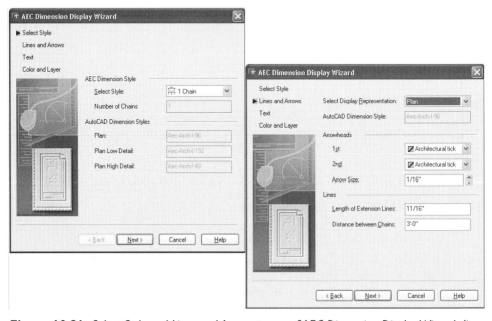

Figure 10.21 Select Style and Lines and Arrows pages of AEC Dimension Display Wizard dialog box

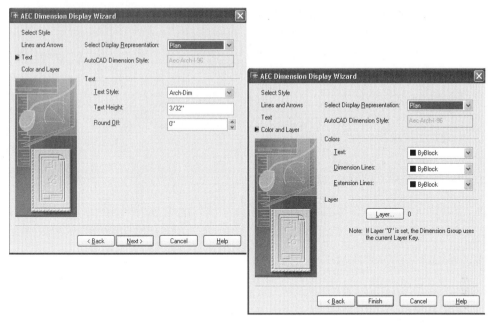

Figure 10.22 Text and Color and Layer pages of AEC Dimension Display Wizard dialog box

Command prompt	DIMANGULAR
Palette	Select Angular from the Annotation palette as shown in Figure 10.1

Table 10.17 DimAngular command access

 Tip: The Angular and Radius tools of the Annotation tool palette are not Aec dimensions; therefore the DimScale of the dimensions does not change when a switch is made from Low Detail, Medium Detail or High Detail.

The dimension shown in Figure 10.23 was placed in the following command line sequence.

    Command: _DimAngular

    Select arc, circle, line, or <specify vertex>: *(Select the wall at p1 as shown in Figure 10.23.)*

    Select second line: *(Select the wall at p2 as shown in Figure 10.23.)*

    Non-associative dimension created.

    Specify dimension arc line location or [Mtext/Text/Angle]: *(Select a point near p3 as shown in Figure 10.23.)*

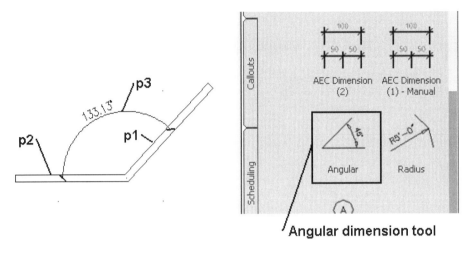

Figure 10.23 Angular dimension placed from Annotation tool palette

Dimension text = 133.13

CREATING RADIAL DIMENSIONS

Radial dimensions can be created by selecting **Radius** from the Annotation tool palette; see Table 10.18 for command access. The dimension is scaled according to the scale specified in the Drawing Window status bar and is placed on the A-Anno-Dims layer. You can edit the arrowhead and other features in the Properties palette. When you select **Radius** from the Annotation tool palette, the **DimRadius** command is selected, and you are prompted to select an arc, circle, or curved building object to dimension. The **Radius** command was used to dimension the curved wall as shown in Figure 10.24.

Command prompt	DIMRADIUS
Tool palette	Select Radius from the Annotation tool palette

Table 10.18 DimRadius command access

Command: _DimRadius

Select arc or circle: *(Select the radial feature at p1 as shown in Figure 10.24 to specify the location of the dimension line.)*

Non-associative dimension created.

Dimension text = 10'-7 3/16» *(Default radial dimension.)*

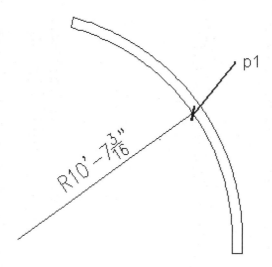

Figure 10.24 *Placing radial dimensions*

> Specify dimension line location or [Mtext/Text/Angle]: ENTER *(Press* ENTER *to locate the dimension as shown in Figure 10.24.)*

 STOP: Do Tutorial 10.1, "Inserting AEC Dimensions" at the end of the chapter.

PLACING STRAIGHT CUT LINES

Straight cut lines are placed in the drawing by selecting **Cut Line (1)** from the Annotation tool palette. When you select **Cut Line (1)**, you are prompted to select the start and end of the straight cut line. After specifying the endpoints of the cut line, you are prompted to select objects to trim as shown in the following command line sequence. The **Cut Line (1)** tool places the cut line using a symbol option of the **AnnoBreakMarkAdd** command. Regardless of the length of the break line, the Z portion remains the same size.

| Tool palette | Select Cut Line (1) from the Annotation tool palette |

Table 10.19 *Cut Line (1) tool command access*

When you select the **Cut Line (1)** tool, the command sequence includes presets that require no user input; therefore only those lines requiring a response are shown below.

Command: _AecAnnoBreakMarkAdd

Specify first point of break line or [Symbol/Type]: *(Select a point near p1 as shown at left in Figure 10.25.)*

Specify second point of break line or [Symbol/Type]: *(Select a point near p2 as shown at left in Figure 10.25.)*

Select objects to trim <None> or [Symbol/Type]: *(Select a point near p3 as shown at left in Figure 10.25.)*

Select objects to trim <None> or [Symbol/Type]: *(Select a point near p4 as shown at left in Figure 10.25.)*

Select objects to trim <None> or [Symbol/Type]: ENTER *(Press ENTER to end the trim.)*

(Walls trimmed as shown at right in Figure 10.25.)

The Architectural Desktop\Imperial\Documentation\Break Marks folder of the DesignCenter includes additional cut and break lines. These break and cut lines can also be accessed from the Break Marks category of the Documentation Tool Catalog-Imperial and the Documentation Tool Catalog-Metric catalogs of the **Content Browser** shown in Figure 10.26.

REVISION CLOUDS

The revision cloud allows you to identify an area of the drawing that has been revised or add a note that applies to this portion of the building. The revision cloud is created free style with a series of arcs connected to enclose an area of the drawing. Access the **AnnoRevisionCloudAdd** command as follows (Table 10.20). Access to this command by typing the command in the command line does not apply the lineweight, attribute, and other settings preset when the command is accessed from the tool palette.

Command prompt	ANNOREVISIONCLOUDADD
Tool Palette	Select Medium Arcs and Tag from the Annotation tool palette

Table 10.20 *Revision Clouds command access*

When you select **Medium Arcs and Tag** from the Annotation tool palette, the **AnnoRevisionCloudAdd** command is executed. This command allows you to draw a series of polyline arcs. The width of the arc is adjusted according to the scale factor defined by **Drawing Setup**. The cloud is drawn with scaled 3/4" radius arcs. Revision clouds are placed on the A-Anno-Revs layer, which has the color 71.

To place a revision cloud, select **Medium Arcs and Tag** from the Annotation tool palette, and then specify the start point. Move the pointer from the start point in a cir-

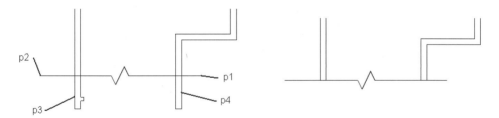

Figure 10.25 *Placing a cut line*

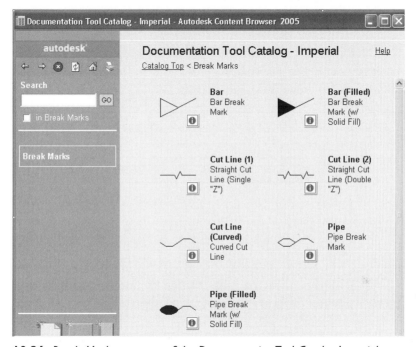

Figure 10.26 *Break Marks category of the Documentation Tool Catalog-Imperial*

cular direction. When the pointer is returned near the start point, the final arc will close, and you will be prompted to specify the callout location.

The command line sequence shown below was used to create the revision cloud shown in Figure 10.27. Because the first lines of command line include presets that do not require user input, only those lines requiring a response are shown below.

>Command: AecAnnoRevisionCloudAdd
>
>Specify cloud starting point or [Symbol block/pline Color/Arc length/pline Width]: *(Select point p1 in Figure 10.27 to start the revision cloud.)*

Figure 10.27 *Creating a revision cloud*

Cloud will close when returned to start point...

Guide crosshairs along cloud path (counter-clockwise) *(Move the mouse in the direction shown to form the enclosure and close the revision cloud.)*

Specify center point of revision tag <None>: *(Select a point near p2 shown in Figure 10.27.)*

(Type the number of the detail in the **Edit Attributes** *dialog box; select* **OK** *to dismiss the dialog box.)*

 Tip: Revision clouds should be drawn counterclockwise. Revision clouds drawn in a clockwise direction will create arc endpoints pointing out from the enclosure.

Additional styles of revision clouds can be inserted from the Documentation Tool Catalog-Imperial shown in Figure 10.28 or the Architectural Desktop\Imperial\Documentation\Revision Clouds folder of the DesignCenter.

INSERTING CHASES

The **Chase** symbol can be used to represent chimneys, ducts, or wall niches. Representations of chases can be placed in the drawing from the Chase category of the Documentation Tool Catalog-Imperial shown in Figure 10.29 or the Architectural Desktop\Imperial\Documentation\Chases folder of the DesignCenter.

DesignCenter	Select Custom tab of the DesignCenter, and select Architectural Desktop\Imperial\Documentation\Chases
Content Browser	Open the Documentation Tool Catalog-Imperial, and choose the Chases category

Table 10.21 *Chases command access*

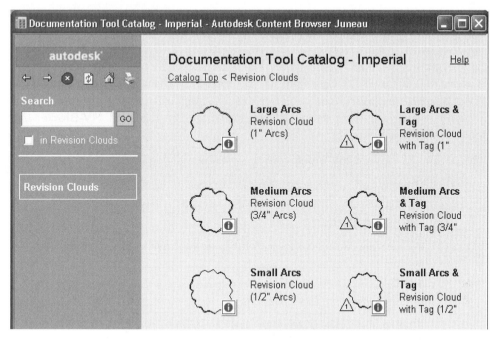

Figure 10.28 *Revision Clouds category of the Documentation Tool Catalog-Imperial*

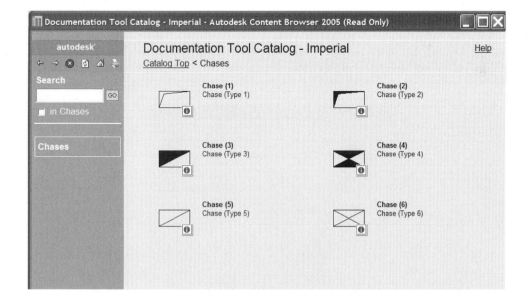

Figure 10.29 *Chase symbols of the Documentation Tool Catalog-Imperial Chase category*

When you **drag** the i-drop for a chase symbol from the **Content Browser**, you are prompted to specify the insertion point, and the **Add MV-Block with Interference** dialog box opens. The interference of the chase with walls, spaces, slabs, and roof slabs can be specified as additive, subtractive, and ignore. The ignore interference condition is active only for walls. The size and description of the chase is specified in the **Add MV-Block with Interference** dialog box as shown in Figure 10.30. The size is specified by editing the scale values of the unit block in the **Add MV-Block with Interference** dialog box.

The options of the **Add MV-Block with Interference** dialog box are described below.

> **Description** – A **Description** can be typed in this field to describe the chase.
>
> **INSERTION POINT**
>
> **Pick Point** – Select the **Pick Point** button to specify on screen the insertion point.
>
> **X** – The **X** absolute coordinate value can be typed to specify the X location of the insertion point of the chase.
>
> **Y** – The **Y** absolute coordinate value can be typed to specify the Y location of the insertion point of the chase.

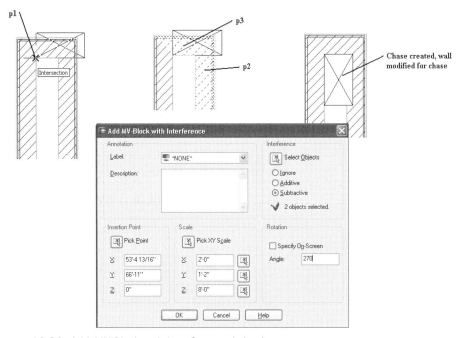

Figure 10.30 *Add MV Block with Interference dialog box*

Z – The **Z** absolute coordinate value can be typed to specify the Z location of the insertion point of the chase.

SCALE

Pick XY Scale – Selecting the **Pick XY Scale** button returns you to the workspace. When you return to the workspace the cursor rubber-bands from the insertion point, and you can select a point diagonally located from the insertion point to size the chase.

X – Specify the **X** scale dimension to size the unit block. Therefore if 24 is typed in this edit field, the chase will be 24 units wide in the X direction.

Y – Specify the **Y** scale factor to size the unit block in the Y direction.

Z – Specify the **Z** scale factor to size the unit block in the Z direction.

Interference Select Objects – The **Select Objects** button of the **Interference** section will return you to the drawing area, allowing you to select AEC objects, which may interfere with the chase. **Ignore**, **Additive**, and **Subtract** Boolean operations can be applied to the interference between the chase and the selected AEC objects. The chase inserted as shown in Figure 10.30 is applied to subtract from the wall, and the shrinkwrap is excluded. The Ignore option will break the wall, and the shrinkwrap is applied without regard to the chase, whereas the Additive option will break the wall and include the shrinkwrap with the chase.

Rotation – The **Angle** edit field allows you to type the rotation angle or check the **Specify on Screen** check box to specify the rotation of the chase.

STEPS TO INSERTING A CHASE USING THE CONTENT BROWSER

1. Drag the i-drop for **Chase (6)** from the **Content Browser**.
2. Select an insertion point at **p1** as shown in Figure 10.30.
3. The **Add MV-Block with Interference** dialog box opens; select the **Select Objects** button and select the walls at **p2** and **p3** as shown in Figure 10.30.
4. Edit the **Add MV-Block with Interference** dialog box: specify rotation = 270 and Y Scale value = 1'-2". Specify Subtractive interference as shown in Figure 10.30.
5. Select **OK** to close the dialog box; chase is created as shown in Figure 10.30.

CONTENTS OF THE MISCELLANEOUS FOLDER

Additional dimensions and other annotation techniques such as Fire Rating Lines, Match Lines, and North Arrows are included in the Miscellaneous folder of the DesignCenter and the Documentation Tool Catalog-Imperial. The explanation regarding annotation inserted from these categories is included in Appendix B of the CD.

CREATING TAGS AND SCHEDULES FOR OBJECTS

Schedules can be created to list doors, windows, walls, or any object in the drawing. The schedule tools allow you to collect data from one or more drawings and to export the schedule data to a Microsoft Excel spreadsheet. The schedule provides detailed information regarding the size and construction of the object. Each object of a schedule is usually assigned a mark, which is placed as a tag near the object in the drawing. Schedules can be used to determine the quantity of a certain type of building component repeated throughout the building.

Tags and schedules can be developed within Construct drawings, or if project-based tags are used, the schedule can be developed based upon all drawings of the project. When you insert a tag for an object, the **properties** associated with the **tag** are inserted in the drawing. The tag is a multi-view block with properties defined in its attributes. The data defined for the **properties** is extracted and placed in the schedule when the schedule is inserted. The schedule data for the object consists of property sets; the properties of a door schedule are shown in Figure 10.31. Schedule table styles are located in the **Schedule Table Styles** folder of Documentation Objects in the **Style Manager**. The **Columns** tab of the **Schedule Table Style Properties** dialog box for a Door Schedule Style is shown in Figure 10.31. The **Style Manager** allows you to view the properties included in the schedule table. Properties consist of data such as height, width, hardware, or materials. Some data for the properties is **Automatic**, extracted from the object in the drawing. However, other data is **Manual**, inserted by editing the **Edit Property Set Data** dialog box. When a schedule is created, the data from the property sets of the drawing is extracted and included in the schedule.

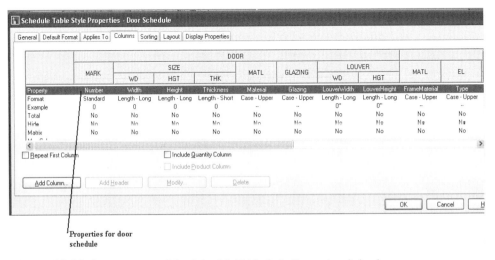

Figure 10.31 *Property sets of the Schedule Table Style Properties dialog box*

Note: The **AEC Content** tab of the **Options** dialog box allows you to toggle ON **Display Edit Property Data Dialog During Tag Insertion**. Therefore, upon insertion of the tag you can enter the data for the schedule in the **Edit Property Set Data** dialog box.

Door and window schedules and tags are located on the Scheduling tool palette shown in Figure 10.32. Fifteen schedule tables and associated tags are located in the Schedule Tables and Schedule Tags categories of the Documentation Tool Catalog-Imperial shown in Figure 10.33. Schedule tags are also located in the Architectural Desktop\Imperial\Documentation\Schedule Tags folder of the DesignCenter. Schedule tables can be imported within the **Style Manager** from the *C:\Documents and Settings\All Users\Application Data\Autodesk\ADT 2005\enu\Styles\Imperial\Schedule Table (Imperial).dwg* directory. The following schedule tables can be imported in the **Style Manager** from the Documentation Objects\Schedule Table Styles folder: Area Grouping-BOMA, Area List BOMA, Door Schedule, Door Schedule Project Based, Equipment Schedule, Furniture Schedule, Room Finish Matrix, Room Finish Matrix Project Based, Room Finish Schedule, Room Finish Schedule Project Based, Room Schedule, Room Schedule Project Based, Space Inventory, Wall Schedule, and Window Schedule. You can also modify these schedule table styles or create new ones to meet your schedule needs.

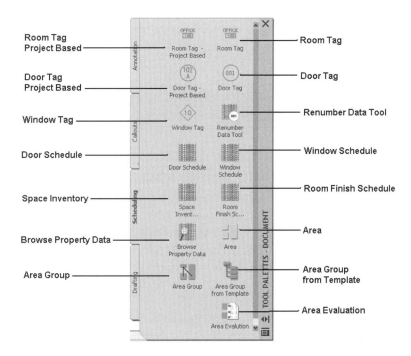

Figure 10.32 *Scheduling palette*

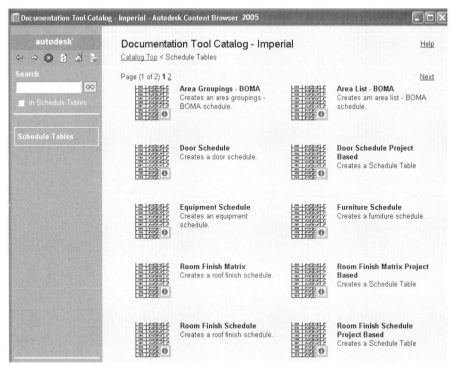

Figure 10.33 Schedule Tables category of the Documentation Tool Catalog-Imperial

In summary, the schedule is created based upon the format defined in schedule table style, which extracts property information from the schedule data associated with the Architectural Desktop object and the object's style. When tags are inserted in the drawing, the properties for the tagged object are inserted. After a schedule is inserted, it can be dynamically updated as the drawing changes.

PROPERTIES OF OBJECTS AND OBJECT STYLES

The properties included in the drawing are inserted when you tag an object, or the properties can be pasted from other drawings into the Documentation\Property Set Definitions folder of the current drawing in the **Style Manager**. When an object is tagged, the properties associated with the tag are inserted into the drawing. When you place a door tag, the **Door Object** property and the **Door Style** property are inserted. The **Door Object** property set is displayed in the **Edit Property Set Data** dialog box of the door as shown in Figure 10.34. The style-based properties associated with a door style are edited by selecting the **Property Sets** button of the **General** tab in the **Door Style Properties** dialog box as shown in Figure 10.35. When you select the **Property Sets** button, the **Edit Property Set Data** dialog box is displayed, which lists the style-based properties of the door.

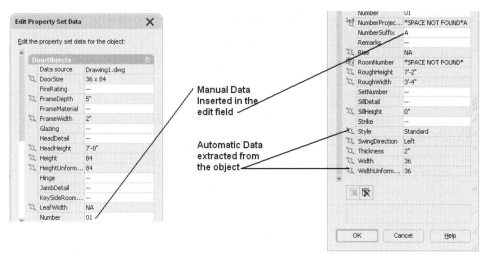

Figure 10.34 *Edit Property Set Data for a Door object*

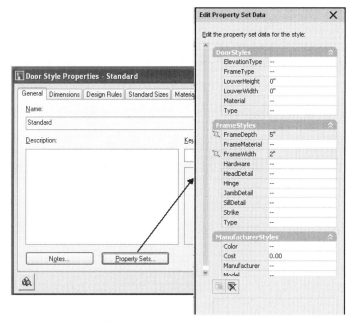

Figure 10.35 *Edit Property Set Data for Property sets of Door style*

If no tags have been inserted, property sets can be pasted into the drawing in the **Style Manager**. If a drawing consists of doors but no tags, the Properties palette will display an empty **Extended Data Property** field as shown on left in Figure 10.36. If you paste the **Door Objects** property set into the drawing, the **Add property sets** button becomes active in the **Extended Data** tab of the Properties palette as shown at right in Figure 10.36.

If you select the **Add property sets** button, the **Add Property Sets** dialog box opens, listing the property sets appropriate for the door object.

The Room Finish Objects, Room Objects, and Space Objects are the only property sets included in the drawing prior to placing tags. There are 30 property sets include in the Schedule Table (Imperial).dwg of the Documentation Objects\Property Set Definitions folder. The Schedule Table (Imperial).dwg file is located in the *C:\Documents and Settings\All Users\Application Data\Autodesk\ADT 2005\enu\Styles\Imperial* directory. The property sets shown in Table 10.22 are inserted in the drawing when a door tag is placed.

PLACING DOOR TAGS

Door tags can be applied to door objects by selecting the **Door Tag** or **Door Tag Project Based** command on the Scheduling palette. The tag is a multi-view block with schedule properties defined as its attributes. The door tag can be applied in the View drawing, although the door exists within the Construct drawings attached as a reference to create the view. When you place a door or window tag, the **AnnoScheduleTagAdd** command is used to place the symbol. Preset symbol, leader,

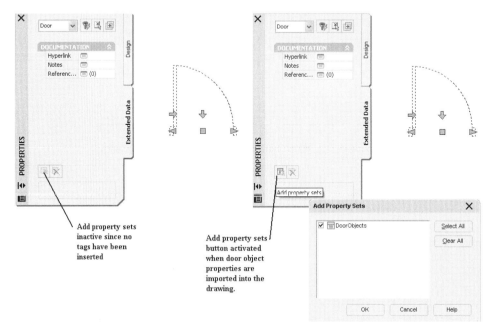

Figure 10.36 *Adding property sets to the object*

Property Sets	Objects/Styles and Definition	Applies to
Door Object	Object	Door, Door/Window Assembly
Door Styles	Styles and Definition	Door Style, Door/Window Assembly
Frame Styles	Styles and Definitions	Door Style, Door/Window Assembly Style, Window Styles
Manufacturer Styles	Styles and Definitions	All Objects

Table 10.22 *Property Sets*

and dimstyle values for the command are selected when the symbol is selected from the Scheduling tool palette or the DesignCenter. Door tags are restricted multi-view blocks; therefore you must select a door object to place a door tag. The steps to placing a door tag are shown below.

STEPS TO PLACING A DOOR TAG

1. Select the **Door Tag** command from the Scheduling tool palette.
2. You are prompted in the command line to select a door; select a door at **p1** as shown in Figure 10.37.
3. Specify the location of the tag near the door at **p2** as shown in Figure 10.37.
4. Make the necessary changes to the **Edit Property Set Data** dialog box as shown in Figure 10.37.
5. Select OK to dismiss the **Edit Property Set Data** dialog box.
6. Door tag is placed in drawing. Press ESC to end the command.

The location of the tag can be selected with the pointer, or you can press ENTER to center the tag about the door. The **Edit Property Set Data** dialog box opens each time you select an additional door. Tags for multiple doors can be placed by selecting the Multiple option of the command. The Multiple option allows you to select additional doors without reopening the **Edit Property Set Data** dialog box for each insertion. If you select Multiple, you can select the entire drawing and all doors will be tagged. The Multiple option can be selected by typing **M** in the command line after you close the **Edit Property Set Data** dialog box for the first tag, as shown in the following command line sequence. The position of the additional tags is defined based upon the position of the first tag. Therefore, if you place the first door tag in the center of the opening, the multiple insertions of the remaining tags will be placed in a similar position. The door tag is placed on the **A-Door-Iden** layer, which has the color 132 (green).

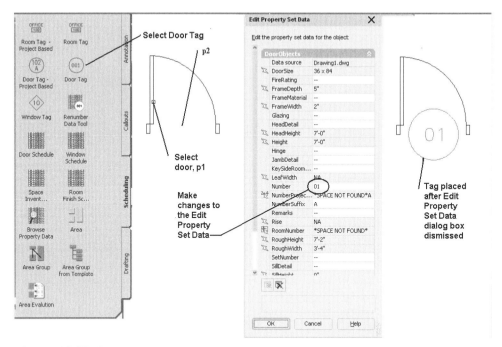

Figure 10.37 Door tag placed in the drawing

Select object to tag [Symbol/Leader/Dimstyle/Edit]: *(Select a door.)*

Specify location of tag <Centered>: *(Select a location for the tag or press ENTER to center the tag about the object.)*

(The **Edit Property Set Data** *dialog box opens; edit the properties, and select* **OK** *to dismiss the dialog box.)*

Select object to tag [Symbol/Leader/Dimstyle/Multiple/Edit]: **M** ENTER *(Press ENTER to continue selecting doors.)*

Select objects to tag: *(Select additional doors; press ENTER to end selection and edit the* **Edit Property Set Data** *dialog box.)*

PLACING WINDOW TAGS

A window tag is also located on the Annotation tool palette as shown in Figure 10.37. Additional door and window tags can be inserted from the Schedule Tags category of the Documentation Tool Catalog-Imperial as shown in Figure 10.38. The size of the Window Tag Scale Dependent tag is controlled by the scale associated with the Display Configuration. (See Table 10.8 for scales associated with Display Configurations.)

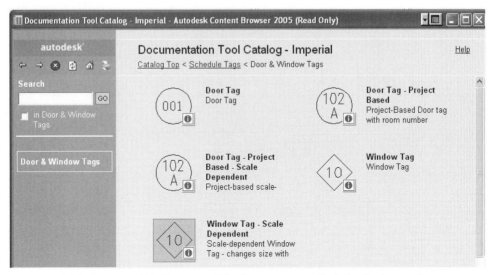

Figure 10.38 *Door and Window tags of the Documentation Tool Catalog-Imperial*

Window tags are multi-view blocks that are restricted to tagging only windows. Therefore you cannot insert a window tag unless a window exists in the drawing. The window tag is placed on the **A-Glaz-Iden** layer, which has the color 152 (a hue of blue).

STEPS TO PLACING A WINDOW TAG

1. Select **Window Tag** from the Scheduling palette.
2. Select a window at **p1** shown at left in Figure 10.39 for the placement of a tag.
3. Select a location for the window tag (**p2** as shown at right in Figure 10.39). The **Edit Property Set Data** dialog box opens as shown in Figure 10.39.
4. Edit the **Edit Property Set Data** dialog box, and select **OK** to close the dialog box and display the window tag as shown at right in Figure 10.39.
5. Continue to select additional windows to place additional tags or press ENTER to end the command.

Tags can also be placed from the AutoCAD DesignCenter. To place window tags from the DesignCenter, double-click on the tag, select a window, specify the location of the tag, and then edit the **Edit Property Set Data** dialog box. Door tags are selected from the DesignCenter using a similar procedure.

ROOM AND FINISH TAGS

Room tags can be placed by selecting **Room Tag** from the Scheduling palette or by selecting other room tags from the Room & Finish Tags folder of the Architectural

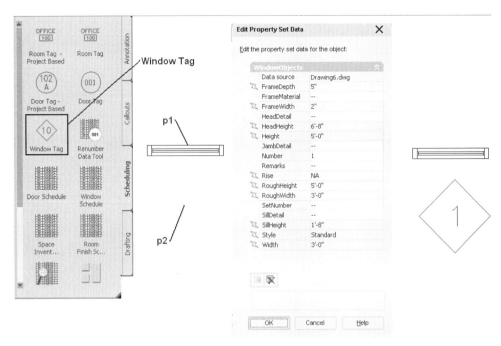

Figure 10.39 *Placing a window tag*

Desktop\Documentation\Schedule Tags\Room & Finish Tags folder of the AutoCAD DesignCenter. The tags located in the Architectural Desktop\Schedule Tags folder of the DesignCenter allow you to specify the room number, finish, and area of a room for the development of a schedule. When you select **Room Tag** from the Scheduling palette, the **AecAnnoScheduleTagAdd** command is used to place the room tag. To place a room tag, you must select a space object to tag the room. The space object includes data regarding the length, width, and height of the room. The room and finish tags are specifically designed for use with Aec Space objects created with the **SpaceAdd** command. When tags are applied to an Aec Space object, the area of the space or room is extracted from the properties of the Aec Space object.

Creating Spaces

A building can be designed based upon spaces, which are discussed in Chapter 12. Designing with Aec Space objects allows you to place space objects, which represent different rooms. Spaces can be assigned the names of the various rooms such as living, dining, and kitchen. Architectural Desktop includes predefined space styles for residential and commercial construction. These predefined spaces can be placed in the drawing to perform space planning.

However, if a floor plan is created without the use of the space planning tools, an Aec Space object can be created from the existing wall geometry. The room tags can then be attached to the Aec Space. The creation of an Aec Space object allows you to quickly determine the area of a room. You can create an Aec Space object from an existing floor plan with the **AecSpaceAutoGenerate** command. Access the **AecSpaceAutoGenerate** command as shown in Table 10.23.

Command prompt	AECSPACEAUTOGENERATE
Tool palette	Select Space Auto Generate tool from the Design palette

Table 10.23 *Space Auto Generate command access*

When you select the **AecSpaceAutoGenerate** command, the **Generate Spaces** dialog box opens as shown in Figure 10.40. The **Generate Spaces** dialog box includes **Automatic** and **Manual** radio buttons.

If you select the **Automatic** radio button with the filter set to **Walls only** and then move the pointer over a room, the boundary enclosed by the walls will be highlighted. If you click inside the room, the highlighted boundary will be confirmed. A

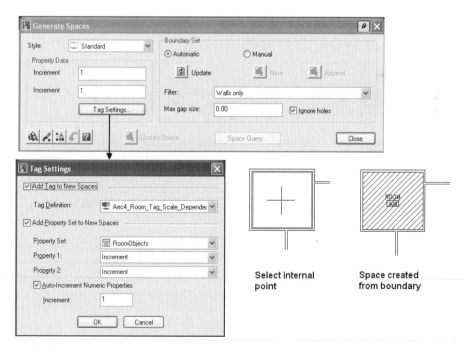

Figure 10.40 *Creating a space with the Space Auto Generate command*

Filter drop-down list allows you to specify which of the following components to contain the boundary: Walls only; Walls lines arcs polyline and circles; and All Linework.

When you select the **Manual** option, the cursor converts to a pick box that allows you to select the geometry to create the boundary. After you select the components of the boundary, you are prompted to select an internal point for the development of the boundary. This command creates a boundary for the space similar to the flood option of **Bhatch** command. The options of the **Generate Spaces** dialog box shown in Figure 10.40 are described below:

> **Style** – The **Style** list allows you to select the space styles included in the drawing. Unless space styles have been pasted into the drawing, only the Standard style is listed.
>
> **Automatic** –The **Automatic** radio button searches for a boundary based upon the objects specified in the filter when you move the pointer over rooms and select an internal point.
>
> **Manual** – The **Manual** option converts the pointer to a pick box, which allows you to select the geometry that bounds the space.
>
> **Append** – The **Append** button is active when the **Manual** method is selected, which allows you to modify the selection set used to create the boundary.
>
> **New** – The New button is active when the **Manual** method is selected, which allows you to create a new selection set for a boundary.
>
> **Filter** – The **Filter** drop-down list allows you to specify which of the following components to contain the boundary: Walls only; Walls lines arcs polyline and circles; and All Linework.
>
> **Update Space** – The **Update Space** button at the bottom of the dialog box allows you to modify an existing space by redefining its boundaries.
>
> **Increment** – The **Increment** option specifies the increment of increase for tags placed with numeric input. The **Property Data** field title depends upon the tag definition specified in the **Tag Settings** dialog box. The **Property Data** fields may include such properties as Base Color, Name, Increment, or Manufacturer depending upon the Property Set specified with the **Tag Definition**.
>
> **Tag Settings** – Select the **Tag Settings** button to open the **Tag Settings** dialog box as shown in Figure 10.40. You can tag each space upon creation if you check the **Add Tag to New Spaces** check box and select a tag from the **Tag Definition** list. The Aec3_RoomTag will allow you to place a room number automatically incremented as you select additional rooms. If your drawing is part of a project, you can select the Aec4_Room_Tag_Scale_Dependent to tag and increment the room numbers based upon levels defined in the project. The Aec4_Room_Tag_Scale_Dependent tag text height is scaled based upon

the scale specified by the **display configuration** (see Table 10.8). Additional tags can be selected from the Room Finish Tags category of the Documentation Tool Catalog-Imperial. The properties of the specified tag definition are listed in the **Tag Settings** dialog box shown in Figure 10.40. If you check the **Add Property Set to New Spaces** check box, you can identify the **Property Set** to attach to the new space from the list. The **Auto-Increment Numeric Properties** check box allows you to automatically increment room or space numbers. Therefore, if you tag a room 101, the next tag will automatically increment to 102. The room numbering tool increments room numbers based upon level defined in the project; therefore, second-level rooms increment from 200.

Tip: If you have inserted a tag into the drawing from the Room and Finish tag category of the Documentation Tool Catalog-Imperial, the tag will be listed in **Tag Definition** drop-down list of the **Tag Settings** dialog box.

Space Query – The **Space Query** button executes the **Space Query** command from within the dialog box. The **Space Query** command summarizes the spaces inserted in the drawing as shown in Figure 10.41.

Base Color – The **Base Color** edit fields shown in Figure 10.42 are displayed when the Property Set = RoomFinishObjects is specified in the **Tag Settings** dialog box.

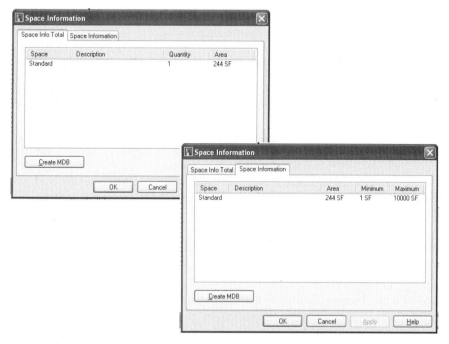

Figure 10.41 *Space Information dialog box*

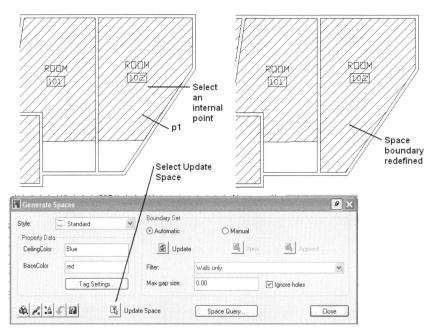

Figure 10.42 *Revising a space with the Update Space option*

Increment – The **Increment** fields allow you to specify the integer to increment the tag. The **Increment** fields are displayed when the **Property Set** = RoomObjects, is specified in the **Tag Settings** dialog box as shown in Figure 10.40.

If you have repositioned walls after a space has been formed, you can select the **Space Auto Generate** Tool and choose the **Update Space** button to revise the space. This option will prompt you to select the internal space to revise the space as shown in the following command sequence.

Select a space object to update: *(Select a space at p1 shown in Figure 10.42.)*

Select internal point: *(Select a point near p2 as shown in Figure 10.42.)*

(The space is redefined as shown at right.)

The Standard space style can be applied, because it can be used for spaces from 1 to 10,000 square feet in area. The Standard space style is the default space style and can be used to label the room with the room and finish tags. Additional space styles can be imported into the drawing from the Spaces category of the Design Tools Catalog-Imperial. This directory includes the following categories, Commercial, Educational, Medical, and Residential. The Residential space styles are shown in Figure 10.43. After

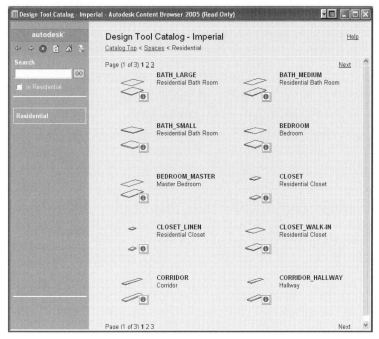

Figure 10.43 *Residential space styles of the Design Tools Catalog-Imperial*

inserting space styles into the drawing, you can assign a space style to an existing space in the Properties palette.

A space style includes a name that describes the space and a range of acceptable width and length dimensions for the space. The space name can be included in the tag of the space.

STEPS TO CREATING A SPACE FROM WALLS WITHIN A PROJECT

1. Verify that the current drawing is assigned to a project in the Drawing Window status bar. Select the **Space Auto Generate Tool** from the Design palette to open the **Generate Spaces** dialog box.

   (If you want to insert the tag during this operation, select the **Tag Settings** button. Check **Add Tag to New Spaces**, **Add Property Set to New Spaces**, and **Auto-Increment Numeric Properties**. Verify that Tag Definition = Aec4_Room_Tag_Scale_Dependent and Property Set = Room Objects, Property 1 = Increment, and Property 2 = Increment. Select **OK** to dismiss the **Tag Settings** dialog box.)

2. Toggle ON the **Automatic** radio button, and then verify that **Walls only** is set as the **Selection Filter** of the **Generate Spaces** dialog box. Set **Property Data Increment** to 1 for each property.

3. Move the pointer over an internal point, and verify that the boundary is displayed red.

4. Click inside the boundary as shown in the following command line sequence.

>   Command: _AecSpaceAutoGenerate
>
>   Select internal point or [Style/UPdate space/Query/Match]: *(Select point p1 as shown in Figure 10.44.)*
>
>   Select internal point or [Style/UPdate space/Query/Undo/Match]: *(Press* ENTER *to end selection.)*

The display of the hatch pattern representing the AEC Space object can be turned off by freezing the **A-Area-Spce** layer. The Presentation display configuration displays the space as a shaded object.

Placing Room and Finish Tags

Room and finish tags can be placed by selecting **Room Tag** from the Scheduling tool palette. Additional tags can be placed from the Room and Finish Tags category of the Documentation Tool Catalog-Imperial or the DesignCenter. The Room Tag shown in Figure 10.45 was inserted with a room number tag as shown in the following command line sequence. The room tag automatically increments the room number if additional spaces are selected.

>   *(Select* **Room Tag** *from the Scheduling tool palette.)*
>
>   Command: _AecAnnoScheduleTagAdd

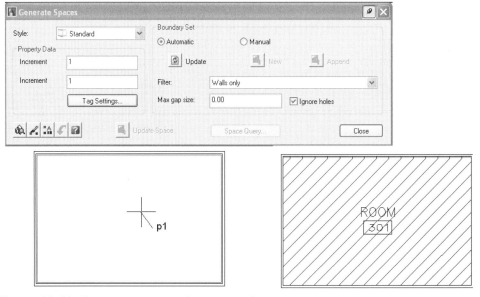

Figure 10.44 *Creating a space tag for a project drawing*

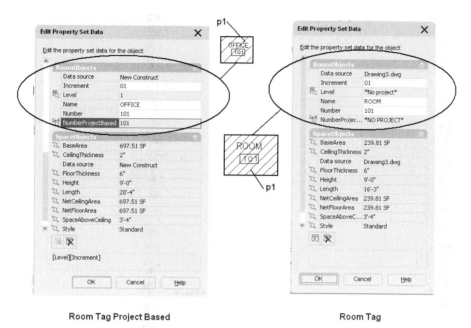

Figure 10.45 Room Tag placed for a space

(Note: Preset values of the command are not shown in the command line because they require no user input.)

Select object to tag [Symbol/Leader/Dimstyle/Edit]: *(Select the space at p1 as shown in Figure 10.45.)*

Specify location of tag <Centered>: ENTER *(Press ENTER to center the tag about the space.)*

(Edit the **Edit Property Set Data** *dialog box as shown in Figure 10.45, and select* **OK** *to dismiss the dialog box.)*

Select object to tag [Symbol/Leader/Dimstyle/Multiple/Edit]: ENTER *(Press ENTER to end selection.)*

The property data for the **Room Tag Project Based** tag included on the Scheduling palette shown at left in Figure 10.45 includes additional data regarding the project. Additional room and space tags can be selected from the Room & Finish Tags category of the Documentation Tool Catalog-Imperial as shown in Figure 10.46. The **Space Tag** of the Room and Finish Tags folder will extract the name of the space style when the tag is placed. The **Room Finish Tag** allows you to type the finish materials for surfaces of the room in the **Edit Property Set Data** dialog box.

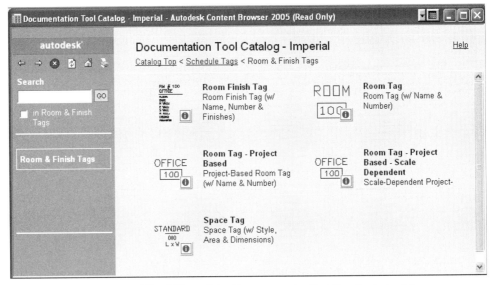

Figure 10.46 *Room and Finish Tags of the Documentation Tool Catalog-Imperial*

The room tags and space tag symbols are placed on the **A-Area-Iden** layer, which has the color 152 (a hue of blue). The Room Finish Tag symbol is placed on the **A-Flor-Iden** layer, which has the color 171 (a hue of blue).

 STOP. Do Tutorial 10.2, "Adding Space Tags for Rooms," at the end of the chapter.

OBJECT TAGS

Tags for structural components, equipment, and furniture can be placed in the drawing to identify a mark for the development of a schedule. The object tags listed in Table 10.24 are located in the Schedule Tags category of the Documentation Tool Catalog-Imperial. These tags are also located in the Architectural Desktop\Imperial\Documentation\Schedule Tags\Object Tags folder of the DesignCenter.

The Object Tags folder includes structural tags for beams, braces, and columns. The structural tags include the StructuralMemberStyles:Style and StucturalMemberObjects: Number properties. Therefore, when you create a structural member style from the Structural Catalog, the name assigned to the member is extracted and included in the tag. Structural tags are restricted to structural members created from the **Structural Catalog** discussed in Chapter 13, "Drawing Commercial Structures." The Equipment tag is designed to tag block references or multi-view block references such as equipment, furniture, or appliances. The **AecAnnoScheduleTagAdd** command is used to place the object tags. Table 10.24 lists each tag, and its layer and color properties. Object tags are placed using the same procedure as other tags; therefore, only objects of the type specified for the tag can be selected.

Tag	Tag Purpose	Layer	Color
Beam Tag	Structural Beam	S-Beam-Iden	131 hue of cyan
Brace Tag	Structural Brace	S-Cols-Brce-Iden	151 hue of blue
Column Tag	Structural Columns	S-Cols-Iden	171 hue of blue
Equipment	Multi-view block	A-Eqpm-Iden	132 hue of cyan
Equipment Leader	Multi-view block	A-Eqpm-Iden	132 hue of cyan
Equipment Tag	Multi-view block	A-Eqpm-Iden	132 hue of cyan
Equipment Tag Scale Dependent	Multi-view block	A-Eqpm-Iden	132 hue of cyan
Furniture	Furniture	I-Furn-Iden	232 hue of magenta
Furniture Leader	Furniture	I-Furn-Iden	232 hue of magenta
Furniture Tag Scale Dependent	Furniture	I-Furn-Iden	232 hue of magenta

Table 10.24 *Object tags of DesignCenter*

WALL TAGS

Wall tags are used to identify different types of wall construction. A wall tag identifies a wall that is detailed in a different location or in a wall schedule. Wall tags are located in the Wall Tags category of the Documentation Tool Catalog-Imperial. Tagging a wall allows you to create a schedule of the walls that lists the height, width, surface area, and volume. There are two wall tags included in the **Wall Tags** folder: Wall Tag and Wall Tag (Leader), as shown in Figure 10.47.

Wall tags can be attached to a wall with a leader or placed near the wall. The wall tag is placed on the **A-Wall-Iden** layer, which has the color 211 (magenta). The **AecAnnoScheduleTagAdd** command is used to place the wall tags symbol. The text or value placed in the tag is extracted from the wall style. Therefore, after you place the tag, you can assign a value for the tag in the **Wall Style Properties** dialog box for the wall. The value is assigned by selecting the **Property Sets** button on the **General** tab of the **Wall Style Properties** dialog box. The **Property Sets** button opens the **Edit**

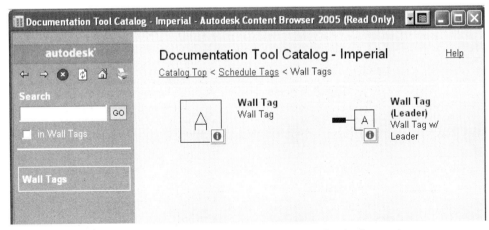

Figure 10.47 *Wall Tags category of the Documentation Tool Catalog-Imperial*

Property Set Data dialog box, allowing you to enter a value for the **Type** property. The procedure to place a Wall Tag (Leader) is shown in the following steps.

STEPS TO PLACING WALL TAGS

1. Drag the i-drop of the **Wall Tag (Leader)** from the Documentation Tool Catalog-Imperial into the workspace. Respond to the command line prompts as shown below to place the tag. (The command line sequence shown includes only required user input content.)

    Command: _AecAnnoScheduleTagAdd

    Select object to tag [Symbol/Leader/Dimstyle/Edit/Constrain/Rotation]: *(Select the wall at point p1 in Figure 10.48.)*

    Specify next point of leader line or [New] : *(Select a point near p2 in Figure 10.48 to end the leader.)*

    Specify next point of leader line or <End leader>: ENTER *(Press ENTER to end leader location.)*

    (Select **OK** *to dismiss the* **Edit Property Set Data** *dialog box.)*

    Select object to tag [Symbol/Leader/Dimstyle/Multiple/Edit/Constrain/Rotation]: ENTER *(Press ENTER to end the command.)*

2. To add text to the tag, select the Wall, right-click, and choose **Edit Wall Style** from the shortcut menu.

3. Select the **General** tab, and select the **Property Sets** button to open the **Edit Property Set Data** dialog box. Type the text for the wall tag in the **Type** edit field as shown at right in Figure 10.48.

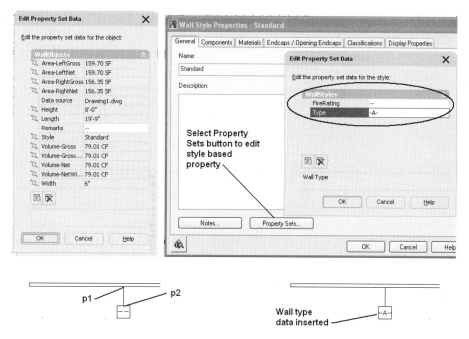

Figure 10.48 Adding a wall tag

4. Select **OK** to dismiss the **Edit Property Set Data** dialog box and select **OK** to dismiss the **Wall Style Properties** dialog box.

EDITING TAGS AND SCHEDULE DATA

When tags are inserted in the drawing, the object based property information for a schedule is typed in the **Edit Property Set Data** dialog box that opens when the tag is placed. The object based property data is also displayed on the **Extended Data** tab of the Properties palette. To add style based data, select the **General** tab of the object's style properties dialog box, and then select the **Property Sets** button to open the **Edit Property Set Data** dialog box. The **Edit Property Set Data** dialog box of the object style can be changed after the tag has been inserted.

ADDING A SCHEDULE TABLE

The Scheduling tool palette includes the Door Schedule, Window Schedule, Space Inventory Schedule, and Room Finish Schedule. Prior to inserting a schedule table, tag the objects related to the table. The process of adding the tag and editing the **Edit Property Set Data** dialog box creates most of the data required for the schedule. Access the **ScheduleAdd** command as shown in Table 10.25. The content of the schedule includes properties and property sets assigned to that object. To add a schedule table, click on one of the schedule tables included on the Scheduling palette as shown in Figure 10.49.

Command prompt	SCHEDULEADD
Tool palette	Select a schedule table from the Scheduling palette

Table 10.25 *TableAdd command access*

When you select a schedule table from the Scheduling palette, you are prompted in the command line to select objects for the schedule or press ENTER to apply the schedule to an external drawing. Only objects of the type designed for the schedule can be selected and applied to the schedule. Therefore you can select all the objects of the drawing, and only the objects of the type designed for the schedule will be selected for the schedule.

The **ScheduleAdd** command is used to add schedules to the drawing. When you select the **Window Schedule** command from the Scheduling palette, the **ScheduleAdd** command is used to insert the schedule. The **ScheduleAdd** command opens the Properties palette, which allows you to control the objects included in the schedule and options regarding the update of the schedule.

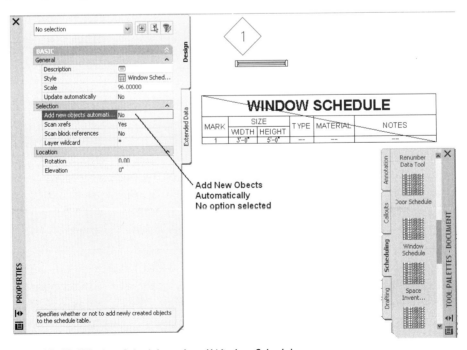

Figure 10.49 *Window Schedule tool and Window Schedule*

The Properties palette of the **ScheduleAdd** command consists of the following options.

Basic

General

Description – The **Description** button opens the **Description** dialog box, which allows you to add text describing the schedule table.

Style – The **Style** list includes all schedule table styles that have been inserted in the drawing.

Scale – The default **Scale** value is equal to the specified value in the **Scale** tab of the **Drawing Setup** dialog box.

Update automatically – The **Yes** option of this property allows changes to the drawing to be automatically reflected in the schedule. If the object is changed after the schedule is inserted when the **No** option of this property is selected, a diagonal line is drawn through the schedule as shown in Figure 10.49. The diagonal line can be removed by updating the schedule.

Selection

Add new objects – If the **Yes** option is specified for **Add new objects**, objects added after the schedule is placed will automatically be included in the schedule.

Scan xrefs – The **Yes** option of **Scan xrefs** allows the search of objects to include the content of external reference files attached to the drawing.

Scan block reference – If the **Scan block reference drawings** option is **Yes**, Architectural Desktop objects included in blocks that are inserted in the current drawing will be included in the schedule selection set.

Layer wildcard – The **Layer wildcard** field allows you to create a layer filter to limit the search for objects that apply to the schedule. The asterisk wildcard will include all layers of the drawing.

Location

Rotation – The **Rotation** edit field is active if you are editing an existing schedule. The angle of rotation of the schedule table is displayed in this edit field.

Elevation – The **Elevation** edit field is active if you are editing an existing schedule. The elevation of the schedule table is displayed in this edit field.

After editing the Properties palette for the schedule table and selecting objects, you are prompted to specify the location for the schedule in the drawing area. When you specify the location of the schedule, you can specify its size by selecting the location of the upper left and lower right corners, or you can accept the default size by pressing ENTER. If you select the default size, the schedule size is scaled to the scale value specified by the **AecDwgSetup** command.

USING SCHEDULE TABLE STYLES

Schedule tables are used to organize and present the data inserted in the **Edit Property Set Data** dialog box in the form of a schedule. Most schedules include a mark column, which is a letter or number identifying such objects as a door or window. The mark is defined when a tag is placed in the drawing. The tables included allow you to create door, window, space, and room finish schedules. Additional tables can be imported in the **Style Manager** from the *C:\Documents and Settings\All Users\Application Data\Autodesk\ADT 2005\enu\Styles\Imperial* folder or from the Documentation Tool Catalog-Imperial of the **Content Browser**. Schedule tables can be developed to display the schedule data in the desired order. Access the schedule table style as shown in Table 10.26.

 Tip: Schedule tables can be placed on the Scheduling palette or on a new palette by opening the **Content Browser** and dragging the i-drop of the schedule table, from the Documentation Tool Catalog-Imperial or the Documentation Tool Catalog-Metric of the **Content Browser**.

Command prompt	SCHEDULESTYLE
Palette	Select a schedule table on the Scheduling tool palette, right-click, and choose Schedule Table Styles

Table 10.26 *Schedule Table Styles command access*

The schedule tables of the Scheduling palette are shown in Figure 10.50. The **EditScheduleStyle** command allows you to edit the format of the schedule according to user preferences.

When you select a schedule table, right-click, and select **Edit Schedule Table Style**, the **Style Manager** opens to the **Documentation Objects\Schedule Table Styles** folder. Double-click on the name of an existing schedule table to open the **Schedule Table Style Properties** dialog box as shown in Figure 10.51. The tabs of the **Schedule Table Style Properties** dialog box allow you to define the content of the schedule. A description of these tabs follows.

General Tab

    Name – The **Name** of the schedule table is displayed in this field.

    Description – A **Description** of the schedule can be typed in this field.

    Notes – The **Notes** button allows you to add notes and files in the **Notes** and **Reference Docs** tabs.

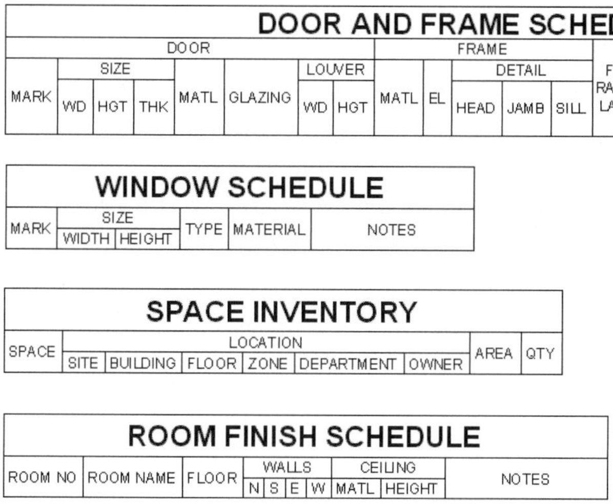

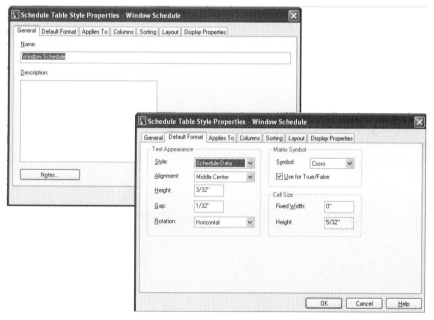

Figure 10.50 *Schedules of the Scheduling tool palette*

Figure 10.51 *General and Default Format tabs of the Schedule Table Style Properties - Window Schedule dialog box*

Default Format Tab

The **Default Format** tab is shown in Figure 10.51.

Text Appearance

    Style – The **Style** drop-down list displays the text styles of the drawing.

    Alignment – The **Alignment** options define the justification of the text in the schedule. The Middle Center alignment will center the text horizontally and vertically within the cell of the schedule.

    Height – The **Height** edit field allows you to define the text height, which will be scaled according to the scale factor of the **Scale** tab of the **Drawing Setup** dialog box.

    Gap – The **Gap** field defines the distance between rows of text and the schedule table lines.

    Rotation – The **Rotation** list includes the Horizontal and Vertical options, which control the orientation of the text.

Matrix Symbol

    Symbol – The **Symbol** list allows you to specify one of the following symbols used in a matrix schedule: Check, Dot, Cross, and Slash.

    Use for True/False – The **Use for True/False** check box, if selected, applies the matrix symbol when the option of the schedule applies.

Cell Size

    Fixed Width – The **Fixed Width** of a cell can be set to a specific distance. If the **Fixed Width** is set to zero, the width varies with the width of text necessary for the cell.

    Height – The **Height** edit field is the vertical dimension of the cell.

Applies To Tab

The **Applies To** tab includes a list of all objects to which the schedule can be applied for the development of the schedule. The Window schedule shown in Figure 10.52 includes a check in the Window box.

Columns Tab

The **Columns** tab allows you to preview the list of properties to be included in the table. It includes buttons for creating and modifying each component. The **Columns** tab shown in Figure 10.53 lists the properties of the Window schedule. This tab allows you to identify the property set and property used in each column of the schedule to extract the data from the object. Therefore the **Mark** column will display the value of the **Number** property for each window object. The window schedule is developed from the WindowObjects and WindowStyles property sets as shown in Figure 10.53.

A description of the **Columns** tab follows:

    Repeat First Column – The **Repeat First Column** check box allows you to insert a copy of the first column on the right side of the schedule.

Annotating and Documenting the Drawing 707

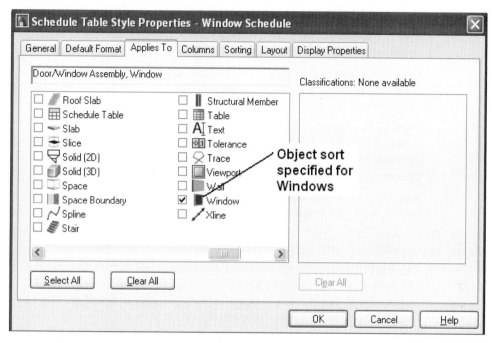

Figure 10.52 *Applies To tab of the Schedule Table Style Properties - Window Schedule dialog box*

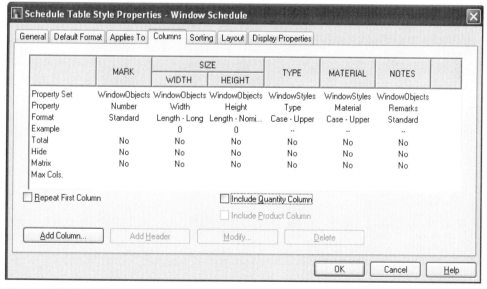

Figure 10.53 *Columns tab of the Schedule Table Style Properties - Window Schedule dialog box*

Repeating the first column assists the reader when the chart is extremely wide. Using **Repeat First Column** can allow the Mark column to be placed on the left and right margins of the schedule.

Include Quantity Column – The **Include Quantity Column** check box allows you to insert a column that lists the number of repetitions of a building component in the schedule.

Include Product Column – The **Include Product Column** check box allows you to insert a column that lists the name of the product identified by the mark. **Include Product Column** can only be selected if the **Include Quantity Column** check box has been selected.

Add Column – The **Add Column** button allows you to add a column to the schedule. The **Add Column** button opens the **Add Column** dialog box shown in Figure 10.54. The **Add Column** dialog box allows you to define the contents of the new column.

The **Add Column** dialog box allows you to select the property to be included in a new column. The **Add Column** dialog box shown in Figure 10.54 includes the **Property Set/Properties** list, **Column Properties**, and **Column Position** sections described below.

> **Property Set/Properties** List – The **Property Set/Properties** list includes a comprehensive list of the properties of the drawing available for the schedule. When you select a property from the list, you are adding that property to the schedule. The position and properties for the new column are defined in the sections located at right in the dialog box.

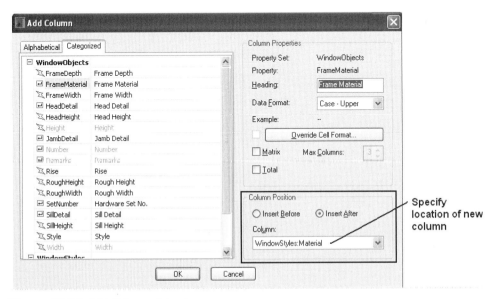

Figure 10.54 *Add Column dialog box*

Column Properties

Heading – The **Heading** edit field specifies the title of the heading for the new column in the schedule.

Data Format – The **Data Format** allows you to specify how the data of the column will be displayed. Data Formats can be selected from the drop-down list.

Override Cell Format – The **Override Cell Format** button opens the **Cell Format Override** dialog box, which allows you to specify the text size and properties as shown in Figure 10.55. The **Cell Format Override** dialog box allows you to specify the text appearance, including style, alignment, height, gap, and rotation for the heading of the column. If a matrix schedule is specified, you can select the symbol used in the matrix. The width of the cell can be specified in the **Cell Size Fixed Width** edit field. If the width is set to zero, the width will vary according to the needs of the text width.

Matrix – The **Matrix** check box allows you to display the data for the property in the matrix format.

Total – The **Total** option allows you to generate a total for the property.

Column Position – This section allows you to define the position of the column relative to other columns. In Figure 10.54, the new column will be inserted after the **WindowStyles:Material** column. The radio

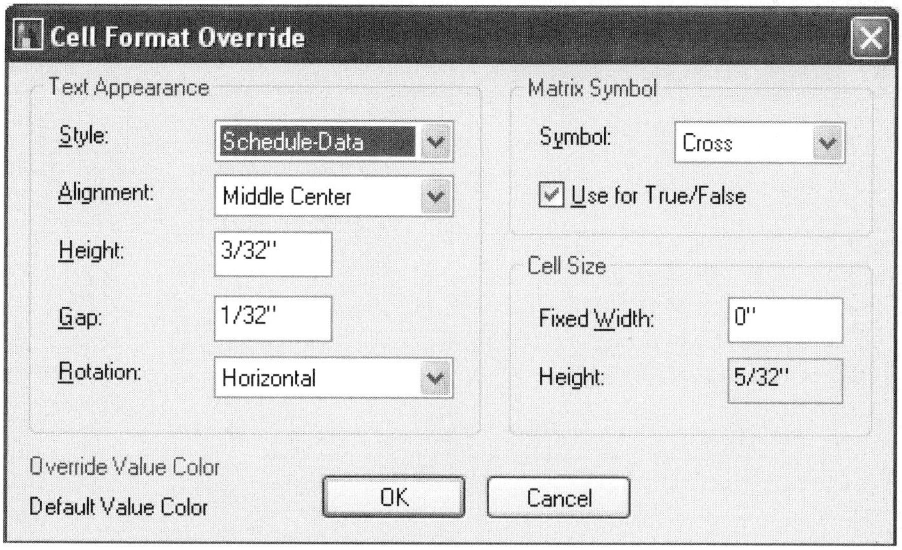

Figure 10.55 *Cell Format Override dialog box*

buttons allow you to position the new column before or after an existing column. The **Column** list shows the properties of the schedule that can be selected to set the relative position of the new column.

Add Header – The **Add Header** button of the **Columns** tab allows you to create a header for a column. The header titled SIZE was created as shown in Figure 10.53 by selecting the **Width** and **Height** column titles and then selecting the **Add Header** button. Holding down CTRL when you select the column titles allows you to select more than one title. The title of the header can then be typed in the header space. You can delete headers by selecting the header and then selecting the **Delete** button.

Modify – If you select the title of a column and then select the **Modify** button, the **Modify Column** dialog box opens, as shown in Figure 10.56. The **Modify Column** dialog box allows you to edit the contents of the column. As shown in Figure 10.56, the **Type** column can be edited to change the heading, data format, or matrix style, or to include a total.

Delete – The **Delete** button allows you to delete a selected column from the schedule.

You can **reposition** a column within the schedule by selecting the **column title** in the **Columns** tab and **dragging** the column title to the left or right. Releasing the mouse button over another column will reposition the selected column within the schedule as shown in Figure 10.57.

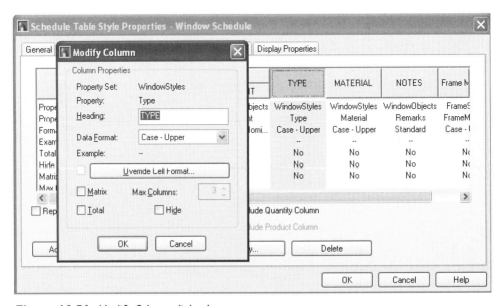

Figure 10.56 *Modify Column dialog box*

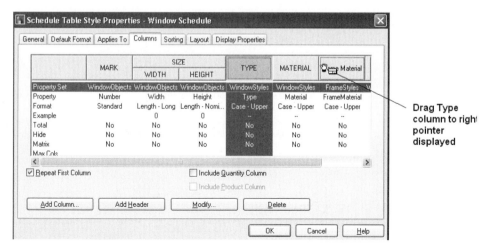

Figure 10.57 *Moving a column in a schedule table*

Sorting Tab

The **Sorting** tab allows you specify how the schedule data is displayed in the schedule. In Figure 10.58, the schedule data is sorted according to the **WindowObjects: Number** property in Ascending order. You can define the sorting of the properties by selecting one of the following four buttons: **Add, Remove, Move Up,** and **Move Down**.

> **Add** – The **Add** button opens the **Select Property** dialog box, which lists other properties that can be defined for sorting.
>
> **Remove** – To remove a property for sorting, select a property from the property list, and then select the **Remove** button.
>
> **Move Up** – If more than one property is used for sorting, you can select a property and then select the **Move Up** button, and that property will move to the top of the list.
>
> **Move Down** – When a property has been moved up in priority in the property list, you can select the property and then select the **Move Down** button, and that property will be moved down to lower priority when the sort is executed.
>
> **Sort Order** – The **Sort Order** can be specified as **Ascending** or **Descending**.

Layout Tab

The **Layout** tab allows you to define the text size for the schedule title, column headers, and matrix headers. The **Layout** tab shown in Figure 10.58 includes a **Table Title** edit field and buttons to define the format of the title, column headers, and matrix headers.

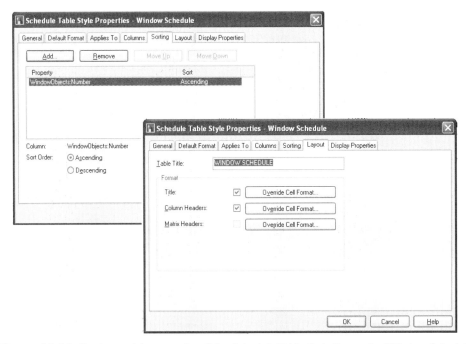

Figure 10.58 *Sorting and Layout tabs of the Schedule Table Style Properties-Window Schedule dialog box*

The **Table Title** edit field allows you to enter the text for the title of the schedule, and the remaining portion of the dialog box allows you to define the text height and other properties of the title and headers. You can edit the appearance of the title, column headers, and matrix headers by selecting the **Override Cell Format** buttons. When you select one of the **Override Cell Format** buttons, the **Cell Format Override** dialog box opens. This dialog box allows you to increase the text height and properties of the title and headers.

Display Properties Tab

The **Display Properties** tab is similar to most object styles; the visibility, color, linetype, lineweight, and Ltscale can be controlled as shown in Figure 10.59.

RENUMBERING TAGS

Tags are placed on a drawing and assigned a mark in a manner that allows you to predict the location of the next number. Placing tag numbers or letters in a drawing in a systematic manner makes it easier for the craftsman to interpret the drawing. As the design develops, additional windows or doors can be inserted in a drawing, which creates disorder in the tag numbering sequence. Therefore, the **Renumber Data Tool**

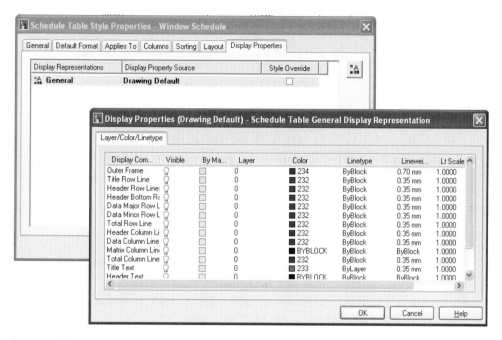

Figure 10.59 *Display Properties tab and Layer/Color/Linetype tab*

located on the Documentation tool palette can be used to renumber existing tags. Access the **Renumber Data Tool** command as shown in Table 10.27.

Command prompt	PROPERTYRENUMBERDATA
Scheduling palette	Renumber Data Tool

Table 10.27 *Renumber Data Tool command access*

When you select the **Renumber Data Tool** command, the **Data Renumber** dialog box opens, as shown in Figure 10.60. This dialog box allows you to edit the property, start number, and increment for changing the tag number. The settings shown in Figure 10.60 would select the first tag and change its value to 1, and each successive tag selected would be increased by one.

The **Data Renumber** dialog box consists of the following options.

> **Property Set** – The **Property Set** drop-down list displays property sets defined for the drawing.
>
> **Property** – The **Property** drop-down list displays the property specified for the renumbering operation.

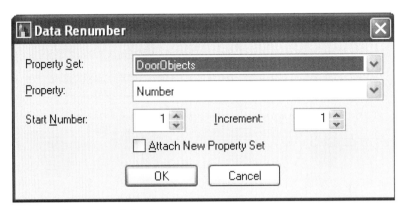

Figure 10.60 *Data Renumber dialog box*

Start Number – The **Start Number** edit field allows you to specify the beginning number of the renumbering operation.

Increment – The **Increment** value increases the tag number by adding the increment number to the previous tag number.

Attach New Property Set – When you check the **Attach New Property Set** check box, the property of the objects in the drawing will change as you renumber them.

UPDATING A SCHEDULE

The content of a drawing often changes after a schedule is developed and objects are tagged. If the **Add New Objects Automatically** option is set to **Yes** in the Properties palette when a schedule is placed, the additional objects added to the drawing will automatically be inserted in a schedule. When the schedule is created, the **Add New Objects Automatically** can be set to **No** as shown in Figure 10.49 in the Properties palette; the objects are not included in the schedule. When additional objects are added to this schedule, a diagonal line is drawn across the schedule to indicate it is not current, as shown in Figure 10.49. This diagonal line indicates that the table needs to be updated; it will be removed when the table is updated with the **Update Schedule Table** command (**ScheduleUpdateNow**). Access the **Update Schedule Table (ScheduleUpdateNow)** command as follows (Table 10.28).

Command prompt	SCHEDULEUPDATENOW
Shortcut menu	Select the schedule, right-click, and choose Update Schedule Table

Table 10.28 *Update Schedule Table command access*

When you select a schedule, right-click, and choose the **Update Schedule Table** command, the table is updated without additional input, and the diagonal line is removed. You can select one or more tables, and the **Update Schedule Table** command will update the contents of each schedule.

 Tip: The diagonal line may not be drawn across schedules placed with **Add New Objects Automatically** set to **No** when additional objects are added. Select **Regenerate Model** on the **Standard** toolbar to refresh the content of the schedule.

EDITING THE CELLS OF A SCHEDULE

The manual data values in a schedule can be changed by editing the **Edit Property Set Data** of the Properties palette of the object or of the style properties dialog box. You can also edit the value of a property for an object by selecting the cell directly in the schedule. The **Edit Table Cell** command (**ScheduleCellEdit**) is used to edit the data directly in the schedule. This command allows you to edit cells with manual data. Cells with Automatic data extracted from the object cannot be edited with this command, because the object must be changed to change the automatic data. Access the **Edit Table Cell** command as shown in Table 10.29.

Command prompt	SCHEDULECELLEDIT
Shortcut menu	Select the schedule, right-click, and choose Edit Table Cell

Table 10.29 *Edit Table Cell command access*

When you select the **ScheduleCellEdit** command, you are prompted to select a cell within a schedule. However, if you move your pointer over a cell, the drawing object represented by the cell will be highlighted. If you press CTRL when you select a cell, the graphic screen will zoom to the object represented by the cell. Selecting the cell in the schedule will open the **Edit Referenced Property Set Data** dialog box, as shown in Figure 10.61.

The **Edit Referenced Property Set Data** dialog box allows you to enter a new value in the schedule. Included in the dialog box are the names of the property and property set that are defined for the cell. To change the value of the cell, overtype a new value and select the **OK** button; the schedule is then changed.

USING BROWSE PROPERTY DATA

The **Browse Property Data** tool located on the Documentation tool palette allows you to view and edit the values for the properties of all objects of a specific type in a single dialog box. The values for the properties can be changed in the **Browse**

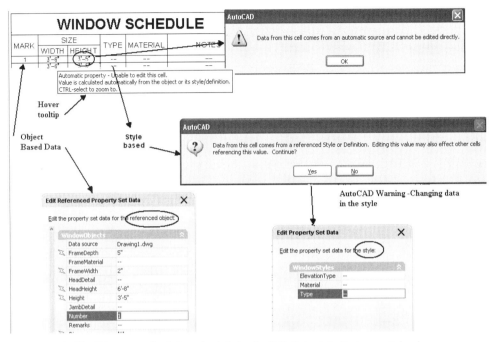

Figure 10.61 *Editing a cell of the schedule in the Edit Schedule Property dialog box*

Property Data dialog box. Access the **Browse Property Data** command as shown in Table 10.30.

Command prompt	PROPERTYDATABROWSE
Palette	Select the Browse Property Data command from the Scheduling tool palette

Table 10.30 *PropertyDataBrowse command access*

When you select the **Browse Property Data** command from the Scheduling palette, the **Browse Property Data** dialog box opens as shown in Figure 10.62. This dialog box has been edited to specify the **WindowObjects** property set definition and filter out objects other than windows. Therefore only windows are listed in the left pane. If you select a window in the left pane, its properties will be displayed in the right pane, as shown at right in Figure 10.63. You can select additional windows if you hold CTRL during selection. If you have selected more than one window in the left pane, you can edit properties in the right pane that will be applied to all windows selected. The remark "Insulated Glass" was added to all windows of the schedule in Figure 10.63.

Annotating and Documenting the Drawing 717

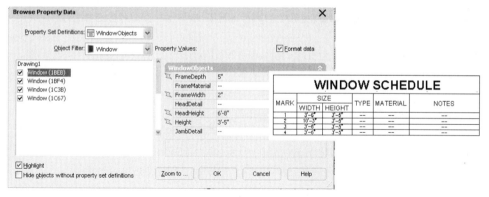

Figure 10.62 *Browse Property Data dialog box*

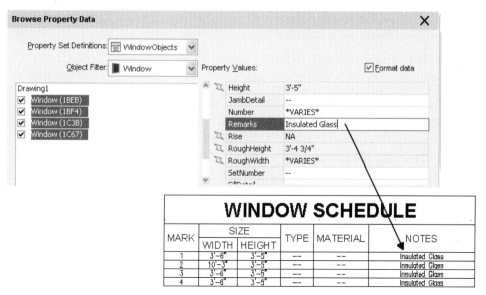

Figure 10.63 *Editing window properties in the Browse Property Data dialog box*

The **Zoom to** button in the bottom center of the dialog box will zoom the drawing to the location of the selected window. The **Highlight** check box in the dialog box will highlight the window in the drawing when it is selected in the list located in the left pane.

 STOP. Do Tutorial 10.3, "Placing Tags for Windows and Creating Window Schedules," at the end of the chapter.

CHANGING THE SELECTION SET OF A SCHEDULE

You can revise a schedule by adding or removing objects from the schedule. **Selection** is an option of the shortcut menu for a selected schedule, which allows you to **Add**, **Remove**, **Select**, and **Show** objects. Each of the options of the **Selection** shortcut menu is described below.

Add – The **Add** option (**ScheduleSelectionAdd**) allows you to select additional objects in the drawing to add to the selected schedule.

Remove – The **Remove** option (**ScheduleSelectionRemove**) allows you to remove from a schedule an object that is listed in the schedule.

Select – The **Select** option (**ScheduleSelectionReselect**) allows you to reselect the objects to be included in the schedule.

Show – The **Show** option (**ScheduleSelectionShow**) allows you to select the text within one or more cells in the schedule, and the associated object in the drawing will be highlighted.

Each of these options is included on the shortcut menu of a selected schedule. If Automatic Update is not being used for a schedule, these commands allow you to selectively revise the schedule.

ADDING OBJECTS TO AN EXISTING SCHEDULE

When you select **Selection>Add** from the shortcut menu of a selected schedule, the **ScheduleSelectionAdd** command is selected as shown in Table 10.31.

Command prompt	SCHEDULESELECTIONADD
Shortcut menu	Select a schedule, right-click, and choose Selection>Add from the shortcut menu

Table 10.31 *ScheduleSelectionAdd command access*

The window schedule shown in Figure 10.64 includes three windows. The following command line sequence will add the fourth window to the existing schedule.

Select the Window Schedule at **p1** as shown at left in Figure 10.64, right-click, and choose **Selection>Add** from the shortcut menu. Respond to the command line prompts as shown below:

Command: ScheduleSelectionAdd

Select objects or ENTER to schedule external drawing: 1 found
(Select the window at p2 as shown at left in Figure 10.64.)

Select objects or ENTER to schedule external drawing: ENTER *(Press* ENTER *to end selection.)*

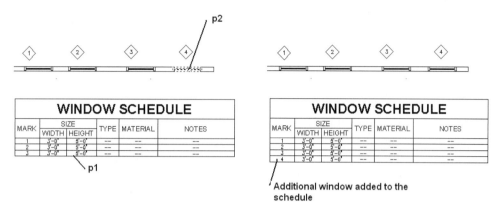

Figure 10.64 *Window Schedule developed after adding to the selection*

> 1 object(s) added.
>
> *(Additional window added to the schedule as shown at right in Figure 10.64.)*

REMOVING OBJECTS FROM AN EXISTING SCHEDULE

When you select **Selection>Remove** from the shortcut menu of a selected schedule, the **ScheduleSelectionRemove** command is selected as shown in Table 10.32.

Command prompt	SCHEDULESELECTIONREMOVE
Shortcut menu	Select the schedule, right-click, and choose Selection>Remove

Table 10.32 *ScheduleSelectionRemove command access*

The window schedule shown in Figure 10.65 includes four windows. The following command line sequence was used to remove windows from the schedule.

> *(Select the Window Schedule at p1 as shown at left in Figure 10.65, right-click, and choose **Selection>Remove** from the shortcut menu.)*
>
> Command: ScheduleSelectionRemove
>
> Select objects: 1 found *(Select window at p2 as shown in Figure 10.65.)*
>
> Select objects: 1 found, 2 total *(Select window at p3 as shown in Figure 10.65.)*
>
> Select objects: ENTER *(Press ENTER to end selection.)*
>
> 2 object(s) removed.
>
> *(Windows removed from the schedule as shown at right in Figure 10.65.)*

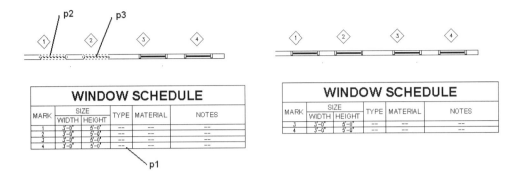

Figure 10.65 Revised schedule based upon remaining objects

RESELECTING OBJECTS FOR AN EXISTING SCHEDULE

The **Reselect** option will allow you to reselect the objects you want included in the schedule. Only those objects selected in the new selection set will be included in the schedule. Objects in the schedule prior to your selecting this command and not included in the reselection will be removed from the schedule. Select a schedule, right-click, and choose the **Selection>Reselect** (**ScheduleSelectionReselect**) command as shown in Table 10.33.

Command prompt	SCHEDULESELECTIONRESELECT
Shortcut menu	Select the schedule, right-click, and choose Selection>Reselect

Table 10.33 ScheduleSelectionReselect command access

The window schedule shown in Figure 10.66 includes four windows. The following command line sequence was used to reselect the windows for the schedule.

(Select the Window Schedule at p1 as shown at left in Figure 10.66, right-click, and choose **Selection>Reselect** *from the shortcut menu.)*

Command: ScheduleSelectionReselect

Select objects or ENTER to schedule external drawing: 1 found
(Select window at p2 as shown at left in Figure 10.66.)

Select objects or ENTER to schedule external drawing: 1 found
(Select window at p3 as shown at left in Figure 10.66.)

Select objects or ENTER to schedule external drawing: 1 found
(Select window at p4 as shown at left in Figure 10.66.)

Select objects or ENTER to schedule external drawing: ENTER *(Press* ENTER *to end selection.)*

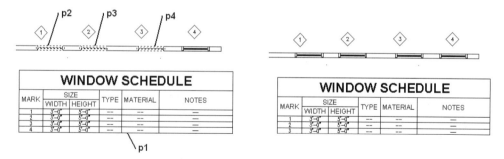

Figure 10.66 Reselect option to change selection set of a schedule

(Revised schedule shown at right in Figure 10.66.)

FINDING OBJECTS IN THE DRAWING LISTED IN THE SCHEDULE

The **Selection>Show** option of the shortcut menu for a schedule allows you to select an object in the schedule, and it will be highlighted in the drawing. This option allows you to quickly locate an object in the drawing that has been included in the schedule. Access the **ScheduleSelectionShow** command as shown in Table 10.34.

Command prompt	SCHEDULESELECTIONSHOW
Shortcut menu	Select a schedule table, right-click, and choose Selection>Show

Table 10.34 Table Selection Show command access

When you select the **Selection>Show** option, you are prompted to select the text in the schedule; the object will be highlighted in the drawing. The command line prompt includes a CTRL-select option that will zoom the display to the object associated with the text selected in the schedule.

The command line prompt also allows you to select the border of the schedule to highlight all items in the drawing associated with the schedule. The following command line sequence was used to zoom to a selected item of the schedule.

> (Select the Window Schedule at p1 as shown in Figure 10.67, right-click, and choose **Selection>Show**.)
> Command: ScheduleSelectionShow
> Select schedule table item (or the border for all items) or CTRL-select to zoom: <Cycle on> <Cycle off> (Hold CTRL down, and select the window mark at p2 as shown in Figure 10.67.)

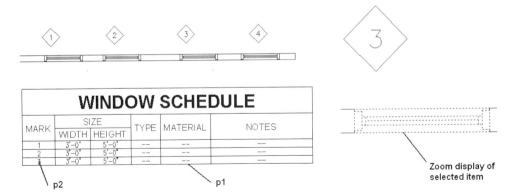

Figure 10.67 *Show selection of the schedule table*

(Zoom display of the window is shown at right in Figure 10.67.)

EXPORTING SCHEDULE DATA

The data of the schedules can be exported from Architectural Desktop to a text file or spreadsheet. If the data is exported to a spreadsheet, the data can then be used in cost estimates. The command to export schedule data is **ScheduleExport**. Select a schedule, right-click, and choose **Export** from the shortcut menu to access the **ScheduleExport** command (see Table 10.35).

Command	SCHEDULEEXPORT
Shortcut menu	Select a schedule, right-click, and choose Export

Table 10.35 *Table Export command access*

When you select the **Export** option from the shortcut menu of a selected schedule, the **Export Schedule Table** dialog box opens, which allows you to define the output and input for exporting the schedule data. The **Export Schedule Table** dialog box shown in Figure 10.68 includes an **Output** and **Input** section. The content of the **Output** and **Input** sections is described below.

Output

        SaveAs Type – The **SaveAs Type** list allows you to select the file format for the exported data. The type of file can be selected from one of the following: Microsoft Excel 97 (*.xls), Text (Tab delimited)(*.txt), and CSV (Comma delimited) (*.csv).

File Name – The **File Name** edit field allows you to type the name of the file for the data. The **Browse** button opens the file dialog box, which allows you to specify the directory for the data file.

Input

Use Existing Table – If the **Use Existing Table** check box is selected, when you select the **OK** button of the **Export Schedule Table** dialog box, you will be prompted to select an existing schedule to export the data. If **Use Existing Table** is not selected, the remainder of the dialog box is activated, and you can create a new schedule.

Schedule Table Style – The **Schedule Table Style** list shows the schedule table styles that have been either created or imported into the drawing. Selecting a table style from the list specifies the table style for the new schedule.

Layer Wildcard – The **Layer Wildcard** edit field allows you to specify the layer filter for searching objects of the drawing.

Scan Xrefs – If the **Scan Xrefs** check box is selected, the objects included in attached external reference files can be included in the new schedule.

Scan Block References – If the **Scan Block References** check box is selected, blocks inserted in the current drawing can be selected for the schedule.

Editing the **Export Schedule Table** dialog box specifies the format for the data file. When you select **OK** to dismiss the dialog box, you are prompted to select a schedule in the drawing. The format of the text in the schedule causes the **Format** dialog box to open, as shown in Figure 10.68. The **Format** dialog box allows you to select the format for the text of cells in an Excel spreadsheet. The options of the **Format** dialog box are as follows.

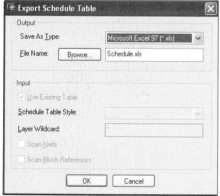

Figure 10.68 *Export Schedule Table dialog box*

Use Unformatted Decimal Value – The **Use Unformatted Decimal Value** option will convert feet and inches to decimal inches.

Convert to Formatted Text – The **Convert to Formatted Text** option will retain the feet and inches format in data.

After you select **OK** in the **Format** dialog box, the file will be created. You can check the file by opening the file in the Excel software program.

STEPS TO EXPORTING SCHEDULES TO DATA FILES

1. Select the schedule, right-click, and choose **Export** from the shortcut menu.
2. Select **Microsoft Excel 97** as the **SaveAs Type** and type the name of the data file in the **File Name** edit box of the **Export Schedule Table** dialog box as shown in Figure 10.69. Select the **Browse** button and specify your student directory.
3. Select **OK** to dismiss the **Export Schedule Table** dialog box.
4. Select the **Convert to Formatted Text** radio button, select **Apply to All Columns**, and then select the **OK** button.
5. You can check the file by launching the Excel program and opening the file.

 STOP. Do Tutorial 10.4, "Placing Alphabetic Tags and Quantities for Door Schedules," at the end of the chapter.

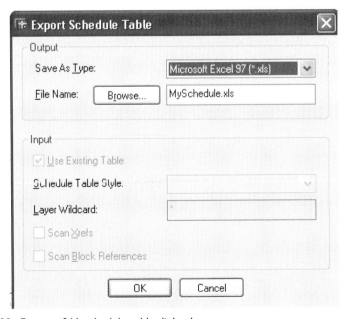

Figure 10.69 *Export of Myschedule table dialog box*

CREATING SCHEDULES FOR PROJECT DATA

The benefits of using project based schedules can be realized if project based tags for spaces and doors are used in drawings included in a project. This technique provides flexibility in maintaining an accurate and up-to-date comprehensive schedule. The project based door tags increment the door number using the room number as a prefix for the door identifier. The Room Tag-Project Based or the Room Tag-Project Based-Scaled creates a room identifier prefixed by the level. Therefore room tags for the first level would increment from 101 to 102, and the second level would increment from 201 to 202. When the Door Tag-Project Based or Door Tag-Project Based Scale Dependent is inserted, the door identifier includes the room number suffixed by a letter. Since no two rooms share the same number across levels, each door is assigned a unique identifier. Therefore, if the door leaf is placed over a space, the door identifier will be numbered based upon the associated space. The Property Data Location grip shown in Figure 10.70 determines which space to link the door. Therefore the leaf of exterior doors that swing out to comply with egress are not linked to a space. You can select the Property Data Location grip and drag the grip to a point over the interior space to link the door identifier to the space as shown at right in Figure 10.70.

The comprehensive schedule is developed in a Sheet drawing from the data located in View drawings. After placing a schedule in the Sheet drawing, toggle ON **Schedule external drawing** and specify the View drawings to included in the **External drawing** field of the **Schedule** properties palette as shown in Figure 10.70. When you update the schedule, the data from the View drawings specified will be displayed in the schedule.

STEPS TO CREATING A COMPREHENSIVE PROJECT-BASED SCHEDULE

1. Open the Construct drawing, and create spaces for each room.
2. Create a View drawing that consists of the Construct drawing. Tag each space using the Room Tag-Project Based or the Room Tag-Project Based-Scale.

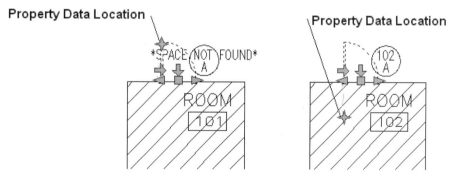

Figure 10.70 Door Tag-Project Based

3. Tag each door in the View drawing using the Door Tag-Project Based or Door Tag-Project Based Scale Dependent.

4. Create a new Sheet drawing in the Schedules and Diagrams category of the **Sheets** tab in the **Project Navigator**. Open the new Sheet drawing.

5. Open the **Content Browser**, navigate to the Documentation Tool Catalog-Imperial catalog, choose Schedule Tables, and i-drop the Door Schedule-Project Based into the new Sheet drawing.

6. Select the schedule, right-click, and edit the **External Source** section of the Properties palette as follows: toggle **Yes** for **Schedule external drawings** and edit **External drawing path** to include the View drawings of the project as shown in Figure 10.71.

7. Select the schedule, right-click, and select **Update Schedule Table** from the shortcut menu.

INSERTING PROPERTIES WITHOUT TAGS

When tags are placed in a drawing, the multi-view block consists of attributes. The tags of the attributes specify the property set and property. Therefore, when an object is tagged, the property is inserted in the drawing with the multi-view block. The properties can then be used to develop schedules. The attribute tags of the Window

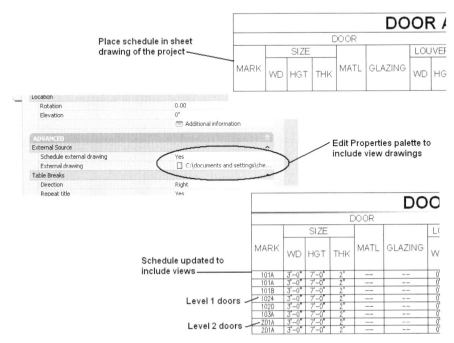

Figure 10.71 *Creating a comprehensive project door schedule*

tag multi-view block are shown in Figure 10.72. Property sets can be imported into a drawing in the **Style Manager**. If the property set and properties are imported in the **Style Manager**, schedules can be developed for any object of the drawing without tagging each object. New tags can be developed that specify the property set and properties. The master set of Property Sets are located in the *C:\Documents and Settings\All Users\Application Data\Autodesk\ADT 2005\enu\Aec Content\Imperial\Documentation\Schedule Tags\PropertySetDefs.dwg*.

PROPERTY SET DEFINITIONS

The Documentation Objects\Property Set Definitions folder of the PropertySetDefs.dwg file includes the comprehensive list of properties, as shown in Figure 10.73. If you open the *PropertySetDefs.dwg* file, the properties can be imported into the current drawing. The content of a property such as the Door Style property shown in Figure 10.73 can be reviewed by selecting the **Edit** button when the property set is selected in the right pane of the **Style Manager**. The **Property Set Definition Properties** dialog box consists of three tabs: **General**, **Applies To**, and **Definition**.

General Tab

The **General** tab shown in Figure 10.74 allows you to type a name and description of the property set.

Applies To Tab

The **Applies To** tab includes options to specify to which type of object the property set will be applied when the schedule is developed. The options of the **Applies To** tab are described below.

> **Applies To Objects** – The **Objects** radio button applies the data to the object without regard to its style.
>
> **Applies To Styles and Definitions** – The **Styles and Definitions** radio button applies the properties to a specific style.

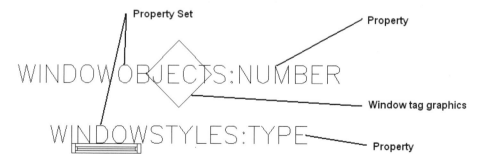

Figure 10.72 *Attribute tags of multi-view block*

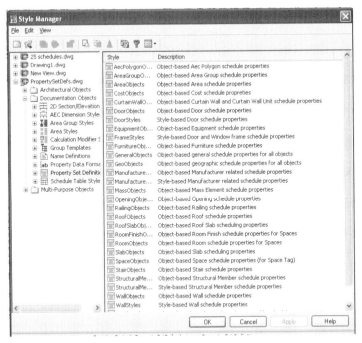

Figure 10.73 *Property Set Definitions of the PropertySetDefs drawing*

Object list – The object list includes all Architectural Desktop objects. The Door Object property can be applied only to doors and door/window assemblies, as shown in Figure 10.74.

Definition Tab

The **Definition** tab lists all the properties of a property set. The **Definition** tab shown in Figure 10.75 lists the properties for a DoorObject. The Door Size property is highlighted in the top window; this property can be edited in the lower **Property Definition** window.

Name – Select the **Name** button to edit the name of the property.

Description – Type a **Description** in the edit field to specify a description for the property.

Edit Source – Select the **Edit Source** button to open the **Automatic Property Source** list of data sources for the property. The **Edit Source** button is active for automatic properties.

Type – The **Type** of property can be auto increment-character, auto increment-integer, integer, real, text, and true/false. Auto increment-character allows the letter of the tag to increase in the alphabet as you tag additional objects.

Annotating and Documenting the Drawing 729

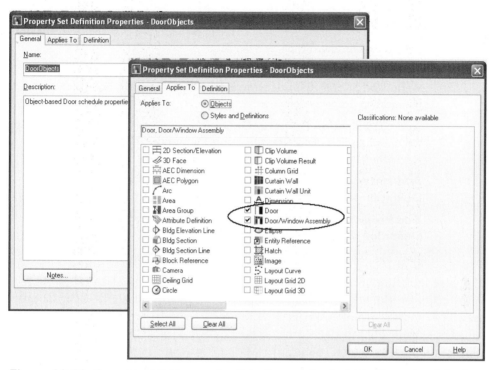

Figure 10.74 General and Applies to tabs of the Property Set Definition Properties dialog box

Default – A **Default** value will be displayed if no value is typed in the dialog box. A data field can be defined as the default value for the schedule.

Format – The **Format** list includes the following **Property Data Format**, which controls whether the values are displayed in such formats as feet and inches, decimal, or whole numbers. You can view the properties of a data format by editing the data format in the **Style Manager**. Data formats are located in the Documentation\Property Data Formats folder. The data format controls whether the data is displayed in all caps and the precision of the data. The Length-long format sets the precision to 1/4" while the Length-short format displays the data with 1/16" precision. Refer to the Property Data Format folder of the **Style Manager**.

The buttons located in the upper right corner are described below:

Add Manual Property Definition – Select **Add Manual Property Definition** to create a new property in the **New Property** dialog box. The **Start With** edit field allows you to copy the attributes of existing properties.

Add Automatic Property Definition – Select **Add Automatic Property Definition** to specify an automatic property for the new property.

Add Formula Property Definition – Select **Add Formula Property Definition** to define a property based upon a formula that applies arithmetic operations on other property definitions.

Add Location Property Definition – Select **Add Location Property Definition** to open the **Location Property Definition** dialog box to define the location property definition.

Add Classification Property Definition – If Classifications are attached to the drawing, select **Add Classification Property Definition** to add reference to a classification.

Add Material Property Definition – Select **Add Material Property Definition** to include materials in the property definition.

Add Project Property Definition – Select **Add Project Property Definition** to reference a project to the property.

Remove – Select **Remove** to remove a property from the property set.

Property sets allow you to control the text of a tag. When you place a door tag, the Number property of the DoorObjects property set is used as the tag. The default property Type for the Number object is Auto Increment-Integer, and the Format is Number-Object. Therefore, you can edit the property Type to **Text** and tag doors with alphanumeric characters. If you set the Type to **Auto Increment-Character**, you can place tags that increment like their numeric counterparts.

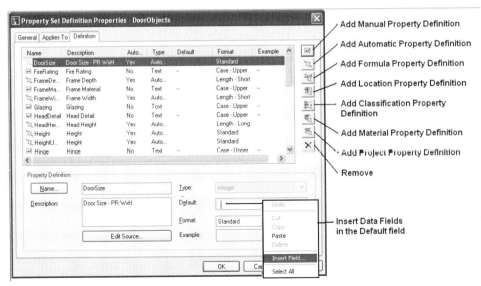

Figure 10.75 *Definition tab of the Property Set Definition Properties dialog box*

INSERTING DATA FIELDS IN A SCHEDULE

You can define a data field as a default value for a property in the property set definition. The shortcut menu of the Default value of a property includes **Insert Field**, as shown in Figure 10.75. When the **Insert Field** option is selected, the **Field** dialog box opens as shown in Figure 10.76. The **Field** dialog box lists field categories and field names. You can select a field name such as File name or date, and this data will be displayed in the schedule as a default value for the property.

 STOP. Do Tutorial 10.5, "Creating Tags and Schedules for a Project" at the end of the chapter.

SUMMARY

1. The scale factor for annotation is set in **Drawing Setup**. However, annotation identified as scale dependent is controlled by the scale linked in the display configuration.
2. Commands to place Aec dimensions, leaders, tags, and revision clouds are located on the Annotation tool palette.
3. AEC dimensions can be placed from the Annotation palette, whereas non-AEC dimensions are selected from the Documentation\Miscellaneous folder of the DesignCenter.

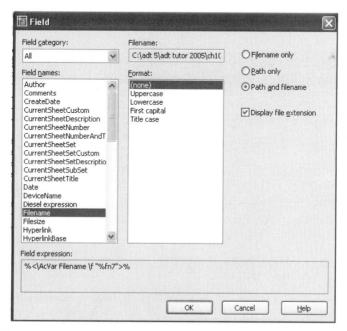

Figure 10.76 *Field dialog box*

4. Match lines, north arrows, and datum elevations are selected from the Miscellaneous folder of the Documentation Tool Catalog-Imperial.

5. Tags for Architectural Desktop objects can be placed in the drawing from the Scheduling tool palette. Additional tags for door, windows, walls, furniture, equipment, and structural members are located in the Schedule Tags category of the Documentation Tools Catalog-Imperial.

6. The property set and properties for the schedule are inserted into the drawing when objects are tagged.

7. The **Space Auto Generate Tool** is selected from the Design tool palette. After spaces are created, room and finish tags can be attached to the space.

8. Residential and other space styles are imported from the Spaces category of the Design Tool Catalog-Imperial of the **Content Browser**.

9. Property Set Definitions for AEC objects can be imported in the **Style Manager** from the *C:\Documents and Settings\All Users\Application Data\Autodesk\ADT 2005\Aec Content\Imperial\Documentation\Schedule Tags\PropertySetDefs.dwg*.

10. The units or format of data in a schedule is displayed according to the Property Data Formats.

11. The **ScheduleStyleEdit** command is used to define the components of a schedule.

12. Additional schedules can be i-dropped from the Schedules category of the Documentation Tool Catalog-Imperial of the **Content Browser**.

REVIEW QUESTIONS

1. The quantity of an item in a schedule can be added to a schedule by _____.

2. Place leaders in a drawing by selecting _____ from the Annotation tool palette.

3. A tag is a _____ used to attach a property to an object.

4. AEC Space objects can include an area property included in the _____ tag.

5. A door schedule can be inserted in a new drawing if _____ exist in the drawing.

6. Object based property values can be added to a drawing in the _____ tab of the Properties palette.

7. Style based property values can be added to a drawing by selecting the **Property Sets** button of the _____ tab of the dialog box to edit the style of the object.

8. A schedule with a diagonal line across the schedule indicates _____.

9. Text data in a schedule is revised by the _____ command.

10. Schedule data can be exported to an _____ spreadsheet.

11. The dimension tool used to place dimensions that will automatically update the changes in the objects is _____.

12. The command used to add points included in an AEC Dimension is _____.

13. Schedule table styles are edited in the _____.

14. Editing the _____ definition of a property used in a tag allows you to insert alphabetic tags.

15. The _____ _____ of a property determines the units and precision of the text displayed in the schedule.

16. Prior to placing a room tag, a _____ must be created.

17. The _____ option of a door tag allows you to tag all the objects of a given type with one selection.

18. The _____ command allows you to change a property used in a schedule for all objects in the drawing.

TUTORIAL 10.1 INSERTING AEC DIMENSIONS

1. Open Autodesk Architectural Desktop 2005, and select **File>Project Browser** from the menu bar. Use the Project Selector drop-down list to navigate to your *ADT Student\ADT Tutor\Ch 10* directory. Double-click on **Ex 10-1** to set this project current. Select **Close** to dismiss the **Project Browser**.

2. Select the Views tab of the **Project Navigator**, and open Floor Plan - 1.

3. Toggle OFF SNAP, GRID, POLAR, OSNAP, OTRACK, and toggle ON ORTHO on the status bar.

4. If the Tool Palettes window is not open, press CTRL +3.

5. To set the view of the dimensions to 1/4" = 1'-0", select the High Detail display configuration from the Drawing Window status bar.

6. Select **AEC Dimension (2)** from the Annotation tool palette, and respond to the command prompts as shown below.

    Command: DimAdd

    Select geometry to dimension or [Style]: *(Create a crossing select, select a point near p1 as shown in Figure 10T.1.)*

Specify opposite corner: 15 found *(Select a point near p2 as shown in Figure 10T.1.)*

1 was filtered out

Select geometry to dimension or [Style]: ENTER *(Press ENTER to end selection.)*

Specify insert point or [Rotation/Align] *(Move the pointer to display a horizontal dimension as shown in Figure 10T.1. Select a point near p3 as shown in Figure 10T.1.)*

13 added

(Aec dimension placed as shown in Figure 10T.1.)

7. Select **AEC Dimension (2)** from the Annotation tool palette, place the dimension as shown in Figure 10T.2, and refer to the following command line sequence to select the walls.

   Command: DimAdd

   Select geometry to dimension or [Style]: *(Select the wall at p1 as shown in Figure 10T.2.)* 1 found

   Select geometry to dimension or [Style]: *(Select the wall at p2 as shown in Figure 10T.2.)* 1 found, 2 total

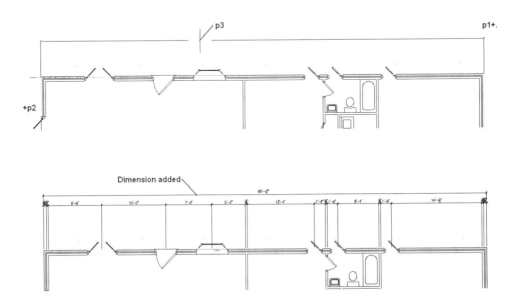

Figure 10T.1 *Crossing selection window to specify objects for dimension*

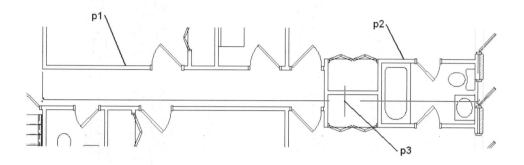

Figure 10T.2 *Placing interior dimension chain*

Select geometry to dimension or [Style]: ENTER *(Press* ENTER *to end selection.)*

Specify insert point or [Rotation/Align] *(Move the pointer to display a horizontal dimension as shown in Figure 10T.2. Select a point near p3 as shown in Figure 10T.2.)*

2 added

(Aec dimension placed as shown in Figure 10T.3.)

8. Select the dimension at **p1** as shown in Figure 10T.3, right-click, and choose **Properties**. Edit the Style to **1 Chain** in the **Design** tab of the Properties palette. Press ESC to clear the selection.

9. Select the dimension at **p2** as shown in Figure 10T.3, right-click, and choose **Edit AEC Dimension Style**. Select the **Display Properties** tab. Verify that the **Plan High Detail** display representation is highlighted, and select the **Edit Display Properties** button to open the **Display Properties (AEC Dimension Style Override – 1 Chain) – AEC Dimension Plan High Detail Display Representation** dialog box. Select the **Contents** tab, and edit the Apply to = **Opening in Wall**; clear the **Center of Opening**, and verify that **Opening Max Width** and **Opening Min Width** are clear as shown in Figure 10T.4. Select **OK** to close the **Display Properties (AEC Dimension Style Override – 1 Chain) – AEC Dimension Plan High Detail Display Representation** dialog box. Select **OK** to close all dialog boxes.

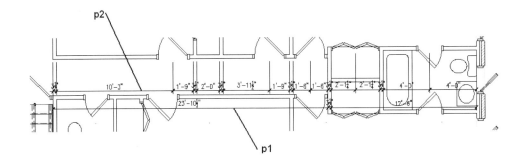

Figure 10T.3 *AEC Dimension 2 Chain placed for interior partitions*

10. Select the revised dimension as shown in Figure 10T.5, right-click, and choose **Remove Dimension Points** from the shortcut menu. Respond to the command line prompts as shown below to delete the dimensions; dimension point extension lines will turn red when selected:

    Command: DimPointsRemove

    Select extension lines to remove dimension points: 1 found
    (Select extension line above dimension line at p1 as shown in Figure 10T.5.)

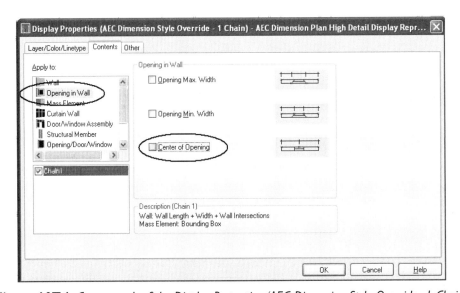

Figure 10T.4 *Contents tab of the Display Properties (AEC Dimension Style Override -1 Chain) - AEC Dimension Plan High Detail Display Representation dialog box*

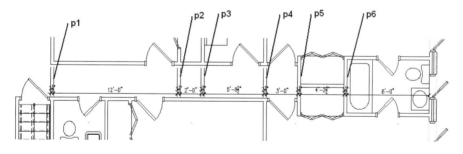

Figure 10T.5 *Remove Dimension Points from dimension string*

Select extension lines to remove dimension points:*Cancel*

Select extension lines to remove dimension points: 1 found
(Select extension line above dimension line at p2 as shown in Figure 10T.5.)

Select extension lines to remove dimension points:*Cancel*

Select extension lines to remove dimension points: 1 found
(Select extension line above dimension line at p3 as shown in Figure 10T.5.)

Select extension lines to remove dimension points:*Cancel*

Select extension lines to remove dimension points: 1 found
(Select extension line above dimension line at p4 as shown in Figure 10T.5.)

Select extension lines to remove dimension points:*Cancel*

Select extension lines to remove dimension points: 1 found
(Select extension line above dimension line at p5 as shown in Figure 10T.5.)

Select extension lines to remove dimension points:*Cancel*

Select extension lines to remove dimension points: 1 found
(Select extension line above dimension line at p6 as shown in Figure 10T.5.)

Select extension lines to remove dimension points:*Cancel*

Select extension lines to remove dimension points: ENTER
(Press ENTER to end the command.)

Override set for 1 Dimension Group(s)

(Dimension string revised as shown in Figure 10T.6.)

11. Select Zoom Extents from the Navigation toolbar.

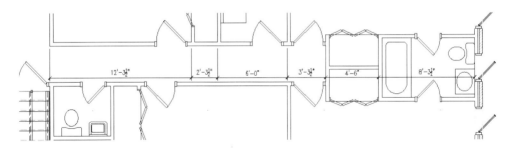

Figure 10T.6 *Revised AEC 2 Chain dimensions*

12. Select **AEC Dimension (2)** from the Annotation tool palette and place the dimensions for each of the remaining exterior walls as shown in Figure 10T.8. Select the upper left dimension string at p1 shown in Figure 10T.7, right-click, and choose **Remove Dimension Points** from the shortcut menu. Respond to the command line prompts as shown below to delete the extension lines at **p1** and **p2**.

   Command: DimPointsRemove

   Select extension lines to remove dimension points: 1 found
   (Select extension line at p1 as shown in Figure 10T.7.)

   Select extension lines to remove dimension points: ENTER *(Press ENTER to end selection.)*

   Override set for 1 Dimension Group (s).

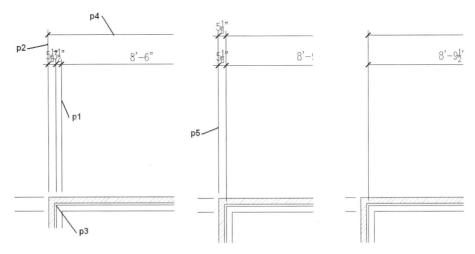

Figure 10T.7 *Aec Dimension 2 Chain created for exterior wall*

13. Select the dimension string at **p2** shown in Figure 10T.7, right-click, and choose **Add Dimension Points** from the shortcut menu. Respond to the command line prompts as shown and add the extension line as shown in the middle figure of Figure 10T.7.

    Command: DimPointsAdd

    Pick dimension points on screen to add: *(SHIFT + right-click and select the Endpoint object snap mode, and select the wall at p3.)*

    Pick dimension points on screen to add: ENTER *(Press ENTER to end selection.)*

    Select Dimension chain: *(Select the dimension string at p4.)*

14. Select the dimension string at **p5** shown in Figure 10T.7, right-click, and choose **Remove Dimension Points** from the shortcut menu. Respond to the command line prompts as shown below to delete the outer extension line.

    Command: DimPointsRemove

    Select extension lines to remove dimension points: 1 found
    (Select extension line at p5 as shown in Figure 10T.7.)

    Select extension lines to remove dimension points: ENTER *(Press ENTER to end selection.)*

    Override set for 1 Dimension Group (s).

15. Edit each of the remaining exterior walls to add dimension points as necessary to locate each dimension string to the outside of the stud, and remove dimension points locating brick veneer and the interior stud surface as shown in Figure 10T.8. Edit the exterior dimensions that locate the interior walls. The location of interior walls should be specified to one surface of the stud as shown in Figure 10T.8.

16. Save and close the drawing.

17. Select the Sheets tab, and double-click on the **A-3 Floor Plan Level 2** drawing of the **Project Navigator**.

18. Select the viewport to display the grips of the viewport. Select each grip and drag the viewport to display the dimensions of the view.

19. To set the view of the dimensions for the viewport, select the viewport and edit the Display Configuration flyout to **High Detail** display configuration in the Drawing Window status bar.

20. Save and close the drawing.

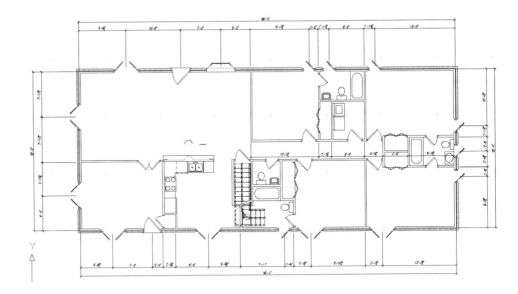

Figure 10T.8 Aec dimensions edited for each of the exterior walls

TUTORIAL 10.2 ADDING SPACE TAGS FOR ROOMS

1. Open Autodesk Architectural Desktop 2005, and select **File>Project Browser** from the menu bar. Use the Project Selector drop-down list to navigate to your *ADT Student\ADT Tutor\Ch 10* directory. Double-click on **Ex 10-2** to set this project current. Select **Close** to dismiss the **Project Browser**.

2. Select the **Views** tab of the **Project Navigator**, and open Floor Plan-1 view drawing.

3. Select **High Detail** from the **Display Configuration** flyout of the Drawing Window status bar. Select the **1/4"=1'-0"** scale from the flyout scale list of the Drawing Window status bar.

4. Toggle OFF SNAP, GRID, POLAR, OTRACK on the status bar. Toggle ON ORTHO and OSNAP on the status bar. Right-click over the OSNAP toggle on the status bar, select **Settings**, and toggle OFF all object snaps except **Endpoint**. Select **OK** to close the dialog box.

5. If tool palettes are not displayed, press CTRL +3. Right-click over the title bar of the tool palette and select the **Design** palette set. Right-click over the title bar of the tool palette, and select **New Palette** from the shortcut menu. Overtype the name **Residential** to name the tool palette.

6. To import space styles to the new palette, choose the **Content Browser** from the **Navigation** toolbar. Double-click to open the **Design Tool Catalog-Imperial**. Choose the **Spaces** category and open the **Residential** subcategory. Drag the following spaces from the **Residential** subcategory onto the new palette: **Bath_Small**, **Bedroom**, **Dining Room**, **Family Room**, **Kitchen**, and **Living Room**. You can select a space, hold down CTRL and select additional spaces, and then drag the i-drop for multiple spaces onto the tool palette. Close the **Content Browser**.

7. Choose the **Space Auto Generate Tool** from the Design tool palette to open the **Generate Spaces** dialog box. Edit the **Generate Spaces** dialog box as follows: Filter list = **Walls only**, toggle ON **Automatic**, and Style = **Standard**. Choose the **Tag Settings** button, verify that all check boxes are clear, and select **OK** to dismiss the **Tag Settings** dialog box. Move the pointer over the bedroom to display the boundary (shown in red) of the room, and then left-click near **p1** to specify the boundary as shown in Figure 10T.9. Move the pointer over the remaining bedrooms, and then left-click at **p2**, **p3**, and **p4** to specify their boundaries. Select **Close** to close the **Generate Spaces** dialog box.

8. Select the Residential palette, right-click on **Bedroom**, and select **Apply Tool Properties to>Space**. Respond the to the command line as follows to edit the spaces.

    Command: ApplyToolToObjects

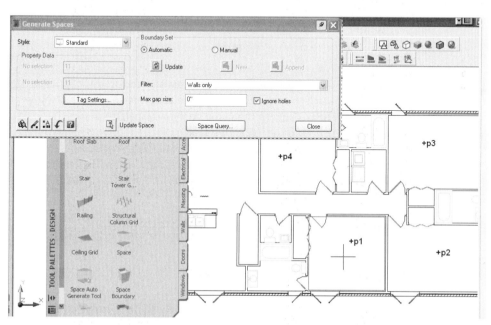

Figure 10T.9 *Creating spaces with Space Auto Generate Tool*

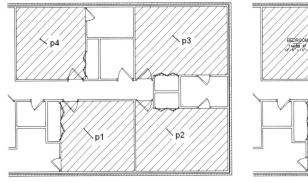 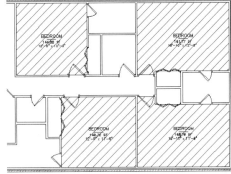

Bedroom Spaces Bedroom Spaces Tagged

Figure 10T.10 *Adding space tags to bedroom spaces*

Select Space(s): 1 found *(Select space at p1 as shown in Figure 10T.10.)*

Select Space(s): 1 found, 2 total *(Select space at p2 as shown in Figure 10T.10.)*

Select Space(s): 1 found, 3 total *(Select space at p3 as shown in Figure 10T.10.)*

Select Space(s): 1 found, 4 total *(Select space at p4 as shown in Figure 10T.10.)*

Select Space(s): ENTER *(Press ENTER to end edit and press ESC to clear selection.)*

9. Right-click over the title bar of the tool palettes and choose **Document** palette set. Right-click over the title bar of the tool palette and select **New Palette**, and overtype **Tags** as the name of the new palette. Select the **Content Browser** from the Windows taskbar. Open the Documentation Tool Catalog-Imperial catalog, select the **Schedule Tags** category in the left pane, and choose the **Room and Finish Tags** subcategory. Click and drag the i-drop for the **Space Tag** from the catalog onto the new Tags palette. Close the **Content Browser**.

10. Choose the **Space Tag** from the new Tags palette, and respond to the following command line prompts to tag each bedroom space as shown in Figure 10T.10.

   Command: ANNOSCHEDULETAGADD

   Select object to tag [Symbol/Leader/Dimstyle/Edit/Constrain/Rotation]: *(Select the space at p1 as shown in Figure 10T.10.)*

Specify location of tag <Centered>: *(Press* ENTER *to place the tag centered about the space and open the* **Edit Property Set Data** *dialog box, press* ENTER *to close the* **Edit Property Set Data** *dialog box.)*

Select object to tag [Symbol/Leader/Dimstyle/Multiple/Edit/Constrain/Rotation *(Select the space at p2 as shown in Figure 10T.10.)*

Specify location of tag <Centered>: ENTER *(Press* ENTER *to place the tag centered about the space and open the* **Edit Property Set Data** *dialog box, press* ENTER *to dismiss this dialog box.)*

Select object to tag [Symbol/Leader/Dimstyle/Multiple/Edit/Constrain/Rotation]: *(Select the space at p3 as shown in Figure 10T.10.)*

Specify location of tag <Centered>: ENTER *(Press* ENTER *to place the tag centered about the space and open the* **Edit Property Set Data** *dialog box, press* ENTER *to dismiss this dialog box.)*

Select object to tag [Symbol/Leader/Dimstyle/Multiple/Edit/Constrain/Rotation]: *(Select the space at p4 as shown in Figure 10T.10.)*

Specify location of tag <Centered>: ENTER *(Press* ENTER *to place the tag centered about the space and open the* **Edit Property Set Data** *dialog box, press* ENTER *to dismiss this dialog box.)*

Select object to tag [Symbol/Leader/Dimstyle/Multiple/Edit/Constrain/Rotation]: *(Press* ENTER *to end the command.)*

11. Save and close the drawing.

TUTORIAL 10.3 PLACING TAGS FOR WINDOWS AND CREATING WINDOW SCHEDULES

1. Open **Ex 10-3.dwg** from your *ADT Student\ADT Tutor\Ch10* directory.

2. Select **File>SaveAs** from the menu bar and save the drawing as **Lab 10-3** in your student directory.

3. Select **Window Tag** from the Scheduling tool palette of the Document palette set. Respond to the command line prompts as shown below to tag the windows.

Note: To view preset values of the command, press F2.

Select object to tag [Symbol/Leader/Dimstyle/Edit/Constrain/Rotation]: *(Select window at p1, the sash of the window, as shown in Figure 10T.11. Note, if you select the lines representing the window that are coincident with the wall lines, the tag will not recognize the window selection and the command will be terminated.)*

Specify location of tag <Centered>: *(Select a point near p2 as shown in Figure 10T.11.)*

*(Verify Number = 1 in the **Edit Property Set Data** dialog box. Select **OK** to dismiss the **Edit Property Set Data** dialog box.)*

Select object to tag [Symbol/Leader/Dimstyle/Multiple/Edit/Constrain/Rotation]: *(Select window at p3 as shown in Figure 10T.11.)*

Specify location of tag <Centered>: *(Select a point near p4 as shown in Figure 10T.11.)*

*(Verify Number = 2 in the **Edit Property Set Data** dialog box. Select **OK** to dismiss the **Edit Property Set Data** dialog box.)*

(Continue to tag the remaining windows as shown in Figure 10T.11.)

4. Select **Window Schedule** from the Scheduling tool palette. Verify the following settings in the Properties palette: Style = Window Schedule, Scale = 96, Update automatically = Yes, Add new objects automatically = Yes, Scan Xrefs = Yes, Scan block references = Yes, and Layer wildcard = *. Respond to the command line prompt as follows.

   Command: ScheduleAdd

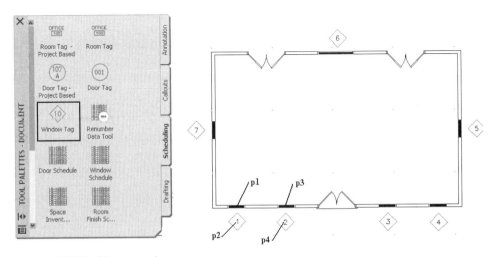

Figure 10T.11 *Placing window tags*

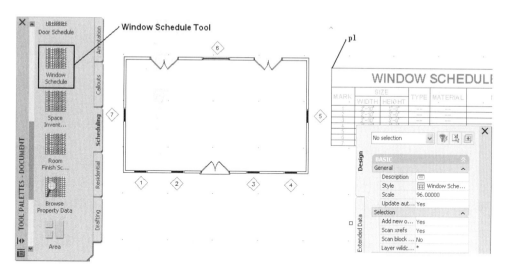

Figure 10T.12 *Window Schedule placed with initial data*

Select objects or ENTER to schedule external drawing: **all**
ENTER *(Type all to select all window objects in the drawing.)*

7 found

3 were not in current space.

18 were filtered out.

Select objects or ENTER to schedule external drawing: ENTER
(Press ENTER *to end selection.)*

Upper left corner of table: *(Select a point near p1 as shown in Figure 10T.12.)*

Lower right corner (or RETURN): ENTER *(Press* ENTER *to insert the schedule scaled to the drawing scale. Schedule inserted as shown in Figure 10T.13.)*

5. Right-click over the title bar of the tool palette and choose **Design** palette set. Select the **Casement** tool from the Windows palette. Edit the Width = 3', Height = 4'-0", Measure to = Outside of Frame, Swing Angle = 45, Position along wall = Offset/Center, Automatic Offset = 36, Vertical alignment = Head, Head height = 6'-8". Add two casement windows on the left wall of the building as shown at p1 and p2 in Figure 10T.13. Note: When casement windows are added to the drawing, the data for the windows is added to the schedule.

6. To remove the question marks of the first two schedule entries, right-click over the title bar of the tool palette, and select the **Document** palette set.

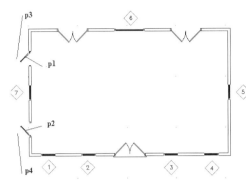

Figure 10T.13 *Casement windows added to floor plan*

Select **Window Tag** from the Scheduling tool palette. Respond to the command line prompts as shown below to tag the windows.

Note: To view preset values of the command, press F2.

Select object to tag [Symbol/Leader/Dimstyle/Edit/Constrain/Rotation]: *(Select window at p1 as shown in Figure 10T.13.)*

Specify location of tag <Centered>: *(Select a point near p3 as shown in Figure 10T.13 to locate the tag.)*

(Verify that 8 is the value of the Number edit field of the Edit Property Set Data dialog box. Select OK to dismiss the Edit Property Set Data dialog box.)

Select object to tag [Symbol/Leader/Dimstyle/Multiple/Edit/Constrain/Rotation]: *(Select window at p2 as shown in Figure 10T.13.)*

Specify location of tag <Centered>: *(Select a point near p4 as shown in Figure 10T.13.)*

(Verify that 9 is the value of the Number edit field of the Edit Property Set Data dialog box. Select OK to dismiss the Edit Property Set Data dialog box.)

(Press ESC to end the command; question marks are removed from schedule as shown in Figure 10T.14.)

7. Select the **Renumber Data Tool** from the Scheduling tool palette. Edit the **Data Renumber** dialog box as follows: Property Set: Window Objects, Property = **Number**, Start Number = **7**, and Increment = **1**. Select **OK** to dismiss the **Data Renumber** dialog box. Select the windows of the left wall as shown in Figure 10T.15.

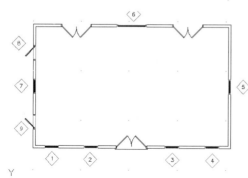

Figure 10T.14 Mark data from window tag automatically updated in window schedule

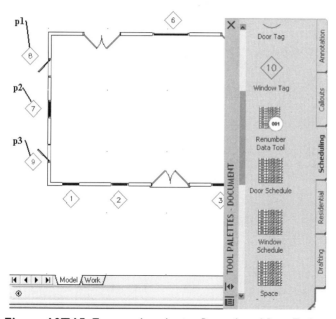
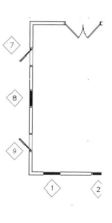

Figure 10T.15 Tags numbered using Renumbered Data Tool

Command: _AecPropertyRenumberData

Select object to renumber to 7: *(Select window at p1.)*

Select object to renumber to 8: *(Select window at p2.)*

Select object to renumber to 9: *(Select window at p3.)*

Select object to renumber to 10: ENTER *(Press* ENTER *to end selection.)*

8. Select the Window Schedule, right-click, and choose **Edit Schedule Table Style** from the shortcut menu. Select the Columns tab of the **Schedule Table Style Properties – Window Schedule** dialog box.

9. Select the **Add Column** button to open the **Add Column** dialog box. Scroll down the list of properties in the left column and select **FrameDepth** from the **WindowObjects** property set. Overtype **FRAME DEPTH** in the **Heading** edit field of the **Column Properties** section at right. Edit **Data Format = Length-Long**. Select the **Insert After** radio button of the **Column Position** section and select **WindowObjects:Remarks** from the drop-down list as shown in Figure 10T.16. Select **OK** to dismiss the **Add Column** dialog box.

10. Select the **Add Column** button to open the **Add Column** dialog box. Scroll down the left column of properties and select **FrameMaterial** from the **WindowObjects** property set. Overtype **FRAME MATERIAL** in the **Column Properties** section. Verify that Data Format = **Case-Upper**. Select the **Insert After** radio button of the **Column Position** section and select **WindowObjects:FrameDepth** from the **Column** drop-down list as shown in Figure 10T.17. Select **OK** to dismiss the **Add Column** dialog box.

11. Select the scroll bar at the bottom of the **Schedule Table Style Properties – Window Schedule** dialog box as shown at **p1** in Figure 10T.18 and move it all the way to the right. Hold down CTRL, and select the **FRAME DEPTH**

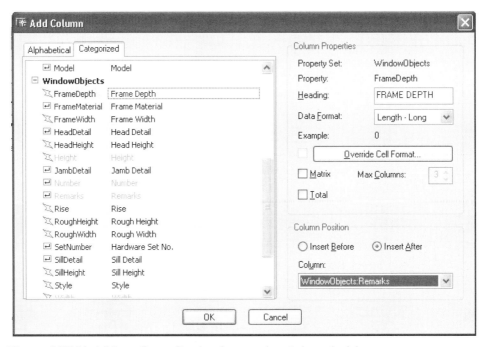

Figure 10T.16 *Adding a Frame Depth column to the window schedule*

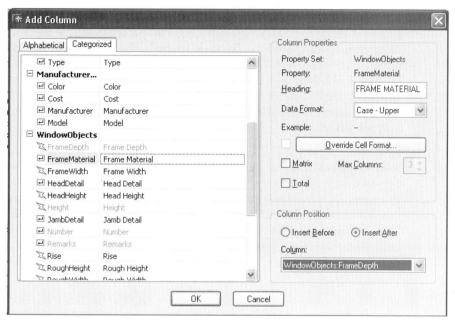

Figure 10T.17 *Adding a Frame Material column to the window schedule*

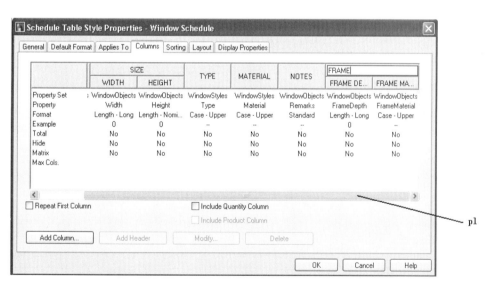

Figure 10T.18 *Adding a header*

and **FRAME MATERIAL** column headings. Select the **Add Header** button, and type **FRAME** to add the header as shown in Figure 10T.18. Select **OK** to dismiss **Schedule Table Style Properties – Window Schedule** dialog box.

12. Manual data for a window style is added in this step. Select the window tagged number **1** in the floor plan, right-click, and choose **Edit Window Style** from the shortcut menu. Select the **General** tab, and select the **Property Sets** button. Scroll down to the Window Styles property set, type **DOUBLE HUNG** in the **Type** edit field of the **WindowStyles** property set, and type **VINYL CLAD** in the **Material** edit field as shown in Figure 10T.19. Select **OK** to dismiss the **Edit Property Set Data** dialog box. Select **OK** to dismiss the **Window Style Properties – Double Hung** dialog box. Style based data is added to schedule.

13. Select the window tagged number **9**, right-click, and choose **Edit Window Style** from the shortcut menu. Select the **General** tab, and select the **Property Sets** button. Scroll down to the **WindowStyles** property set, type **SINGLE CASEMENT** in the **Type** edit field, and type **VINYL CLAD** in the **Material** edit field. Select **OK** to dismiss the **Edit Property Set Data** dialog box. Select **OK** to dismiss the **Window Style Properties – CASEMENT** dialog box. Type and material data are added to the schedule as shown in Figure 10T.20.

14. Select the Window Schedule, right-click, select **Edit Table Cell**. Click the "– –" in the **Type** column for number **6** window at p1 in Figure 10T.21. Select **Yes** to dismiss the AutoCAD warning box and open the **Edit Property Set Data** dialog box. Type **PICTURE ARCHED** in the **Type** edit field of the

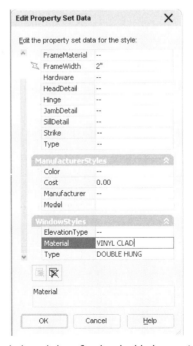

Figure 10T.19 *Inserting style based data for the double hung window*

WINDOW SCHEDULE

MARK	SIZE		TYPE	MATERIAL	NOTES	FRAME	
	WIDTH	HEIGHT				FRAME DEPTH	FRAME MATERIAL
1	2'-10"	5'-0"	DOUBLE HUNG	VINYL CLAD	--	5"	--
2	2'-10"	5'-0"	DOUBLE HUNG	VINYL CLAD	--	5"	--
3	2'-10"	5'-0"	DOUBLE HUNG	VINYL CLAD	--	5"	--
4	2'-10"	5'-0"	DOUBLE HUNG	VINYL CLAD	--	5"	--
5	2'-10"	5'-0"	DOUBLE HUNG	VINYL CLAD	--	5"	--
6	6'-0"	5'-0"	--	--	--	5"	--
7	3'-0"	4'-0"	SINGLE CASEMENT	VINYL CLAD	--	5"	--
8	2'-10"	5'-0"	DOUBLE HUNG	VINYL CLAD	--	5"	--
9	3'-0"	4'-0"	SINGLE CASEMENT	VINYL CLAD	--	5"	--

Style based data added to schedule

Figure 10T.20 *Window Type and Material data entered in schedule from style based properties*

WindowStyles property set. Type **VINYL CLAD** in the **Material** edit field of the **WindowStyles** property set. Select **OK** to dismiss the **Edit Property Set Data** dialog box. Press ESC to end the command.

15. Select **Browse Property Data** from the Scheduling tool palette. Edit the Property Set Definitions = **Window Objects** and Object Filter = **Window** in the **Browse Property Data** dialog box. Hold down CTRL and select all the windows as shown in Figure 10T.22. Type **Vinyl Clad** in the **FrameMaterial** edit field of the **WindowObjects** property set. Select **OK** to dismiss the **Browse Property Data** dialog box.

16. Select the window schedule, right-click, and choose **Edit Schedule Table Style**. Select the **Columns** tab. Select the scroll bar and scroll to the right to display the **Notes** and **Frame** column headings. (If the scroll bar is not displayed, move the pointer over the lower edge of the dialog box to display the

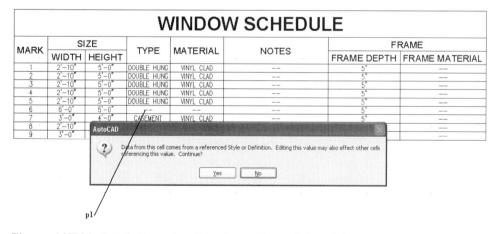

Figure 10T.21 *AutoCAD warning dialog box editing style based data*

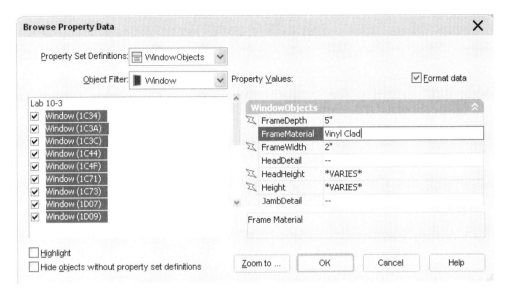

Figure 10T.22 *Editing the Frame Material in the Browse Property Data dialog box*

double arrow, and click and drag down the lower edge to increase the size of the dialog box.) Select the **Notes** heading and drag the heading right as shown in Figure 10T.23. Select **OK** to dismiss the **Schedule Table Style Properties – Window Schedule** dialog box.

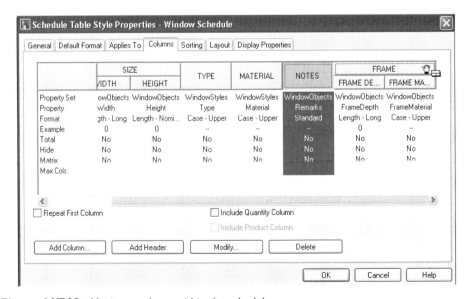

Figure 10T.23 *Moving a column within the schedule*

WINDOW SCHEDULE

MARK	SIZE		TYPE	MATERIAL	FRAME		NOTES
	WIDTH	HEIGHT			FRAME DEPTH	FRAME MATERIAL	
1	2'-10"	5'-0"	DOUBLE HUNG	VINYL CLAD	5"	VINYL CLAD	---
2	2'-10"	5'-0"	DOUBLE HUNG	VINYL CLAD	5"	VINYL CLAD	---
3	2'-10"	5'-0"	DOUBLE HUNG	VINYL CLAD	5"	VINYL CLAD	---
4	2'-10"	5'-0"	DOUBLE HUNG	VINYL CLAD	5"	VINYL CLAD	---
5	2'-10"	5'-0"	DOUBLE HUNG	VINYL CLAD	5"	VINYL CLAD	---
6	6'-0"	5'-0"	PICTURE ARCHED	VINYL CLAD	5"	VINYL CLAD	---
7	3'-0"	4'-0"	CASEMENT	VINYL CLAD	5"	VINYL CLAD	---
8	2'-10"	5'-0"	DOUBLE HUNG	VINYL CLAD	5"	VINYL CLAD	---
9	3'-0"	4'-0"	CASEMENT	VINYL CLAD	5"	VINYL CLAD	---

Figure 10T.24 *Completed window schedule*

Completed window schedule is shown in Figure 10T.24.

17. Save and close the drawing.

TUTORIAL 10.4 PLACING ALPHABETIC TAGS AND QUANTITIES FOR DOOR SCHEDULES

1. Open **Ex 10-4** from your *ADT Student\ADT Tutor\Ch10\directory*.

2. Select **File>SaveAs** from the menu bar and save the drawing as **Lab 10-4** in your student directory.

3. This tutorial will modify the property used in the door tag to place text tags. In the following steps you will insert the door tag from the Scheduling tool palette, insert the schedule, alter the property set definition, place tags, and modify the schedule.

4. If tool palettes are not displayed, select **Window>Tool Palettes** from the menu bar.

5. Insert the **DoorObject** property into the drawing by selecting **Door Tag** from the Scheduling tool palette. Respond to the following command line prompts to tag the front door.

    Select object to tag [Symbol/Leader/Dimstyle/Edit/Constrain/Rotation]: *(Select the front door at p1 shown in Figure 10T.25.)*

    Specify location of tag <Centered>: *(Select a location at p2 as shown in Figure 10T.25.)*

    *(Edit the Number property = **01** in the **Edit Property Set Data** dialog box. Select **OK** to dismiss the **Edit Property Set Data** dialog box.)*

    Select object to tag [Symbol/Leader/Dimstyle/Edit/Constrain/Rotation]: ESC *(Press ESC to end the command.)*

6. Select **Door Schedule** from the Scheduling tool palette. Edit the Properties palette as follows: Style = **Door Schedule**, Scale = **96.00**, Update automatically = **yes**, Add new objects automatically = **yes**, Scan xrefs = **yes**, Scan block references = **yes**, Layer wildcard = *. Respond to the following command line prompts to place the schedule as shown in Figure 10T.25.

   Command: ScheduleAdd

   Select objects or ENTER to schedule external drawing: **all** ENTER

   5 found

   3 were not in current space.

   Select objects or ENTER to schedule external drawing: ENTER *(Press ENTER to end selection.)*

   Upper left corner of table: *(Select a point near p3 as shown in Figure 10T.25.)*

   Lower right corner (or RETURN): ENTER *(Press ENTER to insert the schedule scaled to the drawing scale.)*

7. Select **Format>Style Manager** from the menu bar. Expand **Documentation Objects>Property Set Definitions** in the left pane. Double-click on the **DoorObjects** property to open the **Property Set Definition Properties – Door Objects** dialog box. Select the **Definition** tab, scroll down the list of properties and select **Number**. Edit Type = **Text**, Default = **A**, and Format = **Case-Upper**. Select **OK** to close the **Property Set Definition Properties – Door Objects** dialog box. Select **OK** to close the **Style Manager**.

8. Select **Door Tag** from the Scheduling palette. Respond to the command line prompts as shown below to tag the doors.

Note: To view preset values of the command, press F2.

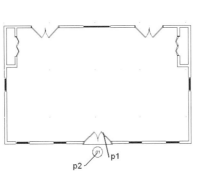

Figure 10T.25 *Door schedule placed in the drawing*

Select object to tag [Symbol/Leader/Dimstyle/Edit/Constrain/Rotation]: *(Select the door at p1 as shown in Figure 10T.26.)*

Specify location of tag <Centered>: *(Select a point near p2 as shown in Figure 10T.26.)*

(Verify **A** *is the Number value of the* **Edit Property Set Data** *dialog box. Select* **OK** *to dismiss the* **Edit Property Set Data** *dialog box.)*

Select object to tag [Symbol/Leader/Dimstyle/Multiple/Edit/Constrain/Rotation]: **M** ENTER *(Use a crossing window to select all the doors of the floor plan as shown in Figure 10T.26.)*

Select objects to tag: *(Select a point near p3 shown in Figure 10T.26.)*

Specify opposite corner: *(Select a point near p4 shown in Figure 10T.26.)* 9 found

6 were filtered out.

Select objects to tag: ENTER *(Press* ENTER *to end selection.)*

(Select **OK** *to dismiss the* **Edit Property Set Data** *dialog box.)*

Select object to tag [Symbol/Leader/Dimstyle/Multiple/Edit/Constrain/Rotation]: ENTER *(Press* ENTER *to end the command.)*

9. Select the tags to display their grips and move tags as necessary to improve clarity.

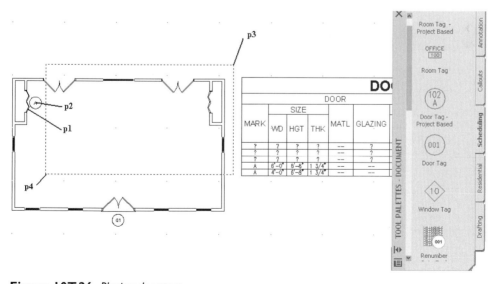

Figure 10T.26 *Placing door tags*

10. The outcome of this tutorial is to create a schedule that lists all doors of the same type with the same mark. Therefore select the doors at **p1** and **p2** as shown in Figure 10T.27, right-click, and choose **Properties** to open the Properties palette, select the **Extended Data** tab, and edit the **Number** property to **B**. Move the pointer over the workspace and press ESC to clear selection.

11. Double-click on the door at **p3** as shown in Figure 10T.27, to open the Properties palette, select the **Extended Data** tab, and edit the **Number** property to **C**. Press ESC to clear selection.

12. Select **Zoom Extents** from the **Zoom** flyout of the **Navigation** toolbar.

13. Select the schedule, right-click, and choose **Edit Schedule Table Style** from the shortcut menu. Select the **Columns** tab, and check **Include Quantity Column**. Hold CTRL down and select the following headings as shown in Figure 10T.28:

    DOOR: MATL

    DOOR: GLAZING

    DOOR: LOUVER: WD, HGT

    FRAME: MATL, EL

    FRAME: DETAIL, HEAD, JAMB, SILL

    FIRE RATING

    HARDWARE: SET NO, KEYSIDE RM NO

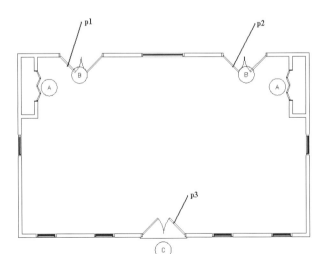

Figure 10T.27 *Door tags edited*

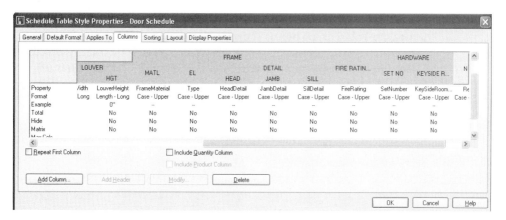

Figure 10T.28 *Deleting columns from schedule*

Select **Delete** to remove the columns. Select **OK** to dismiss the **Remove Columns/Headers** dialog box. Select **OK** to dismiss the **Schedule Table Style Properties – Door Schedule** dialog box.

14. The schedule shown in Figure 10T.29 lists door A with incorrect thickness. Select the schedule, right-click, and choose **Selection>Show**. Select Mark **A** in the schedule to identify where the doors are located in the drawing with incorrect thickness in the drawing. Press ESC to end selection.

15. Door thickness is a style based property; select door **A** at **p1** as shown in Figure 10T.29, right-click, and choose **Edit Door Style** from the shortcut menu. Select the **Dimensions** tab of the **Door Style Properties – Bifold-Double** dialog box, edit Door Thickness = 1-3/8". Select **OK** to dismiss the **Door Style Properties – Bifold-Double** dialog box.

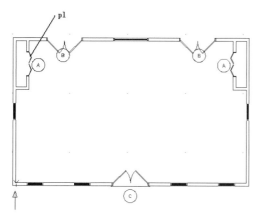

Figure 10T.29 *Door schedule revised by style*

16. Select the schedule, right-click, and choose **Export** to open the **Export Schedule Table** dialog box as shown in Figure 10T.30. Select the **Browse** button and edit the File name = **Lab 10-4** and Save in your student directory. Select **Save** to complete the selection and dismiss the **Create File** dialog box.

17. Select **OK** to dismiss the **Export Schedule Table** dialog box. Verify that the **Convert to Formatted Text** radio button is selected and **Apply to All Columns** is checked in the **Format** dialog box.

18. Save and close the drawing.

TUTORIAL 10.5 CREATING TAGS AND SCHEDULES FOR A PROJECT

1. Open Autodesk Architectural Desktop 2005, and select **File>Project Browser** from the menu bar. Use the Project Selector drop-down list to navigate to your *ADT Student\ADT Tutor\Ch 10* directory. Double-click on **Ex 10-5** to set this project current. Select **Close** to dismiss the **Project Browser**.

2. Select the **Constructs** tab of the **Project Navigator**. Double-click on **Floor 1** to open the Floor 1 drawing.

3. Select **High Detail** display configuration from the **Display Configuration** flyout of the Drawing Window status bar. Select the **1/4"=1'-0"** scale from the flyout scale list of the Drawing Window status bar.

4. Toggle OFF SNAP, GRID and ORTHO on the status bar. Toggle ON POLAR, OSNAP, and OTRACK on the status bar. Right-click over the OSNAP toggle

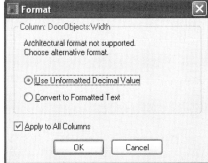

Figure 10T.30 *Door schedule exported to Excel program*

on the status bar, select **Settings**, and toggle OFF all object snaps except **Endpoint**. Select **OK** to close the dialog box.

5. If tool palettes are not displayed, press CTRL +3. Right-click over the title bar of the tool palette and select the **Design** palette set. Right-click over the title bar of the tool palette, and select **New Palette** from the shortcut menu. Overtype the name **Commercial** to name the tool palette.

6. To import space styles to the new palette, choose the **Content Browser** from the **Navigation** toolbar. Double-click to open the Design Tool Catalog-Imperial. Choose the **Spaces** category and open the **Commercial** subcategory. Drag the following spaces from the **Commercial** subcategory onto the new palette: **Corridor**, **Restroom_Men_Small**, **Restroom_Women_Small**, **Office_Medium**, and **Entry_Room**.

7. Choose the **Space Auto Generate Tool** from the Design tool palette to open the **Generate Spaces** dialog box. Edit the **Generate Spaces** dialog box as follows: **Filter** list = **Walls only**, toggle ON **Automatic**, and Style = **Standard**. Choose the **Tag Settings** button and check **Add Tag to New Spaces** and **Add Property Set to New Spaces** check boxes. Edit the Tag Definition = **Aec4_Room_Tag_Scale_Dependent**, Property Set = **Room Objects**, Property 1 = **Increment**, and Property 2 = **Increment**, check **Auto-Increment Numeric Properties**, and **Increment** = **1**. Select **OK** to dismiss the **Tag Settings** dialog box. In the **Property Data** section edit **Increment** = **1** and **Increment** = **1**. Move the pointer over the room at **p1** to display the boundary (shown in red) of the room, and then left-click near **p1** to specify the boundary as shown in Figure 10T.31. Move the pointer over the remaining rooms, and then click at **p2**, **p3**, **p4**, **p5**, **p6**, **p7**, **p8**, **p9**, and **p10** to specify their boundaries. Select **Close** to close the **Generate Spaces** dialog box.

8. Select the **Office_Medium** space on the Commercial palette, right-click, and choose **Apply Tool Properties to>Space**. Select the space in rooms **102**, **103**, **104**, **108**, and **109**. Press ENTER to end the edit and press ESC to clear selection.

9. Select the **Corridor** space on the Commercial palette, right-click, and choose **Apply Tool Properties to>Space**. Select the space in Room **110**. Press ENTER to end the edit and press ESC to clear the selection.

10. Select the **Entry_Room** space on the Commercial palette, right-click, and choose **Apply Tool Properties to>Space**. Select the space in Room **107** and Room **101**. Press ENTER to end the edit and press ESC to clear the selection.

11. Select the **Restroom_Women_Small** space on the **Commercial** palette, right-click, and choose **Apply Tool Properties to>Space**. Select the space in Room **106**. Press ENTER to end the edit and press ESC to clear the selection.

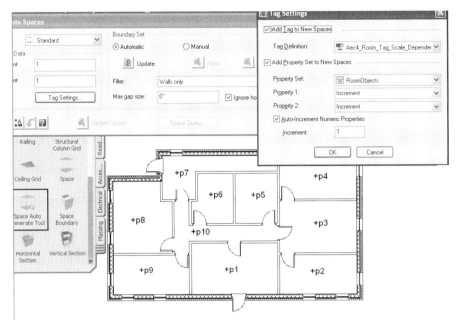

Figure 10T.31 *Creating spaces for level 1*

12. Select the **Restroom_Men_Small** space on the Commercial palette, right-click, and choose **Apply Tool Properties to>Space**. Select the space in Room **105**. Press ENTER to end the edit and press ESC to clear the selection.

13. Save and close the drawing.

14. Select the **Constructs** tab of the **Project Navigator**. Select **Floor 2**, right-click, and choose **Open**. Select 1/4"=1'-0" scale and **High Detail** display configuration from the Drawing Window status bar.

15. Choose the **Space Auto Generate Tool** from the Design tool palette to open the **Generate Spaces** dialog box. Edit the **Generate Spaces** dialog box as follows: **Filter** list = **Walls only**, toggle ON **Automatic**, and Style = **Standard**. Choose the **Tag Settings** button and check **Add Tag to New Spaces** and **Add Property Set to New Spaces** check boxes. Edit the Tag Definition = **Aec4_Room_Tag_Scale_Dependent**, Property Set = **Room Objects**, Property 1 = **Increment**, and Property 2 = **Increment**, check **Auto-Increment Numeric Properties**, and Increment = 1. Select **OK** to dismiss the **Tag Settings** dialog box. In the **Property Data** section, edit Increment = 1 and Increment = 1. Move the pointer over the office at **p1** to display the boundary (shown in red) of the room, and then left-click near **p1** to specify the boundary as shown in Figure 10T.32. Move the pointer over the remaining offices, and then click at **p2, p3, p4, p5, p6, p7, p8, p9** and **p10** to

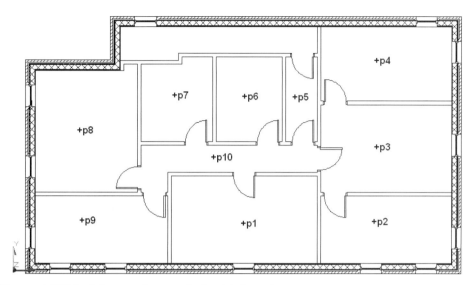

Figure 10T.32 *Selection of spaces for level 2 floor plan*

specify their boundaries. Press ENTER to end selection and end the command.

16. Select the **Office_Medium** space on the Commercial palette, right-click, and choose **Apply Tool Properties to>Space**. Select the space for rooms **201, 202, 203, 204, 206, 207, 208**, and **209**. Press ENTER to end the edit and press ESC to end selection of the spaces.

17. Select the **Corridor** space on the Commercial palette, right-click, and choose **Apply Tool Properties to>Space**. Select the space for Room **210** and Room **205**. Press ENTER to end the edit and press ESC to end selection of the spaces.

18. Save and close the **Floor 2** drawing.

19. Select the **Views** tab of the **Project Navigator**, and double-click on the **Floor Plan 1** to open the drawing. Press F3 to turn off object snaps. Right-click over the title bar of the tool palettes and select **Document** palette set. Select **Door Tag Project Based** from the Scheduling tool palette. Tag each door as shown in Figure 10T.33 and the following prompts.

> Select object to tag [Symbol/Leader/Dimstyle/Edit/Constrain/Rotation]: *(Select the door at p1 as shown in Figure 10T.33.)* (1 found)
>
> Specify location of tag <Centered>: ENTER ENTER *(Press* ENTER *to center tags and dismiss the* **Edit Property Set Data** *dialog box.)* (1 found)

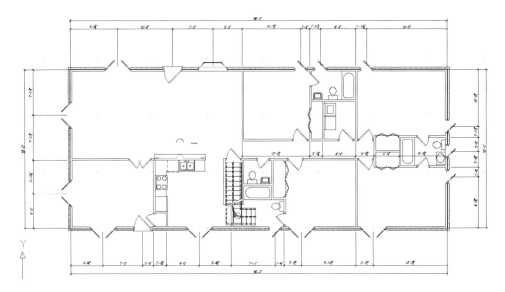

Figure 10T.33 *Selection of doors for tags floor 1*

Select object to tag [Symbol/Leader/Dimstyle/Multiple/Edit/Constrain/Rotation]: **M** ENTER

*(Type **M** to select multiple doors.)*

Select objects to tag: *(Select doors using a crossing window from p2 to p3 as shown in Figure 10T.33.)*

Specify opposite corner: 12 found

*(Choose **No** to dismiss the AutoCAD dialog box. Select **OK** to dismiss the **Edit Property Set Data** dialog box.)*

Select objects to tag: ENTER *(Press ENTER to end selection.)*

Select object to tag [Symbol/Leader/Dimstyle/Multiple/Edit/Constrain/Rotation]: ENTER *(Press ENTER to accept the settings of the **Edit Property Set Data** dialog box.)*

Select object to tag [Symbol/Leader/Dimstyle/Multiple/Edit/Constrain/Rotation]: ENTER *(Press ENTER to end the command.)*

20. To link the exterior doors to a space, select the door at **p4** as shown in Figures 10.33 and 10.34, right-click, and choose **Open Xref**. Select the door at **p4** to display its grips. Drag the **Property Data Location** grip from point **p4** to a point near **p5** as shown in Figure 10.34.

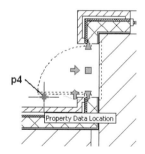 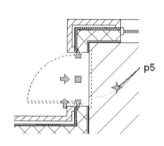 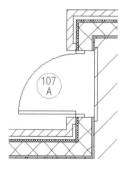 

Property Data Linked to Space

Figure 10T.34 *Editing the Property Data Location*

21. Select the door at **p6** in Figure 10.33 to display its grips. Drag the **Property Data Location** grip to the adjoining space of Room **101**. Save and close the **Floor 1** Construct drawing.

22. Select **Manage Xrefs** from the Drawing Window status bar, choose **Floor 1**, and choose the **Reload** button. Select **OK** to dismiss the **Xref Manager**.

23. Save and close the **Floor Plan 1** View drawing.

24. Verify that the **Views** tab of the **Project Navigator** is current; and double-click on the **Floor Plan 2** to open the drawing.

25. Select the **1/4"=1'-0"** scale from the **Scale** flyout of the Drawing Window status bar. Select **Door Tag Project Based** from the Scheduling tool palette. Tag each door as shown in Figure 10T.35 and the following prompts:

    Select object to tag [Symbol/Leader/Dimstyle/Edit/Constrain/Rotation]: *(Select the door at p1 as shown in Figure 10T.35.)* (1 found)

    Specify location of tag <Centered>: ENTER, ENTER *(Press ENTER twice to center tags and dismiss the Edit Property Set Data dialog box.)* (1 found)

    Select object to tag [Symbol/Leader/Dimstyle/Multiple/Edit/Constrain/Rotation]: **M** ENTER

    *(Type **M** to select multiple doors.)*

    Select objects to tag: *(Select doors using a crossing window from p2 to p3 as shown in Figure 10T.35)*

    Specify opposite corner: 46 found 36 were filtered out

Select objects to tag: ENTER *(Press ENTER to end selection.)*

*(Choose **No** to dismiss the AutoCAD dialog box. Select **OK** to dismiss the **Edit Property Set Data** dialog box.)*

Select object to tag [Symbol/Leader/Dimstyle/Multiple/Edit/Constrain/Rotation]: ENTER

*(Press ENTER to accept the settings of the **Edit Property Set Data** dialog box.)*

26. Save and close the **Floor Plan 2** drawing. Select **QNew** from the Standard toolbar.

27. In the **Views** tab of the **Project Navigator**, select the Views category, right-click, and choose **New View Dwg>General**. Type **Model** as the name of the new drawing, select **Next**, and check Level 1 and Level 2 for Division 1. Select **Next** to verify that **Floor 1** and **Floor 2** constructs are checked. Select **Finish** to close the **Add General View** dialog box.

28. Select the **Sheets** tab, select the **Schedules and Diagrams** category, right-click, and choose **New>Sheet**. Type **A-10** in the **Number** field and type **Schedules** in the **Sheet title** edit field. Select **OK** to close the **New Sheet** dialog box.

29. Select the **Sheets** tab, and double-click on the **A-10 Schedules** file of the Sheet Set View.

30. Select the **Content Browser**, and open the Documentation Tool Catalog-Imperial. Open the **Schedule Tables** category and click and drag the i-drop

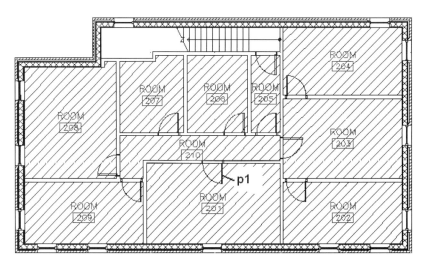

Figure 10T.35 *Selection of doors for the 2nd Floor*

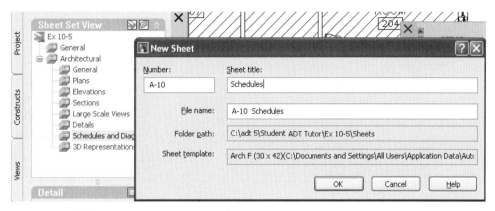

Figure 10T.36 *Create a new sheet for schedules*

of the **Door Schedule Project Based** into the workspace. Respond to the following command line prompts to insert the schedule.

Command: ScheduleAdd

Select objects or ENTER to schedule external drawing: ENTER
(Press ENTER to schedule external drawings.)

Upper left corner of table: *(Select a point near p1 in the drawing as shown in Figure 10T.37.)*

Lower right corner (or RETURN): ENTER *(Press ENTER to accept default size.)*

31. Select the schedule, right-click, and select **Properties** from the shortcut menu. Select the **Design** tab of the Properties palette, scroll down to the

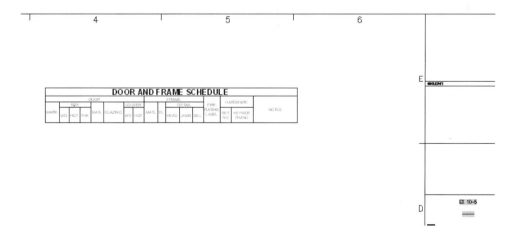

Figure 10T.37 *Specify the insert point of the schedule*

DOOR AND FRAME SCHEDULE														
	DOOR						FRAME					HARDWARE		
MARK	SIZE			MATL	GLAZING	LOUVER		MATL	EL	DETAIL			FIRE RATING LABEL	SET NO	KEYSIDE RM NO	NOTES
	WD	HGT	THK			WD	HGT			HEAD	JAMB	SILL				
101A	3'-0"	7'-0"	1 3/4"	---	---	0"	0"	---	---	---	---	---	---	---	---	---
101A	3'-0"	7'-0"	2"	---	---	0"	0"	---	---	---	---	---	---	---	---	---
102A	3'-0"	7'-0"	2"	---	---	0"	0"	---	---	---	---	---	---	---	---	---
103A	3'-0"	7'-0"	2"	---	---	0"	0"	---	---	---	---	---	---	---	---	---
104A	3'-0"	7'-0"	2"	---	---	0"	0"	---	---	---	---	---	---	---	---	---
105A	3'-0"	7'-0"	2"	---	---	0"	0"	---	---	---	---	---	---	---	---	---
106A	3'-0"	7'-0"	2"	---	---	0"	0"	---	---	---	---	---	---	---	---	---
107A	3'-0"	7'-0"	1 3/4"	---	---	0"	0"	---	---	---	---	---	---	---	---	---
107A	3'-0"	7'-0"	2"	---	---	0"	0"	---	---	---	---	---	---	---	---	---
108A	3'-0"	7'-0"	2"	---	---	0"	0"	---	---	---	---	---	---	---	---	---
109A	3'-0"	7'-0"	2"	---	---	0"	0"	---	---	---	---	---	---	---	---	---
201A	3'-0"	7'-0"	2"	---	---	0"	0"	---	---	---	---	---	---	---	---	---
202A	3'-0"	7'-0"	2"	---	---	0"	0"	---	---	---	---	---	---	---	---	---
203A	3'-0"	7'-0"	2"	---	---	0"	0"	---	---	---	---	---	---	---	---	---
204A	3'-0"	7'-0"	2"	---	---	0"	0"	---	---	---	---	---	---	---	---	---
205A	3'-0"	7'-0"	2"	---	---	0"	0"	---	---	---	---	---	---	---	---	---
205B	3'-0"	7'-0"	2"	---	---	0"	0"	---	---	---	---	---	---	---	---	---
206A	3'-0"	7'-0"	2"	---	---	0"	0"	---	---	---	---	---	---	---	---	---
207A	3'-0"	7'-0"	2"	---	---	0"	0"	---	---	---	---	---	---	---	---	---
208A	3'-0"	7'-0"	2"	---	---	0"	0"	---	---	---	---	---	---	---	---	---
209A	3'-0"	7'-0"	2"	---	---	0"	0"	---	---	---	---	---	---	---	---	---

Figure 10T.38 *Schedule created for all floors of the building*

Advanced section, toggle **Schedule external drawing** to **Yes**, click the **External drawing** field, and select the *\ADT Student\ADT Tutor\Ch 10\Ex 10-5\Views\Model.dwg* drawing.

32. Verify that the schedule is selected, right-click, and select **Update Schedule Table** to create the comprehensive schedule shown in Figure 10T.38.

33. Save and close all drawings.

PROJECT

Ex 10.6 Creating Spaces and Dimension

1. Open Autodesk Architectural Desktop 2005, and select **File>Project Browser** from the menu bar. Use the Project Selector drop-down list to navigate to your *ADT Student\Adt Tutor\Ch10* directory. Double-click on the **Ex 10-6** to set this project current. Select **Close** to dismiss the **Project Browser**.

2. Select **Space Auto Generate Tool** to create additional spaces for each of the rooms in the Floor Plan 1 drawing. Insert additional spaces from the Residential and Medical catalogs of the **Content Browser**. Edit the **Space Style** of each space created according to Figure 10T.39.

3. Insert the **Space Tag** from the Documentation Tool Catalog -Imperial Schedule Tags category, Room and Finish Tags subcategory of the **Content Browser**.

4. Close and save the drawing.

5. Open the Floor Plan-Basement view drawing. Edit the display configuration to High Detail. Edit the **Cut Height** of the Display Configuration to **6'** and

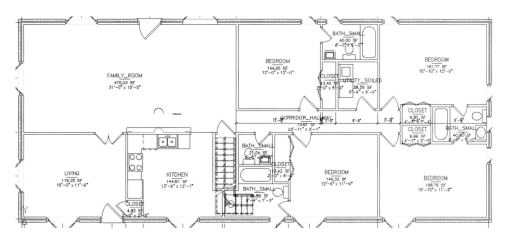

Figure 10T.39 *Spaces added to floor plan*

dimension the plan using **AEC Dimension (2)**. Open the **A-2 Floor Plan Basement Level** sheet and resize the viewport as necessary to display the dimensions.

CHAPTER 11

Creating Elevations, Sections, and Details

INTRODUCTION

This chapter includes the commands to extract elevations and sections from a building model and develop details. The use of the **Project Navigator** to create the building model and the callout tools used to create elevations and sections will be presented. The model is created as a View drawing in the **Project Navigator** by attaching each floor or building component as a construct. The **Project Navigator** provides for flexibility in design because it tracks changes that occur in Construct drawings and displays the latest version of a drawing in the model. Detail callouts, section marks, and elevation marks are used to link callouts to sheet numbers.

OBJECTIVES

Upon completion of this chapter, you will be able to

- Use the **Project Navigator** to create a 3D model of a building
- Create elevations and sections using the elevation and section marks tools located on the Callouts tool palette
- Revise the display of the elevation based upon changes in the model using **Refresh** (**2dSectionResultRefresh**) and **Regenerate** (**2dSectionResultUpdate**) commands
- Edit the dimensions and subdivision of the projection box for elevation or section objects
- Create blocks for the development of elevation and section working drawings with the **Hidden Line Projection** (**CreateHLR**)command
- Create live sections from pictorial views of a model using (**LiveSectionEnable**)

- Modify lines of a section or elevation using the **2dSectionResultEdit**, **2dSectionEditComponent**, and **2dSectionResultAddHatchBoundary** commands
- Create and modify 2d Section/Elevation styles
- Create keynote and dimension details using the **Detail Component Manager**

CREATING THE MODEL FOR ELEVATIONS AND SECTIONS

The elevation or section drawings of a building are developed from a model drawing. The model is constructed in the **Project Navigator** by attaching such Construct drawings as floor plans, decks, porches, or other detached components associated with each level. The levels specified in the **Project Navigator** as shown in Figure 11.1 determine the Z coordinate insertion point for the Construct drawings assigned to each level. The insertion point of the construct is located at the baseline of wall. The siding wall component defined in the wall style of floor 1 extends below the baseline of the wall to cover the floor framing as shown in Figure 11.1. In this example, the **FloorLine** command was used in the construct drawing to extend a wall component of the upper level down below its baseline to cover the floor framing. You can adjust the offset values for the **FloorLine** command based upon distances between the levels determined from elevation views of the model. The floor framing can be represented by a slab object with its thickness set equal to the joist and sill wood framing. The steps to creating a model using a project are listed below.

STEPS TO CREATING A MODEL

1. Open the **Project Browser** (select **File>Project Browser** from the menu bar) and set your project current.
2. Select **Close** to close the **Project Browser** and open the **Project Navigator**.
3. Select the **Levels** button of the **Project** tab as shown in Figure 11.1 to open the Levels dialog box. Specify values for floor elevation and finish floor to finish floor vertical distances for the each level (foundation, floor 1, floor 2).
4. Select the **Views** tab.
5. Select the Views category, right-click, and choose **New View Dwg>General**.
6. Edit the **Add General View** dialog box: type **Model** in the drawing edit field of the **General** page as shown in Figure 11.2. Click **Next** to open the **Context** page, and check Level **1**, Level **2**, and Level **3** for Division **1**. Select **Next** to open the **Content** page, and verify that all constructs are selected for the levels. Select **Finish** to close the **Add General View** dialog box.

Creating Elevations, Sections, and Details 771

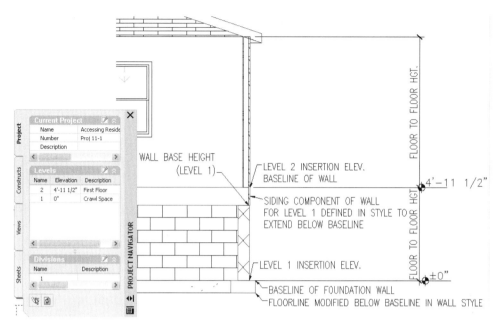

Figure 11.1 Creating levels for a model

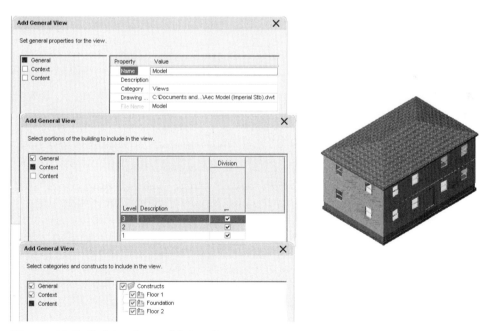

Figure 11.2 Defining the model view drawing

7. Double-click on the **Model** view drawing in the **View** tab to open the drawing as shown at right in Figure 11.2.

CREATING BUILDING ELEVATIONS AND SECTIONS

The model view drawing can include all levels and divisions of the building. After a model view drawing is created, you can open the drawing and place a callout mark. The callouts tools, located on the Callouts palette of the Document palette set, includes tools for creating the elevation or section within the model drawing or creating additional view drawings for the elevations. The elevation tools of the Callouts palette are as follows:

- Elevation Mark A1 Elevation Mark
- Elevation Mark A2 Elevation Mark (w/ Sheet No.)
- Exterior Elevation Mark A3-Exterior Elevation Mark (Entire Building)
- Interior Elevation Mark B1 Interior Elevation Mark (4 way) "1/2/3/4"
- Interior Elevation Mark B2 Interior Elevation Mark (4-Way) "N/E/S/W"

Each of the elevation mark tools of the Callouts palette opens the **Place Callout** dialog box, which includes options to place the mark as shown in Figure 11.5. When you

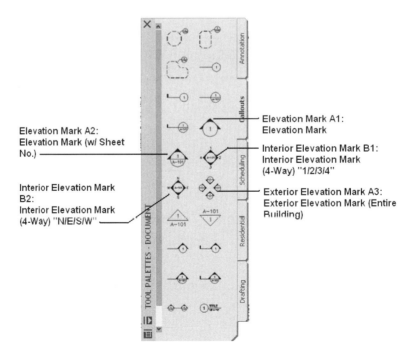

Figure 11.3 *Callouts tab of Document tool palette set*

select the elevation mark tool, you are prompted to specify the location and direction of view for the tag (see Figure 11.4), and then the **Place Callout** dialog box opens. The building elevation line is created during the process of placing the callout if you check **Generate Section/Elevation**. The **Place Callout** dialog box allows you to create a new view drawing, replace an existing view elevation drawing, or add an elevation within the current drawing, as shown in Figure 11.5. After you place the elevation mark, you are prompted to select the corner of the elevation region, which places the elevation building line and defines the view direction of the projection box for the elevation. The elevation or section line and projection box, shown in Figure 11.12, specifies the objects included in the elevation or section view.

The following steps outline the process of placing Elevation Mark A1, developing the elevation as a separate view drawing, and finally placing the view on a sheet. When you place a callout, a tag is inserted from the Callouts folder. The tag and title blocks with associated attributes can be identified in the Properties palette of each callout tool. The attributes of the callout include fields that extract view numbers or sheet names when placed. Therefore, when the callout is placed, a placeholder field is created that is displayed as a question mark until the view is placed on a sheet. The background of the field does not plot; only the view number is plotted. The **Place Callout** dialog box allows you to specify the name, scale, and titlemark of the view. When the view is placed on a sheet the view number will be displayed on the sheet and on the elevation view drawing.

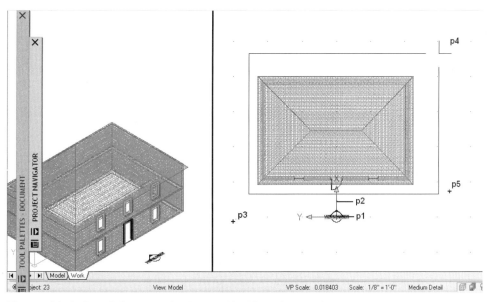

Figure 11.4 *Specify location of callout and building elevation line*

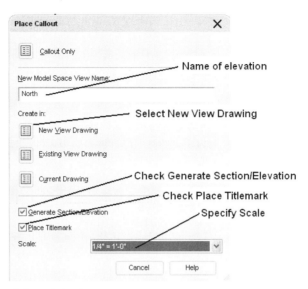

Figure 11.5 *Place Callout dialog box*

Figure 11.6 *Callout Properties palette*

STEPS TO CREATING AN ELEVATION

1. Open the **Project Navigator**, and select the **View** tab.

2. Open the model view drawing.

3. Select **Elevation Mark A1** from the Callouts tool palette shown in Figure 11.3, and respond to the command line prompts as shown below:

    Command: AecCallout

    Specify location of elevation tag: *(Select a point near p1 shown in Figure 11.4 to place the callout.)*

    Specify direction for elevation: *(Select a point near p2 shown in Figure 11.4, to specify the direction of view.)*

4. The **Place Callout** dialog box opens as shown in Figure 11.5; type **North** in the **New Model Space View Name** edit field, verify that **Generate Section/Elevation** and **Place Titlemark** are checked, set scale to **1/4"=1'-0"**, and then check **New View Drawing** to open the **Add Section/Elevation View** dialog box.

5. Edit the **Add Section/Elevation View** dialog box as follows: Type the name = **Exterior Elevations** of the view drawing in the **General** page, click **Next**, and then verify levels and constructs are selected in the **Context** and **Content** pages as shown in Figure 11.7. Select **Finish** to close the dialog box.

6. Respond to the remaining command line prompts:

    Command:

    Specify first corner of elevation region: *(Select a point near p3 shown in Figure 11.4 to define the start location of the projection box.)*

    Specify opposite corner of elevation region: *(Select a point near p4 shown in Figure 11.4 to define the size and content of the projection box.)*

    Specify insertion point for the 2D elevation result: *(Select a point near p5 shown in Figure 11.4 to locate the coordinates of the new elevation view.)*

    (The elevation is created in the new view drawing named North.)

7. To view the elevation, select the **Views** tab of the **Project Navigator**, and double-click on the **Exterior Elevations** view drawing.

8. To place the elevation on a sheet, create the new sheet in the **Sheets** tab, select the **Elevations** sheet category, right-click, and select **New>Sheet** as shown in Figure 11.8. Type a **A-111** for the number and **North Elevation** for the title of the sheet as described in Figure 11.8.

9. Select the **Sheets** tab of the **Project Navigator**, and double-click on the **A-111**, the new drawing, to open the drawing.

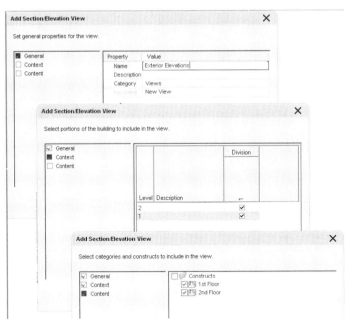

Figure 11.7 Add Section/Elevation View dialog box

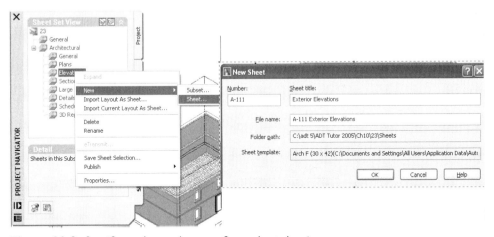

Figure 11.8 Specify number and name of new sheet drawing

10. Select the **Views** tab, select the **North** drawing, right-click, and choose **Place on Drawing** as shown in Figure 11.9. Drag the view to the sheet and left-click to position the view on the sheet. The view number is placed in the title bubble of the elevation view as shown in Figure 11.10.

Creating Elevations, Sections, and Details 777

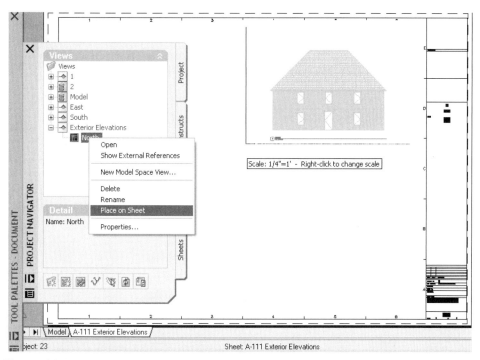

Figure 11.9 Place view drawing on a sheet

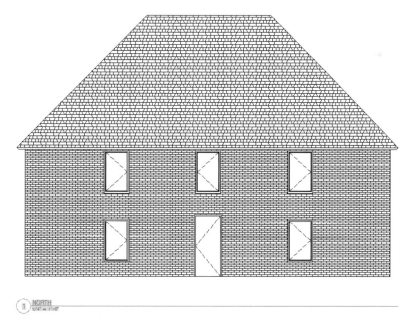

Figure 11.10 Callout number placed in the elevation drawing

When you create an elevation within a View drawing, you can select the elevation to display its view boundary. The grips of the view boundary are shown in Figure 11.11. You can select the grips of the view boundary and drag them to a different location to resize the view boundary. The scale of the elevation view is specified in the **Place Callout** dialog box when the view is created.

Editing the Building Section/Elevation Line Using Grips

After you create the elevation view, you can open the Construct drawings and make revisions. The elevation can be updated to reflect the changes. However, the building elevation line defines the viewing direction and size of the three-dimensional projection box for content of the elevation; therefore only those objects that fall within the box will be included in the elevation. For the elevation to be captured correctly, the changes in the construct may require the size and location of the building elevation line to be changed. You can select the building elevation line in the model view drawing to modify its location and size using grips. The size of the elevation projection box can be changed by dragging the grips located at the base of the projection box shown at left. The default height of the building elevation projection box is set equal to the height of the model space entities within the building line. If you select the trigger grip shown at right in Figure 11.12, the height of the projection box is displayed, and you can select the height and lower extension grips to edit the projection box.

EDITING THE PROPERTIES PALETTE OF THE BUILDING ELEVATION/SECTION LINE

The building elevation line and building section line properties are similar. Editing the properties of the building elevation line allows you to define what is included in the elevation or section and refine the display of the object using graphic subdivisions.

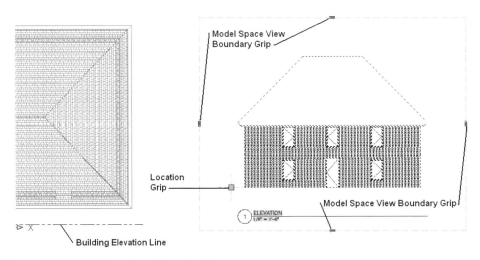

Figure 11.11 *View Boundary of an elevation*

Creating Elevations, Sections, and Details 779

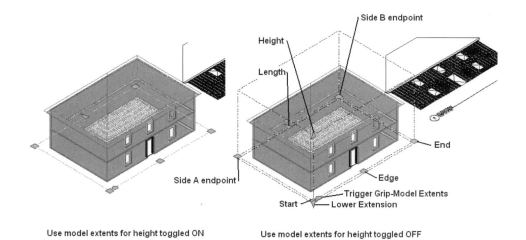

Figure 11.12 *Elevation drawing in the view drawing*

The properties of the building elevation line are displayed in the Properties palette as shown in Figure 11.13, when you select the building elevation or section line. The Properties palette of a Bldg Elevation Line consists of four sections: **General**,

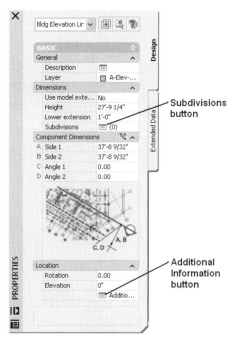

Figure 11.13 *Properties palette of the Building Elevation Line*

Dimensions, **Component Dimensions**, and **Location**. A description of the components of the Properties palette for the elevation line follows:

General

Description – The **Description** button opens the **Description** dialog box, which allows you to type a description.

Layer – The name of the layer of the Bldg Elevation Line is displayed. The Bldg Elevation Line is placed on A-Elev-Line if the AIA 256 Color layer key style is used in the drawing.

Dimensions

Use model extents for height – The **Yes** option of the **Use model extents for height** property will create a projection box inclusive of the height of the objects selected for the elevation. If the **No** option is selected, the **Height** and **Lower** extension edit fields are added to the Properties palette.

Height – The **Height** dimension is the vertical dimension up from the building elevation line plane. The building elevation line is placed at Z=0 coordinate.

Lower Extension – The **Lower Extension** is the vertical dimension below the building elevation line plane of the projection box.

Subdivisions – Selecting the **Subdivisions** button opens the **Subdivisions** dialog box, which allows you to create user-defined subdivisions for controlling elevation object display. The subdivisions allow you to create elevations or sections that emphasize those planes closest to the observer by increasing the line weight or color of the entities within those subdivisions. The layer, color, linetype, and lineweight of subdivisions are controlled in the **Display Properties** dialog box of the **2D Section/Elevation** display representation.

Component Dimensions

A-Side 1 – The **A-Side 1** dimension is the distance from the first point of the building elevation line measured in the view direction of the building elevation line as shown in Figure 11.14.

B-Side 2 – The **B-Side 2** dimension is the distance specified by the opposite corner of the elevation region from the building elevation line measured in the view direction perpendicular to the building elevation line. The depth dimensions of each side can vary, allowing the elevation to include only selected building components.

C-Angle 1 – The **C-Angle 1** dimension is the deflection from a line perpendicular to the building elevation line. Positive angles deflect to the left of the viewing direction, while negative angles deflect to the right. The angle of deflection is relevant to your orientation when facing toward the viewing direction for the elevation as shown in Figure 11.14.

D-Angle 2 – The **D-Angle 2** dimension is the deflection from along side 2 along a line perpendicular to the building elevation line. The positive angles deflect to the left and the negative angles deflect to the right, when oriented from the second point.

Location

Rotation – The **Rotation** edit field allows you to specify the angle of rotation from the 3:00 o'clock position for the building elevation line.

Elevation – The **Elevation** edit field displays the elevation of the building elevation line.

Additional Information – Select the **Additional Information** button to open the **Location** dialog box. The **Location** dialog box specifies the x, y, z coordinate insertion point location of the building elevation line. The **Normal** edit field allows you to specify the extrusion direction for the development of the elevation. Normal values should be set x=1, y=0, and z=0 to develop an elevation view. The **Rotation** angle edit field allows you to enter an angle of rotation for the building elevation line.

REFRESHING AND REGENERATING ELEVATION/SECTIONS

Because elevations and sections are generated from a model drawing that can change as the design develops, the **Refresh** (**2DsectionResultRefresh**) and **Regenerate**

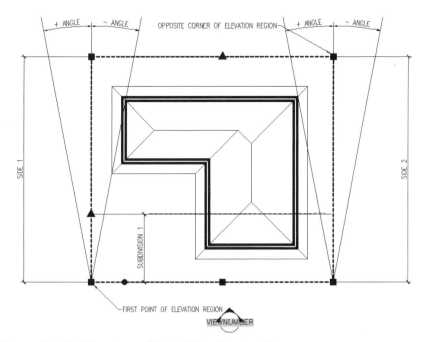

Figure 11.14 *Dimenisons of the Building Elevation Line*

(**2DsectionResultUpdate**) commands are provided to revise the elevation or section to the current state of the model. The **Refresh** (**2DsectionResultRefresh**) command (refer to Table 11.1) updates the elevation based on the original settings of the elevation. If additional objects have been added to the model drawing, the **Refresh** command updates the elevation **without** including the new objects. Therefore to include new objects, the **Regenerate** (**2DsectionResultUpdate**) command should be selected. Refer to Table 11. 2 for **Regenerate** command access. The **Regenerate** command reopens the **Generate Section/Elevation** dialog box, shown in Figure 11.15.

Command prompt	2DSECTIONRESULTREFRESH
Menu bar	Select Document>Sections and Elevations>Refresh
Shortcut menu	Select an elevation, right-click, and choose Refresh

Table 11.1 *Refresh the section or elevation object command access*

Command prompt	2DSECTIONRESULTUPDATE
Shortcut menu	Select an elevation, right-click, and choose Regenerate

Table 11.2 *Regenerate section or elevation objects command access*

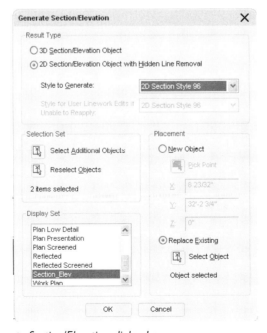

Figure 11.15 *Generate Section/Elevation dialog box*

The **Generate Section/Elevation** dialog box allows you to change the elevation style, selection set of the object, and placement of the elevation.

The options of the **Generate Section/Elevation** dialog box are described below:

Result Type

3D Section/Elevation Object – The **3D Section/Elevation Object** radio button creates a three-dimensional elevation object. The **Hidden Line Removal** (**CreateHLR**) command can be applied to this object to create a 2D representation of a 3D section/elevation object.

2D Section/Elevation Object with Hidden Line Removal – The **2D Section/Elevation Object with Hidden Line Removal** radio button creates a two-dimensional elevation object.

Style to Generate – The **Style to Generate** drop-down list includes elevation styles defined in the drawing. Elevation styles can be created to control display of components of the elevation object.

Style for User Linework Edits if Unable to Reapply – The **Style for User Linework Edits if Unable to Reapply** drop-down list allows you to assign a section/elevation style for linework edits. This option is active if lines of the elevation have been selected for edit.

Selection Set

Select Additional Objects – The **Select Additional Objects** button becomes active if objects have been selected previously using the **Select Objects** button. The **Select Objects** button changes to **Reselect Objects** after an object has been selected.

Select Objects – The **Select Objects** button returns the focus to the workspace, allowing you to select objects to be included in the elevation.

Display Set

Display Set – The **Display Set** lists the display representation sets. Select the **Section_Elev** display representation set to display objects appropriate for elevations or sections.

Placement

New Object – The **New Object** radio button allows you to create a new elevation object when objects are selected. Existing elevations can be updated if you select the **Replace Existing** radio button.

Pick Point – The **Pick Point** button returns you to the workspace and prompts you to specify the location for the elevation object.

X, Y, Z – The **X**, **Y**, **Z** edit fields allow you to specify the location of the elevation using absolute coordinates.

Replace Existing – The **Replace Existing** radio button allows you to select and update an existing elevation object.

Select Object – The **Select Object** button returns you to the workspace to select an existing elevation.

Object not selected – The **Object not selected** message field indicates if an elevation object has not been selected for update.

Generating Elevations from a Building Section/Elevation Line

You can also update an elevation or create additional elevation views if you select a building elevation line, right-click, and choose **Generate Elevation**. The **Generate Elevation** command (see Table 11.3 for **BldgElevationLineGenerate** command access) opens the **Generate Section/Elevation** dialog box, shown in Figure 11.15, which allows you to select new objects for the elevation, specify a new location, and select the elevation style for an elevation or section. The **BldgElevationLineGenerate** command will develop an elevation from all entities selected. The objects selected should be represented by the SECTION_ELEV display representation set because it includes the typical representation of architectural objects for elevation views. The Elevation object is created on the **A-Elev** layer.

Command prompt	BLDGELEVATIONLINEGENERATE
Shortcut menu	Select the building elevation line, right-click, and choose Generate Elevation

Table 11.3 *BldgElevationLineGenerate command access*

Reversing the Building Section/Elevation Line

The **BldgElevationLineReverse** command can be used to change the viewing direction of an existing building elevation line. Access the **BldgElevationLineReverse** command as shown in Table 11.4. When you select the building elevation line, right-click, and choose **Reverse** from the shortcut menu, the viewing direction changes 180 degrees.

Command prompt	BLDGELEVATIONLINEREVERSE
Shortcut menu	Select the building elevation line, right-click, and choose Reverse

Table 11.4 *BldgElevationLineReverse command access*

After placing the building elevation line, you can edit the text in the elevation bubble by the **EAttEedit** command. Access the **EAttEdit** command as shown in Table 11.5.

Command prompt	EATTEDIT
Shortcut menu	Select the elevation bubble, right-click, and choose Edit Attributes or double-click on the elevation bubble

Table 11.5 *Enhanced Attribute Editor command access*

If you edit the attributes and save the drawing, when the Sheet drawing is reloaded, the revised text will be displayed. In addition, the attributes can be edited by selecting the block and editing them in the **Attributes** section of the Properties palette.

CONTROLLING DISPLAY OF AN ELEVATION OR SECTION

The display of the elevation can be controlled by its Object Display or by creating elevation styles. The Object Display and elevation styles control display by assigning display components that govern the appearance of the elevation. The material hatching displayed in the final elevation is extracted from the material assignments of the objects selected from the model. Therefore, if a wall style includes brick veneer with brick assigned as a material, the resulting elevation will be hatched in brick. Since the materials are usually assigned in a Construct drawing, to change the elevation you must edit the Construct drawing and update the elevation to implement a material hatch change. The display of material hatching in the elevation is controlled by the **Surface Hatch Linework** component of the elevation, as shown in the **Display Properties** dialog box of Figure 11.16. If the Surface Hatch Linework component is turned off, the visibility of the Surface Hatch Linework in the elevation will be removed and only the defining lines of the elevation will be displayed.

Therefore you can control the display of the components within the elevation by editing its **Object Display** dialog box or creating elevation styles that customize the display properties. Editing an elevation using Object Display applies the changes to only the selected elevation. The Object Display of an elevation can be edited if you select an elevation, right-click, and choose **Edit Object Display**. Selecting **Edit Object Display** from the shortcut menu opens the **Object Display** dialog box as shown in Figure 11.16. The **Display Properties** tab of the **Object Display** dialog box lists the display representations of the elevation. When you select the **Edit Display Properties** button, the **Display Properties** dialog box opens for the selected display representation as shown in Figure 11.16.

The display components are listed in the far left column of the **Layer/Color/Linetype** tab. When the elevation is created, the edges of objects in the model drawing are projected to create lines placed on the **Defining Line** component. The subdivisions listed in Figure 11.16 are assigned by creating subdivisions in the building elevation line as shown in Figure 11.14. Therefore you could create a subdivision within 10 feet of the building elevation line. All objects projected from this zone are placed in the ele-

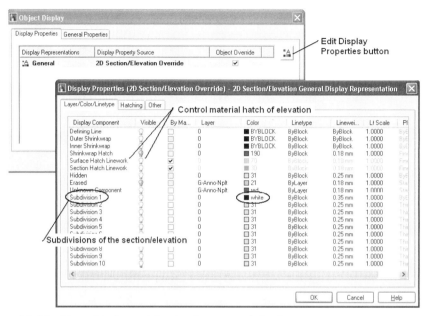

Figure 11.16 *Object Display and Display Properties dialog box*

vation object with the color, layer, linetype, and lineweight as defined by the Subdivision 1 in the **Layer/Color/Linetype** tab shown in Figure 11.16. The color of the objects within Subdivision 1 will be displayed white as shown in Figure 11.16. The lineweight of Subdivision 1 could be set wider to display the objects within Subdivision 1 with greater emphasis. The color, layer, linetype, and lineweight of each component listed can be changed in the **Layer/Color/Linetype** tab. The **Other** tab allows you to create additional custom display components.

 STOP. Do Tutorial 11.1, "Creating a Model View Using the Project Navigator," at the end of the chapter.

CREATING ELEVATION STYLES TO CONTROL ELEVATION DISPLAY

Elevation styles can be created to apply the display settings for all elevations of the elevation style. When you create a 2D elevation using the elevation marks on the Callouts palette, the **2D Section Style 96** elevation style is used to develop the elevation. Additional styles can be created to alter the display of the elevation by creating custom **display components** that have special layer, color, linetype, and lineweight properties. Elevation styles can be created to identify objects with specific colors within the model drawing and to assign the representation of the entities using the new elevation display components. When the elevation is developed with the new elevation style, all objects from the model with a specified color can be assigned to a new elevation display component. The display component can specify color, linetype, and line weight. Therefore an elevation style could be created that represents all objects with blue color in the model drawing by hidden lines in the elevation. The Standard elevation style will create an elevation without any recognition of colors within in the model drawing.

You can create a new elevation or section style in the Documentation Objects\2D Section/Elevation Styles folder of the **Style Manager**. You can create an elevation style by selecting the 2D Section/Elevation Style folder in the left pane of the **Style Manager**, right-clicking, and choosing **New**. A New Style name is displayed in the right pane; overtype the name of the new style. The specific properties of the new style are identified by selecting the new style, right-clicking, and choosing **Edit**. When you select the **Edit** option from the shortcut menu, the **2DsectionResultStyleEdit** command is selected and the **2D Section/Elevation Style Properties** dialog box opens, as shown in Figure 11.17.

Existing elevation styles can be modified by selecting the elevation, right-clicking and selecting **Edit 2D Section/Elevation Style** (see Table 11.6). When you select the **Edit 2D Section/Elevation Style** command, the **2D Section/Elevation Style Properties** dialog box opens to display the content for its style as shown in Figure 11.17. Refer to Table 11.6 for **2DsectionResultStyleEdit** command access.

Command prompt	2DSECTIONRESULTSTYLEEDIT
Shortcut menu	Select an elevation or section, right-click, and choose Edit 2D Elevation/Section Style

Table 11.6 *2DsectionResultStyleEdit command access*

A description of the tabs of the **2D Section/Elevation Style Properties** dialog box follows:

General Tab

> **Name** – The **Name** edit field allows you to change the name of the style.
>
> **Description** – The **Description** edit field allows you to add a description to the Section/Elevation style.
>
> **Notes** – The **Notes** button opens the **Notes** dialog box, which is identical to other styles, consisting of **Reference Docs** and **Text Notes** tabs for adding additional descriptions regarding the style.

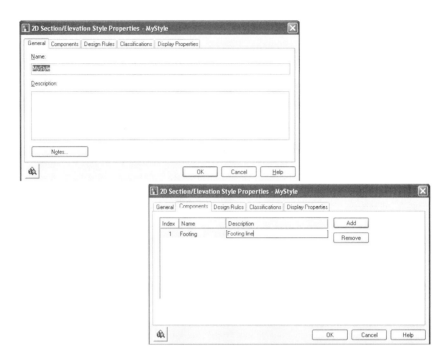

Figure 11.17 *General and Components tabs of an Elevation Style Properties dialog box*

Components Tab

The **Components** tab is shown in Figure 11.17.

> **Index** – The **Index** specifies a number for the new component used in the display of the elevation.
>
> **Name** – The **Name** specifies the name of the new component used to display the elevation.
>
> **Description** – The **Description** edit field allows you to add a description to the component name.
>
> **Add/Remove** – Select the **Add** or **Remove** button to create components for the list or delete them from the list.

Design Rules Tab

The **Design Rules** tab is shown in Figure 11.18.

> **Rule** – The **Rule** column specifies the number of the rule used to define color and display method.
>
> **Color** – The **Color** column specifies the color number/name of objects in the model, which will be represented in the display method of the elevation.

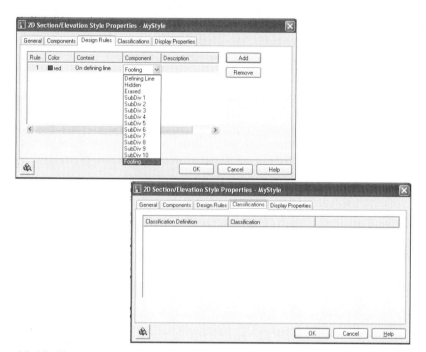

Figure 11.18 *Design Rules and Classification tabs of the Elevation Style Properties dialog box*

>**Context** – The **Context** specifies which component within the elevation display properties to apply the rule. The drop-down list of the **Context** column includes such components as on defining lines, within any subdivision, any visible, hidden, within any of the subdivisions defined in the elevation.
>
>**Component** – The **Component** drop-down list of the column specifies the display components to assign the lines based upon the color specified.
>
>**Description** – Select in the **Description** edit field to add a description for each rule.

Classifications Tab

The **Classifications** tab includes a list of classification definitions included in the drawing and the classification content of the classification definitions.

Display Properties Tab

The **Display Properties** tab is shown in Figure 11.19. It lists the display representations available for the section or elevation object.

>**Display Representation** – The name of the display representation used in the elevation object is listed in the **Display Representation** column.

Display Property Source – The current display property source is listed for the display representation. The property source can either be Drawing Default or 2D Section/Elevation Style Override.

Edit Display Properties – The **Edit Display Properties** button opens the **Display Properties (2D Section/Elevation Style Override – style name)** dialog box. The **Display Properties (2D Section/Elevation Style Override – style name)** dialog box consists of the **Layer/Color/Linetype**, **Hatching**, and **Other** tabs, which allow you to define the properties of the components. The **Other** tab allows you to create custom display components for the style.

STEPS TO CREATING AND EDITING AN ELEVATION STYLE

The following steps outline the procedure to edit the **2D Section/Elevation Style Properties** dialog box to create an elevation style that controls object display based upon the color of the object. The following steps create a style that will identify any red line in the model and assign it to the Footing custom display component. The Footing display component is shown in the elevation with hidden lines.

1. Select **Format>Style Manager** from the menu bar. Expand the Documentation Objects>2D Section/Elevation folder. Select the 2D Section/Elevation folder, right-click, and choose **New**. Overtype **MyStyle** as the name of the new style.

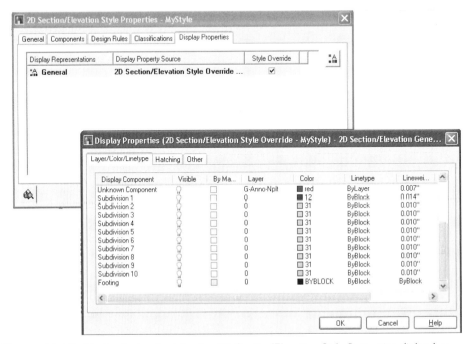

Figure 11.19 *Display Properties tab of a 2D Section/Elevation Style Properties dialog box*

2. Double-click on MyStyle to open the **2D Section/Elevation Style Properties – MyStyle** dialog box.

3. Select the **Components** tab, choose the **Add** button, and overtype **Footing** as the name of the custom component.

4. Select the **Display Properties** tab and choose the **Edit Display Properties** button to open the **Display Properties (2D Section/Elevation Style Override-MyStyle)** dialog box. Select the **Other** tab, choose the **Add** button, and type **Footing** as the name of the custom display component. Select **OK** to close the **Display Properties** dialog box.

5. Select the **Design Rules** tab (shown in Figure 11.18), select the **Add** button, and specify the color name/number. Select in the **Context** column and select a context such as **Any visible** or **Hidden** of the object from the model that will be assigned unique display. Select in the **Component** column and specify the name of the display component such as **Footing**, created in step 3 above.

6. Select the **Display Properties** tab, verify that **Style Override** is checked, and select the **Edit Display Properties** button to open the **Display Properties (2D Section/Elevation Style Override – MyStyle)** dialog box. Select the **Layer/Color/Linetype** tab, and edit the properties such as color or linetype of the Footing display component created in step 2. Select **OK** to dismiss the **Display Properties (2D Section/Elevation Style Override-MyStyle)** dialog box. Select **OK** to dismiss the **2D Section/Elevation Style Properties – MyStyle** dialog box.

EDITING THE LINEWORK OF THE ELEVATION

The lines of an elevation are displayed according to the display component definition within the elevation style. If a custom display component has been created in the elevation style, the elevation can be opened for edit and any entity can be edited to assume the display defined by that custom display component. The line shown at **p1** in the elevation at A in Figure 11.21 is displayed as an object line because it is assigned to a display component with continuous linetype. This line can be changed to a hidden line as shown at B in Figure 11.21 if you create a display component with the hidden linetype property. The first step in changing the linetype of a line is to create a display component with the linetype property in the elevation style. You can create a new component by selecting the style override on the **Display Properties** tab of the **2D Section/Elevation Style Properties** dialog box shown in Figure 11.20. When you check the **Style Override** check box, the **Display Properties (2D Section/Elevation Style Override – style name)** dialog box opens. Select the **Other** tab of the **Display Properties (2D Section/Elevation Style Override – style name)** dialog box as shown in Figure 11.20. Select the **Add** button and type the name of the new display component. New components are listed, added, and removed in the **Other** tab of the

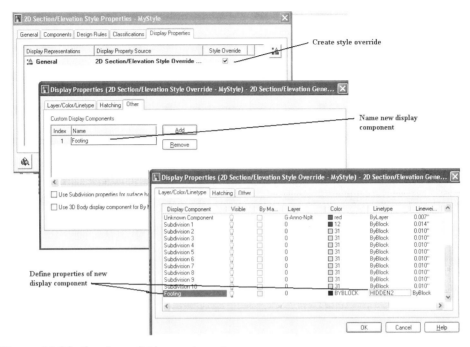

Figure 11.20 *Creating a display component*

Display Properties (2D Section/Elevation Style Override – style name) dialog box. Display components created for display by object color are also listed in that tab.

After creating a name for the display component, select the **Layer/Color/Linetype** tab shown in Figure 11.20, and edit the properties of the new display component. The **Layer/Color/Linetype** tab of the **Display Properties (2D Section/Elevation Style Override – style name)** dialog box, shown in Figure 11.20, includes the **Footing** component with hidden linetype.

You can open the elevation for edit by selecting the elevation, right-clicking, and choosing **Linework**, and then **Edit** from the shortcut menu (**2DsectionResultEdit** command, see Table 11.7). When the elevation is opened for edit, entities can be erased or modified. In addition, the hatch pattern is not displayed when the elevation is opened for edit. Upon completion of the changes, you save the changes, and the display state is saved back to the elevation. This command allows you to change linetype, color, or lineweight of selected entities. After opening the elevation for edit, you can erase any entity in the elevation

The entities can be assigned to any of the display components or custom display components defined in the elevation style. After opening the elevation for edit, you can select an entity shown at **p1** in Figure 11.21 of the elevation, right-click, and choose the **Modify Component** command from the shortcut menu (**2dSectionEdit Component**, see Table 11.8). The **Modify Component** command opens the **Select Linework Component** dialog box, shown at left in Figure 11.21. The **Select Linework Component** dialog box allows you to select the **Footing** display component, and the line will be displayed as a hidden line in the elevation. When you have finished editing the lines, select any entity of the elevation, right-click, and choose **Edit in Place>Save Changes** or **Edit in Place>Discard Changes** to complete the change.

Command prompt	2DSECTIONRESULTEDIT
Shortcut menu	Select an elevation, right-click, and choose Linework>Edit

Table 11.7 *Edit Linework command access*

Command prompt	2DSECTIONEDITCOMPONENT
Shortcut menu	Select an elevation entity, right-click, and choose Modify Component

Table 11.8 *Modify Component command access*

The entities of an elevation object can be changed to different display components of elevation style. If you assign an entity to the Erased Vectors linework component, it will not be displayed in the elevation.

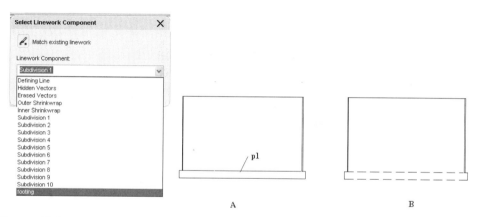

Figure 11.21 *Editing linework of an elevation*

MERGING LINES TO THE ELEVATION

Lines and hatch patterns can be merged to the elevation object using the **Linework>Merge** command. This command allows you to add reference lines for the finish floor and finish ceiling of an elevation. The lines are merged to the elevation and assigned to a display component. Access the **Linework>Merge** (**2dSectionResult Merge**) command as follows.

Command prompt	2DSECTIONRESULTMERGE
Shortcut	Select an elevation, right-click, and choose Linework>Merge

Table 11.9 *Merge Linework command access*

When you select the elevation, right-click, and choose **Linework>Merge**, you are prompted to select the entities to merge with the elevation object as shown in the command sequence below. In the sequence shown, a centerline was added to the elevation. Note that prior to beginning the **Linework>Merge** command shown below, a display component named Floor Line was created in the **Components** tab. The component was assigned center linetype in the **Layer/Color/Linetype** tab of the **Display Properties (2D Section/Elevation Style Override – style name)** dialog box.

> (Select the elevation at p1, right-click, and choose **Linework>Merge** as shown at left in Figure 11.22.)
>
> Command: _Aec2dSectionResultMerge
>
> Select entities to merge: 1 found *(Select the line at p2 to merge it with the elevation object.)*
>
> Select entities to merge: ENTER *(Press ENTER to end the selection.)*
>
> *(Select the Floor Line display component from the drop-down list of the* **Select Linework Component** *dialog box as shown in Figure 11.22.)*
>
> *(Line merged to elevation as shown at right in Figure 11.22.)*

MODIFYING THE MATERIAL HATCH PATTERN

The **Material Boundary>Add** command (**2dSectionResultAddHatchBoundary**, refer to Table 11.10) can be used to limit the area of material hatch displayed in an elevation. The material represented in the elevation is defined in the material definition of the objects selected for the elevation. Therefore, if you change a wall from brick to CMU, when the elevation is updated, the material displayed in the elevation will change. To limit the material hatch display, draw a polyline boundary where you want to highlight or mask the hatch in the elevation. The material display can be limited

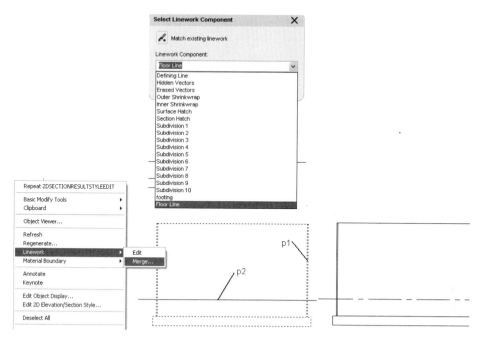

Figure 11.22 Merging a line into an elevation

to within a closed polyline or erased within the closed polyline. The **2dSectionResultAddHatchBoundary** command allows you to specify which materials to limit or erase in the elevation.

Command prompt	2DSECTIONRESULTADDHATCHBOUNDARY
Shortcut menu	Select the elevation, right-click, and choose MaterialBoundary>Add

Table 11.10 Material Boundary>Add (2dSectionResultAddHatchBoundary) command access

A boundary was applied to the material representation of the concrete masonry unit in the following command line sequence. The **Material Boundary>Add** was set to **Limit**, and therefore the hatch was limited to within the polyline as shown at right in Figure 11.23.

> (Select the elevation, right-click, and choose **Material Boundary>Add** from the shortcut menu.)
>
> Command: 2dSectionResultAddHatchBoundary

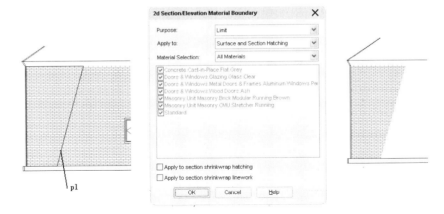

Figure 11.23 *Adding a material boundary to a hatch*

Select a closed polyline for boundary: *(Select the polyline at p1 as shown in Figure 11.23.)*

Erase selected linework? [Yes/No] <No>: **y** *(Type y to erase the polyline; the* **2dSection/Elevation Material Boundary** *dialog box opens, select the* **Erase** *or* **Limit** *purpose and specify material to apply the boundary.)*

Command: o*Cancel*

(Material pattern is limited to within the polyline as shown at right in Figure 11.23.)

The boundary can be edited after it is applied by selecting the elevation, right-clicking, and choosing **Boundary>Edit In Place** (**2dSectionResultEditHatchBoundaryInPlace** command). When you edit the boundary in place, the grips are displayed at each vertex. You can select the grips or select from the shortcut menu to edit the vertex of the polyline.

 STOP. Do Tutorial 11.2, "Creating Elevations and Elevation Styles," at the end of the chapter.

CREATING SECTIONS OF THE MODEL USING CALLOUTS

The procedure to create a section is similar to creating an elevation. The section marks are located on the Callouts tool palette as shown in Figure 11.24. The procedure to place the **Section Mark A1** is shown below:

STEPS TO CREATING A SECTION

1. Open the **Project Navigator**, and select the **Views** tab.

Creating Elevations, Sections, and Details 797

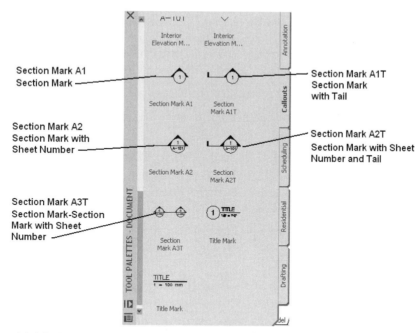

Figure 11.24 *Section marks located on the Callouts palette*

2. Open the model drawing.

3. Select **Section Mark A1** from the Callouts tool palette shown in Figure 11.24, and respond to the command line prompts as shown below:

   Command: AecCallout

   Specify first point of section line: *(Select a point near p1 shown in Figure 11.25 to specify the start point of the cutting plane.)*

   Specify next point of line: *(Select a point near p2 to specify the endpoint of the cutting plane as shown in Figure 11.25.)*

   Specify next point of line or [Break]: ENTER *(Press ENTER to end the cutting plane line.)*

   Specify side for Arrow: *(Select a point near p3, to open the **Place Callout** dialog box as shown in Figure 11.25.)*

   *(The **Place Callout** dialog box opens as shown in Figure 11.25.)*

4. Type **Section A-A** in the **New Model Space View Name** edit field, verify that **Generate Section/Elevation** and **Place Titlemark** are checked, set scale to **1/4"=1'-0"**, and then check **New View Drawing** to open the **Add Section/Elevation View** dialog box.

5. Edit the Add Section/Elevation View dialog box, **General** page: type **Transverse** in the **Name** edit field. Choose **Next** to open the **Context** page, verify that level **1** and level **2** are checked for division **1**. Choose **Next** to open the **Content** page, and verify that all constructs are checked. Choose **Finish** to close the **Add Section/Elevation View** dialog box.
6. Continue responding to the command line as follows:

    Specify insertion point for the 2D section result: *(Select a point near p4 as shown in Figure 11.25.)*

    The section is created in the View drawing named Section.
7. To view the section, select the **Views** tab of the **Project Navigator**, and double-click on the **Transverse** view drawing. A question mark is placed in the mark to identify the section until it is placed on the sheet.
8. Select the **Sheets** tab, select the Sections sheet category, right-click, and select **New>Sheet**. Type **A-112** in the number edit field and **Sections** in the title field of the sheet. Select **OK** to close the **New Sheet** dialog box.
9. Select the **Sheets** tab of the **Project Navigator**, and double-click on **A-112**, the new drawing.
10. Select the **Views** tab, expand the **Transverse** drawing, select the Section A view, right-click, and choose **Place on Sheet**. Drag the view to the sheet and left-click to position the view on the sheet.

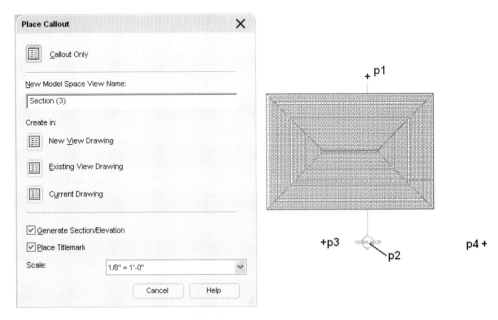

Figure 11.25 *Placing a section building line*

The view number is placed in the title bubble of the elevation view on the Sheet and View drawings. The Section View drawing is shown at left in Figure 11.26.

When you place the section mark, you are prompted to specify the cutting plane line, view direction, and location of the section. Unlike placing an elevation, you do not specify the region or objects to include in the section. The section can be created in the current drawing, an existing view drawing, or a new view drawing.

The display components of a 2D Section/Elevation object allow you to display material hatch of a wall by turning ON Section Hatch Linework. The Section Hatch Linework can be displayed By Material. If the **By Material** option is checked in the **Display Properties** dialog box, the material definition of the objects will govern the hatch pattern. The Shrinkwrap and Surface Hatch are turned off in the elevation style of the section shown at right in Figure 11.26. The display of each object cut in the section can be controlled if you edit the display properties of each material definition used in the object style. In the section shown at right in Figure 11.26, the section hatch is turned on in the elevation style. The section hatch can be turned off in the wall style of the Construct drawing, which will display a section of the wall without material hatch representation. In Figure 11.27 the section hatch has been turned off for the standard wall on the left, whereas the section hatch for the brick masonry wall is turned on at right.

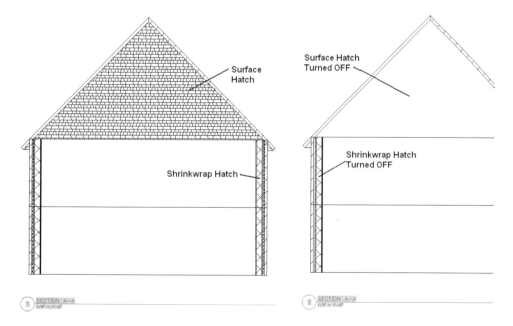

Figure 11.26 *Section View drawing*

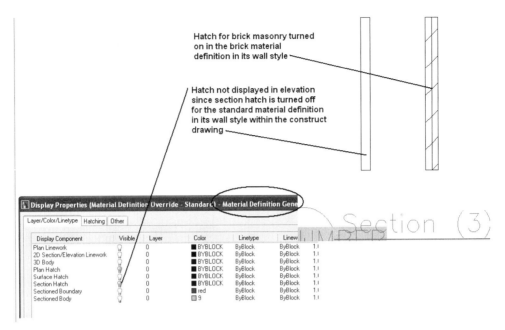

Figure 11.27 Turning off section hatch in the material definition

CREATING A HORIZONTAL AND VERTICAL SECTIONS

During the design of a drawing, you can develop a horizontal or vertical section. Access the **AecHorizontalSection** command as shown in Table 11.11. The command prompts shown below prompt you to specify the region for the section, cutting plane height, and depth of the section. The result of the command creates a building section line. The section is created when you select the building section line, right-click and choose **Generate Section** or **Enable Live Section** as shown in Figure 11.29.

Command prompt	AECHORIZONTALSECTION
Palette	Select the Horizontal Section tool from Design tool palette

Table 11.11 Horizontal Section tool access

STEPS TO CREATING A HORIZONTAL SECTION

1. Select the **Horizontal Section** tool from the Design tool palette.

    Command: AecHorizontalSection

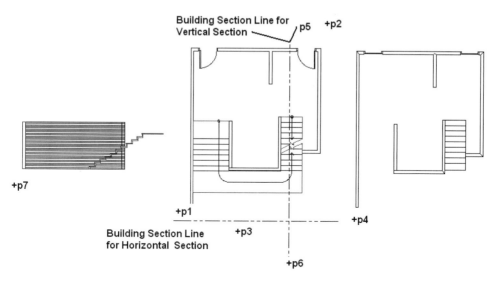

Figure 11.28 Horizontal and Vertical Sections

> Select corner for horizontal section: *(Select the drawing area at p1 as shown in Figure 11.28.)*
>
> Select corner for horizontal section: *(Select the drawing area at p2 as shown in Figure 11.28.)*
>
> Enter elevation of section cutting plane<5'-0">: *(Press* ENTER *to accept default height.)*
>
> Enter depth of section <8'-0">: *(Press* ENTER *to accept default depth and create the Building Section Line.)*

2. Select the building section line, right-click, and choose **Generate Section**.
3. Choose the **Select Objects** button in the **Selection Set** section then create a crossing selection set by selecting a point near **p2** and **p1**.
4. Choose the **New Object** radio button, choose the **Pick Point** button and select a point near p4. Select **OK** to dismiss the **Generate Section Elevation** dialog box.

The **Vertical Section** tool allows you to create a section from a vertical cutting plane. The command prompts for the tool allow you to specify the location of the cutting plane and the depth of the projection box. Access the **Vertical Section** tool as shown in Table 11.12. A 2D section or live section can be developed from the vertical building section line.

Command prompt	AECVERTICALSECTION
Palette	Select the Vertical Section tool from Design tool palette

Table 11.12 *Vertical Section tool access*

STEPS TO CREATING A VERTICAL SECTION

1. Select the **Vertical Section** tool from the Design tool palette.

    Command: _AecVertical Section

    Create polyline for section

    Specify Start point: *(Select a point near p5 as shown in Figure 11.28.)*

    Specify next point: *(Select a point near p6 as shown in Figure 11.28.)*

    Specify next point: *(Press ENTER to end selection.)*

    (Enter length <20'-0">: (Press ENTER to accept default depth.)

2. Select the vertical building section line at **p5**, right-click, and choose **Generate Section**.
3. Choose **Select Objects** button in the **Selection Set** section, and then create a crossing selection set by selecting a point near **p2** and **p1**.
4. Choose the **New Object** radio button, choose the **Pick Point** button, and select a point near p7. Vertical section created as shown at left in Figure 11.28.

CREATING A LIVE SECTION

A live section creates a view of the model that removes the display of the objects in front of the cutting plane line. A live section can be created from any building section line. Additional objects can be added to the drawing while the live section is active. The live section does not create a 2D section object. The live section is a tool to enable you to work on the model by removing the display of some objects. Prior to creating the live section, select a pictorial view of the building as shown in Figure 11.29, select the building section line, right-click, and choose **Enable Live Section**, (see Table 11.13). The building components in front of the section line are removed from display, as shown in Figure 11.29.

Command prompt	LIVESECTIONENABLE
Shortcut menu	Select the building section line, right-click, and choose Enable Live Section.

Table 11.13 *LiveSectionEnable command access*

Creating Elevations, Sections, and Details 803

If a live section has been enabled, you can toggle **Sectioned Body Display** to display the outline of the components, turned off as shown at right in Figure 11.29. Access **Sectioned Body Display** as shown in Table 11.14.

Command prompt	AECTOGGLESECTIONEDBODY
Shortcut menu	Select the building section line of a live section, right-click, and choose Toggle Sectioned Body Display.

Table 11.14 *Toggle Section Body Display command access*

The live section can be turned off by selecting the section line, right-clicking, and choosing **Disable Live Section** (see Table 11.15).

Command prompt	LIVESECTIONDISABLE
Shortcut menu	Select the building section line of a live section, right-click, and choose Disable Live Section.

Table 11.15 *LiveSectionDisable command access*

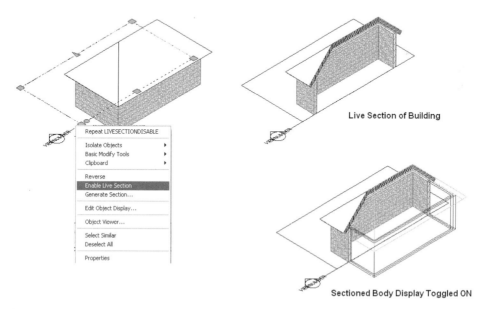

Figure 11.29 *Enabling a Live Section*

CREATING 2D SECTIONS WITH HIDDEN LINE PROJECTION

The **Hidden Line Projection** (**CreateHLR**) command will create a two-dimensional drawing of a view. This command allows you to view the model in an orthographic or pictorial view and create a two-dimensional block from the view. Therefore, you can create elevation views, 2D floor plans, and 2D views of isometric views. You can use the **Live Section** command to create a cut-away view of the model and capture the pictorial view of the model as a 2D drawing.

Access the **Hidden Line Projection** command as shown in Table 11.16. The **Hidden Line Projection** command is located in the Architectural Desktop Stock Tool Catalog>Helper Tools. You can drag the tool to a current tool palette as shown in the following steps:

1. Select **Content Browser** from the **Navigation** toolbar.
2. Double-click on the Architectural Desktop Stock Tool Catalog, and select Helper Tools in the left pane to display the **Hidden Line Projection** command as shown in Figure 11.30.
3. Click on the **Hidden Line Projection** command, select the i-drop of **Hidden Line Projection** and drag the command to a tool palette.

Command prompt	CREATEHLR
Menu bar	Select Document>Hidden Line Projection
Tool palette	Open Content Browser, select Autodesk Architectural Desktop Stock Tool Catalog\Helper Tools folder, and drag the Hidden Line Projection command from the Helper Tools folder

Table 11.16 *CreateHLR - Hidden Line Projection command access*

Prior to selecting the **Hidden Line Projection** command, set the view of the object. You can select **SW Isometric** view and then select the **Hidden Line Projection** command to capture the view. When you select **Hidden Line Projection** from the tool palette, you are prompted to select objects to create the content of the block, as shown in the following command line sequence. After selecting the objects, you are prompted to specify the insertion point for the block. The insertion point can be at 0,0 or any X,Y location. The "Insert in plan view" prompt allows you to insert the block in plan view or in the view that is current when the **Hidden Line Projection** command is executed. Creating the block in plan view allows you to easily add annotation and dimensions to the block.

_AecCreateHLR

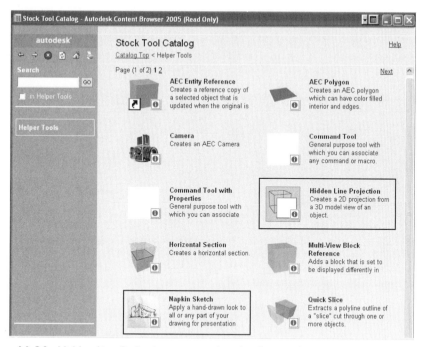

Figure 11.30 Hidden Line Projection command in the Content Browser

Select objects: *(Select objects using a window selection from p1 to p2 as shown in Figure 11.31.)*

Select objects: ENTER *(Press* ENTER *to end selection.)*

Block insertion point: *(Select a point near p3 as shown in Figure 11.31.)*

Insert in plan view [Yes/No]<Yes>: **y** *(Press* **Y** *to enter the drawing projected to the top view; press N to insert the drawing in the current view. Block is created as shown at right in Figure 11.31.)*

(Select **Top View** *from the* **View** *flyout of the* **Navigation** *toolbar to view the block as shown in Figure 11.32.)*

CREATING A NAPKIN SKETCH VIEW

You can display the objects of the model or the two-dimensional drawings with a sketch appearance for presentation purposes as shown in Figure 11.33. The napkin sketch creates a sketch line pattern over the existing lines of the 3D object or 2D drawings. The napkin sketch can be applied to orthographic or pictorial views of the object. A napkin sketch increases the drawing size because it traces each line to apply several sketch lines. Therefore, to reduce the geometry you should apply the **Napkin Sketch** procedure to Hidden views of the model, elevations, sections, or 2D drawings creat-

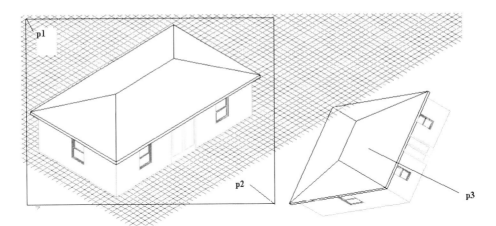

Figure 11.31 *Creating a 2d block using Hidden Line Projection CreateHLR command*

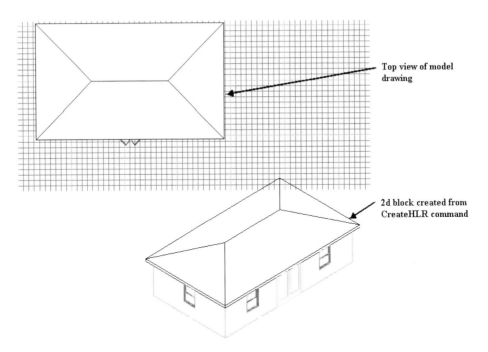

Figure 11.32 *Top view of the 2d elevation created with CreateHLR*

ed by the **CreateHLR** command. Access the **Napkin Sketch** (**AECNapkin**) command as shown in Table 11.17. The **AECNapkin** command is located in the Architectural Desktop Stock Tool Catalog>Helper Tools as shown in Figure 11.30.

Creating Elevations, Sections, and Details 807

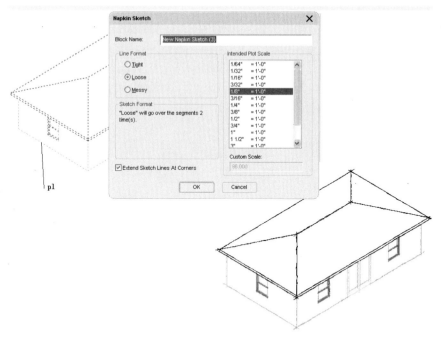

Figure 11.33 *Napkin Sketch procedure applied to straight line drawing*

Command prompt	AECNAPKIN
Menu bar	Select Document>Napkin Sketch
Tool Palette	Open Content Browser, select Autodesk Architectural Desktop Stock Tool Catalog\Helper Tools folder, and drag the Napkin Sketch command from the Helper Tools folder

Table 11.17 *AecNapkin command access*

When you apply the **AecNapkin** command, the sketch lines are applied over the straight lines and are created as a block. Therefore you can move the napkin sketch block to a different location in the drawing or **Wblock** the content to create separate drawings. The following command line sequence applied the napkin sketch to the elevation created using the **AecNapkin** command.

    _AecNapkin

    Select objects: *(Select the block at p1 as shown in Figure 11.33.)*

    Select objects: ENTER *(Press ENTER to end selection.)*

*(Edit the **Napkin Sketch** dialog box shown in the center of Figure 11.33 and select **OK** to dismiss the **Napkin Sketch** dialog box.)*

After applying the **AecNapkin** command, use the **Move** command to move the block to a different location. Since the outcome of the **Napkin Sketch** procedure is a block, select the block, right-click, and choose **Edit Block in place**. The Refedit procedures can be used to edit the sketch.

 STOP. Do Tutorial 11.3, "Creating a 2D Section and a Live Section," at the end of the chapter.

USING THE DETAIL COMPONENT MANAGER

The **Detail Component Manager** provides access to 2D drawings that represent building components. The drawings can consist of fasteners, building units, and assemblies. Depending on the detail component, it can be inserted as an AutoCAD block, entity, or a parametric block. The detail components may include top, elevation, or section views of building components with appropriate material hatch representation. The detail components are not AEC objects and do not interact with other detail components or AEC objects. All detail components can be accessed from the **Detail Component Manager** or the Detailing palettes. Access the **Detail Component Manager** as shown in Table 11.18.

Command prompt	AECDTLCOMPMANAGER
Menu bar	Select Insert>Detail Component Manager
Toolbar	Select Detail Component Manager from the Navigation toolbar
Tool Palette	Select a detail tool from one of the palettes of the Detailing palette set, right-click, and choose Detail Component Manager
Content Browser	Select a tool from the palettes of the Sample Palette Catalog>Detailing category

Table 11.18 *Accessing the Detail Component Manager*

The **Detail Component Manager**, shown in Figure 11.34, includes components categorized in fifteen divisions. The fifteen divisions are represented on the **Basic** palette of the Detailing palette set as shown in Figure 11.35. You can open the **Detail Component Manager** to a division by selecting a division tool on the Basic

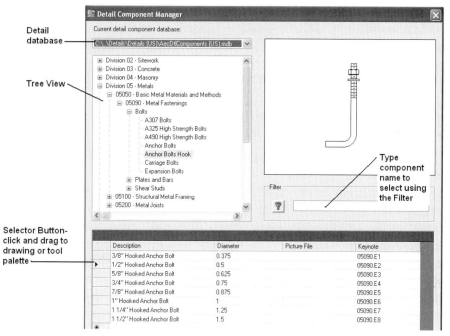

Figure 11.34 *Detail Component Manager*

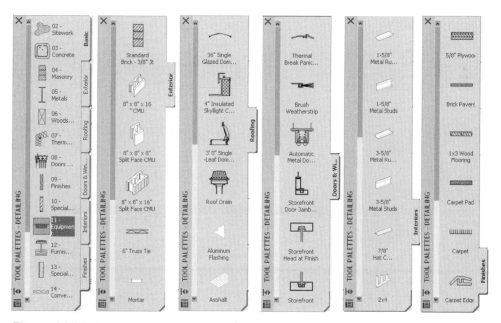

Figure 11.35 *Detailing tool palettes*

palette, right-clicking and choosing **Detail Component Manager**. Division 15 of the **Detail Component Manager** is shown expanded in Figure 11.36. The properties of the selected component are displayed in the lower pane. The keynote number based upon Construction Specification Institute format is listed for each component. The keynote number can be identified with the keynote tools located on the Annotation tool palette. Details can be inserted from **Detail Component Manager** directly if you double-click on the **Selector** at **p1** or select the component and then choose the **Insert Component** button shown in Figure 11.36.

When you select a tool from the **Detail Component Manager**, you can edit the properties of the tool in the Properties palette. You can edit such properties as view direction, content type, and other features of the detail. The properties of a tool can be preset if you right-click on detailing tool located on a palette and select **Properties** to open the **Tool Properties** dialog box. If you drag a detail component from the **Detail Component Manager** to a tool palette, you can set the properties in its **Tool Properties** dialog box for future applications. The **Properties** dialog box of the Anchor Bolt tool shown in Figure 11.38 allows you to specify bolt type, view, and length. After specifying the settings, close the dialog box and insert the anchor bolt as specified. The settings in the Properties palette should be reviewed prior to inserting a detail since it controls the detail component. The procedure to insert components

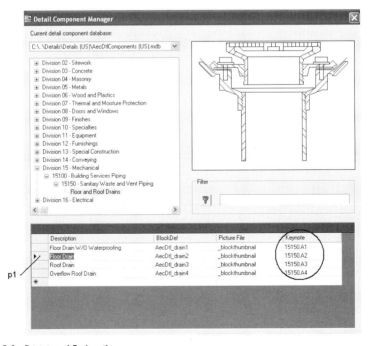

Figure 11.36 *Division 15 details*

Creating Elevations, Sections, and Details 811

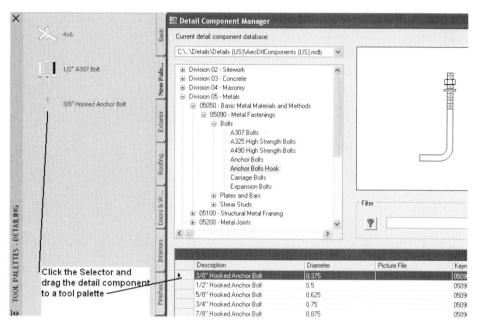

Figure 11.37 *Dragging a tool from the Detail Component Manager to a tool palette*

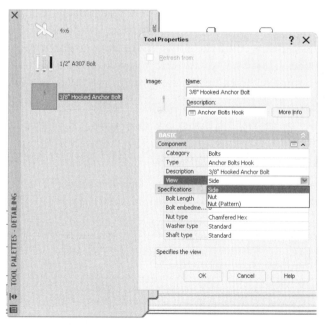

Figure 11.38 *Properties of a detail tool*

is dependent upon the setting in the Properties palette. The prompts to insert a detail differ according to the view and detail type. Detail components can consist of blocks, linework entities, or parametric drawing routines. Linework entities are used to represent components of variable length such as structural members or reinforcement, while parametric drawing routines are often applied to specify the shape and size of such components as bolts. The types of detail components are classified based upon the insertion methods as follows:

- Stamp-Type (Multiple Insertion) Components
- Linear Repeating Pattern Components
- Boundary Repeating Filling Components
- Bookends-Type Components
- Countable Linear Repeating Pattern Components
- Rectangular Predefined-Depth Surface Components
- Dynamically-Sized Rectangular Components
- Bolt Components

Inserting a Stamp-Type Component

The procedure to insert a stamp type component such as a 2 × 4 as shown in section is accessed by specifying a Section view in the Properties of the tool as shown in Figure 11.39. The Section view option of this tool inserts the detail as a stamp type component. The **Description** edit field shown in the Properties palette includes a drop-down list of lumber sizes. The tool can also insert lumber in plan or elevation views. The command prompts shown below allow you to rotate or flip the 2 × 4 about the x or y axis to set its orientation. The block is inserted on the A-Detl-Wide layer.

> *(Select the 06-Wood & Plastics tool, review settings in the Properties palette.)*
>
> Command: _AecDtlCompManager
>
> Insert point or [Base point/Rotate/Xflip/Yflip]: *(Select a point near p1 to insert the block.)*
>
> Insert point or [Base point/Rotate/Xflip/Yflip/Undo]: ENTER *(Press ENTER to end the command.)*

Linear Repeating Pattern Components

A detail can also consist of a collection of repeating entities placed based on your response to the command line. A detail for welded wire fabric consists of a series of "X"s placed along a line as shown in Figure 11.40. To place this detail, open the **Detail Component Manager** to 03200 Concrete Reinforcement>Wiremesh Reinforcing and

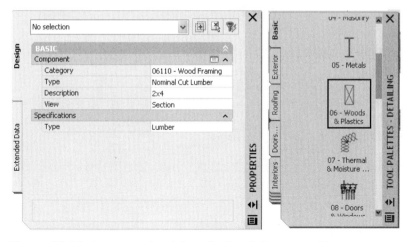

Figure 11.39 Inserting a 2 x 4 from the Detail Component Manager

drag the tool onto a tool palette or into the workspace. Edit the Properties palette to place the detail as described below.

STEPS TO CREATING WELDED WIRE FABRIC

1. From the Basic tool palette of the Detailing palette set, select 03-Concrete, right-click, and select **Detail Component Manager**. Expand the Division 3 in the right pane to 03200 Concrete Reinforcement>03220 Welded Wire Fabric>Wiremesh Reinforcing as shown in Figure 11.40.

2. Double-click the **Selector** button for 6x6-W1.4xW1.4 Welded Wire Mesh, review the Properties palette, and respond to the command line as shown below to place the detail component.

    Command: AECDTLCOMPADD

    Start point: *(Select the start point of the wire fabric at p1 as shown in Figure 11.40.)*

    End point: *(Select the end point of the wire fabric at p2 as shown in Figure 11.40.)*

    Start point: *(Press ENTER to end the command.)*

Boundary Filling Components

The Boundary Filling Components detail component applies a hatch pattern that represents material to a selected boundary. This type of detail component will prompt you to select a closed polyline that represents the boundary for the hatch. The 02-Sitework tool of the Basic tool palette includes tools for a boundary filling component. The Properties palette of the 02-Sitework tool, shown in Figure 11.41, allows you to

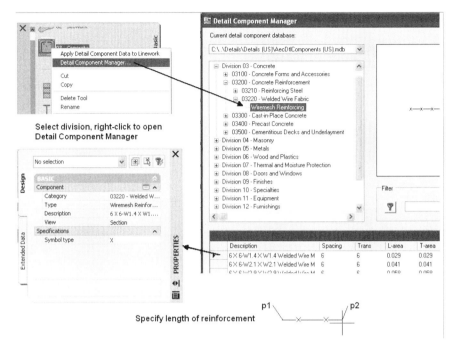

Figure 11.40 *Insert welded wire fabric from the Detail Component Manager*

select the fill options such as gravel or sand. The steps to insert a boundary filling component are shown below.

STEPS TO CREATING GRAVEL BOUNDARY FILLING COMPONENT

1. Draw a polyline boundary to enclose the gravel hatch.
2. From the Basic tool palette select 02-Sitework. Edit the **Description** field to **Gravel** in the Properties palette. Select **OK** to close the Properties palette. Respond to the command line prompts as shown below.

   Command: _AecDtlCompManager

   Select objects to form the backfill boundary: 1 found *(Select the polyline at p1 as shown in Figure 11.41.)*

   Select objects to form the backfill boundary: ENTER *(Press ENTER to end selection.)*

   Select a point within a boundary to backfill: *(Select inside the boundary at p2 as shown in Figure 11.41.)*

   Select objects to form the backfill boundary: ENTER *(Press ENTER to end the command.)*

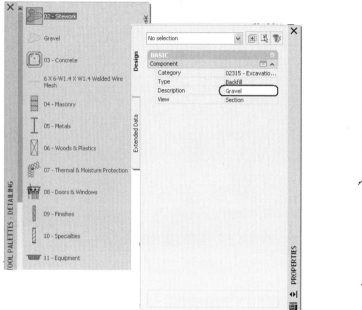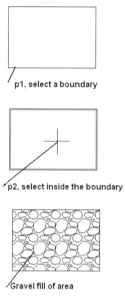

Figure 11.41 Inserting a boundary fill detail component

Bookends-Type Components

The Bookends-Type component starts and ends with unique geometry at each end. A repeated pattern fills the span between the two ends as shown in Figure 11.42. Therefore objects such as vents and louvers are inserted by specifying the start and end points of the vent. The command line options shown below include the options to flip the component about the x or y axis of the component.

> *(Select* **Insert>Detail Component Manager** *and expand the Division 10 Specialties>10200 Louvers and Vents>10210 Wall Louvers>Fixed Aluminum Louvers. Select the 4" Fixed Aluminum Louver detail, and then click* **Insert Component**.*)*
>
> Command:
>
> Start point or [Xflip/Yflip]: *(Select the start of the louver at p1 shown in Figure 11.42.)*
>
> End point or [Xflip/Yflip]: *(Select the start of the louver at p2 shown in Figure 11.42.)*
>
> Start point or [Xflip/Yflip]: ENTER *(Press* ENTER *to end the command.)*

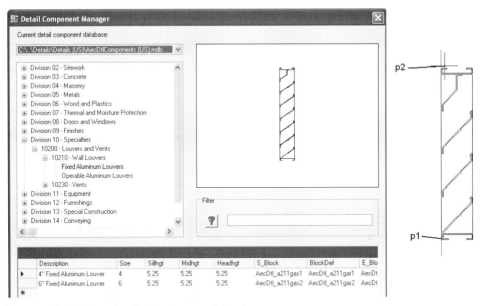

Figure 11.42 *Inserting Bookends type of detail component*

Countable Linear Repeating Pattern Components

The section view of masonry units, located on the 04-Masonry palette, is an example of the Countable Linear Repeating Pattern. This section view of the masonry unit can be specified from the drop-down list in the Properties palette or the **Tool Properties** dialog box of the tool as shown in Figure 11.43. This type of detail allows you to specify the number of courses of brick or block to insert in the detail. The following steps outline the procedure to place multiple courses of CMUs.

STEPS TO CREATING MASONRY UNITS

1. Select **04-Masonry** from the Basic tool palette.

2. Edit the Properties palette as follows: Category = Concrete Masonry Units, Type = 3 Core CMU, Description = 8"x8"x16" CMU, and View = Section.

3. Respond to the command line as follows to place the concrete masonry units:

    Command: _AecDtlCompManager

    Start point or [Xflip/Yflip/Count]: **C** *(Type **C** to specify the course count.)*

    Number of courses <1>: **6** ENTER *(Type the quantity of courses.)*

    Start point or [Xflip/Yflip/Count]: *(Select a point near p1 to specify the insertion point as shown in Figure 11.43.)*

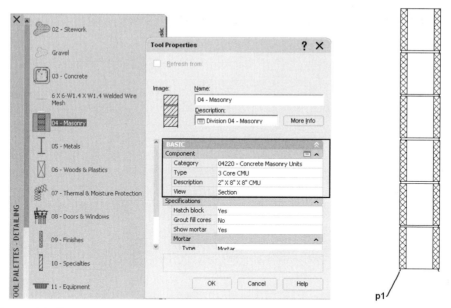

Figure 11.43 *Insert a countable repeating pattern component*

Pick point for direction or [Xflip/Yflip/Count]: *(Select a point above p1 shown In Figure 11.43 to specify the direction.)*

Start point or [Xflip/Yflip/Count]: *(Press* ENTER *to end the insertion.)*

Rectangular Predefined-Depth Surface Components

Rectangular Predefined-Depth Surface components allow you to dynamically size a rectangular surface pattern about a baseline. The rectangular shape can be centered, left, or right about the baseline. The detail created in Figure 11.44 is developed about the baseline from **p1** to **p2**. The rectangle pattern is developed right justified about the baseline. The component is accessed from Division 02-Sitework>02700 Bases, Ballast, Pavements, and Appurtenances> 02720 Unbound Base Courses and Ballasts>Base Courses. This detail component could be added to a palette by selecting the button at p3 and dragging the cursor to detailing tool palette as shown in Figure 11.44.

(Select **Insert>Detail Component Manager**, *expand the Division 02 Sitework>02700 Bases, Ballasts, Pavements, and Appurtenances> 02720 Unbound Base Courses and Ballasts>Base Courses>4" Base Course, and then click* **Insert Component**.*)*

Command: _AecDtlCompManager

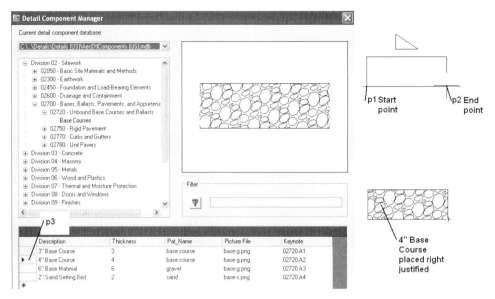

Figure 11.44 *Inserting Rectangular Predefined-Depth Surface Components*

Start point or [Xflip/Yflip/Left/Center/Right]: *(Select the start of baseline at p1 as shown in Figure 11.44.)*

End point or [Xflip/Yflip/Left/Center/Right]: *(Select the end of the baseline at p2 as shown in Figure 11.44.)*

Start point or [Xflip/Yflip/Left/Center/Right]: *Cancel* *(Press* ENTER *to end the command.)*

Bolt Components

Bolt components are located in the Division 05-Metals>Basic Metal Materials and Methods of the **Detail Component Manager**. There are seven categories of bolts in the 05090 Metal Fastenings>Bolts division. In Figure 11.45 the Anchor Bolt Hooks category is expanded; click on the **Selector** button of a specific size listed and drag it to a tool palette. You can edit the Properties palette or the **Tool Properties** dialog box of the tool to specify view, length, bolt embedment, head type, nut type, and other features as shown in Figure 11.46. After editing the properties, insert the tool as shown in the following command line sequence.

Command: _AecDtlCompManager

Projection point: *(Select point p1 as shown in Figure 11.45.)*

Nut location: *(Select point p2 as shown in Figure 11.45.)*

Hook location: *(Select point p3 as shown in Figure 11.45.)*

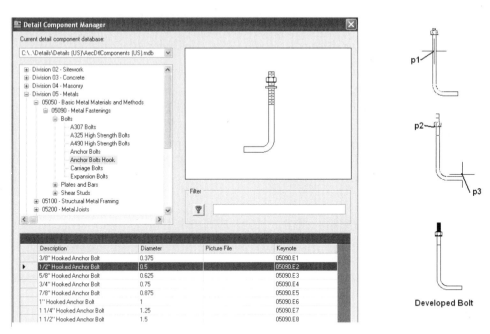

Figure 11.45 *Inserting an anchor bolt*

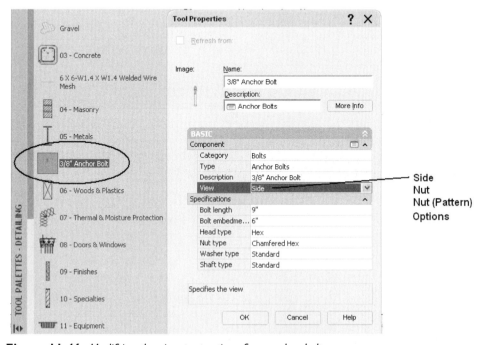

Figure 11.46 *Modifying the view properties of an anchor bolt*

Insert point or [Base point/Rotate/Xflip/Yflip]:

Insert point or [Base point/Rotate/Xflip/Yflip/Undo]: *Cancel*

Editing Details Using AEC Modify

Details can be edited as AutoCAD entities with the commands of the **Basic Modify Tools** shortcut menu. In addition, the **Extended** tab of the Properties palette of each detail component lists the properties of the detail component as shown in Figure 11.47. Additional editing tools are accessed from the shortcut menu of a detail component within the following options of **AEC Modify Tools**: **Trim**, **Divide**, **Subtract**, **Obscure**, **Merge**, and **Crop**.

Trim

The **LineworkTrim** command will trim an Aec object and detail components, including the associated hatch patterns. Access the **LineworkTrim** command as shown in Table 11.19.

Command prompt	LINEWORKTRIM
Shortcut	Select the object for edit, right-click, and choose AEC Modify Tools>Trim

Table 11.19 *Accessing LineworkTrim command*

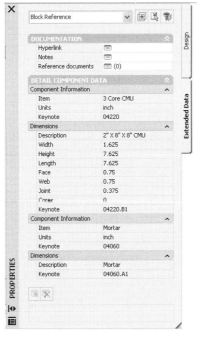

Figure 11.47 *Extended Data tab of the Properties palette*

Creating Elevations, Sections, and Details   821

The CMU is trimmed by the horizontal line in the following command line sequence.

(Select the CMU at p1, right-click, and choose **AEC Modify Tools>Trim** *as shown in Figure 11.48.)*

Command: LineworkTrim

Select the first point of the trim line or ENTER to pick on screen: *(Select a point near p2 as shown in Figure 11.48.)*

Select the second point of the trim line: *(Select a point near p3 as shown in Figure 11.48.)*

Select the side to trim: *(Select a point near p4 as shown in Figure 11.48.)*

Divide

The **LineworkDivide** command will divide an object or detail component, allowing each section to be modified independently. Access **LineworkDivide** as shown in Table 11.20.

Command prompt	LINEWORKDIVIDE
Shortcut menu	Select an entity, right-click, and choose AEC Modify Tools>Divide

Table 11.20 *Accessing LineworkDivide command*

(Select the brick component at p1, right-click, and choose **AEC Modify Tools>Divide**.*)*

Command: LineworkDivide

Select the first point of the dividing line or ENTER to pick on screen: *(Select a point near p1 as shown in Figure 11.49.)*

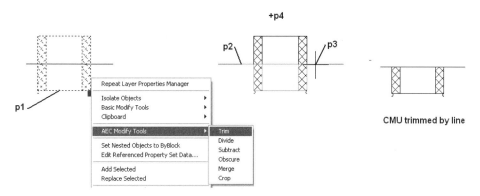

Figure 11.48 *Using AEC Modify Trim to trim CMU*

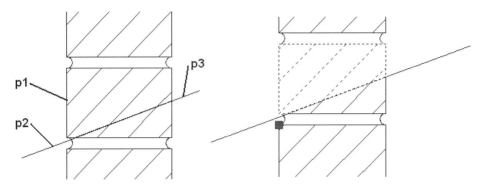

Figure 11.49 Using LineworkDivide to divide a brick

Select the second point of the dividing line: *(Select a point near p2 as shown in Figure 11.49.)*

Subtract

The **LineworkSubtract** command will erase the content of the object which is enclosed within a selected polyline. Access **LineworkSubtract** as shown in Table 11.21.

Command prompt	LINEWORKSUBTRACT
Shortcut menu	Select an entity, right-click, and choose AEC Modify Tools>Subtract

Table 11.21 Accessing LineworkSubtract command

(Select CMU at p1, right-click, and choose **Aec Modify Tools>Subtract***.)*

Command: LineworkSubtract

Select object(s) to subtract: *(Select polyline at p2 as shown in Figure 11.50.)*

1 found

Select object(s) to subtract: ENTER *(Press* ENTER *to end selection.)*

Erase selected linework? [Yes/No] <No>: y

(Slot cut in CMU as shown at right in Figure 11.50.)

Obscure

The **LineworkObscure** command will mask or hide an object or detail component. This command can be used to hide a hatch pattern that is coincident with hatch pat-

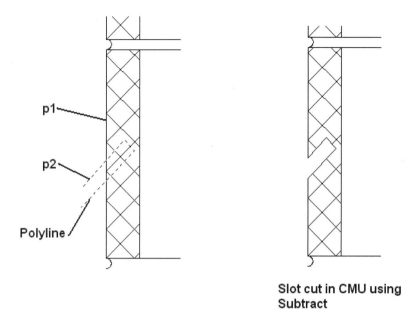

Figure 11.50 *Cutting a slot in the CMU using LineworkSubtract*

terns of other materials. Content that is hidden is placed on the A-Detl-Hide layer. Access the **LineworkObscure** command as shown in Table 11.22.

Command prompt	LINEWORKOBSCURE
Shortcut menu	Select an entity, right-click, and choose AEC Modify Tools>Obscure

Table 11.22 *Accessing LineworkObscure command*

(Select the concrete hatch shown at p1 in Figure 11.51, right-click, and choose **AEC Modify Tools>Obscure**.*)*

Command: LineworkObscure

Select obscuring object(s): 1 found *(Select ridge insulation hatch pattern at p2 as shown in Figure 11.51.)*

Select obscuring object(s): 1 found, 2 total *(Select ridge insulation hatch at p3 as shown in Figure 11.51.)*

Select obscuring object(s): ENTER *(Press ENTER to end selection.)*

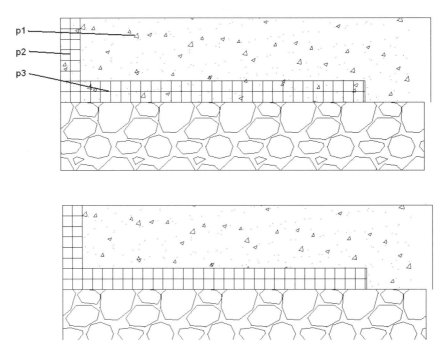

Figure 11.51 *Edit area of display for hatch using the LineworkObscure command*

Merge

The **LineworkMerge** command will merge closed polylines, hatches, or mass element extrusions. Access **LineworkMerge** as shown in Table 11.23.

Command prompt	LINEWORKMERGE
Shortcut menu	Select an entity, right-click, and choose AEC Modify Tools>Merge

Table 11.23 *Accessing LineworkMerge command*

(Select a wall at p1 as shown in Figure 11.52, right-click, and choose **AEC Modify Tools>Merge**.)

Command: LineworkMerge

Select object(s) to merge: *(Select coping at p2 as shown in Figure 11.52.)* 1 found

Select object(s) to merge: ENTER *(Press ENTER to end selection.)*

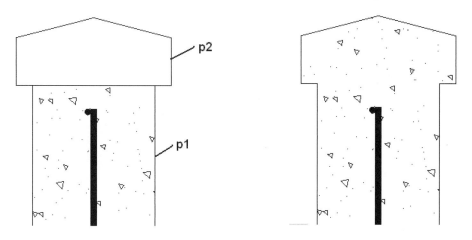

Figure 11.52 *Creating a detail using Linework Merge*

Erase selected linework? [Yes/No] <No>: **y** *(Select Yes option to erase line work and merge hatch as shown at right in Figure 11.52.)*

Select obscuring object(s): ENTER *(Press ENTER to end the command.)*

Crop

The **LineworkCrop** command will screen out portions of objects and entities that extend outside a boundary selected for the crop operation. Access **LineworkCrop** as shown in Table 11.24.

Command prompt	LINEWORKCROP
Shortcut menu	Select an entity, right-click, and choose AEC Modify Tools>Crop

Table 11.24 *Accessing LineworkCrop command*

(Select the wall shown at p1 in Figure 11.53, right-click, and choose **AEC Modify Tools>Crop**.*)*

Command: LineworkCrop

Select object(s) to form crop boundary: 1 found *(Select the rectangle at p2 as shown in Figure 11.53.)*

Select object(s) to form crop boundary: ENTER *(Press ENTER to end boundary selection.)*

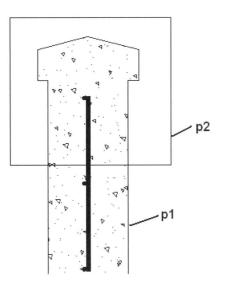

Figure 11.53 *Modifying a detail using LineworkCrop*

Erase selected linework? [Yes/No] <No>: y *(Press* ENTER *to remove geometry outside of boundary.)*

Tip: Details that have been modified can be pasted onto a detailing palette. You can create a block that consists of multiple detail components and drag it onto a detailing palette. In addition, you can convert any entity to assume the keynote properties of division by selecting the division tool, right-clicking, and selecting **Apply Detail Component Data to Linework**.

Linear dimensions can be added to details from Documentation Tool Catalog>Miscellaneous>Dimensions category of the **Content Browser**. The dimensions from the Miscellaneous folder are scaled based on **Drawing Setup**.

KEYNOTING

Keynotes annotate each component of a drawing with an identifying number that can refer to a specification document. Keynotes that refer to a specification document are reference keynotes. The specification document defines the attributes of the product or required assembly techniques. The identifying number for detail components is based upon the Master Format standard of the Construction Specification Institute. However, the identifying number for assemblies is based upon the CSI Uniformat standard. The master list of identifying numbers is included in an Access database file specified in the **AEC Content** tab of the **Options** dialog box. The keynote number embedded in the detail component, object, 2D linework, or material definition is accessed from the **AecKeynote** field in the drawing. Select **Insert>Fields** to view the Aec Keynoting fields as shown in Figure 11.54. The keynoting tools extract the field data from the object and display it as Mtext or the attribute text within a block of a leader. The text-based tools used for reference keynoting extract the keynote data using Mtext. The **Field** picks up the key from the **AecKeynote** field. In addition to reference keynoting, tools for sheet keynoting are also located on the Annotation palette. Sheet keynoting inserts a key number based upon a sheet by sheet number system. The sheet keynote number is assigned after the keynotes are placed on a sheet. The keynoting tools shown on the Annotation tool palette of the Document palette set are shown in Figure 11.55. The keynote tools are also located in the Documentation Tool Catalog of the **Content Browser**. The keynoting tools extract the keynote information using text-based or block-based tools.

REFERENCE KEYNOTING.

The **Reference Keynote (Straight Leader)** tool (see Table 11.25) extracts the keynote from the selected object as shown in the following command line sequence. The keynote data for the CMU selected in the command sequence is extracted from the detail component. The text and leader geometry is inserted in the drawing based upon the scale specified in the **Drawing Setup** dialog box.

(Select **Reference Keynote (Straight Leader)** from the Annotation tool palette.)

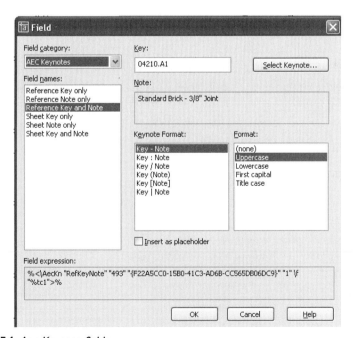

Figure 11.54 *Aec Keynote fields*

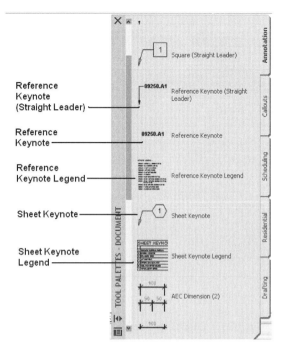

Figure 11.55 *Keynoting tools of the Annotation tool palette*

Command: AecDtlAnnoLeaderAdd

Select object to keynote or ENTER to select keynote manually: *(Select the masonry CMU at p1 as shown in Figure 11.56.)*

Select first point of leader: _mid of *(Select the location of the leader at p2 shown in Figure 11.56.)*

Specify next point of leader line: *(Select the end of the leader line at p3 shown in Figure 11.56.)*

Specify next point of leader line: ENTER *(Press ENTER to end the command.)*

Select text width<0"> ENTER *(Press ENTER to accept the default.)*

Tool palette	Select Reference Keynote (Straight Leader) from the Annotation tool palette

Table 11.25 *Accessing Reference Keynote (Straight) Leader command*

If you press ENTER rather than select an object, the **Reference Keynote (Straight Leader)** tool opens the **Select Keynotes** dialog box shown in Figure 11.57. You can

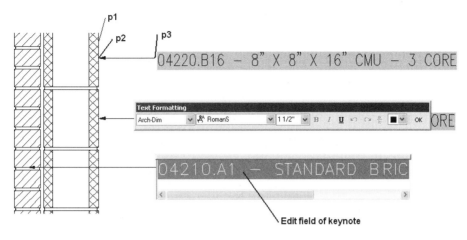

Figure 11.56 *Adding a reference keynote*

expand the Master Format divisions and select a specific keynote in the **Select Keynotes** dialog box without selecting an entity or object.

The text of the leader is created in Mtext, which includes the **AecKeynoting** data field. You can double-click on the text of the leader to edit the text. The text consists of a field as shown in Figure 11.56. If you select the field in the Mtext editor, right-click, and select **Edit Field**, the **Field** dialog box opens, which allows you to edit the format of the field as shown in Figure 11.56. If the keynote is inserted as a placeholder, the keynote information is extracted from the selected object.

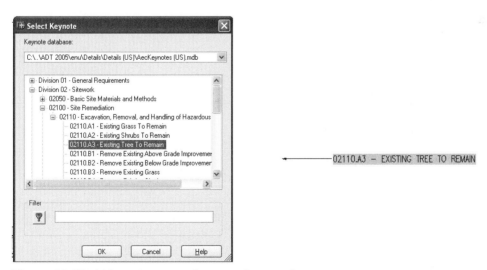

Figure 11.57 *Adding a keynote without specifying an object*

Creating a Keynote Legend

The **Reference Keynote Legend** tool (**KeynoteLegendAdd** command) can be selected from the Annotation tool palette to develop a legend of reference keynotes from the drawing. Refer to Table 11.26 to access the **KeynoteLegendAdd** command.

| Tool palette | Select the Reference Keynote Legend tool from the Annotation palette of the Document tool palette set |

Table 11.26 *KeynoteLegendAdd command access table*

When you select the **Reference Keynote Legend** tool from the Annotation palette, you can create a legend of the keynotes selected, as shown in the following command line sequence.

>Command: AecKeynoteLegendAdd
>
>Select keynotes to include in the keynote legend or [Sheets/from Database]: *(Create a crossing selection window, select a point near p1 then select a point near p2 as shown in Figure 11.58.)*
>
>Specify opposite corner: 3 found
>
>Select keynotes to include in the keynote legend or [Sheets/from Database]: ENTER *(Press ENTER to end selection.)*
>
>Insertion point of table: *(Select a point near p3 to specify the location of the table.)*

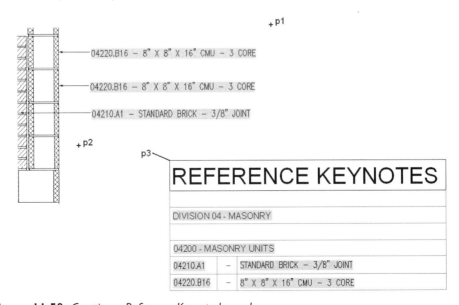

Figure 11.58 *Creating a Reference Keynote Legend*

If you select the Database option in the command line rather than select the keynotes, the **Select Keynote** dialog box opens. You can expand the categories of the keynotes. Hold down CTRL when selecting keynotes to develop a legend that includes the selected keynotes.

The Sheets option opens **Sheets** dialog box, which list the sheets of the current project. You can select the sheets in the right pane, select **Add** to add the keynotes from the added sheets, and develop a legend that includes all the keynotes from the sheets as shown in Figure 11.59.

SHEET KEYNOTES

The **Sheet Keynote**, accessed as shown in Table 11.27, inserts a leader and a specification number, which is linked to a specification listed on the current sheet. The identifying number is not inserted when the keynote is placed. A question mark is displayed to hold the place of the keynote number as shown at right in Figure 11.60 until the sheet keynote legend is generated.

Tool palette	Select the Sheet Keynote tool from the Annotation tool palette

Table 11.27 *Accessing Sheet Keynote command*

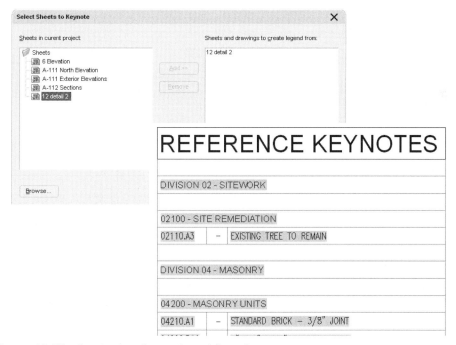

Figure 11.59 *Creating a reference legend from sheets*

Command: AecDtlAnnoLeaderAdd

Select object to keynote or ENTER to select keynote manually: *(Select the detail component at p1 as shown in Figure 11.60.)*

Select first point of leader: *(Select the start of the leader at p2 as shown in Figure 11.60.)*

Specify next point of leader line: *(Specify the endpoint of the leader at p3 as shown in Figure 11.60.)*

Specify next point of leader line: ENTER *(Press ENTER to end the command.)*

Sheet Keynote Legend

The identifying number of the key is placed when you select the sheet keys to generate the sheet keynote legend. Access the **Sheet Keynote Legend** tool (**AecKeynoteLegendAdd**) from the Annotation tool palette as shown in Table 11.28. If two keys are selected for the same type of detail component, the command recognizes the duplication and will repeat the key number.

Tool Palette	Select the Sheet Keynote Legend from the Annotation tool palette

Table 11.28 *Accessing the Sheet Keynote Legend command*

*(Select **Sheet Keynote Legend** from the Annotation tool palette.)*

Command: AecKeynoteLegendAdd

Select keynotes to include in the keynote legend or [from Database]: *(Create a crossing selection window, select point p4 then p5 as shown in Figure 11.60.)*

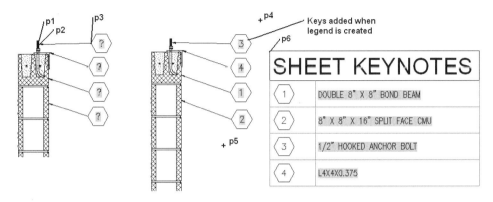

Figure 11.60 *Sheet Keynote and Sheet Keynote Legend*

Creating Elevations, Sections, and Details 833

Specify opposite corner: 15 found

10 were filtered out

Select keynotes to include in the keynote legend or [from Database]: ENTER *(Press* ENTER *to end the command.)*

Insertion point of: *(Select a point near p6 as shown in Figure 11.60.)*

After a legend has been added, additional sheet keys can be added and the legend applied to the new keys. The **KeynoteLegendApplyKeys** command allows can be used to assign an identifying key number to the new keys. Access the **KeyNoteLegendApplyKeys** command as shown in Table 11.29.

| Shortcut menu | Select the Sheet Keynote Legend, right-click, and choose Selection>Apply Keys |

Table 11.29 *Keynote Legend Selection Apply Keys command access*

The **KeynoteLegendApplyKeys** command will assign the key identifying number to new or duplicated detail components as shown in the following command sequence.

Command: KeynoteLegendApplyKeys

Select sheet keynotes to key: 1 found *(Select the new key shown at left in Figure 11.61.)*

Select sheet keynotes to key: *(Press* ENTER *to end selection.)*

EDITING LEGENDS

The selection set of a Reference or Sheet legend can be revised by selecting commands from the shortcut menu. The shortcut menu of a sheet keynote legend includes **Add**, **Reselect**, and **Show** options. The **KeynoteLegendSelectionAdd** command (see table 11.30) allows you to add additional keynotes to a table.

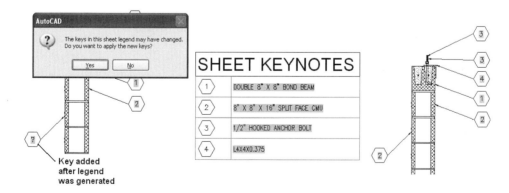

Figure 11.61 *Updating a sheet keynote*

Shortcut menu	Select the Sheet Keynote Legend, right-click, and choose Selection>Add

Table 11.30 *Sheet Keynote Legend Selection Add command access*

    Command: KeynoteLegendSelectionAdd

    Select keynotes to add to the keynote legend or [from Database]: 1 found *(Select a new sheet keynote.)*

The **Selection>Reselect** command (see Table 11.31) located on the shortcut menu of a keynote legend allows you to reselect the keynotes included in the schedule.

Shortcut menu	Select a Reference Keynote Legend or a Sheet Keynote Legend, right-click, and choose Selection>Reselect

Table 11.31 *Reselecting the keys of Reference Keynote Legend*

    (Select a legend, right-click, and choose **Selection>Reselect**.*)*

    Command: KeynoteLegendSelectionReselect

    Select keynotes to include in the keynote legend or [Sheets/from Database]: *(Select the keys using a crossing window.)*

    Specify opposite corner: 15 found

    15 were filtered out

The **Locate Keynotes** command allows you to locate the location of all keys referenced in the table. Access **Locate Keynotes** (**KeynoteLegendSelectionShow**) as shown in Table 11.32.

    (Select a Keynote Legend, right-click, and choose **Selection>Show**.*)*

    Command: KeynoteLegendSelectionShow

    Hover over keynote legend row to highlight corresponding keynotes or

    CTRL-select to zoom: *(Move pointer over keynote 2 shown in Figure 11.62.)*

Command prompt	KEYNOTELEGENDSELECTIONSHOW
Shortcut menu	Select a Keynote Legend, right-click, and choose Selection>Show

Table 11.32 *Show Selection of a Legend command access*

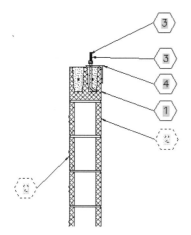
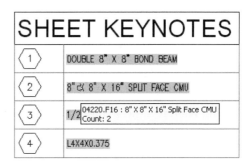

Figure 11.62 *Editing a Sheet Keynote Legend*

 STOP: Do Tutorial 11.4, "Creating Details and Keynoting," at the end of the chapter.

SUMMARY

1. The **Project Navigator** allows you attach construct drawings as reference files to create a model as a View drawing.

2. Floor elevation and floor to floor heights are defined in the Levels of the **Project** tab of the **Project Navigator**.

3. Sections and elevations should be developed from a model that consists of the floor plans of the building attached as external reference files.

4. Building elevation lines are created with the **Building Elevation Mark A2** located on the Callouts tool palette. Additional elevation marks are located in the Documentation>Callouts folder of the **Content Browser**.

5. Building section lines are created when you insert the **Section Mark A2T** from the Callouts tool palette. Additional section marks are located in the **Content Browser**.

6. Changes made in the model can be reflected in the elevation and section objects with the **Refresh** (**2DsectionResultRefresh**) and **Regenerate** (**2DsectionResultUpdate**) commands.

7. Editing the Object Display of the elevation and section object allows you to add and modify linetype, color, and lineweight of the lines of the object.

8. Elevation styles can be created to modify the display content of an elevation.

9. Lines of the elevation or section can be assigned to different display components when you select **2DSectionResultEdit** and **2DsectionEditComponent** commands.
10. Subdivisions can be created in the building elevation line and building section line to add emphasis to objects within the subdivision.
11. The surface hatch can be limited by a closed polyline when you select the **Material Boundary>Add** (2dSectionResultAddHatchBoundary) command.
12. Live Sections can be created from orthographic and pictorial views of the model.
13. The **Hidden Line Projection** (**CreateHLR**) command creates a 2D block from a view of the model.

REVIEW QUESTIONS

1. Drawings that consist of simple components repeated throughout the design are _____.
2. Drawings that consist of unique components of the building are _____.
3. The floor plans for each level of a building are inserted in a View drawing as _____.
4. The insertion point Z coordinate is specified in the _____ definition of the project.
5. Each floor plan inserted in the model should be inserted at a Z coordinate equal to the _____.
6. The projection box of an elevation is created when you insert a _____ _____ _____.
7. The _____ command will change the viewing direction 180°.
8. The subdivisions of a building elevation line are created in the _____ palette.
9. Components of an elevation object displayed based upon the object color are defined in the _____ _____ _____ _____ dialog box.
10. The command that allows you to add new objects to an elevation or section is the _____ command.
11. The _____ command will redisplay an elevation based upon the original set of objects included in the elevation or section.
12. The _____ _____ _____ command allows you to add objects to a 3D section view.

13. Tools for creating elevations and sections are located on the _____ palette of the _____ palette set.

14. The properties of detailing tools can be set in the _____ palette or on the _____ _____ of the tool.

TUTORIAL 11.1 CREATING A MODEL VIEW USING THE PROJECT NAVIGATOR

1. Open Autodesk Architectural Desktop 2005, and select **File>Project Browser** from the menu bar. Use the Project Selector drop-down list to navigate to your *ADT Student\ADT Tutor\Ch11* directory. Double-click on **Ex 11-1** to set this project current. Select **Close** to dismiss the **Project Browser**.

2. If the **Project Navigator** is not displayed, select **Window>Project Navigator Palette** from the menu bar. Select the **Projects** tab. Select **Edit Levels** button and edit the levels are as shown in Table 11T.1. Clear the **Auto-Adjust Elevation** check box. Select **OK** to dismiss the **Levels** dialog box. Select **Yes** to respond to the AutoCAD dialog box and regenerate all views from the level changes.

Name	Floor Elevation	Floor to Floor Height	ID	Description
2	8'-11-1/2"	8'	2	First Floor
1	0	8'-11-1/2"	1	Basement

Table 11T.1 *Level definitions*

3. Select the **Views** tab, and select the **Add Category** command from the toolbar located at the bottom of the **Views** tab. Overtype **Exterior** as the name of the category. Select the **Exterior** category, and then select the **Add View** command from the **Views** tab toolbar. Check the **Section/Elevation View** radio button, select **OK** to close the **Add View** dialog box, and open the **Add Section/Elevation View** dialog box to the **General** page. Edit the following on the **General** page: Name = Model, Description = Model View, and select **OK** to dismiss the **Description** dialog box. Select **Next** and edit the **Context** page as follows: check Division 1 for **Level 1** and **Level 2**. Select **Next** to display the **Content** page, and verify that Floor 1, Basement, and Stair constructs are listed as shown in Figure 11T.1. Select **Finish** to close the dialog box.

4. Select the **Model** drawing in the Exterior category of the **Views** tab, right-click, and choose **Open** from the shortcut menu. Select the Model tab. Select

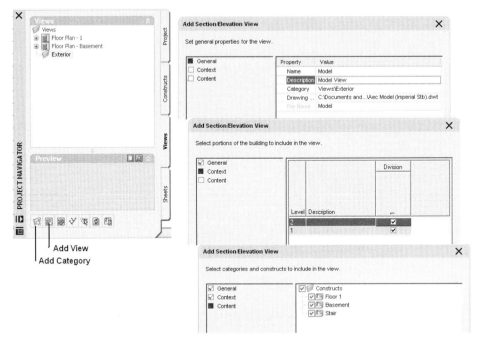

Figure 11T.1 *Creating the Model view drawing*

Drawing Setup from the Drawing Window status bar. Select the **Scale** tab and set the scale to **1/4" = 1'-0"**.

5. Select **Zoom Extents** from the **Zoom** flyout of the **Navigation** toolbar. Select **SW Isometric** from the **View flyout** of the **Navigation** toolbar to verify that level **1** and level **2** are displayed. Select **Top View** from the **View** flyout of the **Navigation** toolbar.

6. Right-click over the title bar of the tool palette, and choose **Document**. Select **Elevation Mark A2** from the Callouts palette. Respond to the command line prompts as follows:

   Command: AecCallout

   Specify location of elevation tag: *(Select a point near p1 as shown in Figure 11T.2.)*

   Specify direction for elevation: *(Select a point near p2 as shown in Figure 11T.2.)*

   (Edit the **Place Callout** *dialog box shown in Figure 11T.3 as follows: Check* **Generate Section/Elevation** *and* **Place Titlemark**, *set scale to* **1/4"=1'-0"**, *type* **Front** *in the* **New Model Space View Name** *and then choose the* **New View**

Creating Elevations, Sections, and Details 839

Drawing *button.)*

(Edit the **Add Section/Elevation View** *dialog box as follows and as shown in Figure 11T.4: type* **Front** *in the* **Name** *edit field, select* **Next**, *check level* **1** *and level* **2** *for the division* **1** *of the* **Context** *page as shown in Figure 11T.4, select* **Next** *to open the* **Content** *page, and check* **Floor 1**, **Basement**, *and* **Stair**. *Select* **Finish** *to close the dialog box.)*

Specify first corner of elevation region: *(Select a point near p3 as shown in Figure 11T.2.)*

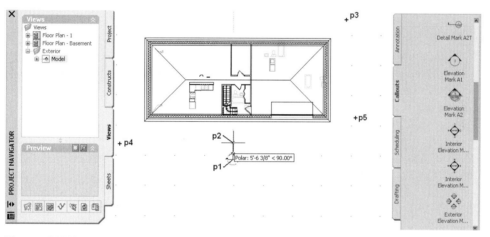

Figure 11T.2 *Views tab of the Project Navigator*

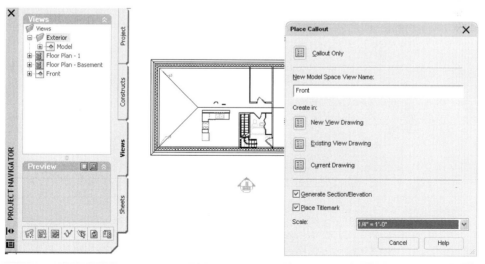

Figure 11T.3 *Editing Place Callout dialog box*

Specify opposite corner of elevation region: *(Select a point near p4 as shown in Figure 11T.2.)*

Specify insertion point for the 2D elevation result: *(Select a point near p5 as shown in Figure 11T.2.)*

7. Select the **Sheets** tab. Select the **Elevations** category, right-click, and choose **New>Sheet**. Edit the **New Sheet** dialog box, type Number = **A-12** and Sheet Title = **Front Elevation**. Select **OK** to dismiss the **New Sheet** dialog box.

8. Select the **A-12 Front Elevation** sheet listed on the **Sheets** tab of the **Project Navigator**, right-click, and choose **Open**. The A-12 Front Elevation drawing opens, including the display of the title block and border.

9. Select the **Views** tab, select and drag the **Front** view into the **A-12 Front Elevation** sheet, and select a point near **p1** as shown in Figure 11T.5.

10. Save and close all drawings. Choose **File>Project Browser** and close the project.

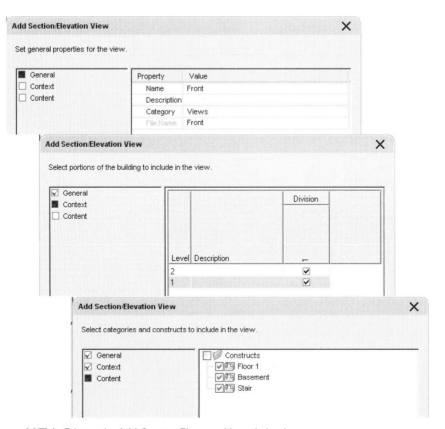

Figure 11T.4 *Editing the Add Section Elevation View dialog box*

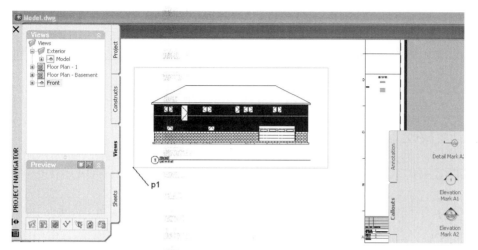

Figure 11T.5 *Front elevation view inserted into sheet*

TUTORIAL 11.2 CREATING ELEVATIONS AND ELEVATION STYLES

1. Open Autodesk Architectural Desktop 2005, and select **File>Project Browser** from the menu bar. Use the Project Selector drop-down list to navigate to your *ADT Student\ADT Tutor\Ch11* directory. Double-click on **Ex 11-2** to set this project current. Select **Close** to dismiss the **Project Browser**.

2. If the **Project Navigator** is not displayed, select **Window>Project Navigator Palette** from the menu bar. Select the **Levels** button and verify that the level is as shown in Table 11T.2.

Name	Floor Elevation	Floor to Floor Height	ID	Description
1	0	8'	1	Floor 1

Table 11T.2 *Level definitions*

3. Select the **Views** tab, and select the **Add Category** command from the toolbar located at the bottom of the **Views** tab. Overtype **Exterior** as the name of the category. Select the **Exterior** category, and then select the **Add View** command from the **Views** tab toolbar. Check the **Section/Elevation View** radio button as shown in Figure 11T.6, and select **OK** to close the **Add View** dialog box and open the **Add Section/Elevation View** dialog box to the **General** page. Edit the following in the **General** page: Name = **Model**, Description = **Model View**, and select **OK** to dismiss the

Description dialog box. Select **Next** and edit the **Context** page as follows: check Division 1 for **Level 1**. Select **Next** to display the **Content** page, and verify that **Floor 1** construct is checked. Select **Finish** to close the dialog box.

4. Select the **Model** drawing in the **Exterior** category of the **Views** tab, right-click, and choose **Open** from the shortcut menu. Select the Model tab. Select **Zoom Extents** from the **Zoom** flyout of the **Navigation** toolbar.

5. Select **Exterior Elevation Mark A3** from the Callouts palette of the Document palette set. Respond to the command line prompts as follows:

   Command: AecCallout

   Specify first corner of elevation region: *(Select a point near p1 as shown in Figure 11T.7.)*

   Specify opposite corner of elevation region: *(Select a point near p2 as shown in Figure 11T.7.)*

   (Edit the **Place Callout** *dialog box as follows: check* **Generate Section/Elevation** *and check* **Place Titlemark**, *set scale to* **1/4"=1'-0"**, *check* **Current Drawing**.*)*

   Specify insertion point for the 2D elevation result: *(Select a point near p3 as shown in Figure 11T.7.)*

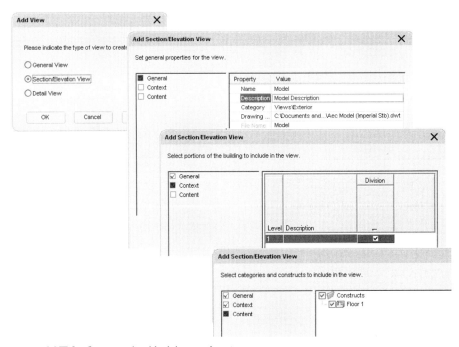

Figure 11T.6 *Creating the Model view drawing*

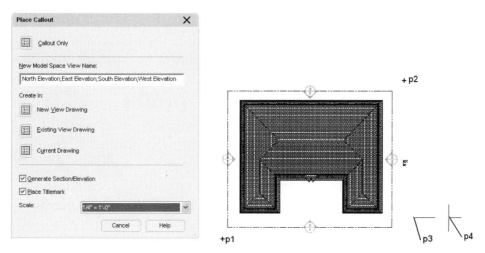

Figure 11T.7 *Creating the Elevation view*

Pick a point to specify the spacing and direction of elevations: *(Move the pointer right, select a point near p4 as shown in Figure 11T.7.)*

6. Select the building elevation line at **p1** shown in Figure 11T.8, right-click, and choose **Properties** to open the Properties palette. Select the **Subdivisions** button in the Properties palette. Select the **Add** button of the **Subdivisions** dialog box, and edit the subdivision distance to **15'** for the building elevation line as shown in Figure 11T.8. Select **OK** to dismiss the **Subdivisions** dialog box.

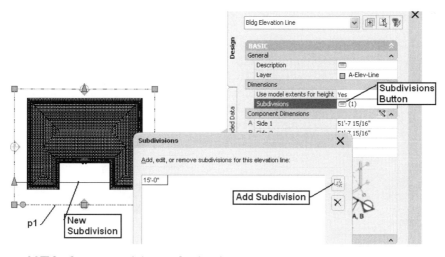

Figure 11T.8 *Creating a subdivision for the elevation*

7. To view the elevation, select **Zoom Extents** from the **Navigation** toolbar.
8. Select **Zoom Window** from the **Zoom** flyout of the **Navigation** toolbar. Respond to the command line as shown below to view the South Elevation.

   Command: '_zoom

   Specify corner of window, enter a scale factor (nX or nXP), or

   [All/Center/Dynamic/Extents/Previous/Scale/Window/Object] <real time>: _w

   Specify first corner: *(Select a point near p1 in Figure 11T.9.)*

   Specify opposite corner: *(Select a point near p2 in Figure 11T.9.)*

9. To customize the display of the elevation using elevation styles, select the South Elevation, right-click, and choose **Copy 2D Elevation/Section Style and Assign**. Edit the new style, choose the **Display Properties** tab, and select the **Edit Display Properties** button to open the **Display Properties (2D Section/Elevation Style Override-2D Section Style 96(2))** dialog box. Select the **Layer/Color/Linetype** tab and then select Subdivision 2. Click in the **Linetype** column for Subdivision 2 to open the **Select Linetype** dialog box. Choose **Hidden2** linetype. Choose **OK** to dismiss all dialog boxes.
10. To view the changes of the elevation style modifications for subdivision 2, select the elevation, right-click, and choose **Refresh**.
11. The elevation style can be modified to display a floor line with centerline type. The style will represent the slab object based upon its color number, color 192 with centerlines. Select the south elevation, right-click, and choose **Edit 2D Elevation/Section Style**.
12. Select the **Components** tab, and select the **Add** button. Edit for Index 3 the Name to **Slab**, and the Description to **slab line**.
13. Select the **Design Rules** tab. Select the **Add** button to add Rule 14. Select the **Color** column to open the **Select Color** dialog box, type **192** in the color edit field, and select **OK** to dismiss the **Select Color** dialog box. Select the **Context** column to display the drop-down list, and select **Any visible** or **hidden**. Select the **Component** column to display the drop-down list, and select the **Slab** component as shown in Figure 11T.10.

Figure 11T.9 *Specifying points for the Zoom Window command*

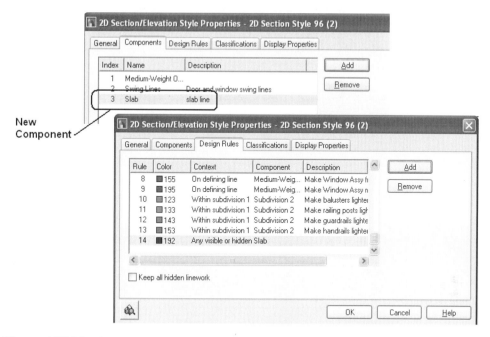

Figure 11T.10 *Creating a new display component*

14. Select the **Display Properties** tab. Select the **Edit Display Properties** button. Select the **Layer/Color/Lineytpe** tab, and scroll down the list of components and select **Slab**. Click in the **Linetype** column for the Slab component, and select **Center2** from the **Select Linetype** dialog box. Select **OK** to dismiss the **Select Linetype** dialog box. Select in the **Color** column, and edit the color to color **7**. Select **OK** to dismiss the **Select Color** dialog box. Select **OK** to dismiss the **Display Properties (2D Section/Elevation Style Override – 2D Section Style 96 (2))** dialog box. Select **OK** to dismiss the **2D Section/Elevation Style Properties – 2D Section Style 96 (2)** dialog box.

15. To apply the elevation style select the elevation, right-click, and choose Refresh to display the elevation as shown in Figure 11T.11.

16. Select **Zoom Window** from the **Navigation** toolbar, and respond to the command line prompts as shown below:

    Specify corner of window, enter a scale factor (nX or nXP), or [All/Center/Dynamic/Extents/Previous/Scale/Window]<real time>:_w

    Specify first corner: *(Select a point near p1 as shown in Figure 11T.11.)*

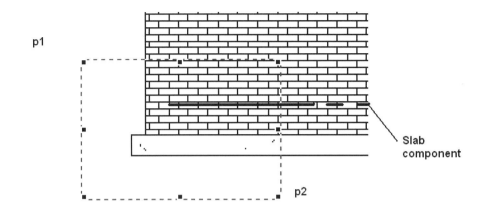

Figure 11T.11 *Slab component displayed in elevation*

Specify opposite corner: *(Select a point near p2 as shown in Figure 11T.11.)*

17. Select **Line** from the **Shapes** toolbar, and respond to the following command line prompts to draw a line as shown in Figure 11T.12.

    Command:_line Specify first point: *(Hold SHIFT down, right-click and choose **Endpoint**.)*

    _endp of *(Select the end of the top of the slab at p1 as shown in Figure 11T.12.)*

    (Move the pointer left to set the direction.)

    Specify next point or [Undo]: **6'** ENTER *(Line drawn 6' long using direct distance entry.)*

    Specify next point or [Undo]: ENTER *(Press ENTER to end the command.)*

18. Select the **Offset** command from the **Modify** toolbar, and respond to the command line as shown below.

    Command:_offset

    Specify offset distance or [Through]<Through>:8' enter

    Select object to offset or <exit>: *(Select the line at p4 in Figure 11T.12.)*

    (Move the pointer up and select a point near p5 as shown in Figure 11T.12)

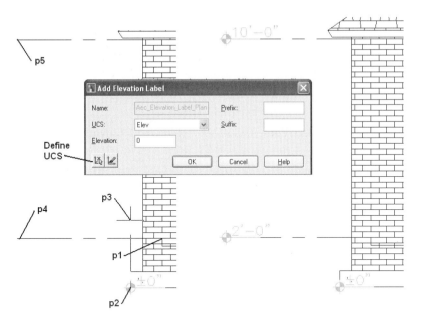

Figure 11T.12 *Adding lines to an elevation*

Select object to offset or <exit>: ENTER *(Press ENTER to end the command.)*

19. Select the elevation, right-click, and choose **Linework>Merge**. Respond to the command line prompts as shown below.

    2dSectionResultMerge

    Select objects to merge: *(Select line at p4 as shown in Figure 11T.12.)*

    Select objects to merge: *(Select line at p5 as shown in Figure 11T.12.)*

    Select objects to merge: ENTER *(Press ENTER to end selection.)*

    *(Select the **Slab** component from the drop-down list of the **Select Linework Component** dialog box. Select **OK** to dismiss the **Select Linework Component** dialog box.)*

20. Select **1/4"=1'-0"** from the **Scale** flyout located on the Drawing Window status bar.

21. Right-click over the title bar of the tool palettes, and choose **Document** palette set. Select the Annotation palette, and scroll down the tools to **Plan Elevation Label (1)**. Respond to the command line prompts as shown below.

Command: _.aecannoelevationlabeladd

Enter multi view block name or [?]: Aec_Elevation_Label_Plan_1_1

Specify insertion point [Leader]: _endp of (SHIFT + *right-click and choose* **Endpoint** *object snap, and then select the end of the footing at p2 as shown in Figure 11T.12.*)

(Select the **Define UCS** *button of the* **Add Elevation Label** *dialog box and respond to the remaining command line prompts.)*

Specify base point: @ ENTER *(Type* @ ENTER *to select the last point entered, p2 as shown in Figure 11T.12.)*

Specify z-direction: *(Select a point near p3 as shown in Figure 11T.12.)*

Enter name for UCS or [?]: **Elev** ENTER *(Name of UCS defined and* **Add Elevation Label** *dialog box reopened.)*

(Verify elevation value = **0** *and then click* **OK** *to dismiss the* **Add Elevation Label** *dialog box.)*

22. Select the **Plan Elevation Label (1)** tool. Respond to the command line prompts as shown below.

    Specify insertion point [Leader]: _endp of (SHIFT + *right-click and choose* **Endpoint** *object snap, and then select the end of floor line at p4 as shown in Figure 11T.12.)*

    (Verify the UCS=Elev and the elevation value is 2'-0", and then select **OK** *to dismiss the dialog box.)*

23. Select the Plan Elevation Label (1) tool. Respond to the command line prompts as shown below.

    Specify insertion point [Leader]: _endp of (SHIFT + *right-click and choose* **Endpoint** *object snap, and then select the end of floor line at p5 as shown in Figure 11T.12)*

    (Verify that the UCS = Elev and the elevation value is **10'-0"**, *and then select* **OK** *to dismiss the dialog box.)*

24. Select the Elevation, right-click, and choose **Linework>Edit**. Select the footing lines shown at **p1, p2, p3,** and **p4** in Figure 11T.13, right-click, and choose **Modify Component**. Select **Subdivision 2** from the list of display components. Select **OK** to dismiss the **Select Linework Component** dialog box.

25. Select the lines representing bottom of the slab using a crossing window from **p1–p2, p3–p4,** and **p5–p6** as shown in Figure 11T.14, and then select **Erase** from the **Modify** toolbar. Select **Save All Changes** from the **Refedit** toolbar.

Creating Elevations, Sections, and Details 849

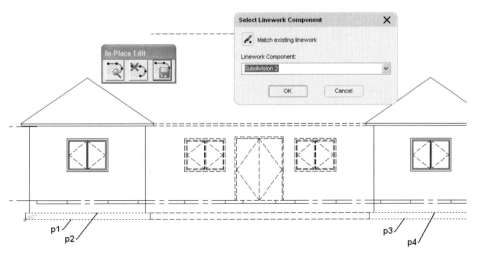

Figure 11T.13 *Footing lines selected for edit*

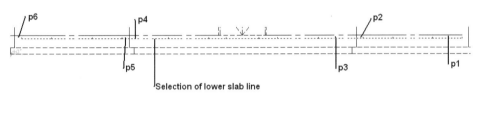

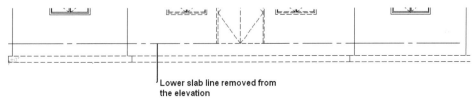

Figure 11T.14 *Lower slab lines removed using Linework>Edit*

26. Select the **Polyline** command from the **Shapes** toolbar, and draw a closed polyline as shown in Figure 11T.15. Select the elevation, right-click, and choose **Material Boundary>Add**. Respond to the command line as shown below to modify the hatch pattern.

    Command: 2dSectionResultAddHatchBoundary

    Select a closed polyline for boundary: *(Select the polyline as shown at p1 in Figure 11.15.)*

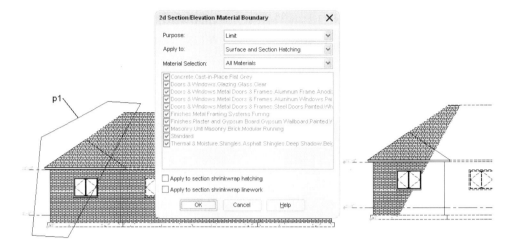

Figure 11T.15 *Editing hatch of the elevation*

Erase selected linework? [Yes/No] <No>: y

*(Verify Purpose=***Limit***; Apply to=* **Surface & Section Hatching***; Material Selection =* **All Materials***. (Select* **OK** *to dismiss the dialog box.)*

27. Select the **Constructs** tab of the Navigation tool palette. Double-click to open the **Floor 1** drawing. Select **SW Isometric** from the **Navigation** toolbar. Select the two windows shown in Figure 11T.16, right-click, and choose **Properties**. Edit the Height of the windows to **5'**. Save the drawing.

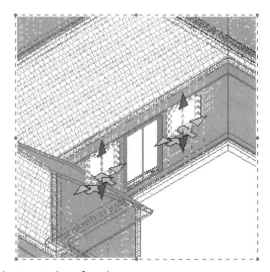

Figure 11T.16 *Selecting windows for edit*

28. Click **Views** tab, and double-click on the **Model** drawing. Select the **Manage Xrefs** button located on the right of the Drawing Window status bar as shown in Figure 11T.17. Select **Floor1** and then check **Reload**. Select **OK** to close the dialog box.

29. Select the **Views** tab, expand the Model, and double-click on the **South Elevation**. Select the elevation object, right-click, and choose **Refresh** from the shortcut menu. Elevation is updated as shown in Figure 11T.18. Save the drawing.

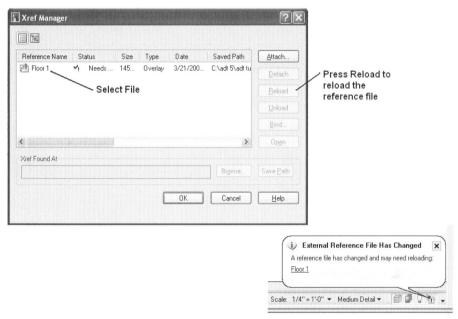

Figure 11T.17 *Reloading reference file*

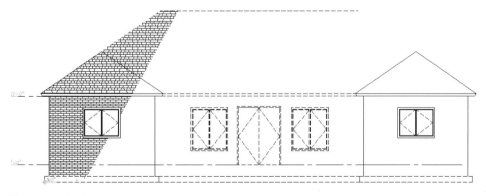

Figure 11T.18 *Window height changed in updated elevation*

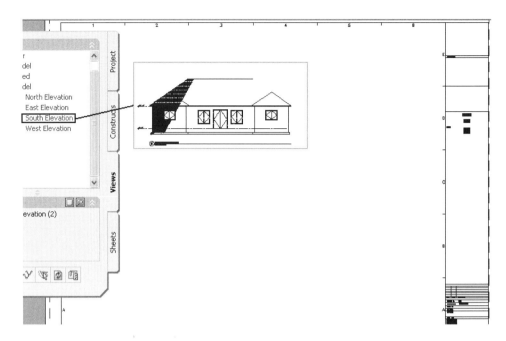

Figure 11T.19 *Elevation placed on sheet*

30. Select the **Sheets** tab of the **Project Navigator**, select the **Elevations** category, right-click and choose **New>Sheet**. Add a new sheet, type **A-13** in the **Number** field and type **South and North Elevations** in the **Sheet Title** field. Select **OK** to dismiss the dialog box. Double-click on the **A-13** sheet drawing in the **Project Navigator** to open the file.
31. Select the **Views** tab of the **Project Navigator**, and expand the **Model** drawing. Click and drag the **South Elevation** onto Sheet A-13 as shown in Figure 11T.19.
32. Save and close the drawing.
33. Select **File>Project Browser** to open the **Project Browser**. Select the **Ex 11-2** project, right-click, and choose **Close Current Project**.

TUTORIAL 11.3 CREATING A 2D SECTION AND A LIVE SECTION

1. Open Autodesk Architectural Desktop 2005, and select **File>Project Browser** from the menu bar. Use the Project Selector drop-down list to navigate to your *ADT Student ADT Tutor\Ch11* directory. Double-click on Ex 11-3 to set this project current. Select **Close** to dismiss the **Project Browser**.

Creating Elevations, Sections, and Details 853

2. If the **Project Navigator** is not displayed, select **Window>Project Navigator Palette** from the menu bar. Select the **Views** tab of the **Project Navigator**.

3. Select the **Add Category** command of the **Views** tab toolbar. Overtype **Sections**. Select the **Sections** category, and then select **Add View** from the **Views** tab toolbar. Select **Section/Elevation View** from the **Add View** dialog box. Select **OK** to close the dialog box. Edit the **General** page of the **Add Section/Elevation View** dialog box as follows: Name = **Transverse-1**, Description = **Transverse**. Select **OK** to close the **Description** dialog box. Select **Next** and edit the **Context** page as follows: for Division 1 check 1 for Level 1 and Level **2**. Select **Next** to display the **Content** page, and verify **Floor 1**, **Basement**, and **Stair** constructs are checked. Select **Finish** to close the dialog box.

4. Select the **Transverse-1** view drawing in the **View** tab of the **Project Navigator**, right-click, and choose **Open**.

5. Select **Zoom Extents** from the **Navigation** toolbar. Toggle ORTHO ON on the status bar. Select **High Detail** from the **Display Configuration** flyout of the Drawing Window status bar.

6. Verify **Document** is the current palette set, select **Section Mark A2T** from the Callouts tool palette, and respond to the command line prompts as shown below.

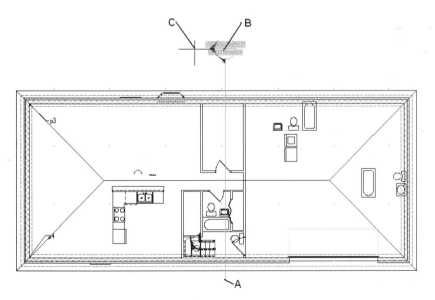

Figure 11T.20 Section line placement

Specify first point of section line: *(Select near point A as shown in Figure 11T.20.)*

Specify next point of line: *(Select near point B as shown in Figure 11T.20.)*

Specify next point of line or [Break]: ENTER *(Press* ENTER *to end line.)*

Specify side for arrow: *(Select a point near C as shown in Figure 11T.20 and then edit the* **Place Callout** *dialog box as described in the next step.)*

7. Edit the **Place Callout** dialog box as follows: type **Section A-A** in the **New Model Space View Name** field, verify that **Generate Section/Elevation** and **Place Titlemark** are checked, select **1/4"=1'-0"** in the **Scale** drop-down list, and check **New View Drawing**.

8. Edit the **Add Section Elevation View** dialog box as follows: type **Section A-A** in the name field, click **Next**, and check Level **1** and Level **2** for division **1** on the **Context** page. Click **Next**, and verify that **Basement**, **Floor 1**, and **Stair** are checked. Click **Finish** to close the dialog box. Specify the location of the section in the command line as follows:

    Specify insertion point for the 2D section result: *(Select a point near p1 as shown in Figure 11T.20.)*

9. Save and close the Transverse-1 drawing.

10. Double-click on the **Section A-A** drawing in the **Views** tab of the **Project Navigator** to open the drawing as shown in Figure 11T.21.

11. Note the hatch patterns assigned in the section include hatch of the wood walls and roof objects. To refine the section, you can turn off the section hatch component of the Standard material definition. Select **Format>Style Manager** from the menu bar. Select the **Section A-A** drawing in the left pane, and expand Multi-Purpose Objects>Material Definitions. Double-click on **Standard** material definition to open the dialog box. Select the **Edit Display Properties Material Definition Properties-Standard** button, and select the **Layer/Color/Linetype** tab toggle **Off** the Section Hatch component as shown in Figure 11T.22. Select **OK** to dismiss all dialog boxes.

12. Select the section, right-click, and choose **Edit 2D Elevation/Section Style**. Select the **Display Properties** tab, and select the **Edit Display Properties** button. Check **By Material** for the Section Hatch Linework and toggle OFF **Shrinkwrap hatch** as shown in Figure 11T.22. Select **OK** to dismiss all dialog boxes. Hatch is removed from components with Standard material definition as shown in Figure 11T.23.

Creating Elevations, Sections, and Details   855

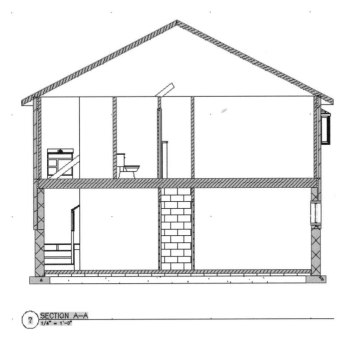

Figure 11T.21 *Section A-A of the building*

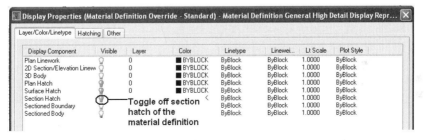

Edit Material Definition

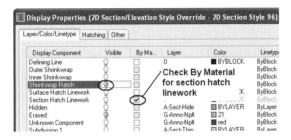

Edit 2D Section/Elevation Style

Figure 11T.22 *Material definition and elevation style edit*

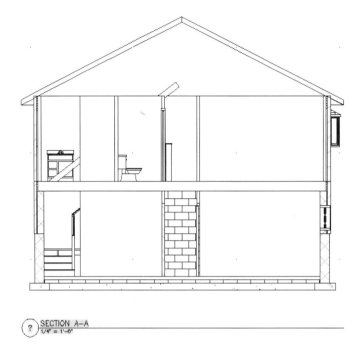

Figure 11T.23 *Hatch pattern removed from objects with Standard material definition*

14. Select **SE Isometric** from the **View** flyout of the **Navigation** toolbar.

15. Select the building section line, right-click, and choose **Enable Live Section**. Select **Hidden** from the **Shade** flyout of the **Navigation** toolbar. Toggle OFF GRID on the status bar to view the model as shown in Figure 11T.24.

16. Save and close the Section A-A.

17. Select **File>Project Browser** to open the **Project Browser**. Select the **Ex 11-3** project, right-click, and choose **Close Current Project**.

TUTORIAL 11.4 CREATING DETAILS AND KEYNOTING

1. Select **File >Project Browser** from the menu bar. Verify that the Project Selector folder location is set to your *ADT Student\ADT Tutor\Ch11*. Select the **New Project** button located in the lower left corner of the **Project Browser** to open the **Add Project** dialog box. Type **Ex 11-4** in the number edit field and **Ex 11-4** in the **Name** edit field. Select **OK** to dismiss the **Add Project** dialog box. Select **Close** to close the **Project Browser**.

2. If the **Project Navigator** is not displayed, select **Window>Project Navigator** from the menu bar. Select the **Views** tab of the **Project**

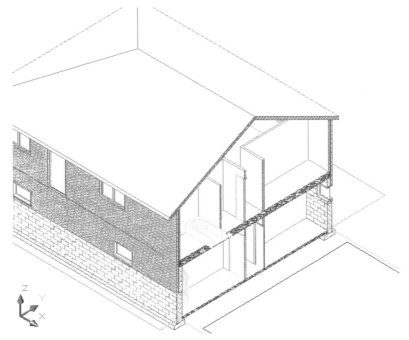

Figure 11T.24 *Live section enabled*

Navigator. Select the **Views** category, right-click, and choose **New View Dwg>Detail**. Edit the **General** page as shown in Figure 11T.25: set Name = **Masonry Wall**. Click **Next** to edit the **Context** page: check **Level 1** for **Division 1**. Click **Next** to open the **Content** page and verify that **Constructs** is checked. Select **Finish** to close the **Add Detail View** dialog box.

3. Select the **Views** tab of the **Project Navigator**. Double-click on **Masonry Wall** to open the file.

4. Right-click over the titlebar of tool palettes, and choose **Detailing** palette set. Toggle ON POLAR on the status bar.

5. Select **04-Masonry** from the Basic palette of the Detailing palette set. Click in the edit fields of the Properties palette and specify the following: Component: Category = **04220 Concrete**, Type = **3 Core CMU**, Description = **8"x8"x16" CMU**, View = **Section**; Specifications: Hatch block = **yes**, Grout file cores = **No**, Show mortar = **yes**; Mortar: Type= **Mortar**, Description = **Type "S" Mortar**, Left joint type = **Rodded**, Right joint type = **Rodded**, and Hatch mortar = **yes** as shown in Figure 11T.26. Respond to the command line as follows to place 14 courses.

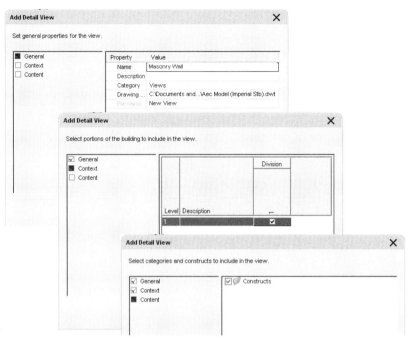

Figure 11T.25 *Add Detail View dialog box*

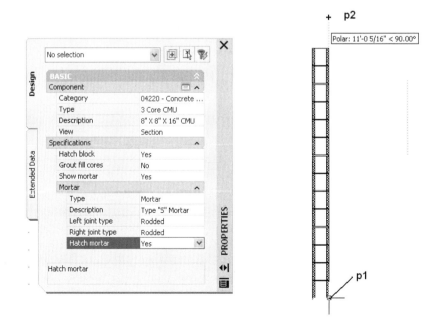

Figure 11T.26 *Start point of wall*

Command:

Start point or [Xflip/Yflip/Count]: **c** *(Type **C** in the command line to select the Count option.)*

Number of courses <1>: **14** ENTER *(Enter the number of courses.)*

Start point or [Xflip/Yflip/Count]: *(Select a point near p1 as shown in Figure 11T.26.)*

Pick point for direction or [Xflip/Yflip/Count]: *(Select a point above p1 near p2, shown in Figure 11T.26, with the polar tooltip angle equal to 90.)*

Start point or [Xflip/Yflip/Count]: ENTER *(Press ENTER to end the command.)*

6. Right-click over the title bar, select **New Palette**, and overtype **My Detail** in the palette field. Select **Insert>Detail Component Manager** from the menu bar. Expand Division 03-Concrete, 03300 Cast-in-Place Concrete>03310 Structural Concrete>Strip Footings. Select the **Selector** of the **18 x 12 Concrete Footing** tool and drag it to the My Detail tool palette. Close the **Detail Component Manager**.

7. Select the **18 x 12 Concrete Footing** tool on the tool palette, right-click, and choose **Properties**. Edit and verify the following tool properties as shown in Figure 11T.27: Component: Category = **03310- Structural Concrete**, Type = **Strip Footings**, Description = **18" x 12" Concrete Footing**, View = **Section**, Specifications: Hatch footing = **Yes**, Show reinforcing = **Yes**, Rebar Longitudinal Type = **ACI Reinforcing Bar**, Description = **#4 Rebar**, Draw type = **Solid**, Bars = **2**, Bars on outside = **No**, Rebar-Lateral: Type = **ACI Reinforcing Bar**, Description = **#4 Rebar**, Draw type = **Solid**, Edges= **Center**, and Edge offset = **1 1/2"**.

8. Select **OK** to close the **Tool Properties** dialog box.

9. Right-click over the OSNAP button of the status bar, select **Settings**, check **Endpoint** object snap, and toggle off all other object snap modes. Select the **Options** button in the lower left corner, and verify that **Display polar tracking vector**, **Display full-screen tracking vector**, and **Display Auto Track tooltip** are checked in the **AutoTrack Settings** section of the **Drafting** tab. Verify that the **Display AutoSnap tooltip** is checked in the **AutoSnap Settings** section. Select **OK** to dismiss the **Options** dialog box. Select **OK** to dismiss the **Drafting Settings** dialog box. Toggle **ON** Otrack and Polar in the status bar.

10. Select the **18 x 12 Concrete Footing** tool from the My Detail palette and place the footing below the CMU wall as shown in the following command line sequence.

Command: _DTLCOMPTOOL

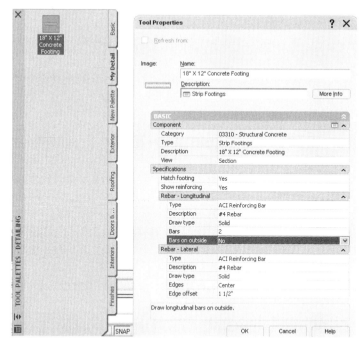

Figure 11T.27 *Settings of a detail tool*

Insert point or [Base point/Rotate/Xflip/Yflip]: *(Hover over the end of the arc representing the mortar joint at p1 in Figure 11T.28, move the cursor left toward p2 and type **5** in the command line.)* **5** ENTER

Insert point or [Base point/Rotate/Xflip/Yflip/Undo]: ENTER *(Press ENTER to end the command.)*

11. Select the footing, right-click, and choose **Edit Block in-place** to open the **Reference Edit** dialog box. Verify that **Prompt to select nested objects** is checked. Select **OK** to dismiss the **Reference Edit** dialog box. Respond to the command line prompt as follows:

Command. _refedit

Select nested objects: *(Create a window selection from p3 as shown in Figure 11T.28.)*

Specify opposite corner: *(Select a point near p4 as shown in Figure 11T.28.)*

5 entities added

Select nested objects: ENTER *(Press ENTER to end selection and open the **Refedit** toolbar.)*

Creating Elevations, Sections, and Details   861

> 5 items selected
>
> Use REFCLOSE or the Refedit toolbar to end reference editing session.

12. Select the **Move** command from the **Modify** toolbar and respond to the command line prompts as shown below.

    > Command: _Move
    >
    > Select objects: *(Select the rebar using a window selection from p3 to p4 as shown in Figure 11T.28.)* 5 found
    >
    > Select objects: *(Press* ENTER *to end selection.)*
    >
    > Specify base point or displacement: *(Select a point near p5 as shown in Figure 11T.28.)*
    >
    > Specify second point of displacement or <use first point as displacement>: **3** *(Move the cursor down, verify the polar tracking angle =>270 and then type* **3**.*)*

13. Select the **Save back changes to reference** button on the **Refedit** toolbar. Select **OK** to dismiss the AutoCAD warning dialog box and save changes from the edit.

14. Select the **Rectang** command from the **Shapes** toolbar to create a boundary for the gravel as shown in following command line prompts.

    > Command: _rectang

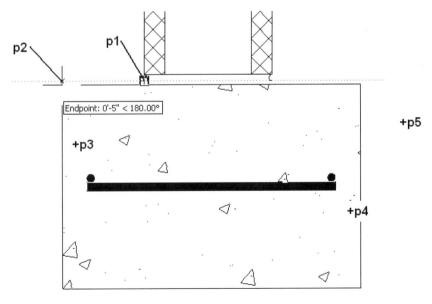

Figure 11T.28 *Creating the footing component*

Specify first corner point or [Chamfer/Elevation/Fillet/Thickness/Width]: *(Select point p1 shown in Figure 11T.29 using the Endpoint object snap mode.)*

Specify other corner point or [Dimensions]: **d** *(Type **d** to choose the Dimensions option.)*

Specify length for rectangles <0'-0">: **5'** ENTER

Specify width for rectangles <0'-0">: **4** ENTER

Specify other corner point or [Dimensions] *(Specify a point near p2 as shown in Figure 11T.29.)*

15. Select **Insert>Detail Component Manager** from the menu bar, and expand Division 02- Sitework>02300 Earthwork>02315 Excavation & Fill>Backfill. Drag the **Selector** for **Gravel** to the My Detail tool palette. Select **Close** to close the **Detail Component Manager**. Select the **Gravel** tool from the My Detail tool palette. Verify Description = **Gravel** in the Properties palette and place the gravel as shown in the following command prompts.

    Command: _DTLCOMPTOOL

    Select objects to form the backfill boundary: 1 found *(Select the rectangle at p3 as shown in Figure 11T.29.)*

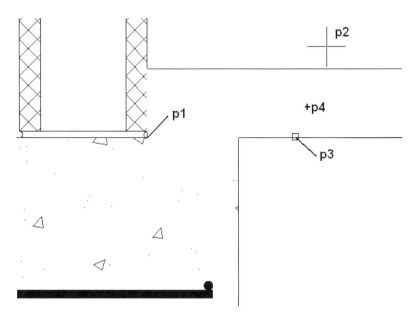

Figure 11T.29 *Creating the boundary for the gravel*

Select objects to form the backfill boundary: ENTER

Select a point within a boundary to backfill: *(Select a point near p4 inside the rectangle as shown in Figure 11T.29. The border of the rectangle will display red prior to selecting the interior point.)*

Select objects to form the backfill boundary: ENTER *(Press ENTER to end command.)*

16. Select the **18 x 12 Concrete Footing** tool of the My Detail tool palette, right-click, and choose **Detail Component Manager**. Expand **Division 03-Concrete>03300 Cast in Place Concrete>03310 Structural Concrete>Slabs with Optional Haunch** of the **Detail Component Manager**. Choose the **Selector** of the **4" Slab with Haunch** and drag to the My Detail tool palette. Select **Close** to close the **Detail Component Manager**.

17. Select the **4" Slab with Haunch** tool, right-click, and choose **Properties** as shown in Figure 11T.30. Edit the **Tool Properties** as follows: Component: Category = **03310 Structural Concrete**, Type = **Slabs with Optional Haunch**, Description = **4" Slab with Haunch**, View = **Section**; Specifications: Use custom size = **No**, Hatch beam = **Yes**, Show reinforcing = **No**, and Draw haunch = **None**. Select **OK** to dismiss the **Tool Properties** dialog box.

18. Select the **4" Slab with Haunch** tool from the My Detail tool palette and respond to the command line prompts as shown below to place the slab.

    Command: _DTLCOMPTOOL

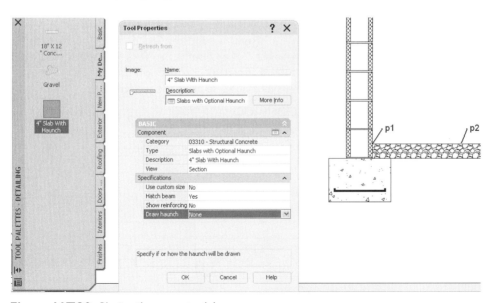

Figure 11T.30 *Placing the concrete slab*

Start point or [Xflip/Yflip]: *(Select a point at p1 as shown in Figure 11T.30.)*

End point or [Xflip/Yflip]: **5'** *(Move the cursor right polar tracking tip >0, then type 5'.)*

Start point or [Xflip/Yflip]: ENTER *(Press ENTER to end the command.)*

19. Right-click the **4" Slab with Haunch** tool of the My Detail tool palette and choose **Detail Component Manager**. Expand Division 03-Concrete>03200 Concrete Reinforcement>03320 Welded Wire Fabric>Wiremesh Reinforcing. Choose **6x6 x W1.4 x W1.4 Welded Wire Mesh** and drag this tool to My Detail tool palette. Select **Close** to close the **Detail Component Manager**.

20. Toggle OFF Osnap. Select **6x6 x W1.4 x W1.4 Welded Wire Mesh** from the My Detail tool palette. Respond to the command line prompt shown below to place the welded wire fabric in the slab.

    Command:

    DTLCOMPTOOL

    Start point: *(Select a point at p1, move the cursor right, SHIFT + right-click, choose **Perpendicular** and select a point near p2 as shown in Figure 11T.31.)*

    Start point: ENTER *(Press ENTER to end the command.)*

21. Toggle ON Osnap, and select the Basic tool palette of the Detailing palette set. Right-click **07-Thermal and Moisture Protection** and choose **Properties**. As shown in Figure 11T.32, edit the **Tool Properties** as follows: Component: Category = **07210-Building Insulation**, Type = **Rigid Insulation**; Description = **1" Rigid Insulation**; Specifications: Hatch item = **Yes**. Verify View=Section. Select **OK** to dismiss the **Tool Properties** dialog box. Select the **Thermal and Moisture Protection** tool and respond to the command line prompts as follows to place the insulation.

    Command: _DTLCOMPTOOL

    Start point or [Xflip/Yflip/Left/Center/Right]: *(Select point p1 shown in Figure 11T.32 using the endpoint object snap.)*

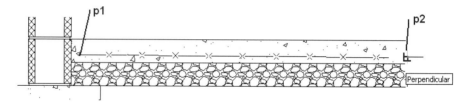

Figure 11T.31 *Placing the welded wire fabric*

End point or [Xflip/Yflip/Left/Center/Right]: *(Select point p2 shown in Figure 11T.32 using the endpoint object snap.)*

Start point or [Xflip/Yflip/Left/Center/Right]: *(Select point p2 as shown in Figure 11T.32 using the endpoint object snap.)*

End point or [Xflip/Yflip/Left/Center/Right]: *(Move the cursor to the right, type* **24** ENTER.*)*

Start point or [Xflip/Yflip/Left/Center/Right]: ENTER *(Press* ENTER *to end the command.)*

22. Select the 4" concrete slab, right-click, and select **Edit Block in-place**. Verify that **Prompt to select nested objects** is checked. Select **OK** to dismiss the **Reference Edit** dialog box. Select the concrete hatch material to open the **Refedit** toolbar. Select the concrete hatch material, right-click, choose **Hatch Edit**, and change the hatch scale to **0.5**. Select **OK** to dismiss the **Hatch Edit** dialog box. Select **Save back changes to reference** button on the **Refedit** toolbar. Select **OK** to dismiss the AutoCAD dialog box and save reference changes.

23. Select the concrete slab, right-click, and choose **Aec Modify Tools>Obscure**. Respond to the command line prompts as follows:

    Command: LineworkObscure

    Select obscuring object(s): 1 found *(Select the rigid insulation at p1 in Figure 11T.33.)*

    Select obscuring object(s): 1 found, 2 total *(Select the rigid insulation at p2 in Figure 11T.33.)*

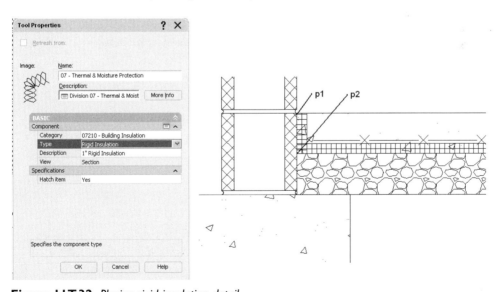

Figure 11T.32 *Placing rigid insulation detail*

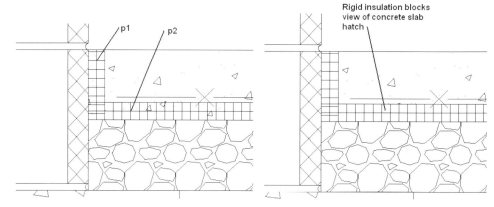

Figure 11T.33 *Obscuring concrete hatch*

> Select obscuring object(s): ENTER *(Press* ENTER *to end the command; concrete hatch is removed from the background of the rigid insulation.)*

24. Select **3/4"=1'-0"** from the **Scale** flyout of the Drawing Window status bar.
25. Right-click over the title bar of the tool palette, and select the **Document** palette set. Choose the Annotation tool palette and select **Reference Keynote (Straight Leader)**.

> Command: AecDtlAnnoLeaderAdd
>
> Select object to keynote or ENTER to select keynote manually: *(Select the 8" CMU at p1 as shown in Figure 11T.34.)*
>
> Select first point of leader: *(Select a point near p1 as shown in Figure 11T.34.)*
>
> Specify next point of leader line: *(Select a point near p2 as shown in Figure 11T.34.)*
>
> Specify next point of leader line: ENTER *(Press* ENTER *to end the leader.)*
>
> Select text width<0">: ENTER *(Press ENTER to accept default text width.)*
>
> *(Keynote information is placed as shown in Figure 11T.34.)*

26. Repeat the **Reference Keynote (Straight Leader)** command to place additional reference keynotes as shown in Figure 11T.34.
27. Select **Reference Keynote Legend** from the Annotation tool palette. Respond to the command line as shown below create the legend.

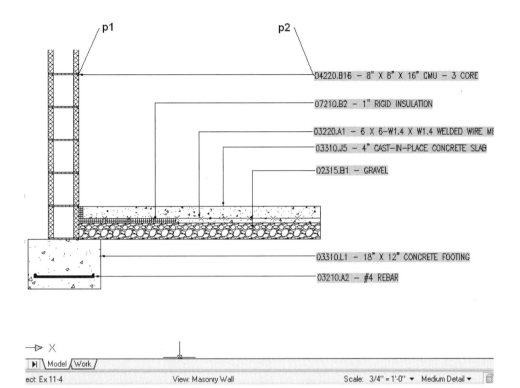

Figure 11T.34 *Reference keynotes placed*

Command: AecKeynoteLegendAdd

Select keynotes to include in the keynote legend or [Sheets/from Database]: *(Select a point at p1 as shown in Figure 11T.35.)*

Specify opposite corner: 24 found *(Select a point at p2 as shown in Figure 11T.35.)*

17 were filtered out

Select keynotes to include in the keynote legend or [Sheets/from Database]: ENTER

Insertion point of table: *(Select a point near p3 as shown in Figure 11T.35.)*

28. Save and close the drawing.
29. Select **File>Project Browser**, select **Ex 11-4**, and close the project.

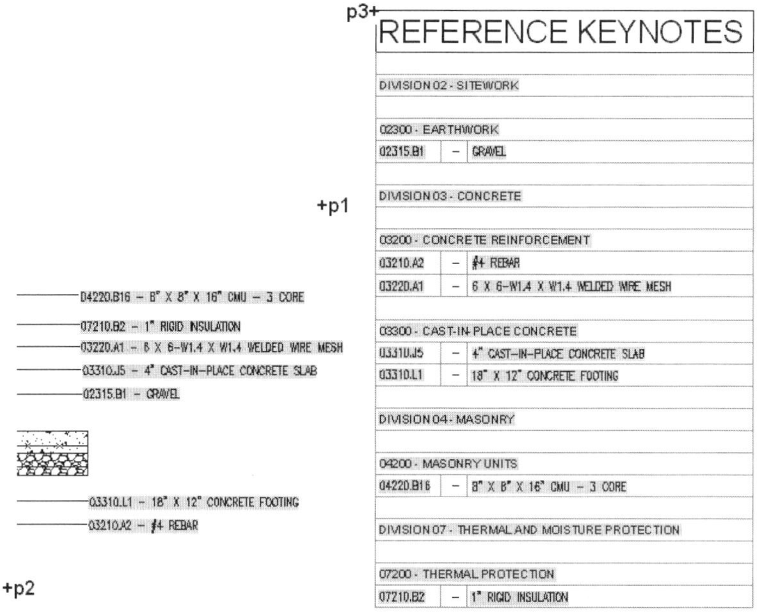

Figure 11T.35 *Placing a reference keynote legend*

PROJECT

Ex 11.5 Creating Projects, Elevations, and Sections

1. Create a project numbered Ex 11-5 and named Accessing Residence 2 in your *ADT Student\ADT Tutor\Ch 11* folder. Select **Yes** to Prefix the file name with the project number.

2. Open the **Project Navigator**, and create a division Name = 1, ID = 1, and Description = 1. Create two levels as follows:

Name	Floor Elevation	Floor to Floor Height	ID	Description
1	0	4'-11 3/4"	1	Crawl Space
2	4'-11 3/4"	8'	2	First Floor

Table 11T.3 *Elevations of levels in the project*

3. Open the *Ex 11-5* drawing from your *ADT Student\ADT Tutor\Ch 11* folder and convert it to a Construct named Crawl. Crawl will be assigned to level 1.

Open the *Ex 11-6* drawing from your *ADT Student\ADT Tutor\Ch 11* folder, convert it to a Construct named Floor 1, and assign Floor 1 to level 2.

4. Create a new General view named Model that includes level 1 and level 2.

5. Create an elevation named Front in the Model drawing. Create a napkin sketch from the Front elevation. Name the napkin sketch Front-Presentation. Set the following properties of the Napkin Sketch: Line Format = Loose, extend sketch lines at corners and Intended Plot Scale = 1/4"=1'-0".

6. Create a transverse section in the Model drawing.

7. Create a sheet named Front. Select the **Views** tab, expand the Model drawing, and drag the Elevation onto the Front sheet. Insert the Front elevation drawing at 1/4"=1'-0" scale.

8. Create a live section from the transverse section line as shown in Figure 11T.36.

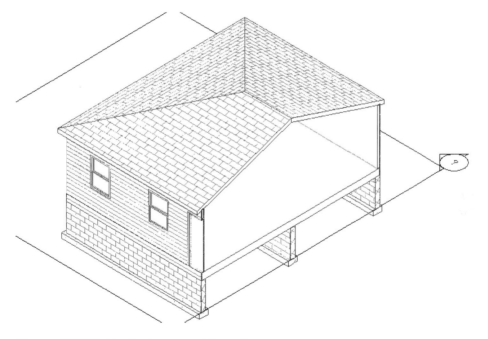

Figure 11T.36 *Live Section created for project*

CHAPTER 12

Creating Mass Models, Spaces, and Boundaries

INTRODUCTION

Designers create physical models to refine their design and to communicate to clients the form or shape of the proposed building. Computer models can serve the same purpose as physical models, except that the computer model is easily edited to meet the intent of the designer. Computer models also allow the designer to perform space planning as the model is developed. Architectural Desktop includes the tools for creating mass models of building components or of the building. Computer models are created by the joining together of a combination of mass elements in the form of twelve possible prism shapes, extrusions of profiles, drapes, or revolutions of profiles. The mass elements can be joined together through Boolean operations of addition, subtraction, or intersection. The mass elements can also be used to simulate such building components as columns, floors, or fireplaces in other drawings. Once a mass model is created, you can slice the model to create polylines that can be used for the development of floor plans. Polylines can be converted to spaces and space boundaries. Space objects include size restrictions, ceilings, and floors to represent the space. Spaces can be enhanced by adding space boundaries to simulate walls. Doors and windows can be placed in space boundaries to separate different spaces.

Additional material regarding Spaces and Boundaries is included in the Chapter 12 Supplement located on the CD. The material included on the CD also includes additional Chapter 12 tutorials.

OBJECTIVES

Upon completion of this chapter, you will be able to

- Create mass elements using the **MassElementAdd** command
- Edit the size of a mass element using its grips and the Properties palette
- Extrude or revolve an AEC Profile and create a Drape mass element

- Use the Model Explorer to create and edit mass elements and groups
- Create and modify a mass element style
- Create and modify a space style to include ceiling and floor components
- Create a space from a polyline using the **Apply Tool Properties to>Polyline** tool of the **Space** tool
- Divide and combine spaces using the **LineworkDivide** and **LineworkMerge** commands
- Create space boundaries with specific height, width, and justification properties using the **Add Edges** command (**SpacePBoundaryAdd**)
- Combine independent boundaries with the **Merge Boundaries** (**SpaceBoundaryMerge**) command
- Attach a space to a boundary using the **Attach Spaces to Boundary** command (**SpaceBoundaryMergeSpace**) and detach a space from a boundary using the **Split Boundary** command (**SpaceBoundarySplit**)
- Convert spaces, slices, and linework to space boundaries using the **Apply Tool Properties to>Slice**, **Apply Tool Properties to>Space**, and **Apply Tool Properties to>Edges** commands
- Insert space boundaries in an existing space boundary using the **Add Edges** command (**SpaceBoundaryAddEdges**)
- Modify each space boundary segment using the **Edit Edges** command (**SpaceBoundaryEdge**)
- Delete a space boundary segement using the **Remove Edges** command (**SpaceBoundaryRemoveEdges**)
- Release or attach objects to the anchor of the space boundary using the **Anchor to Boundary** command (**SpaceBoundaryAnchor**)
- Summarize the spaces of the drawing with the **SpaceQuery** command
- Generate walls from space boundaries using the **Generate Walls** command (**SpaceBoundaryGenerateWalls**)

CREATING MASS MODELS

Mass models are created from mass elements. The elements can be joined together in a group. The group operations allow you to add, subtract, and intersect the mass of an element to create complex mass models. The tools for creating mass elements are located on the Massing tool palette shown in Figure 12.1. When a mass element tool is selected, the Properties palette opens, which displays the dimensions and other properties of the element. The **Shape** drop-down list of the Properties palette includes the following shapes: Arch, Barrel Vault, Box, Cone, Cylinder, Dome, Doric, Gable, Pyramid, Sphere, Isosceles Triangle, Right Triangle, Extrusion, Revolution, and Free Form. Custom mass elements can also be created as extrusions or revolutions from a closed polyline. The prisms of the **Shape** list are shown in Figure 12.3. You can select a shape from the drop-down list of the Properties palette or select the tool from the Massing tool palette. Mass elements are created on the **A-Area-Mass** layer, which has color 70.

INSERTING MASS ELEMENTS

The dimensions of the mass element are defined in the Properties palette. The **MassElementAdd** command (refer to Table 12.1) is used to add the mass element tools shown on the Massing tool palette.

Command prompt	MASSELEMENTADD
Palette	Select any of the following from the Massing tool palette: Arch, Barrel Vault, Box, Pyramid, Isosceles Triangle, Right Triangle, Cone, Cylinder, Dome, Sphere, Gable, Drape, Extrusion, and Revolution

Table 12.1 *MassElementAdd command access*

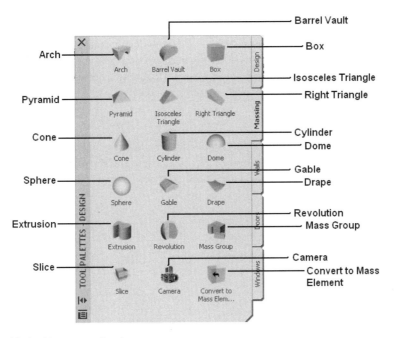

Figure 12.1 *Massing tool palette*

When a mass element tool is selected, the Properties palette opens, allowing you to define the dimensions of the element. The options of the Properties palette vary according to the mass element you are adding. The options of the Properties palette for a mass element, shown in Figure 12.2, are described below.

General

Description – The **Description** button opens the **Description** dialog box, which allows you to add a description of the mass element.

Layer – The **Layer** edit field displays the name of the layer of existing mass elements. Mass elements are placed on the **A-Area-Mass** layer if the AIA 256 Color layer key style is used.

Style – The **Style** edit fields lists the element styles included in the drawing. Element styles allow you to attach materials to the element.

Shape – The **Shape** list includes each of the following prisms: Arch, Barrel Vault, Box, Cone, Cylinder, Dome, Doric, Gable, Pyramid, Sphere, Isosceles Triangle, Right Triangle, Extrusion, Revolution, and Free Form.

Attached to – The **Attached to** drop-down list displays the mass groups. Mass group markers can be placed in the drawing. If an element is attached to a group, the add, subtract, and intersect Boolean operations can be applied to members of the group. Once a group name is defined, this edit box allows you to set a group name current and then assign mass elements to the group.

Operation – The **Operation** drop-down list is active if a group is defined for the mass element. The **Operation** edit field has the options of **Add**, **Subtract**, and **Intersect**. Once a group has been created, mass elements can be attached and defined to add, subtract, or intersect their mass with other mass elements. If the operation is set to **Subtract**, the mass element being inserted will subtract its shape from the existing mass elements of the group. The default setting for operations is **Add**; therefore a new mass element would be added to previous mass elements of the group.

Profile – The **Profile** option allows you to select predefined AEC Profiles for the Extrusion or Revolution to create a shape. The **Profile** option is displayed only when extrusion and revolution tools are selected.

Dimensions

Width – The **Width** edit field allows you to insert the width dimension of the mass element.

Depth – The **Depth** edit field allows you to insert the depth dimension of the mass element.

Height – The **Height** edit field allows you to insert the height dimension of the mass element.

Radius – The **Radius** edit field is active if the mass element includes a radial shape. The radius of the radial shape is defined in this field.

Rise – The **Rise** edit field allows you to specify the rise of the Gable mass element.

Location

Rotation – The **Rotation** edit field displays the angle of rotation from the 3:00 o'clock position.

Elevation – The **Elevation** edit field lists the vertical distance from Z=0.

Add Information – The **Add Information** button opens the **Location** dialog box, which allows you to specify the insertion point, normal, and rotation data.

SELECTING MASS ELEMENT SHAPES

The **Shape** drop-down list of the Properties palette includes twelve mass elements, and extrusion and rotation options. Each shape has a specified insertion point or handle in which the mass element is inserted in the drawing.

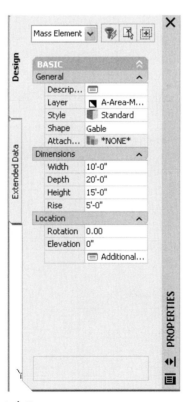

Figure 12.2 *The Properties palette*

Arch

Define the Arch mass element by setting the width, depth, height, and radius. It consists of a box prism, which has the semi-circular prism subtracted from the box. The radius dimension of the Arch determines the size of the semi-circular shape. The center for the Arch is located in the bottom plane. The grips and the dimensions of the Arch are shown in Figure 12.3. There are points at the center of the bottom and top planes that can be snapped to with the Node object snap. The point located in the center of the bottom plane is used by default as the insertion point and the Location grip. If other mass elements are placed on top of the Arch, the top node can be snapped to with the Node object snap to align the mass elements together. The Radius grip shown in Figure 12.3 can be used to edit the radius of the arch. The other grips, when edited, will expand the prism symmetrically, allowing the arch to remain centered in the prism as the prism dimensions change.

Barrel Vault

The Barrel Vault is a prism consisting of a semi-circle that has been extruded. Setting the width and radial dimensions creates a prism that is one-half of a cylinder. The grips of the Barrel Vault are shown in Figure 12.3. The insertion point for placing the Barrel Vault is located at the point centered on the bottom plane. If you edit the grip for the radial dimension, the radius of the semi-circle will change. To edit the width and radius, select any of the four grips located at the corners of the base, and the Barrel Vault will stretch in the direction the mouse is moved. Select the triangular grip located at the midpoint of the sides to change the radius, and select the triangular grip located at each end to edit the width.

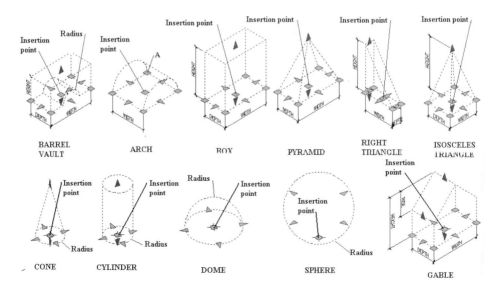

Figure 12.3 *Mass element prisms*

Box

A Box prism is a shape frequently used in buildings, and its simplicity makes the union of multiple boxes easy to visualize. Width, depth, and height are the dimensions controlled in forming various sizes of rectangular prisms. The grip points for the Box are shown in Figure 12.3. The insertion point for the Box is located in the center of the bottom plane. To edit the depth or width dimensions of the Box, select the grips located on the corners and stretch the Box. Edit the height of the Box by selecting the grip located at the top of the Box. You can snap to the grips at the top and bottom of the Box with the Node object snap if other mass elements are stacked together.

Cone

Create the Cone prism by establishing the radius and height dimensions. Cones are seldom used in building designs except for roof forms. The insertion point for the cone is located at its center on the bottom plane. The grips of a Cone are shown in Figure 12.3. When locating other mass elements relative to the Cone, you can use the Node object snap to select the point at the center of the bottom plane of the Cone. The Center object snap mode can be used to locate the center of the cone if the cone is viewed in isometric.

Cylinder

Design the Cylinder prism by establishing the radius and height. When you insert the Cylinder, the insertion point is located at the center of the bottom plane. The grips of the Cylinder are shown in Figure 12.3.

Dome

The Dome mass element is one-half a sphere. Only the radius dimension is set to create the size of the Dome. The point located in the center of the bottom plane is used to place the insertion point of the Dome. The four grips at the quadrant points of the Dome, shown in Figure 12.3, can be used to increase the radius.

Gable

The Gable shape is frequently used for residential and commercial buildings because it includes the Box shape and the gable roof. Create the Gable mass element by establishing the depth, width, height, and rise as shown in Figure 12.3. When designing a Gable, you need to determine the distance from the top ridge down to the cornice or rise. The grips on the base of the Gable allow you to stretch the depth and width dimensions of the building. Gable mass elements are inserted in the drawing with the insertion point centered on the bottom of the base plane. The height is edited by the grip located at the top of the ridge of the gable, while the rise can be edited with the grip at the cornice as shown in Figure 12.3.

Pyramid

Create the Pyramid mass element by setting the depth, height, and width dimensions. The insertion point for the Pyramid is located on the center of the bottom plane of the Pyramid. The grips for the Pyramid are shown in Figure 12.3. Editing the grips located on the corners of the Pyramid mass element will change the depth and width dimensions. The grip at the apex can be used to edit the height, while the base will remain stationary. The bottom central grip can be used to move the Pyramid rather than edit its dimensions. Pyramid mass elements are usually used for roof elements, which necessitates that the insertion point be moved up in the Z direction to the height of the base of the roof plane.

Isosceles Triangle

Create the Isosceles Triangle mass element by setting the width, depth, and height dimensions. This mass element is also a shape used for roofs, and therefore, when it is used as a roof, its insertion point should be established at the Z elevation of the bottom of the roof plane. The grips and the dimensions of the Isosceles Triangle are shown in Figure 12.3. The grips of the Isosceles Triangle edit the width, depth, and height of the mass element similarly to the Pyramid. The insertion point for the Isosceles Triangle is located in the center of the bottom plane.

Right Triangle

Establishing the height and width of the mass element sets the dimensions of the legs of the Right Triangle. Right Triangles can be used as mass elements to form roofs or walls. The grips and dimensions of the Right Triangle are shown in Figure 12.3. The insertion point for this mass element is located in the center of the bottom plane.

Sphere

The Sphere has one dimension: its radius. The grips are located at the four quadrants of the Sphere parallel to the XY plane, as shown in Figure 12.3. Editing these grips will cause the radius of the Sphere to change. The grip located at the bottom of the Sphere is at the zero Z axis location and can be used move the sphere. The insertion point for the Sphere is located on the bottom of the sphere. Although a Sphere is seldom used as a wall or roof element, it can be used to edit other mass elements or for ornament.

Doric Column

The Doric Column is not a mass element on the Massing tool palette. However, it can be selected from the **Shape** drop-down list of the Properties palette. To create a Doric column, select any mass element from the Massing palette and choose the Doric shape from the drop-down list in the Properties palette prior to specifying an insertion point. Set the height and radius in the Properties palette or toggle **Specify on Screen** to **Yes** and specify the dimensions in the workspace. The radius dimension is set at the top of the column. The taper of the column is fixed and increases from the radius dimension set for the top of the column. Other columns are available in the Structural Catalog. The Doric Column is not extruded as a profile. The Doric Column is shown in Figure 12.4.

Revolution and Extrusions

Revolution and Extrusions are methods in the **Shape** list in which mass elements are created from profiles. Any closed polyline can be converted to an AEC Profile and extruded or revolved to create a mass element.

STEPS TO INSERTING A MASS ELEMENT (GABLE)

1. Select **SW Isometric** from the **View** flyout of the **Navigation** toolbar.
2. Select the **Gable** mass element of the Massing tool palette.
3. Move the pointer to the workspace; the Gable mass element is phantom displayed with the pointer. Dimensions of the mass element can be specified on screen or in the Properties palette.
4. Edit the Properties palette as follows: Dimensions-Specify on screen = **No**, Width = **10'**, Depth = **20'**, Height = **15'**, and Rise = **5'**.
5. Move the pointer to the workspace, and specify the insert point and rotation as shown in the following command line sequence.

    Command: MassElementAdd

Figure 12.4 *Doric Column*

Insert point or [SHape/WIdth/Depth/Height/RIse/Match]: *(Edit Width in Properties palette.)*

Insert point or [SHape/WIdth/Depth/Height/RIse/Match]: *(Edit Depth in Properties palette.)*

Insert point or [SHape/WIdth/Depth/Height/RIse/Match]: *(Edit Height in Properties palette.)*

Insert point or [SHape/WIdth/Depth/Height/RIse/Match]: *(Edit Rise in Properties palette.)*

Insert point or [SHape/WIdth/Depth/Height/RIse/Match]: *(Select a point near p1 in the workspace as shown in Figure 12.5.)*

Rotation or [SHape/WIdth/Depth/Height/RIse/Match/Undo] <0.00>: ENTER *(Press* ENTER *to accept 0° rotation.)*

Insert point or [SHape/WIdth/Depth/Height/RIse/Match]: ENTER *(Press* ENTER *to end the command.)*

The top view of the mass element is hatched as shown in Figure 12.6. Therefore place mass elements in a pictorial view, because the hatch is not displayed in model view.

CREATING MASS ELEMENTS USING EXTRUSION AND REVOLUTION

In addition to the twelve mass elements prisms, you can also create a mass element from an extrusion or revolution of a profile. The **ExtrudeLinework** command can be

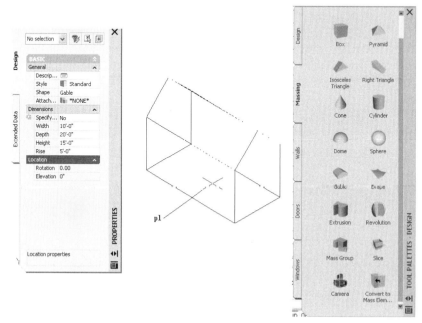

Figure 12.5 *Placing the gable mass element*

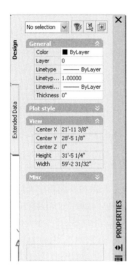  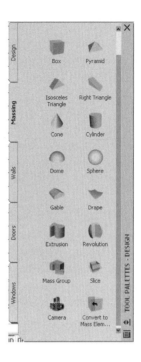

Figure 12.6 *Plan view of gable mass element*

used to create an extrusion from a closed polyline. The outline of a building component or the entire building can be drawn using a polyline. Select a polyline, right-click, and choose **Convert to>Mass element** (**ExtrudeLinework**). The height of the extrusion is specified in the command line as shown below.

*(Select a closed polyline, right-click, and choose **Convert to>Mass Element**.)*

Command: ExtrudeLinework

Erase selected linework? [Yes/No] <No>: **y** *(Specify Yes to erase polyline.)*

Specify extrusion height <20'-0">: *(Specify height of mass element.)*

(1) new mass element(s) created.

Command prompt	EXTRUDELINEWORK
Shortcut	Select a closed polyline, right-click, and choose Convert to>Mass Element

Table 12.2 *ExtrudeLinework command access*

Extrusions and Revolutions can be created by selecting the **Extrusion** or **Revolution** tool from the Massing tool palette, and then specifying height and or radius and the name of a profile in the Properties palette. Profiles are created in the **Style Manager** as presented in Chapter 3. The revolution or extrusion is placed by specifying the insert point and rotation.

 STOP. Do Tutorial 12.1, "Creating Mass Element Components for a Fireplace," at the end of the chapter.

CREATING MASS ELEMENTS USING DRAPE

The Drape mass element is used to create irregular 3D surfaces that drape polylines. The mesh density specified in the command specifies the number of faces created to define the surface. The polylines create the elevation points for the mass elements. The **Drape** command is very useful in creating simulation of land contours. The polylines used in the **Drape** command should vary in the Z coordinate direction according to the elevation of the land. The right viewport shown in Figure 12.7 is a right side view of the contour lines, and the left viewport is the top view. Access the **Drape** tool from the Massing tool palette as shown in Table 12.3.

Command prompt	DRAPE
Palette	Select Drape from the Massing tool palette

Table 12.3 *Drape command access*

The **Drape** command was applied to the polylines shown in Figure 12.7 and the following command line sequence.

> Command: _AecDrape
>
> Select objects representing contours: *(Select the contour lines using a crossing window, select p1 as shown in Figure 12.7.)*
>
> Specify opposite corner: *(Select a point at p2 as shown in Figure 12.7.)*
> 13 found
>
> Select objects representing contours: ENTER *(Press ENTER to end the command.)*
>
> Select mesh corner: *(Select a point near p3 as shown in Figure 12.7)*
>
> Select opposite mesh corner: *(Select a point near p4 as shown in Figure 12.7.)*
>
> Enter mesh size <30>: ENTER *(Press ENTER to accept the default mesh size.)*

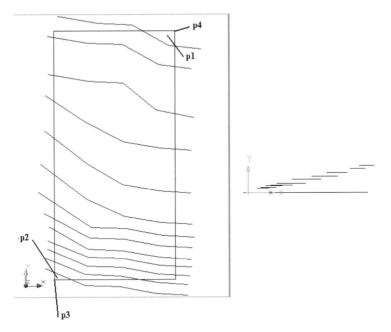

Figure 12.7 *Selection of points for Drape*

Enter base thickness<6">: **6** ENTER *(Type **6** to specify the base thickness.)*

*(Select the Model tab, and select **SW Isometric** from the **View** flyout of the **Navigation** toolbar. Select **Flat Shaded** from the **Shade** flyout of the **Navigation** toolbar to view the drape as shown in Figure 12.8.)*

MODIFYING MASS ELEMENTS

To modify mass elements, you can edit the grips of the mass element or edit the dimensions in the Properties palette. The functions of the grips of the Gable mass element are shown in Figure 12.9. A mass element can be moved, mirrored, rotated, scaled, or stretched by its grips. Using the stretch operation with grips allows you to modify mass elements to fit within existing mass elements. When the grips of the mass element shown in Figure 12.9 are edited, the mass element retains its gable shape; only the dimensions of the mass element change. When you insert a mass element, it is inserted at Z=0; however, the height grip located on the base can be selected and the elevation of the base changed. Selecting a wedge-shaped grip changes one dimension of the mass element, whereas the square grips change two dimensions or the location.

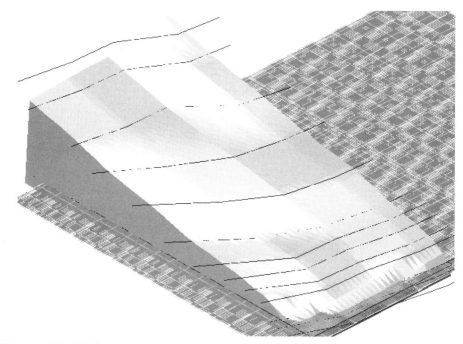

Figure 12.8 SW Isometric view of drape

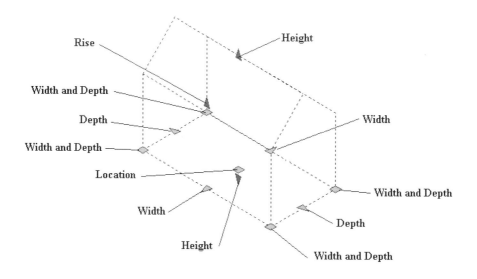

Figure 12.9 Grips of the gable mass element

The grips of mass elements allow you to directly manipulate the dimensions as shown in Figure 12.10. The width of the Gable mass element is resized in Figure 12.10 to 18'.

TRIMMING MASS ELEMENTS

Mass elements can be trimmed to create custom mass elements. The command **Trim by Plane** (**MassElementTrim**) creates a trim plane that cuts across the element. You can then select which side of the trim plane to retain. Access the **Trim by Plane** command from the shortcut menu of the mass element as shown in Table 12.4.

Command prompt	MASSELEMENTTRIM
Shortcut menu	Select a mass element, right-click, and choose Trim by Plane

Table 12.4 *Mass Element Trim by Plane command access*

The **Trim by Plane** command is used in the following command line sequence to trim the Doric column.

> Command: MassElementTrim
>
> Specify trim plane start point or [3points]: *(Select p1 using Endpoint object snap as shown in Figure 12.11.)*

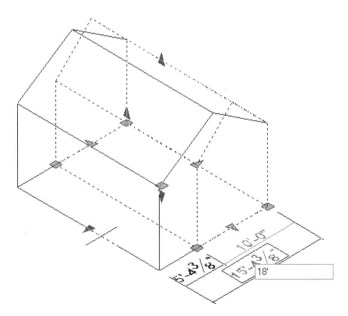

Figure 12.10 *Direct manipulation and grip editing*

Specify trim plane end point: *(Select p2 using Endpoint object snap as shown in Figure 12.11.)*

Select mass element side to remove: *(Select p3 using Endpoint object snap as shown in Figure 12.11.)*

The plane defined to trim the mass element is extruded from a line specified by two points or from a plane defined by three points. After you specify the first point, the trim plane is dynamically shown red as you move your pointer to create an additional point that crosses the mass element and defines a line on the plane. The points selected for the trim plane can be located on or off the mass element.

SPLITTING MASS ELEMENTS

Mass elements can be divided by a cutting plane to create two mass elements. The **Split by Plane** command (**MassElementDivide**) creates a cutting plane that cuts the mass element. Access the **Split by Plane** command as shown in Table 12.5.

Command prompt	MASSELEMENTDIVIDE
Shortcut menu	Select the mass element, right-click, and choose Split by Plane

Table 12.5 *Command access for the Split by Plane command*

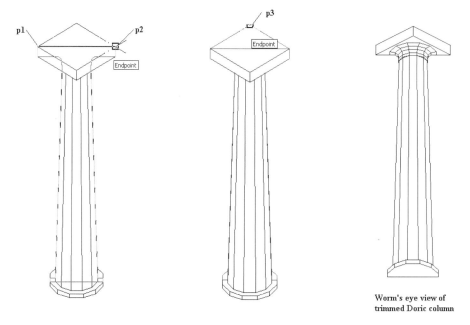

Figure 12.11 *Trimming a mass element*

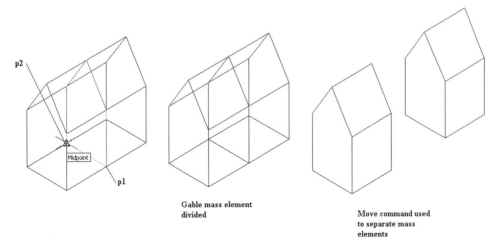

Gable mass element divided

Move command used to separate mass elements

Figure 12.12 Splitting a mass element

The **Split by Plane** command is used in the following command line sequence to divide the Gable mass element. The **Move** command can be used after the split to separate the mass elements.

> Command: MassElementDivide
>
> Specify divide plane start point or [3points]: _mid of *(Select point p1 using the Midpoint object snap as shown in Figure 12.12.)*
>
> Specify divide plane end point: _mid of *(Select point p2 using the Midpoint object snap as shown in Figure 12.12.)*

EDITING FACES

The mass element can be modified by editing its faces. When you split one of the faces of the mass element, you can stretch a face to a new position. The **Split Face** (**MassElementFaceDivide**) command (refer to Table 12.6) creates a free form mass element; therefore the split face and the remaining faces of the mass element can be edited. After you split the face of the mass element, additional grips are displayed on each face. Select the grips and drag the face to a new position as shown in Figure 12.13.

Command prompt	MASSELEMENTFACEDIVIDE
Shortcut menu	Select a mass element, right-click, and choose Split Face

Table 12.6 Split Face command access

The **Split Face** command was applied to the gable shown in Figure 12.13 in the following command line sequence.

> *(Select the mass element, right-click, and choose **Split Face**.)*
>
> Command: MassElementFaceDivide
>
> Select first point on face: _mid of *(Select point p1 using the Midpoint object snap as shown in Figure 12.13.)*
>
> Select second point on the same face: _mid of *(Select point p2 using the Midpoint object snap as shown in Figure 12.13.)*
>
> Gable converted to Free Form.
>
> Select first point on face: ENTER *(Press ENTER to end the command.)*

After you split the face, the mass element is in free form, and therefore you can stretch any face. The grip shown at **p3** was stretched using the **Move orthogonally** option of the grip to create the shape shown in the middle of Figure 12.13. The figure shown at far right was created by again selecting the grip shown at **p3** and pressing CTRL three times to toggle to the **Pull Face orthogonally** option and stretch the face. After a face is split, the face can be rejoined, if **it remains coplanar**, by selecting the **Join Faces** (**MassElementFaceJoin**) command from the shortcut menu of a selected mass element.

> *(Select a coplanar face, right-click, and choose **Join Faces**.)*
>
> Command: MassElementFaceJoin
>
> Select edges of coplanar faces: 1 found *(Select dividing face line at p1 as shown in Figure 12.14.)*
>
> Select edges of coplanar faces: ENTER *(Press ENTER to end selection.)*
>
> Free Form converted to Gable.

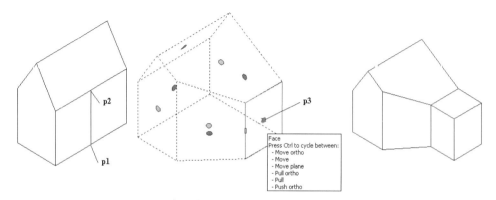

Figure 12.13 *Split Face command applied to Gable mass element*

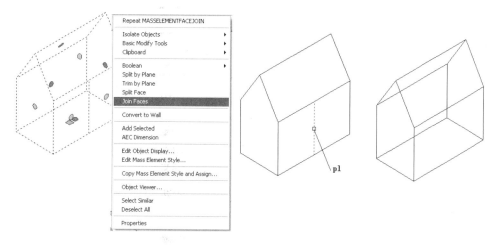

Figure 12.14 *Faces joined with Join Faces*

CREATING STYLE DEFINITIONS FOR MASS ELEMENTS

A style can be created for the mass element. A benefit of creating a mass element style is that you can assign materials to the style. Therefore mass elements assigned to the style will be displayed with materials representation. Masonry fireplaces and masonry wall projections can be modeled with mass elements and assigned brick or concrete masonry unit materials. Mass element styles are created and edited in the Multi-Purpose Objects\Mass Element Styles folder of the drawing in the **Style Manager**. To create a new style, open the **Style Manager**, expand the Multi-Purpose Objects\Mass Element Styles folder, select **New** from the **Style Manager** toolbar, and overtype the name of the new style in the right pane. Select the new style, right-click, and choose **Edit** to open the **Mass Element Style Properties** dialog box as shown in Figure 12.15. The content of the tabs of the **Mass Element Style Properties** dialog box defines the style. The options of the **Mass Element Style Properties** dialog box are described below.

General Tab

The **General** tab allows you to edit the name and add a description to the style.

Materials Tab

The **Materials** tab lists the body component and the assigned material definition. Material definitions must be imported into the drawing from the other drawings or the *C:\Documents and Settings\All Users\Application Data\Autodesk\ADT 2005\Styles\Imperial* directory prior to assigning materials to the mass element style. Click in the **Material Definitions** column to assign materials to the mass element. The materials

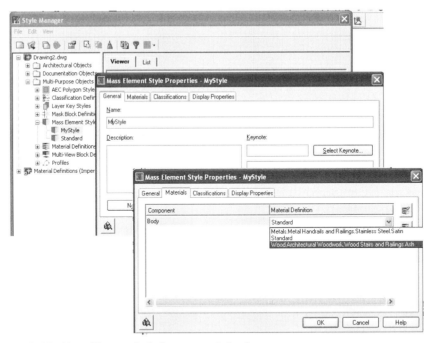

Figure 12.15 *Mass Element Style Properties dialog box*

listed in the drop-down list shown in Figure 12.15 include the material definitions imported into the drawing.

Classifications Tab

The **Classifications** tab includes a list of classification definitions included in the drawing. Figure 12.16 does not include any classifications because no classifications have been defined for the drawing.

Display Properties Tab

The **Display Properties** tab, shown in Figure 12.16, lists the display representations and display property sources of the drawing. Mass elements are simple objects and consist only of a limited number of components. The hatch component of the mass element can be turned off in the **Layer/Color/Linetype** tab of the **Display Properties** dialog box for the Plan Presentation display representation.

 STOP. Do Tutorial 12.2, "Creating Mass Elements for a Terrain," at the end of the chapter.

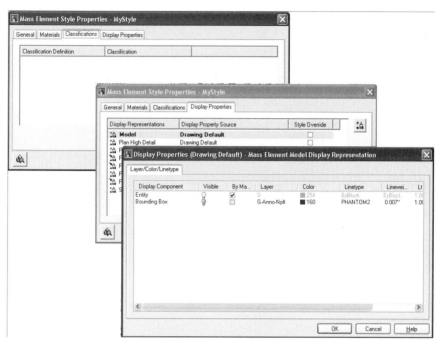

Figure 12.16 *Classifications and Display Properties tabs*

CREATING COMPLEX MODELS

Most mass models usually include several mass elements that create the model when combined. Mass elements are combined when attached to a mass group using Boolean operations. The purpose of each mass element inserted is to add, subtract, or intersect with other mass elements. The mass elements carve out or project from a base mass element to create the model. Each element of the model is created on the XY plane and extruded up in the positive Z direction. The model shown in Figure 12.17 consists of two Gable mass elements, a Box element, and a Barrel Vault. The Barrel Vault mass element was moved up in the positive Z direction to create the model. The Box mass element was subtracted from the remaining elements of the model.

CREATING GROUPS FOR MASS ELEMENTS

Complex models are created by defining a group and attaching elements to the group. The **Mass Group** tool of the Massing tool palette (refer to Table 12.7) allows you to create a group. The **Mass Group** tool inserts a group marker, which represents the group. Mass elements are attached to the group when the group marker is established.

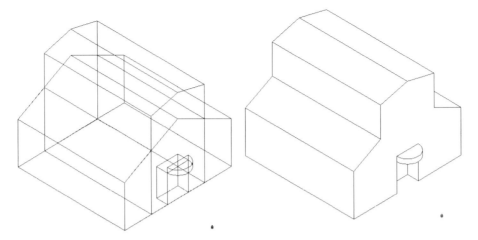

Figure 12.17 *Mass element and mass groups used to create mass model*

Command	MASSGROUPADD
Tool palette	Select Mass Group from the Massing tool palette

Table 12.7 *Command access table for MassGroup*

When you select the **Mass Group** tool, you are prompted to select elements to attach and a location as shown in the following command sequence. Grips of the mass group marker are displayed when an element is selected. A name for the mass group is assigned when the marker is placed. The name will be listed in the Attached to drop-down list in the Properties palette.

>*(Select **Mass Group** from the Massing palette.)*
>
>Command: MassGroupAdd
>
>Select elements to attach: 1 found
>
>Select elements to attach:
>
>Location: *(Select a location in the workspace.)*
>
>Attaching elements…
>
>1 element(s) attached.
>
>*(The grips of the mass group marker are displayed when an element attached to the group is selected as shown in Figure 12.18.)*

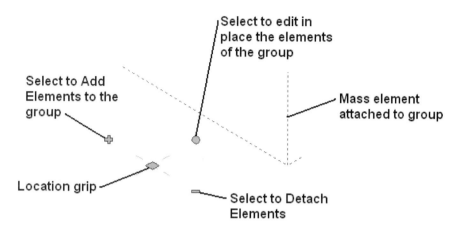

Figure 12.18 Mass Group trigger grips

The mass group marker is used to graphically display the presence of a group, and you can select the (+) to add element to the group or select the (-) to remove elements from the group. The mass group marker and members of the group are placed on the A-Area-Mass-Grp layer (color 13). The command sequence above does not prompt you to assign a name for the group. Architectural Desktop will assign a name to the mass group, such as 4A8. The assigned name is the AutoCAD handle name. AutoCAD handle names exist for all entities created in AutoCAD; they are not intended to describe the entity. To aid in modeling, names for mass elements or groups can be assigned in the Model Explorer, discussed later in this chapter.

ADDING MASS ELEMENTS TO A GROUP

You can assign mass elements to a group when they are created, or you can assign their group definition after insertion by modifying the properties of the mass element. If a mass group has been established in the drawing, when you select a mass element tool from the Massing tool palette, you can edit the **Attached to** field to assign a group upon insertion. The Boolean Add operation is assigned by default to mass elements defined to a group upon insertion.

ATTACHING ELEMENTS TO A GROUP

If a mass group exists, you can assign mass elements to a group by selecting the mass group marker or any component of the group, right-clicking, and choosing **Attach Element** (refer to Table 12.8). The mass elements combine to form one object, the mass group, by adding their volume and geometry to the mass of the group. The mass group can be viewed by turning off the A-Area-Mass layer and viewing only the A-Area-Mass-Grps layer. Mass elements can also be defined to subtract or intersect their mass with other mass elements of a group. These Boolean operations can be changed as the model develops. Several mass groups can be defined for the drawing according to the needs of the model.

Command prompt	MASSGROUPATTACH
Shortcut menu	Select any member of the group to display the mass group marker, right-click, and choose Attach Elements
Properties palette	Select the element, right-click, and choose Properties; edit the Attached to group in the Properties palette

Table 12.8 *Attach Elements to group command access*

The following command line sequence adds the mass element to the selected mass group as shown in Figure 12.19.

(Select a mass element of the group at p1 in Figure 12.19, right-click, and choose **Attach Elements***.)*

Command: MassGroupAttach

Select elements to attach: 1 found *(Select a mass element at p2 as shown in Figure 12.19.)*

Select elements to attach: ENTER *(Press* ENTER *to end selection.)*

Select elements to attach: ENTER *(Press* ENTER *to end the command.)*

Attaching elements...

1 element(s) attached.

All mass elements that are attached to a group are highlighted when you select the group marker or any member of the group as shown at right in Figure 12.19.

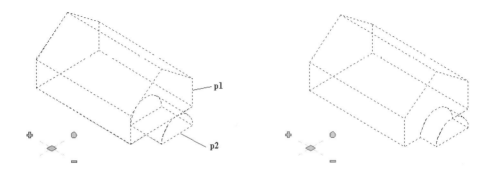

Figure 12.19 *Mass elements attached to mass group*

DETACHING ELEMENTS FROM A MASS GROUP

Elements that have been attached to a group can be detached from the group. Detached mass elements are not erased from the drawing. The mass element no longer functions in the group to add, subtract, or intersect its mass with other elements of the group. To detach an element from a group, select a mass element of the group, right-click, and choose **Detach Elements** from the shortcut menu (refer to Table 12.9). The members of the mass group are displayed in green (color 70) when you select **Detach Elements**, allowing you to select the elements to detach.

Command prompt	MASSGROUPDETACH
Shortcut menu	Select the mass group, right-click, and choose Detach Elements

Table 12.9 *MassGroupDetach command access*

The procedure to detach an element from a group is shown in the following command line sequence.

(Select a mass element of the group, right-click, and choose **Detach Elements**.*)*

Command: MassGroupDetach

Select elements to detach: *(Select mass elements to detach.)* 1 found

Select elements to detach: ENTER *(Press* ENTER *to end the command.)*

Detaching elements...

1 element(s) removed.

In addition, a mass element can be edited by freezing the A-Area-Mass-Grps layer. When the A-Area-Mass-Grps layer is frozen, you can easily select the mass element, right-click, and choose **Properties** from the shortcut menu. Edit the **Attached to** field to **None** in the Properties palette to detach the element from the group. Mass elements can be attached to and detached from the mass element group when you select the (+) and (-) grips of the mass group marker.

BOOLEAN OPERATIONS WITH MASS ELEMENTS

Mass elements that are attached to a group can be defined to add, subtract, or intersect their mass with the other elements. If a mass group exists in the drawing, the mass group and operation can be specified in the Properties palette when new elements are added. After a mass element becomes a member of the group, when you select a mass element, all elements of the group and the group marker are displayed. To isolate and edit a mass element that belongs to the group, select a member to display the group marker grip and choose the **Edit in Place** grip as shown in Figure 12.18. The Edit in

Place option turns off the display of the group and allows you to select the mass elements. After editing the mass elements select **Save changes** from the **Edit in Place** toolbar. When the element is isolated, the operation of a mass element within a group can be edited in Properties palette or by selecting the Boolean operation from the shortcut menu.

When mass elements are attached to a group, the default operation is additive. The Additive operation can be assigned to an element using the **MassElementOpAdd** command; refer to Table 12.10.

Command prompt	MASSELEMENTOPADD
Shortcut menu	Select the Edit in Place grip, select the mass element, right-click, and choose Mass Group Operation>Additive
Properties palette	Select the Edit in Place grip, select the mass element, right-click, and choose Add from the Operation drop-down list of the Properties palette

Table 12.10 *Mass Group Operation>Additive command access*

Addition is similar to performing a union of the mass elements. Figure 12.20 shows two mass elements shown as **p1** and **p2** that are attached to a group using the additive operation. The Barrel Vault and the Gable are displayed in the right viewport as a group.

SUBTRACTING MASS ELEMENTS

When a mass element is defined to subtract, its volume is cut from the group as shown in Figure 12.21. The **MassElementOpSubtract** command (refer to Table 12.11) is

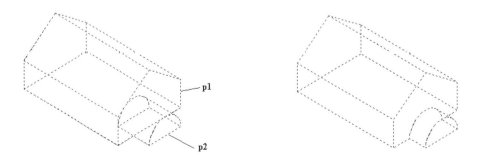

Figure 12.20 *Barrel Vault added to Gable mass element*

used to define the Subtract operation of a mass element. The element selected for the subtraction must be attached to the group after other elements from which it is subtracting its mass. The Model Explorer, discussed later in this chapter, allows you to change the sequence of attachment.

Command prompt	MASSELEMENTOPSUBTRACT
Shortcut menu	Select the Edit in Place grip, select the mass element, right-click, and choose Mass Group Operation>Subtract
Properties palette	Select the Edit in Place grip, select the mass element, right-click, and choose Subtract from the Operation drop-down list of the Properties palette

Table 12.11 *Subtract mass element from mass group command access*

USING THE INTERSECT OPERATION

The **MassElementOpIntersect** command (refer to Table 12.12) joins the two mass elements and creates a mass group, which consists of the mass or volume that is common to the two mass elements.

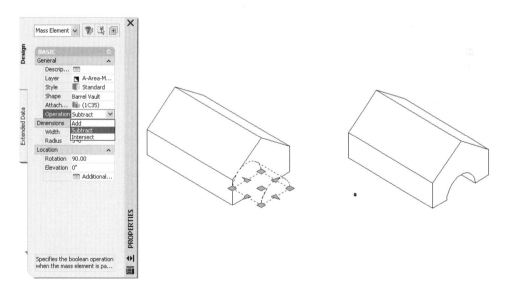

Figure 12.21 *Barrel Vault subtracted from the Gable mass element*

Command prompt	MASSELEMENTOPINTERSECT
Shortcut menu	Select the Edit in Place grip, select the mass element, right-click, and choose Mass Group Operation>Intersect
Properties palette	Select the Edit in Place grip, select the mass element, right-click, and choose Intersect from the Operation drop-down list of Properties palette

Table 12.12 *Make Element Intersection command access*

The Intersect operation allows you to create custom-shaped mass elements that combine the two mass elements and retain the volume and shape that is common. The mass elements shown at left in Figure 12.22 consist of a Gable and an Isosceles Triangle. The mass elements are combined in the mass group. The Isosceles Triangle is defined to intersect with the Gable to form a gambrel roof shape, as the mass group shows on the right.

USING THE MODEL EXPLORER

The Model Explorer is a window that can be used to better visualize and edit mass groups and mass elements. Open the Model Explorer by selecting a mass group, right-clicking, and choosing **Show Model Explorer** from the shortcut menu. See Table 12.13 for **Show Model Explorer** command access.

Command prompt	MODELEXPLORER
Shortcut menu	Select a mass group, right-click, and choose Show Model Explorer

Table 12.13 *Show Model Explorer command access*

The Model Explorer window includes a tree view of the model in the left pane and an Object Viewer area on the right. Located at the top of the Model Explorer are the **Tree View** toolbar and **Object Viewer** toolbar, which include view controls and commands for creating and editing the mass elements and groups. As the model is developed, the graphics area displays the group and the mass elements.

If you select the plus sign of a mass group, the tree will expand and display the mass elements of the mass group. The graphics area shown in Figure 12.23 displays the selected mass group in the Object Viewer. Selecting the box or cylinder in the Tree View will display the box or cylinder in the Object Viewer. If you select a mass element and then right-click, the shortcut menu allows you to edit the mass element, as shown in Figure 12.24.

Creating Mass Models, Spaces, and Boundaries 899

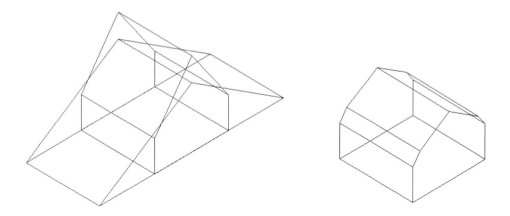

Figure 12.22 *Gambrel roof created with intersect operation between mass elements*

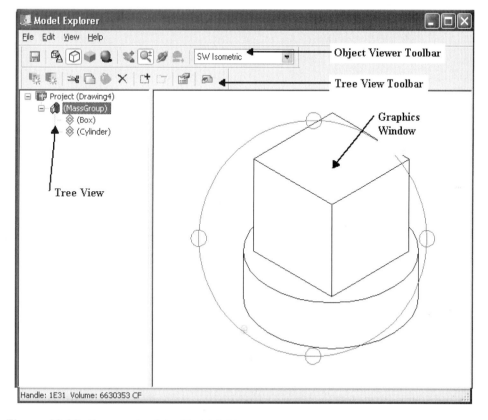

Figure 12.23 *Commands of the Model Explorer*

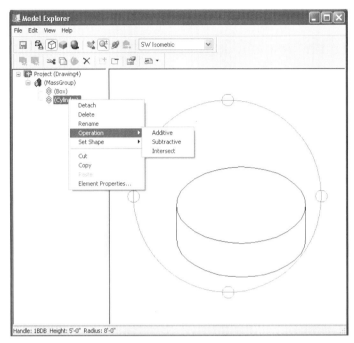

Figure 12.24 *Shortcut menu for a mass element in the Tree*

The shape and operation of the mass element can be edited. If you select a mass element within the Tree View and right-click, you can choose **Element Properties** from the shortcut menu. The **Element Properties** option opens the Properties palette, which allows you to change the dimensions of the mass element.

The shortcut menu of the Object Viewer, shown in Figure 12.26, allows you to view the mass element or mass group in perspective, change the zoom, pan the display, and copy the display as a bitmap image. The menu bar of the Model Explorer also provides access to commands for viewing and changing the mass elements and mass groups. The commands of the **Tree View** toolbar are shown in Figure 12.25 and summarized below.

>**New Grouping** – Creates a new group at a location specified; the group will be listed at the end of the Tree.
>
>**New Element** – Creates a new mass element. The new element will be attached to the group highlighted in the Tree View.
>
>**Cut** – Places a selected element in the clipboard.
>
>**Copy** – Copies the selected object to the clipboard.
>
>**Paste** – Inserts the selected object from the clipboard into the Tree View.

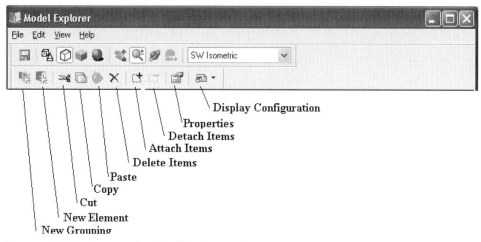

Figure 12.25 *Commands of the Tree View toolbar*

Delete – Removes the selected element or group from the tree.

Attach Items – Select a mass group and select **Attach Items** to attach an element to a group

Detach – Removes an element from the group in the Tree View.

Properties – Opens the Properties palette for the selected element.

Display Configuration – The **Display Configuration** flyout displays configurations for the graphics area.

MOUSE OPERATIONS IN THE MODEL EXPLORER

Using the mouse in the Model Explorer provides an extension of the command menu. When the Model Explorer is opened and the mouse is positioned over the Tree View, the mouse pointer is displayed as a pointer, allowing you to select mass elements or groups from the tree. When you select a mass element or group in the Tree View, you can right-click and display the shortcut menu, as shown in Figure 12.24.

USING THE OBJECT VIEWER SHORTCUT MENU

The shortcut menu of the Object Viewer is shown in Figure 12.26. This shortcut menu provides quick access to commands to control the view in the graphics area.

The options of this shortcut menu are described below.

Close – Closes the Model Explorer.

Pan – Executes the real time **Pan** command.

Zoom – Executes real time **Zoom** command.

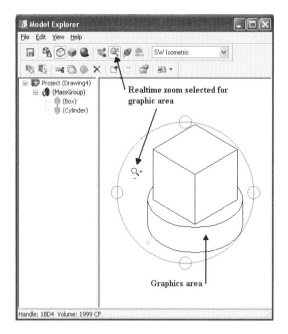

Figure 12.26 *Shortcut menu of the Object Viewer*

Orbit – Executes the **3DOrbit** command in the Object Viewer.

More – The **More** flyout includes **Zoom Window**, **Zoom Extents**, **Zoom Center**, **Zoom In**, and **Zoom Out**.

Projection – The **Projection** flyout includes **Parallel** and **Perspective** options. The **Perspective** option allows you to view the model in a perspective view.

Shading Modes – The **Shading Modes** flyout consists of **Wireframe**, **Hidden**, **Flat Shaded**, and **Gouraud Shaded** modes.

Visual Aids – The **Visual Aids** flyout consists of the **Compass** and **Grids** options.

Reset View – Resets the view to the top view.

Preset Views – The **Preset Views** flyout includes the **Top**, **Bottom**, **Left**, **Right**, **Front**, **Back**, **Southwest**, **Southeast**, **Northeast**, **Northwest** view directions.

Undo View – Performs an **Undo** on previous view settings; returns the view of the model to the previous settings.

Redo View – Performs a **Redo** on last view of the model, returning the view of the model to the settings prior to the **Undo**.

Set View – **Set View** is inactive; however, in other applications this option sets the view in the Object Viewer equal to the workspace.

Save Image – Allows you to save the content of the graphic area as an image file.

USING THE TREE VIEW SHORTCUT MENUS

There are two shortcut menus available in the project tree. If you select a group and then right-click, the shortcut menu includes edit operations for a group. The shortcut menu for the group is shown at left in Figure 12.27.

If you select a mass element and then right-click, the shortcut menu includes options to edit the selected mass element. The shortcut menu for a mass element is shown at right in Figure 12.27. The shortcut menu for the mass element allows you to change the shape of the mass element or the Boolean operation assigned to it. The **Rename** option allows you to append names to the groups and each mass element. You can also change the names of groups and mass elements by clicking twice on the name in the Tree. Naming the mass elements and groups greatly enhances model development in the Model Explorer.

CREATING MASS ELEMENTS WITH THE MODEL EXPLORER

You can create a mass element for a model by selecting the **New Element** command from the **Tree View** toolbar of the Model Explorer after selecting the group name in

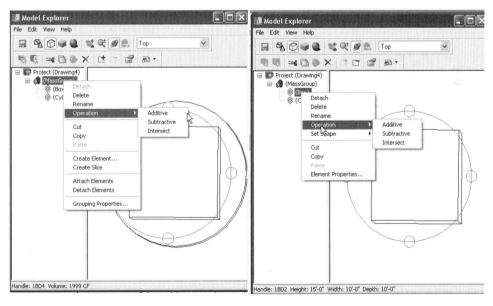

Shortcut menu of Mass Group   Shortcut menu of Mass Element

Figure 12.27 *Shortcut menu of mass group and mass element*

the left pane. The **New Element** and **New Group** commands are active when a group or a drawing is selected in the left pane. You can create a new mass group by selecting the **New Grouping** command from the **Model Explorer** toolbar. When you select the **New Grouping** command, the Model Explorer temporarily closes, and you can specify the location of the mass group marker in the AutoCAD workspace. After selecting the location and rotation, reselect the Model Explorer from the Windows status bar.

To add mass elements to the Tree View, select a mass group in the Model Explorer, right-click and choose **Create Element** from the shortcut menu. The mass element is added to the group selected in the Tree View. The Additive, Subtractive, or Intersect Boolean operation symbols associated with a mass element will prefix the name of each mass element in the Tree View, as shown in Figure 12.28. The Boolean operation can be changed by selecting the element in the Tree View, right-clicking, and choosing the operation from the **Operation** shortcut menu option.

Using the Model Explorer to Reposition Mass Elements in the Tree View

Mass elements are added to the Tree View as they are created in the drawing. The subtractive and intersect operations may require you to reposition the order of the mass elements in the Tree View. If you are subtracting mass elements to remove part of a mass element, the smaller mass element must occur below the larger mass element in

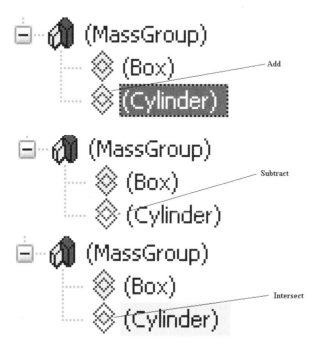

Figure 12.28 *Boolean operation markers*

the Tree. To move a mass element within the Tree, select the mass element with the mouse, continue holding down the mouse button, and drag the mass element name up or down within the Tree. The Tree allows you to copy mass elements to the clipboard and paste them in one or more mass groups.

 STOP. Do Tutorial 12.3, "Using Mass Groups to Create a Fireplace," at the end of the Chapter 12 Supplement located on the CD.

CREATING FLOORPLATE SLICES AND BOUNDARIES

After you develop a mass model of a building, you can create a floor plan from the model. The **Slice** command creates a horizontal slice of the model for the development of floorplates. The **Slice** command allows you to slice through the model at elevations equivalent to floor elevations. The **AecSliceToPline** command can convert the slice to a polyline. The polyline can be converted to walls (**Wall>Apply Tool Properties to>Wall**), or the slice can be converted to space boundaries (**Space Boundaries>Apply Tool Properties to>Slice**).

The **Slice** (**AecSliceCreate**) command is located on the Massing tool palette. The **Slice** command creates slice markers. See Table 12.14 for **Slice** command access. The slice markers are attached to mass elements or mass groups to create a slice of the mass elements. The slice should be taken at the finish floor elevation. A multi-story building would be sliced at the elevations of each finish floor. The exact elevation of finish floors may have to be adjusted later in the design, and therefore the slice elevation can be edited and the cut moved on the mass elements. The floorplate becomes a horizontal section from which floor plans can be developed, or the geometry can be used to develop space plans.

Command prompt	AECSLICECREATE
Palette	Select Slice from the Massing tool palette

Table 12.14 *Slice command access*

When you select the **Slice** command, you are prompted to specify the number of slices, the location for the slice marker, rotation, the elevation of the first slice, and the distance between slices. The slice marker is a graphic marker indicating the slice; the actual slice is executed when the slice marker is attached to the mass elements or mass group. The slice marker and the geometry of the slice through mass elements is created on the **A-Area-Mass-Slce** layer, which has the color 51 (a hue of yellow) and Dashed2 linetype. Shown below is the command line sequence that was used for creating the slice shown in Figure 12.29.

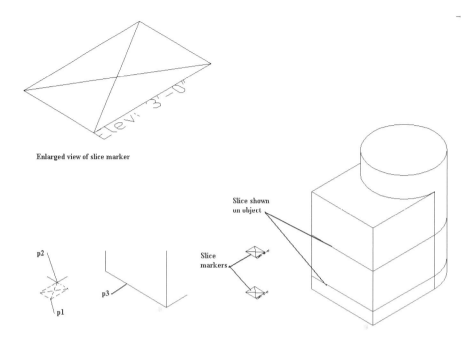

Figure 12.29 *Placing a slice*

> *(Select **Slice** from the Massing tool palette.)*
> Number of slices <1>: **2** ENTER
> First corner: *(Select a point near point p1 in Figure 12.29.)*
> Second Corner or [Width]: *(Select a point near point p2 in Figure 12.29.)*
> Rotation <0.00>: ENTER *(Press ENTER to accept the 0° rotation.)*
> Starting height <0">: **3'** ENTER
> Distance between slices <1'-0">: **9'** ENTER
> *(Slice markers are created for each of the slices, as shown at right in Figure 12.29.)*

The slice marker that is created by the **Slice** command includes the text describing the elevation of each slice as shown at the top in Figure 12.29. The location, size, and rotation of the marker can be anywhere in the drawing.

After placing the marker, select the marker, right-click, and choose **Attach Objects** from the shortcut menu. The **Attach Objects** command (**AecSliceAttach**) is used to assign the objects to be sliced. See Table 12.15 for **Attach Objects** command access.

Command prompt	AECSLICEATTACH
Shortcut menu	Select a marker, right-click, and choose Attach Objects

Table 12.15 *Attach Objects to slice command access*

When you choose **Attach Objects** command (**AecSliceAttach**) from the shortcut menu, you are prompted to select elements or mass groups to attach for the slice. The command line prompts for attaching the mass elements to a slice marker are shown in the command line sequence and in Figure 12.29.

> *(Select the slice markers, right-click, and choose* **Attach Objects**.*)*
>
> Command: AecSliceAttach
>
> Select elements to attach: *(Select the mass element at p3 in Figure 12.29.)* 1 found
>
> Select elements to attach: ENTER *(Press* ENTER *to end the selection of mass elements or mass groups.)*
>
> Adding elements...
>
> 1 element(s) added.
>
> *(Slices created as shown at right in Figure 12.29.)*

Mass elements and mass groups that have been attached to a slice can be detached from the slice through the **Detach Objects** command (**AecSliceDetach**). See Table 12.16 for **Detach Objects** command access.

Command prompt	AECSLICEDETACH
Shortcut menu	Select a slice, right-click, and choose Detach Objects

Table 12.16 *Detach Objects command access*

When you select the **Detach Objects** command, you are prompted to select the slice or slice marker and then to specify the mass elements or mass group to remove from the slice. The command line prompts for detaching a mass element from a slice are shown below.

> *(Select a slice, right-click, and choose Detach Objects.)*
>
> Command: _AecSliceDetach
>
> Select elements to detach: 1 found *(Select the mass element to detach.)*

Select elements to detach: *(Press* ENTER *to end the selection of mass elements or mass groups.)*

Removing elements...

1 element(s) removed.

The elevation of the slice can be revised after the **Slice** command has been executed. Changing the elevation of the slice allows you to adjust the finish floor elevation as the design changes. To change the elevation of the slice, select the slice or the slice marker, right-click, and choose the **Set Elevation** command (**AecSliceElevation**). See Table 12.17 for **Set Elevation** command access.

Command prompt	AECSLICEELEVATION
Shortcut menu	Select a slice marker or slice, right-click, and choose Set Elevation

Table 12.17 *Set Slice Elevation command access*

When you select the **Set Elevation** command, you are prompted in the command line to specify a new elevation. If you type a different elevation in response to the command prompt, the slice will move to the specified elevation. The geometry of the slice cannot be edited. The AutoCAD **Move** command can be used to move the slice marker in the drawing. When the slice marker is moved, the slice location on the mass element moves.

CONVERTING THE SLICE TO A POLYLINE

The slice can be converted to a polyline and the polyline edited to create the shape of the floor plan. The polyline can then be used to create a space boundary or a wall. To convert a slice to a polyline, select the **Convert to Polyline** command (**AecSliceToPline**). See Table 12.18 for **Convert to Polyline** command access.

Command prompt	AECSLICETOPLINE
Shortcut menu	Select a slice, right-click, and choose Convert to Polyline

Table 12.18 *Convert to Polyline command access*

When you select the **AecSliceToPline** command, the slice geometry shown on each mass element is converted to a polyline, which can be edited according to the desired shape. The polyline is created on layer 0 and is displayed with continuous linetype. The command line prompts for the **Convert to Polyline** command are shown below.

(Select the Slice, right-click, and choose **Convert to Polyline**.*)*

Command: AecSliceToPline

1 polyline(s) created.

(Polyline created from the slice. Polyline selected to display grips as shown in Figure 12.30.)

CONVERTING AEC OBJECTS TO MASS ELEMENTS

AEC objects such as walls, doors, and windows can be converted to mass elements. Converting AEC objects to a mass element allows you to perform Boolean operations on a mass element that is the same shape and size as the AEC object. The **Convert to Mass Element Tool** (**AecMassElementConvert**) command, (refer to Table 12.19) can be used to convert AEC objects to mass elements. When you convert an AEC object to a mass element, you can apply the mass element editing tools to create custom mass elements. The AEC objects are converted to free form. Free form objects include additional grips, as shown in Figure 12.31, and increase the drawing file size. Custom shapes can be created from the free form mass elements.

Command	AECMASSELEMENTCONVERT
Palette	Select Convert to Mass Element from the Massing tool palette of the Design palette set

Table 12.19 *MassElementConvert command access*

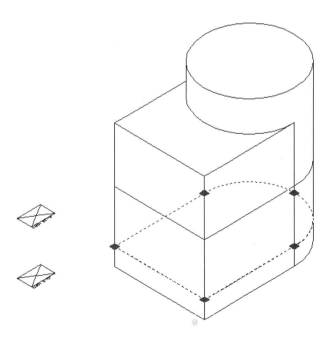

Figure 12.30 *Polyline created from the slice as shown with continuous line*

When you select the **Convert to Mass Element Tool** from the **Content Browser**, you are prompted to select the AEC objects and name the free form mass element. The command line sequence shown below converted a door object to a free form mass element.

> Command: _AecMassElementConvert
>
> Select objects to convert: *(Select a door AEC object at p1 in Figure 12.31.)* 1 found
>
> Select objects to convert: ENTER *(Press ENTER to end selection.)*
>
> Erase layout geometry?[Yes/No]<No>: ENTER *(Press ENTER to retain the geometry.)*
>
> Mass element [Name/Show bounding box]: **N**
>
> Name: DoorMass enter *(Type a name for the free form object.)*
>
> Mass element [Name/Show bounding box]: ENTER *(Press ENTER to end the command.)*
>
> Converting AEC Objects to 3D Solids
>
> *(Door object converted to free form mass element as shown at right in Figure 12.31 at p2.)*

The **AEC Objects to 3D Solids** command located in the Architectural Desktop Stock Tool Catalog\ Modeling Tools\Mass Element Tools can be used to convert AEC

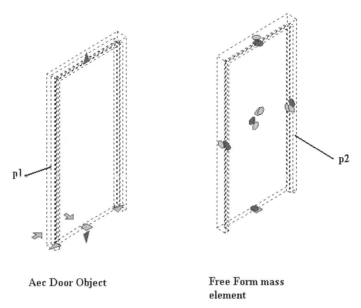

Aec Door Object   Free Form mass element

Figure 12.31 *AEC Door and Free Form Door*

objects to 3D solids. When an AEC object is converted to a 3D solid, you can apply AutoCAD 3D Solid Modeling editing tools to create custom objects.

 STOP: Refer to the Chapter 12 Supplement located on the CD for additional information regarding Spaces, Boundaries, and additional tutorials. Read the material located in the Chapter 12 Supplement and complete the tutorials prior to answering the questions at the end of the chapter.

SUMMARY

1. Create mass models using mass groups, which include mass elements modeled from regular prisms, extrusions, revolutions, and drapes.
2. Create mass element styles in the **Style Manager** that include material definitions.
3. Select the mass group marker tool from the Massing tool palette, and attach mass elements to create a mass model.
4. Select a mass group, right-click, and choose **Show Model Explorer**. The Model Explorer allows you to create and edit mass elements and mass groups.
5. Select the **Slice** tool from the Massing tool palette to create a slice through a mass model. Aec slices can be converted to polylines, which can be developed as spaces, space boundaries, or walls.
6. Select the **Spaces** tool (**SpaceAdd**) from the Design tool palette to add spaces.
7. Create Space Styles in the **Style Manager** to include length, width, and area parameters as well as material definitions for the ceiling and floor.
8. To convert polylines to spaces, select **Spaces** from the Design palette, right-click, and choose **Apply Tool Properties to>Polyline**.
9. Divide spaces with the **Divide Spaces** command (**SpaceDivide**).
10. Combine adjacent spaces with the **Join Spaces** command (**SpaceJoin**).
11. Convert slice, space, or edge to space boundaries by selecting the **Space Boundary** tool on the Design tool palette, right-click, and choose **Apply Tool Properties to>Slice**, **Apply Tool Properties to>Space**, or **Apply Tool Properties to>Edge** respectively.
12. To create space boundaries, select **Space Boundary** (**SpaceBoundaryAdd**) from the Design tool palette.
13. Change the height, width, justification, ceiling condition, and floor condition in the Properties palette of the selected space boundary.
14. Combine independent space boundaries using the **Merge Boundaries** command (**SpaceBoundaryMerge**) located on the shortcut menu of a boundary.

15. Attach a space to a boundary using the **Attach Spaces to Boundary** command (**SpaceBoundaryMergeSpace**) located on the shortcut menu of a boundary.
16. Add space boundaries to existing space boundaries using the **Add Edges** command (**SpaceBoundaryAddEdges**) located on the shortcut menu of a boundary.
17. Change the ceiling and floor boundary conditions of a space boundary segment using the **Edit Edges** command (**SpaceBoundaryEdge**) located on the shortcut menu of a boundary.
18. Delete a boundary segment using the **Remove Edges** command (**SpaceBoundaryRemoveEdges**) selected from the shortcut menu for the boundary.
19. Release or attach the anchor of objects in a space boundary using the **Anchor to Boundary** command (**SpaceBoundaryAnchor**).
20. Type **SpaceQuery** in the command line to determine the areas of the spaces in a drawing.
21. Create walls from space boundaries using the **Generate Walls** command (**SpaceBoundaryGenerateWalls**) located on the shortcut menu for the boundary.

REVIEW QUESTIONS

1. The _____ _____ tool is selected from the _____ tool palette to create a mass group.
2. Elements attached to a mass group marker are by default attached with the _____ Boolean operation.
3. The _____ _____ can be used to assign names to mass elements.
4. A Doric Column can be created by _____.
5. The _____ tool selected from the _____ tool palette is used to create a slice through a mass element or mass group.
6. The _____ layer is created for mass elements in the AIA 256 Color layer key style.
7. Use the _____ object snap mode to place spaces by snapping to points on the gross or net boundary.
8. The _____ _____ option of the command line of a space moves the insertion point in a counterclockwise direction from its default position.
9. The default insertion point of a space is located in the _____ _____ _____ of a space.

10. The distance between the Net and Gross space boundaries is specified in the _____ _____ _____ _____ _____ option of the **Space Style Properties** dialog box.

11. Materials can be assigned to the _____ and _____ components of a space style.

12. The **Divide Spaces** command requires that the points specified _____ _____ _____ on the space.

13. The **Join Spaces** command can be used to _____ adjoining spaces.

14. Space boundaries with a width dimension greater than zero are _____.

15. Specify how the floor and ceiling of a space boundary meet the wall component by selecting the _____ _____ command.

16. The _____ _____ command can be used to trim a space boundary that extends beyond other space boundaries.

17. To detach a space from its boundary, use the _____ _____ command.

18. Space boundaries can be converted to walls by the _____ _____ command.

19. Summarize the area per space by typing _____ in the command line.

20. Insert additional space boundaries attached to the adjacent spaces by selecting the _____ _____ command.

21. List six mass element shapes that you can create.

TUTORIAL 12.1 CREATING MASS ELEMENT COMPONENTS FOR A FIREPLACE

1. Open *Ex 12-1* from your *ADT Student\ADT Tutor\Ch12* folder.

2. Choose **File>SaveAs** from the menu bar and save the drawing as **Lab 12-1** in your student directory.

3. If tool palettes are not displayed, select **Window>Tool Palettes** from the menu bar.

4. Select the **Gable** tool from the Massing tool palette.

5. Edit the Properties palette as follows: Style = **Standard**, Shape = **Gable**, Attach to = **None**, Specify on Screen = **No**, Width = **4'-4"**, Depth = **3'-0"**, Height = **8'** and Rise = **8"**. Specify the Insert point and Rotation in the command line as shown in the following command line sequence.

    Command: MassElementAdd

Insert point or [SHape/WIdth/Depth/Height/Match]: **15', 20'**
ENTER *(Specify the insert point for the mass element.)*

Rotation or [SHape/WIdth/Depth/Height/Match/Undo] <0.00>:
0 ENTER *(Specify the rotation of the mass element.)*

6. Edit the Properties palette as follows: Style = **Standard**, Shape = **Box**, Attach to = **None**, Specify on Screen = **No**, Width = **5'-0"**, Depth = **1'-6"**, Height = **1'-2"**. Specify the Insert point and Rotation as shown in the following command line sequence:

    Command: MassElementAdd

    Insert point or [SHape/WIdth/Depth/Height/Match]: **24', 10'**
    ENTER *(Specify the insert point for the mass element.)*

    Rotation or [SHape/WIdth/Depth/Height/Match/Undo] <0.00>:
    0 ENTER *(Specify the rotation of the mass element.)*

7. Edit the Properties palette as follows: Style = **Standard**, Shape = **Box**, Attach to = **None**, Specify on Screen = **No**, Width = **2'-0"**, Depth = **2'-0"**, and Height = **24'**. Specify the Insert point and Rotation as shown in the following command line sequence.

    Command: MassElementAdd

    Insert point or [SHape/WIdth/Depth/Height/Match]: **6', 20'**
    ENTER *(Specify the insert point for the mass element.)*

    Rotation or [SHape/WIdth/Depth/Height/Match/Undo] <0.00>:
    0 ENTER *(Specify the rotation of the mass element.)*

    Insert point or [SHape/WIdth/Depth/Height/Match]: ENTER
    (Press ENTER to end the command.)

8. Select the polyline at **p1** as shown in Figure 12T.1, right-click, and choose **Convert To>Mass Element**. Respond to the following command line prompts to specify the mass element.

    Command: ExtrudeLinework

    Erase selected linework? [Yes/No] <No>: **N** *(Select No to retain the polyline.)*

    Specify extrusion height <1'-0">: **2'-5"**

9. Select the mass element and verify properties in the Properties palette as follows: Style = **Standard**, Profile = **EMBEDDED**, Attach to = **None**, Width = **3'**, Depth = **1'-11"**, Height = **2'-5"**. Mass elements are developed as shown in Figure 12T.2.

10. Save and close the drawing. (The mass elements developed in this tutorial will be combined to create a fireplace in Tutorial 12.3, "Using Mass Groups to Create a Fireplace" located in the Chapter 12 Supplement of the CD.)

pl

Figure 12T.1 Selection of polyline to create Profile

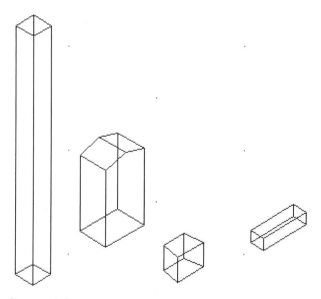

Figure 12T.2 Profile extruded

TUTORIAL 12.2 CREATING MASS ELEMENTS FOR A TERRAIN

1. Open your *ADT Student\ADT Tutor\Ch12\Ex 12-2.dwg*.

2. Choose **File>SaveAs** from the menu bar and save the drawing as **Lab 12-2** in your student directory.

3. If the tool palette is not displayed, select **Window>Tool Palettes** from the menu bar. Toggle OFF OSNAPS in the status bar.

4. Verify that the left viewport is current, and select **Drape** from the Massing tool palette. Respond to the command line prompts as shown below:

   Command: _AecDrape

   Select objects representing contours: *(Select a point near p1 as shown in Figure 12T.3)*

   Specify opposite corner: *(Select a point near p2 as shown in Figure 12T.3)* 11 found

   Select objects representing contours: ENTER *(Press ENTER to end selection.)*

   Select mesh corner: *(Use SHIFT + right-click to select Endpoint and select a point near p3 as shown in Figure 12T.3.)*

   Select opposite mesh corner: *(Use shift and right-click to select **Endpoint** and select a point near p4 as shown in Figure 12T.3.)*

   Enter mesh size <30>: ENTER *(Specify mesh size equal to 30.)*

   Enter base thickness <6>: **6** ENTER *(Specify base thickness.)*

5. Select **Format>Style Manager** from the menu bar. Select **Open Drawing** from the **Style Manager** toolbar. Select Content from the **Places** panel, select the Styles>Imperial folder, choose *Material Definitions (Imperial).dwg*, and select the **Open** button to complete the selection.

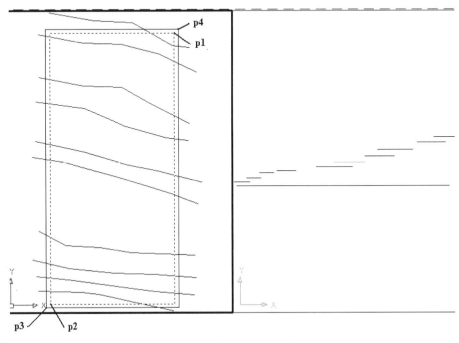

Figure 12T.3 *Selection of contour lines for Drape mass element*

6. Expand the Multi-Purpose Objects>Material Definitions folder of the Material Definitions (Imperial).dwg in the left pane. Scroll down the list of materials and select **Site Construction.Planting.Groundcover.Grass.Short**, right-click, and choose **Copy** from the shortcut menu. Select the current drawing **Lab 12-2**, right-click, and choose **Paste**.

7. Expand Multi-Purpose Objects>Mass Element Styles of the Lab 12-2.dwg drawing in the left pane. Select **New Style** from the toolbar. Click on the **New Style** name and overtype the name **Grass** in the right pane. Double-click on the **Grass** name to open the **Mass Element Style Properties – Grass** dialog box.

8. Select the **Materials** tab of the **Mass Element Style Properties – Grass** dialog box. Click in the Material Definition column for the Body component, and select **Site Construction.Planting. Goundcover.Grass.Short** as shown in Figure 12T.4.

9. Select **OK** to close the **Mass Element Style Properties – Grass** dialog box. Select **OK** to close the **Style Manager**.

10. Select the Drape mass element, right-click, and choose **Properties**. Edit Style = **Grass** in the Properties palette.

11. Set the Model tab current, and select **SW Isometric** from the **View** flyout of the **Navigation** toolbar. Select **Flat Shaded** from the **Shade** flyout of the **Navigation** toolbar.

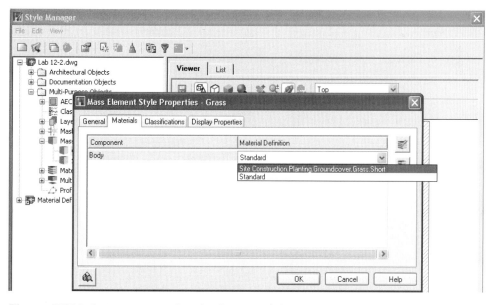

Figure 12T.4 *Assigning material to the Grass mass element style*

12. Select the Work layout tab, click to set the left viewport current, and select **Gable** from the Massing tool palette. Edit the Properties palette as follows: Style = **Standard**, Shape = **Gable**, Specify on Screen = **No**, Width = **28'**, Depth =**68'**, Height = **22'**, and Rise = **5'**.

    Command: MassElementAdd

    Insert point or [SHape/WIdth/Depth/Height/Match]: **80', 70', 8'** ENTER *(Specify the insert point for the mass element.)*

    Rotation or [SHape/WIdth/Depth/Height/Match/Undo] <0.00>: **90** ENTER *(Specify the rotation of the mass element.)*

    Insert point or [SHape/WIdth/Depth/Height/Match]: ENTER *(Press ENTER to end the command.)*

    (Right side view of Gable mass element and drape shown in Figure 12T.5.)

13. Save and close the drawing.

PROJECTS

Ex 12.5 Creating a Mass Group and Slicing the Model

1. Open **Ex 12-5** from your *ADT Student\ADT Tutor\Ch12* Directory.
2. Save the file as **Lab 12-5** in your student directory.
3. Move each of the mass elements in the drawing together to create a model as shown in Figure 12T.6.
4. Create slices at **1'** and **12'** elevations.
5. Create a mass group that includes all mass elements.
6. Slice the mass model at each of the slice elevations.
7. Convert the slices to a polyline.
8. Create a perspective view of the model in the Model Explorer as shown at left in Figure 12T.6.

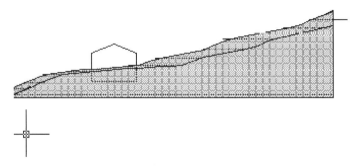

Figure 12T.5 *Right side view of Gable mass element*

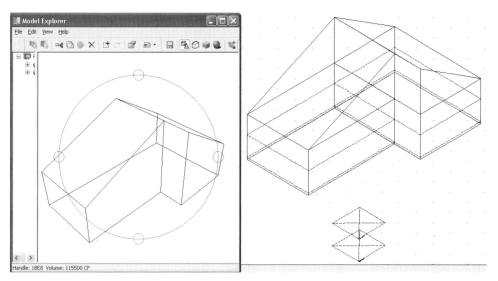

Figure 12T.6 *Model and slices used in Exercise 12.5*

9. Save and close the drawing.

Ex 12.6 Positioning Spaces to Create Space Boundaries and Walls

1. Open **Ex 12-6** from your *ADT Student\ADT Tutor\Ch12* directory and save the drawing in your student directory as **Lab 12-6**.

2. Import the Space styles from the Design Tool Catalog, Commercial category.

3. Resize the space in area 4 as shown in Figure 12T.7 to complete the plan as shown in Figure 12T.8.

4. Convert the spaces to space boundaries. Edit the space boundaries to solid type with 6" width as shown in Figure 12T.8.

5. Use **LineworkDivide** to divide space 5 using a horizontal line that equally divides the spaces. The upper space will be 5a and the lower space will be 5b.

6. Edit the space style for each of the spaces using the space styles of the Spaces-Commercial (Imperial).dwg. 1 = **Office_Large**, 2 = **Library**, 3 = **Conference_Large**, 4 = **Entry_Room**, 5a = **Restroom_Men-Small**, 5b = **Restroom_Women-Small**.

7. Open the Documentation Tool Catalog of the Content Browser and insert the Space Tag for each of the spaces.

8. Type **spacequery** in the command line and list the names of the spaces that have areas that exceed the maximum area defined for the space style.

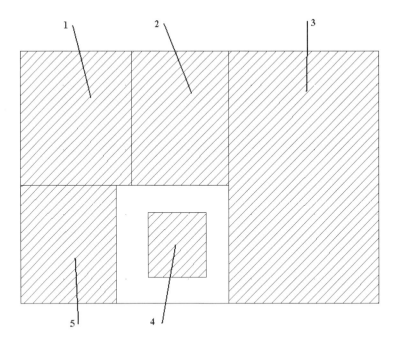

Figure 12T.7 *Existing spaces inserted for edit*

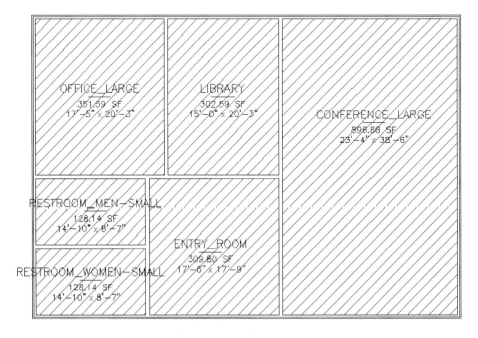

Figure 12T.8 *Spaces resized and space boundaries created*

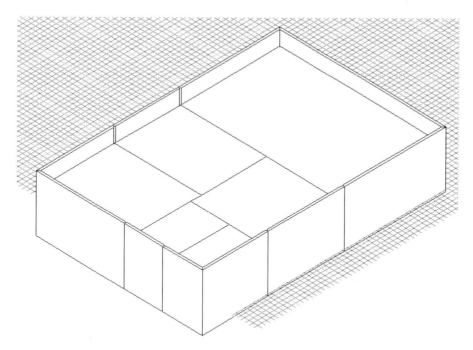

Figure 12.T.9 *Space boundaries shown in pictorial view*

9. Select **SW Isometric** to view the space boundaries as shown in Figure 12T.9. Select **Hidden** from the **Shade** flyout of the **Navigation** toolbar.
10. Save and close the drawing.

CHAPTER 13

Drawing Commercial Structures

INTRODUCTION

The development of drawings for commercial building requires the insertion of structural grids and ceiling grids. Layout curves can be developed to insert columns or other AEC objects spaced uniformly along a layout curve. Structural grids, layout curves, and ceiling grids are presented in this chapter. In addition, the development of views and walkthroughs using a camera is presented.

OBJECTIVES

Upon completion of this chapter, you will be able to

- Create column, beam, and brace styles from the **Structural Member Catalog**
- Create a structural column grid using the **ColumnGridAdd** command
- Add beams, columns, and braces to create a structural frame
- Dynamically size a structural grid using the **Specify on Screen** option of the **ColumnGridAdd** command
- Add, move, remove, mask, and clip column grid lines
- Adjust the location of the column grid using **Start offset** and **End offset** options
- Change the size of the column grid and add AEC Dimensions to the column grid
- Label the column grid using the **ColumnGridLabel** command
- Insert columns in the column grid
- Create a ceiling grid with boundaries using the **CeilingGridAdd** command

- Add and remove ceiling grid lines using the **CeilingGridXAdd/CeilingGridYAdd** and **CeilingGridXRemove/CeilingGridYRemove** commands
- Use the **CeilingGridClip** command to trim the ceiling grid
- Use the **LayoutCurveAdd** command to place columns and other AEC objects symmetrically along a layout curve
- Place a camera in the drawing using the **AecCameraAdd** command
- Create a video of a walkthrough using the **AecCameraVideo** command

INSERTING STRUCTURAL MEMBERS

Structural members including steel, concrete, and wood shapes can be inserted in a drawing as columns, beams, and braces. The components of the structural framework are inserted as 3D objects but can be viewed from the top view to create 2D plan views. Lines, arcs, or polylines can be converted to structural shapes by extruding the shape along the geometry. Structural components are selected from the **Structural Member Catalog** and saved as a style. The style extracts the properties of the component from the catalog and assigns a name. The style is then inserted into the drawing as a structural component. The design rules of a structural member style allow you to add custom shapes to the style. Two structural components can be combined to create a composite column or beam. Beams can be created that are tapered with different dimensional properties at each end of the beam and used as a rigid frame. The physical properties of the structural member are defined in the **Structural Member Catalog**. Styles of structural members can be defined for structural grids and anchored to the structural grid or inserted independent of a grid. Although structural members are styles, they should be inserted in the drawing from the **Structural Member Catalog** because its window provides a detailed description of the structural member physical properties.

The tools for inserting columns, beams, and braces are located on the Design tool palette. The member inserted as a column, beam, or brace is a style of a component developed from the **Structural Member Catalog**.

USING THE STRUCTURAL MEMBER CATALOG

The **Structural Member Catalog** is a library of structural components, which includes Imperial and Metric shapes of concrete, steel, and wood materials. Access the **Structural Member Catalog** as shown in Table 13.1.

Menu bar	Format>Structural Member Catalog
Command prompt	AECSMEMBERCATALOG

Table 13.1 *Structural Member Catalog command access*

When you select the **Structural Member Catalog** command (**AecSMemberCatalog**), the **Structural Member Catalog** window opens, allowing you to select a structural component to create a style.

The left pane of the **Structural Member Catalog** lists a hierarchical tree view of the components available as shown in Figure 13.1. If you select a structural member in the left pane, a graphic preview of the category of members is shown in the upper right pane and the text descriptions of the physical properties of the members are shown in the lower right pane. The columns describing the physical properties of the beam usually exceed the width of the lower right pane; therefore you can use the scroll bar at the bottom of the pane to view the additional properties.

The contents of the Structural Member Catalog Imperial folder are summarized in Table 13.2.

In addition, open web steel bar joists can be added to the drawing by importing their styles from the Architectural Desktop Design Tool Catalog-Imperial\Structural catalog of the **Content Browser**. This catalog consists of Steel Bar Joists shown in Figure 13.2 and other Members shown in Figure 13.3. The tools of the catalog are i-dropped into the drawing from the **Content Browser** to a tool palette or the workspace for immediate use.

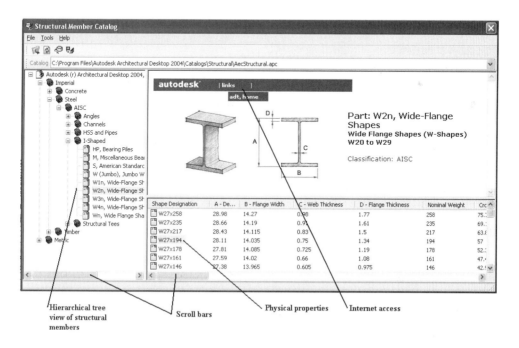

Figure 13.1 *Structural Member Catalog*

Material	Form	Components
Concrete	Cast-in-Place	Beams, Circular columns, Rectangular columns
	Precast	Inverted Tee Beams, Joist Double T, Joist T, L Shaped Beams, Rectangular Beams, Rectangular Columns
Steel	AISC	Angles, Channels, HSS and Pipes, I Shaped, Structural Tees
Timber	Glue Laminated Beams	Beams
	Lumber	Heavy Timber, Nominal Cut Lumber, Rough Cut Lumber
	Plywood Web Wood Joist	Plywood Joist

Table 13.2 *Structural member styles*

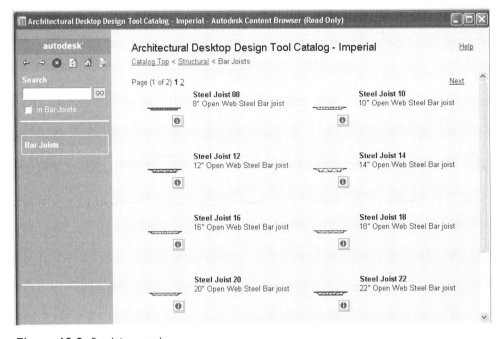

Figure 13.2 *Bar Joists catalog*

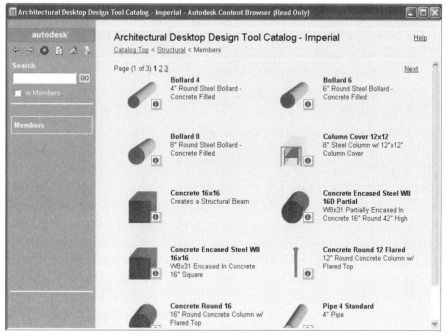

Figure 13.3 Members catalog

CREATING A STRUCTURAL MEMBER STYLE

Structural member styles are created from the components in the **Structural Member Catalog**. When you create a structural member style, you select a component from the **Structural Member Catalog** and assign a name to the style. The style has the physical properties of the component selected from the **Structural Member Catalog**. The styles created can be used as a column, beam, or brace.

STEPS TO CREATING A STRUCTURAL MEMBER STYLE

1. Select **Format>Structural Member Catalog** from the menu bar to open the **Structural Member Catalog** window.
2. Expand a folder in the left pane of the desired structural member category.
3. Select the file in the left pane to display the contents in the lower right pane.
4. Select the structural member in the lower right pane.
5. Create a style name by one of the following five methods:

   Double-click on the component name in the lower right pane.

   Select **Tools>Generate Style** from the menu bar of the **Structural Member Catalog**.

Select a member, and then type CTRL+G.

Select the member, right-click, and choose **Generate Member Style**.

Select a member, and then select the **Generate Member Style** icon from the **Structural Member Catalog** toolbar.

6. Type a name in the **Structural Member Style** dialog box as shown in Figure 13.4.
7. Select **OK** to dismiss the **Structural Member Style** dialog box.

ADDING STRUCTURAL MEMBERS

The commands to add a beam, column, or brace are located on the Design tool palette as shown in Figure 13.5. The styles created from the **Structural Member Catalog** are listed in the **Style** list of the Properties palette when you add a column, beam, or brace. When a structural member is added to the drawing, the associated cross-section of the structural member is extruded along an axis line. The length of the axis line is specified in the **Logical length** edit field of the Properties palette. The axis line includes grips at the start, end, and midpoint. The **Node** object snap is active at the start, end, and midpoint of this axis line. Therefore **Start offset**, **End offset**, **Roll**, and **Justify** are all relative to the axis line.

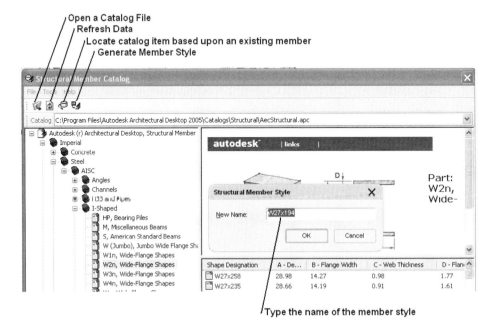

Figure 13.4 *Creating a structural member style*

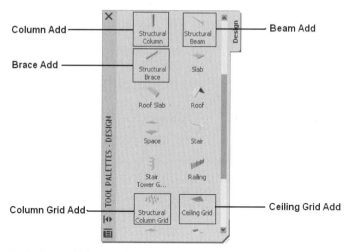

Figure 13.5 *Tools to add beams, braces, columns, and column grids*

INSERTING COLUMNS

The **Structural Column** tool is used to place columns in a drawing. Access the **Structural Column** command as follows.

Command prompt	COLUMNADD
Palette	Select Structural Column from the Design Tool palette as shown in Figure 13.5

Table 13.3 *ColumnAdd command access*

When the **Structural Column** tool is selected from the Design tool palette, the Properties palette opens, which allows you to specify the style and other properties of the column. The structural shape of the column is extruded along the axis line. The Properties palette for a column shown in Figure 13.6 is described below.

General

> **Description** – The **Description** button opens a Description dialog box, which allows you to add text describing the column.
>
> **Layer** – The **Layer** field displays the layer of the column. Columns are placed on the **S-Cols** layer (color 52, mustard) if the AIA 256 Color layer key style is used in the drawing. This option is only displayed in the Properties palette when you are editing existing columns.

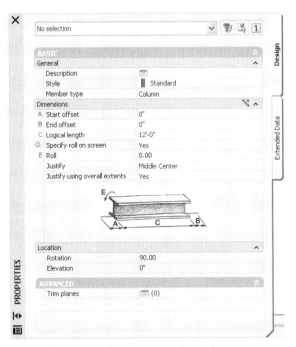

Figure 13.6 *Properties palette of a column structural member*

Style – The **Style** drop-down list consists of the styles created in the drawing from the **Structural Member Catalog**.

Member Type – **Column**.

Dimensions

A-Start offset – The **Start offset** edit field allows you to specify the distance to adjust the bottom of the column relative to start position (Z=0) of the axis line. A positive distance starts the column shape above the axis start point, as shown in Figure 13.7, while a negative distance extends the column shape below the axis start point.

B-End offset – The **End offset** edit field allows you to specify the distance to adjust the column above the top of the axis line. A positive offset extends the shape above the axis line and a negative offset will shorten the column from the end of the axis line.

C-Logical length – The **Logical length** edit field allows you to specify the length of the column axis line.

Specify roll on screen – The **Specify roll on screen Yes** option allows you to specify the **Roll** in the workspace or in the Properties palette.

E-Roll – The **Roll** edit field allows you to specify the rotation angle about the axis line. Positive angles rotate the shape about the axis line in a counterclockwise direction.

Justify – The **Justify** list allows you to specify the location of the axis line relative to the geometry of the shape as shown in Figure 13.8. In addition to the nine positions shown in Figure 13.8, the **Baseline** option is located on the centroid of the column.

Justify using overall extents – The **Justify using overall extents** options apply to structural members with multiple shapes throughout the length of the member. If the Yes option is selected, the justification is applied at each vertex of the member based upon the largest cross-section. If the No option is selected, the justification is based upon the priorities of the shaped defined in the style.

Advanced

Trim planes – A trim plane can be specified for the ends of the column. The trim plane cuts the end of the member at an angle relative to the baseline of the column. Selecting the **Trim Planes** button opens the **Column Trim Planes** dialog box to specify angle and location of the plane.

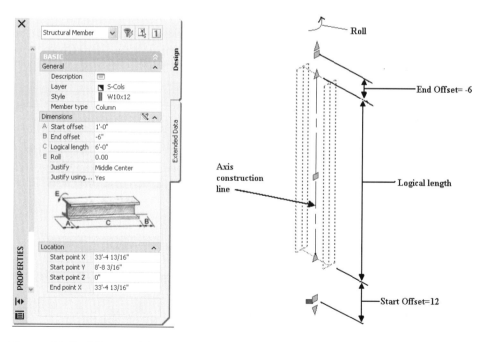

Figure 13.7 *Offsets and logical length of the column*

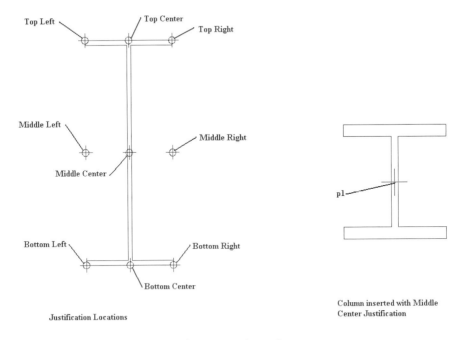

Figure 13.8 *Justification locations of a structural member*

STEPS TO ADDING A COLUMN

1. Select **Generate Member Style** from the toolbar of the **Structural Member Catalog** to create a style. (Structural member styles can also be imported into the drawing in the **Style Manager**.)
2. Select **Structural Column** from the Design tool palette.
3. Specify the Style, Logical length, Start offset, End offset, Justify, Roll, and trim planes of the column in the Properties palette.
4. Specify the location of the column by either selecting a node on a column grid or by pressing ENTER to specify a freestanding location as shown in the following command sequence.

    Command: AECSCOLUMNADD

    Select grid or RETURN: ENTER *(Press ENTER to specify a freestanding column.)*

    Insert point or [STyle/Length/STArt offset/ENd offset/Justify/Roll/Match]: *(Specify a location p1 as shown in Figure 13.8.)*

    Insert point or [STyle/Length/STArt offset/ENd offset/Justify/Roll/Match/Undo]: ENTER *(Press ENTER to end the command.)*

INSERTING BEAMS

Beams are placed in the drawing by selecting the **Structural Beam** tool from the Design tool palette (refer to Table 13.4 for command access).

Command prompt	BEAMADD
Palette	Select Structural Beam from the Design tool palette as shown in Figure 13.5

Table 13.4 *BeamAdd command access*

When you select the **Structural Beam** tool from the Design palette, the Properties palette opens, allowing you to specify the offsets, justification, and roll of the beam. The Properties palette of a beam is shown in Figure 13.9.

The options of the Properties palette of a **Structural Beam** are described below.

General

    Description – The **Description** button opens the **Description** dialog box, allowing you to add text.

    Layer – The **Layer** edit field lists the layer name of existing beams. Beams are placed on the **S-Beam** layer, (color 12, red) if the AIA 256 Color layer key style is used.

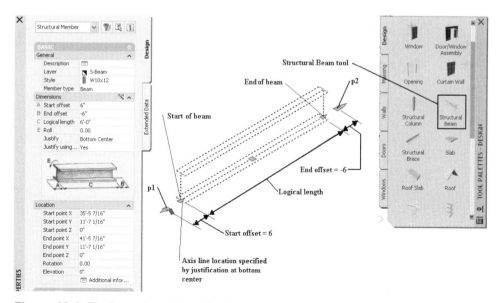

Figure 13.9 *The Properties palette of a beam*

Style – The **Style** list consists of the styles created in the drawing from the **Structural Member Catalog**.

Member type – **Beam**.

Dimensions

A-Start offset – The **Start offset** edit field allows you to specify the distance to adjust the start of the beam relative to the start position of the axis line for the beam. A positive distance starts the beam short of the axis start point, while a negative offset increases the length of the beam from the axis start point. A +6 offset is applied to the start point of the beam as shown in Figure 13.9. The start offset shown in Figure 13.10 creates clearance for assembly and attachment of connecting angles.

B-End offset – The **End offset** edit field allows you to specify the distance to adjust the beam relative to the end of the axis line. A positive offset extends the shape beyond axis line, and a negative offset will shorten the beam from the axis line. A –6 offset is applied to the beam shown in Figure 13.9.

C-Logical length – The dynamic display of the **Logical length** is the length of the axis line of the beam.

E-Roll – The **Roll** edit field allows you to specify the rotation angle about the axis line. Positive angles rotate the shape about the axis line in a counterclockwise direction. The counterclockwise direction is from the orientation of the observer positioned at the start point of the beam and facing the endpoint of the beam.

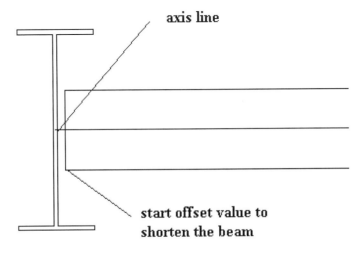

Figure 13.10 *Start offset adjusted to enable assembly*

Justify – The **Justify** list allows you to specify the location of the axis line relative to the geometry of the shape as shown in Figure 13.8. In addition to the nine positions shown, the **Baseline** option is located on the centroid of the beam.

Justify using overall extents – The **Justify using overall extents** applies to justification throughout the length of the member based upon the largest cross-section. If the **Yes** option is selected, the justification is applied at each vertex of the member based upon the largest cross-section. If the **No** option is selected, the justification is based upon the priorities of the shape defined in the style.

Advanced

Trim planes – The **Trim planes** button opens the **Beam Trim Planes** dialog box, which allows you to specify an angle to trim the beam relative to the axis line.

STEPS TO ADDING A BEAM

1. Create a style from the components of the **Structural Member Catalog**.
2. Select **Structural Beam** from the Design tool palette.
3. Specify the Style, Start offset, End offset, Justify, Roll, and trim planes of the beam in the Properties palette as shown in Figure 13.9.
4. Specify the location of the beam by specifying the start point and end point of the axis line as shown in the following command sequence.

   Command: BeamAdd

   Start point or [STyle/STArt offset/ENd offset/Justify/Roll/Match]: *(Specify the start point of the beam at p1 as shown in Figure 13.9.)*

   End point or [STyle/STArt offset/ENd offset/Justify/Roll/Match]: *(Specify the end of the beam at p2 as shown in Figure 13.9.)*

   End point or [STyle/STArt offset/ENd offset/Justify/Roll/Match/Undo]: ENTER *(Press ENTER to end the command.)*

 STOP. Do Tutorial 13.1, "Creating Deck Floor Framing," at the end of the chapter.

INSERTING A BRACE

Braces are placed in the drawing by selecting the **Structural Brace** tool in the Design tool palette. The brace is added on the **S-Cols-Brce** layer, which has color 32 (a hue of brown) and dashed linetype. Access the **Structural Brace** tool as shown in Table 13.5.

Command prompt	BRACEADD
Palette	Select Structural Brace from the Design Tool palette as shown in Figure 3.5.

Table 13.5 *BraceAdd command access*

If you select the **Structural Brace** tool from Design tool palette, the Properties palette opens as shown in Figure 13.11. This palette allows you to specify the style and position of the structural member about the axis line.

The options of a **Structural Brace** in the Properties palette are described below.

General

Description – The **Description** button opens the **Description** dialog box for adding text.

Layer – The **Layer** edit field lists the layer name of existing braces.

Style – The **Style** list consists of the styles created in the drawing from the **Structural Member Catalog**.

Member type – **Brace**.

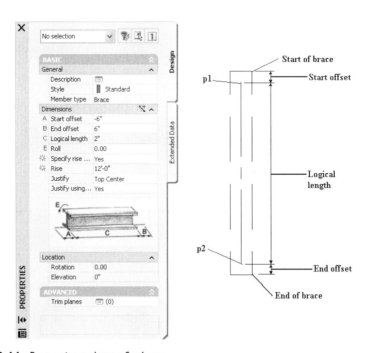

Figure 13.11 *Properties palette of a brace*

Dimensions

A-Start offset – The **Start offset** edit field allows you to specify the distance to adjust the start of the brace relative to the start position of the axis line for the brace. A positive distance starts the brace short of the axis start point, while a negative offset increases the length of the brace from the axis start point. A zero start offset will start the brace at the same location as the start of the axis line.

B-End offset – The **End offset** edit field allows you to specify the distance to adjust the brace relative to the axis line. A positive offset extends the shape beyond the axis line, and a negative offset will shorten the brace from the axis line. A zero end offset will end the brace at the same location as the end of the axis line.

C-Logical length – The **Logical length** readout displays the length of the brace dynamically as the end point of the brace is specified.

E-Roll – The **Roll** edit field allows you to specify the rotation angle about the axis line. Positive angles rotate the shape about the axis line in a counterclockwise direction. The counterclockwise direction is from the orientation of the observer positioned at the start point of the brace and facing the end point of the brace.

Specify rise on screen – The **Specify rise on screen** includes **Yes** and **No** options. The **Yes** option allows you to specify the rise in the z direction in the workspace. The **No** option activates the **Rise** edit field of the Properties palette.

Rise – The **Rise** value is the z direction displacement of the end of the brace.

Justify – The **Justify** list allows you to specify the location of the axis line relative to the geometry of the shape as shown in Figure 13.8. In addition to the nine positions shown, the **Baseline** option is located on the centroid of the brace.

Justify using overall extents – The **Justify using overall extents** applies to structural members that apply multiple shapes throughout the length of the member. If the **Yes** option is selected, the justification is applied at each vertex of the member based upon the largest cross-section. If the **No** option is selected, the justification is based upon the priorities of the shaped defined in the style.

STEPS TO ADDING A BRACE

1. Create a style from the components of the **Structural Member Catalog**.
2. Select **Structural Brace** from the Design tool palette.
3. Specify the Style, Start offset, End offset, Justify, Roll, and Rise of the brace in the Properties palette as shown in Figure 13.11.

4. Specify the location of the brace by specifying the start point and end point of the axis line as shown in the following command sequence.

Command: _AecsBraceAdd

Start point or [STyle/STArt offset/ENd offset/Justify/Roll/RIse/Match]: *(Specify the start point of the brace at p1, as shown in Figure 13.11.)*

End point or [STyle/STArt offset/ENd offset/Justify/Roll/RIse/Match]: *(Specify the end of the brace at p2, as shown in Figure 13.11.)*

Start point or [STyle/STArt offset/ENd offset/Justify/Roll/RIse/Match/Undo]: ENTER

(Brace is added as shown in Figure 13.11.)

A brace that is added from the top view with a z coordinate above the cut plane elevation will be displayed as shown in Figure 13.12. The rectangle shown in Figure 13.12 represents the point where the brace passes through the cutting plane. The center line is the axis line of the brace. A horizontal brace would be displayed as shown in Figure 13.11. The cutting plane can be controlled in the Object Display of the brace, the style of the brace, or in the display configuration definition. The display configuration definition of Medium Detail includes a cutting plane elevation of 3'-6" as shown in the **Cut Plane** tab of Figure 13.13.

MODIFYING ENDS WITH TRIM PLANES

Trim planes can be created to modify the start and end of structural components. Trim planes are specified by selecting the **Trim planes** button located in the **Advanced** section of the Properties palette. The brace shown in Figure 13.14 includes four trim planes. The **Brace Trim Planes** dialog box shown in Figure 13.14 is similar to **Column Trim Planes** and **Beam Trim Planes** dialog boxes. The **Add**, **Copy**, and **Remove** buttons located at the right of the **Brace Trim Planes** dialog box allow you to create, copy, and delete trim planes. The location of the trim plane can be positioned relative to the start or end of the brace. The offset values position the trim plane relative to the end of the axis line. The trim plan is rotated in the x, y, and z directions.

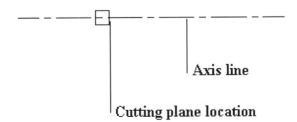

Figure 13.12 *Brace added with end point above the cutting plane line*

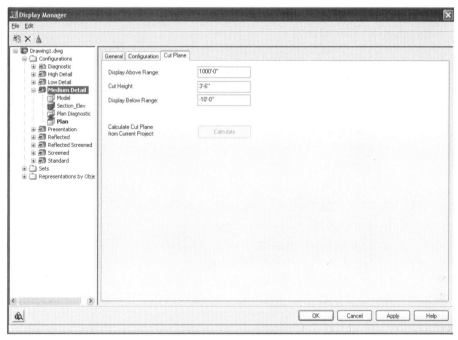

Figure 13.13 *Cut Plane location specified in the Display Manager*

The x axis is aligned with the axis line of the structural member. Therefore, in the example shown in Figure 13.14, Trim Plane 1 cuts the brace with a horizontal trim plane located 1" short of the start of the axis line. The brace is trimmed 1" short of its total length to provide clearance with the web of the column.

CUSTOMIZING STRUCTURAL MEMBERS USING STYLES

Styles created from the **Structural Member Catalog** can be customized in the style definition. The **Edit Member Style** command edits the properties of a structural member style. A style created from the **Structural Member Catalog** assigns a shape for the axis line. Materials, classifications, and display properties can be edited in the member style. You can assign a shape for each end of the structural component or combine two shapes for a structural component. Access the **Edit Member Style** command as shown in Table 13.6.

Command prompt	MEMBERSTYLEEDIT
Shortcut menu	Select a structural member, right-click, and choose Edit Member Style

Table 13.6 *MemberStyleEdit command access*

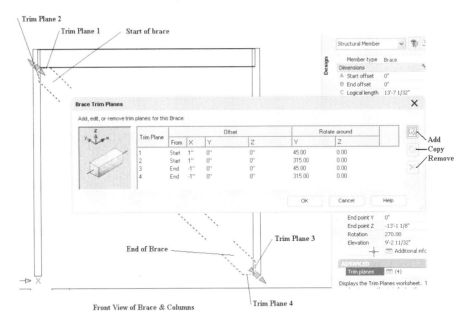

Figure 13.14 *Trim planes of a brace*

When you select a structural member, right-click, and choose **Edit Member Style**, the **Structural Member Style Properties** dialog box opens as shown in Figure 13.15.

General Tab

The **General** tab shown in Figure 13.15 lists the name of the style, description, and keynotes.

Design Rules Tab

The **Design Rules** tab consists of columns that define the shape name for the start and end segments of the structural member. The content of the tab is displayed when you select the **Show Details** button. The content of the **Design Rules** tab is described below.

> **Component** – The **Component** name allows you to create a component name for the shape that is developed along the axis line.
>
> **START AND END SHAPES**
>
> > **Name** – The **Name** drop-down list displays the structural member styles included in the drawing. The W10x12 structural style is the shape assigned to the start of the structural component as shown in Figure 13.15.

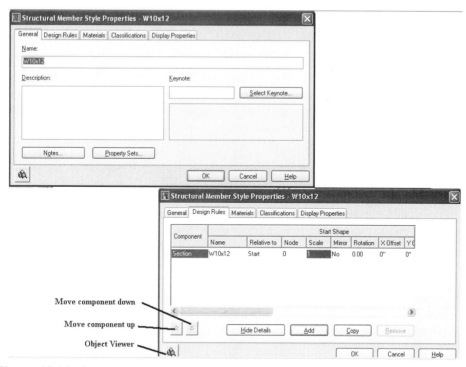

Figure 13.15 *Structural Member Style Properties dialog box*

Relative to – The **Relative to** options include Start and End. The shape can be defined relative to the start or end of the axis line. The component shapes for the start or end of the axis line can be different.

Node – The **Node** column allows you to divide the length of the structural member by placing nodes at each vertex of the member path defined by a polyline. Node 0 is assigned to the start of the path. A different shape can be defined for each node.

Scale – The **Scale** option allows you to scale the shape. A scale of 1 inserts the shape equal to the size defined by the structural member style.

Mirror – The **Mirror** option will develop a mirrored image of the shape.

Rotation – The **Rotation** angle will rotate the shape about the axis line.

X Offset – The **X Offset** will start the shape an offset distance along the X axis. The X axis is aligned with the axis line of the structural member.

Y Offset – The **Y Offset** will move the shape in the Y direction relative to the axis line.

Z Offset – The **Z Offset** will move the shape in Z direction relative to the axis line.

Priority – Components that share a common node are mitered if they have the same Priority number. Clear the **Justify Overall Extents** check box on the **Dimensions** tab of the member properties dialog box to create the miter about the justification line for the lowest priority component.

Hide Details/Show Details – The **Hide Details/Show Details** button expands or contracts the component information. When the details are hidden, only the start shape and component name are displayed. The details are shown in Figure 13.15; select the scroll bar to view the End Shape data hidden on the right.

Add – Select the **Add** button to create a new component.

Copy – Select the **Copy** button to create a new component as a copy of an existing component.

Remove – Select the **Remove** button to remove a selected component.

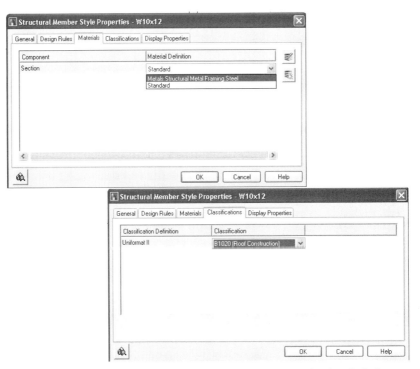

Figure 13.16 *Materials and Classifications tabs of the Structural Member Style Properties dialog box*

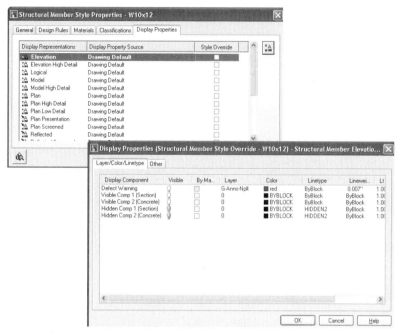

Figure 13.17 Display Properties tab and Layer/Color/Linetype tab of Display Properties dialog box

Materials Tab

The **Materials** tab consists of a list of the components of the member. Select the Material Definition to assign materials to a component as shown in Figure 13.16.

Classifications Tab

The **Classifications** tab lists the classifications that are defined in the drawing. Classifications are imported in the **Style Manager**.

Display Properties Tab

The **Display Properties** tab lists the display representations and specifies the display property source for each display representation. If you check the **Style Override** for a display representation, you can specify the display properties for each component as shown in Figure 13.17.

CONVERTING LINEWORK TO STRUCTURAL MEMBERS

The **Structural Column**, **Structural Beam**, and **Structural Brace** tools located on the Design tool palette all include a shortcut menu option to convert linework to the structural member. Therefore you can draw an arc, line, or polyline and convert the geometry to a structural member. The linework becomes the axis line for the structural member. Access the **ColumnToolToLinework** command as shown in Table 13.7.

Command prompt	COLUMNTOOLTOLINEWORK
Palette	Select Structural Column from the Design Tool palette, right-click, and choose Apply Tool Properties to>Linework

Table 13.7 *ColumnToolToLinework command access*

The **ColumnToolToLinework** command allows you to draw in elevation or pictorial views lines, arcs, or polylines and convert the geometry to a column. The linework is converted to a column by selecting the **Structural Column** tool of the Design tool palette, right-clicking, and choosing **Apply Tool Properties to>Linework**. The following command line sequence converted a line shown in front view to a column.

Command: ColumnToolToLinework

Select lines, arcs, or open polylines to convert into members: 1 found *(Select linework as shown in Figure 13.18.)*

Select lines, arcs, or open polylines to convert into members: ENTER *(Press ENTER to end selection.)*

Erase layout geometry? [Yes/No] <No>: ENTER *(Press ENTER to retain the linework.)*

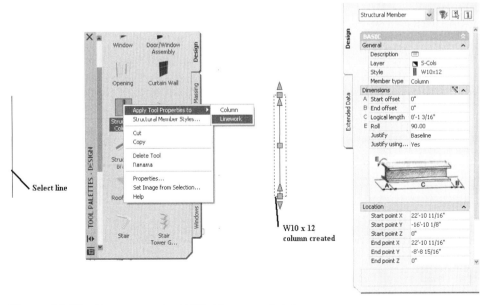

Figure 13.18 *Line converted to W10x12 structural style*

1 new member(s) created

(Edit the Properties palette to specify the column type as shown in Figure 13.18.)

If you draw a polyline to represent the column, you can assign various structural shapes to each segment of the polyline. The polyline drawn in Figure 13.19 consists of a lower segment 12" long and an upper segment 4' long. The lower segment is converted at right to a 12 × 12 concrete shape in the **Design Rules** tab of the **Structural Member Style Properties** dialog box shown in Figure 13.19. The upper segment is converted to a 4 × 4 wood shape. A node has been assigned to the vertex of the polyline. A shape can be defined to start and end at the node in the **Design Rules** tab. The **ColumnToolToLinework** tool allows you to convert the polyline to a column that applies the shapes defined for the nodes.

The **BeamToolToLinework** tool allows you to draw linework such as arcs to create beams for canopies. Access the **BeamToolToLinework** tool as shown in Table 13.8.

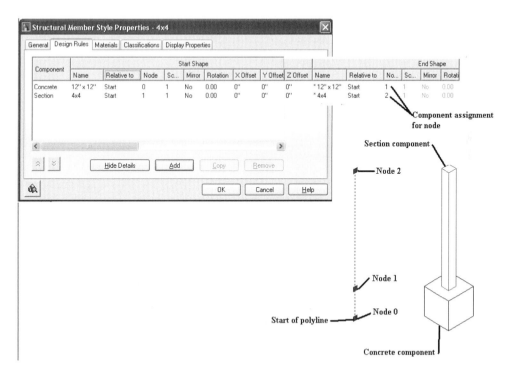

Figure 13.19 *Polylines converted to members and the Structural Member Style Properties dialog box*

Command prompt	BEAMTOOLTOLINEWORK
Palette	Select Structural Beam from the Design Tool palette, right-click, and choose Apply Tool Properties to>Linework

Table 13.8 *BeamToolToLinework command access*

The **BeamToolToLinework** command was applied to the arc drawn in elevation view to create the curved beam shown in Figure 13.20.

> *(Select the **Structural Beam** tool of the Design tool palette, right-click, and choose **Apply Tool Properties to>Linework**.)*
>
> Command: BeamToolToLinework
>
> Select lines, arcs, or open polylines to convert into members: 1 found *(Select the arc at p1 shown in Figure 13.20.)*
>
> Select lines, arcs, or open polylines to convert into members: ENTER *(Press ENTER to end selection.)*
>
> Erase layout geometry? [Yes/No] <No>: ENTER *(Press ENTER to retain the geometry.)*
>
> 1 new member(s) created.
>
> *(Edit the Properties palette to assign structural style.)*

Braces can also be created from linework. Access the **BraceToolToLinework** command as shown in Table 13.9.

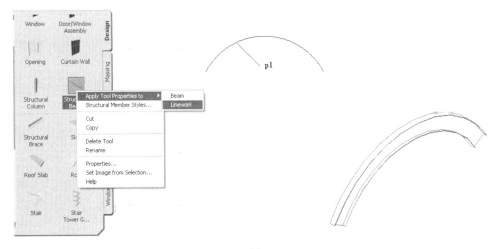

Figure 13.20 *Converting an arc to a structural beam*

Command	BRACETOOLTOLINEWORK
Palette	Select Structural Brace from the Design Tool palette, right-click, and choose Apply Tool Properties to>Linework

Table 13.9 *BraceToolToLinework command access*

CREATING A RIGID FRAME

A rigid frame can be created by drawing a polyline in the shape of the rigid frame as shown in Figure 13.21. The polyline becomes the path for the development of the rigid frame. The structural shapes used in the rigid frame are then developed as styles from the **Structural Member Catalog**. The styles are applied to each segment of the polyline by defining a shape to the nodes in the **Design Rules** tab of the **Structural Member Style Properties** dialog box.

The polyline is converted to a structural shape by selecting **Structural Column**, right-clicking, and choosing **Apply Tool Properties to>Linework** from the Design tool palette. When the polyline is converted to a structural shape, the Standard style is applied. The Standard style applies a 12 × 12 square shape to the path defined by the polyline. Therefore, to create a rigid frame that includes wide flange shapes, you must create styles from the **Structural Member Catalog** and apply them to the style. The wide flange styles can be applied to the nodes of the rigid frame as shown in Figure 13.22.

The **Vertical left** component shown in Figure 13.22 starts and ends with the W14×53 shape; however, the **Horizontal left** component transitions from W14×53 to W10×12 at node 2. The **Design Rules** tab allows you to assign the shapes necessary to create the rigid frame.

Figure 13.21 *Polyline drawn to create path for rigid frame*

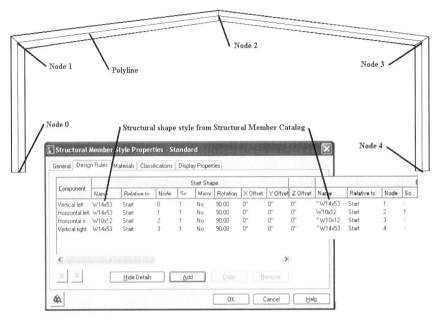

Figure 13.22 *Rigid frame developed from polyline*

CREATING COMPLEX COLUMNS

A structural member style can be created to provide encasement materials for a column. The Architectural Desktop Design Tool Catalog-Imperial>Structural>Members catalog in the **Content Browser** includes examples of encased columns. The tools provided are shown in Figure 13.3 and can be used as a base to develop encased columns for other column sizes. The **Design Rules** tab of the **Structural Member Style Properties** dialog box is shown in Figure 13.23. The Standard and W10×12 shapes are applied throughout the path to create the encasement.

CREATING CUSTOM SHAPES

Custom shapes not included in the **Structural Member Catalog** can be created. The **-AecsMemberShapeDefine** command allows you to create a shape from a closed polyline. Access the **-AecsMemberShapeDefine** command as shown in Table 13.10. The command allows you to assign three levels of display representation: plan **Low Detail**, **Plan**, and plan **High detail**. Separate geometry can be defined for each of the display representations, or the same geometry can be used for each of the display representations. The shape name that is created can be assigned to a structural style in the **Design Rules** tab of the **Structural Member Style Properties** dialog box as shown in Figure 13.24. The options of the command also allow you to edit existing defin-

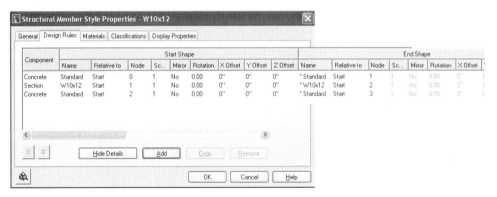

Figure 13.23 *Encased columns*

itions and purge the shape definitions from the drawing. The Copy option allows you to copy shapes included in the drawing for the development of new shapes.

Command prompt	-AECSMEMBERSHAPEDEFINE
Menu bar	Format>Define Custom Member Shape

Table 13.10 *-AecMemberShapeDefine command access*

STEPS TO CREATING A CUSTOM SHAPE

1. Draw a closed polyline that represents the shape.
2. Select **Format>Define Custom Member Shape** from the menu bar, and respond to the command line as follows:

    Command: -AECSMEMBERSHAPEDEFINE

    Shape [New/Copy/Edit/Purge/?]: **n** ENTER *(Type **N** to create a new shape.)*

    New style name or [?]: **brick** *(Type the name of the new shape.)*

    New style brick created.

    Shape definition [Name/Description/Graphics]: **g** *(Type **g** to define the graphics for the new shape.)*

    Shape [plan Low detail/Plan/plan High detail]: **L** *(Type **L** to define plan Low detail display representation.)*

    Base Point for Sketch Representation: *(Select a point near p1 as shown in Figure 13.24.)*

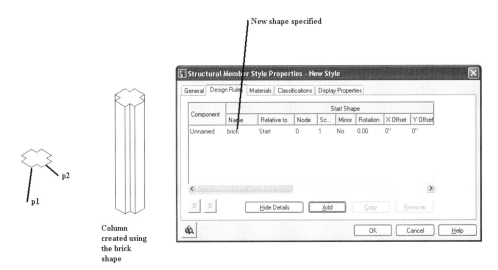

Figure 13.24 *Creating a custom shape from a closed polyline*

Select lines, arcs, or polylines for Sketch Representation: 1 found *(Select the closed polyline near p2 as shown in Figure 13.24.)*

Select lines, arcs, or polylines for Sketch Representation: ENTER *(Press ENTER to end selection.)*

Shape [plan Low detail/Plan/plan High detail]: **p** *(Type **P** to specify the Plan display representation.)*

Erase polyline? [Yes/No] <No>: ENTER *(Press ENTER to retain the polyline.)*

Select a closed polyline: *(Select the polyline at p2 as shown in Figure 13.24.)*

Add another ring? [Yes/No] <No>: ENTER *(Press ENTER to reject adding another ring.)*

Insertion Point or <Centroid>: ENTER *(Press ENTER to define the centroid as the insertion point.)*

Shape [plan Low detail/Plan/plan High detail]: **h** *(Type **h** to specify the plan High detail display representation.)*

Erase polyline? [Yes/No] <No>: ENTER *(Press ENTER to retain the polyline.)*

Select a closed polyline: *(Select the polyline at p2 as shown in Figure 13.24.)*

Add another ring? [Yes/No] <No>: ENTER *(Press* ENTER *to reject adding another ring.)*

Insertion Point or <Centroid>: ENTER *(Press* ENTER *to accept the centroid as the insertion point for the plan high detail.)*

Shape [plan Low detail/Plan/plan High detail]: ENTER *(Press* ENTER *specifies null shape to end the shape prompt.)*

Shape definition [Name/Description/Graphics]: ENTER *(Press* ENTER *to end shape definition prompt.)*

Shape [New/Copy/Edit/Purge/?]: ENTER *(Press* ENTER *to end the command.)*

3. Select **Format>Style Manager** from the menu bar. Expand the Architectural Objects>Structural Member Styles. Select Structural Member Styles in the left pane of the **Style Manager**, and select **New Style** from the **Style Manager** toolbar. Select **New Style** in the right pane, right-click, and choose **Edit** to open the **Structural Member Style Properties – New Style** dialog box.

4. Select the **Design Rules** tab of the **Structural Member Style Properties – New Style** dialog box, and select **brick** from the drop-down list in the **Name** column. This step defines the new brick shape created in step 2 above, for the structural component.

CREATING STRUCTURAL MEMBERS IN THE STYLE WIZARD

Custom sizes for the shapes included in the **Structural Member Catalog** can be developed in the **Structural Member Style Wizard**. Access the **Structural Member Style Wizard** as shown in Table 13.11. The **Structural Member Style Wizard** shown in Figure 13.25 includes a tree listing of the shapes used in the **Structural Member Catalog**. The first page of **Structural Member Style Wizard** allows you to select a shape from the tree list. Select the **Next** button at the bottom of the first page to open the next page and define the sizes for the shape as shown in Figure 13.26. The final page of the wizard allows you to specify the name for the style. A summary of properties of the style is listed as shown in Figure 13.26. The new shape can be specified as a shape for a component in the **Structural Member Style Properties** dialog box.

Command prompt	MEMBERSTYLEWIZARD
Menu bar	Format>Structural Member Style Wizard

Table 13.11 *MemberStyleWizard command access*

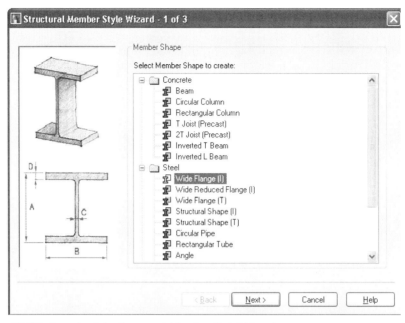

Figure 13.25 *Page 1 of the Structural Member Style Wizard*

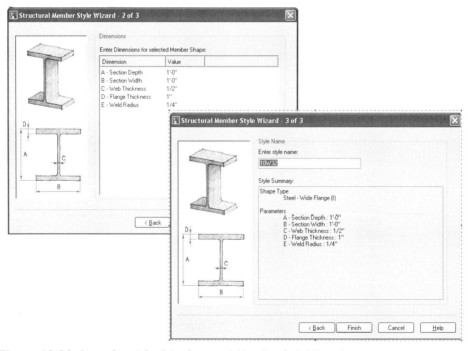

Figure 13.26 *Pages 2 and 3 of the Structural Member Style Wizard*

CREATING BAR JOISTS

The tools for creating open web joists are located in the Bar Joists catalog. This catalog is accessed from the **Content Browser** in the Architectural Desktop Design Tool Catalog-Imperial>Structural catalog. The bar joist tools can be i-dropped from the catalog to the current drawing. When a bar joist tool is i-dropped into the drawing, the style will be included in the **Style** list of the Properties palette. Therefore, additional joists can be added using the **Structural Beam** tool with the Bar Joists style assigned in the Properties palette. The **Baseline** justification should be used when placing bar joists, because the baseline is located at the bearing surface of the bar joist as shown in Figure 13.27.

STEPS TO CREATING A BAR JOIST

1. Select **Content Browser** from the **Navigation** toolbar.
2. Double-click on the **Architectural Desktop Design Tool Catalog-Imperial**.
3. Select the **Structural >Bar Joists** catalog as shown in Figure 13.28.
4. Click and drag the steel joist from the catalog to the drawing as shown in Figure 13.29.
5. The **BeamAdd** command is current, and the bar joist style is the current style. Verify that the justification is Baseline. Select the start and end point locations of the joist

Note: The **Bar Joist** tool can be i-dropped from the catalog to a tool palette for future use.

CREATING STRUCTURAL GRIDS

Column grids are developed to uniformly place structural columns in the plan view. The size and orientation of the structural grid are dependent upon the floor plan and maximum span of the structural components. The column grid is created on the S-Grid layer, which has the color 191 (a hue of purple) with Center2 linetype.

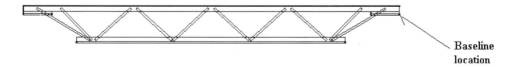

Figure 13.27 *Baseline location of a bar joist*

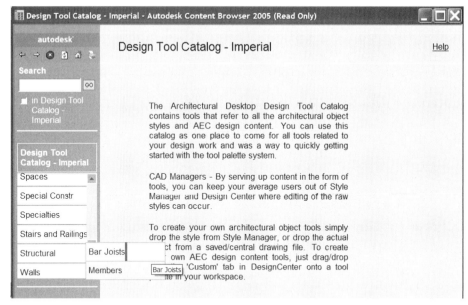

Figure 13.28 *Architectural Desktop Design Tool Catalog-Imperial*

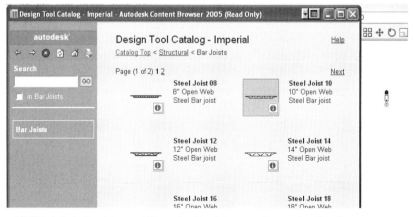

Figure 13.29 *i-drop Bar Joist tool from the Bar Joists catalog to the drawing*

The column grid can be created with structural components from the **Structural Member Catalog**. The structural component is assigned in the Style edit field of the Properties palette as shown in Figure 13.30. The column shape is inserted at each grid intersection. The **Structural Column Grid** command (**ColumnGridAdd**) is used to create a column grid. The grid can be rectangular or radial in shape. See Table 13.12 for **Structural Column Grid** command access.

Command	COLUMNGRIDADD
Palette	Select Structural Column Grid from the Design tool palette

Table 13.12 *Structural Column Grid command access*

When you select the **Structural Column Grid** tool, the Properties palette opens, allowing you to specify the parameters of the column grid, as shown in Figure 13.30.

The rectangular column grid allows you to specify the overall size of the grid according to the **X-Width** and **Y-Depth** dimensions. The **X-Width** and **Y-Depth** dimensions can be spaced evenly or repeated by a bay size dimension. The column grid is developed relative to its insertion point, which is in the lower left corner, as shown in Figure 13.30. The options of the Properties palette of a structural column grid are described below.

Basic

General

> **Description** – The **Description** button opens the **Description** dialog box, allowing you to add additional text that describes the grid.
>
> **Shape** – The **Shape** list includes radial and rectangular grid shapes. The options of the dialog box change according to the shape selected.

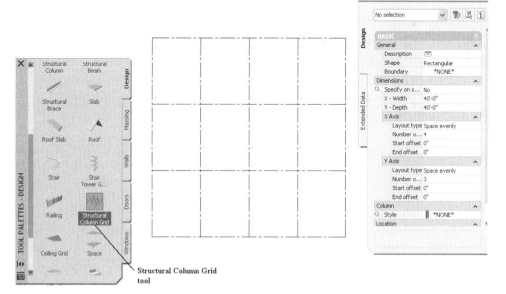

Figure 13.30 *Column Grid Properties palette*

Boundary – The **Boundary** option includes a list of closed polylines applied to the grid to mask the grid beyond the boundary of the polyline.

Dimensions

X-Width – The **X-Width** dimension of the grid is the overall horizontal dimension of the grid.

Y-Depth – The **Y-Depth** dimension of the grid is the vertical dimension of the grid.

X AXIS

Layout type – The **Layout type** options include Manual, Space evenly, and Repeat.

> **Manual** – The **Manual** option is only available for editing existing column grids. The **Manual** option creates a **Bays** button in the Properties palette. The **Bays** button opens a **Bays along the X Axis** dialog box, which lists the distance to line for each bay of the grid.
>
> **Space evenly** – The **Space evenly** option expands the X-Axis options, allowing you to specify the Number of bays, Bay size, Start offset, and End offset.
>
> **Repeat** – The **Repeat** option allows you to specify the bay distance.

Number of Bays – The **Number of Bays** specifies the number of bays to divide the overall X-Width dimension to create the grid. The **Number of Bays** option is active if the **Space evenly** layout method is used.

Bay size – The **Bay size** displays the bay distance in the X direction when the Repeat layout type is specified.

Start offset – The **Start offset** allows you to specify the distance from the insertion point to start the column grid.

End offset – The **End offset** edit field allows you to specify the distance from the diagonal corner of the overall column grid to end the grid.

Y AXIS

Layout type – The **Layout type** options include Manual, Space evenly, and Repeat.

> **Manual** – The **Manual** option is only available for editing existing column grids. The Manual option creates a **Bays** button in the Properties palette. The **Bays** button opens a **Bays along the Y Axis** dialog box, which lists the distance to line for each bay of the grid.
>
> **Space evenly** – The **Space evenly** option expands the Y-Axis, allowing you to specify the Number of bays, Bay size, Start offset, and End offset.
>
> **Repeat** – The **Repeat** option allows you to specify the bay distance.

Number of Bays – The **Number of Bays** specifies the number of bays to divide the overall **Y-Width** dimension to create the grid. The **Number of Bays** option is active if the **Space evenly** layout method is used.

Bay size – The **Bay size** lists the bay distance in the Y direction when the **Repeat** layout type is specified. The **Bay size** also includes **Pick two points to set a distance** button in the edit field, which allows you specify the distance by selecting two points in the workspace.

Start Offset – The **Start Offset** allows you to specify the distance from the insertion point to start the column grid.

End Offset – The **End Offset** edit field allows you to specify the distance from the diagonal corner of the overall column grid to end the grid.

Column

Style – The **Style** edit field allows you to specify the structural member style. The specified structural member style is inserted at each of the intersections of the column grid. If a style is specified, the **Dimensions** section of the Properties palette is added.

DIMENSIONS

Start offset – The **Start offset** edit field allows you to specify the distance to adjust the bottom of the column relative to start position (Z=0) of the axis line. A positive distance starts the column shape above the axis start point, while a negative distance extends the column shape below the axis start point.

End offset – The **End offset** edit field allows you to specify the distance to adjust the column above the top of the axis line. A positive offset extends the shape above the axis line, and a negative offset will shorten the column from the end of the axis line.

Logical length – The **Logical length** edit field allows you to specify the length of the column axis line.

Justify – The **Justify** list allows you to specify the location of the axis line relative to the geometry of the shape as shown in Figure 13.8.

Justify using overall extents – The **Justify using overall extents** options apply to structural members with multiple shapes throughout the length of the member. If the **Yes** option is selected, the justification is applied at each vertex of the member based upon the largest cross-section.

Location

Rotation – The **Rotation** edit field specifies the angle of rotation of the grid from the three o'clock position

Elevation – The **Elevation** edit field lists the elevation in the z direction of the grid.

Additional Information – The **Additional Information** button opens the **Location** dialog box, which allows you to edit the insertion point, normal, and rotation fields of the grid.

Advanced

Trim planes – The **Trim planes** button opens the **Column Trim Planes** dialog box for defining planes to cut the end of the column at an angle.

Creating a Radial Column Grid

A radial column grid is created by selecting the **Radial** shape from the **Shape** list of the Properties palette. The radius of the column grid specifies the overall size and the angle specifies the included angle of the grid, as shown in Figure 13.31. The insertion point of the grid is at the center of the radius. The grid can be created by the **Repeat** or **Space evenly** layout. In Figure 13.31, the radial dimension has been divided into four equal spaces, while the **A-Angle** has been divided into three equal spaces.

The options of the Properties palette unique to a radial column grid are described below.

X-Width – The **X-Width** edit field defines the length of the radial grid lines.

A-Angle – The **A-Angle** edit field specifies the included angle of the column grid. The column grid is developed counterclockwise from the three o'clock direction.

Baysize – The **Baysize** is the distance between curved column grid lines. The **Baysize** specifies the grid size when the **Repeat** layout type is specified.

Number of bays – The **Number of bays** specifies the number of divisions to divide the **X-Width**. The **Number of bays** option is available when **Space evenly** is the layout type.

Inside Radius – The **Inside Radius** specifies the distance from the insertion point to the inner most radial column grid.

End offset
Y AXIS

Bay angle – The **Bay angle** is the angular measure between the radial column grid lines. The **Bay angle** specifies the angle between radial column grids when the Repeat layout type is specified. Three bays would be created if the **A-Angle** were set to 90 and the **Bay angle** were set to 30. If the **A-Angle** measure is not divisible by the **Bay angle** measure, the column grid is created equal to the **Bay angle** measure. In Figure 13.31, if the **A-Angle** is set to 90 and the **Bay angle** is set to 50, then a 50-degree column grid would be created.

Start angle – The **Start angle** is the angle of measure from the three o'clock position to start the radius.

End angle – The **End angle** reduces the total angle of the grid by a specified angle.

Number of bays – The **Number of bays** specifies the number of divisions to divide the **A-Angle**. The **Number of bays** option is available when **Space evenly** is the layout type.

SPECIFYING THE SIZE OF THE COLUMN GRID

The size and number of divisions of the rectangular column grid can be specified in the **X-Width** and **Y-Depth** fields of the Properties palette. The **Repeat** and **Space evenly** layout types allow you to type the number of bays or the bay size in the Properties palette. The bay size and number of bays can also be determined dynamically on screen. When the **Specify on screen** option is selected, you can specify the overall size of the column grid by selecting two diagonal points. Moving the pointer from the last diagonal point dynamically sets the number of divisions. As you move the pointer back **toward the direction of the insertion point,** the **number of bays increases**. Movement of the pointer in a horizontal direction toward the insertion point increases the number of divisions along the horizontal grid line. Vertical movement of the pointer in the direction toward the insertion point will increase the number of divisions along the vertical grid line.

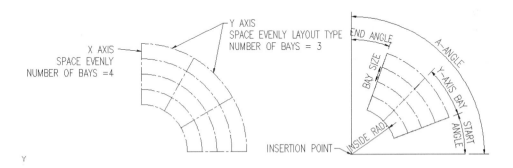

Figure 13.31 *Radial column grid*

STEPS TO DYNAMICALLY SIZING A RECTANGULAR COLUMN GRID

1. Select the **Structural Column Grid** tool from the Design palette.
2. Select the **Rectangular** shape from the **Shape** list.
3. Select **Yes** to the **Specify on screen** option of the Properties palette.
4. Select **Space** evenly Layout type for the **X-Axis** and **Y-Axis** dimensions.
5. Respond to the command line prompts as follows to place the column grid.

    Command: ColumnGridAdd

    Insertion point or [WIdth/Depth/XCount/YCount/XDivide by toggle/YDivide by toggle/Match]: *(Select the insertion point at p1 as shown in Figure 13.32.)*

    New size or [WIdth/Depth/XCount/YCount/XDivide by toggle/YDivide by toggle/Match]: *(Select the diagonal corner of the column grid at p2 in Figure 13.32.)*

    New size for cell or [WIdth/Depth/XCount/YCount/XDivide by toggle/YDivide by toggle/Match]: *(Move the pointer horizontally toward point p3 as shown in Figure 13.32 and click to specify the number of **X-Width** divisions and **Y-Depth** divisions.)*

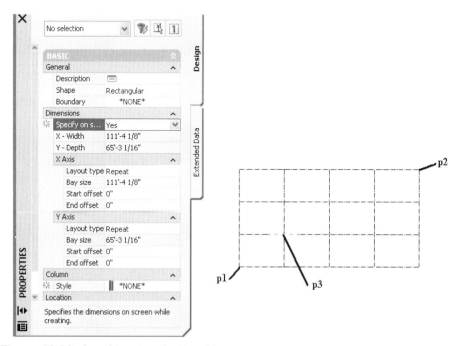

Figure 13.32 *Specifying size of grid and bay size on screen*

Rotation or [WIdth/Depth/XCount/YCount/XDivide by toggle/YDivide by toggle/Match] <0.00>: **0** ENTER *(Rotation specified.)*

Insertion point or [WIdth/Depth/XCount/YCount/XDivide by toggle/YDivide by toggle/Match/Undo]: ENTER *(Press* ENTER *to end the command.)*

Radial column grids can also be specified on screen. Selecting points in the drawing area specifies the insertion point, outer radius, and inner radius. Select the number of divisions by moving the pointer toward the inside radius and down.

STEPS TO DYNAMICALLY SIZING A RADIAL COLUMN GRID

1. Select the **Structural Column Grid** command from the Design tool palette.
2. Select the **Radial** shape from the **Shape** list.
3. Select **Yes** for **Specify on screen**.
4. Select **Space** evenly for both the **X-Axis** and **Y-Axis** dimensions.
5. Respond to the following command line prompts to specify the insertion point, inner radius, and the number of bays.

    Command: COLUMNGRIDADD

    Insertion point or [Radius/Angle/Inside radius/Xbay/Bay angle/XDivide by toggle/YDivide by toggle/Match]: *(Select point p1 in Figure 13.33 to specify the insertion point.)*

    New size or [Radius/Angle/Inside radius/Xbay/Bay angle/XDivide by toggle/YDivide by toggle/Match]: *(Select point p2 in Figure 13.33 to specify the radial dimension.)*

    New inside radius or [Radius/Angle/Inside radius/Xbay/Bay angle/XDivide by toggle/YDivide by toggle/Match] <0">: *(Select point p3 in Figure 13.33 to specify the inner radius.)*

    New size for cell or [Radius/Angle/Inside radius/Xbay/Bay angle/XDivide by toggle/YDivide by toggle/Match]: *(Move the mouse to point p4 as shown in Figure 13.34 to specify the number of bays.)*

    Rotation or [Radius/Angle/Inside radius/Xbay/Bay angle/XDivide by toggle/YDivide by toggle/Match] <0.00>: ENTER *(Press* ENTER *to accept 0° rotation.)*

    Insertion point or [Radius/Angle/Inside radius/Xbay/Bay angle/XDivide by toggle/YDivide by toggle/Match/Undo]: ENTER *(Press* ENTER *to end the command.)*

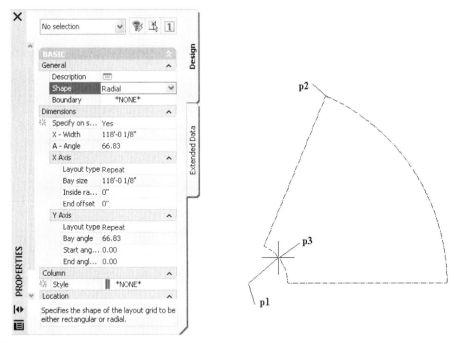

Figure 13.33 Specfying the insertion point of a radial column grid

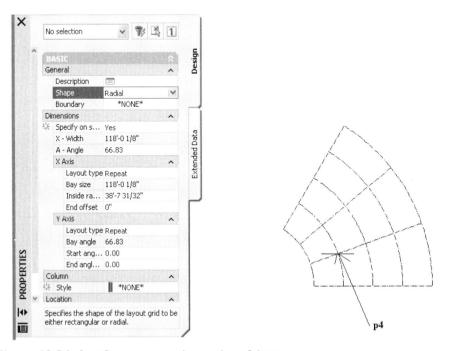

Figure 13.34 Specifying on screen the number of divisions

STOP. Do Tutorial 13.2, "Creating Structural Plans," at the end of the chapter.

CLIPPING THE COLUMN GRID

The column grid can be clipped by a closed polyline. Clipping a column grid allows you to see only that part of the grid inside the closed polyline. The clip can be applied when the grid is inserted or added to an existing grid. To clip a structural grid upon insertion, select the **Boundary** edit field of the Properties palette, choose the **Select object** option, and select a closed polyline at **p1** as shown in Figure 13.35. The column grid will be restricted to inside the closed polyline as shown at **p2** in Figure 13.35. The clip can be removed by selecting the column grid and editing the **Boundary** edit field to **None** in the Properties palette.

Holes can be placed in column grid with the **LayoutGridClip** command. Refer to Table 13.13 for **LayoutGridClip** command access.

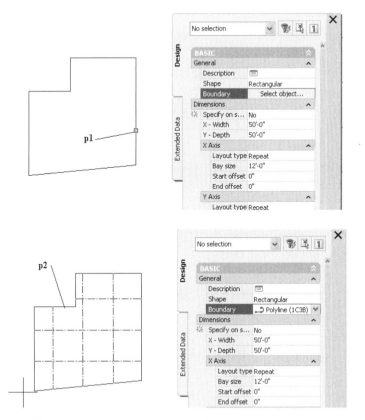

Figure 13.35 *Clipping a column grid with a boundary*

Command prompt	LAYOUTGRIDCLIP
Shortcut menu	Select a column grid, right-click, and choose Clip

Table 13.13 *LayoutGridClip command access*

When you select a grid, right-click, and choose **Clip**, you are prompted in the command line to Set boundary/Add hole/Remove hole. The Set boundary option will mask the grid to only include the column grid inside the closed polyline. The Set boundary option clips the grid in the same manner as selecting a **Boundary** in the Properties palette. The Add hole option removes the grid from within the closed polyline. The Remove hole option removes the clip from the column grid. A hole was placed in the column grid shown in Figure 13.36 as shown in the following command line sequence.

*(Select a column grid, right-click, and choose **Clip**.)*

Command: LayoutGridClip

Layout grid clip [Set boundary/Add hole/Remove hole]: **a** ENTER
*(Type **a** to add a hole.)*

Select layout grids: 1 found *(Select the column grid at p1 shown in Figure 13.36.)*

Select layout grids: ENTER *(Press ENTER to end the command.)*

Select a closed polyline for hole: *(Select the polyline at p2 as shown in Figure 13.36.)*

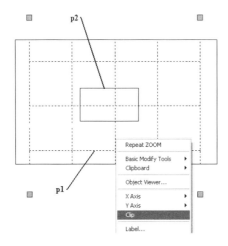 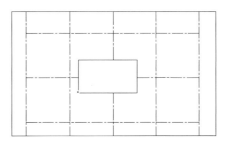

Figure 13.36 *Hole clipped in column grid*

Layout grid clip [Set boundary/Add hole/Remove hole]: *Cancel*
(Press ESC to end the command and view the column grid as shown at right in Figure 13.36.)

DEFINING THE COLUMN FOR THE GRID

Columns can be inserted at each intersection of the column grid by selecting a style from the **Style** drop-down list in the **Column** section of the Properties palette. The content of the drop-down list includes the structural member styles created from the **Structural Member Catalog** and used in the drawing. The column style will be inserted at each intersection of the column grid; however, you can erase selected columns from the grid. The properties such as **Logical length**, **Start offset**, **End offset**, **Justify**, and **Justify using overall extents** are set in the **Dimensions** section of the Properties palette of the column grid. Columns inserted in the column grid can be edited independent of the remaining columns by selecting the column and changing the style or other properties in the Properties palette. The Properties palette can be used to globally edit all columns of a column grid by editing the filter at the top of the Properties palette. If you select all elements of a drawing, you can isolate the **Structural Member** objects by editing the filter of the Properties palette as shown in Figure 13.37.

The Properties palette of an existing column grid does not include a column style edit field. Therefore, if this edit field is not specified when the column grid is placed, you must use the **Structural Column** tool to insert columns in the grid. The **Node** and **Intersection** object snap modes can be used to insert the column at the intersection of the column grids.

REFINING THE COLUMN GRID

The Properties palette allows you to edit most features of an existing column grid. The **Layout type** of the column grid determines how the grid will be edited. Column grids placed with **Space evenly** and **Repeat** include grips at the four corners. The **X-**

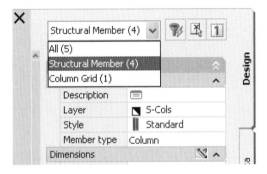 

Figure 13.37 *Filtering objects in the Properties palette*

Width and **Y-Depth** dimensions of the column grid can be changed if you select one of the four corner grips and stretch the grip to a new location. Column grids created with **Space evenly** or **Repeat** layouts can be converted to **Manual** by selecting the **Manual** layout type in the Properties palette. Column grids converted to **Manual** have grips for each grid line, as shown on right in Figure 13.38. A grid line can be relocated by selecting one of the grips and stretching it to a new location.

When a grid is converted to **Manual** layout, a **Bays** button is added in the Properties palette. Selecting the **Bays** button opens the **Bays along the X Axis** or the **Bays along the Y axis** dialog box. This dialog box lists each bay of the grid. The **Distance to line** is the distance from the insertion point. The **Spacing** distance is the distance between grid lines. Click in the **Distance to line** or the **Spacing** column to edit the dimensions of the bay. The **Add** and **Remove** buttons located to the right of the dialog box allow you to add or delete grid lines. A grid line can also be added if you double-click in the **Bay** column.

ADDING AND REMOVING COLUMN GRID LINES

The commands to add, remove, and change the layout type of a column grid are located on the shortcut menu of the column grid. The shortcut menu of the grid includes **X-Axis** and **Y-Axis** cascade menus. The cascade menu includes the following options: **Add Grid Line**, **Remove Grid Line**, and **Layout Mode**. These options allow you to quickly change the column grid mode and add or remove column grid lines for the X axis or the Y axis.

The **Layout Mode** command on the shortcut menu allows you to define the spacing system used in the column grid. See Table 13.14 for **Layout Mode** command access.

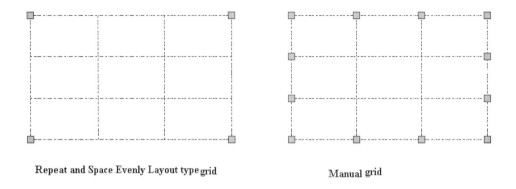

Repeat and Space Evenly Layout type grid   Manual grid

Figure 13.38 *Grips of the rectangular column grid*

Command prompt	COLUMNGRIDXMODE or COLUMNGRIDYMODE
Shortcut menu	Select the column grid; then right-click and choose X-Axis>Layout Mode or Y-Axis>Layout Mode from the shortcut menu

Table 13.14 *ColumnGridXMode or ColumnGridYMode command access*

Choosing the **Layout Mode** command from the shortcut menu allows you to select one of the following modes for the column grid: **Manual**, **Repeat**, or **Space Evenly**. Selecting the **Manual** option allows you to add or remove individual column grid lines at specific locations. The **Repeat** option allows you to specify in the command line the bay size to be repeated in the column grid. The **Space Evenly** option allows you to specify the number of equal spaces in the column grid. These options allow you to edit the column grid by entering values in the command line. The layout mode can also be changed in the **Layout type** edit field of the Properties palette for existing column grids.

Adding Column Grid Lines

Additional column grid lines can be added to the grid at specific locations if the column grid is set to the **Manual** mode. The **Space evenly** layout type allows you to add grid lines uniformly spaced within the grid. However, **Repeat** layout does not allow you to add or remove grid lines with the **ColumnGridXAdd** or **ColumnGridYAdd** command.

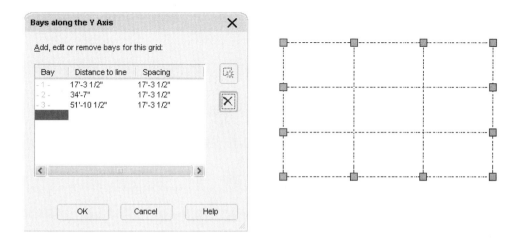

Figure 13.39 *Bays along the Y Axis dialog box for a manual grid*

You can add column grid lines for the X axis or the Y axis by choosing the column grid, right-clicking, and choosing **Add Grid Line** from the X axis or Y axis shortcut menu as shown in Table 13.15.

Command prompt	COLUMNGRIDXADD or COLUMNGRIDYADD
Shortcut menu	Select the column grid, right-click, and choose X Axis>Add Grid Line or select the column grid, right-click, and choose Y Axis>Add Grid Line

Table 13.15 *ColumnGridXAdd and ColumnGridYAdd command access*

When you select a grid, right-click, and choose **X Axis>Add Grid Line** or **Y Axis>Add Grid Line**, you are prompted to specify the location for the new line. The new line can be specified in the command line or selected with the pointer. The distance inserted in the command line is relative to the insertion point of the column grid. The new column grid line is combined with the remainder of the column grid. The procedure for adding a new column grid line is shown in the following command line sequence.

(Select the column grid with Manual layout type, as shown in Figure 13.40; then right-click and choose **X-Axis>Add Grid Line**.*)*

Command: ColumnGridXAdd

Enter X length <10'-0">: **15'** ENTER *(Column grid line added 15' from the insertion point along the x axis as shown at left in Figure 13.40.)*

Removing Column Grid Lines

You can remove column grid lines by choosing the **ColumnGridXRemove** or **ColumnGridYRemove** command. See Table 13.16 for command access.

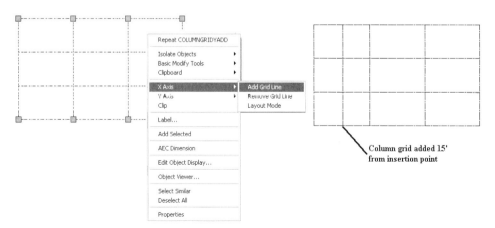

Figure 13.40 *Column grid prior to adding a grid line*

Command prompt	COLUMNGRIDXREMOVE or COLUMN-GRIDYREMOVE
Shortcut menu	Select the column grid, right-click, and choose X Axis>Remove Grid Line or Y Axis>Remove Grid Line

Table 13.16 *ColumnGridXRemove or ColumnGridYRemove command access*

When you select a column grid, right-click, and choose **X Axis>Remove Grid Line** or **Y Axis>Remove Grid Line**, you are prompted in the command line to specify the location of the grid line to remove. Selecting near the line with the pointer or typing the location as the distance from the insertion point specifies the location. The procedure for removing the column grid line is shown in the following command line sequence.

(Select the column grid; then right-click and choose **X-Axis>Remove Grid Line**.)

Command: ColumnGridXRemove

Enter approximate X length to remove <10'-0">: (Select the location of the column grid line at point p1 in Figure 13.41.)

(Column grid removed as shown in Figure 13.41.)

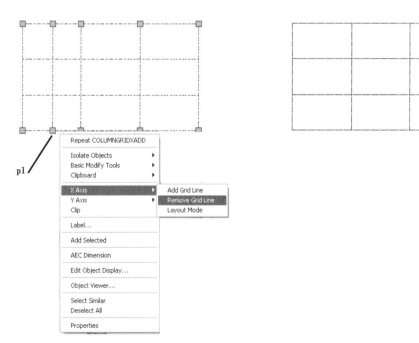

Figure 13.41 *Selected column grid line removed*

CONVERTING EXISTING LINES TO COLUMN GRIDS

Existing lines can be converted to column grid lines by selecting **CustomColumnGridAdd**. The **CustomColumnGridAdd** command allows you to convert existing center lines to a column grid object. The column grid is placed on the S-Grid layer. Access **CustomColumnGridAdd** as shown in Table 13.17.

Command prompt	CUSTOMCOLUMNGRIDADD
Palette	Select the Structural Column Grid tool of the Design tool palette, right-click, and choose Apply Tool Properties to>Linework

Table 13.17 *CustomColumnGridAdd command access*

The CustomColumnGridAdd command was used to convert the geometry shown in Figure 13.42 to a column grid in the following command line sequence.

Command: CustomColumnGridAdd

Select linework: Specify opposite corner: 2 found *(Select from p1 to p2 to select lines for the column grid as shown in Figure 13.42.)*

Select linework: ENTER *(Press ENTER to end the selection.)*

Erase selected linework? [Yes/No] <No>: ENTER *(Press ENTER to retain the geometry.)*

(Lines converted to column grid as shown on right in Figure 13.42.)

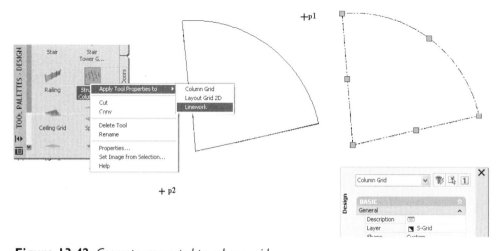

Figure 13.42 *Geometry converted to column grid*

Labeling the Column Grid

The column grid is labeled with the **ColumnGridLabel** command. The column grid can be labeled automatically with numbers or letters, or each column grid line can be assigned a unique label. If the column grid labels are automatically generated, the numbers or letters can be assigned in an ascending or descending pattern generated from the column grid line passing through the insertion point. See Table 13.18 for **ColumnGridLabel** command access.

Command prompt	COLUMNGRIDLABEL
Shortcut menu	Select a column grid, right-click, and choose Label

Table 13.18 *ColumnGridLabel command access*

When you select a column grid, right-click, and choose **Label**, the **Column Grid Labeling** dialog box opens, as shown in Figure 13.43.

The **Column Grid Labeling** dialog box consists of the **X - Labeling** and **Y - Labeling** tabs. The column grid lines for each grid direction are labeled in the tab. The **Automatically Calculate Values for Labels** check box allows the numbering of the

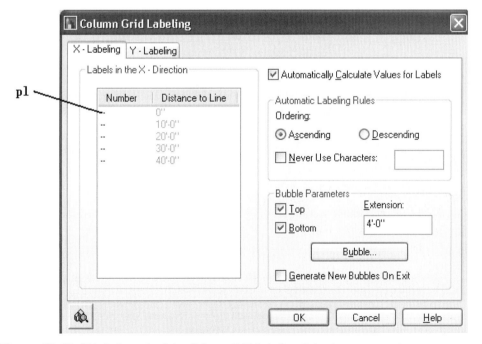

Figure 13.43 *X-Labeling tab of the Column Grid Labeling dialog box*

column grid lines to be assigned in ascending or descending order from the first column grid line. The number assigned to the first column grid line as shown at **p1** in Figure 13.43 defines the beginning character of the sequence. In the figure, if 1 is assigned at **p1** and **Ascending** order is toggled ON, the columns will be labeled 1, 2, 3. The options of the **X - Labeling** and **Y - Labeling** tabs are described below.

Labels in the X - Direction
The **Labels in the X - Direction** list box shows each grid label and the distance from the insertion point.

>**Number** – The **Number** column allows you to assign an alphanumeric character for each column grid line. If **Automatically Calculate Values for Labels** has been selected, the labels are generated from the label specified in the top field of the **Number** column. You can override the column grid labels generated with the **Automatically Calculate Values for Labels** check box by selecting the grid label and typing the character for the column grid in the **Number** column.
>
>**Distance to Line** – The **Distance to Line** column lists the distance for each column grid line from the insertion point of the column grid.
>
>**Automatically Calculate Values for Labels** – The **Automatically Calculate Values for Labels** check box will automatically assign the labels to the column grid from the beginning column grid in ascending or descending order.

Automatic Labeling Rules
>**Ascending** – The **Ascending** radio button assigns the column grid label in increasing numeric or alphabetic order.
>
>**Descending** – The **Descending** radio button assigns the column grid label in decreasing numeric or alphabetic order.
>
>**Never Use Characters** – The **Never Use Characters** check box allows you to identify characters to exclude when the labels are automatically generated.

Bubble Parameters
The **Bubble Parameters** section allows you to define the size and location of the bubbles used to identify the column grid.

>**Top** – The **Top** check box will create bubbles at the top of the column grid.
>
>**Bottom** – The **Bottom** check box will create bubbles at the bottom of the column grid.
>
>**Extension** – The **Extension** edit field defines the distance from the grid to extend the grid centerline and place the bubble.

Bubble – The **Bubble** button opens the **MvBlockDef Select** dialog box, which allows you to select the multi-view block for the bubble.

Generate New Bubbles On Exit – The **Generate New Bubbles On Exit** check box redisplays the position and size of the bubbles according to the changes. The size of the column bubbles is determined in the **Drawing Setup** dialog box (**AecDwgSetup** command).

Note: The **Y-Labeling** tab includes similar options for specifying the labeling of the column grid in the Y direction.

DIMENSIONING THE COLUMN GRID

The column grid can be automatically dimensioned with the **AECDimension** command (**DimAdd**). The dimensions will be updated to reflect any changes in the column grid. Access the **AECDimension** command as shown in Table 13.19.

Command prompt	DIMADD
Shortcut menu	Select the column grid, right-click, and choose AecDimension

Table 13.19 *AEC Dimension for column grids command access*

When you select the column grid, right-click, and choose **AecDimension**, the direction in which you move the pointer determines whether you create a horizontal or vertical dimension. In the example shown in Figure 13.44, the pointer was moved up from the grid. In the Properties palette, specify a dimension style for the Aec Dimension. Verify that an Aec Dimension style is current, and then select a location with the pointer to specify the location of the dimension string. Shown below is the command sequence for dimensioning a column grid.

(*Select the column grid at p1 as shown in Figure 13.44; then right-click and choose* **AecDimension** *from the shortcut menu.*)

Command: DimAdd

Specify insert point or [Style/Rotation/Align] (*Edit the Style = 2 Chain in the Properties palette, and then specify the location of the dimension string at p2 as shown in Figure 13.44.*)

1 added

(*Column grid is dimensioned as shown on right in Figure 13.44.*)

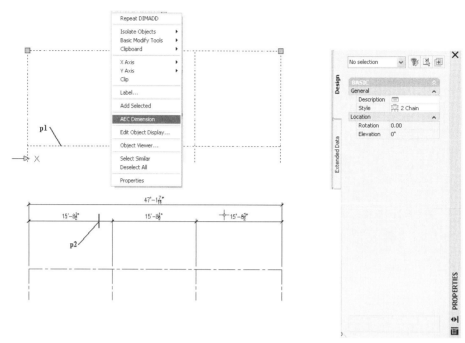

Figure 13.44 *Dimensioned column grid*

 STOP. Do Tutorial 13.3, "Dimensioning the Column Grid," at the end of the chapter.

CREATING CEILING GRIDS

Ceiling grids represent the building components of the suspended ceiling. The ceiling grid is created as an object with properties similar to the column grid. The ceiling grid is developed in a rectangular shape; when it is inserted, its size should exceed the maximum dimensions of the room. If the ceiling grid is attached to a space or closed polyline as a boundary, only that part of the ceiling grid inside the room will be displayed. The ceiling grids are displayed in the Reflected and Reflected Screened display configurations. The ceiling grid should be anchored to a space or a polyline; however, it can be placed free standing. The command to place a ceiling grid is **CeilingGridAdd**. See Table 13.20 for **CeilingGridAdd** command access.

Command prompt	CEILINGGRIDADD
Palette	Select Ceiling Grid from the Design Tool palette

Table 13.20 *CeilingGridAdd command access*

Ceiling grids are displayed in Reflected and Reflected Screened display configurations. The display configuration also displays the space. Therefore, select the **Reflected** or **Reflected Screened** display configuration prior to selecting the **Ceiling Grid** tool from the Design tool palette. When you select the **Ceiling Grid** command, the Properties palette opens, allowing you to specify the size and properties of the ceiling grid, as shown in Figure 13.45.

The Properties palette for adding a ceiling grid is described below.

General

  Description – The **Description** button opens the **Description** dialog box, allowing you to add a description.

  Boundary – The **Boundary** lists existing boundaries, **Select object**, and **None**. If a closed polyline or space is selected, the ceiling grid will be masked by the boundary.

Dimensions

  SPECIFY ON SCREEN

  X-Width – The **X-Width** value is the dimension parallel to the X axis of the ceiling grid.

  Y-Depth – The **Y-Depth** value is the dimension parallel to the Y axis of the ceiling grid.

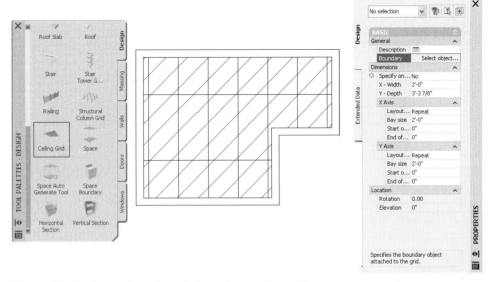

Figure 13.45 *Properties palette for inserting a ceiling grid*

X AXIS

Layout type – The **Layout type** options include **Repeat** and **Space evenly**. The **Repeat** option divides the total X-Width with grids a specific distance apart. The **Space evenly** option divides the X-Width by a specified number of divisions.

Bay size – If **Repeat** layout is used, the bay size is the distance between ceiling grids.

Number of bays – The **Number of bays** value is divided into the the total X-Width to determine grid spacing

Start offset – The **Start offset** is the distance from the insertion point to the start of the first grid on the X axis.

End offset – The **End offset** distance reduces the width of the ceiling grid from the total X-Width.

Y AXIS

Layout type – The **Layout type** options include **Repeat** and **Space evenly**. The **Repeat** option divides the total Y-Depth with grids a specific distance apart. The **Space evenly** option divides the Y-Depth by a specified number of divisions.

Bay size – If **Repeat** layout is used, the bay size is the distance between ceiling grids.

Number of bays – The **Number of bays** value is divided into the total Y-Depth to determine grid spacing.

Start offset – The **Start offset** is the distance from the insertion point to the start of the first grid on the Y axis

End offset – The **End offset** distance reduces the depth of the ceiling grid from the total Y-Depth.

Location

Rotation – The **Rotation** edit field is inactive during placement of the ceiling grid. The rotation for existing ceiling grids is displayed and can be edited.

Elevation – The **Elevation** is inactive during placement of ceiling grids. The elevation of existing grids is displayed and can be modified in the Properties palette.

The ceiling grid is placed as a rectangle regardless of the shape of the room. If the ceiling grid is attached to a boundary, the ceiling grid outside the boundary is not displayed. The **Boundary** field of the Properties palette includes a **Select object** option, which allows you to select a closed polyline or space to attach to the ceiling grid. The closed polyline or space creates a boundary for the ceiling grid, which turns off the display of the ceiling grid outside the boundary. The ceiling grid shown in Figure 13.45

was attached to a Space object. Ceiling grids attached to any closed polyline will be bound by the polyline and placed at the elevation of the polyline. Using polylines and spaces as the boundaries for the ceiling grid creates ceiling grids that fit irregularly shaped spaces.

 Tip: When you attach a ceiling grid to a space, the grid can be attached to the floor or ceiling boundaries. If a ceiling grid has been attached to the polyline of the floor, view the drawing in pictorial, choose **Select object** for the **Boundary** option in the Properties palette, and select the ceiling object.

When the **Ceiling Grid** command is selected, the default width and depth are retained from the last ceiling grid insertion. The grid size can be determined in the workspace if you select **Yes** to the **Specify on screen** option in the Properties palette. The **Specify on screen** option allows you to select a location for the insertion point (**p1**) and a diagonal point (**p2**) that is inclusive of the room geometry as shown in Figure 13.46. Selecting two points equal to or greater than the room size creates a ceiling grid that will cover the room. When **Specify on screen** is selected, you are prompted in the command line to select an insertion point, and the new size is determined by selecting the diagonal location as shown in the following command line sequence and Figure 13.46. The ceiling grid shown in Figure 13.46 and in the following command line sequence is **not attached to a boundary**; therefore the ceiling grid is displayed beyond the walls of the room.

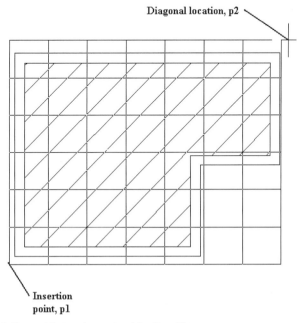

Figure 13.46 Ceiling grid size determined by Specify on screen

Command: CeilingGridAdd

Insertion point or [WIdth/Depth/XSpacing/YSpacing/XDivide by toggle/YDivide by toggle/Set boundary/SNap to center/Match]: *(Select a point near p1 as shown in Figure 13.46.)*

New size or [WIdth/Depth/XSpacing/YSpacing/XDivide by toggle/YDivide by toggle/Set boundary/SNap to center/Match]: *(Prior to selecting the 2nd point, verify that* **Specify On screen** *in the Properties palette is set to* **Yes***, and then select a point near p2 as shown in Figure 13.46.)*

Rotation or [WIdth/Depth/XSpacing/YSpacing/XDivide by toggle/YDivide by toggle/Set boundary/SNap to center/Match] <0.00>: ENTER *(Press* ENTER *to accept 0° rotation.)*

Insertion point or [WIdth/Depth/XSpacing/YSpacing/XDivide by toggle/YDivide by toggle/Set boundary/SNap to center/Match/Undo]: ESC *(Press* ESC *to exit the command.)*

If the ceiling grid is attached to a boundary, the ceiling grid can be centered about the boundary by selecting the SNap to center option in the command line in response to the Rotation prompt. The SNap to center option centers the ceiling grid about the boundary as shown in the following steps.

Note: You can use the **Space Auto Generate Tool** of the Design palette to create spaces for ceiling grids of floor plans that were developed without space planning.

STEPS FOR ATTACHING A CEILING GRID TO A SPACE OBJECT

1. Select **Reflected** or **Reflected Screened** display configuration from the Drawing Window status bar.
2. Select the **Ceiling Grid** tool from the Design palette.
3. Choose **Select object** from the **Boundary** edit field of the Properties palette. Select a space as shown at p1 in Figure 13.47.
4. Select **Specify on screen**, move the pointer to the workspace, and respond to the following command line prompts:

    Command: CeilingGridAdd

    Insertion point or [WIdth/Depth/XSpacing/YSpacing/XDivide by toggle/YDivide by toggle/Set boundary/SNap to center/Match]: *(Select the* **Select Object** *option of the* **Boundary** *edit field of the Properties palette.)*

    >>Select a space or closed polyline for boundary: *(Select the space at p1 as shown in Figure 13.47.)*

Resuming CEILINGGRIDADD command.

Insertion point or [WIdth/Depth/XSpacing/YSpacing/XDivide by toggle/YDivide by toggle/Set boundary/SNap to center/Match]: *(Select a point near p2 as shown in Figure 13.47.)*

New size or [WIdth/Depth/XSpacing/YSpacing/XDivide by toggle/YDivide by toggle/Set boundary/SNap to center/Match]: *(Select a point near p3 to specify the diagonal location for ceiling grid size.)*

Rotation or [WIdth/Depth/XSpacing/YSpacing/XDivide by toggle/YDivide by toggle/Set boundary/SNap to center/Match] <0.00>: **SN** *(Type SN to select Snap to center option.)*

Insertion point or [WIdth/Depth/XSpacing/YSpacing/XDivide by toggle/YDivide by toggle/Set boundary/SNap to center/Match]: ENTER *(Press ENTER to end the command.)*

CHANGING THE CEILING GRID

After the ceiling grid is placed, the layout type, bay size, start offset, and end offset can be changed in the Properties palette. The grips of the Repeat and Space evenly layouts are located at the four corners of the ceiling grid as shown at left in Figure 13.48. The grips located at the corners of the ceiling grid allow you to resize the X-Width and Y-Depth of the grid. If you select a ceiling grid and edit the **Layout type** to **Manual** for the X axis and the Y axis in the Properties palette, the grips are locat-

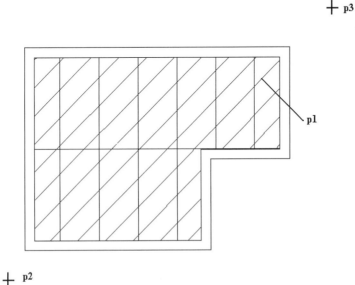

Figure 13.47 *Ceiling grid sized and attached to a boundary*

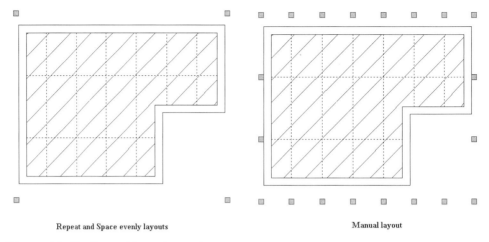

Repeat and Space evenly layouts   Manual layout

Figure 13.48 *Grips of the ceiling grid*

ed on each ceiling grid line. The grips of the Manual layout are shown at right in Figure 13.48.

The shortcut menu displayed when you select a ceiling grids includes the following commands for modifying the ceiling grid: **Add Grid Line**, **Remove Grid Line** (for both the X and Y axes), **Layout Mode**, **Clip**, and **Resize**. These commands allow you to change the dimensions of the ceiling grid and add or remove sections of the grid as needed.

ADDING AND REMOVING CEILING GRID LINES

The **Layout Mode** command of the shortcut menu allows the ceiling grid to be set to **Manual**, and ceiling grid lines can then be added or removed from the grid. The **Layout Mode** command (**CeilingGridXMode** or **CeilingGridYMode**) allows editing of each ceiling grid line. You can edit grid lines in the X axis direction by selecting the **Layout Mode** command for the X axis (**CeilingGridXMode**). Edit the ceiling grid lines in the Y direction by selecting the **Layout Mode** command for the Y axis (**CeilingGridYMode**). See Table 13.21 for **Layout Mode** command access.

Command prompt	CEILINGGRIDXMODE or CEILINGGRIDYMODE
Shortcut menu	Select the ceiling grid; then right-click and choose X-Axis>Layout Mode or Y-Axis>Layout Mode

Table 13.21 *CeilingGridXMode or CeilingGridYMode command access*

When you select the **Layout Mode** command for either the X or Y axis, the ceiling grid is released from the automatic mode, and additional ceiling grid lines can be added or removed. Selecting the **Manual** option is equivalent to selecting **Manual** in the Properties palette to edit an existing ceiling grid.

Adding Ceiling Grid Lines

After the layout mode has been toggled to **Manual**, you can add ceiling grid lines with the **Add Grid Line** command (**CeilingGridXAdd** or **CeilingGridYAdd**). See Table 13.22 for **Add Grid Line** command access.

Command prompt	CEILINGGRIDXADD or CEILINGGRIDYADD
Shortcut menu	Select the ceiling grid; then right-click and choose X-Axis>Add Grid Line or Y-Axis>Add Grid Line

Table 13.22 *CeilingGridXAdd or CeilingGridYAdd command access*

When you select the **Add Grid Line** command (**CeilingGridXAdd** or **CeilingGridYAdd**), you are prompted to specify the location for the new line. You can specify the location by typing a distance in the command line, or you can left-click with the mouse to select the location. If a distance is entered in the command line, it is measured relative to the insertion point of the grid. The procedure for adding a ceiling grid line is shown in the following command line sequence.

Note: Layout mode must be set to **Manual** prior to grid lines being added.

*(Verify that ORTHO is toggled ON in the status bar and grid layout mode is **Manual**.)*

*(Select the ceiling grid; then right-click and choose **X-Axis>Add Grid Line**.)*

Command: CeilingGridXAdd

Enter X length <10'-0">: *(Select a point near p1 in Figure 13.49.)*

(Ceiling grid line added as shown in Figure 13.49.)

Removing Ceiling Grid Lines

Ceiling grid lines can be removed if the layout mode has been set to **Manual** for the axis of the ceiling grid and you select the **CeilingGridXRemove** or **CeilingGridY Remove** command. See Table 13.23 for **Remove Grid Line** command access.

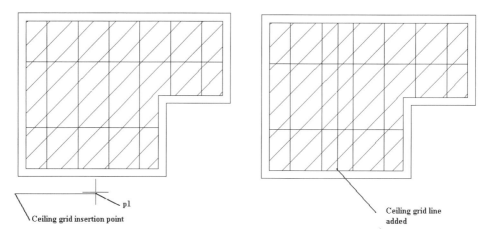

Figure 13.49 *Specifying the location to add a ceiling grid line*

Command prompt	CEILINGGRIDXREMOVE or CEILING-GRIDYREMOVE
Shortcut menu	Select the ceiling grid; then right-click and choose X-Axis>Remove Grid Line or Y-Axis>Remove Grid Line

Table 13.23 *CeilingGridXRemove or CeilingGridYRemove command access*

When the layout mode of the ceiling grid is set to **Manual**, you can remove selected ceiling grid lines. The following command line sequence removes a ceiling grid line.

(Verify that ORTHO is toggled ON on the status bar.)

(Select the ceiling grid at p1 in Figure 13.50; then right-click and choose **X-Axis>Remove Grid Line** *from the shortcut menu.)*

Command: CeilingGridXRemove

Enter approximate X length to remove <10'-0">: *(Select near the grid line at p2 in Figure 13.50 to remove the grid line. Grid line is removed as shown in Figure 13.50.)*

ADDING BOUNDARIES AND HOLES

Boundaries and holes can be added to existing ceiling grids with the **CeilingGridClip** command. This command has three options: Set boundary, Add hole, and Remove hole. The command can be used to create a boundary for a ceiling that has not been bounded to a space or polyline. The Add hole and Remove hole options allow you to carve out an area from the ceiling grid for penetrations through the ceiling, such as

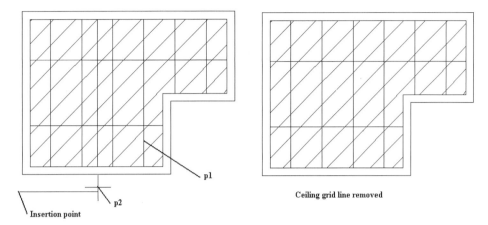

Figure 13.50 *Ceiling grid line removed from the ceiling grid*

columns, shafts, and plumbing chases. See Table 13.24 for **Attach Clipping Boundary/Holes** command access.

Command prompt	CEILINGGRIDCLIP
Shortcut menu	Select a ceiling grid, right-click, and choose Clip

Table 13.24 *CeilingGridClip command access*

When you select a ceiling grid, right-click, and choose **Clip**, you are prompted in the command line to select one of three options: Set boundary/Add hole/Remove hole. If a grid has been placed without being attached to a boundary, the Set boundary option can be used to assign a boundary, which can be a space or closed polyline. The command sequence for assigning a boundary to an existing ceiling grid is shown below.

> *(Select a ceiling grid at p1 as shown at left in Figure 13.51, right-click, and choose* **Clip***.)*
>
> Command: CeilingGridClip
>
> Ceiling grid clip [Set boundary/Add hole/Remove hole]: **S** ENTER *(Select the Set boundary option.)*
>
> Select a closed polyline or space object for boundary: *(Select the space at p2 as shown at left in Figure 13.51.)*
>
> Ceiling grid clip [Set boundary/Add hole/Remove hole]: ENTER *(Press* ENTER *to end the command and display the grid as shown at right in Figure 13.51.)*

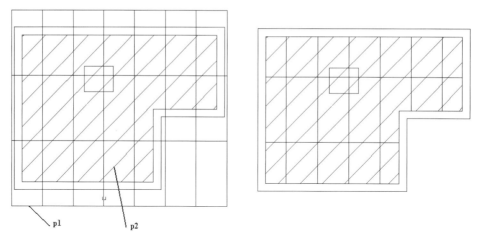

Figure 13.51 Ceiling grid attached to boundary

If a space is selected as the boundary of the ceiling grid, the ceiling grid is restricted to the boundary limits of the space. If the space dimensions are changed later, the ceiling grid does not stretch to the new dimensions of the space. However, the grips of the ceiling grid can be stretched to include the new dimensions of the space, and the space will continue to serve as the boundary for the ceiling grid.

Creating Holes in the Ceiling Grid

The Add hole option of the **CeilingGridClip** command allows you to carve from the ceiling grid the shape of any closed polyline or AEC object. The polyline does not have to have the same elevation or intersect with the plane of the ceiling grid. The ceiling grid shown at right in Figure 13.52 is trimmed by the closed polyline. The following command sequence was used to edit the ceiling grid as shown in Figure 13.52.

> *(Select ceiling grid at p1 as shown at left in Figure 13.52, right-click, and choose* **Clip**.*)*
>
> Command: _AecCeilingGridClip
>
> Ceiling grid clip [Set boundary/Add hole/Remove hole]: **A** ENTER *(Selects the Add hole option.)*
>
> Select a closed polyline or AEC entity for hole: *(Select the column at p2 as shown at left in Figure 13.52.)*
>
> Ceiling grid clip [Set boundary/Add hole/Remove hole]: ENTER *(Press ENTER to end the command.)*
>
> *(Ceiling grid modified as shown at right in Figure 13.52.)*

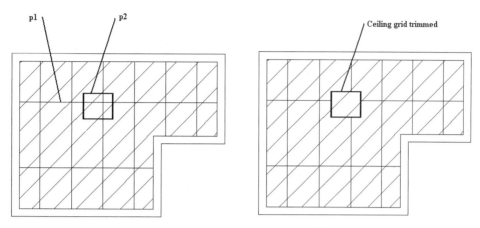

Figure 13.52 Hole added to ceiling grid

The Remove hole option allows you to edit the grid and remove any holes that have been created. The following command line sequence resulted in removing the hole from the ceiling grid.

>(Select ceiling grid at *p1* as shown at left in Figure 13.52, right-click, and choose **Clip**.)
>
>Ceiling grid clip [Set boundary/Add hole/Remove hole]: **R** ENTER
>*(Type **R** to remove existing hole.)*
>
>Select object for hole: *(Select polyline that was used to create the hole.)*
>
>Ceiling grid clip [Set boundary/Add hole/Remove hole]: ENTER *(Press ENTER to remove the clip boundary that created the hole.)*

 STOP. Do Tutorial 13.4, "Creating and Modifying a Ceiling Grid," at the end of the chapter.

Changing the Size of the Ceiling Grid

The **Resize** option of the ceiling grid shortcut menu allows you to change the overall size of the ceiling grid. The **Resize** option allows you to type values in the command line for the X-Width and Y-Depth dimensions of the ceiling grid. Access the **Resize** option as shown in Table 13.25.

Command prompt	CEILINGGRIDDIM
Shortcut menu	Select the ceiling grid, right-click, and choose Resize

Table 13.25 *CeilingGridDim command access*

When you select a ceiling grid, right-click, and choose **Resize**, you are prompted in the command line to specify the X-Width and Y-Depth dimensions as shown in the following command line sequence.

    Command: CeilingGridDim

    Enter X length <10'-0">: **25'** ENTER *(Specify new size the horizontal direction.)*

    Enter Y length <10'-0">: **35'** ENTER *(Specify new size in the vertical direction.)*

USING LAYOUT CURVES TO PLACE COLUMNS

Placing columns spaced evenly along a center line does not require the creation of a column grid line. Columns and other AEC objects can be placed along a layout curve. Isolated pier footings used in the foundations of residences can be placed on a layout curve. When a layout curve is created, nodes are defined on the curve, and AEC objects will snap to the node when placed in the drawing. The AEC objects can be spaced evenly or at a specified distance apart. Layout curves are similar to the AutoCAD **Divide** and **Measure** commands; however, the layout curve provides additional enhancements.

Access the **Layout Curve** tool from the **Content Browser**; select **Stock Tool catalog>Parametric Layout and Anchoring**. See Table 13.26 for **LayoutCurveAdd** command access.

Command prompt	LAYOUTCURVEADD
Palette	Access the Content Browser, select Stock Tool catalog>Parametric Layout and Anchoring, select Layout Curve

Table 13.26 *LayoutCurveAdd command access*

When you select the **LayoutCurveAdd** command, you are prompted to select a curve. The curve can be a straight line, arc, or other AutoCAD entity. Walls and boundaries can be used as the layout curve. The **LayoutCurveAdd** command allows you to use layout nodes equally spaced or at a specified distance apart. Architectural

Desktop objects can be inserted at the nodes to create columns along an arc or irregular curve. After selecting the entity to be used as a layout curve, you are prompted in the command line to select the mode for the layout. The options for the layout mode are described below.

>**Manual** – Requires that you specify the location of each node relative to the start point of the layout curve.
>
>**Repeat** – Requires that you specify the starting offset distance, ending offset distance, and the spacing value along the curve to place the nodes.
>
>**Space evenly** – Requires that you specify the starting offset distance, ending offset distance, and the number of nodes you want placed along the curve.

The **LayoutCurveAdd** command is used in the following command sequence to place nodes along a curved wall. The nodes can then be used to locate the columns, as shown in Figure 13.53.

>Command: LayoutCurveAdd
>
>Select a curve: *(Select the arc at p1 in Figure 13.53.)*
>
>Select node layout mode [Manual/Repeat/Space evenly]<Manual>: **S** ENTER *(Select the Space evenly option.)*
>
>Start offset <0">: **2'** ENTER
>
>End offset <0">: **2'** ENTER
>
>Number of nodes <3>: **5** ENTER
>
>*(The Structural Column command can then be used to place columns along the arc using the Node object snap.)*

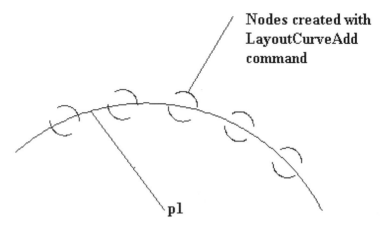

Figure 13.53 *Nodes placed on the arc with the LayoutCurveAdd command*

VIEWING THE MODEL

Most frequently, the designs of commercial buildings are presented to clients in the form of perspective drawings or a walkthrough. Architectural Desktop includes a camera to facilitate the view of the building model. The camera provides a quick visualization tool for presenting the model. The model can be viewed through one or more cameras placed in the plan view of the drawing. The still view or walkthrough presentation can be created with the camera; the camera is inserted in the floor plan and pointed toward objects of interest. Access the camera from the Massing tool palette or the **Content Browser** as shown in Figure 13.54. See Table 13.27 for **Camera** command access.

Command prompt	AECCAMERAADD
Palette	Select Camera from Massing tool palette of the Design palette set
Content Browser	Select the Content Browser, select Architectural Desktop Stock Tool Catalog>Helper Tools, and select Camera

Table 13.27 *AecCameraAdd command access*

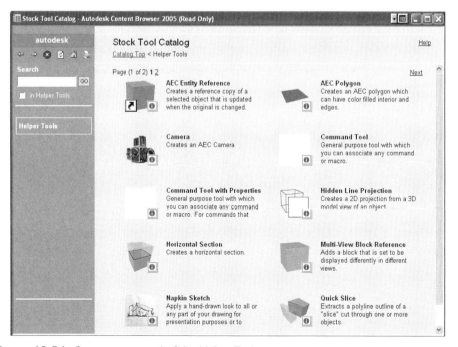

Figure 13.54 *Camera command of the Helper Tools category*

When the **AecCameraAdd** command is selected, the Camera Properties palette opens, and you are prompted to specify the insertion point and target point of the camera as shown in the following command line sequence.

>*(Select Camera from the Massing tool palette.)*
>
>Command: AecCameraAdd
>
>Insertion point or [Name/Zoom/Eye level/Generate view after add/Match]: *(Select a point near p1 as shown in Figure 13.55.)*
>
>Target point or [Name/Zoom/Eye level/Generate view after add/Match]: *(Specify the target select a point near p2 as shown in Figure 13.55.)*

The Properties palette includes the following options:

General

>**Name** – The **Name** edit field allows you to specify a name for the camera placed in the drawing.
>
>**Description** – The **Description** button allows you to add text to describe the camera.
>
>**Layer** – The **Layer** of the camera is G-Anno-Nplt if the AIA 256 Color layer key style is current. This field is not displayed during insertion. However, it is displayed when you edit with the Properties palette.
>
>**Zoom** – The **Zoom** factor of the camera can be modified in this field.

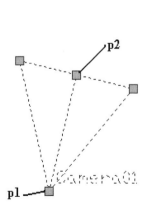
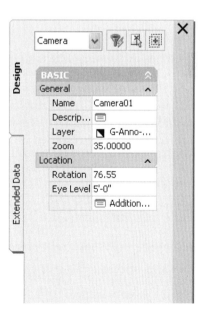

Figure 13.55 *Insertion point and target of camera*

Generate view after add – The **Yes** option of **Generate view after add** will display the view of the camera immediately after the camera is placed. This field is displayed during insertion only. The field is not shown in Figure 13.55 because the screen capture is taken from an existing camera.

Location

Rotation – The angle of **Rotation** from the three o'clock position of the camera.

Eye Level – The **Eye Level** or vertical displacement from Z=0 of the camera.

After setting the eye level and zoom of the camera in the Properties palette, you can select the insertion point and target of the camera. A camera symbol and the camera name are placed in the drawing. The target can be adjusted with grips after you insert it in the drawing.

You can obtain the view of the camera later by selecting the camera, right-clicking, and choosing **Create View** from the shortcut menu. When you select the **Create View** command, the workspace displays the view through the camera. See Table 13.28 for **Create View** command access.

Command prompt	AECCAMERAVIEW
Shortcut menu	Select the camera, right-click, and choose Create View

Table 13.28 *Create View command access*

After the view is created, select **Top View** from the **View** flyout of the **Navigation** toolbar to return to the view of the drawing.

CREATING A WALKTHROUGH

Create a walkthrough by inserting a camera and defining the path and target for the camera to follow. The camera follows the path looking toward the target to create the walkthrough, which is recorded as a video file. The video is created by selecting the **AecCameraVideo** command. See Table 13.29 for **AecCameraVideo** command access.

Command prompt	AECCAMERAVIDEO
Shortcut menu	Select the camera, right-click, and choose Create Video

Table 13.29 *AecCameraVideo command access*

When you select a camera, right-click, and choose **Create Video**, the **Create Video** dialog box opens as shown in Figure 13.56.

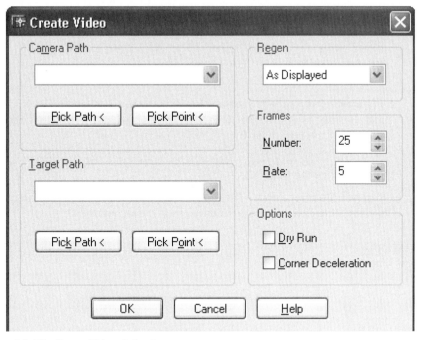

Figure 13.56 *Create Video dialog box*

The **Create Video** dialog box allows you to name the path and target. The path and target can be assigned to a polyline path or a point. The camera will move along the path of the polyline to create the video. The options of the **Create Video** dialog box are as follows:

> **Camera Path** – The **Camera Path** list shows the paths defined for the cameras in the drawing.
>
> **Pick Path<** – The **Pick Path<** button of the **Camera Path** section prompts you to select a polyline to use as the path for the camera. The **Path Name** dialog box opens if the selected polyline has not previously been defined as a path. Create a new name by typing the name in the **Path Name** dialog box.
>
> **Pick Point<** – The **Pick Point<** button of the **Camera Path** section allows you to pick a location in the drawing for the camera to point.
>
> **Target Path** – The **Target Path** list shows the targets defined for the camera.
>
> **Pick Path<** – The **Pick Path<** button of the **Target Path** section allows you to select a polyline as the target for the camera to follow.

Pick Point< – The **Pick Point<** button of the **Target Path** section allows you to select a point in the drawing for the camera to point to as it travels the path.

Regen – The **Regen** list includes the following options: As Displayed, None, Hide, Shade 256 Color, Shade 256 Edge, and Shade Filled.

Frames – The **Frames** section allows you to specify the number of frames and the rate. Increasing the number of frames creates a smoother video, while decreasing the rate slows the pace of the walkthrough.

Dry Run – The **Dry Run** check box provides a dry run or animation of the camera following its path.

Corner Deceleration – The **Corner Deceleration** check box adjusts the speed of the camera as it turns the corner of the path.

STOP. Do Tutorial 13.5, "Creating a Walkthrough," located in the Chapter 13 Supplement on the CD.

SUMMARY

1. Styles of structural components are created in the **Structural Member Catalog**.
2. The logical length of a structural member is the length of its axis line.
3. Start offset and end offset values modify the extension of the column about the end of the axis line.
4. **ColumnToolToLinework**, **BeamToolToLinework**, and **BraceToolToLinework** commands convert linework to the structural shape.
5. Additional structural shapes, including bar joist tools, are located in the Architectural Desktop Design Tool Catalog-Imperial>Structural catalog of the **Content Browser**.
6. Columns, beams, and braces are inserted in the drawing using the **Structural Column**, **Structural Beam**, and **Structural Brace** commands of the Design palette.
7. Column grids are placed in the drawing with the **ColumnGridAdd** command of the Design tool palette.
8. Set the Layout Mode to **Manual** for an axis to add column grid lines at specific locations along the axis.
9. The start offset and end offset distances of the column grid are relative to the insertion point of the column grid.
10. The column grid can be dimensioned with the AEC Dimensions.
11. Ceiling grids are added to the drawing with the CeilingGridAdd command.

12. The **Clip** option for ceiling grids and column grids allows you to trim the grid by a polyline or an AEC object.

13. The **LayoutCurveAdd** command can be used to create nodes that are evenly spaced along a line, arc, or polyline.

14. Obtain views of the building by placing a camera with the **AecCameraAdd** command.

15. A video of a walkthrough in the building is developed with the **AecCameraVideo** command.

REVIEW QUESTIONS

1. The columns of a column grid are placed on the _____ layer, and the column grid lines are placed on the _____ layer.

2. The distance measured between column grids horizontally is the _____ axis baysize.

3. The angle of the radial column grid is measured relative to the _____ position.

4. Column grid lines can be added at specific locations if the _____ layout type is assigned to the column grid.

5. Columns can be added to a column grid after the grid is placed using the _____ _____ _____ of the Design palette.

6. The size of the text used in the column label bubble is specified in the _____ _____ dialog box.

7. Ceiling grids only have four grips when the _____ or _____ _____ layout type is selected.

8. A _____ _____ can be used to place columns spaced evenly without the column grid.

9. You can combine two columns to create one column by assigning two _____ in the _____ _____ tab of the column style.

10. Ceiling grids should always be sized _____ than the room size.

TUTORIAL 13.1 CREATING DECK FLOOR FRAMING

1. Open **Ex 13-1** from your ADT Student\ADT Tutor\Ch13 directory.

2. Select **Save As** from the menu bar and save the drawing as **Lab 13-1** in your student directory.

3. Select **Format>Structural Member Catalog** from the menu bar.

4. Expand the tree in the left pane to **Imperial>Timber>Lumber>Nominal Cut Lumber** as shown in Figure 13T.1.

5. Select the **2x8** in the right pane, right-click, and choose **Generate Member Style** from the shortcut menu.

6. Type/verify the name **2x8** in the **Structural Member Style** dialog box. Select **OK** to dismiss the dialog box.

7. Select **File>Close** from the **Structural Member Catalog** menu.

8. Select the **Structural Beam** tool from the Design tool palette.

9. Edit the Properties palette as follows: Style = **2x8**, Start offset = **0**, End offset = **0**, Roll = **0**, and Justify = **Top Right**. Move the pointer to the workspace and respond to the command line as shown below:

    Command: BeamAdd

    Start point or [STyle/STArt offset/ENd offset/Justify/Roll/Match]: *(Select the top of the post at p1 as shown in Figure 13T.2, using the Endpoint object snap.)*

    End point or [STyle/STArt offset/ENd offset/Justify/Roll/Match]: *(Select the top of the post at p2 as shown in Figure 13T.2, using the Endpoint object snap.)*

    End point or [STyle/STArt offset/ENd offset/Justify/Roll/Match]: ENTER *(Press ENTER to end placing the first beam.)*

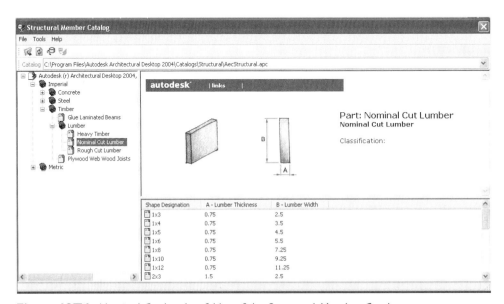

Figure 13T.1 *Nominal Cut Lumber folder of the Structural Member Catalog*

Command: ENTER *(Press ENTER to start the BeamAdd command.)*

Start point or [STyle/STArt offset/ENd offset/Justify/Roll/Match]: *(Select the top of the post at p3 as shown in Figure 13T.2, using the Endpoint object snap.)*

End point or [STyle/STArt offset/ENd offset/Justify/Roll/Match]: *(Select the top of the post at p4 as shown in Figure 13T.2, using the Endpoint object snap.)*

End point or [STyle/STArt offset/ENd offset/Justify/Roll/Match/Undo]: ENTER *(Press ENTER to end the command.)*

10. Select **Hidden** from the **Shade** flyout of the **Navigation** toolbar.
11. Select the **Structural Beam** tool from the Design tool palette.
12. Edit the Properties palette as follows: Style = **2x8**, Start offset = **0**, End offset = **0**, Roll = **0**, Justify = **Top Right**. Move the pointer to the workspace and respond to the following command line sequence shown below:

    Command: BeamAdd

    Start point or [STyle/STArt offset/ENd offset/Justify/Roll/Match]: *(Select the top of the 2x8 at p1, as shown in Figure 13T.3 using the Endpoint object snap.)*

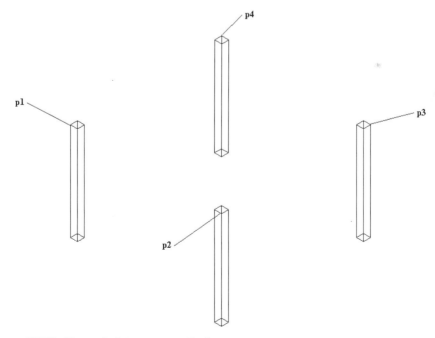

Figure 13T.2 *Placing 2x8 joist at top of column*

End point or [STyle/STArt offset/ENd offset/Justify/Roll/Match]: *(Select the top of the post at p2, as shown in Figure 13T.3 using the Endpoint object snap.)*

End point or [STyle/STArt offset/ENd offset/Justify/Roll/Match]: ENTER *(Press ENTER to end placing the first beam.)*

Command: ENTER *(Press ENTER to start the BeamAdd command.)*

Start point or [STyle/STArt offset/ENd offset/Justify/Roll/Match]: *(Select the top of the post at p3, as shown in Figure 13T.3 using the Endpoint object snap.)*

End point or [STyle/STArt offset/ENd offset/Justify/Roll/Match]: *(Select the top of the post at p4, as shown in Figure 13T.3 using the Endpoint object snap.)*

End point or [STyle/STArt offset/ENd offset/Justify/Roll/Match/Undo]: ENTER *(Press ENTER to end the command.)*

13. Select **Array** from the **Modify** toolbar. Select the **Select objects** button and select the joist at p1 as shown in Figure 13T.4. Edit the **Array** dialog box as follows: Rows = **1**, Column = **5**, Row offset = **1**, Column offset = **16**, Angle of array = **0**. Select **Preview** to view the array, and select the **Accept** button of the **Array** message box to complete the operation.

14. Select **Top View**; the cut plane height of the Display Configuration has created a section view of the joist. To remove the cutting plane display, select

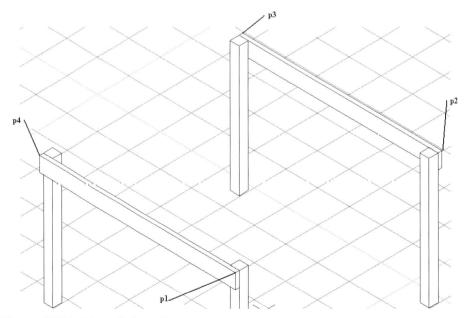

Figure 13T.3 *Adding 2x8 header*

Format>Display Manager. Select Medium Detail Configuration in the left pane, select the Cut Plane tab in the right pane, and edit the Cut Height = 5'. Select OK to close the Display Manager dialog box.

15. Select SW Isometric to view the joists as shown in Figure 13T.5.
16. Save and close the drawing.

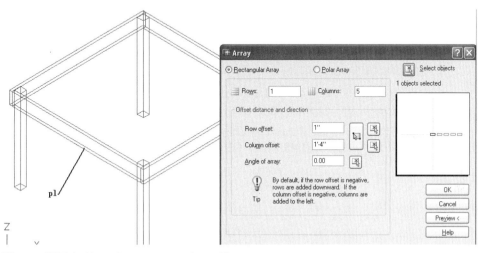

Figure 13T.4 *Using Array command to add joists*

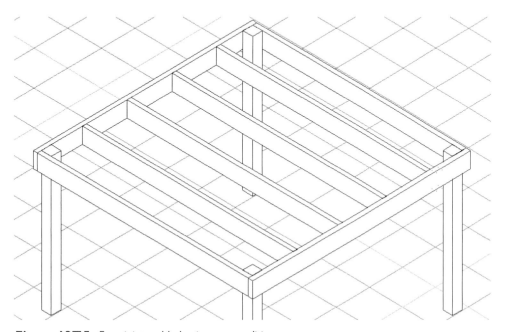

Figure 13T.5 *Four joists added using array editing*

TUTORIAL 13.2 CREATING STRUCTURAL PLANS

1. Open **Ex 13-2** from your *ADT Student\ADT Tutor\Ch13* directory.

2. Select **Save As** from the menu bar and save the drawing as **Lab 13-2** in your student directory.

3. Select **Format>Structural Member Catalog** from the menu bar.

4. Expand the tree in the left pane to **Imperial>Steel>AISC>I Shaped>W1n, Wide Flange Shapes** as shown in Figure 13T.6.

5. Select the **W14x109** beam in the right pane, right-click, and select **Generate Member Style** from the shortcut menu.

6. Type or verify the name **W14x109** in the **Structural Member Style** dialog box. Select **OK** to dismiss the dialog box.

7. Scroll down the list of beam sizes and double-click on the **W10x77** shape to open the **Structural Member Style** dialog box.

8. Type or verify the name **W10x77** in the **Structural Member Style** dialog box. Select **OK** to dismiss the dialog box.

9. Select **File>Close** to close the **Structural Member Catalog**.

10. Select the **Structural Column Grid** from the Design tool palette.

11. Edit the Properties palette as follows: Shape = **Rectangular**, Boundary = **None**, Specify on screen = **No**, X-Width = **72'**, Y-Depth = **36'**, X Axis: Layout type = **Repeat**, Bay size = **18'**, Start offset = **0**, End offset = **0**, Y Axis: Layout

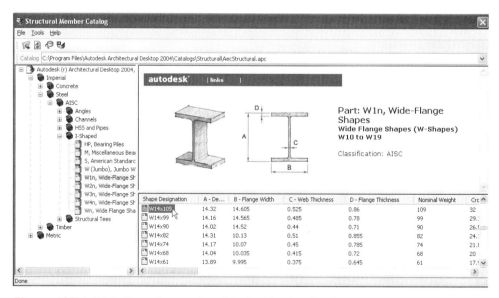

Figure 13T.6 *Wide Flange beam selected in the Member Catalog*

type = **Repeat**, Bay size = **18'**, Start offset = **0**, End offset = **0**, Column Style = **W10x77**, Column Dimensions: Start offset = **0**, End offset = **0**, Logical length = **12'**, Justify = **Middle Center**, and Justify using overall extents = **Yes** as shown in Figure 13T.7. Move the pointer to the workspace and type the insertion point in the command line as shown below:

Command: _AecColumnGridAdd

Insertion point or [WIdth/Depth/Xspacing/Yspacing/XDivide by toggle/YDivide by toggle/Match]: **15',15'** ENTER *(Specify the insertion point location.)*

Rotation or [WIdth/Depth/Xbay/Ybay/XDivide by toggle/YDivide by toggle/Match] <0.00>: ENTER *(Press ENTER to accept 0 rotation.)*

Insertion point or [WIdth/Depth/Xbay/Ybay/XDivide by toggle/YDivide by toggle/Match/Undo]: ENTER *(Press ENTER to end the command.)*

12. Right-click over the OSNAP toggle of the status bar, turn OFF all object snap modes, and toggle ON the **Node** object snap. Check **Object Snap On (F3)** and click **OK** to close the **Drafting Settings** dialog box.

Figure 13T.7 *Column grid Properties palette*

13. Verify that the Model tab is current and select the **SW Isometric** view from the **View** flyout of the **Navigation** toolbar.
14. Select **Structural Beam** from the Design palette as shown in Figure 13T.8.
15. Edit the Properties palette for the Structural Beam as follows: Style = **W14x109**, Start Offset = **5.5**, End Offset = **-5.5**, Roll = **0**, Justify = **Top Center** and Justify using overall extents = **Yes** as shown in Figure 13T.8. (Note that start offset and end offset will shorten the beam to fit between the columns.)
16. Respond to the command line prompts to place the beam as shown below.

    Command: BeamAdd

    Start point or [STyle/STArt offset/ENd offset/Justify/Roll/Match]: *(Select the node of the column at p1 as shown in Figure 13T.8.)*

    End point or [STyle/STArt offset/ENd offset/Justify/Roll/Match]: *(Select the node of the column at p2 as shown in Figure 13T.9.)*

    End point or [STyle/STArt offset/ENd offset/Justify/Roll/Match/Undo]: *(Select the node of the column at p3 as shown in Figure 13T.9.)*

    End point or [STyle/STArt offset/ENd offset/Justify/Roll/Match/Undo]: ENTER *(Press ENTER to end the command.)*

17. Continue to place the beams as shown in Figure 13T.10.

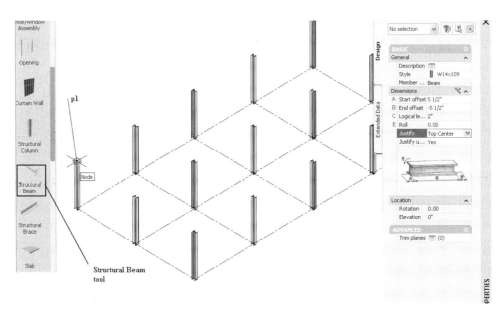

Figure 13T.8 *Placing the start point of the beam*

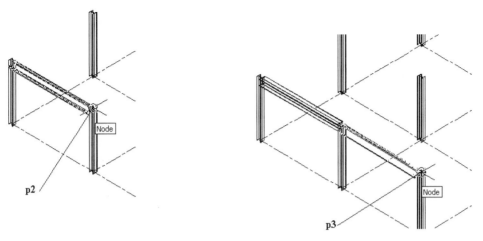

Figure 13T.9 *Placing the end point of the beam*

18. Select the **Content Browser** from the **Navigation** toolbar. Select the Architectural Desktop Design Tool Catalog>Structural>Bar Joist.
19. Select the i-drop for the **Steel Joist 20** and drag it from the catalog to the workspace of the current drawing as shown in Figure 13T.11.

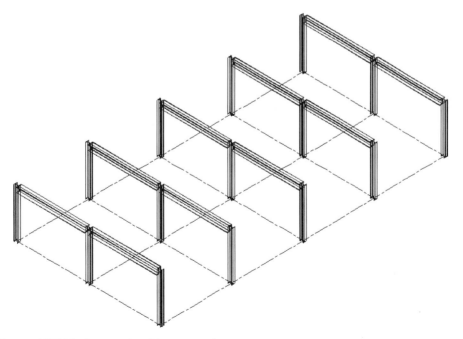

Figure 13T.10 *Beams placed between columns*

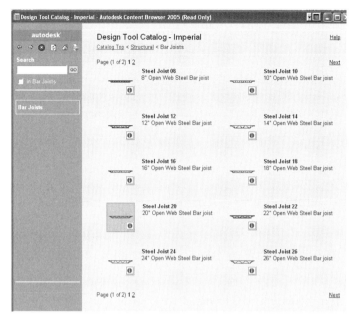

Figure 13T.11 *Steel Joist 20 i-dropped into the drawing*

20. Verify the settings of the Properties palette as follows: Style = **Steel Joist 20**, Start Offset = **1**, End Offset = **−1**, Roll = **0**, and Justify = **Baseline**. Note that Baseline justification will place the joist on top of the beam or column.
21. Select the node at **p1** of the column shown in Figure 13T.12 and draw the joist to the node at **p2** of the column shown in Figure 13T.13. Press ENTER to end the command.

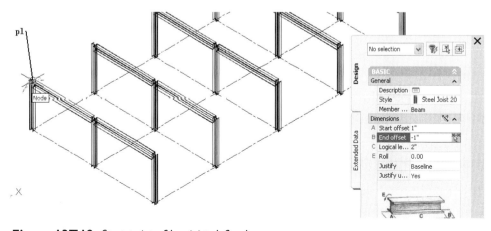

Figure 13T.12 *Start point of bar joist defined*

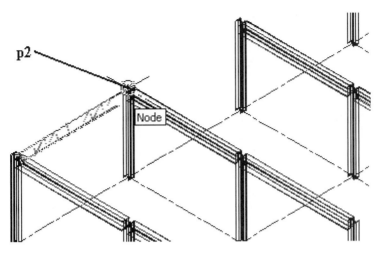

Figure 13T.13 *Endpoint of bar joist defined*

22. Verify that ORTHO is turned on. Select the bar joist, and then select the grip at **p1**. Move the pointer right, and press TAB to toggle the dynamic dimension. Type **12"** in the command line as shown below and in Figure 13T.14.

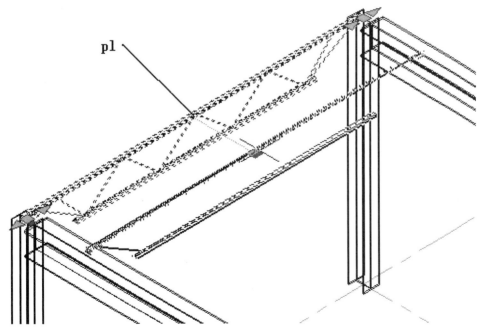

Figure 13T.14 *Joist moved using grip editing*

Command

STRETCH

Specify stretch point or [Base point/Copy/Undo/exit]: **12"**
ENTER *(Specify stretch distance in the command line.)*

(Press ESC *to end the command.)*

(Note the temporary position of the bar joist in Figure 13T.14 exceeds 12" in order to enter a direct distance entry of 12".)

23. Select the **Array** command from the **Modify** toolbar. Edit the **Array** dialog box as shown in Figure 13T.15. Toggle ON the **Rectangular Array**, set Rows = **12**, Columns = **4**, Row offset = **-36**, Column offset = **18'**, Angle of array = **0**, and select the **Select objects** button to select the joists shown at **p1** in Figure 13T.15. Select **Preview** to view the array, and select the **Accept** button of the **Array** message box to complete the operation.

24. Joist placed as shown in Figure 13T.16; save and close the drawing.

TUTORIAL 13.3 DIMENSIONING THE COLUMN GRID

1. Open *Ex 13-3* from your *ADT Student\ADT Tutor\Ch13* directory.

2. Select **Save As** from the menu bar and save the drawing as **Lab 13-3** in your student directory.

3. Select **Top View** from the **View** flyout of the **Navigation** toolbar. Select the column grid line, right-click, and choose **Label**. Edit the **X-Labeling** tab: Check **Automatically Calculate Values for Labels**, select the **Descending** radio button, for Bubble parameters check **Top** and check **Bottom**, Extension = **12'-0"**, and select **Generate New Bubbles on Exit**.

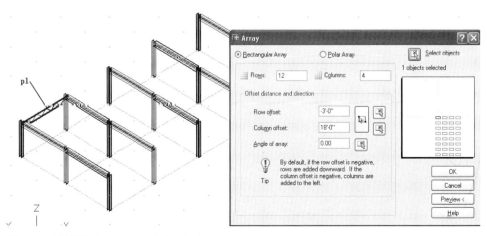

Figure 13T.15 *Joist selected for the Array command*

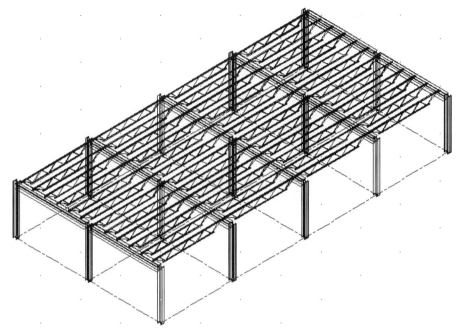

Figure 13T.16 *Joist placed using the Array command*

Select in the **Number** column, and type **E** in the field as shown at p1 in Figure 13T.17.

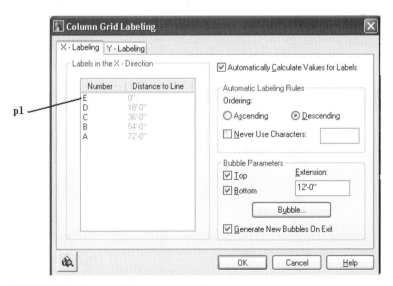

Figure 13T.17 *X-Labeling of the column grid*

4. Select the **Y-Labeling** tab; then check **Automatically Calculate Values for Labels**, select the **Descending** radio button, verify that the **Bubble Parameter Left** is checked and clear the **Right** bubble, set the extension to **12'-0"**, and select **Generate New Bubbles on Exit**. Type **3** in the top row of the **Number** column as shown at **p1** in Figure 13T.18. Select **OK** to dismiss the **Column Grid Labeling** dialog box.

5. Select the column grid, right-click, and choose **AEC Dimension**. Edit the Properties palette: Style = 2 Chain. Select a point near **p1** as shown in Figure 13T.19.

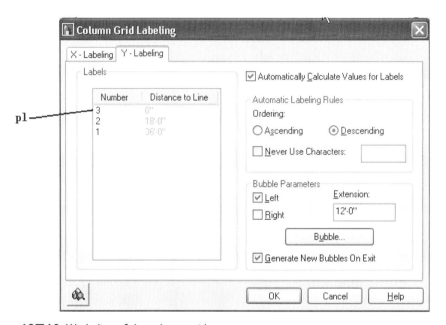

Figure 13T.18 *Y-Labeling of the column grid*

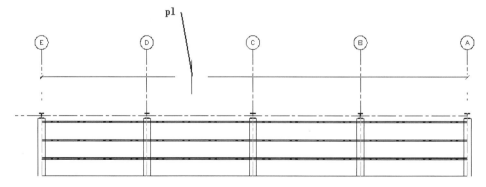

Figure 13T.19 *Location of horizontal dimension*

6. Select the column grid, right-click, and choose **AEC Dimension**. Verify the Style = 2 Chain in the Properties palette. Add a vertical dimension as shown in Figure 13T.20.

7. Select the **Content Browser** of the **Navigation** toolbar, and double-click to select the Architectural Desktop Documentation Tool Catalog-Imperial > Schedule Tags>Object Tags. Select the i-drop of the **Beam Tag** and drag the tool to the workspace as shown in Figure 13T.21. Respond to the command line as shown below to place the tag.

    Command: ContentTool

    Select object to tag [Symbol/Leader/Dimstyle/Edit] *(Select the beam at p1 as shown in Figure 13T.22.)*

    Specify location of tag <Centered>: ENTER *(Press ENTER to center the tag about the beam.)*

    *(Select **OK** to dismiss the **Edit Property Set Data** dialog box.)*

    (Press ESC to exit the command.)

8. Save and close the drawing.

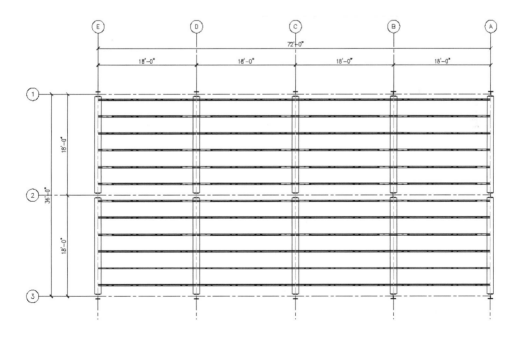

Figure 13T.20 *Column grid dimensioned*

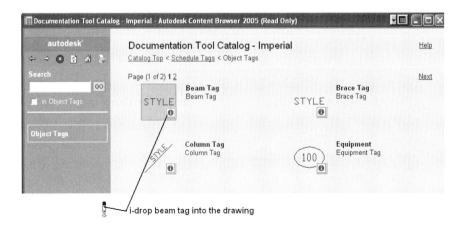

Figure 13T.21 *Beam tag i-dropped from the Architectural Desktop Documentation Tool Catalog-Imperial*

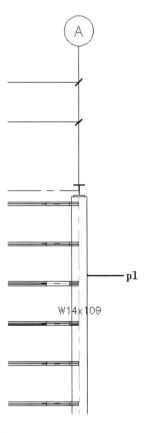

Figure 13T.22 *Beam tag placed*

TUTORIAL 13.4 CREATING AND MODIFYING A CEILING GRID

1. Open **Ex13-4.dwg** from your *ADT Student\ADT_Tutor\Ch13* directory.
2. Save the file to your student directory as **Lab 13-4**.
3. Select **Reflected** display configuration from the Drawing Window status bar. Toggle OFF Snap, Grid, Ortho, Polar, Osnap, Otrack in the status bar.
4. Select **Ceiling Grid** from the Design palette.
5. Edit the Properties palette as follows: Specify on screen = **No**, Layout type = **Repeat**, Bay size = **2'-0"**, Start offset = **0**, End offset = **0**, Y Axis: Layout type = **Repeat**, Bay size = **2'-0"**, Start offset = **0**, End offset = **0**. To set the remainder of the ceiling grid dimensions select Specify on screen = **Yes**. Choose **Select object** from the **Boundary** list, and select the edge of the Aec Space for the room at **p1** as shown in Figure 13T.23. Move the pointer to the workspace and respond to the command line prompts as follows:

   Insertion point or [WIdth/Depth/XSpacing/YSpacing/XDivide by toggle/YDivide by toggle/Set boundary/SNap to center/Match]: *(Select a point near p1 as shown in Figure 13T.24.)*

   New size or [WIdth/Depth/XSpacing/YSpacing/XDivide by toggle/YDivide by toggle/Set boundary/SNap to center/Match]: *(Select a point near p2 as shown in Figure 13T.24.)*

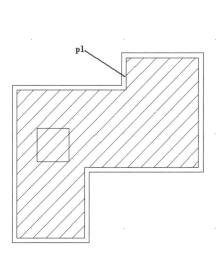

Figure 13T.23 *Selecting the AEC Space boundary for the ceiling grid*

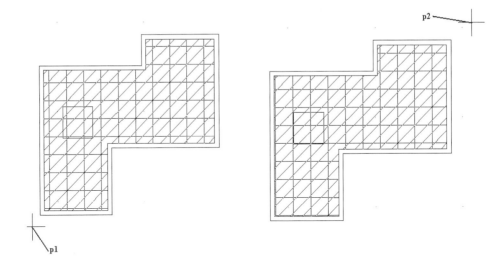

Figure 13T.24 *Specifying the size of the ceiling grid*

Rotation or [WIdth/Depth/XSpacing/YSpacing/XDivide by toggle/YDivide by toggle/Set boundary/SNap to center/Match] <0.00>: **sn** ENTER *(Type **SN** to center the grid in the room.)*

Insertion point or [WIdth/Depth/XSpacing/YSpacing/XDivide by toggle/YDivide by toggle/Set boundary/SNap to center/Match]: ESC *(Press ESC to end the command.)*

6. Select **SW Isometric** view from the **View** flyout of the **Navigation** toolbar. Select the ceiling grid, select **Select object** from the **Boundary** field of the Properties palette, and then select the ceiling boundary at **p1** as shown in Figure 13T.25.

7. Select the **Top View** from the **View** flyout of **Navigation** toolbar.

8. Select the ceiling grid, right-click, and choose **Clip**. Respond to the command line prompts as shown below to remove the ceiling grid from inside the rectangle shown in Figure 13T.26.

*(Select the ceiling grid, right-click, and choose **Clip**.)*

Command: CeilingGridClip

Ceiling grid clip [Set boundary/Add hole/Remove hole]: **a** ENTER *(Type a to add a hole in the ceiling grid.)*

Select a closed polyline or AEC object for hole: *(Select the rectangle at p1, as shown in Figure 13T.26.)*

Ceiling grid clip [Set boundary/Add hole/Remove hole]: *Cancel* ESC *(Press ESC to end the command.)*

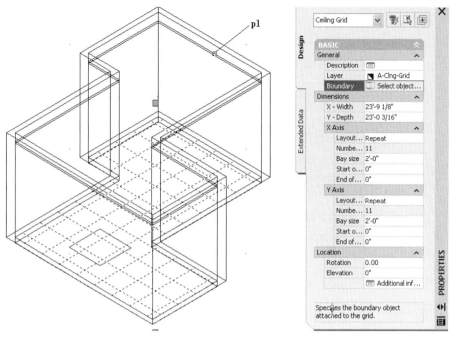

Figure 13T.25 *Select the ceiling polyline boundary*

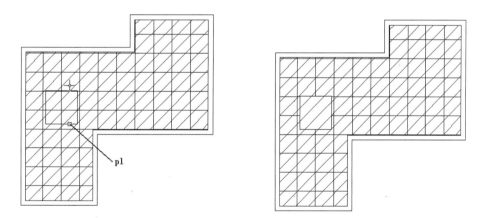

Figure 13T.26 *Creating a hole in the ceiling grid*

9. Select the Freeze icon for **A-Area-Space** layer from the **Layer** drop-down list in the **Layer Properties** toolbar.
10. Save changes and close the drawing.

PROJECT

Ex 13.6 Creating Columns and Column Grids

1. Open **EX 13-6** from your *ADT Student\ADT_Tutor\Ch13* directory.

2. Save the drawing as **Proj 13-6** in your student directory.

3. Open the **Structural Member Catalog** and create a member style for the HSS9x5x3/16 from the Imperial\Steel\AISC\HSS and Pipes\HSS, Small Hollow Structural Sections folder.

4. Create a column grid using the **Specify on screen** option. Specify the column grid at the inner surface of the diagonal corners of the building. Specify HSS9x5x3/16 columns that are 9' high.

5. Specify the drawing scale as 1/8"=1'-0".

6. Use **ColumnGridLabel** to label the columns descending from the C in the upper left corner and 1 at the top grid, using automatic labeling to label the columns as shown in Figure 13T.27.

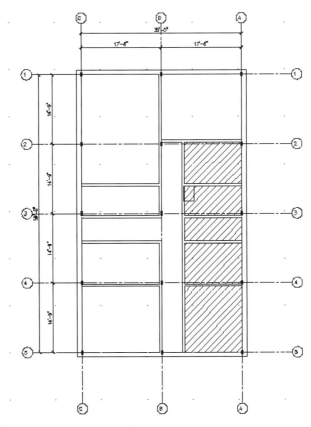

Figure 13T.27 *Column grid placed in the drawing*

7. Dimension the column grid as shown in Figure 13T.27.
8. Create suspended ceilings in the rooms on the right side of the building, and attach the ceiling to the ceiling boundary using the boundary option of Ceiling Grid.
9. Specify ceiling panels equal to 2' × 2'.
10. Trim the ceiling grid by the chase in the corner as shown at **p1** in Figure 13T.28.
11. The column bubbles are not displayed in the Reflected display configuration. Select the **Medium Detail** display configuration, select a column bubble, right-click, and choose **Edit Multi-View Block Definition**. Select the **View Blocks** tab. Select **Reflected** display representation and select the **Add** button to choose the **BubbleDef** view block for this display representation.
12. Save and close the drawing.

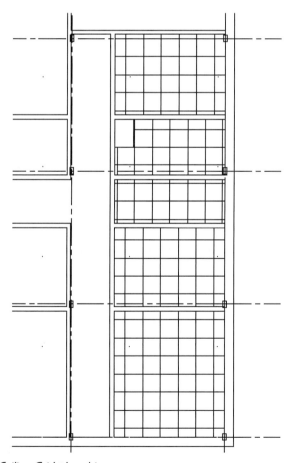

Figure 13T.28 *Ceiling Grid placed in rooms*

INDEX

(A) Height column, 384-85
(B) Overhang column, 384-85
(C) Eave
　column, 385
　value, 386
04-Masonry option, 816, 857
07-Thermal and Moisture Protection option, 864
1/2 Landing turn type, 509
1/2 turn Turn type, 509-10
1 Chain style, 662, 735
10_bull_nose wall style, 225-26
12diffuser view block, 640
18 x 12 Concrete Footing tool, 859, 863
2D
　Section
　　Rules, 436
　　Style 96 elevation style, 786
　Section/Elevation
　　display representation, 780
　　Object with Hidden Line Removal radio button, 783
　　Style Override-2D Section Style 96(2) dialog box, 844
　　Style Properties dialog box, 787-91
　　Style Properties – MyStyle dialog box, 791
　Wireframe, 42, 238, 451, 454
2dSection/Elevation Material Boundary dialog box, 796
2dSectionEditComponent command, 770, 793, 836
2dSectionResultAddHatchBoundary command, 770, 794-95, 836
2dSectionResultEdit command, 770, 792-93, 836
2dSectionResultEditHatchBoundaryIn Place command, 795
2dsectionResultRefresh command, 781-82, 835
2dSectionResultMerge command, 794
2DsectionResultStyleEdit command, 787
2DsectionResultUpdate command, 782, 835
3D
　Orbit
　　command, 902
　　flyout, 24-25
　　toolbar, 24
　Section/Elevation Object radio button, 783
　Solids Pulldown menu, 6
4" Slab with Haunch tool, 863-64

A

A-10 Schedules file, 764
A-12 Front Elevation sheet, 840
A-Angle edit field, 958-59
A-Anno_Dims layer, 672
A-Anno-Note layer, 652
A-Area-Iden layer, 698
A-Area-Mass layer, 873-74
A-Area-Mass-Grps layer, 424
A-Area-Mass-Slce layer, 905
A-Area-Space layer, 1011
A-Area-Spce layer, 62, 695
A-Auto-Adjust to Edge Height check box, 433
Above Cut Plane, 166
Absolute coordinate entry method, 82
Accept button, 996, 1004
ADCENTER command, 591
Add
　a New Opening Endcap Style button, 161
　another ring? prompt, 618
　Automatic Property Definition, 729
　Body Modifier dialog box, 204-5
　button, 150-51, 278, 297, 302, 305, 327, 373, 526, 572, 601, 605, 622, 638-40, 711, 788, 791, 830, 843-44, 938, 942, 966
　Category
　　button, 52
　　command, 49, 837, 841, 853
　Classification Property Definition button, 730
　Column
　　button, 708, 748
　　dialog box, 708, 748
　Component button, 151, 153, 218
　Construct
　　command, 49, 51
　　dialog box, 50-52, 66, 562, 567, 571
　Detail View dialog box, 857
　Dimension Points option, 666, 668, 739
　Division button, 49
　Edge Profiles
　　command, 438, 441-42, 479-81
　　option, 439, 480
　Edges command, 872, 912
　Element
　　command, 49
　　dialog box, 49-50, 52
　Elevation Label dialog box, 848
　Fascia/Soffit Profiles dialog box, 439-40. 480
　Formula Property Definition button, 730
　Frame Profile dialog box, 359-60
　General View dialog box, 52-53, 66-68, 251, 764, 770
　Grid Line, 966, 968, 980-81
　Gridline option, 369
　Header button, 710, 749
　Information button, 875
Level
　Above level name, 48
　Below level name, 48
　button, 48, 66
Location Property Definition button, 730
Manual Property definition button, 729
Material Property Definition button, 730
Model Space View dialog box, 54
MV-Block with Interference dialog box, 680-81
New Material button, 39-40, 160, 171-72, 279-80, 299, 343, 428, 436, 530, 547
New Objects Automatically option, 714-15
new objects option, 703
Opening Profile dialog box, 310
operation, 874
option, 188, 418, 551, 718, 833
Post Profile dialog box, 557
Priority Override button, 187
Profile
　command, 359, 556-57
　option, 310, 557
Project
　dialog box, 43-45, 65-66, 856
　Property Definition button, 730
Property
　Set to New Spaces check box, 693, 695, 759-60
　Sets dialog box, 686
　property sets button, 600-1, 619, 686
　Ring option, 203, 310, 360, 362
Section/Elevation View dialog box, 775, 797-98, 839, 841, 853-54
Selected option, 72, 74, 138, 640, 644
Standard Size dialog box, 278-79, 297-98, 327
Tag to New Spaces check box, 692, 695, 759-60
Tread Depth check box, 528-29, 572
Vertex option, 183, 194, 203, 310, 360-62, 413, 439, 441, 558
View
　command, 52, 837, 841, 853
　dialog box, 52, 837, 841, 853
Wall Modifier
　button, 190
　dialog box, 193, 198, 231-33
Wall Sweep dialog box, 200-2
Additional information button, 77, 383, 386, 406, 473, 508, 597, 615
Additional Information button, 781, 958
Additional Width value, 527-28
Additive – Cut Openings operation, 205
Additive Boolean operation, 681
AddWindow command, 325
A-Distance to First Tread DOWN edit field, 528-29
A-Door-Iden layer, 687
ADT
　menu group toolbars, 23-25, 57
　tool palettes, 4
Advanced section, 529, 931, 935, 938, 958
Advanced-Worksheets section, 529
AEC Content
　tab, 588, 593, 626, 683, 826
　Wizard, 585, 627
AEC Dimension
　Group, 660
　Marker, 660-61
　option, 1006
　Settings section, 662-63
　Style
　　dialog box, 664
　　Override – 2 Chain dialog box, 660-63
　Properties dialog box, 658-59
AEC Dimension Style Wizard
　command, 671
　dialog box, 671-73
AEC Dimension (1) – Manual, 654, 656-57
AEC Dimension (2), 654-57, 733-34, 738
AEC dimensions
　attaching building objects to, 666
　editing, 664-72
　overriding text of, 670-71
Aec MassElementConvert command, 909-10
Aec Model (Imperial Stb).dwt template, 2-3, 58, 87, 117
Aec Model (Metric Ctb).dwt template, 3
Aec Model (Metric Stb).dwt template, 3
Aec Model (Imperial Ctb).dwt template, 3, 58, 87
AEC Modify
　editing details using, 820-26
　Tools, 820-26
Aec Modify Tools>Crop option, 825
Aec Modify Tools>Divide option, 821
Aec Modify Tools>Merge option, 824
Aec Modify Tools>Obscure option, 823, 865
Aec Modify Tools>Subtract option, 822
Aec Modify Tools>Trim option, 821
AEC Object Settings tab, 93, 127-28, 563

1015

Aec Objects, converting to Mass Elements, 909-11
Aec Polygon option, 347
Aec Profile, creating the, 199
AEC Project Defaults tab, 45
Aec2dSectionResultMerge command, 794
Aec4_Room_Tag_Scale_Dependent tag definition, 759
AecAnnoRevisionCloudAdd command, 677-78
AecAnnoScheduleTagAdd command, 690, 696, 698-700
AecCallout command, 838842
AecCameraAdd command, 924, 988-89, 993
AecCameraVideo command, 924, 990, 993
AecCameraView command, 990
AECCOPYANDASSIGNSTYLE command, 149
AECDimension command, 973
AecDimStyle command, 658-59
AecDimStyleEdit command, 659
AecDimWizard command, 671
AecDisplayManager command, 31
AECDISPLAYMANAGER command, 31
AecDrape command, 916
AecDtlAnnoLeaderAdd command, 650-52, 828, 866
AECDTLCOMPADD command, 813
AecDtlCompManager command, 21, 808, 812, 816-18
AECDWGSETUP command, 25
AecDwgSetup command, 586, 626, 650, 654, 703, 973
AecHorizontalSection command, 800
AecKeynote field, 827
AecKeynoteLegendAdd command, 830, 832, 867
AecKeynoting data field, 829
AecLayerStd file, 27, 113, 117
AECNapkin command, 806-8
AECPROFILEDEFINE command, 199
AecPropertyRenumberData command, 747
AecSliceAttach command, 906-7
AecSliceCreate command, 905
AecSliceDetach command, 907
AecSliceElevation command, 908
AecSliceToPline command, 905, 908-9
AecSMemberCatalog command, 21, 924
-AecsMemberShapeDefine command, 948-49
AecSpaceAutoGenerate command, 691
AECSTYLEMANAGER command, 18, 143, 603
AecToggleSectionedBody command, 803
AECWALLBODY command, 204
AecWallStyle command, 144
AED Objects to 3D Solids command, 910-11
A-Extension of ALL Posts from Top Railing edit field, 542
A-Flor-Iden layer, 698
A-Flor-Strs layer, 508
A-Glaz-Iden layer, 689
A-Graphline position option, 77

AIA (256 color) Layer Key Style, 73, 113, 117
A-Length option, 337
Align command, 110
Alignment
  column, 525-26, 572
  options, 348, 706
All
  Materials setting, 850
  Palettes option, 59
  radio button, 373-74
All Programs>Autodesk>Autodesk Architectural Desktop 2005, 2
Allow
  docking, 10
  Each Railing to Vary check box, 537, 541, 543, 546, 577, 579
  Each Stair to Vary check box, 506, 527-29, 572
A-Maximum Slope, 524
Anchor
  button, 270-71, 291, 308
  dialog box, 270-71, 291, 308
  Railing command, 555
  to Boundary command, 872, 912
  Type drop-down list, 623
AnchorRelease command, 534, 556
Anchors, wall, 73
AnchorToStairLanding command, 533-34
Andersen_Casement style, 326-27
Angle
  edit field, 433, 532, 534, 681
  property, 405
Angular
  command, 672
  option, 672
  section (of Drawing Setup dialog box), 26
  tool, 672
AnnoBreakMarkAdd command, 675-76
AnnoRevisionCloudAdd command, 675-76
AnnoScheduleTagAdd command, 686, 742
Annotation
  option, 624
  Plot Size factor, 624
  Scale menu flyout, 4-5
  template, 3-4
  tool palette, 651, 674, 676
Any visible context, 791, 844
A-Overhang option, 404, 432, 438, 477
Append button, 692
Application status bar, 6
Applies To
  Objects radio button, 727
  Styles and Definitions radio button, 727
  tab, 706-7, 727-28
Apply
  'L' Cleanup command, 100
  Detail Component Data to Linework option, 826
  Roof/Floor Lines to Sweeps check box, 201
  to
    All Columns check box, 724, 758
    drop-down list, 59, 190, 533-34
    window list, 661-62

Tool Properties to
  command, 75-76, 311-12, 392-93, 442, 470-71
  Door option, 312
  option, 378
Apply Tool Properties to>Edges command, 872, 911
Apply Tool Properties to>Linework and Walls, 392-93, 450, 470-71
Apply Tool Properties to>Linework option, 944, 946-47
Apply Tool Properties to>Linework, Walls and Roof, 392-93, 458
Apply Tool Properties to>Polyline tool, 872, 911
Apply Tool Properties to>Roof command, 442
Apply Tool Properties to>Slice command, 872, 911
Apply Tool Properties to>Space, 741, 759-61, 872, 911
ApplyToolToObjects command, 741
Arc
  Angle edit field, 512
  Constraint drop-down list, 512
  option, 76, 111
  segment type, 81-82
Arch mass element shape, 876
Architectural
  Desktop
    Design Tool Catalog – Imperial, 953
    installation options, 2
    screen layout, 4-25
    starting, 2-4
  Objects folder, 217, 229, 235, 245
  Objects\Roof Slab Styles folder, 425
  Archive option, 57
  Area section (of Drawing Setup dialog box), 26
A-Roof layer, 401
A-Roof-Slab layer, 401, 404
Array
  command, 110, 996, 1004
  dialog box, 996, 1004
  message box, 996, 1004
Arrow section, 533-34
As Inserted option, 624
Ascending
  radio button, 972
  sort order, 711
A-Side 1 dimension, 780
A-Slab layer, 470, 472
Assemblies
  adding door/window, 335-40
  trimming using Interference Add, 366
A-Start offset edit field, 930, 934, 937
A-Straight Length option, 502
At
  Floor Levels section, 577-78
  Landings section, 577-78
A-Thickness
  edit field, 426
  option, 396, 467
A-Top of Entire Stair edit fields, 547
A-Tread Thickness edit field, 527-28
Attach
  Clipping Boundary/Holes command, 983
  Elements option, 423, 893-94
  New Property Set check box, 704

Objects
  command, 906-7
  option, 616, 644, 666, 906-7
  to Mask command, 615
Spaces to Boundary command, 872, 912
to option, 537-39
Attached
  Items command, 901
  to drop-down list, 537-38, 874, 893, 895
Attribute Text
  Angle toggle, 624
  Style toggle, 624
Attributes
  button, 597, 644
  section, 785
Auto
  Calculate Edge Offset check box, 159
  Increment-Character type, 730
  option, 261, 290, 308, 339
  project option, 176
  Track
    Settings section, 859
    tooltip, 74
Auto-Adjust
  Elevation check box, 49, 66, 837
  Profiles options, 350
  to Width of Wall check box, 275, 295-96
AutoCAD Dimension Settings section, 662-63
AutoCAPS option, 651
Autodesk Architectural Desktop 2005 icon, 2
Autodesk Render Material Catalog, 160
Auto-hide toggle, 9-11
Auto-Increment Numeric Properties check box, 693, 695, 759-60
Automatic
  data, 682
  Landings, 529
  Offset edit field, 258, 289, 309
  Offset/Center option, 287
  Placement options, 538
  Property Source list, 728
  radio button, 691-92, 695
  toggle, 502, 504, 514, 564, 574, 741, 759-60
Automatically
  Apply to Other Display Representations check box, 373
  Calculate Values for Labels check box, 971-72, 1006
  Choose Above and Below Cut Plane Heights check box, 168-69, 237
  Create Dimstyle Override check box, 59
  select all nested objects, 565
Autosnap Grip Edited Wall Baselines check box, 93, 127
Autosnap New Wall Baselines check box, 93, 127
Autosnap Radius = 6" check box, 127
AutoSnap Settings section, 859
A-Wall layer, 73
A-Wall-Iden layer, 699
A-Width
  column, 525-26
  edit field, 274, 291-92, 295, 307, 502, 513
  option, 76, 258, 288

B

Back
  arrow, 13-14
  button, 621
  option, 902
Back/Forward toggle, 592
Balanced winder, 510
Balcony element check box, 54
Balusters check box, 543, 577, 579
Bar Joists
  catalog, 926
  creating, 953
  tool, 953
Bar_pilaster component, 229
Barrel Vault mass element shape, 876, 897
Base
  Color edit fields, 693-94
  height
    option, 397, 467, 468
    wall, 471
  Width value, 154-56, 248
Baseline, 155, 220, 226, 468
  Edge edit field, 397
  justification, 953
  Offset
    edit field, 86
    option, 76
  option, 92, 111, 931, 935, 937
Basement
  check box, 839, 853-54
  wall style, 246, 582
Basement.dwg, 235
Basement_Int style name, 241-42
Basic
  Legend tool, 15-16
  Modify Tools shortcut menu, 820
  palette, 808
Basic Modify Tools>Erase option, 240, 245, 425
Basic Modify Tools>Extend option, 136, 238, 241, 249
Basic Modify Tools>Trim option, 239
Bath_Small space, 741
B-Auto-Adjust to Overhang Depth check box, 433
Bay
  angle, 958
  column, 966
  size, 956-57, 976
  window style, 325-26
Bays
  along the
    X Axis dialog box, 956, 966
    Y Axis dialog box, 956, 966
  button, 956, 966
Baysize, 958
B-Base Height (Design - Basic Dimensions) option, 76
B-Bottom of Entire Stair edit fields, 547
B-Cleanup radius (Design - Advanced Cleanups) option, 77
B-Depth edit field, 274-75, 295
B-Distance to First Tread UP edit field, 528-29
Beam
  Tag, 1007
  Trim Planes dialog box, 935, 938
BeamAdd command, 933, 953, 994, 1000

Beams, inserting, 933-35
BeamToolToLinework
  command, 946, 992
  tool, 945-46
B-Edge Cut option, 405, 422-23, 438, 477
Bedroom space, 741
Below Cut Plane, 166, 223, 237, 248
B-End offset edit field, 930, 934, 937
B-Extension of ALL Posts from Floor Level edit field, 543
Bhatch command, 692
B-Height
  edit field, 258, 288, 291-92, 307, 502, 504
  option, 337
Bifold-Double tool, 312
Bitmap image files, 2
BldgElevationLineReverse command, 784
BldgElevationLineGenerate command, 784
Block
  content type, 622
  Definition dialog box, 639
  list, 533
Blocks
  masking
    attaching objects to, 615-17
    creating, 613-14, 617-20
    detaching, 616
    inserting new, 620-21
    using Properties to edit existing, 615
  multi-view, inserting new, 620-21
  new mask, setting definition properties of a, 618-20
  properties of the, 614-15
Body Modifiers>Add, 204-5
B-Offset column, 526
Bolt components, 818, 820
Boolean editing option, 205-6
Boolean>Add option, 418-19
Boolean>Detach option, 420
Boolean>Subtract option, 419, 506
B-Optimum Slope, 524
Bottom
  check box, 972, 1004
  Elevation column, 156-58
  of Entire Stair edit fields, 547
  of Flight edit field, 547
  Offset value, 504
  option, 396-98, 467, 483-84, 487, 491, 902
Bottomrail check box, 542, 577
Boundaries and holes, adding, 982-86
Boundary
  edit field, 963, 976, 978, 1009-10
  option, 166, 956, 964, 975-76
Boundary (2) FTG component, 224
Boundary>Edit In Place command, 796
Box mass element shape, 877
Brace
  inserting a, 935-38
  Trim Planes dialog box, 938
BraceAdd command, 936
BraceToolToLinework command, 946-47, 992
Break
  command, 110
  Mark section, 533-34

marks, 20
Brick subcategory. (*See* Walls category)
BRICKFND, 246
B-Riser
  Count option, 502
  Thickness edit field, 527-28
Browse
  button, 65, 210-11, 252, 593, 625, 723-24, 758
  Project button, 45
  Property Data
    command, 716
    dialog box, 715-17, 751
B-Side 2 dimension, 780
Bubble
  button, 973
  Parameter Left check box, 1006
  Parameters section, 972
Building
  Elevation Mark A2 option, 835
  elevation/section line, editing the properties palette of the, 778-81
  Object Dimension Points check boxes, 662
  objects, detaching, 666-68
  section/elevation line, reversing the, 784-85
Bull_Nose style, 225
B-Vertical offset option, 396, 467
By
  adjusting
    baseline height option, 405, 473
    overhang option, 405, 473
  Material check box, 39, 41, 166, 799, 854
ByLayer lineweight, 41

C

C:\Documents & Settings\All Users\Application Data\Autodesk\ADT 2005\enu\Styles directory, 142, 170
C:\Documents and Settings\All Users\Application Data\Autodesk\ADT 2005\enu\Template folder, 3
CAD
  Manager Pulldown menu, 6
  Standards feature, 2
Calculate Automatically
  command, 185-87
  option, 183
Calculation Rules, 497, 512, 514, 561, 564, 568, 574
  button, 499, 502, 563
  edit field, 502-3
  dialog box, 499, 502-4, 563-64, 568, 574
Calculator Rule section, 524-25
Callout Properties palette, 774
Callouts, 20
Camera
  command, 988
  Path
    list, 991
    section, 991
C-Angle
  1 dimension, 780
  option, 405, 433, 438, 477
Casement tool, 745

Casement-Double
  style, 325
  tool, 324
Case-Upper value, 748, 754
Casework subcategory. (*See* Walls category)
Ceiling
  Grid, 60, 62, 1009
    changing the, 979-86
    command, 975, 977
    creating, 974-79
    lines, adding, 981
    lines, removing, 981*82
    tool, 975, 978
  multi-view block, 640
  symbol, 637
CeilingGridAdd command, 923, 974, 978-79
CeilingGridClip command, 924, 982-84
CeilingGridDim command, 986
CeilingGridXAdd command, 924, 981
CeilingGridXMode command, 980
CeilingGridXRemove command, 924, 981-82
CeilingGridYAdd command, 924, 981
CeilingGridYMode command, 980
CeilingGridYRemove command, 924, 981-82
Cell
  Assignments, 347, 371
  Format Override dialog box, 709, 712
  Location Assignment dialog box, 371
  Size section, 706
Center
  alignment, 371, 502
  of Opening check box, 662, 735
  option, 396-98, 467, 572, 639
  Side for Offset value, 577
Center2 option, 845
Chamfer
  command, 110, 112
  profile, 329
  style, 330
  value, 330
Chain list, 661
Chains tab, 660
Chase
  (6), 681
  symbols, 678-79
Chases, inserting, 678-81
C-Horizontal offset option, 396, 467
Circle grip, 406
Circle option, 639
Classification
  column, 299, 428, 548
  Definition column, 163, 299, 428, 530-31, 548, 601, 619
  Definitions, 162
  drop-down list, 548, 602, 619
Classifications
  drop-down list, 530-31
  tab, 149, 162-63, 273, 280, 282, 294, 299-300, 341, 344-45, 428, 530-31, 540, 548, 601-2, 618, 660, 789, 890-91, 942-43
Clean
  Screen Mode toggle, 6-7
  Up Joints toggle, 304

Cleanup
  Automatically option, 76, 94
  column, 526
  controlling using priority, 96
  drop-down list, 526
  group definition option, 76, 96-97
    creating, 96-97
  properties of a wall, 94
  Radius, setting the, 94-95, 112
  value, 526
Cleanup 'L' tool, 71
Cleanup 'T' tool, 71
Cleanups shortcut menu, 99
Cleanups>Add Wall Merge Condition option, 101, 135
Cleanups>Apply 'L' Cleanup, 100, 133-34
Cleanups>Apply 'T' Cleanup, 100, 133-34
Cleanups>Remove All Wall Merge Conditions option, 102
Clear All button, 628
C-Length (Design - Basic *Dimensions*) option, 76
Clip option, 964, 983-85, 993, 1010
C-Logical length edit field, 930, 934, 937
Close
  button, 253, 315, 323
  Current Project, 68, 129, 243, 329, 566, 570, 635, 852, 856
  menu option, 37-38, 66, 122, 216, 235, 244, 251
  option, 83, 111, 229, 252-53, 399, 443, 562, 567, 571, 628, 733, 740-41, 758-59, 770, 837, 841, 852, 856, 862-63, 901
  toggle, 9-10, 16-19
Close Refedit Session>Save Reference Edits option, 566
C-Maximum Center to Center Spacing edit field, 543
C-Minimum Slope, 524
CMU subcategory. (*See* Walls category)
C-Nosing Length edit field, 527-28
Code Limits section, 524
Color
  and Layer page, 672-73
  column, 39, 41, 788, 844-45
Column
  drop-down list, 710, 748
  grid
    clipping the, 963-65
    converting existing lines to, 970-73
    creating a radial, 958-59
    dimension the, 973-74
    labeling the, 971-73
    lines, adding, 967-68
    lines, removing, 968-69
    refining the, 965-74
    specifying the size, 959
    Grid Labeling dialog box, 971-73, 1006
    plan modifier, 233
    Position sections, 708-10, 748
    Properties section, 708-10, 748
    section, 957, 965
    Trim Planes dialog box, 931, 938, 958
ColumnAdd command, 929
ColumnGridAdd command, 923, 954-55, 960-61, 992

ColumnGridLabel command, 923, 971
ColumnGridXAdd command, 967-68
ColumnGridXMode command, 967
ColumnGridXRemove command, 968-69
ColumnGridYAdd command, 967-68
ColumnGridYMode command, 967
ColumnGridYRemove command, 968-69
Columns
  complex, creating, 948
  encased, 949
  inserting, 929-32
  tab, 682, 706-11, 751, 756
ColumnToolToLinework command, 943-45, 992
Command String edit field, 623
Commands, using toolbars to access, 23-25
Commercial
  palette, 759
  subcategory, 759
Compass option, 902
Component
  column, 39-40, 427, 434, 529-30, 544-45, 547, 791, 844
  Dimensions section, 780-81
  drop-down list, 789
  field, 159-60, 547
  Name
    column, 940
    drop-down list, 190
  Offset dialog box, 236
  stamp-type, inserting a, 812
Components, 2
  bolt, 818, 820
  bookends-type, 815-16
  boundary filling, 813-14
  button, 506
  countable linear repeating pattern, 816-17
  gravel boundary filling, creating, 814
  linear repeating pattern, 812-13
  rectangular predefined-depth surface, 817-18
  structural, inserting, 21-22
  tab, 149, 151-59, 218, 221, 223, 236, 242, 246, 248, 506, 527-28, 540, 544-45, 573, 577, 633, 788, 791, 794, 844
Concrete subcategory. (*See* Walls category)
Concrete.Cast In-Place.Flat Grey material definition, 235
Cone mass element shape, 877
Configuration
  list, 64
  tab, 35, 60
Configurations
  folder, 31-32
  tab, 34-35
Construct
  folder, 49
  subdirectory, 46
  tab, 51, 498, 562
Construction
  documents, using components to develop, 47-51
  Line grip, 514
Constructs
  check box, 857
  creating, 49, 51
  definition, 49

tab, 46, 49-52, 66, 122, 206-7, 216, 229, 235, 244, 315, 323, 443, 498, 562, 567, 571, 628, 758, 760, 850
Content
  Browser™, 1, 9, 11, 13-15, 17-20, 24, 57, 59, 160, 170-72, 256, 313, 325, 327, 336, 425, 475, 497, 522, 569, 571, 576, 585, 587-89, 593, 595, 610-11, 628-29, 632-33, 636, 649-50, 652, 654, 676, 680-81, 704, 726, 732, 741-42, 759, 764, 804, 826-27, 835, 910, 925, 948, 953, 986, 988, 992, 1001, 1007
    accessing the, 13
    customizing with the, 11-16
  File list, 622
  option, 216, 313, 318, 425, 489, 539-40
  page, 570, 582, 775, 798, 839, 842, 854, 857
  using the, 587-91
CONTENTBROWSER command, 13
Contents tab, 661-62, 735
Context
  column, 789, 791, 844
  page, 52-54, 68, 251, 570, 582, 770, 775, 798, 837, 842, 853
  Type
    page, 621-23
    section, 622
Continue Editing check box, 201
Convert Polyline to Wall Modifier command, 197
  option, 188
Convert to
  Body check box, 304
  Formatted Text radio button, 724, 758
  Mass Element
    option, 914
    Tool command, 909-10
  Polyline
    command, 908
    option, 908
  Roof Slabs option, 402-3, 442
Convert to>Mass element option, 881
Copy
  2D Elevation/Section Style and Assign option, 844
  button, 173-74, 527, 938, 942
  command, 110, 143, 173, 217, 235, 900
Construct to Levels
  dialog box, 235
  option, 235, 244
Level
  and Contents option, 49
  option, 48-49
  option, 89-90, 150, 314, 319, 491, 540, 579, 606-7, 917
Wall Style and Assign option, 221, 241, 248
CopyAndAssignStyle command, 148-49
Corner Deceleration check box, 992
Corridor space, 759, 761
Counterclockwise setting, 512
Create
  AEC Content
    command, 621

Wizard, 621-23
  button, 45
  Element option, 904
  File dialog box, 758
  New Drawing dialog box, 3
  Sheet Set-Begin dialog box, 45
  Video command, 990-91
  View command, 990
CreateContent command, 621
CreateHLR command, 783, 804, 806, 836
C-Riser option, 502
Crop option, 820, 825-26
C-Start Miter option, 337
C-Top of Flight edit field, 547
CTRL + 1 command, 16. (*See also* Properties palette)
CTRL + 2 command, 18. (*See also* Style Manager)
CTRL + 3 command, 7. (*See also* Tool Palettes)
CTRL + 5 command 46. (*See also* Window>Project Navigator Palette command)
Current Drawing
  check box, 842
  list, 622
Curtain Wall Unit option, 347
Custom
  Block dialog box, 305
  Command content type, 622
  installation, 2
  Plan Components tab, 345
  selection, 307
  shape selection, 533
  tab, 593
  view, 591, 593, 626
Custom Applications>Design folders, 18
CustomColumnGridAdd command, 970
Customize, 11, 13
  dialog box, 11, 13, 24
  Edge, 498, 517-20
  menu, 6
  palette dialog box, 13
Customize Edge>Generate Polyline command, 520-21
Customize Edge>Offset command, 517-18
Customize Edge>Project command, 519
Customize Edge>Remove Customization command, 520
Cut
  command, 900
  Height, 242, 246
  Line (1)
    option, 675
    tool, 675
  lines, placing straight, 675-76
  option, 150, 412, 424
  Plane
    Height edit field, 168-69
    section, 532, 534
    tab, 35, 166-70, 237, 242, 246, 245, 582, 938997
C-Waist column, 526
C-Width dimension, 275, 296
Cycle measure to option, 261, 264, 287, 290, 308-9, 339
CYcle measure to property, 261, 290, 309
Cylinder mass element shape, 877

D

D Plot, 252-53
D-Angle 2 dimension, 781
Data
　Format drop-down list, 709, 748
　Renumber dialog box, 713-14, 746
Database feature, 2
D-Bottom of Flight edit field, 547
DC Online tab, 593
D-Depth dimension, 275, 296
Default
　Format tab, 705-6
　Frame assignment, 349
　Layer Standard Layer Key Style list, 27. (*See also* Layering tab)
　Orientation option, 348
　output device option, 213, 252-53
　value, 729
Defaults tab, 431-33
Defect Warning, 98, 166
Define
　UCS button, 848
　View Window button, 54-55
Defining Line component, 785
Definition
　edit field, 596, 614, 640, 644
　option, 596
　tab, 728-30, 754
Degrees per Tread arc constraint, 512
Delete
　button, 30, 150-51, 710, 757
　command, 901
　option, 150
　Vertex option, 362
D-End Miter option, 338
Depth
　column, 545
　dimension, 545
　edit field, 191-92, 304, 349-50, 374, 874
Descending
　radio button, 972, 1004, 1006
　sort order, 711
Description
　button, 258, 288, 306, 337, 379, 395-96, 466, 501, 537, 592, 596, 614, 703, 780, 874, 929, 933, 936, 955, 975, 989
　column, 49
　dialog box, 76, 258, 288, 306, 337, 379, 395, 466, 501, 537, 596, 614, 703, 780, 837, 842, 853, 874, 929, 933, 936, 955, 975
　edit field, 149-50, 217, 221, 236, 248, 273, 295, 326, 341, 369, 426, 524, 541, 600, 618-19, 660, 680, 704, 728, 787-89, 812, 814
　option, 76, 395-96, 501, 537, 596, 874
Design
　Component Manager, 1
　directories, 19
　palette, 4-5, 9, 81-82, 740, 745, 759
　Pulldown menu, 6
　resources, accessing, 17-18
　Rules tab, 161-62, 273, 276-77, 294, 296-97, 304, 327, 330, 340-41, 345-46, 349, 352, 358, 369, 374, 426-27, 431, 433-34, 476, 479-80, 502-4, 514, 524-25, 571-72, 788-89, 791, 844, 940-42, 945, 947-48, 951

defining the components using, 345-52
　tab, 16-17, 291, 352, 383, 386, 735, 765
　Tool
　　Catalog, 20, 587-88, 593-95
　　Catalog-Imperial catalog, 19, 142, 741
　　Catalog-Metric catalog, 19, 142
　tool palette, 73, 136, 466
Design Rules>Save to Style option, 356
Design Rules>Transfer to Object option, 356
DesignCenter™, 1-2, 9, 17-120, 592-94, 643
　button, 589-90
　command, 24
　creating symbols for the, 621-26
　Documentation folder, 19
　inserting content from the, 19
　toolbar, 591
　using the, 591-93
Detach
　button, 566, 570
　command, 901
　Elements option, 424, 895
　Objects
　　command, 616, 907
　　option, 616, 666, 668, 907
　option, 418, 498, 582
Detail Component
　creating, 20-21
　Manager, 17, 20, 24-25, 770, 808-14, 818, 859, 862-64
　accessing the, 21
　using the, 808-13
　tool palette, 20
Detailed Description edit field, 625
Detailing palette, 9, 857
Details command, 144
D-Extension of Balusters from Floor Level edit field, 544
D-Horizontal Offset from Roof Slab Baseline edit field, 433
Diagnostic display configuration, 95, 98, 127-28
Dictionaries feature, 2
Dim Style edit field, 533-34
DimAdd command, 654, 973
DimAngular command, 673
DimAttach command, 666
DimDetach command, 666-68
Dimension
　points
　　adding, 668
　　removing AEC, 668-70
　Style Manager, 662
　toolbar, 24
Dimensioning template, 3
Dimensions, 20
　creating, 653-75
　dynamic, 263
　placing in a drawing, 19
　radial, 674-75
　section, 780, 874-75, 930-31, 934-35, 937, 956-57, 965
　tab, 273-76, 294-97, 318, 327, 373, 757, 942, 975
DimManAdd command, 656-57
DimPointsAdd command, 668
DimPointsRemove command, 669

DimRadius command, 674
DimRemoveOverrideExtLines command, 665
DimRemoveOverrideTextOffsets command, 665
DimScale, 654
DimTextOverride command, 670
Dining Room space, 741
Direct
　distance entry method, 82
　Distance Entry, 81-82
　mode, 395-96, 466, 468, 487
Direction
　property, 399
　toggle, 397, 467
Disable Live Section option, 803
Discard All Changes
　button, 353
　command, 185, 353
　option, 184, 195, 203, 311, 360, 362, 439, 441
Display
　Auto Track tooltip toggle, 74, 113, 118, 123, 130, 136, 175, 859
　AutoSnap tooltip, 105, 113, 118, 123, 130, 136, 859
　Below Range, 582
　Component column, 39, 41
　Configuration flyout, 4-6, 60, 120, 127, 129, 612, 637, 639, 740, 758, 853, 901
　configurations, 58
　control, defining for objects and viewports, 31-37
　Edit Property Data Dialog During Tag Insertion toggle, 683
　full-screen tracking vector toggle, 113, 118, 123, 130, 136, 859
　Inner Lines Below check box, 169-70
　Manager
　　Configurations folder, 605
　　dialog box, 1, 31-33, 62, 166-67, 242, 246, 997
　of objects, updating the, 42
　Options page, 621, 625-26
　Path drop-down list, 533-34
　polar tracking vector toggle, 113, 118, 123, 130, 136, 859
　Properties
　　button, 320, 581
　　dialog box, 40, 164-65, 170, 224, 237, 248, 283, 300-2, 345-46, 427, 429-30, 531, 535, 548-50, 559, 561, 620, 660-63, 670, 780, 785-86, 791, 799, 844, 890
　　tab, 39-41, 149, 163-69, 171, 223, 237, 248, 273, 279-80, 282-84, 294, 299-300, 319-20, 341, 345-46, 373, 428-30, 434, 443, 445, 459, 531-33, 535, 540, 548-51, 569, 618, 620, 635, 660-61, 712-13, 735, 785, 789-91, 844-45, 854, 890-91, 943
　Properties – Door Display Representation dialog box, 282-83
　Properties – Door High Display Representation dialog box, 283
　Properties – Door Threshold Plan Display Representation dialog box, 284

Properties – Material Definition dialog box, 299, 343-44, 434-36
Properties – Material Display dialog box, 279-31, 299
Properties – Window Elevation Display Representation dialog box, 305
Properties – Window Model Display Representation dialog box, 305, 373
Properties – Window Plan Display Representation dialog box, 300
Properties (2D Section/Elevation Style Override – 2D Section Style 96(2)) dialog box, 845
Properties (2D Section/Elevation Style Override – MyStyle) dialog box, 791
Properties (2D Section/Elevation Style Override – style name) dialog box, 791-92, 794
Properties (AEC Dimension Style Override – 1 Chain) – AEC Dimension Plan High Detail Display Representation dialog box, 735
Properties (Door Style Override-Hinged-Single) – Door Threshold Plan Display Representation dialog box, 320, 322
Properties (Drawing Default) – Material Definition General Medium Detail Display Representation dialog box, 171-72
Properties (Drawing Default) – Stair Plan Display Representation dialog box, 569
Properties (Drawing Default) Roof Slab Plan Display Representation dialog box, 459
Properties (Material Definition Override-Doors&Windows.WoodDoors.Ash) – Material Definition General Medium Detail Display Representation dialog box, 320-21
Properties (Wall Style Override-Frame Brick) – Wall Plan Display Representation dialog box, 443-44
Property
　Set Data dialog box, 683
　Source column, 39-40, 164, 429, 434, 531-32, 549, 620, 660, 790
Representation, 58
　column, 789
　Control tab, 33-34, 60
　High detail, 948
　Low detail, 948
　Plan, 948
　Set column, 35-36
Representations
　column, 39-40, 428, 434, 531-32, 549, 620, 660
　list, 601
　option, 27, 428
　selecting, 31-32
Riser check box, 527-28, 572

Set
  list, 783
  section, 783
  tab, 25-27, 60
  Tread check box, 527-28, 572
Distance
  edit field, 532, 534
  to Line column, 972
  to line, 966
Divide
  command, 986
  option, 820-22
  Spaces command, 911
Division
  Assignment Override dialog box, 364
  assignment, creating an Override of, 363-64
  Types drop-down list, 347
Division>Override Assignment command, 363-64
Divisions
  dialog box, 49
  element, 340, 369
  section, 49
D-Landing Thickness edit field, 527-28
Do not show a Startup dialog, 3
Document
  option, 838
  palette, 9, 742, 745, 761, 847, 853, 866
  Pulldown menu, 6
Documentation
  folder, 19
  inserting, 19-20
  menu, 6
  Objects\AEC Dimension Styles folder, 658
  Objects\Schedule Table Styles folder, 704
  Objects>Property Set Definitions, 754
  Tool
    Catalog-Imperial, 19, 653
    Catalog-Metric, 19
Dome mass element shape, 877
Door
  insertion point, locating the, 262-69
  modifying using grips, 265-69
  Object property, 684
  Objects property set, 686
  option, 313, 316
  Override source, 40
  placing in walls, 256-72
  Plan High Detail display representation, 282
  properties of a, 257-62
    defining, 269-72
  Schedule
    option, 754
    Project Based, 765
  shifting within a wall, 284-85
  Style
    option, 347
    Properties – Bifold-Double dialog box, 757
    Properties - Hinged Single Exterior dialog box, 319-20, 322
    Properties dialog box, 273-75, 282, 318, 330, 373, 684-85
    property, 684

Styles
  (Imperial) option, 313
  command, 272-73
  option, 313, 329
  swing, setting the, 262
Tag
  command, 687
  option, 686, 753-54
  Project Based command, 686, 761, 763
  tool, 256-57
Type
  list, 276
  section (of Design Rules tab), 276
Door/Window Assembly
  modifying, 352-67
  option, 336-38, 371
  setting the miter angle of, 364-65
  Style
    components of, 340-52
    Properties dialog box, 339-43, 345, 358, 371, 374
    style in the drawing, changing the, 352-55
    Styles option, 340-41
DoorAdd command, 74, 255-56, 260, 263-64, 272, 287, 314, 316, 330
DoorObject property, 753
DoorObjects property, 754
Doors &
  Windows.Glazing.Glass.Clear option, 327
Doors & Windows.Metal Doors & Frames.Aluminum Windows.Painted.White option, 327
Doors & Windows.Metal Doors & Frames.Steel.Doors.Painted.White option, 319
Doors & Windows.Wood Doors.Ash option, 319-20
Doors, editing properties of multiple, 272
DoorWinAssemblyAdd command, 335-36, 339, 372
DoorWinAssemblyStyle command, 340
Doric Column mass element shape, 878-79
Double
  Casement
    Windows, 326
    window type, 327
  Door selection, 351-52
  slope
    option, 379-80
    shape, 391, 450
DOUBLE HUNG property set value, 750
D-Radius field, 120
Drafting
  Settings dialog box, 113, 118, 123, 130, 136, 226, 229, 443, 446, 450, 455, 483, 489, 562, 568, 571, 595, 628, 636, 642, 859, 999
  tab, 74, 105, 175, 859
Drape
  command, 882, 916
  tool, 882
Draw toolbar, 25

Drawing
  annotation, placing on a, 650-53
  content type, 622
  Default
    Display Configuration option, 27, 60
    display property source, 531, 549
    global settings, 164, 429, 620
    source, 40
  Encryption feature, 2
  menu
    button, 59, 117
    flyout, 4-6, 113
  Menu option, 595
  Properties, 5-6
  Scale flyout, 26, 636
  scale factor, 624
  Setup
    command, 5-6, 25, 59, 113, 117, 595, 650-51, 654, 672, 676, 730, 826, 837
    configuring, 25-27
    controlling display in, 27
    dialog box, 25-26, 57, 60, 113-14, 117, 586, 595, 626, 652, 703, 706, 827, 973
  Template edit field, 49-50
  Units drop-down list, 26
  walls, 72-76
  Window status bar, 4-6, 26, 34, 57, 60, 68, 95, 98, 113, 117, 120
Dry Run check box, 992
D-Total column, 526
D-Tread option, 502
Duplex Receipt symbol, 642
Dynamic Posts check box, 543, 577, 579

E
E - "Y" Direction edit field, 433-34
E-Additional Width edit field, 527-28
EAttEedit command, 784-85
Eave property, 385
Edge
  Column, 384
  Cut, 379, 405, 432
  Grip, 203, 406, 515, 551
  Offset
    column, 218
    edit field, 156, 218
    grip, 406
  overhang, using grips to edit, 406
  profile, adding to the workspace, 479-81
  Style
    column, 438, 477
    drop down list, 404, 438, 477
Edges button, 403-4, 436, 472, 478
Edges/Faces button, 383
Edit
  2D Section/Elevation Style command, 787
  AEC Dimension Style option, 666, 735
  Attribute Locations option, 609
  Attributes dialog box, 652-53, 678
  Block in place option, 808, 860, 865
  Block in-Place option, 610
  button, 30, 148, 150-51, 278, 297, 727
  Display Properties button, 171, 279-80, 299, 320, 343, 345, 373, 429,

434, 459, 531-32, 535, 549, 569, 581, 620, 635, 660, 735, 785, 790-91, 844-45, 854
Display Properties Material Definition Properties-Standard button, 854
Divisions button, 48-49
Door Style option, 318-19, 373, 757
Door/Window Assembly Style option, 358, 373
Edge grips toggle, 513
Edge Profile In Place
  command, 441-42, 481-82
  option, 441
Edges command, 872, 912
Edges/Faces
  button, 388
  command, 388
  option, 388-89, 452-53
Elevation/Section Style, 844, 854
Field option, 829
field, 149-50
In Place
  command, 183, 194, 360
  option, 183-84, 194, 205
  toolbar, 896
  trigger grips, 664-65, 893, 895-96
  using with Wall Sweep, 202
In-Place option, 666
Justification command, 91, 130-31
Levels button, 48, 66, 244, 837
Material
  button, 39-40, 160, 171-72, 279, 299, 319, 343, 427, 434, 530, 547
  Definition Properties dialog box, 427
  dialog box, 436
Member Style
  command, 939
  option, 940
menu, 6, 8
Multi-View Block Definition option, 599, 637
Object Display
  command, 300
  option, 39, 58, 459, 785
  option, 326, 330, 369, 425, 540, 787, 792, 889
Profile In Place
  command, 359-60, 557-58
  option, 331
Project button, 48
Property
  Set Data command, 649, 715
  Set Data dialog box, 682, 684, 687-89, 697, 699-701, 704, 743-44, 746, 750-52, 753, 755, 761-62, 764, 1007
  Sets Data dialog box, 150-51, 295, 426, 524, 600-1, 619
Railing Style
  command, 539-40
  option, 540
Referenced Property Set Data dialog box, 715-16
Roof Slab Edges
  command, 437
  dialog box, 437-39, 477
  option, 437
Schedule
  Property dialog box, 716

Index

Table Style option, 704, 748, 751, 756
Slab Edge Style option, 478-79
Slab Edges
  command, 476, 478
  dialog box, 477-78
  option, 476-77
Slab Style
  command, 475
  option, 476
Source button, 728
Stair Style
  command, 522-23, 569
  option, 581
Standard Size dialog box, 278, 297
Style
  button, 147-49, 217, 223, 236, 273, 523, 604
  command, 143-44, 273
Table Cell command, 715, 750
the Selected Opening Endcap Style button, 161
View Block
  Offset command, 608
  Offsets option, 608
Wall Style option, 169, 182, 443, 445, 635, 700
Window Style option, 372, 750
Xref in place option, 568, 580
Edit in Place>Discard Changes option, 793
Edit in Place>Save Changes option, 793
EditScheduleStyle command, 704
E-Door Thickness dimension, 275
E-Glass Thickness dimension, 296
Element
  column, 351-52, 371
  Definitions, 345
    defining, 347-49
  folder, 49
  Properties option, 900
  subdirectory, 46
Elements, 49
Elevation
  controlling display of, 785-86
  edit field, 406, 473, 508, 532, 534, 596, 614, 635, 703, 781, 875, 957, 976
  field, 380, 468
  Mark A1, 775
  Mark A2, 838
  merging lines to the, 794-96
  option, 77
  styles, creating to control elevation display, 785-91
Elevations
  category, 840, 852
  generating from a building section/elevation line, 784
  sheet category, 775
E-Maximum Center to Center Spacing edit field, 544
Emergency Square 1 symbol, 644, 646
Enable
  AEC Unit Scaling check box, 624
  Live Section option, 800, 802, 856
End
  angle, 959
  check box, 371
  Elevation Offset section, 191
  grip, 551
  Miter angle, 338

object snap, 264
offset, 923, 928, 930, 934, 937, 956-57, 976
  property, 965
Shape columns, 940-42
Endcap Styles command, 179, 215
Endcaps
  creating wall, 178-83
  feature, 308
  modifying, 182-83
  shortcut menu, 183, 291
Endcaps/Opening Endcaps tab, 149, 160-62, 182, 185, 225, 633
Endcaps>Calculate Automatically option, 185
Endcaps>Edit In Place, 183, 226
Endcaps>Override Endcap Style option, 185
Ending endcap (Design - Advanced *Style Overrides*) option, 78
Endpoint
  entry methods, establishing, 82
  Object Snap mode, 86, 113, 118, 123, 130, 136, 226, 568, 628, 740, 759, 848, 859, 916
  option, 846
Enhanced Attribute Editor command, 785
Entry_Room space, 759
Erase
  command, 110
  option, 461, 848
  Selected Object check box, 205
E-Roll edit field, 931, 934, 937
E-Roof line offset from base height option, 76, 79
eTransmit, 5-6
E-Waist column, 526
Exit Edit
  Attribute Locations option, 609
  In Place trigger grips, 664
  View Block Offsets option, 609
Expand button, 623
Explode
  check box, 610
  command, 111
  on Insert check box, 623
EXPLODE command, 623
Export
  button, 30
  option, 722, 724, 758
  Schedule Table dialog box, 722-24, 758
Express Tools, 28
Extend
  command, 110, 112, 137
  Landings to
    Merge Flight Stringers with Landing Stringers check box, 528-29, 572
    Prevent Risers and Treads Sitting Under Landings check box, 528-29, 572
  option, 409
Extended Data
  Property field, 685-86
  tab, 16-17, 271, 291, 383, 386, 686, 701, 756, 820
Extension
  Distances section, 528-29
  edit field, 972
  of ALL Posts from Floor Level edit field, 543

Top Railing edit field, 542
of Balusters from Floor Level edit field, 544
Extensions tab, 540, 546-47, 577-78
Exterior
  category, 837, 841-42
  Elevation Mark A3 option, 842
  Elevations, 775
External
  drawing
    field, 725, 766
    path, 726
  Source section, 726
ExtrudeLinework command, 880-81
Extrusion tool, 882
Eye Level edit field, 990

F

F - "" Direction edit field, 433-34
Face column, 385
Faces, editing, 887-89
Family Room space, 741
Fascia check box, 433
Favorites
  folder, 592
  toggle, 592
F-Floor line offset from baseline option, 77, 79
Field
  dialog box, 731, 827, 829
  Verify, 670
File
  menu, 4, 6, 8
  Name edit field, 625, 723-24
  name edit field, 60
File>Close All option, 235, 245
File>Close, 116, 121, 994, 998
File>New, 2, 117
File>Project Browser, 43, 65, 122, 129, 216, 229, 235, 244, 251, 315, 323, 329, 443, 562, 567, 571, 566, 570, 628, 635, 733, 740, 758, 770, 837, 840-41, 852, 856, 867
File>Save, 65
File>SaveAs, 60, 113, 117, 130, 136, 329, 369, 483, 489, 635, 642, 743, 753, 913, 915
Files tab, 588
Fillet command, 110, 112
Filter
  drop-down list 692, 759-60
  Style Type command, 144-45
Finish
  option, 68, 251, 764, 770, 775, 798, 837, 839, 842, 854, 857
  Tags, 689-90
    placing, 696-98
Finished Floor to Floor selection, 563
First Riser Number edit field, 533-34
Fixed
  Cell Dimension division, 345, 347, 355
  Manual division, 345, 347-48
  Number of Cells division, 345, 347-48, 354
  Posts
    at Railing Corners check box, 543, 577, 579
    check box, 542, 577, 579
  View Direction list, 35-36
  Width edit field, 706

Flat Shaded mode, 322, 493, 883, 902, 917
Flight
  & Landing Corners, 563
  columns, 526
  Dimensions section, 527-28
  End grip, 514
  Height value, 529
  Start grip, 515
  Taper grip, 514-15
  Width grip, 514
Flip
  grips, 515
  Hinge trigger grip, 267-68, 314
  Swing trigger grip, 267-68, 314
Floor
  1 check box, 839, 850-51, 853-54
  creating additional, 206-9
  Elevation value, 49
  Level 3 drawing, 570
  Plan 1 option, 761, 763
  Plan 2 option, 763-64
  Plan-Basement Level, 251-52, 582
  to Floor Height value, 49
FloorLine command, 158, 176-78, 207, 215, 238, 248, 770
Floorplate slices and boundaries, creating, 905-8
FND level name, 244-45
F-Number per Tread edit field, 544
Folders tab, 593
Fonts feature, 2
Footing component name, 791, 793
Force Horizontal option, 624
Format
  dialog box, 723-24, 758
  list, 729
  menu, 6, 8
Format>AEC Dimension Styles, 658
Format>Blocks>Block Definition option, 639
Format>Define Custom Member Shape option, 949
Format>Display Manager, 34-35, 60, 242, 246, 582, 997
Format>Multi-View Block>Multi-View Block Definitions option, 639
Format>Structural Member Catalog option, 994, 998
Format>Style Manager, 97, 143, 148, 171, 174, 181, 196, 199, 216, 223, 229, 245, 318, 326, 329, 369, 489, 523, 540, 571, 576, 579, 604, 606, 617, 633, 642, 754, 790, 854, 916, 951
Foundation plan, creating a, 209
Frame
  Assignment Override dialog box, 362, 374
  Brick, 217, 220
  column, 751
  Display tab, 282-83
  editing the profile of, 359
  Material column, adding a, 749
  rigid, creating, 947-48
  section, 319
FRAME
  command, 749
DEPTH
  column heading, 748-49
  value, 748

MATERIAL
    column heading, 749
    value, 748
FrameDepth option, 748
FrameMaterial
    edit field, 751
    option, 748
Frame/Mullion>Add Profile command, 359-60
Frame/Mullion>Edit Profile in Place option, 361
Frame/Mullion>Override Assignment option, 362, 374
Frames
    component, 340, 371
    creating custom, 349
    option, 374
    section, 992
Free arc constraint, 512
from list, 190-91
Front
    Elevation property, 840
    option, 902
    View option, 373, 493, 840
F-Total column, 526

G

Gable
    mass element shape, 877, 879
    option, 379, 387, 917
    roof, creating a, 387
    Shape value, 387, 391, 442, 447-48, 456-58
    tool, 913
Gap edit field, 706
General
    display representation, 638-40
    Medium Detail value, 319
    Page, 52, 67, 251, 570, 582, 770-71, 775, 798, 837, 841, 853, 857
    Properties tab, 39, 41-42
    section, 779-80, 874, 929, 933-34, 936, 975, 989
    Spaces dialog box, 692-94
    tab, 34-35, 149-51, 161-62, 217, 221, 236, 241, 248, 273-74, 294-95, 326, 341, 369, 425, 431-32, 434, 523-24, 540-41, 600-1, 618, 660, 684, 699-701, 704, 727, 729, 750, 787-88, 889, 940-41, 955-56
Generate
    Elevation command, 784
    Member Style
        option, 928, 932, 994, 998
        toolbar icon, 928
    New Bubbles on Exit check box, 973, 1004, 1006
    Polyline option, 176, 517, 520
    Section
        Elevation dialog box, 801
        option, 800-2
    Section/Elevation
        check box, 773, 775, 797, 838, 843, 854
        dialog box, 782-84
    Spaces dialog box, 691, 695, 741, 759-60
    Stair Tower option, 535
    View after add edit field, 990
    Walls command, 872, 912
Global option, 167

Gouraud Shaded mode, 902
Graphic Path grip, 514
Graphline position option, 94
Grass, 917
Gravel property, 814, 862
Grid
    defining the column for the, 965
    editing in the workspace, 353-55
    infill, 347
    nested, 347
    template, 3-4
GRID, setting, 90
GridAssemblyAddCellOverride command, 358
GridAssemblyAddDivisionOverride command, 364
GridAssemblyAddEdgeOverride command, 362
GridAssemblyAddProfileOverride command, 359
GridAssemblyCopyFromStyle command, 355-56
GridAssemblyInterferenceAdd command, 365-66
GridAssemblyInterferenceRemove command, 367
GridAssemblyMakeStyleBased command, 356
GridAssemblyMergeCells command, 357, 368
GridAssemblySaveChanges command, 356
GridAssemblySetEditDepthAll command, 357
GridAssemblySetMiterAngles command, 365
Grids option, 902
Grips, 71, 73, 80, 474
    circle, 406
    Construction Line, 514
    copying walls using, 90
    edge offset, 406
    edge, 406
    Edge, 515, 551
    Edit in Place, 893, 895
    editing
        a stair using, 514
        AEC Dimensions, 664-65
        walls using, 89
        windows using, 292-93
    End, 551
    Fixed Post Position, 551
    Flight
        End, 514-15
        Start, 515
        Taper, 514-15
        Width, 514-15
    Flip, 515-16
    gable mass element, 884-85
    Graphic Path, 514
    Landing Width, 515
    Lengthen Flight, 514
    Location, 514, 517
    Mass Group trigger, 893
    Move Flight, 515
    negative symbol, 406
    Property Data Location, 762-63
    Radius, 517
    rhombus, 406
    Start, 551
    Turn Points, 515
    using to edit a

building section/elevation line, 778
    roof, 390-91
    roof slab, 406
    wall, 88-89, 112
    wedge shaped, 406
Group property, 75
Groups function, 28-29
Guardrail check box, 542, 577
Guard-Rect Balusters, 569

H

H command, 397, 467
Handrail
    check box, 542, 577
    railing style, 576
Has Fixed Thickness check box, 427
Hatch
    component, 166, 429
    Edit dialog box, 865
    Pattern dialog box, 167
    tab, 345
Hatching tab, 166-67, 170, 280-81, 429-30, 436, 790
Head
    Height option, 259, 289, 338-39, 505
    height option, 261, 290, 308
    option, 259, 307, 318, 325
HEAd height property, 261, 290, 308, 339
Header, adding a, 749
Heading edit field, 709, 748
Headroom edit field, 505
Height
    column, 710
    edit field, 258, 278, 288, 292, 337, 502, 504, 706, 780, 874
    option, 290, 308, 339
    property, 75, 111, 261, 384-85
HEight option, 261, 290, 308, 339
Help
    command, 24
    menu, 6, 8
Hide
    Details button, 942
    Edge option, 184, 195
    Lines Below Openings at Cut Plane check box, 169-70
    option, 551
Hidden
    context, 791
    Line
        Projection command, 804-5, 836
        Projection, creating 2D sections with, 804-5
        Removal command, 783
    mode, 902
    option, 222, 242, 330, 419, 445, 449, 454, 462, 582-83, 635, 844, 856, 995
Hidden2 linetype, 223-24, 237, 248, 844
High Detail display, 330, 635, 739-40, 758, 760, 853
Highlight check box, 717
Hinged-Single-Exterior option, 318
Hip roof, creating a, 383
History tab, 593
Hold facia elevation edit field, 405, 473
Hole command, 416
Hole>Add option, 416, 487
Hole>Remove option, 417

Holes
    and Boundaries, adding, 982-86
    creating in the ceiling grid, 984-85
Home button, 592
Horizontal
    Height column, 542
    left component, 947
    offset, 396, 398, 467
        editing with grips, 406
    Orientation edit field, 501, 510, 512
    Section tool, 800-1
Hot Grips, 89. (*See also* Grips)
Hub Style setting, 304
Hyperlink button, 271, 291, 383-84, 386
Hyperlinks button, 598, 615

I

Icon
    only, 11-12
    section, 625
    with text, 11-12, 59
ID column, 49
i-drop, 15-16, 19
Ignore Boolean operation, 681
Illustration toggle, 16-17
Image Size, 11-12
Imperial content installation option, 2
Imperial>Timber>Lumber>Nominal Cut Lumber, 994
Import button, 30
Import/Export – Duplicate Names Found dialog box, 173-74, 606
In Place
    Edit changed, saving to the style, 355-58
    option, 92
Include
    Product Column check box, 708
    Quantity Column check box, 708, 756
Increment
    fields, 694, 714, 759-60
    option, 692
Index
    column, 218, 788
    number, 152-53
Infill
    Assignment Override dialog box, 358-59
    modifying the, 356-57
    overriding the, 358-59
    specifying, 348-49
    Type options, 348, 369, 371
Infill>Merge command, 357
Infill>Override Assignment command, 358-59
Infill>Show Markers command, 357-58
Infills element, 340, 358, 369
In-Place Edit toolbar, 185, 226, 228, 310-11, 331, 353, 360-62, 439-41, 480-82, 557-58
InplaceEditAddRing command, 203, 310
InplaceEditAddVertex command, 183, 194, 203, 310
InplaceEditHideEdge command, 184, 195
InplaceEditRemoveEndcap command, 184, 194
InplaceEditRemoveModifier command, 195

InplaceEditRemoveVertex command, 183, 203, 310
InplaceEditReplaceEndcap command, 184
InplaceEditReplaceModifier command, 195
InplaceEditReplaceRing command, 203, 311
InplaceEditSave command, 195
InplaceEditSaveAll command, 184
InplaceEditShowEdge command, 184, 195
Input section, 722-23
Inquiry toolbar, 24
Insert
  After radio button, 748
  as Block option, 52
  Component button, 810, 815, 817
  dialog box, 610
  Field option, 731
  Hyperlink dialog box, 291, 386
  Hyperlinks dialog box, 598, 615
  menu, 6, 8
  Options page, 621, 623
INSERT command, 607
Insert>Detail Component Manager, 815, 817, 859, 862
Insertion
  offsets button, 596, 598
  Point
    section, 598, 680
    value, 386
    X edit field, 598
    Y edit field, 598
    Z edit field, 598
Inside
  of frame option, 258-59, 288, 330
  Radius, 958
Inspect command, 175
Install Supplemental Tools, 28
Interference (Select Objects) section, 681
Interference Add, applying to the assembly, 365-67
Interference>Add command, 365-66
Interference>Remove command, 367
Interior, 221
  partitions, 221
Intersect operation, 874, 897-98
Intersection object snap mode, 381-82, 424, 446, 450, 455, 562, 628, 642, 965
Isolate Objects>End Objects Isolation, 23
Isolate Objects>Hide Objects, 23
Isolate Objects>Isolate Objects, 23
Isosceles Triangle mass element shape, 878

J

J command, 396, 467
Join
  command, 71, 99, 103
  Face editing option, 205-6
  Faces
    command, 888
    option, 888
  option, 133
  Spaces command, 911
Joint type, 857
Justification
  column, 545
  option, 545

setting, 90-91, 111
tool, 71
types of, 80-81
Justify
  drop-down list, 502, 931, 935, 937, 957
  property, 75, 928, 965
  option (Design - Basic *Dimensions*), 76, 396-98, 467
  Overall Extents check box, 942
  using overall extents, 931, 935, 937, 957, 965

K

Keynote
  button, 295, 524
  edit field, 149-50, 341
  Legend, creating a, 830-31
  option, 600
KeynoteLegendAdd command, 830
KeynoteLegendApplyKeys command, 833
KeynoteLegendSelectionAdd command, 833-34
KeynoteLegendSelectionShow command, 834
Keynotes, 20
  button, 274
Keynoting, 826-35
  Reference, 827-31
Kitchen space, 741

L

Label option, 971, 1004
Labels in the X- Direction list box, 972
Landing
  columns, 526-27
  Dimensions section, 527-28
  Extensions
    button, 506, 529
    dialog box, 506-7
    tab, 506, 528-29, 572, 574
  Length edit field, 529
  Location drop-down list, 529
  Thickness value, 527-28
  Width grip, 515
Launch Project Browser button, 46-47, 68, 243
Layer
  button, 41
  column, 39, 41
  command, 28
  drop-down list, 269, 780, 1011
  edit field, 269, 291, 308, 404, 472, 508, 537, 597, 615, 874, 929, 933, 936, 989
  Key Styles, 28-29, 60
  Manager, 25, 28-29, 57, 64, 459, 635
    accessing the, 28
    dialog box, 29
    toolbar, 62-64
    tools of the, 28-30
  Name column, 624
  option, 508
  Properties
    Manager command, 24-25, 29, 62, 64
    toolbar, 4, 23, 25, 28-29, 62, 64, 1011
  Snapshots, 28
    creating, 29-30
    standard, defining the, 27
  standards template, 3-4

Standards/Key File to Auto Import edit field, 27, 60
Wildcard edit field, 723-24
wildcard field, 703
LAYER command, 28
Layer/Color/Linetype tab, 39-41, 165-66, 169, 223-24, 237, 248, 280, 282-84, 320, 345, 429, 434, 444, 531-32, 548-50, 559, 620, 635, 660-61, 713, 785-86, 790-92, 794, 844-45, 854, 890
Layering tab, 25-27, 60, 117, 626
Layout
  Curve tool, 986
  curves, placing columns using, 986-87
  Mode command, 966-67, 980-81
  setting, 213, 252-53
  tab, 711-12
  type
    edit field, 967
    option, 956, 965, 976, 979
LayoutCurveAdd command, 986-87, 992
LayoutGridClip command, 963-64
Layouts
  inserting with multiple fixtures, 610-13
  using, 30
Leaders, placing in a drawing, 651-53
Leaf edit field, 278
Leave Existing option, 173, 606
Left
  justification, 502
  option, 902
Legends, editing, 833-35
Length
  edit field, 74, 191-92, 337, 339
  of Wall check box, 662
  overall, 504
LEngth option, 339
Lengthen Flight grip, 514
Length-Long value, 748
Levels
  button, 770, 841
  dialog box, 48, 66, 244, 837
Lights
  High edit field, 304
  Wide edit field, 304
Limit property, 850
Limits template, 3-4
Line
  command, 83
  option, 76, 846
  segment type, 81
Linear section, 26
Lines and Arrows page, 672
Linetype
  button, 41
  column, 39, 41, 224, 237, 248, 844-45
  Scale edit field, 41
Lineweight
  button, 41
  column, 39, 41
  dialog box, 41
  option, 223
Linework
  and Walls option, 470
  converting to
    structural members, 943-47
    walls, 75-76
  option, 76, 792

Linework>Edit command, 848
Linework>Merge command, 794, 847
LineworkCrop command, 825
LineworkDivide command, 821, 872
LineworkMerge command, 824, 872
LineworkObscure command, 822-23, 865
LineworkSubtract command, 822
LineworkTrim command, 820
Link to Autodesk VIZ Render, 5-6
List
  tab, 146-47
  View, 11-12
LIST command, 616
Live Section
  command, 804
  Rendering option, 436
LiveSectionDisable command, 803
LiveSectionEnable command, 802
Living Room space, 741
Load
  command, 591
  dialog box, 591
Locate Keynotes command, 834
Location
  dialog box, 77, 386, 597-99, 615, 781, 875, 958
  grip, 514, 517
  Property Definition dialog box, 730
  section, 533-34, 780-81, 875, 957, 975, 990
Logical length
  edit field, 928, 930, 934, 957
  property, 965
  readout, 517
Look in directory, 245
Low Detail, 948
Lower
  Extension edit field, 780
  Slope angle, 380
LtScale column, 39, 41

M

M command line input. (*See* Match option)
MA option, 398-99
Manage Xrefs button, 498, 566, 570, 582, 763, 851
Manual
  Above and Below Cut Plane Heights check box, 168-69, 237
  data, 682
  division, 355
  layout, 966-67, 979-82, 992
  mode, 967
  option, 692, 956, 987
  radio button, 691
Mark column, 706
Mask Block
  Definition Properties dialog box, 618-19
  Definitions
    folder, 617
    option, 642
  tool, 621
MaskAdd command, 621
MaskDetach command, 616
Masking Block content type, 622
Masonry.
  Unit
    Masonry.CMU.Stretcher.Running material definition, 235
  Wall file name, 857

Mass
  Element Style Properties
    - Grass dialog box, 917
    dialog box, 889-90
  Elements
    adding to a group, 893-911
    adding using Body Modifiers, 204-6
    attaching to a group, 893-94
    Boolean operations with, 895-96
    creating style definitions for, 889-91
    creating using Drape, 882-83
    creating using extrusion and revolution, 880-82
    creating with the Model Explorer, 903-5
    detaching from a group, 895
    groups for, creating, 891-93
    inserting, 873-75
    modifying, 883-89
    shapes, selecting, 875-79
    splitting, 886-87
    tools catalog, 13-15
    trimming, 885-86
    using Boolean add/subtract/detach to combine, 418-20
  Group
    command, 422
    Operation>Additive command, 896
    option, 424, 893
    tool, 891-93
  models, creating, 872-83
MassAttach command, 615-16
MassElementAdd command, 871, 873, 879
MassElementDivide command, 886-87
MassElementFaceDivide command, 206, 887-88
MassElementFaceJoin command, 206, 888-89
MassElementOpAdd command, 896
MassElementOpIntersect command, 897-98
MassElementOpSubtract command, 896-97
MassElementTrim command, 206, 885
MassGroupAdd command, 422, 892
MassGroupAttach command, 423-24, 894
MassGroupDetach command, 424, 895
Massing toolbar, 422, 424
Match option, 75, 137, 261, 290, 309, 339, 398-99, 687-88
Material Boundary>Add command, 794-95, 836, 849
Material Definition
  column, 39-40, 159-60, 218, 237, 278, 298, 319, 343, 427, 434, 530, 547, 889-90
  Properties dialog box, 160, 279, 299, 343, 434-35, 530, 547
  Material Definition Properties - Stone dialog box, 171
  Material Definition Properties-Doors & Windows.Wood Doors.Ash dialog box, 320
Material edit field, 750-51
MaterialList command, 171
Materials
  assigning to wall components, 170

Definitions (Imperial).dwg file, 216, 235, 318
list, 917
tab, 39-40, 149, 159-60, 170-71, 218, 237, 248, 273, 279-80, 294, 298-99, 319, 327, 341, 343-44, 427-28, 431, 434-36, 476, 529-30, 540, 547-48, 889-90, 917, 942-43
Matrix
  check box, 709
  Symbol section, 706
Maximize
  option, 64
  Viewport toggle, 6-7, 122
Maximum
  Center to Center Spacing edit field, 543
  Height edit field, 505
  Limit Type options, 505
  Risers edit field, 505
  Slope edit field, 524
Measure to
  Center check box, 192
  command, 986
  edit field, 258-59, 288
  to
    Inside of frame, 288, 291-92
    Outside of frame, 288, 291-92
Medium
  Arcs and Tag option, 676
  Detail display configuration, 35, 60, 87, 97, 129, 242, 322, 582, 639, 997
Member type – Beam, 934
Member type – Brace, 936
Member type – Column, 930
Members catalog, 925, 927
MemberStyleEdit command, 939
MemberStyleWizard command, 951
Merge
  Boundaries command, 872, 911
  option, 820, 824-25
Metric
  content installation option, 2
  D A CH content, 2
Microsoft Excel 97, 724
Mid option, 414
Migrate Custom Settings feature, 2
Minimize Viewport toggle, 6
Minimum
  Height edit field, 505
  Limit Type options, 505
  Maximum Limits, 525
  Risers edit field, 505
  Slope edit field, 524
Mirror
  column, 941
  command, 110
  In options, 350
  option, 941
Mirror X check box, 192
Mirror Y check box, 192
Miscellaneous folder, contents of, 681
Miter
  option, 410-11, 485
  Selected Walls check box, 201
MO command, 396, 467
Mode options, 395-96, 466, 468
Model
  complex, creating, 891
  display representation set, 35

drawing, 837, 842, 851-52
Explorer, 897
  mouse operations in the, 901
  toolbar, 904
  using the, 898-901
  using to reposition Mass Elements in the Tree View, 904-5
tab, 4, 30, 59, 117, 242
value, 770, 841
view
  description, 841
  drawing, 772
  viewing the, 988-92
MODEL/PAPER status bar toggle, 30
Modifier Style drop-down list, 190
Modify
  button, 710
  Column dialog box, 710
  Component command, 793, 848
  General View dialog box, 582
  Project dialog box, 45, 48
  toolbar, 24-25, 461, 569, 846, 848, 861, 996, 1004
More
  flyout , 902
  menu option, 37-38
Move, 10
  All Chains trigger grips, 664-65
  Component
    Down In List button, 152-53
    Up In List button, 152-53
  command, 110, 510, 861, 887, 908
  Down button, 711
  Flight grip, 515
  Frame radio button, 374
  option, 569
  orthogonally option, 888
  Up button, 711
  vertically option, 266
Mullions
  components, 340
  creating custom, 350-52
  editing the profile of, 359, 371
  option, 350
Multi-landing stair shape, 501-2
Multi-Purpose Objects folder, 604, 606, 617, 642
Multi-View Block
  Definition Properties
    - 12 Diffuser dialog box, 640
    dialog box, 603-5, 637, 639
  Definitions folder, 604, 606
  Reference tool, 607
Multi-view
  Block
    Attributes dialog boxes, 597, 644, 646
    Definition folder, 639
    content type, 622
    Reference Properties dialog box, 626
    Blocks Offsets dialog box, 596-599
  blocks
    creating, 602-6
    editing using Properties, 597-99
    importing and exporting, 606-7
    inserting new, 607-8
    new, defining properties of, 604-6
    properties of, 595-99
    viewing with reflected display representation, 612
Muntin section, 373

Muntins
  Block
    Definition Properties dialog box, 599-600
    dialog box, 302-3, 373-74
    creating for windows, 300-5
    tab, 300, 302, 373
MvBlockAdd command, 607
MvBlockDefEdit command, 599
MvBlockEditAttributeLocations command, 609
MvBlockEditViewBlockOffsets command, 608
My Endcap style, 181
My Tool Catalog, 15-16
MyMaskBlock, 617
Mystyle option, 97
MyStyle style name, 790

N

Name
  button, 728
  column, 49, 788, 940
  division, new, 355
  drop-down list, 940
  edit field, 65, 67, 153, 171, 218, 248, 273, 294, 302, 341, 347-50, 426, 524, 541, 562, 600, 618-19, 639, 704, 787, 798, 839, 856, 989
Napkin Sketch
  command, 806
  dialog box, 808
  procedure, 805-8
  view, creating a, 805-8
Navigation toolbar, 4, 13, 18, 23-25, 42, 59, 113, 117, 120, 122, 124-25, 139, 171, 220-23, 226, 228, 235, 238, 242, 245, 315, 317, 322-25, 328, 331, 372-74, 382, 419, 443, 445, 449, 451, 454, 458, 459, 462, 485-86, 488, 493, 565, 567, 569, 571, 582-83, 587-89, 595, 628, 630, 632-33, 635-36, 643, 741, 756, 759, 804-5, 838, 842, 844-45, 850, 853, 856, 879, 883, 917, 990, 995, 1000-1, 1004, 1007, 1010
NE Isometric option, 233, 328, 582-83
Nearest option, 574, 636
Negative symbol grip, 406
Never Use Characters check box, 972
New
  button, 30, 62, 213, 229, 252-53, 353, 356, 692
  Cell Assignment button, 351-52, 371
  Construct option, 198
  division name, 355
  Drawing
    button, 173
    command, 143-44, 173
    dialog box, 173
  drawing, creating a new, 59-60
  Element command, 900, 903-4
  Endcap Style dialog box, 184
  Features Workshop feature, 2
  Frame
    Assignment button, 351-52
    button, 3450
  Group
    command, 904
    Filter, 28-29

Grouping command, 900, 904
Icon button, 625
Material dialog box, 171, 279, 299, 428, 436, 530, 547
Model Space View
   Name edit field, 775, 797, 838, 854
   option, 54
Modifier Style dialog box, 195
Mullion button, 351-52
Name edit field, 233
Object radio button, 783, 801-2
   option, 329, 358, 369, 371, 374, 617, 639, 642, 787, 790, 889
Page Setup dialog box, 213-14, 252-53
page setup edit field, 252-53
Palette, 11, 59, 628, 636, 740, 742, 759, 859
Profile Name edit field, 201, 310
Project button, 43, 65-66
Property dialog box, 729
Sheet dialog box, 54, 68, 211-12, 252, 764, 798, 840
Style
   button, 146-48, 196, 199, 217, 223, 225, 236, 273, 294, 523, 604
   command, 97, 143-44, 273, 917
   creating from an existing style, 148-49
   default name, 273, 523
   edit field, 371
   name, 927
   option, 951
View Drawing
   button, 838-39
   check box, 775, 797, 854
View Dwg>Detail option, 857
View Dwg>General option, 66, 251, 764, 770
Wall Modifier Style Name dialog box, 198, 233
Window, 13
New>Construct option, 66, 562, 567, 571
New>Sheet option, 54, 68, 211, 252, 764, 775, 798, 840, 852
Next button, 67-68, 251, 621, 764, 770, 775, 798, 837, 839, 843, 853-54, 857, 951
No option, 405, 473
Node
   column, 941
   object snap mode, 87-88, 113, 118, 123, 130, 136, 381, 483, 489, 642, 644, 928, 965, 999
None option, 538, 624, 895, 963, 975
Normal
   edit field, 781
   X edit field, 598
   Y edit field, 598
   Z edit field, 598
   value, 386
North Elevation, 775
Northeast option, 902
Northwest option, 902
Nosing Length value, 527-28
Notes
   button, 150, 271, 273, 291, 295, 383, 386, 426, 524, 541, 598, 600, 615, 618-19, 704, 787
   column, 751-52

   dialog box, 150-51, 271, 273, 291, 295, 386, 426, 524, 541, 598, 600, 615, 618, 787
   tab, 150-51, 704
Number
   column, 972, 1005-6
   edit field, 65, 68, 252, 746, 764, 852
   Final Riser check box, 533-34
   of Bays option, 956-59, 976
   of Rails edit field, 542
   per Tread edit field, 544
   property, 706, 746, 754, 756

O

Object
   Display dialog box, 39, 459, 785-86
   list, 728
   not selected message field, 784
   orientation of hatch, 167-68
   Override check box, 39-40
   Snap
      On (F3) check box, 443, 446, 450, 455, 483, 489, 562, 568, 571, 999
      tab, 113, 118, 123, 130, 136, 443, 446, 483, 489, 562, 568, 571
      toolbar, 24
      Tracking On (F11) check box, 443, 483, 489, 562, 568, 571
   Tags, 698-99
   Type
      technology, 1-2
      list, 27
      option, 27
   Viewer, 2, 37-39, 58, 64, 322
      options, 901-3
      toolbar, 38, 898
ObjectARX™, 2
ObjectDisplay, 2
   accessing, 39
   editing objects with, 39-42
OBJECTDISPLAY command, 39
Objects, 1
   and object styles, properties of, 684-86
   creating tags and schedules for, 682-701
ObjRelUpdate command, 42, 57
Obscure option, 820, 822-24
OF command. (See OFfset option)
Office_Medium space, 759, 761
Offset
OFFSET command, 103, 110-12
   command, 846
   Distance edit field, 307, 338
   edit field, 533-34
   from Post column, 542
   Increment text box, 159
   option, 176, 261, 290, 309, 348, 517, 519, 526
   Set From
      command, 108-9, 132
      tool, 71
OFfset option, 79, 261, 290, 309
Offset/Center option, 257-58, 289-90, 307, 325, 330, 338
Offset>Copy command, 103-5, 124
Offset>Move command, 107
Offset>Set From command, 108, 131
Offsets edit field, 350
ON POLAR toggle, 74, 331
Open
   button, 174, 235, 313, 606, 916

Drawing
   button, 173-74, 216, 235, 245, 252, 313, 489, 539-40, 916
   command, 143-44, 173, 425
   dialog box, 143, 173, 216, 235, 313, 318, 489-90, 540, 606
Drawings tab, 593
   option, 51-52, 117, 130, 136, 159, 318, 760, 837, 840, 837, 841, 853, 856-57
   toggle, 16-17
   Xref option, 762
Opening
   adding with precision, 308-9
   Endcap Style dialog box, 161-62
   Endcaps option, 271, 291
   in Wall, 662, 735
   Max Width check box, 735
   Min Width check box, 735
   modifying, 309
   Percent, 289
OpeningAdd command, 256, 305-6
OpeningAddProfile command, 310
OpeningFlipHinge command, 268
OpeningFlipSwing command, 269
OpeningProfileEdit command, 311
Openings
   creating, 305-11
   properties, defining, 308
Operation
   drop-down list, 874
   edit field, 874
   options, 205
   shortcut menu option, 904
Optimum Slope edit field, 524
Options
   button, 113, 118, 123, 130, 136, 859
   dialog box, 3, 7, 24, 45, 74, 93, 105, 113, 118, 123, 127, 130, 136, 175, 563, 588, 593, 626, 826, 859
   option, 24, 127, 563
   selecting from the command line, 79
OR command line entry, 84. (See also Ortho close option)
Orbit menu option, 37-38, 902
Orientation buttons, 345, 347
Ortho close option, 84-86, 111, 120, 124, 397, 399, 467, 485
OSNAP button, 113, 118, 123, 130, 136
Osnap shortcut menu, 574
Other
   radio button, 302, 304, 373
   tab, 39-40, 169, 171-72, 280-84, 299, 305, 320, 427, 429-30, 436, 459, 531-32, 535, 549-51, 559, 569, 581, 662-63, 786, 790-91
OTRACK button, 113, 118, 123, 130, 136
Output section, 722-23
Outside of frame selection, 258-59, 288, 325
OV command, 397, 467
Overall
   check box, 662
   length, 504
Overhang
   column, 389
   edit field, 397, 432, 467
   option, 379
   property, 384-85, 404

Override
   Assignment command, 358-59, 362
   column, 187-88
   Display Configuration Cut Plane check box, 168-69, 237, 459, 532, 535, 559, 569, 581
   end cleanup radius option, 78, 95
   Endcap Style command, 183, 185
   Override Cell Format button, 709, 712
   start cleanup radius option, 78, 95
   Text & Lines
      dialog box, 670-71
      Marker, 660-61, 670
      option, 666
   View Direction check box, 35-36
Overrides tab, 341-43
Oversize:ANSI D (landscape) paper size, 252-53
Overwrite Existing option, 173

P

Page Setup
   dialog box, 213-14, 252-53
   Manager
      dialog box, 252-53
      option, 213-14, 252-53
Palette Properties dialog box, 11
Palettes, tool, 9
Pan
   command, 24, 382, 901
   menu option, 37-38, 901
Panel, 319
Parallel option, 902
Paste
   button, 173-74, 217, 235
   command, 143-44, 173, 900
   option, 150, 314, 491, 540, 606, 917
Path Name dialog box, 991
Pattern column, 166-67
Patterns drop-down list, 304
Perimeter edge edit field, 397, 467
Perpendicular option, 864
Perspective option, 902
PH command, 380
Pick
   Path< button, 991
   Point button, 680, 783, 801-2
   Point< button, 991-92
   XY Scale button, 681
PICTURE ARCHED property set value, 750
Pivot Point
   editing with grips, 406
   X value, 405, 473
   Y value, 405, 473
   Z value, 405, 473
Place
   Callout dialog box, 772-75, 778, 797, 838, 842, 854
   on
      Drawing option, 776-77
      Sheet option, 798
   Titlemark check box, 775, 797, 838, 842, 854
Placement section, 783-84
Places panel, 216, 313, 318, 425, 489, 539-40, 916
Plan
   display representation, 33, 428, 443, 445, 581
   Elevation Label (1) option, 847-48

High Detail display representation, 735
Modifiers button, 78, 189, 198, 233
set, 34
Plan Modifier>Convert Polyline to Wall Modifier option, 197, 233
Plan Modifiers>Add, 192, 231
Plans subset, 251-52
Plate Height, 379-80
Plot, 5-6
and Publish Details dialog box, 6
Preview, 5-6
Style
button, 42
column, 39, 41
Plotting Status button, 6
Plumb option, 379
POINT command, 602
POLAR button, 113, 118, 123, 130, 136
Polar Tracking
check box, 229
tab, 229, 483, 489
Polar Tracking (F10), 483
Polar coordinate entry method, 82
Polyline
command, 849
converting to roof, 392-93
creating from the edge of the stair, 520-21
option, 232, 424
Portable License Utility feature, 2
Position along wall property, 257-58, 289-90, 307, 309, 338
Post
hiding, 553
Location option, 551
Locations
dialog box, 537
editing, 551-54
tab, 540, 542-44, 577-79
locations button, 537
Placement option, 551, 553
redistributing, 554
removing, 552-53
showing, 553-54
Post Placement>Add option, 551
Post Placement>Hide option, 553, 580
Post Placement>Remove option, 551
PR command, 380
Predefined
radio button, 296, 341
Rectangular shape, 327
shapes, 341
Presentation Format (No Cut Lines or Path) check box, 563
Preset
Elevation edit field, 623
Views menu option, 37-38, 902
Preview toggle, 592, 996, 1004
Primary Grid option, 350-51, 371
Priority
column, 188, 218, 942
drop-down list, 308
edit field, 153-54, 308
Grid, 345
Overrides
button, 187
dialog box, 187, 189
overrides option, 78
overriding in a wall style, 187-88
ing to control cleanup, 96

Profile
changing with Edit Edge Profile In Place, 441, 481-82
In Place, 557-58
Definition
command, 198-99
edit field, 310
option, 200
drop-down list, 350
feature, 307
grips, editing, 203
Name
column, 544-45
drop-down list, 544-45
option, 874
tab, 7, 24
Profiles, 24
Project
Browser, 1, 42-46, 57, 65-66, 68, 122, 216, 229, 244, 251, 315, 323, 443, 562, 567, 571, 628, 635, 733, 740, 758, 770, 837, 841, 852, 856
data, creating schedules for, 725-26
defining divisions and levels of the, 48-49
file, 46
Navigator, 1, 7, 24, 30, 42-43, 46-49, 51-52, 54-55, 57, 66, 68, 122, 141, 206-7, 209-10, 213, 215-16, 229, 235, 243-44, 251-52, 315, 323, 443, 498-99, 562, 566-67, 571, 582, 628, 726, 733, 739-40, 758, 760-61, 763-64, 769-70, 775, 796, 798, 835, 837, 840-41, 852-54, 856-57
creating content with, 46-47
keeping up to date, 55, 57
option, 517-19
Properties command, 45
tab, 46, 48, 66, 68, 207, 244, 770-71, 835
to polyline option, 176
PROJECTBROWSER command, 43
Projected mode, 395-96, 466-67, 468, 483-84, 491
Projection menu option, 37-38, 902
Projection>Perspective option, 64
PROJECTNAVIGATORTOGGLE command, 46
Projects
creating, 42-57
tab, 243, 837
Prompt
column, 597
for template check box, 251-52
option, 580
to select nested objects check box, 860, 865
Properties, 11
button, 40
command, 16, 24
dialog box, 210, 582, 810
inserting without tags, 726-27
option, 54, 210-11, 220, 251-52, 570, 582, 597, 635, 657, 735, 756, 765, 810, 843, 850, 859, 863-64, 895, 917
palette, 4, 7, 9, 16-17, 59, 73-76, 79, 81, 94-95, 97-98, 111, 123, 141-42, 188, 242, 257, 336-37, 403-4, 444, 466, 874-75
Advanced section, 77-79

Basic section, 76-77
toggle, 9-10
toolbar, 24, 635
Property
Data
fields, 692
Format, 729
Increment, 695
Location grip, 762-63
section, 759
Definition window, 728
drop-down list, 713
Set, 693-94, 713
Definitions, 728
Definition Properties – Door Objects dialog box, 754
Set/Properties list, 708
Sets
button, 150, 295, 426, 524, 541, 600, 619, 684, 699-701, 750
dialog box, 541
list, 271, 291
PropertyRenumberData command, 713
PropertySetDefs drawing, 728
Pyramid mass element shape, 878
Publish to
Architectural Studio, 5-6
Web, 5-6
Publish>Publish to Plotter, 213, 253
Pull Face orthogonally option, 888
Purge
options, 174
Styles command, 144

Q

QNew, 2, 59, 65, 113, 122, 764

R

Radial shape, 958, 961
Radius
column, 385
command, 674
edit field, 512, 875
grip, 517
option, 304, 639, 674
property, 385
Rail
locations
button, 537
dialog box, 537
tab, 537
Locations tab, 540-43, 577
Railing
anchoring, 555-56
command, 498, 506, 537-39
creating, 536-38
modifying the, 551-58
option, 538, 577, 579
reversing, 554-55
style, components of a, 540-51
Styles
- Handrail dialog box, 576-79 (Imperial).dwg, 540
defining, 538-51
dialog box, 537, 540-41, 543
tool, 536
Railing Anchor>Anchor to Object option, 555
Railing Anchor>Release option, 556
RailingAdd command, 498, 536, 561
RailingAddComponentProfile command, 556-57

RailingAnchorToObjects command, 555
RailingPostAdd command, 551-52
RailingPostHide command, 553
RailingPostRemove command, 552-53
RailingPostShow command, 553-
RailingProfileEdit command, 557-58
RailingRedistributePosts command, 554
RailingReverse command, 555
Realtime
Pan command, 38
Zoom command, 38
Rectang command, 861
Rectangular shape, 960
Redistribute option, 551, 554
Redo
command, 902
View menu option, 37-38, 902
Refedit toolbar, 24, 581, 848, 860-61, 865
Reference
Docs
dialog box, 598
tab, 150-51, 273, 704, 787
Documents
button, 271, 291, 383, 386, 598, 615
dialog box, 291, 615
Edit dialog box, 565, 580, 860, 865
Keynote
Legend tool, 830, 866
tool, 827-28, 866
Manager feature, 2
point on option, 262-65, 287, 290, 308-9, 339
REference point on
option, 290, 309, 339
property, 262
Reflected
display configuration option, 60-61, 64, 637, 975, 978, 1009
Screened display configuration, 975, 978
Refresh
command, 781-82, 835, 844
option, 851
Project command, 57
Regen list, 992
Regenerate
command, 781-82, 835
Model option, 329, 715
Relative
coordinate entry method, 82
to column, 941
Relaxed Calculator Limits property, 563
Reload button, 763, 851
Remove
button, 278, 297, 351-52, 527, 601, 622, 711, 730, 788, 938, 942, 966
Columns/Headers dialog box, 757
Component button, 151, 153
Customization option, 517, 519
Dimension Points, 666, 669, 736-39
Edges command, 872, 912
Endcap option, 184
Frame button, 349-50
Grid Line, 966, 980-81
Modifier option, 195
option, 188, 205, 310, 551, 718
Priority Override button, 187

Index

property sets button, 601, 619
Vertex option, 183, 194, 203, 226, 360, 362, 439, 441, 558
Wall Modifier button, 190
Removed Points Marker, 660
Rename
   option, 235, 244, 246, 571, 576, 579, 903
   Palette Set, 11
   to Unique option, 173
Render Material
   Catalog, 171-72
   list, 320
Renumber Data Tool command, 712-13, 746
Repathing Xref command, 57
Repeat
   First Column check box, 706-8
   layout, 966-67
   mode, 967
   option, 956, 958-59, 965, 976, 987, 1009
Replace
   Endcap option, 184
   Existing radio button, 783-84
   Modifier option, 195
   Ring option, 203, 311, 360, 362, 439, 441
Reposition
   Along Wall command, 286-87
   Within Wall command, 285
RepositionAlong command, 255, 265, 286-87
RepositionWithin
   command, 255, 267, 284, 287
   Wall option, 322
Representations by Objects folder, 31-33, 62
Reselect option, 720-21, 833
Reset
   cascade menu, 664
   Extension Lines
      command, 665
      option, 664-65
   option, 176, 664, 666
   Text to Original Position
      command, 664-65
      option, 664-65
   View menu option, 37-38, 902
Residential
   stair style, 571
   subcategory, 741
Resize
   arrow, 10
   option, 980, 985-86
Restore button, 30, 64
Restroom_Men_Small space, 759-60
Restroom_Women_Small space, 759
Result Type section, 783
Reverse
   Baseline command, 92-93
   command, 72, 92, 554
   In Place command, 92-93
   option, 92, 554, 784
Reverse>Baseline option, 131
Revolution
   and Extrusions mass element shape, 879
   tool, 882
Rhombus grip, 406
Right
   justification, 502, 512

option, 902
Reading option, 624
Triangle mass element shape, 878
Rise
   edit field, 278, 289, 337, 380, 397, 467, 875, 937
   option, 337, 348, 383
   property, 386
   value, 937
Riser
   Count control option, 502-4
   dimension, 504
   number of, 504
   Numbering tab, 531, 533
   setting, 512
   Thickness value, 527-28
Risers option, 505
Riser/Tread Calculation edit field, 504
Rodded joint type, 857
Roll edit field, 928, 930-31, 934, 937
Roof
   and Floor Line dialog box, 76-77, 79
   command, 394, 442-43, 446, 455, 457
   creating holes in a, 416-17
   Dormer
      command, 420, 422
      option, 421, 459
   dormer, creating a, 420-21
   Edges
      list box, 384
      section, 389-90, 452-54
   Edges and Faces dialog box, 384, 388-90, 452-54
   existing, editing to create gables on an, 388-91
   Faces (by Edge)
      list box, 385-86
      section, 384-85, 389
   hip, creating a, 443-46
   intersections, determining, 422-25
   properties, defining, 383-86
   removing holes from a, 417-18
   Slab
      and Roof Slab Edge Styles (Imperial).dwg, 425
      command, 458
      Edge Styles dialog box, 431-32, 434
      Edges dialog box, 403-5, 436
      option, 394-95
      Styles command, 425
      Styles dialog box, 425-26
      Styles option, 425
      tool, 400, 425
   slabs
      creating, 393-403
      cutting, 412-13
      extending, 409-10
      mitering, 410-12
      modifying, 403-6
      tools for editing, 406, 408-21
      trimming, 406, 408-9
      vertices, adding, 413-14
      vertices, removing, 415
   Slab
      Edge Styles dialog box, 479
      Styles dialog box, 476
      tool, 378, 387, 393, 450
   Roof/Floor Line (Design - Advanced *Worksheets*) option, 79

Roof/Floor Line>Modify Floor Line option, 176-77, 238, 246
Roof/Floor Line>Modify Roof Line option, 176-77, 238, 248, 454, 460, 566, 580
RoofAdd command, 377-78, 381, 441, 443
RoofEditEdges command, 388
RoofLine command, 158, 176, 207, 215, 441
RoofSlabAdd command, 394-95, 397
RoofSlabAddEdgeProfiles command, 438
RoofSlabAddHole command, 416-17
RoofSlabAddVertex command, 413-14
RoofSlabBoolean command, 418-20
RoofSlabBooleanAdd command, 418-19
RoofSlabBooleanDetach command, 420
RoofSlabBooleanSubtract command, 419-20
RoofSlabConvert command, 402-3
RoofSlabCut command, 412, 424
RoofSlabDormer command, 420-21
RoofSlabEdgeEdit command, 437
RoofSlabEdgeProfileEdit command, 441
RoofSlabEdgeStyleEdit command, 429
RoofSlabExtend command, 409
RoofSlabMiter command, 410-11
RoofSlabRemoveHole command, 417-18
RoofSlabRemoveVertex command, 415
RoofSlabStyleEdit command, 425
RoofSlabToolToLinework command, 400
RoofSlabTrim command, 406, 408
RoofToolToLinework command, 392, 401, 403, 450
Room
   and Finish Tags, 742
   Finish Tag, 697
   Objects property set, 759-60
   Tag
      option, 689-90, 695-97
      placing, 696-98
      Project Based, 697
Rotate command, 110
Rotation
   column, 545, 941
   dimension, 545
   drop-down list, 706
   edit field, 308, 338, 350, 380, 508, 628-29, 703, 781, 875, 957, 976, 990
   option, 77, 259, 289
   section, 598, 681
   value, 386, 468
RoughHeight property value, 291-92
RoughWidth property value, 291-92
Rule column, 788
Run
   base value, 380
   property, 386
   value, 397, 467

S

Sample Palette
   Catalog-Imperial, 15
   Catalog-Metric, 15
Samples feature, 2

Save
   All Changes
      command, 185, 353, 360, 362
      option, 184, 228, 310, 331, 353-54, 439, 441, 848
   As New
      Endcap Style option, 184
      Modifier Style option, 195
      Profile option, 203, 311, 360, 362, 439, 441
   As Type drop-down list, 722-24
   back changes to reference button, 581, 861, 865
   button, 354
   Changes
      dialog box, 353-54, 356
      option, 203, 205, 311, 360, 362, 896
   Content File dialog box, 625
   Current Dwg As
      Construct option, 52, 207
      Element option, 51-52
   Drawing As dialog box, 60
   Image
      dialog box, 39
      menu option, 37, 39, 903
   In list, 60
   option, 60, 758
   Preview Graphics check box, 625
SaveAs option, 4, 998, 1004
S-Beam layer, 933
Scale
   column, 545, 941
   command, 110
   drop-down list, 545, 854
   field, 703
   flyout, 642, 763, 847, 866
   option, 941
   section, 681
   setting, 26
   tab, 25-26, 59, 586, 595, 703, 706
Scan
   block
      reference drawings option, 703
      references, 703
   Block References check box, 723-24
   styles, using, 704-12
   Xrefs check box, 723-24
   xrefs option, 703
Schedule
   Data, exporting, 722-30
   external drawing toggle, 725-26, 766
   properties palette, 725
   Tags category, 742
Schedule Table
   adding a, 701-3
   Style
      drop-down list, 723-24
      Properties dialog box, 682, 704-5, 707, 712
      Properties – Door Schedule dialog box, 757
      Properties – Window Schedule dialog box, 748-49, 752
   Styles
      command, 704
      folder, 682
Schedule Tables category, 764
ScheduleAdd command, 650, 702-3, 754, 765
ScheduleCellEdit command, 715
ScheduleExport command, 722

Schedules, 20
  and Diagrams category, 764
  editing the cells of, 715
  placing in a drawing, 19
  selection set, changing the, 718-22
  updating, 714-15
ScheduleSelectionAdd command, 718
ScheduleSelectionRemove command, 718-19
ScheduleSelectionReselect command, 718, 720
ScheduleSelectionShow command, 718, 721
ScheduleStyle command, 704
ScheduleStyleEdit command, 732
ScheduleUpdateNow command, 714
S-Cols layer, 929
S-Cols-Brce layer, 935
SE Isometric, 449, 451, 454, 488, 565, 635, 856
Search button, 592
Section
  A-A type, 854-55
  controlling display of, 785-86
  creating a live, 802-3
  Mark A2T, 853
Section Mark
  A1, 796-797
  A2T, 835
Section/Elevation View radio button, 841
Section_Elev display representation set, 35, 783
Sectioned Body Display toggle, 803
Sections category, 798, 853
Segment
  edit field, 81-82
  property, 385
  type (Design - Basic *General*) option, 76, 81
Segments column, 385
Select
  A Block dialog box, 305, 601, 639-40
  Additional Objects button, 783
  All button, 150
  an Endcap Style dialog box, 185
  Block button, 305
  Color dialog box, 41, 844-45
  Display Representation dialog box, 616, 644
  Drawing dialog box, 252
  File dialog box, 130, 136
  Keynote
    button, 149-50, 341, 369, 426, 541, 600
    dialog box, 149, 273, 295, 341, 426, 524, 541, 600, 828-29, 831
  Layer
    as Sheet Template dialog box, 211
    dialog box, 41
    Key button, 624
    Key dialog box, 624
    Layout as Sheet Template dialog box, 54, 211
  Levels dialog box, 535, 570
  Linetype dialog box, 41, 223-24, 237, 248, 844-45
  Lineweight dialog box, 41
  Linework Component dialog box, 793-94, 847-48
  Object button, 784

object list option, 1009-10
Object option, 963, 975-78
Objects button, 639, 681, 783, 801-2, 996, 1004
  option, 718
Plot Style dialog box, 41-42
Reference Document dialog box, 150
Reselect Objects button, 783
Style page, 672
Template dialog box, 2-3, 117, 212
Wall Style dialog box, 144, 174-75
Select Layers>Add, 28
Selection
  Filter, 695
  option, 718
  Set section, 783, 801-2
Selection/Elevation View radio button, 837, 853
Selection>Add command, 718
Selection>Remove command, 719
Selection>Reselect command, 720, 834
Selection>Show command, 721, 757, 834
Selector button, 810, 813, 818, 859, 862-63
Set
  Current button, 252-53
  Elevation command, 908
  Fixed Cell Dimension Rules dialog box, 353-55
  From
    button, 199, 225, 229
    command, 144, 197
    option, 329, 617, 642
  Interference Block button, 601
  Miter Angle command, 364-65
  Template dialog box, 210-11
  View menu option, 37-39, 903
Sets folder, 31, 33-34
Settings option, 113, 118, 123, 130, 136, 226, 443, 446, 450, 455, 483, 489, 562, 568, 571, 595, 628, 636, 642, 740, 759, 859
S-Grid layer, 953-54
Shade flyout, 24-25, 222, 238, 242, 445, 449, 454, 462, 582-83, 635, 856, 883, 917, 995
Shading
  Modes menu option, 37-38, 902
  toolbar, 24, 493
Shading Modes>Flat Shaded option, 65
Shape
  drop-down list, 306, 308, 379, 501-2, 533-34, 872-75, 878-79, 955, 958, 960-61
  edit field, 508
  flyout, 419, 451
  option, 308, 380, 383, 501
  section (of Design Rules tab), 276
  tab, 341-42
Shapes
  Predefined, 276
  toolbar, 4, 23, 25, 232, 424, 639, 849, 861
Sheet
  Keynote
    Legend tool, 832-33
    option, 831-33
  Set, 54-55
    Properties dialog box, 54-55
    View toggle, 210, 251

Title edit field, 68, 252, 764, 852
Sheets
  dialog box, 831
  subdirectory, 46
  tab, 46-47, 68, 210, 215, 251-53, 726, 764, 775, 798, 840, 852
Shift
  Baseline arrows, 159
  to Add, 89. (*See also* Grips)
Shortcut menus, accessing, 22-23
Show
  Details button, 940, 942
  Edge option, 184, 195
  hidden files and folders, 18
  Model Explorer, 898, 911
  option, 551, 553, 718, 833
  Startup dialog box, 3
Shrink Wrap component, 166
Shrinkwrap hatch toggle, 854
Side
  Clearance edit field, 505
  for Offset column, 542
  offset edit field, 538
Sidelight, 351-52
Sill Height option, 339
Sill
  height
    edit field, 308
    option, 308, 339
  Height option, 289-90, 338
  option, 307-8
Single
  Door Type value, 330
  Pole switch
    symbol, 636
    tool, 636
  radio button, 302, 304
  slope
    option, 379, 382-83
    Shape value, 387, 446-49, 456-57
SINGLE CASEMENT property set value, 750
Site
  Construction.Planting.Groundcover.Grass.Short option, 917
Size edit field, 10, 533-34
SL command, 397, 467
Slab
  and Slab Edge Styles (Imperial) drawing, 489-91
  attaching a slab edge style to, 476-81
  baseline, 471
  command, 468
  component, 845, 847
  Edge Styles dialog box, 479-80
  Edges dialog box, 472, 479
  edges, assigning styles to, 436-38
  modifying using properties and grips, 472-74
  option, 483-84, 487
  Styles dialog box, 475-76
  styles, accessing, 475-76
  tool, 465, 468, 470-71
SlabAdd command, 465-66, 468-69, 483, 491
SlabAddHole command, 474
SlabAddVertex command, 474
SlabBooleanAdd command, 474
SlabBooleanDetach command, 474
SlabBooleanSubtract command, 474
SlabCut command, 474
SlabEdgeEdit command, 476

SlabEdgeProfileEdit command, 481
SlabEdgeStyleEdit command, 479
SlabExtend command, 474
SlabMiter command, 474
SlabRemoveHole command, 474
SlabRemoveVertex command, 474
Slabs with Optional Haunch type, 863
SlabStyleEdit command, 476
SlabToolToLinework command, 471
SlabTrim command, 474
Slice
  command, 905-6, 908
  converting to a polyline, 908-9
  option, 906
  tool, 911
Slope
  column, 389, 452-53
  option, 380
  property, 385
  value, 397, 468
Slopeline option, 396-98, 467
Sloping
  Height column, 542
  Riser check box, 528, 572
Snap template, 3-4
SNAP, setting, 90
Snapshot
  dialog box, 62-63
  Edit dialog box, 30
Snapshots
  button, 29, 62
  dialog box, 29-30, 62-64
Snowflake icon, 62
Soffit check box, 433
Sorting
  Order, 711
  tab, 711-12
Source column, 39-40
South Elevation option, 851
Southeast option, 902
Southwest option, 902
Space
  Auto Generate Tool, 694-95, 732, 741, 759-60, 978
  Boundary
    command, 911
    tool, 911
    evenly, 987
    layout type, 956-61, 965-67, 976
  Query
    button, 693
    command, 693, 872
  Tag, 697, 742
  tool, 872
Space Boundary>Apply Tool Properties to>Slice, 905
SpaceAdd command, 690, 911
SpaceBoundaryAdd command, 911
SpaceBoundaryAddEdges command, 872, 912
SpaceBoundaryAnchor command, 872, 912
SpaceBoundaryEdge command, 872, 912
SpaceBoundaryGenerateWalls command, 872, 912
SpaceBoundaryMerge command, 872, 911
SpaceBoundaryMergeSpace command, 872, 912
SpaceBoundaryRemoveEdges command, 872, 912

Index

SpaceBoundarySplit command, 872
SpaceDivide command, 911
SpaceJoin command, 911
SpacePBoundaryAdd command, 872
SpaceQuery command, 912
Spaces
   category, 741, 759
   creating, 690-96
   option, 911
   tool, 911
Spacing
   column, 966
   distance, 966
Speaker symbol, 644
Specify
   on Screen
      check box, 681, 878
      drop-down list, 512, 596
      option, 923, 930, 959-62, 977-78
      section, 975
   rise on screen option, 937
   Rotation on Screen edit field, 596, 614
   rotation on screen option, 596, 628-29, 636
   scale on screen edit field, 614, 640
   Scale on Screen, 628
Sphere mass element shape, 878
Spiral stair shape, 501-2, 510
Split
   Boundary command, 872
   by Plane command, 886-87
   Face
      command, 887-88
      editing option, 205-6, 888
Spokes option, 304
Square option, 379
ST command, 395, 466
Stair
   anchoring to a landing, 533-34
   check box, 839, 854
   command, 497, 499-506, 563-64
   Components dialog box, 506-7
   construct. 562, 571, 853
      creating the, 498-99
   customization, removing, 519-20
   displaying in multiple levels, 558-60
   edge, customizing the, 517-21
   Flight option, 538
   Line section, 533-34
   multi-landing, creating a, 508-9
   option, 537-38, 567, 574
   Plan Display Representation, 559
   properties of, 501-8
   setting, 538, 567, 569, 571
   Style Override, 559
   Styles
      command, 522
      dialog box, 502-4, 506, 523, 527, 530-31, 535, 573-74
   styles, 499
      creating, 521-33
      Residential, 571
      Wood-Saddle, 571
   Styles – Residential dialog box, 571-72, 581
   Styles – Standard dialog box, 569
   tower, creating a, 534-36
   Tread Length Override check box, 544
   U-shaped, creating a, 509-11
   using Properties to modify, 506, 508

Stair Landing Anchor>Anchor Object option, 534
StairAdd command, 497, 499-500, 508-9, 512, 561, 563-64, 569
StairGeneratePolyline command, 520-21
StairOffset command, 513, 517-18
StairProject command, 518-19
StairRemoveCustomization command, 519-20
StairTowerGenerate command, 534-35, 570
Stairs
   creating spiral, 512-14
   projecting to walls, 518-19
StairStyleEdit command, 522-23
Standard
   Cleanup Group Definition, 94
   installation, 19
   material definition, 854
   option, 369, 371
   Sizes
      section, 297
      tab, 273, 276, 278-79, 294, 297-98, 327
   sizes edit field, 258, 288
   style, 439, 483-84, 487, 512, 564, 741, 759
   toolbar, 2, 4, 23-24, 59, 65, 113, 130, 136, 329, 715
Start
   angle, 959
   button, 2
   check box, 371
   Elevation Offset edit box, 190
   from scratch option, 200-1, 360, 480
   from Scratch option, 359
   grip, 551
   Miter angle, 337
   Number edit field, 714
   offset
      edit field, 930, 934, 937, 956-57, 976
      option, 923, 928
      property, 965
   Position Offset edit box, 190
   Shape columns, 940-42
   With
      edit field, 729
      window, 213
Starting Endcap option, 78
Startup
   dialog box, 3
   option, 3
Stock Tool catalog, 13-16
Stock Tool catalog>Parametric Layout and Anchoring, 986
Stop, 319
Straight
   Leader, 651, 827-28, 866
   Length edit field, 504, 514
   stair shape, 501-2, 564
Stretch command, 110, 112
Stringer Resolution section, 528-29
Stringers
   column, 525-26, 573
   tab, 525-27, 572-73
Structural
   Beam tool, 933, 935, 943, 946, 953, 992, 994-95, 1000
   Brace tool, 935-37, 943, 992
   Catalog, 17, 698

Column
   command, 929, 932, 992
   Grid command, 954-55, 960-61, 998
   tool, 929, 943-44, 947, 965
grids, creating, 953-65
Member Catalog, 21-22, 923-25, 927-28, 930, 932, 934-35, 937, 939, 947-48, 951, 954, 965, 992, 994, 998
   command, 924-25
   dialog box, 928
   menu bar, 927
   toolbar, 928
Member objects, 965
Member Style dialog box, 994, 998
Member Style Properties - New Style dialog box, 951
   dialog box, 940, 945, 947-48, 951
Member Style Wizard, 951-52
member styles, 926
members
   adding, 928-48
   customizing using styles, 939-43
   inserting, 924-28
Structural>Bar Joists option, 953
Stud subcategory. (See Walls category)
Style
   box, 258, 288
   drop-down list, 141-42, 395-96, 466, 501, 537, 692, 703, 706, 928, 934, 936, 953, 965
   edit field, 238, 327, 874, 957
   for User Linework Edits if Unable to Reapply drop-down list, 783
   Manager, 1, 9, 17-18, 97, 142-43, 146-47, 149, 162-63, 170, 173-74, 187, 194, 196-97, 199, 215-17, 220, 225-26, 229, 235, 238, 245, 247, 256, 272-73, 276, 293-94, 296, 298, 304, 313-14, 319, 327, 329-30, 340, 343, 371, 404, 425, 429, 431, 433, 438, 475, 479, 497, 510, 522, 530, 539-4, 561, 572, 579, 595, 601-4, 606, 617-20, 633, 640, 643, 658, 682-85, 704, 727, 729, 732, 754, 787, 882, 889, 911, 917, 932, 943, 951
   accessing, 143
   copying files from the, 19
   creating Wall Modifier Styles with, 196-97
   dialog box, 143-49, 179-82, 491
   drop-down list, 930
   resources of the, 18
   structural member style, creating a, 927-28
   toolbar, 97, 147-48, 159, 173-74, 181, 197, 199-200, 216-17, 223, 225, 235-36, 245-46, 273, 294, 313, 318, 425, 489, 523, 539-40, 604, 606, 616, 889, 916, 951
   option, 76, 261, 290, 304, 337, 339, 369, 371, 395-96, 501, 703
Override
   check box, 164-65, 169, 223, 248, 282, 300, 302, 320, 343, 345, 429, 434, 443, 445, 531-32, 549, 620, 660, 791, 943
   creating a, 302
section, 533-34

to Generate drop-down list, 783
STyle property, 75, 261, 279, 339
Styles, importing and exporting door and window, 313-14
Subdivision 2 display component, 848
Subdivisions
   button, 780, 843
   dialog box, 780, 843
Subset Properties dialog box, 211, 213, 252
Subtract
   Boolean operation, 681
   operation, 874
   option, 418, 820, 822
Sunshine icon, 62
Surface
   & Section Hatching setting, 850
   Hatch
      Linework component, 785
      Placement option, 436
   Rendering option, 436
SW Isometric option, 64, 139, 330-31, 372, 374, 424, 445, 458-59, 462, 486, 493, 569, 804, 838, 850, 879, 883-84, 917, 997, 1000, 1010
Sweep Profile command, 198
Sweeps>Add, 200-1
Sweeps>Miter, 202
Swing angle, 258, 289
Symbol drop-down list, 706
Symbols
   inserting, 593-95
   placing in a drawing, 19
System tab, 3

T

T command, 396, 467
Table
   Export command, 722
   Selection Show command, 721
   Title edit field, 711-12
TableAdd command, 702
Tag Settings
   button, 692, 695, 741, 759-60
   Definition drop-down list, 692-93
   dialog box, 692-95, 741, 759-60
Tags, 20
   properties, 682
   renumbering, 712-14
   window, placing, 688-89
Target Path list, 991-92
Templates, selecting, 3-4
Terminate with option, 502
Text
   Appearance section, 706
   Docs tab, 273
   edit type,
   Formatting toolbar, 651
   Notes tab, 787
   Override edit field, 670
   page, 672
   placing, 650-51
   Tool, 650-51
   type, 730
Thickness
   edit field, 379
   Offset option, 427
   option, 396, 426, 467
   using *wedge* shaped grips to edit, 406
Threshold
   A component, 282, 284

B component, 282, 284, 320
height option, 261
option, 259
Plan, 320
profile, 374
THreshold height property, 261
Tile-Mode system variable, 30
Toggle View command, 144
Tool
  catalog, inserting tools from the, 19
  package, 12
  Palettes, 7, 8, 11, 24
    creating, 588-89
  Path File Locations, 587-88
  Properties dialog box, 810, 816, 818, 859, 863-64
  properties, applying, 311-12
TOOLBAR command, 24, 57
Toolbars, using to access commands, 23-25
Top
  check box, 972
  Depth setting, 504
  Elevation column, 156-58
  of Entire Stair edit fields, 547
  of Flight edit field, 547
  Offset value, 504
  option, 396-98, 467, 902, 1004
  radio button, 302
  View option, 235, 245, 443, 451, 459, 485, 805, 838, 990, 996, 1004, 1010
Total
  check box, 709
  Degrees arc constraint, 512
  value, 526
Transfer to Object option, 355-56
Transparency, 11
  dialog box, 11-12
Transverse, 1, 798, 853
Tread
  dimension, 504
  Thickness value, 527-28
Tree View
  shortcut menus, using the, 903
  toggle, 592
  toolbar, 898, 900-1, 903
Trim
  by Plane command, 885
  command, 110, 112
  editing option, 205-6, 408
  option, 820-21
  Planes
    button, 931, 935, 938
    modifying ends with, 938-39
  planes button, 958
Truncate value, 526
Turn
  Points grip, 515
  type
    drop-down list, 501
    edit field, 508-10
Tutorials feature, 2
Type
  column, 525-26, 710, 750-51
  drop-down list, 167, 533-34
  edit field, 750
  property, 700, 728

U

U command, 382

nd flyout, 24-25

toolbar, 24
UCS II toolbar, 24
UH command, 380
Unconstrained option, 258, 289-90, 318
Undo
  button, 383
  command, 38, 382, 902
  option, 82, 150, 261, 290, 309
  View menu option, 37-38, 902
Units
  setting, 26
  tab, 25-26, 59
  template, 3-4
Unnamed stringer value, 572
Up command, 592
Update
  Schedule Table
    command, 704-15
    option, 726, 766
  Space button, 692, 694
Upper
  Height edit field, 380
  height option, 380
  Rails section, 577
  Slope, 380
UR command, 380
US command, 380
Use
  Drawn Size check box, 191, 197
  Existing Table check box, 723
  for True/False check box, 706
  model extents for height option, 780
  Profile
    button, 276, 296, 330, 341-42, 374
    option, 342, 350
  Rule Based Calculator check box, 524-25, 571
  Stair Landing Extension check box, 546-47, 577
  Unformatted Decimal Value option, 724
Used In column, 371
User Single, 166-67
U-shaped stair
  grips, 515
  shape, 501-2

V

Value
  column, 597
  field, 644
Varies description, 272
VBA Support feature, 2
Vertex Grip, 203
Vertical
  Alignment, 259, 289, 307, 338
  left component, 947
  offset
    option, 396, 398, 467
    using *wedge* shaped grips to edit, 406
  Orientation drop-down list, 501, 564
  Section tool, 801-2
  Up setting, 512
Vertices, changing, 406
View
  Block Offsets, In-Place editing of, 608-10
  Blocks
    list, 601
    tab, 600-1, 603, 637-38, 640
  Direction, 35-36

Directional Dependent, 35-36
Directions check box, 601
directions, 146
drawing, creating model space views within a, 54-55
flyout, 64, 139, 233, 235, 238, 245, 331, 372-74, 424, 443, 445, 449, 451, 454, 459, 462, 485, 493, 569, 582-83, 635, 805, 837, 856, 879, 883, 917, 990, 1000, 1004, 1010
menu, 6, 8
  Options, 11, 59
  dialog box, 11-12, 59
  subdirectory, 46
  tab, 54, 207, 499, 566, 772, 853
Viewer, 151-52
  button, 151
  tab, 146, 330
Viewing toolbar, 146
Viewport
  applying display configurations to, 34-37
  Floating, 6
  Scale menu flyout, 4-5
Views
  category, 857
  creating, 52-54
  icon, 251
  tab, 46-47, 52-54, 66, 251-52, 570, 582, 740, 761, 763-64, 770, 775-76, 796, 798, 837, 840-42, 851-54, 856-57
  toggle, 592
  toolbar, 24
View>SE Isometric option, 322
View>SE Isometric View option, 120-21
VINYL CLAD value, 750-51
Vinyl Clad value, 751
Visible (light bulb) option, 39, 41, 166
Visual Aids menu option, 37-38, 902
VIZ Render, 6
Volume section (of Drawing Setup dialog box), 26
VP Scale flyout, 116, 226

W

Waist value, 526
Walkthrough, creating a, 990-92
Wall
  Center Line, 94
  cleanup, using, 94
  Component list, 200-1
  defect markers, managing, 97-99
  direction, changing with WallReverse, 92
  Endcap Style list, 182, 225-26, 633
  endcaps, creating custom, 179-82
  handle, moving using the Offset option, 86-87
  Interference command, 166
  Justification Line, 94
  justification, 471
  Merge command, 71, 99
  Modifier
    List Window, 190
    Styles folder, 229
    Styles, creating, 196-97
  Modifiers
    dialog box, 78-79, 188-90, 197-98, 233
    inserting and editing in the work-

  space, 192-95
  Style folder, 196
Offset command, 103
Offset Set From command, 99
option, 76, 136, 226, 229
Plan Display Representation, 164-65
Properties
  dialog box, 169
  palette, 116, 119, 121
Style
  Browser button, 152-53
  command, 215
  Components Browser dialog box, 152-53
  folder, 173
  Properties - Frame Brick dialog box, 217, 220, 443, 445
  Properties - Standard dialog box, 182, 221
  Properties BrickFND dialog box, 246
  Properties dialog box, 143, 149-50, 163, 171, 185, 221, 226, 223-24, 238, 242, 248, 443, 445, 633, 699, 701
  Properties-10_bull_nose dialog box, 225
style
  defining the properties of a, 149-69
  name, creating a new, 146-48
  Standard, 80-81
Styles, 182
  – Base_Int dialog box, 242-43
  category, 148
  dialog box, 148
  folder, 217, 223, 235
styles
  accessing, 141-42
  creating and editing, 142-75
  exporting and importing, 173-74
  identifying using Inspect, 175
  purging, 174-75
Sweeps, creating using profiles, 198-200
Tag (Leader), 700
Tags, 699-701
  tool, 73, 75-76, 81-82, 113, 118, 123
Wall>ApplyToolPropertiesTo>Wall, 905
WallAdd command, 71-73, 75, 79-83, 94, 111, 113, 118, 123, 136, 142, 229, 634-35
WallAddSelected command, 74, 138
WALLAPPLYENDCAP command, 185
WALLAUTOENDCAP command, 185-86
WallBody command, 204
WallCleanupL command, 99
WallCleanupT command, 99-101
WALLEDITJUSTIFICATION command, 91
WallEndCap command, 182
WALLENDCAPEDIT command, 183
WALLENDCAPSTYLE command, 179, 215
WALLGRAPHDISPLAYTOGGLE command, 98
WallJoin command, 103, 133
WALLMERGE command, 102
WallMergeAdd command, 101-2, 135

WallMergeRemove command, 101-2
WallModifierAdd command, 188-89
WALLMODIFIEREDIT command, 194
WallModifierRemove command, 193-94
WallModifierStyle command, 196
WallOffsetCopy
  command, 99, 103-5, 124-26, 129
  tool, 71
WallOffsetMove
  command, 99, 103, 106-7
  tool, 71
WallOffsetSet command, 103, 107-10, 131
Walls
  category, 142
  connecting with Autosnap, 93-99
  converting linework to, 75-76
  converting to roof, 392-93
  creating
    curved, 82-86
    straight, 81-82
    with WallAdd command, 73-74
  drawing, 72-76
  editing
    tools, 71
    with AutoCAD editing commands, 110-11
  extending to the roof, 441
  joining, 102-3
  modifying, 99-111
  swept, 202-4
  only filter, 691, 695, 741, 759-60
  tool palette, 141-42, 173
WALLSTYLE command, 143-44
WallStyle command, 215
WallSweep command, 200
WallSweepMiterAngles command, 202
Warm Grips, 89. (*See also* Grips)
Wblock command, 470, 560, 807
Wedge shaped grip, 406
Welded wire fabric, creating, 813-14
What to plot field, 213, 252-53
Width
  check box, 374
  column, 218, 710
  dimension, 274, 545
    column, 545
  drop-down list, 154-55

edit field, 218, 248, 258, 260-61, 288, 292, 304, 349-50, 502, 525, 874
  option, 290
WIdth property, 75, 111, 261, 290
Winder
  balanced, 510
  property, 510
Window
  editing with grips, 292-93
  menu, 6-8
  muntins, creating for, 300-5
  object filter, 751
  Objects value, 751
  Pane section, 373
  properties, defining, 290-92
  Schedule
    command, 702
    option, 744
  Style
    option, 347
    Properties dialog box, 294-95, 297-98, 300, 302-4, 327, 373
    Properties – CASEMENT dialog box, 750
    Properties – Double Hung dialog box, 750
  style, Bay, 325-26
  Styles option, 294
  Tag option, 689, 743, 746
    placing a 689
  tool, 294
  Type list box, 297
Window>Project Navigator Palette command, 46, 837, 841, 853, 856
Window>Tool Palettes command, 59, 113, 118, 122, 130, 136, 222, 315, 752, 913, 915
WindowAdd command, 255, 287, 293
WindowObjects property set definition, 716, 748, 751
WindowObjects:FrameDepth property, 748
WindowObjects:Number property, 711
WindowObjects:Remarks property, 748
Windows>Properties Palette, 59
Windows>Pulldowns cascade menu, 6, 8

WindowStyle command, 293-94
WindowStyles property set, 750-51
WindowStyles:Material column, 709
Wireframe mode, 902
Wood-Saddle stair style, 571
Work tab, 4-5, 30, 87, 136, 222, 242, 245, 250
World Coordinate System/Current Coordinate System radio button, 598

X

X
  absolute coordinate value, 680
  Axis section, 955-56, 976
  edit field, 598, 614, 783
  Offset column, 941
  offset value, 598
  Scale edit field, 623
  scale factor, 681
X – Labeling tab, 971-72, 1004
X Axis>Add Grid Line option, 968
X Axis>Remove Grid Line option, 969
X-Axis
  cascade menu, 966
  dimension, 960-61
X-Axis>Add Grid Line option, 981
X-Axis>Remove Grid Line option, 982
Xref
  Attach option, 51-52
  Manager
    dialog box, 566, 570, 582, 763
    icon, 57
  Overlay option, 51-52, 498, 562, 567, 571
X-Width
  dimension, 955-56, 958-60, 965-66
  edit field, 958, 975

Y

Y
  absolute coordinate value, 680
  Axis section, 956-57, 976
  edit field, 598, 614, 783
  Offset column, 942
  offset value, 598
  Scale edit field, 623
  scale factor, 681
Y – Labeling tab, 971-72
Y Axis>Add Grid Line option, 968

Y Axis>Remove Grid Line option, 969
Y-Axis
  cascade menu, 966
  dimension, 960-61
Y-Depth dimension, 955-56, 959-60, 966, 975
Y-Labeling tab, 973, 1006

Z

Z
  absolute coordinate value, 681
  edit field, 598, 614, 783
  Offset column, 942
  offset value, 598
  Scale edit field, 623
  scale factor, 681
Zoom
  All option, 228
  camera factor, 989
  Center command, 38, 902
  command, 24, 113, 117, 362, 382, 901
  Extents
    command, 38, 116, 120, 122, 125, 902
    option, 222-23, 226, 238, 571, 643, 756, 837, 842, 844, 853
  flyout, 24-25, 113, 117, 120, 124-25, 220-23, 226, 228, 238, 493, 565, 567, 571, 630, 643, 756, 837, 842, 844
  In command, 38, 902
  menu option, 37-38, 220-22, 238, 382, 901
  Out command, 38, 902
  Previous option, 221
  scale factor, 6
  To
    command, 185, 226, 353, 439, 441, 482
    option, 360-62, 439
  to button, 717
  toolbar, 24
  Window
    command, 38, 113-14, 117, 124, 126, 902
    option, 220-21, 315, 317, 322, 382, 493, 565, 567, 628, 630, 844-45

LICENSE AGREEMENT FOR AUTODESK PRESS
A Thomson Learning Company

Educational Software/Data

You the customer, and Autodesk Press incur certain benefits, rights, and obligations to each other when you open this package and use the software/data it contains. BE SURE YOU READ THE LICENSE AGREEMENT CAREFULLY, SINCE BY USING THE SOFTWARE/DATA YOU INDICATE YOU HAVE READ, UNDERSTOOD, AND ACCEPTED THE TERMS OF THIS AGREEMENT.

Your rights:

1. You enjoy a non-exclusive license to use the enclosed software/data on a single microcomputer that is not part of a network or multi-machine system in consideration for payment of the required license fee, (which may be included in the purchase price of an accompanying print component), or receipt of this software/data, and your acceptance of the terms and conditions of this agreement.
2. You own the media on which the software/data is recorded, but you acknowledge that you do not own the software/data recorded on them. You also acknowledge that the software/data is furnished "as is," and contains copyrighted and/or proprietary and confidential information of Autodesk Press or its licensors.
3. If you do not accept the terms of this license agreement you may return the media within 30 days. However, you may not use the software during this period.

There are limitations on your rights:

1. You may not copy or print the software/data for any reason whatsoever, except to install it on a hard drive on a single microcomputer and to make one archival copy, unless copying or printing is expressly permitted in writing or statements recorded on the diskette(s).
2. You may not revise, translate, convert, disassemble or otherwise reverse engineer the software/data except that you may add to or rearrange any data recorded on the media as part of the normal use of the software/data.
3. You may not sell, license, lease, rent, loan, or otherwise distribute or network the software/data except that you may give the software/data to a student or and instructor for use at school or, temporarily at home.

Should you fail to abide by the Copyright Law of the United States as it applies to this software/data your license to use it will become invalid. You agree to erase or otherwise destroy the software/data immediately after receiving note of Autodesk Press' termination of this agreement for violation of its provisions.

Autodesk Press gives you a LIMITED WARRANTY covering the enclosed software/data. The LIMITED WARRANTY can be found in this product and/or the instructor's manual that accompanies it.

This license is the entire agreement between you and Autodesk Press interpreted and enforced under New York law.

Limited Warranty

Autodesk Press warrants to the original licensee/ purchaser of this copy of microcomputer software/ data and the media on which it is recorded that the media will be free from defects in material and workmanship for ninety (90) days from the date of original purchase. All implied warranties are limited in duration to this ninety (90) day period. THEREAFTER, ANY IMPLIED WARRANTIES, INCLUDING IMPLIED WARRANTIES OF MERCHANTABILITY AND FITNESS FOR A PARTICULAR PURPOSE ARE EXCLUDED. THIS WARRANTY IS IN LIEU OF ALL OTHER WARRANTIES, WHETHER ORAL OR WRITTEN, EXPRESSED OR IMPLIED.

If you believe the media is defective, please return it during the ninety day period to the address shown below. A defective diskette will be replaced without charge provided that it has not been subjected to misuse or damage.

This warranty does not extend to the software or information recorded on the media. The software and information are provided "AS IS." Any statements made about the utility of the software or information are not to be considered as express or implied warranties. Delmar will not be liable for incidental or consequential damages of any kind incurred by you, the consumer, or any other user.

Some states do not allow the exclusion or limitation of incidental or consequential damages, or limitations on the duration of implied warranties, so the above limitation or exclusion may not apply to you. This warranty gives you specific legal rights, and you may also have other rights which vary from state to state. Address all correspondence to:

AutodeskPress
Executive Woods
5 Maxwell Drive
Clifton Park, NY 12065